Methods in Molecular Biology

For further volumes:
http://www.springer.com/series/7651

Human Fertility

Methods and Protocols

Edited by

Zev Rosenwaks

The Ronald O. Perelman and Claudia Cohen Center for Reproductive Medicine, Weill Cornell Medical College, New York, NY, USA

Paul M. Wassarman

Department of Developmental & Regenerative Biology, Ichan School of Medicine at Mount Sinai, New York, NY, USA

Humana Press

Editors
Zev Rosenwaks
The Ronald O. Perelman and Claudia Cohen
Center for Reproductive Medicine
Weill Cornell Medical College
New York, NY, USA

Paul M. Wassarman
Department of Developmental
& Regenerative Biology
Ichan School of Medicine at Mount Sinai
New York, NY, USA

ISSN 1064-3745 ISSN 1940-6029 (electronic)
ISBN 978-1-4939-4117-9 ISBN 978-1-4939-0659-8 (eBook)
DOI 10.1007/978-1-4939-0659-8
Springer New York Heidelberg Dordrecht London

Printed on acid-free paper

Humana Press is a brand of Springer
Springer is part of Springer Science+Business Media (www.springer.com)

Preface

In July 2013, Louise Brown, the first IVF baby, celebrated her 35th birthday. During her lifetime more than five million IVF babies have been born worldwide. In the past three-and-a-half decades we have witnessed a virtual explosion in the clinical application of assisted reproductive technologies, or ART, to help couples experiencing fertility barriers achieve pregnancy. It is estimated that one in six couples worldwide will experience some sort of fertility barrier during their reproductive lifetime. Today more than one percent of all babies born in the United States are conceived by ART practised in more than five hundred fertility clinics around the country.

Optimization of ART, perhaps more than any other clinical discipline, has relied enormously on scientific breakthroughs and methodological innovations in both the embryology laboratory and the clinical arena. Robert Edwards' steadfast adherence to stringent scientific principles, principles that made the clinical application of IVF a reality, continues to propel the field.

This volume is intended for all practitioners of reproductive medicine and ART, as well as for reproductive biologists and embryologists, cell and molecular biologists, and others in the biomedical sciences. Its goal is to present in a straightforward manner best practice approaches for overcoming a host of fertility challenges.

Methods in Human Fertility is grounded in the belief that good medical practice of ART relies on a thorough understanding of the physiologic and genetic basis of male and female reproduction. Accordingly, chapters on the scientific fundamentals of human reproduction (Chaps. 1 and 8), genetics of male and female infertility (Chaps. 2–4), spermatozoal function (Chap. 5), markers of male infertility (Chap. 9), and menstrual cycle physiology (Chap. 7) are followed by detailed presentations of clinical aspects of ART (Chaps. 10, 12–15). Descriptions of oocyte and sperm retrieval techniques (Chaps. 16 and 17) are complemented by a thorough presentation of contemporary approaches for diagnosing and treating male infertility (Chap. 18). Ovarian tissue cryopreservation (Chap. 21) along with traditional and novel approaches to oocyte and embryo cryopreservation (Chaps. 11, 19, and 20) and markers of embryo quality (Chaps. 23 and 24) are described. The dynamic technology known as preimplantation genetic diagnosis or PGD is reviewed (Chap. 22), as is the long-term well-being of children conceived following intracytoplasmic sperm injection, or ICSI (Chap. 26). Finally, embryo transfer techniques and technology involved in human embryonic stem cell derivation are also described in some detail (Chaps. 6, 25, and 27).

We wish to express our gratitude to the many authors included in the volume for their diligence and patience and for generously sharing their knowledge and expertise. We are also very grateful to Daniel Pepper who provided considerable editorial expertise and kept the project pretty much on track.

New York, NY, USA *Zev Rosenwaks*
Paul M. Wassarman

Contents

Contributors

ROBERT L. BARBIERI • *Department of Obstetrics and Gynecology, Harvard Medical School, Brigham and Women's Hospital, Boston, MA, USA*

BARRY BEHR • *Division of Reproductive Endocrinology and Infertility, Department of OB/GYN, Stanford University, Stanford, CA, USA*

DIANA PATRICIA BERNA • *Reproductive Biology Associates, Atlanta, GA, USA*

HANNAH BIDERMAN • *Fertility Preservation Center and IVF Unit, Sheba Medical Center Tel-Hashomer, Sackler School of Medicine, Tel-Aviv University, Tel-Aviv, Israel*

ANATE AELION BRAUER • *Greenwich Fertility and Medical Group, Tuckahoe, NY, USA*

F.J. BROEKMANS • *Department of Reproductive Medicine and Gynaecology, University Medical Centre Utrecht, CS Utrecht, The Netherlands*

YUN-XIA CAO • *Department of Obstetrics and Gynecology, Center for Reproductive Medicine, The First Affiliated Hospital of Anhui Medical University, Hefei, People's Republic of China*

LIVIO CASARINI • *Department of Biomedical, Metabolic and Neural Sciences, University of Modena and Reggio Emilia, Modena, Italy*

CHING-CHIEN CHANG • *Reproductive Biology Associates, Atlanta, GA, USA*

RI-CHENG CHIAN • *Department of Obstetrics and Gynecology, McGill University, Royal Victoria Hospital, Montreal, QC, Canada*

MARK A. DAMARIO • *Department of Obstetrics, Gynecology and Women's Health, University of Minnesota, Minneapolis, MN, USA*

OWEN K. DAVIS • *Center of Reproductive Medicine and Infertility, Weill Medical College of Cornell University, New York, NY, USA*

B.C.J.M. FAUSER • *Department of Reproductive Medicine and Gynaecology, University Medical Centre Utrecht, CS Utrecht, The Netherlands*

MICHAEL FUNARO • *Department of Urology and Reproductive Medicine, Weill Cornell Medical College, New York, NY, USA*

MARC GOLDSTEIN • *Department of Urology, James Buchanan Brady Foundation, Weill Cornell Medical College, New York, NY, USA*

ROGER G. GOSDEN • *Weill Cornell Medical College, New York, NY, USA*

LUCINDA VEECK GOSDEN • *Weill Cornell Medical College, New York, NY, USA*

O. HAMDINE • *Department of Reproductive Medicine and Gynaecology, University Medical Centre Utrecht, CS Utrecht, The Netherlands*

APARNA HEGDE • *Division of Reproductive Endocrinology and Infertility, Department of OB/GYN, Stanford University, Stanford, CA, USA*

JENNIFER HU • *The Ronald O. Perelman and Claudia Cohen Center for Reproductive Medicine, Weill Cornell Medical College, New York, NY, USA*

JACK YU JEN HUANG • *Stanford Fertility Center, Palo Alto, CA, USA*

JUSTIN KOCENT • *The Ronald O. Perelman and Claudia Cohen Center for Reproductive Medicine, Weill Cornell Medical College, New York, NY, USA*

YI-NAN LIN • *Institute of Biomedical Sciences, Academia Sinica, Taipei, Taiwan*

MARTIN M. MATZUK • *Department of Pathology & Immunology, Baylor College of Medicine, Houston, TX, USA; Department of Molecular and Cellular Biology, Baylor College of Medicine, Houston, TX, USA; Department of Molecular and Human Genetics, Baylor College of Medicine, Houston, TX, USA*

JOSE V. MEDRANO • *CIPF & Fundación Instituto Valenciano de Infertilidad (FIVI), Valencia University, INCLIVA, Valencia, Spain*

DROR MEIROW • *Fertility Preservation Center and IVF Unit, Sheba Medical Center Tel-Hashomer, Sackler School of Medicine, Tel-Aviv University, Tel-Aviv, Israel*

DEVIN MONAHAN • *The Ronald O. Perelman and Claudia Cohen Center for Reproductive Medicine, Weill Cornell Medical College, New York, NY, USA*

ZSOLT PETER NAGY • *Reproductive Biology Associates, Atlanta, GA, USA*

LIESL NEL-THEMAAT • *Reproductive Biology Associates, Atlanta, GA, USA*

QUEENIE V. NERI • *The Ronald O. Perelman and Claudia Cohen Center for Reproductive Medicine, Weill Cornell Medical College, New York, NY, USA*

YUVAL OR • *Department of Obstetrics and Gynecology, Kaplan Medical Center, Rehovot, Israel; Hebrew University of Jerusalem and Hadassah School of Medicine, Jerusalem, Israel*

DARIUS A. PADUCH • *Department of Urology and Reproductive Medicine, Weill Cornell Medical College, New York, NY, USA*

GIANPIERO D. PALERMO • *The Ronald O. Perelman and Claudia Cohen Center for Reproductive Medicine, Weill Cornell Medical College, New York, NY, USA*

RENEE REIJO PERA • *Department of Obstetrics and Gynecology, Institute for Stem Cell Biology and Regenerative Medicine, Stanford University School of Medicine, Stanford University, Stanford, CA, USA*

HILA RA'ANANI • *Fertility Preservation Center and IVF Unit, Sheba Medical Center Tel-Hashomer, Sackler School of Medicine, Tel-Aviv University, Tel-Aviv, Israel*

DAVID REICHMAN • *Center of Reproductive Medicine and Infertility, Weill Medical College of Cornell University, New York, NY, USA*

ZEV ROSENWAKS • *The Ronald O. Perelman and Claudia Cohen Center for Reproductive Medicine, Weill Cornell Medical College, New York, NY, USA*

D. SAKKAS • *Boston IVF, Waltham, MA, USA; Department of Obstetrics, Gynecology, and Reproductive Sciences, Yale University School of Medicine, New Haven, CT, USA*

DANIEL B. SHAPIRO • *Reproductive Biology Associates, Atlanta, GA, USA*

GLENN SCHATTMAN • *Center of Reproductive Medicine and Infertility, Weill Medical College of Cornell University, New York, NY, USA*

PETER N. SCHLEGEL • *Department of Urology, James Buchanan Brady Foundation, Weill Cornell Medical College, New York, NY, USA*

ZEEV SHOHAM • *Department of Obstetrics and Gynecology, Kaplan Medical Center, Rehovot, Israel; Hebrew University of Jerusalem and Hadassah School of Medicine, Jerusalem, Israel*

CARLOS SIMON • *CIPF & Fundación Instituto Valenciano de Infertilidad (FIVI), Valencia University, INCLIVA, Valencia, Spain*

MANUELA SIMONI • *Department of Biomedical, Metabolic and Neural Sciences, University of Modena and Reggio Emilia, Modena, Italy*

JOE LEIGH SIMPSON • *Department of Human and Molecular Genetics, Herbert Wertheim College of Medicine, Florida International University, Miami, FL, USA; Department of Obstetrics and Gynecology, Herbert Wertheim College of Medicine, Florida International University, Miami, FL, USA*

PETER J. STAHL • *Department of Urology, Columbia University Medical Center, New York, NY, USA*

EDI VAISBUCH • *Department of Obstetrics and Gynecology, Kaplan Medical Center, Rehovot, Israel; Hebrew University of Jerusalem and Hadassah School of Medicine, Jerusalem, Israel*

KANGPU XU • *Center of Reproductive Medicine and Infertility, Weill Medical College of Cornell University, New York, NY, USA*

NIKICA ZANINOVIC • *Center for Reproductive Medicine, Weill Cornell Medical College, New York, NY, USA*

QIANSHENG ZHAN • *Center for Reproductive Medicine, Weill Cornell Medical College, New York, NY, USA*

Part I

Overview

Chapter 1

General Aspects of Fertility and Infertility

Mark A. Damario

Abstract

Fertility rates have been declining in most Western nations over the past several decades, although it is not entirely clear if an increased rate of infertility substantially contributes to this. As compared to other species, the reproductive efficiency of humans is relatively low. Factors related to fertility include age, exposure to sexually transmitted diseases, frequency of intercourse, coital timing, as well as diet and lifestyle habits. Infertility is considered a disease due to its major disruption of major organ systems and life functions. An infertility evaluation is recommended after 12 months or more of regular, unprotected intercourse and may be considered after 6 months for those female patients over the age of 35 or with other known abnormalities. A proper infertility evaluation is a comprehensive examination of possibly identifiable infertility factors of both female and male partners, lending itself to the most appropriate and potentially effective treatment.

Key words Reproductive age, Infertility, Causes of infertility, Ovulatory function, Ovarian reserve, Cervical factor, Peritoneal factor, Uterine factor, Male factor, Tubal factor

1 Introduction

Fertility is the ability to produce a child. In most Westernized societies, the general fertility rate (births per 1,000 women aged 15–44) has declined over the past several decades [1]. Some of this decline is attributable to intentional factors (purposeful desire for smaller family size), although some of this decline is due to unintentional factors (inability to conceive). Sociodemographic trends in most Westernized nations occurring over this time period include a greater interest in advanced education and career development among women, later age of marriage, more frequent divorce, delayed childbearing, and improved contraceptive methods as well as greater access to family planning. In the United States, the general fertility rate (births per 1,000 women aged 15–44) in 2010 was 64.1, which represented a rate that was approximately 27 and 37 % lower than in 1970 and 1950, respectively [2].

As typical in most Westernized societies, attitudes in the United States among women and towards women have significantly

Zev Rosenwaks and Paul M. Wassarman (eds.), *Human Fertility: Methods and Protocols*, Methods in Molecular Biology, vol. 1154, DOI 10.1007/978-1-4939-0659-8_1, © Springer Science+Business Media New York 2014

changed over the past several decades. Census data shows that more women have completed 4 or more years of college (from 8.2 % in 1970 to 30.5 % in 2012) [3]. Women now represent the majority of college students. Many professional schools (graduate schools, law schools, medical schools) have high proportions of women students.

The greater focus on education and careers among women has contributed to a number of other trends. These include less frequent and later marriage. First marriage rates in the United States peaked just after World War II, between 1945 and 1947 (143 per 1,000 single women), and have declined approximately 50 % in the more than six decades since [4]. The median age of first marriage has also increased over this same time period (to a high of 25.8 years in 2006–2010) [5]. In addition, divorce rates among women of reproductive age more than doubled between 1960 and 1980 (to about 40 per 1,000 married women aged 15–44) and have remained relatively stable since then [4]. Today, approximately 50 % of marriages still end in divorce in the United States. The probability of remarriage is also proportionately related to the age of the woman. Remarriage rates have also been noted to have declined in the United States over the past several decades in parallel with first marriage rates [6].

The availability of expanding and more effective contraceptive methods as well as access to family planning services and legalized abortion have improved the means for women to safely and effectively control their fertility. These options have significantly contributed to the decline in the general fertility rate in the United States over the past several decades. Their effect on the general fertility rate has been both direct (by reducing the number of unplanned pregnancies and births) as well as indirect (allowing for the postponement of pregnancies and births).

The net result of all of these sociodemographic trends has been a trend towards delayed childbearing in the United States and in many Westernized nations. The mean age of first live birth has risen steadily in the United States, from 21.4 years in 1970 to 25.4 years in 2010 [1]. Mean age for all subsequent live births has also increased. Increasing age at first birth and declining fertility rates have contributed to significantly fewer births per woman and the aforementioned decline in general fertility rates.

2 Normal Reproductive Efficiency

When compared to other species, including nonhuman primates, human reproductive efficiency is relatively inefficient. Utilizing clinically recognized pregnancies as the reference, reproductive efficiency in normally fertile couples averages about 20 % per menstrual cycle. Given the approximate 20 % cycle fecundability rate, early studies estimated the time required to conceive in couples who achieved pregnancy and found that approximately 85 % of couples conceived within 1 year of exposure [7]. Zinaman et al. noted that 82 % of 200 couples who desired pregnancy conceived over a 12-menstrual-cycle observation period [8].

Normal sperm can survive for up to 3–5 days in the female reproductive tract and fertilize an egg. An oocyte has the potential to be fertilized for only 12–24 h post-ovulation [9]. As a result, in all conception cycles intercourse occurs within 5 days prior to or on the day of ovulation [10]. The probability of pregnancy after the day of ovulation falls close to zero. As a result, additional focus on coital timing through the use of basal body temperature (BBT) charting, cervical mucus assessment, or urinary LH surge detection methods may result in modest improvements in cycle fecundability. Even with well-timed coitus, however, cycle fecundability still does not appear to exceed 35 % [11].

Gnoth and associates reported on a long-term prospective cohort study regarding the use of natural family planning on the time to conception [12]. In this report, women received teaching of natural family planning methodologies, focusing on BBT and cervical mucus patterns as well as calculation rules. Cumulative probabilities of conception for all couples were 38, 68, 81, and 92 % following their first, third, sixth, and twelfth cycles.

3 Factors Impacting Fertility

3.1 Age

Age alone has a significant impact on female fertility. The biology of female fertility is such that the process of gametogenesis does not occur after birth. Women, therefore, have a discrete reproductive life-span in which reasonable rates of fecundity occur. In general, women have maximal fecundity potential in their late teens, twenties, and early thirties.

Unlike their male partners, women are endowed with a finite and non-replenishable complement of germ cells. The maximum number of germ cells occurs at fetal mid-gestation when a total of 6–7 million are present. Thereafter, the number of germ cells irretrievably declines and no further de novo gametogenesis occurs. At birth, the number of germ cells is estimated to be approximately 1–2 million. At the onset of puberty, the germ cell number is typically reduced to approximately 300,000. Thereafter, during the reproductive years, a number of oocytes begin to develop with only one or a few becoming dominant while the others undergoing a process of atresia [13]. The absolute number of oocytes continues to decline with age irrespective of whether the woman has ovulatory cycles. At approximately the age of 37–38, there is often an accelerated rate of follicular loss which occurs when the number of follicles reaches about 25,000 [14]. This accelerated loss is correlated with a subtle increase in serum follicle-stimulating hormone (FSH) and a decrease in inhibin production. The functional capacity of the remaining germ cells and follicles has been termed "ovarian reserve" or ovarian age. The subtle changes which indicate diminished ovarian reserve are associated with a significantly lowered fertility potential, often without apparent changes in clinically identifiable characteristics or menstrual cyclicity. At the time of menopause, fewer than 1,000 follicles remain.

Historical data shows that populations who do not practice contraception have declining fertility rates with increasing age of the woman. Tietze performed a comprehensive analysis of the fertility rates of the Hutterite sect of the Western United States and Canada during the 1950s [15]. This sect originated in Switzerland and settled in the upper Great Plains in the 1870s. Because of the communal nature of the sect and lack of contraceptive methods, there were no incentives to limit family size. As a result, the birth rate of Hutterite women is one of the highest recorded with an average of 11 children per married woman. Only 5 out of 209 women studied failed to have children (infertility rate of 2.4 %). As the women aged, however, their fertility rates fell. After the age of 34 years, 11 % of women bore no further children, 33 % of women bore no more children after the age of 40, and 87 % of women bore no more children after the age of 45.

Although there is an apparent decrease in sexual activity as women age, this does not entirely explain the observed decrease in fertility. A French study of couples treated with donor insemination also revealed a proportionate decline in pregnancy rates associated with advancing female age [16]. In this study, the pregnancy rate (after up to 12 inseminations) for women <30 years was 73 %, for women aged 31–35 years was 62 %, and for women aged 36–40 years was 54 %.

As demonstrated by numerous reports and registries, success rates achieved with the assisted reproductive technologies (ART) also decline as female age increases. The number of oocytes retrieved, and embryos available are lower. However, since many older patients still have comparable number of embryos transferred, it is not the embryo number per se that is the determinant of the reduced pregnancy rate seen with advancing age. Older patients undergoing ART with their own oocytes experience significant declines in the implantation rate (as assessed by the presence of a fetal heartbeat per embryo transferred) proportionately associated with advancing age [17].

As female age increases, the risks of other disorders that may adversely affect fertility, such as fibroids and endometriosis, also increase. In addition, the age-related decline in female fertility is accompanied by a significant increase in the rate of embryonic aneuploidy and spontaneous abortion. Autosomal trisomies are the most frequent findings and are related to changes in the oocyte meiotic spindle, predisposing to nondisjunction. Even embryos selected for transfer in in vitro fertilization (IVF) based on favorable morphologic features have a high rate of aneuploidy [18]. The fetal loss rate remains significantly higher even after the detection of a fetal heartbeat by transvaginal ultrasonography after IVF [19]. For women <33 years who conceive after fresh IVF, 9.9 % experience a pregnancy loss after a fetal heartbeat is detected. This rate of pregnancy loss after a fetal heartbeat progressively increases with advancing female age, such that it was seen in 11.4 % of women aged 33–34 years, 13.7 % of women aged 35–37 years, 19.8 % of women aged 38–40 years, 29.9 % of women aged 41–42 years, and 36.6 % of women aged >42 years. An additional clear relationship between advancing female age and chromosomal abnormalities in live-born offspring is also observed.

In contrast, the relationship between advancing male age and declining fertility is more difficult to define. This is partly due to the fundamental differences in gametogenesis among men and women. As opposed to women, who functionally exhaust the supply of available oocytes as age advances, mitotic divisions in spermatogonia occur throughout the life-span of men. This results in the continual replenishment of germ cells and allows men to reproduce even very late in life. Fertility in men, however, does appear to subtly decline as age increases, and the issue may be growing in importance as a significantly higher number of men are choosing to father children at older ages. In the United States, birth rates for men between the ages of 35 and 54 increased 46.7 % between 1980 and (68.2 per 1,000 men) and 2010 (100.1 per 1,000 men) [2].

Changes in semen characteristics observed as male age increases include gradual decreases in semen volume, sperm motility, and proportion of morphologically normal sperm but not sperm concentration [20]. One study examined the relationship between male age and semen quality among couples pursuing donor oocyte IVF and observed a decline in total motile sperm count of approximately 2.5 million per year [21]. As semen characteristics do not entirely predict fertilization capacity, the significance of semen quality changes with advancing male age is not entirely certain. On the other hand, however, there is some available evidence that the time to conception and pregnancy rates are subtly decreased with advancing male age, even factoring in the confounders of advancing female age and decreased coital frequency. Results from one British study indicate that the time to pregnancy is five times longer for men older than age 45 than for men younger than age 25, even when the analysis was restricted to men with relatively younger partners [22]. On the other hand, results from oocyte donor IVF (in which all oocytes are derived from young and healthy individuals) fail to demonstrate a clear relationship between advancing male age and diminished live birth rates [21].

It has recently been reported that de novo mutation single-nucleotide polymorphism (SNP) rates in offspring are correlated with the age of the father at the time of conception [23]. The effect shown reflects an increase of about two mutations per year, from which an exponential model predicts a doubling of paternal mutations in 16.5 years. Men appear to transmit a higher number of mutations to their children than women, suggesting a relationship between increasing male age and diseases known to be linked to *de novo* mutations, such as schizophrenia and autism.

3.2 Sexually Transmitted Diseases

Tubal infertility affects 18 % of couples trying to overcome infertility by using ART in the United States [24]. Tubal scarring and occlusions are typically the consequence of chronic pelvic inflammatory disease (PID), the latter of which is associated with sexually transmitted diseases such as chlamydia and gonorrhea. More than one million chlamydia cases are reported to the Centers for Disease Control and Prevention (CDC) annually; although since most chlamydia cases are undiagnosed this fails to reflect the true incidence, which is estimated to be approximately 2.8 million cases per year [25].

Up to 40 % of females with inadequately treated chlamydial infections develop PID, and 20 % of those are estimated to become infertile [26]. Untreated chlamydial infections can result in sterility for men as well, although this is relatively rare. The chlamydia case rate for females is approximately three times that of males (likely reflecting increased screening). In 2006, young females aged 15–19 (2,863 cases per 100,000 population) and aged 20–24 (2,797 per 100,000 population) had the highest incidence [25]. Reported chlamydia rates were more than seven times higher in African Americans than among whites, highlighting a large racial disparity.

The CDC currently recommends annual chlamydia screening for all sexually active women under age 26 as well as older women with particular risk factors. The implementation of global screening has been noted to reduce the incidence of PID by over 50 % in a managed care setting [27]. Although there have been identified barriers to global screening, such a preventive care initiative may have the potential to limit the serious health consequences of chlamydia, including infertility.

3.3 Frequency of Intercourse

Information is emerging that may better define an optimal frequency of intercourse. Based on early studies and current World Health Organization (WHO) guidelines suggesting a 2–7-day abstinence interval before semen evaluation [28], there is a widely held misconception that frequent ejaculation decreases male fertility. An analysis of almost 10,000 semen specimens observed that in normozoospermic men, sperm concentration and motility remain normal even with daily ejaculation [29]. In oligozoospermic men, there was actually an inverse relationship between the mean percentage of motile sperm and increasing abstinence length with both no abstinence and abstinence of just 1 day related to peak sperm quality. Abstinence intervals also generally did not appear to affect sperm morphology, as judged by "strict" criteria. Semen parameters did appear to deteriorate, however, after longer abstinence intervals (of 10 days or more).

Evidence currently suggests that daily intercourse may offer a slight fertility advantage, although clinicians should be mindful about the potential of coital recommendations causing unnecessary stress. In a study involving 221 presumably fertile couples attempting to conceive, statistical modeling demonstrated that the highest cycle fecundability (37 % per cycle) was associated with daily intercourse [10]. Comparable fecundability (33 % per cycle) was seen with every other day intercourse, but the likelihood for success decreased to 15 % per cycle when intercourse occurred only once weekly. The stress of infertility can potentially negatively impact sexual function, particularly when the timing of intercourse is linked to ovulation predictor methods or follows a strict schedule [30]. Couples should be informed, therefore, that reproductive efficiency increases with the frequency of intercourse and is perhaps highest when intercourse occurs every 1–2 days but that they should determine the coital frequency that works best for them.

3.4 Coital Timing

The generally accepted "fertile window" comprises the 6 days of the menstrual cycle consisting of the day of ovulation and the 5 preceding days. Couples should be counseled that their highest chance for conception occurs when coitus occurs during this time period. In a study of 221 healthy women trying to conceive, Wilcox et al. found conceptions occurring only during a 6-day period that ended on the day of ovulation [10]. The probability of conception ranged from 0.10 five days prior to ovulation to 0.33 on the day of ovulation (Fig. 1). From a further analysis of cycles in which only one intercourse event occurs over the span of the "fertile period," it was found that most pregnancies occur within 3 days of ovulation, with only 6 % of pregnancies occurring with sperm that was three or more days old. Irrespective of ovulation timing, it was demonstrated that the probability of pregnancy from a single act of coitus in women with regular cycles was 0.069 on cycle day 10, 0.094 on cycle day 12, 0.085 on cycle day 14, and 0.059 on cycle day 16 [31]. Whereas aging does not impact the overall timing of the "fertile window," the relative probability of pregnancy diminishes with increasing age [32]. Increased coital frequency during the "fertile window" also increases the overall cycle fecundability [33]. Methods to determine ovulation timing and the resultant "fertile window" include analysis of menstrual cycle calendars, BBT recordings, urinary luteinizing hormone determinations, and cervical mucus scores. Subjective perception of ovulation timing based on symptoms alone reveals a low rate of concordance with actual ovulation [34].

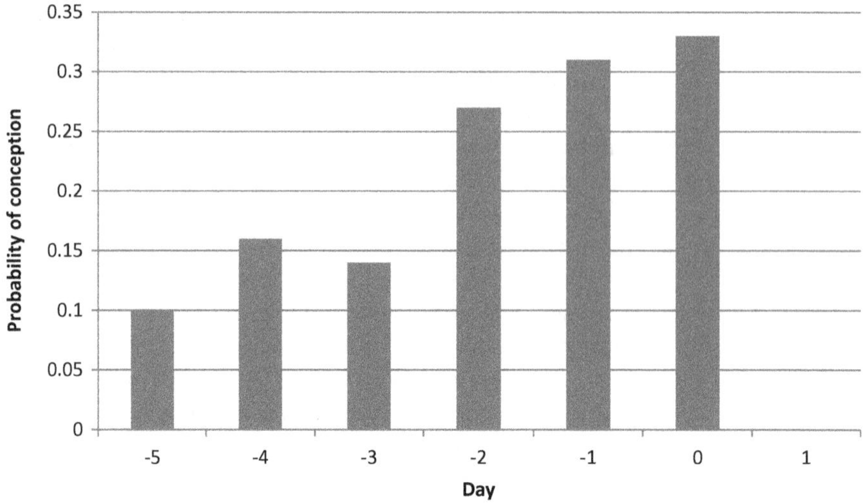

Fig. 1 Probability of conception on specific days relative to the day of ovulation (day 0). Data from Wilcox et al. [10]

The volume of cervical mucus increases over the 5–6 days preceding ovulation in proportion to serum estradiol concentrations and reaches its peak 2–3 days prior to ovulation. The viscosity of cervical mucus also lessens during this time period such that it is classically described as “slippery, clear” just prior to ovulation. Assessing cervical mucus involves ascertaining secretions present at the vaginal introitus. Studies show that perceived cervical mucus changes correlate well with BBT charting in predicting the time of peak fertility [35]. In a prospective cohort study, Scarpa and associates reported the probability of conception to range from 0.003 for days with no noticeable secretions to 0.29 for days with the most optimal mucus [36].

Ovulation predictor kits and monitors provide patients an ability to screen for urinary luteinizing hormone (LH) excretion and may also assist in determining peak fertility. Numerous studies have verified the accuracy of these methods in detecting the midcycle LH surge, although the precise timing of subsequent ovulation is still somewhat variable (up to 2 days) [37]. False-positive home urine luteinizing hormone tests may also occur, with a study by the National Institute of Child Health and Human Development (NICHD) National Cooperative Reproductive Medicine Network showing that more than 7 % of endometrial biopsies performed 7–13 days after a positive home urine LH test failed to demonstrate secretory endometrium [38]. On the other hand, other studies have shown a reduced time to conception with the use of home urinary LH testing. Robinson and coauthors reported a cumulative pregnancy rate of 22.7 % over two cycles as compared to 14.4 % for control patients in a cohort study [39].

3.5 Diet and Lifestyle Habits

Fertility rates are reduced in women who are both underweight and overweight. Whereas diet and lifestyle modification-associated weight loss have improved fertility potential for women with ovulation dysfunction [40], there is little evidence for dietary modifications (including low-fat, vegetarian, vitamin-enriched, gluten-free diets) impacting fertility in normal-weight women. Elevated blood mercury concentrations from frequent seafood consumption have been associated with lowered fertility [41]. Women who are planning to conceive should take at least 0.4 mg of folic acid daily as this has been shown to lower the incidence of neural tube defects in offspring if taken at the time of conception [42].

Smoking has multiple adverse effects on fertility, including gamete production and function, ovulatory cyclicity, fertilization, early embryonic cleavage, embryo transport, and implantation [43]. Numerous studies also demonstrate that smoking is associated with increased risks of spontaneous abortions as well as preterm birth [44, 45]. Smoking has also been shown to cause a mean birth weight decrease of 150–300 g in offspring [44]. A systematic review and meta-analysis of 12 studies revealed an overall odds ratio for risk of infertility in female smokers of 1.60 (1.34–1.91, 95 % confidence intervals) as compared to nonsmokers [46]. In this report, female smokers receiving IVF treatment also had a lower odds ratio of 0.66 (0.49–0.88, 95 % confidence intervals) for pregnancies per cycle as compared to nonsmokers. Female smokers transition to menopause, on average, 1–4 years earlier than nonsmokers suggesting that smoking also accelerates follicular depletion [47]. Although semen abnormalities have been found in men who smoke, a similar relationship with lowered fertility in men has not yet been conclusively demonstrated.

The effects of alcohol on female fertility have produced contradictory results with some studies purporting a relationship with infertility whereas others actually reporting a shortened time to conception [48, 49]. On the other hand, in a prospective cohort study, Rossi et al. found that women drinking at least four drinks per week had a lower odds ratio for live birth following IVF of 0.84 (0.71–0.99, 95 % confidence intervals) than women who drank fewer than four drinks per week [50]. It appears prudent to advise avoidance of high levels of alcohol consumption while attempting to conceive as well as no alcohol at all during pregnancy due to its well-documented detrimental effects on fetal development.

In a large European multicenter study, patients who consumed high levels of daily caffeine (>500 mg) intake had an increased odds ratio of 1.45 (1.03–2.04, 95 % confidence intervals) for subfecundity in the first pregnancy [51]. Reports vary on the relationship between caffeine consumption and the risk of miscarriage, although a few studies have suggested that daily caffeine consumption greater than 200–300 mg during pregnancy is associated with an increased risk of spontaneous abortion [52]. Overall, caffeine consumption in moderation (one to two cups of coffee per day or equivalent) does not appear to be associated with lowered fecundity.

4 Infertility

The American Society for Reproductive Medicine defines infertility as "a disease defined by the failure to achieve a successful pregnancy after 12 months or more of appropriate, timed unprotected intercourse or therapeutic donor insemination" [53]. On this basis, most advise the start of an infertility evaluation if a patient has met this threshold. Based on medical history and physical findings, an earlier evaluation may be warranted in certain circumstances and is advised for women >35 years of age who have been attempting pregnancy for at least 6 months.

4.1 Incidence of Infertility

Infertility is estimated to affect 10–15 % of couples. Hull and co-workers reported that out of 708 couples in an English health district, at least 1 in 6 needed specialist care at some point in their lives because of an inability to conceive or to conceive the number of children they desired [54]. Snick and co-workers noted that 9.9 % of women aged 15–44 years in the Walcheren area of the Netherlands needed specialist fertility care at some point in their lives [55]. According to the National Survey of Family Growth, 7.3 million American women aged 15–44 years (approximately 12 % of all women aged 15–44 years) reported having ever used infertility services [56].

Many infertility couples are subfertile, not truly sterile. A proportion of subfertile couples will eventually conceive even without treatment [57]. The likelihood of achieving a live birth without treatment decreases with increasing age of the female as well as duration of infertility [58]. The vast majority of spontaneous pregnancies occur within the first 3 years of attempting pregnancy, with a relatively rare spontaneous pregnancy occurrence occurring beyond this time.

The use of ART has continued to increase since its inception in 1978 [59]. In 2009, 146,244 ART procedures were performed in the United States resulting in 45,780 live-birth deliveries and 60,190 infants [60]. Overall, ART contributed to 1.4 % of the US births.

4.2 Causes of Infertility

The major causes of infertility include ovulatory dysfunction, tubal and peritoneal factors (including endometriosis), uterine factor, male factor, diminished ovarian reserve, and unexplained. The proportion of patients with a particular factor depends on the patient age as well as the duration of infertility. The composition of

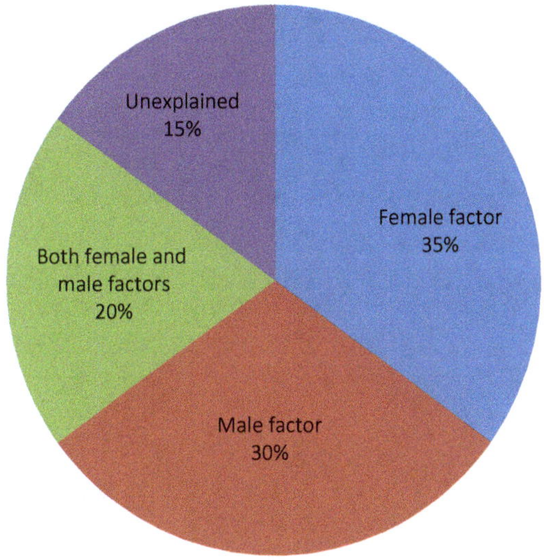

Fig. 2 Distribution of causes of infertility. Data from Forti and Krausz [61]

infertility factors also depends on the level of care (primary, secondary, tertiary). The mean duration of infertility is longer in groups that provide tertiary care (42 months) [55] than those that provide primary care (21 months) [58], resulting in somewhat differing proportions of patients with particular factors. In general, approximately 35 % of infertility is attributable to a female factor alone, 30 % to a male factor alone, and 20 % to both, and in 15 % of cases it is essentially unexplained (Fig. 2) [61].

5 Evaluation of Infertility

The evaluation of infertility starts with a comprehensive history and physical examination of the female, which may uncover clues of a possible etiology for reproductive failure. Based on the history or subfertile semen profile, men may also be referred for a comprehensive evaluation. Following the initial findings, more specific testing may be indicated, although it remains important for both partners to undergo general testing to determine whether there are indicators for the common etiologies of infertility.

5.1 Ovulatory Function

Menstrual cycle history and characteristics are typically sufficient in determining whether there is normal ovulatory function. Patients with normal ovulatory cycles generally have regular menstrual cyclicity with intervals of 25–35 days, consistent flow characteristics, and accompanying moliminal symptoms associated with varying ovarian hormones. Many patients, however, do have some degree of intermenstrual interval variability. A study of more than 1,000 cycles demonstrated at least one intermenstrual interval

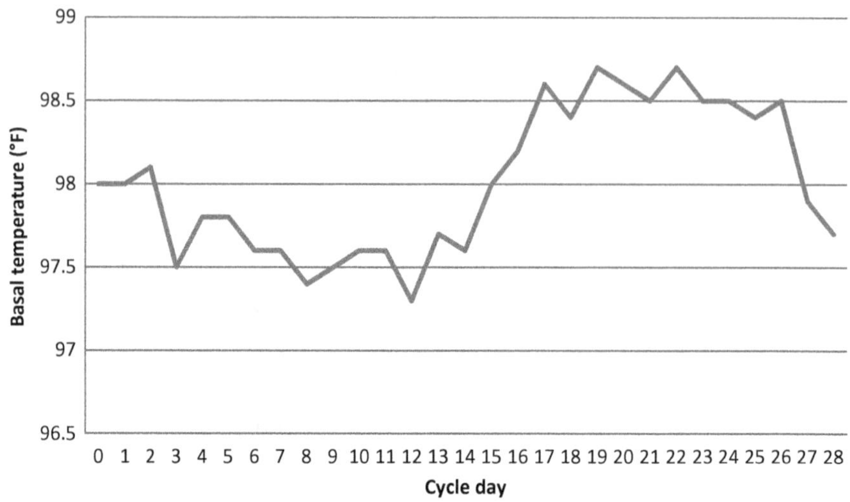

Fig. 3 Basal body temperature chart with a biphasic pattern

variation >5 days in 75 % of patients followed for 1 year [62]. Although history is a strong predictor of ovulatory cyclicity, it remains prudent to further confirm ovulatory function by additional corroborative objective information.

Serial BBT charting provides a simple means for patients to monitor their menstrual cycles. In order to obtain reliable results, temperatures need to be obtained daily at similar times and under similar conditions with an accurate thermometer. In response to the thermogenic properties of progesterone during the luteal phase, there is typically an increase in temperature of 0.4–0.8 °F. Most ovulatory cycles therefore show a biphasic pattern (Fig. 3), although some women with ovulatory cycles have difficulty identifying clear temperature shifts.

Over the counter urinary ovulation predictor kits and monitors may be used to identify the mid-cycle surge of luteinizing hormone occurring 1–2 days preceding ovulation. Urinary LH detection provides presumptive evidence of ovulation and may also assist couples in appropriately timing coitus. Urine LH assessment generally correlates with serum concentrations with reasonable accuracy, particularly when performed during the midday or evening [63]. On the other hand, some patients with ovulatory cycles still have difficulty recording positive LH surges whereas others may experience false positives [38].

Mid-luteal phase serum progesterone determinations provide an objective assessment of luteal function. Due to menstrual variability, serum progesterone determinations should be scheduled approximately 1 week prior to the expected ensuing menses. For patients with 28-day cycles, the test would ordinarily be obtained on cycle day 21. A progesterone concentration >3 ng/ml provides presumptive evidence of ovulation [64]. Importantly, serum progesterone determinations are best used as a qualitative rather than a quantitative test (demonstrating evidence that ovulation has occurred). Although levels >10 ng/ml are typically seen in the mid-luteal phase and reflect appropriate luteal function, issues related to cycle variability as well as pulsatile nature of progesterone secretion limit its utility as a determinant of luteal adequacy [65]. Some practitioners measure progesterone levels every other day in the luteal phase to better assess luteal phase adequacy (personal communication).

Endometrial biopsy with histologic analysis can infer an ovulatory cycle through the demonstration of secretory endometrial changes. In most cases, this evaluation will not be required to determine normal ovulatory function. The endometrium progresses through a sequence of changes in the secretory phase, the "dating" of which was previously used as a standard for the quality of luteal function and the diagnosis of luteal phase deficiency [66]. However, well-performed studies have demonstrated that histologic endometrial dating lacks both accuracy and reproducibility as well as fails to demonstrate differences between fertile and infertile women [67, 68].

5.2 Ovarian Reserve

A number of tests have been devised to indirectly assess ovarian reserve (i.e., reproductive potential as reflected by the number of remaining oocytes). Diminished ovarian reserve (DOR) occurs in women who still have regular menstrual cyclicity but who exhibit diminished gonadotropin responsiveness and lower fecundity potential than other women of similar age. This testing is of particular importance for women over the age of 35, those with a single ovary, those with a family history of early menopause, those with unexplained infertility, those with a prior exposure to chemotherapy, and those planning to undergo ART [69]. In addition to lowered gonadotropin responsiveness, women with DOR have a significantly lower chance to conceive through ART [70].

Ovarian reserve may be assessed by the determinations of basal FSH and estradiol early in the follicular phase (between cycle days 2 through 4). Serum FSH concentrations in the early follicular phase appear to rise several years prior to the menopause, the subtle rise of which is felt to likely indicate reduced secretion of inhibin and other inhibitory substances from the germ cell-depleted ovary. These subtle elevations are associated with poor responses to gonadotropin stimulation and a diminished chance for success after IVF and related therapies [71, 72]. Elevations in serum estradiol in the early follicular phase may also portend a poorer prognosis in some cases, [73, 74] although elevations in serum estradiol might reflect physiology other than lowered ovarian reserve in other instances.

The clomiphene challenge test was designed to evaluate fecundity potential prospectively [75]. In this test, patients have basal FSH and estradiol determinations in the early follicular phase (day 3) and then receive 100 mg of clomiphene citrate daily during days 5–9 of the cycle. Serum FSH is again repeated on cycle day 10. Elevated FSH levels on day 10 are a particularly concerning indicator of DOR and have higher sensitivity but lower specificity as compared to day-3 concentrations [76].

Antral follicle count (AFC) is the quantitation of the number of basal follicles (generally 2–10 mm) seen in both ovaries utilizing high-resolution transvaginal ultrasonography prior to treatment. The number of antral follicles has been noted to decrease with increasing age [14]. A number of reports have demonstrated a low antral follicle count (generally ≤4–6) to also be associated with poor response to ovarian stimulation and failure to achieve pregnancy after IVF [77].

Serum concentrations of antimullerian hormone (AMH), produced by granulosa cells, are gonadotropin independent and relatively constant throughout the menstrual cycle, making determinations reliable at any time [78]. Low levels of AMH (<1 ng/ml) have also been associated with poor responses to gonadotropins and lower embryo quality following IVF, although in general AMH is not considered a marker of embryo quality [79, 80].

5.3 Cervical Factor

Abnormalities of cervical mucus or mucus–sperm interaction are a likely rare cause of infertility. Examination of the cervical mucus following coitus (i.e., postcoital test) had been a widely performed procedure to evaluate both the quality of cervical mucus and mucus–sperm interaction. However, due to patient inconvenience, poor reproducibility, and its inaccuracy in predicting lowered fecundity and then change clinical management, the postcoital test is rarely currently utilized [81, 82].

5.4 Uterine Factor

Abnormalities of the uterine cavity can be detected by various means. Transvaginal ultrasonography provides excellent identification of uterine wall abnormalities, such as leiomyomas, and has moderate sensitivity in the identification of endometrial polyps. Hysterosalpingography (HSG) remains a relatively standard test for both the evaluation of the uterine cavity and fallopian tubes. This examination is able to relatively readily demonstrate significant endometrial polyps or submucous leiomyomata, intrauterine synechiae, and mullerian anomalies (i.e., septate, bicornuate, unicornuate uterus). However, HSG has only moderate sensitivity (50 %) and positive predictive value (28.6 %) for all polypoid lesions [83].

Sonohysterography involves imaging the uterine cavity with high-resolution transvaginal ultrasound following cannulation of the cervix and injecting the uterine cavity with a saline solution. Sonohysterography allows for a higher discernment of subtle lesions within the endometrial cavity, showing an overall sensitivity of 87 %, specificity of 91 %, positive predictive value of 92 %, and negative predictive value of 86 % for all uterine abnormalities [84].

Hysteroscopy remains the gold standard for evaluation of the uterine cavity, but due to its invasiveness and costs, it is usually reserved for further evaluation and treatment of identified intrauterine abnormalities.

5.5 Tubal Factor

HSG remains a relatively commonly performed evaluation test of the uterine cavity and fallopian tubes. HSG can reveal proximal or distal tubal occlusion, salpingitis isthmica nodosa, a pattern of diminished tubal rugae and possible tubal phimosis or peritubal adhesions (through the demonstration of delayed and loculated dye patency, respectively). HSG can be performed utilizing both oil- and water-soluble contrast media, although the former results in some obscuring of subtle findings and has some inherent risk for granuloma formation and embolism. Although it was generally presumed that oil-soluble media was associated with a higher fecundability potential post-procedure than water-soluble media, this advantage was not demonstrated in a systematic review [85]. HSG has a demonstrated sensitivity of only 60 % and specificity of 95 % as a test for tubal patency, with some instances of suspected tubal occlusion attributable to uterine or tubal "spasm" and not true blockage [86].

Laparoscopy with chromopertubation utilizing either methylene blue or indigo carmine dye can further be used to evaluate tubal patency or corroborate prior HSG findings. At the time of laparoscopy, restorative surgery such as lysis of peritubal adhesions, fimbrioplasty, or hysteroscopic tubal cannulation can be performed in order to reestablish tubal patency.

5.6 Peritoneal Factor

Although transvaginal ultrasonography may reveal endometriomas with high sensitivity and specificity, transvaginal ultrasonography often lacks the ability to characterize adhesions or early-stage endometriosis [87]. In these instances, the only currently available method to assess for peritoneal adhesions or early endometriosis is through diagnostic laparoscopy. Laparoscopy allows for the direct visual inspection of the peritoneal cavity and pelvic reproductive anatomy. On the other hand, it appears that the relative impact of minimal and mild endometriosis on fecundity is only modest and the number of cases that must be treated with laparoscopy in order to demonstrate a clinical advantage can be significant [88, 89]. Laparoscopy is most indicated for those patients with pelvic pain symptoms or risk factors, abnormal transvaginal ultrasonography or HSG, and no other clear indications for ART. Laparoscopy may also have a relatively high yield for patients with long-standing infertility (>3 years) which is otherwise unexplained [90].

Table 1
Lower limits of the reference values for semen analysis

Semen parameter	Reference value
Ejaculate volume	1.5 ml
Sperm concentration	15 million/ml
Total sperm number	39 million/ejaculate
Percent motility	40 %
Forward progressive motility	32 %
Normal morphology	4 % normal

Data from World Health Organization, 2010 [28]

5.7 Male Factor

The semen analysis remains the primary means of laboratory evaluation of the male factor contribution to infertility. Patients should be properly instructed for specimen collection, including a recommended abstinence interval of 2–7 days. Specimens should be collected by either masturbation in a sterile cup or intercourse utilizing a suitable collection condom. Although specimen collection is most optimal in the clinic, for those men who are uncomfortable collecting in this setting, specimens may be collected at home provided they are transported promptly, kept at least at room temperature in transit, and able to be evaluated in the laboratory within 1 h of collection. Guidelines for semen parameters have been established by the WHO (Table 1) [28]. The present criteria are based on a population study of almost 2,000 men from 8 countries whose partners conceived within the preceding 12 months [91]. The WHO established the lower limits of the normal range at approximately the fifth centiles of this fertile population, representing a sperm concentration of 15 million/ml, progressive motility of 32 %, and normal sperm morphology of 4 % utilizing "strict" criteria.

Sperm morphology utilizing "strict" criteria involves a quantitative analysis of sperm and was initially reported by Kruger et al. [92]. Laboratory technicians performing "strict" criteria require specialized training and a relatively high lab throughput in order to maintain proficiency. The finding of low sperm morphology by strict criteria was found to correlate with poor oocyte fertilization rates in vitro and represents a frequent indication for intracytoplasmic sperm injection (ICSI) for patients undergoing IVF. The management of patients with isolated low sperm morphology with otherwise normal semen parameters is less clear [93].

Other evaluation tests for the infertile male depend on the clinical circumstances and may include an endocrine evaluation (FSH and total testosterone, at minimum) for men with oligospermia, sexual dysfunction, or other findings suggesting an endocrinopathy. Transrectal or scrotal ultrasonography may be helpful in certain clinical circumstances to rule out ejaculatory duct obstruction or subtle varicocele, respectively. Antisperm antibodies may be considered for men with known risk factors, including trauma, torsion, orchitis, testicular surgery, and vasectomy.

Tests for sperm DNA integrity have also been developed, including the sperm chromatin structure assay (SCSA), terminal deoxynucleotidyl transferase-mediated dUTP—biotin end-labeling assay (TUNEL), modified alkali single-cell gel electrophoresis assay (COMET), and sperm chromatin dispersion test (SCD) [94–97]. Although numerous studies purport a relationship between sperm DNA integrity and pregnancy outcome in multiple contexts (natural conception, intrauterine insemination, ART), the current methods for assessing sperm DNA integrity do not appear to have sufficient reliability in the prediction of clinical outcomes to presently justify their routine use [98].

Certain infertile men should have genetic counseling and/or screening, including men with congenital bilateral absence of the vas deferens (CBAVD), who are presumed to have an abnormality of the cystic fibrosis transmembrane conductance regulator (CFFR) gene and a high proportion demonstrating an abnormality with currently available cystic fibrosis gene mutation panels [99]. Men with non-obstructive azoospermia and severe oligospermia (<5 million/ml) are at increased risk for karyotypic abnormalities and Y chromosome microdeletions and should be offered serum chromosome analysis and Y chromosome microdeletion testing [100, 101].

References

1. Martin JA, Hamilton BE, Ventura SJ, Osterman MJ, Kirmeyer S, Mathews TJ, Wilson EC (2011) Births: final data for 2009. Natl Vital Stat Rep 60(1):1–70
2. Martin JA, Hamilton BE, Ventura SJ, Osterman MJ, Wilson EC, Mathews TJ (2012) Births: final data for 2010. Natl Vital Stat Rep 61(1):1–71
3. U. S. Census Bureau (2012) Years of school completed by people 25 years and over, by age and sex: selected years 1940 to 2012. U.S. Dept. of Commerce Economics and Statistics Administration, U. S. Census Bureau, Washington, DC
4. Norton AJ, Miller LF (1992) Marriage, divorce, and remarriage in the 1990s. Curr Popul Rep Popul Charact (180):i-1–21
5. Copen CE, Daniels K, Vespa J, Mosher WD (2012) First marriages in the United States: data from the 2006-2010 National Survey of Family Growth. Nat Health Stat Report 49:1–21
6. Bramlett MD, Mosher WD (2002) Cohabitation, marriage, divorce and remarriage in the United States. National Center for Health Statistics. Vital Health Stat 23(22): 1–103
7. Guttmacher AF (1956) Factors affecting normal expectancy of conception. JAMA 161:855
8. Zinaman MJ, Clegg ED, Brown CC et al (1996) Estimates of human fertility and pregnancy loss. Fertil Steril 65:503–509
9. Wilcox AJ, Weinberg CR, Baird DD (1998) Post-ovulatory ageing of the human oocyte and embryo failure. Hum Reprod 13:394–397
10. Wilcox AJ, Weinberg CR, Baird DD (1995) Timing of sexual intercourse in relation to ovulation. N Engl J Med 333:1517–1521
11. Dunson DB, Baird DD, Wilcox AJ, Weinberg CR (1999) Day-specific probabilities of clinical pregnancy based on two studies with imperfect measures of ovulation. Hum Reprod 14:1835–1839

12. Gnoth C, Godehardt D, Godehardt E et al (2003) Time to pregnancy: results of the German prospective study and impact on the management of infertility. Hum Reprod 18:1959–1966
13. Faddy MJ, Gosden RG, Gougeon A et al (1992) Accelerated disappearance of ovarian follicles in midlife: implications for forecasting menopause. Hum Reprod 7:1342–1346
14. Gougeon A, Echochard R, Thalabard JC (1994) Age-related changes of the population of human ovarian follicles: increase in the disappearance rate of non-growing and early-growing follicles in aging women. Biol Reprod 50(3):653–663
15. Tietze C (1957) Reproductive span and rate of reproduction among Hutterite women. Fertil Steril 8:89–97
16. Federation CECOS, Schwartz D, Mayaux JM (1982) Female fecundity as a function of age: results from artificial insemination in 2193 nulliparous women with azoospermic husbands. N Engl J Med 306:404–406
17. Rosenwaks Z, Davis OK, Damario MA (1995) The role of maternal age in assisted reproduction. Hum Reprod 10(Suppl 1): 165–173
18. Munne S, Alikani M, Tomkin G et al (1995) Embryo morphology, developmental rates, and maternal age are correlated with chromosome abnormalities. Fertil Steril 64:382–391
19. Farr SL, Schieve LA, Jamieson DJ (2007) Pregnancy loss among pregnancies conceived through assisted reproductive technology, United States, 1999–2002. Am J Epidemiol 165:1380–1388
20. Eskenazi B, Wyrobek AJ, Sloter E et al (2003) The association of age and semen quality in healthy men. Hum Reprod 18:447–454
21. Paulson RJ, Milligan RC, Sokol RZ (2001) The lack of influence of age on male fertility. Am J Obstet Gynecol 184:818–824
22. Hassan MAM, Killick SR (2003) Effect of male age on fertility: evidence for the decline in male fertility with increasing age. Fertil Steril 79(Suppl 3):1520–1527
23. Kong A, Frigge ML, Masson G et al (2012) Rate of *de novo* mutations and the importance of father's age to disease risk. Nature 488:471–475
24. Macaluso M, Wright-Schnapp TJ, Chandra A et al (2010) A public health focus on infertility prevention, detection and management. Fertil Steril 93:16.e1–e10
25. Centers for Disease Control and Prevention (2007) Trends in reportable sexually transmitted diseases in the United States, 2006. U.S. Department of Health and Human Services, Centers for Disease Control and Prevention, Atlanta
26. Hillis SD, Wasserheit JN (1996) Screening for chlamydia—a key to the prevention of pelvic inflammatory disease. N Engl J Med 334:1399–1401
27. Scholes D, Stergachis A, Heidrich FE et al (1996) Prevention of pelvic inflammatory disease by screening for cervical chlamydial infection. N Engl J Med 334:1362–1366
28. World Health Organization (2010) WHO laboratory manual for the examination and processing of human semen, 5th edn. World Health Organization, Geneva
29. Levitas E, Lunenfeld E, Weiss N et al (2005) Relationship between the duration of sexual abstinence and semen quality: analysis of 9,489 semen samples. Fertil Steril 83:1680–1686
30. Lenzi A, Lombardo F, Salacone P et al (2003) Stress, sexual dysfunctions, and male infertility. J Endocrinol Invest 26(Suppl 3):72–76
31. Wilcox AJ, Dunson DB, Weinberg CR et al (2001) Likelihood of conception with a single act of intercourse: providing benchmark rates for assessment of post-coital contraceptives. Contraception 63:211–215
32. Dunson DB, Colombo B, Baird DD (2002) Changes with age in the level and duration of fertility in the menstrual cycle. Hum Reprod 17:1399–1403
33. Standford JB, Dunson DB (2007) Effects of sexual intercourse patterns in time to pregnancy studies. Am J Epidemiol 165: 1088–1095
34. Sievert LL, Dubois CA (2005) Validating signals of ovulation: do women who think they know, really know? Am J Hum Biol 17:310–320
35. Bigelow JL, Dunson DB, Stanford JB et al (2004) Mucus observations in the fertile window: a better predictor of conception than timing of intercourse. Hum Reprod 19:889–892
36. Scarpa B, Dunson DB, Colombo B (2006) Cervical mucus secretions on the day of intercourse: an accurate marker of highly fertile days. Eur J Obstet Gynecol Reprod Biol 125:72–78
37. Miller PB, Soules MR (1996) The usefulness of a urinary LH kit for ovulation prediction during menstrual cycles of normal women. Obstet Gynecol 87:13–17
38. McGovern PG, Myers ER, Silva S et al (2004) Absence of secretory endometrium after false-positive home urine luteinizing hormone testing. Fertil Steril 82:1273–1277

39. Robinson JE, Wakelin M, Ellis JE (2007) Increased pregnancy rate with use of the Clearblue Easy fertility monitor. Fertil Steril 87:329–334
40. Clark AM, Thornley B, Tomlinson L et al (1998) Weight loss in obese infertile women results in improvement in reproductive outcome for all forms of fertility treatment. Hum Reprod 13:1502–1505
41. Choy CM, Lam CW, Cheung LT et al (2002) Infertility, blood mercury concentrations and dietary seafood consumption: a case-control study. BJOG 109:1121–1125
42. De-Regil LM, Fernandez-Gaxiola AC, Dowswell T, et al (2010) Effects and safety of periconceptional folate supplementation for preventing birth defects. Cochrane Database Syst Rev (10):CD007950
43. Stillman RJ, Rosenberg MJ, Sachs BP (1986) Smoking and reproduction. Fertil Steril 46:545–566
44. Ness RB, Grisso JA, Hirschinger N et al (1999) Cocaine and tobacco use and the risk of spontaneous abortion. N Engl J Med 340: 333–339
45. Winter E, Wang J, Davies MJ, Norman R (2002) Early pregnancy loss following assisted reproductive technology treatment. Hum Reprod 17:3220–3223
46. Augood C, Duckitt K, Templeton AA (1998) Smoking and female infertility: a systematic review and meta-analysis. Hum Reprod 13: 1532–1539
47. Adena MA, Gallagher HG (1982) Cigarette smoking and the age at menopause. Ann Hum Biol 9:121–130
48. Eggert J, Theobald H, Engfeldt P (2004) Effects of alcohol consumption on female fertility during an 18-year period. Fertil Steril 81:379–383
49. Tolstrup JS, Kjaer SK, Holst C et al (2003) Alcohol use as a predictor for infertility in a representative population of Danish women. Acta Obstet Gynecol Scand 82:744–749
50. Rossi BV, Berry KF, Hornstein MD et al (2011) Effect of alcohol consumption on in vitro fertilization. Obstet Gynecol 117:136–142
51. Bolumar F, Olsen J, Rebagliato M, Bisanti L (1997) Caffeine intake and delayed conception: a European multicenter study on infertility and subfertility. Am J Epidemiol 15:324–334
52. Signorello LB, McLaughlin JK (2004) Maternal caffeine consumption and spontaneous abortion: a review of the epidemiologic evidence. Epidemiology 15:229–239
53. Practice Committee of the American Society for Reproductive Medicine (2013) Definitions of infertility and recurrent pregnancy loss. Fertil Steril 99:63
54. Hull MGR, Glazener CMA, Kelly NT et al (1985) Population study of causes, treatment, and outcome of infertility. BMJ 291: 1693–1697
55. Snick HKA, Snick TS, Evers JLH, Collins JA (1997) The spontaneous pregnancy prognosis in untreated subfertile couples: the Walcheren primary care study. Hum Reprod 12: 1582–1588
56. Chandra A, Stephen EH (2010) Infertility service use among U.S. women: 1995 and 2002. Fertil Steril 93:725–736
57. Collins JA, Wrixon W, Janes LB, Wilson EH (1983) Treatment-independent pregnancy among infertile couples. N Engl J Med 309: 1201–1206
58. Collins JA, Burrows EA, Wilan AR (1995) The prognosis for live birth among untreated infertile couples. Fertil Steril 64:22–28
59. Steptoe PC, Edwards RG (1978) Birth after the reimplantation of a human embryo. Lancet 2:366
60. Centers for Disease Control and Prevention (2012) Assisted reproductive technology surveillance—United States, 2009. MMWR 61(SS-7):1–24
61. Forti G, Krausz C (1998) Evaluation and treatment of the infertile couple. J Clin Endocrinol Metab 83:4177–4188
62. McCarthy JJ Jr, Rockette HE (1986) Prediction of ovulation with basal body temperature. J Reprod Med 31:742–747
63. Luciano AA, Peluso J, Koch EI et al (1990) Temporal relationship and reliability of the clinical, hormonal, and ultrasonographic indices of ovulation in infertile women. Obstet Gynecol 75:412–416
64. Wathen NC, Perry L, Lilford RJ, Chard T (1984) Interpretation of single progesterone measurement in diagnosis of anovulation and defective luteal phase: observations on analysis of the normal range. Br Med J 288:7–9
65. Filicori M, Butler JP, Crowley WF Jr (1984) Neuroendocrine regulation of the corpus luteum in the human. Evidence for pulsatile progesterone secretion. J Clin Invest 73: 1638–1647
66. Noyes RW, Hertig AT, Rock J (1950) Dating the endometrial biopsy. Fertil Steril 1:3–25
67. Murray MJ, Meyer WR, Zaino RJ et al (2004) A critical analysis of the accuracy, reproducibility, and clinical utility of histologic endometrial

dating in fertile women. Fertil Steril 81: 1333–1343

68. Coutifaris C, Myers ER, Guzick DS et al (2004) Histological dating of timed endometrial biopsy tissue is not related to fertility status. Fertil Steril 82:1264–1272
69. Sharara FI, Scott RT Jr, Seifer DB (1998) The detection of diminished ovarian reserve in infertile women. Am J Obstet Gynecol 179: 804–812
70. Broekmans RJ, Kwee J, Hendriks DJ et al (2006) A systematic review of tests predicting ovarian reserve and IVF outcome. Hum Reprod Update 12:685–718
71. Muasher SJ, Oehninger S, Simonetti S et al (1988) The value of basal and/or stimulated serum gonadotropin levels in prediction of stimulation response and *in vitro* fertilization outcome. Fertil Steril 50:298–307
72. Toner JP, Philput C, Jones GS, Muasher SJ (1991) Basal follicle stimulating hormone (FSH) level is a better predictor of *in vitro* fertilization (IVF) performance than age. Fertil Steril 55:784–791
73. Licciardi FL, Liu HC, Rosenwaks Z (1995) Day 3 estradiol serum concentrations as prognosticator of ovarian stimulation response and pregnancy outcome in patients undergoing in vitro fertilization. Fertil Steril 64:1136–1140
74. Smotrich DB, Widra EA, Gindoff PR et al (1995) Prognostic value of day 3 estradiol on in vitro fertilization outcome. Fertil Steril 64:1136–1140
75. Navot D, Rosenwaks Z, Margalioth EJ (1987) Prognostic assessment of female fecundity. Lancet 2(8560):645–647
76. Hendriks DJ, Mol BW, Bancsi LF et al (2006) The clomiphene challenge test for the prediction of poor ovarian response and non-pregnancy in patients undergoing in vitro fertilization: a systematic review. Fertil Steril 86:807–818
77. Hendriks DJ, Mol BW, Bancsi LF et al (2005) Antral follicle count in the prediction of poor ovarian response and pregnancy after in vitro fertilization: a meta-analysis and comparison with basal follicle-stimulating hormone level. Fertil Steril 83:291–301
78. Hehenkamp WJ, Looman CW, Themmen AP et al (2006) Anti-mullerian hormone levels in the spontaneous menstrual cycle do not show substantial fluctuation. J Clin Endocrinol Metab 91:4057–4063
79. Muttukrishna S, McGarrigle H, Wakim R et al (2004) Antral follicle count, anti-mullerian hormone and inhibin B: predictors of ovarian response in assisted reproductive technology? BJOG 111:1248–1253
80. Silberstein T, MacLaughlin DT, Shai I et al (2006) Mullerian inhibiting substance levels at the time of hCG administration in IVF cycles predict both ovarian reserve and embryo morphology. Hum Reprod 21:159–163
81. Griffith CS, Grimes DA (1990) The validity of the postcoital test. Am J Obstet Gynecol 162:615–620
82. Oei SG, Helmerhorst FM, Bloemenkamp KW et al (1998) Effectiveness of the post-coital test: randomized controlled trial. BMJ 317:502–505
83. Soares SR, Barbosa dos Reis MM, Camargos AF (2000) Diagnostic accuracy of sonohysterography, transvaginal sonography, and hysterosalpingography in patients with uterine cavity diseases. Fertil Steril 73:406–411
84. Schwarzler P, Concin H, Bosch H et al (1998) An evaluation of sonohysterography and diagnostic hysteroscopy for the assessment of intrauterine pathology. Ultrasound Obstet Gynecol 11:337–342
85. Luttjeboer F, Harada T, Hughes E et al (2007) Tubal flushing for subfertility. Cochrane Database Syst Rev 3:CD003718
86. Evers JLH, Land JA, Mol BW (2003) Evidence-based medicine for diagnostic questions. Semin Reprod Med 21:9–15
87. Ubaldi F, Wisanto A, Camus M et al (1998) The role of transvaginal ultrasonography in the detection of pelvic pathologies in the infertility workup. Hum Reprod 13: 330–333
88. Marcoux S, Maheux R, Berube S, Canadian Collaborative Group on Endometriosis (1997) Laparoscopic surgery in infertile women with minimal or mild endometriosis. N Engl J Med 337:217–222
89. Jacobson TZ, Duffy JM, Barlow D et al (2010) Laparoscopic surgery for subfertility associated with endometriosis. Cochrane Database Syst Rev 20(1):CD001398
90. Practice Committee of the American Society for Reproductive Medicine (2012) Diagnostic evaluation of the infertile female: a committee opinion. Fertil Steril 98:302–307
91. Cooper TG, Noonan E, von Eckardstein S et al (2010) World Health Organization reference values for human semen characteristics. Hum Reprod Update 16:231–245
92. Kruger TF, Acosta AA, Simmons KF et al (1988) Predictive value of abnormal sperm morphology in in vitro fertilization. Fertil Steril 49:112–117

93. Keegan BR, Barton S, Sanchez X et al (2007) Isolated teratozoospermia does not affect in vitro fertilization outcome and is not an indication for intracytoplasmic sperm injection. Fertil Steril 88:1583–1588
94. Evenson DP, Jost LK, Marshall D et al (1999) Utility of the sperm chromatin assay as a diagnostic and prognostic tool in the human fertility clinic. Hum Reprod 14:1039–1049
95. Sun JG, Jurisicova A, Casper RF (1997) Detection of deoxyribonucleic acid fragmentation in human sperm: correlation with fertilization in vitro. Biol Reprod 56: 602–607
96. Hughes CM, Lewis SEM, McKelvey-Martin VJ (1996) A comparison of baseline and induced DNA damage in human spermatozoa from fertile and infertile men, using a modified comet assay. Mol Hum Reprod 2: 613–619
97. Fernandez JL, Muriel L, Rivero MT et al (2003) The sperm chromatin dispersion test: a simple method for the determination of sperm DNA fragmentation. J Androl 24:59–66
98. Practice Committee of the American Society for Reproductive Medicine (2013) The clinical utility of sperm DNA integrity testing: a guideline. Fertil Steril. pii: S0015–0282(13)00011–3. doi: 19.1016/j.fertnstert.2012.12.049
99. Anguiano A, Oates RD, Amos JA et al (1992) Congenital bilateral absence of the vas deferens. A primarily genital form of cystic fibrosis. JAMA 267:1794–1797
100. Van Assche E, Bonduelle M, Tournaye H et al (1996) Cytogenetics of infertile men. Hum Reprod 11(Suppl 4):1–26
101. Pryor JL, Kent-First M, Muallem A et al (1997) Microdeletions in the Y chromosome of infertile men. N Engl J Med 336:534–539

Chapter 2

Genetics of Male Fertility

Yi-Nan Lin and Martin M. Matzuk

Abstract

Early in embryogenesis, cells that are destined to become germ cells take on a different destiny from other cells in the embryo. The germ cells are not programmed to perform "vital" functions but to perpetuate the species through the transfer of genetic materials to the next generation. To fulfill their destiny, male germ cells undergo meiosis and extensive morphogenesis that transforms the round-shaped cells into freely motile sperm propelled by a beating flagellum to seek out their missing half. Apparently, extra genes and additional regulatory mechanisms are required to achieve all these unique features, and an estimated 11 % of genes are involved in fertility in Drosophila (Hackstein et al., Trends Genet 16(12):565–572, 2000). If comparative numbers of male fertility genes are needed in mammals, extra risks of male fertility problems are associated with disruptive mutations in those genes. Among human male infertility cases, approximately 22 % were classified as "idiopathic," a term used to describe diseases of unknown causes, with idiopathic oligozoospermia being the most common semen abnormality (11.2 %) (Comhaire et al., Int J Androl (Suppl 7):1–53, 1987). "Idiopathic" is a widely used adjective that is used to reflect our lack of understanding of the genetics of male fertility. Fortunately, after more than two decades of phenotypic studies using knockout mice and identifying genes disrupted in spontaneous mutant mice, we have unveiled new and unexpected aspects of crucial gene functions for fertility. Other efforts to categorize genes involved in male fertility in mammals have suggested a total of 1,188 genes (Hermo et al., Microsc Res Tech 73(4):241–494, 2010). Although intracytoplasmic sperm injection (ICSI) can be used to bypass many fertilization obstacles to achieve fertilization with only a few extracted sperm, the widespread use of ICSI without proper knowledge for genetic testing and counseling could still potentially propagate pleiotropic gene mutations associated with male infertility and other genetic diseases (Alukal and Lamb, Urol Clin North Am 35(2):277–288, 2008). In this chapter, we give a brief account of major events during the development of male germ cells and focus on the functions of several crucial genes that have been studied in mutant mouse models and are potential causes of human male infertility.

Key words Fertility, Sperm, Germ cells, Male fertility, Intracytoplasmic sperm injection

Yi-Nan Lin and Martin M. Matzuk have equally contributed to this work.

Zev Rosenwaks and Paul M. Wassarman (eds.), *Human Fertility: Methods and Protocols*, Methods in Molecular Biology, vol. 1154, DOI 10.1007/978-1-4939-0659-8_2, © Springer Science+Business Media New York 2014

1 Primordial Germ Cell Specification

The development of male germ cells initiates in the early embryo with the specification of primordial germ cells (PGCs) that are detected as alkaline phosphatase-positive cells at the base of the yolk sac before formation of the allantois by embryonic day 7.25 in mice (reviewed in ref. [1]). The critical roles of transforming growth factor-β (TGF-β) superfamily signaling pathways in regulating the development of PGCs have been shown in knockout mice lacking bone morphogenetic protein-4 (BMP-4) or BMP-8b or the downstream signaling proteins, SMAD1 and SMAD5 [2–5]. At midgestation, the PGCs start migrating along the hind gut to reach the genital ridge. Although the involvement of several adhesion molecules or G-protein-coupled receptor–ligand pairs in PGC migration has been studied in flies and zebra fishes (reviewed in ref. [6]), the factors required for PGC migration in mammals are less clear. The cytokine ligand–receptor pair of Kit ligand (KITL) and Kit receptor (KIT) were originally identified as two spontaneous mutant loci, steel and W, respectively, in mice for the infertility in both sexes caused by PGC loss [7, 8], and their binding is required specifically for PGC motility [9]. Additionally, two transcriptional repressors, BLIMP1/PRDM1 and PRDM14, may play critical roles in regulating proliferation and migration while repressing the somatic differentiation program in PGCs [10–12]. A mutation in the putative RNA-binding protein gene, dead-end homolog (*Dnd1*), also depletes PGCs [13, 14]. At this early embryonic stage, PGCs are still pluripotent.

2 Sex Determination

After the PGCs enter the genital ridge, the major gene implicated in human sex determination is sex-determining region Y (*Sry*), the male-determining gene located on the short arm of Y chromosome. The earliest event of male sex determination is the expression of SRY in the pre-Sertoli cells of bipotential gonads. SRY upregulates Sry-related HMG box-containing gene 9 (*Sox9*) expression by binding to multiple gonad-specific enhancer elements in the *Sox9* promoter [15]. It was shown that haploinsufficiency of SOX9 causes 46,XY male-to-female sex reversal [16, 17], and activating mutations of SOX9 result in 46,XX,SRY-negative female-to-male sex reversal [18–20]. During testis differentiation, SOX9 upregulates anti-Müllerian hormone (AMH) to act on its receptor, AMHR2, to induce the regression of the Müllerian duct, whereas testosterone drives the development of the Wolffian duct [21]. Additionally, SRY or SOX9 may act to downregulate the ovarian pathway by suppressing RSPO1 in early development [22] and also suppressing the levels of β catenin directly [1]. Surprisingly, another analysis of *Sox3* transgenic mice and three 46,XX males showed that ectopic expression of SOX3, the closest homolog of SRY [23], in the bipotential gonad can lead to complete sex reversal as XX males [24], suggesting that SOX3 and SRY are functionally interchangeable in sex determination. Future studies on the regulatory roles of SOX3 on SOX9 and other downstream sex determination effectors and more detailed spatiotemporal expression pattern comparison between SRY and SOX3 will help us elucidate the reason why the X-linked SOX3 itself cannot initiate male development.

3 Hypothalamic–Pituitary–Gonadal Axis

Masculinization during fetal development is dependent on the secretion of AMH and androgens by the fetal testis. At the onset of puberty in both sexes, the well-established hypothalamic–pituitary–gonadal (HPG) axis works as follows: the hypothalamus produces pulsatile gonadotropin-releasing hormone (GnRH) secretion, the gonadotropes in the anterior pituitary respond by secreting luteinizing hormone (LH) and follicle-stimulating hormone (FSH), and the gonads are activated to produce estrogen and testosterone. In males, FSH activates Sertoli cells and LH stimulates the steroidogenesis in the Leydig cells, resulting in testosterone production that signals back to the pituitary and hypothalamus. However, the signals that initiate the onset of puberty were only discovered more recently. The deficiency of kisspeptin (KISS1) and its receptor KISS1R/GPR54 both disrupted the axis causing hypogonadotropic hypogonadism, and the exogenous kisspeptin injection was able to restore robust LH secretion in *Kiss1* knockout mice, proving that the axis is still intact [25, 26]. The levels of *Gpr54* mRNA in GnRH neurons are similar in both juvenile and adult mice, but the expression of *Kiss1* mRNA increased dramatically during the transition from juvenile to adult life in the anteroventral periventricular nucleus [27]. Although an additional feedback circuitry of estrogen on KISS1 neurons for preovulatory LH surge has been described [28], there is no similar report on the involvement of androgen in feedback regulation yet.

4 Meiotic Defects Usually Cause Azoospermia or Severe Oligozoospermia

In males, spermatogenesis can be divided into three specific phases: proliferation, meiosis, and spermiogenesis. One recent elegant imaging study showed that in mouse, A_{single} and A_{paired} spermatogonia are the GDNF receptor-positive spermatogonial stem cells, whereas $A_{aligned}$ spermatogonia are formed by mitosis without cytokinesis into interconnected NGN3-positive cysts of 4–16 cells linked through intercellular bridges [29]. More differentiated A_1–A_4 spermatogonia are dependent on membrane-bound KITL from Sertoli cells [30, 31] and subsequently undergo several mitotic and two meiotic divisions to generate haploid spermatids with diverse combinations of genes on independently assorted chromosomes. After the puberty signal is relayed through the hypothalamic–pituitary–gonadal axis into elevated blood testosterone, meiosis in males is initiated to generate sperm. Unlike mitosis, meiosis consists of two cell divisions following a single-DNA replication phase, thereby generating four haploid cells (i.e., round spermatids in the male). Crucial functions during meiosis I include those involved in the formation of synaptonemal complex and the DNA recombination process that helped to form and resolve the chiasmata for the cosegregation of homologous chromosomes (reviewed in ref. [32] with updated references below). Initially, telomere clustering on the nuclear membrane, which requires SUN1/UNC84A, assists homologous chromosome alignment for their subsequent pairing. The axial elements of synaptonemal complex begin to form on chromosomes from SYCP2, SYCP3, and meiosis-specific cohesion proteins, like REC8, STAG3, and SMC1β. After homologous pairing, the axial elements come together to form lateral elements. Homologous recombination is initiated by the double-strand breaks formed by the activity of SPO11 to recruit recA-homologs, RAD51 and DMC1, for strand invasion to form double-Holliday junctions at recombination nodules for meiotic crossovers. During this period, TEX11 binds to SYCP2 to promote synapsis and crossover [33], and TEX15 assists the recruitment of RAD51 and DMC1 [34]. Later, SYCP1 is loaded onto the central region of synapsed chromosomes along with SYCE1, SYCE2, and TEX12. Meiotic MutS homologs, MSH4 and MSH5, form complex at double-Holiday junctions and subsequently recruit MutL homologs, MLH1 and MLH3, to complete homologous recombination. Homologous chromosomes undergo desynapsis and condense after the completion of recombination, and at least one chiasma (crossover site) per homologous chromosome pair is required for correct segregation at anaphase I. Among the knockout studies that unveiled critical functions of cell cycle genes, the viable but infertile phenotypes caused by the disruption of cell cycle kinase, CDK2, came out as a total surprise [35], and a recent follow-up study showed that CDK2 is required for multiple meiotic functions [36]. The incomplete chromosome pairing, nonhomologous synapsis, and unrepaired double-strand breaks found in *Cdk2* knockout spermatocytes suggest that while its roles during mitosis may be replaced by other mitotic CDK–cyclin pairs, CDK2 has irreplaceable roles as a master meiotic regulator. Interestingly, disruption of some of those genes caused infertility in males but only subfertility or no fertility phenotypes in females, reflecting some substantial differences for meiosis between the two sexes. Male-specific infertility has been reported in knockouts of *Sycp2* [37], *Sycp3* [38], *Tex11* [33], and *Tex15* [34]. The dimorphism for the male-specific infertility phenotypes has been attributed to a more stringent control mechanism during meiosis in males [39].

5 Sperm Morphogenesis Can Go Awry

Teratozoospermia in patients is diagnosed based on the prevalence of misshapen sperm [40]. Although paddle-shaped human sperm show a different range of abnormal shapes when compared to hook-shaped mouse sperm, several mutant mice models with abnormally shaped sperm have been studied in details with electron microscopy to provide insights into teratozoospermia. The acrosome–acroplaxome–manchette (AAM) complex was coined by Kierszenbaum and Tres to describe the sperm head shaping mechanism in the elongating spermatids [41]. Golgi-derived proacrosomal vesicles need to attach and fuse along the actin–keratin-containing acroplaxome to anchor the developing acrosome to the nuclear membrane on one end and the longitudinally pulling manchette microtubules need to attach to the sperm nuclear membrane on the other end. The tight association between acrosome and the Sertoli cell thus allows the pulling forces from the manchette to be transduced correctly into forces that stretches and flattens sperm nucleus into shape. Disruption of the acrosome–acroplaxome–manchette complex and other functionally related structures is likely to abrogate the shaping force, resulting in misshapen sperm head. Because proacrosomal vesicles are derived from the endoplasmic reticulum (ER) and Golgi apparatus, defects in ER or Golgi functions also cause acrosome dysfunction. To date, the disruption of *Hrb* [42], *Gopc* [43], *Csnk2a2* [44], *Zpbp1* and *Zpbp2* [45], *Pick1* [46], *Hook1* [47, 48], and *RIM-BP3* [49] in knockout mice have generated abnormally shaped sperm. Because the HRB protein is an important factor for proacrosomal vesicle fusion, proacrosomal vesicles fail to fuse into an acrosome in *Hrb* knockout spermatids [42]. Zona pellucida-binding protein 1 (ZPBP1), in contrast, functions as a critical acrosomal matrix protein for acrosome compaction, and its absence results in a dilated acrosome easily undergoing fragmentation during sperm elongation [45]. Although the functions of GOPC, a Golgi-associated protein, and CSNK2A2, a subunit of casein kinase 2, are largely unknown, the recent identification of PICK1 as a binding partner of both GOPC and CSNK2A2 helped to elucidate their collective involvement in the trafficking of proacrosomal vesicles from the Golgi apparatus to the acrosome [50]. HOOK1 was originally identified as the mutated gene in abnormal spermatozoon head shape (*azh*) mutant mice [48], and it is involved in attaching the manchette microtubules to the nuclear membrane [47]. RIM-BP3, the interacting partner of HOOK1, is also a manchette-associated protein, and the inability of RIM-BP3 to interact with the mutant HOOK1 protein found in *azh* mutant mice may help to explain its structural roles in the manchette for sperm head morphogenesis [49]. A recent study that used *Vasa-Cre* to knock out *Hsp90β1*, an ER chaperone gene, specifically in male germ cells also disrupted acrosome formation and caused a similar teratozoospermic infertility phenotype [51]. These observations strengthen the important roles of acrosome–acroplaxome–manchette complex in shaping the sperm head.

6 Sperm Motility and Hyperactivated Motility Are Required for Male Fertility In Vivo

Halfway during spermiogenesis, the formation of the acrosome marks the anterior end of the sperm head while the centrosomes move to the anterior end of sperm nucleus to function as the basal bodies for flagellar growth. The nine triplets of centriolar microtubules in the basal bodies then extend into the nine doublets of microtubules with the associated force-generating dynein arms plus two central microtubules (central apparatus) in the axoneme, in which many structural proteins are shared between flagella and cilia [52]. It is not surprising that disruptions of DNAHC1 (an inner arm dynein heavy-chain protein [53]), TCTE3 (a putative outer dynein arm light-chain protein [54]), TEKTIN-2 (required for the integrity of the inner dynein arm [55]), SPAG6 (associated with the central apparatus [56]), and PCDP1 (likely also associated with the central apparatus [57]) impair fertility via flagellar dysfunctions and cause phenotypes related to primary ciliary dyskinesia (PCD) in mice. The reduction of sperm motility in asthenozoospermia has also been attributed to genes involved in the formation of functional midpiece from tightly wrapped mitochondria that powers the flagellar beating. Recently, two knockout studies of murine glutathione peroxidase 4 (GPX4, a selenoprotein) that utilized either the deletion of mitochondria-specific leader sequence [58] or a spermatocyte-specific knockout strategy [59] established the links between selenium and essential mitochondrial functions in male fertility. Besides mitochondria, the sperm also rely heavily on the glycolytic pathway to power the flagella and to achieve hyperactivated motility, and knockouts of sperm-specific glyceraldehyde-3-phosphate dehydrogenase (GAPDHS [60]), lactate dehydrogenase C (LDHC [61]), and phosphoglycerate kinase 2 (PGK2 [62]) have resulted in male infertility. Flagellar accessory structures including outer dense fibers and fibrous sheath are also important for structural integrity as shown in the infertile *Akap4* knockout male mice [63], although the preimplantation lethality caused by disruption of *Odf2* (potentially via disruption of the essential ciliary functions of its alternative spliced product, Cenexin [64]) prevented the analysis of its roles in the outer dense fiber [65]. Another instance of a sperm motility defect is caused by the failure to remove the excess residual cytoplasm to achieve a streamlined sperm shape and to fully extend the flagellum, such as the *Spem1*-deficient sperm with bent neck and tail trapped in the cytoplasm remnants [66]; however, the roles of SPEM1 in cytoplasm removal during spermiation remain to be determined. In addition to the requirement of structural components and energy for sperm, it was known that mammalian sperm require capacitation for hyperactivated motility and acrosome reaction prior to fertilization. Although many of the molecular mechanisms involved in capacitation remain to be elucidated [67], the interplay among bicarbonate ions, cyclic AMP (cAMP) regulatory pathway activation, and elevation of cytosolic calcium levels through membrane ion channels have been crucial to capacitation [68]. For the cAMP-dependent sperm functions, the bicarbonate- and calcium-responsive soluble adenylate cyclase (ADCY10/sAC) functions as the major source of cAMP required for activating protein tyrosine phosphorylation and hyperactivated motility during capacitation [69, 70]. Subsequently, the elevated cytosolic cAMP levels activate the alkalization-activated potassium channel, SLO3/KSper [71, 72], to cause membrane hyperpolarization that increases calcium influx through the sperm-specific alkalization-activated calcium-selective CatSper channel (composed of a tetramer of CatSper subunits 1–4 and auxiliary subunits, CatSper β, γ, δ [73–81]). The increased cytosolic calcium levels then hyperactivate sperm motility. Interestingly, progesterone [82] and prostaglandin E1 [83] have long been reported to act on some uncharacterized sperm membrane receptors to activate calcium influx, and two recent reports identified CatSper channel as the target non-genomic receptor for both progesterone and prostaglandin E1, further demystifying their activating effects in capacitated sperm [82, 83]. These discoveries will surely help to unveil more previously unresolved aspects of sperm capacitation.

7 Sperm Passage Through the Epididymis and Vas Deferens

Once spermiogenesis is completed and the sperm are flushed out of the seminiferous tubules by the secretion of Sertoli cells, they undergo maturation in the epididymis and are transported through the vas deferens for ejaculation. Obstruction of the passage results in the absence of sperm in the ejaculate. One major form of obstructive azoospermia is the congenital bilateral absence of the vas deferens (CBAVD), and CBAVD is frequently associated with human mutations in cystic fibrosis transmembrane conductance regulator (CFTR), a cAMP-regulated chloride ion channel [84, 85]. Studies of CFTR mutations in patients show that the development of vas deferens is sensitive to CFTR mutations [86]. Examination of two different *Cftr* mutant mouse models shows age-dependent obstructive azoospermia caused by a collapse of the lumen of the vas deferens [87]. Recently, tissue-specific ablation of androgen receptor from the caput epididymal epithelium in two studies shows that the inactivation of androgen receptor causes epithelial hypoplasia and gradual efferent duct obstruction; this defect results in fluid back pressure that disrupts the seminiferous epithelium and causes male infertility [88, 89]. These mouse models could be used for future clinical studies on the involvement of androgen action in the non-CFTR forms of obstructive azoospermia.

8 New Technologies for Male Fertility Research and Diagnosis

With several emerging new technologies in the postgenomic era, the knowledge of more than 1,000 male fertility-related genes could be used to advance our research and diagnosis of male fertility. In the past few years, genome-wide association studies (GWAS or GWA studies) have been widely used to identify loci associated with diseases or traits. Making use of the catalogued single-nucleotide polymorphisms (SNPs) in the genome, SNP microarray-based GWA studies are used to find associations between SNP-tagged genomic blocks (haplotypes) with diseases in question. Although the effectiveness of these studies is limited by the number of participants, a pilot GWA study of 92 azoospermia and severe oligozoospermia patients has utilized a candidate gene approach to identify some potential genes associated with spermatogenesis dysfunctions [90, 91]. Previously, many studies have attempted to identify causal mutations in candidate genes derived from male sterile knockout mice models by resequencing the corresponding genomic sequences of human infertile patients in order to detect potentially disruptive nucleotide sequence changes. However, due to the technical limitations that permit the examination of only a handful of genes, many have failed to identify specific mutations for future diagnostic applications. Therefore, the ability to generate sequencing data effectively and selectively from interested targets by combining hybridization capture and next-generation sequencing (NGS or massive parallel sequencing) technology (reviewed in ref. [92]) has the potential to revolutionize infertility research and diagnosis. With the continuous reduction of error rates and cost in NGS, exome sequencing (currently estimated to cost about \$3,000) may be the most effective way to generate sequencing data of most of the known genes for familial pedigree analysis to identify disease genes [93]. The detectable mutations by NGS range from point mutations and small indels (insertions/deletions) to copy number alterations [94]. When combining NGS with RNA sequencing (RNA-seq, which sequences cDNAs reverse-transcribed from RNA) or direct RNA sequencing (reviewed in ref. [95]) to analyze the transcriptomes, it is possible to detect splicing disorders and gene translocation events, making it possible to directly and noninvasively examine the spermatozoal RNA pool for molecular defects in sperm [96, 97]. With these new technologies, we could accelerate our research by shifting the focus back to understanding naturally occurring mutations that impair fertility in men. The identification of hypomorphic alleles for crucial fertility genes in humans could complement our functional characterization of null alleles generated in knockout mice (Table 1) to better understand the whole spectrum of disease phenotypes. With better understanding in male fertility, we should have more effective treatment for male infertility.

Table 1
List of genes involved in the different developmental stages of male fertility in this chapter

Primordial germ cell specification	*Bmp4*	*Bmp8b*	*Dnd1*	*Kit*	*Kitl*	*Prdm1*	*Prdm14*	*Smad1*	*Smad5*											
Sex determination	*Amh*	*Amhr2*	*Rspo1*	*Sox3*	*Sox9*	*Sry*														
HPG axis	*Fshb*	*Gnrh1*	*Kiss1*	*Kiss1r*	*Lhb*															
Meiosis	*Cdk2*	*Dmc1*	*Mlh1*	*Mlh3*	*Msh4*	*Msh5*	*Rad51c*	*Rec8*	*Smc1b*	*Spo11*	***Stag3***	***Sun1***	*Sycp1*	*Sycp2*	*Sycp3*	***Syce1***	***Syce2***	*Tex11*	***Tex12***	*Tex15*
AAM complex and morphogenesis	*Csnk2a2*	*Gopc*	*Hook1*	*Hrb*	***Pick1***	***Rimbp3***	*Zpbp*													
Motility and hyperactivated motility	*Adcy10*	*Akap4*	*Catsper1*	*Catsper2*	*Catsper3*	*Catsper4*	***Catsperb***	***Catsperg***	***Catsperd/ Tmem146***	*Dnahc1*	*Gapdhs*	***Gpx4***	*Ldhc*	*Odf2*	*Pcdp1/ Gm101*	***Pgk2***	***Slo3***	*Spag6*	*Spem1*	***Tcte3***	*Tekt2*
Obstructive azoospermia	*Ar*	***Cftr***																		

Genes that were not included in the list of male fertility mouse models (Fig. 5) in the *Nature Medicine* review article by Matzuk and Lamb [98] are underlined in this table

References

1. Edson MA, Nagaraja AK, Matzuk MM (2009) The mammalian ovary from genesis to revelation. Endocr Rev 30(6):624–712
2. Lawson KA et al (1999) Bmp4 is required for the generation of primordial germ cells in the mouse embryo. Genes Dev 13(4):424–436
3. Ying Y et al (2000) Requirement of Bmp8b for the generation of primordial germ cells in the mouse. Mol Endocrinol 14(7):1053–1063
4. Chang H, Matzuk MM (2001) Smad5 is required for mouse primordial germ cell development. Mech Dev 104(1–2):61–67
5. Tremblay KD, Dunn NR, Robertson EJ (2001) Mouse embryos lacking Smad1 signals display defects in extra-embryonic tissues and germ cell formation. Development 128(18):3609–3621
6. Richardson BE, Lehmann R (2010) Mechanisms guiding primordial germ cell migration: strategies from different organisms. Nat Rev Mol Cell Biol 11(1):37–49
7. Manova K et al (1990) Gonadal expression of c-kit encoded at the W locus of the mouse. Development 110(4):1057–1069
8. Matsui Y, Zsebo KM, Hogan BL (1990) Embryonic expression of a haematopoietic growth factor encoded by the Sl locus and the ligand for c-kit. Nature 347(6294):667–669
9. Gu Y et al (2009) Steel factor controls primordial germ cell survival and motility from the time of their specification in the allantois, and provides a continuous niche throughout their migration. Development 136(8):1295–1303
10. Ohinata Y et al (2005) Blimp1 is a critical determinant of the germ cell lineage in mice. Nature 436(7048):207–213
11. Kurimoto K et al (2008) Specification of the germ cell lineage in mice: a process orchestrated by the PR-domain proteins, Blimp1 and Prdm14. Cell Cycle 7(22):3514–3518
12. Yamaji M et al (2008) Critical function of Prdm14 for the establishment of the germ cell lineage in mice. Nat Genet 40(8):1016–1022
13. Sakurai T et al (1995) The ter mutation first causes primordial germ cell deficiency in ter/ter mouse embryos at 8 days of gestation. Dev Growth Differ 37(3):293–302
14. Youngren KK et al (2005) The Ter mutation in the dead end gene causes germ cell loss and testicular germ cell tumours. Nature 435(7040):360–364
15. Harley VR, Clarkson MJ, Argentaro A (2003) The molecular action and regulation of the testis-determining factors, SRY (sex-determining region on the Y chromosome) and SOX9 [SRY-related high-mobility group (HMG) box 9]. Endocr Rev 24(4):466–487
16. Foster JW et al (1994) Campomelic dysplasia and autosomal sex reversal caused by mutations in an SRY-related gene. Nature 372(6506):525–530
17. Wagner T et al (1994) Autosomal sex reversal and campomelic dysplasia are caused by mutations in and around the SRY-related gene SOX9. Cell 79(6):1111–1120
18. Huang B et al (1999) Autosomal XX sex reversal caused by duplication of SOX9. Am J Med Genet 87(4):349–353
19. Cox JJ et al (2011) A SOX9 duplication and familial 46, XX developmental testicular disorder. N Engl J Med 364(1):91–93
20. Vetro A et al (2011) XX males SRY negative: a confirmed cause of infertility. J Med Genet 48(10):710–712
21. Barsoum I, Yao HH (2006) The road to maleness: from testis to Wolffian duct. Trends Endocrinol Metab 17(6):223–228
22. Capel B (2006) R-spondin1 tips the balance in sex determination. Nat Genet 38(11):1233–1234
23. Collignon J et al (1996) A comparison of the properties of Sox-3 with Sry and two related genes, Sox-1 and Sox-2. Development 122(2):509–520
24. Sutton E et al (2011) Identification of SOX3 as an XX male sex reversal gene in mice and humans. J Clin Invest 121(1):328–341
25. d'Anglemont de Tassigny X et al (2007) Hypogonadotropic hypogonadism in mice lacking a functional Kiss1 gene. Proc Natl Acad Sci U S A 104(25):10714–10719
26. Lapatto R et al (2007) Kiss1−/− mice exhibit more variable hypogonadism than Gpr54−/− mice. Endocrinology 148(10):4927–4936
27. Han SK et al (2005) Activation of gonadotropin-releasing hormone neurons by kisspeptin as a neuroendocrine switch for the onset of puberty. J Neurosci 25(49):11349–11356
28. Pineda R et al (2010) Physiological roles of the kisspeptin/GPR54 system in the neuroendocrine control of reproduction. Prog Brain Res 181:55–77
29. Nakagawa T et al (2010) Functional hierarchy and reversibility within the murine spermatogenic stem cell compartment. Science 328(5974):62–67
30. Mauduit C, Hamamah S, Benahmed M (1999) Stem cell factor/c-kit system in spermatogenesis. Hum Reprod Update 5(5):535–545

31. Yoshinaga K et al (1991) Role of c-kit in mouse spermatogenesis: identification of spermatogonia as a specific site of c-kit expression and function. Development 113(2):689–699
32. Handel MA, Schimenti JC (2010) Genetics of mammalian meiosis: regulation, dynamics and impact on fertility. Nat Rev Genet 11(2):124–136
33. Yang F et al (2008) Meiotic failure in male mice lacking an X-linked factor. Genes Dev 22(5):682–691
34. Yang F et al (2008) Mouse TEX15 is essential for DNA double-strand break repair and chromosomal synapsis during male meiosis. J Cell Biol 180(4):673–679
35. Berthet C et al (2003) Cdk2 knockout mice are viable. Curr Biol 13(20):1775–1785
36. Viera A et al (2009) CDK2 is required for proper homologous pairing, recombination and sex-body formation during male mouse meiosis. J Cell Sci 122(Pt 12):2149–2159
37. Yang F et al (2006) Mouse SYCP2 is required for synaptonemal complex assembly and chromosomal synapsis during male meiosis. J Cell Biol 173(4):497–507
38. Yuan L et al (2000) The murine SCP3 gene is required for synaptonemal complex assembly, chromosome synapsis, and male fertility. Mol Cell 5(1):73–83
39. Hunt PA, Hassold TJ (2002) Sex matters in meiosis. Science 296(5576):2181–2183
40. World Health Organization (1999) WHO Laboratory manual for the examination human semen and sperm-cervical mucus interaction. Cambridge University Press, Cambridge
41. Kierszenbaum AL, Tres LL (2004) The acrosome-acroplaxome-manchette complex and the shaping of the spermatid head. Arch Histol Cytol 67(4):271–284
42. Kang-Decker N et al (2001) Lack of acrosome formation in Hrb-deficient mice. Science 294(5546):1531–1533
43. Yao R et al (2002) Lack of acrosome formation in mice lacking a Golgi protein, GOPC. Proc Natl Acad Sci U S A 99(17): 11211–11216
44. Xu X et al (1999) Globozoospermia in mice lacking the casein kinase II alpha' catalytic subunit. Nat Genet 23(1):118–121
45. Lin YN et al (2007) Loss of zona pellucida binding proteins in the acrosomal matrix disrupts acrosome biogenesis and sperm morphogenesis. Mol Cell Biol 27(19):6794–6805
46. Xiao N et al (2009) PICK1 deficiency causes male infertility in mice by disrupting acrosome formation. J Clin Invest 119(4):802–812
47. Meistrich ML, Trostle-Weige PK, Russell LD (1990) Abnormal manchette development in spermatids of azh/azh mutant mice. Am J Anat 188(1):74–86
48. Mendoza-Lujambio I et al (2002) The Hook1 gene is non-functional in the abnormal spermatozoon head shape (azh) mutant mouse. Hum Mol Genet 11(14):1647–1658
49. Zhou J et al (2009) RIM-BP3 is a manchette-associated protein essential for spermiogenesis. Development 136(3):373–382
50. Xiao N et al (2009) PICK1 deficiency causes male infertility in mice by disrupting acrosome formation. J Clin Invest 119(4):802
51. Audouard C, Christians E (2011) Hsp90beta1 knockout targeted to male germline: a mouse model for globozoospermia. Fertil steril 95(4):1475–7 e1-4
52. Pazour GJ et al (2005) Proteomic analysis of a eukaryotic cilium. J Cell Biol 170(1):103–113
53. Neesen J et al (2001) Disruption of an inner arm dynein heavy chain gene results in asthenozoospermia and reduced ciliary beat frequency. Hum Mol Genet 10(11):1117–1128
54. Rashid S et al (2010) Disruption of the murine dynein light chain gene Tcte3-3 results in asthenozoospermia. Reproduction 139(1):99–111
55. Tanaka H et al (2004) Mice deficient in the axonemal protein Tektin-t exhibit male infertility and immotile-cilium syndrome due to impaired inner arm dynein function. Mol Cell Biol 24(18):7958–7964
56. Sapiro R et al (2002) Male infertility, impaired sperm motility, and hydrocephalus in mice deficient in sperm-associated antigen 6. Mol Cell Biol 22(17):6298–6305
57. Lee L et al (2008) Primary ciliary dyskinesia in mice lacking the novel ciliary protein Pcdp1. Mol Cell Biol 28(3):949–957
58. Schneider M et al (2009) Mitochondrial glutathione peroxidase 4 disruption causes male infertility. FASEB J 23(9):3233–3242
59. Imai H et al (2009) Depletion of selenoprotein GPx4 in spermatocytes causes male infertility in mice. J Biol Chem 284(47): 32522–32532
60. Miki K et al (2004) Glyceraldehyde 3-phosphate dehydrogenase-S, a sperm-specific glycolytic enzyme, is required for sperm motility and male fertility. Proc Natl Acad Sci U S A 101(47):16501–16506
61. Odet F et al (2008) Expression of the gene for mouse lactate dehydrogenase C (Ldhc) is required for male fertility. Biol Reprod 79(1):26–34

62. Danshina PV et al (2010) Phosphoglycerate kinase 2 (PGK2) is essential for sperm function and male fertility in mice. Biol Reprod 82(1):136–145
63. Miki K et al (2002) Targeted disruption of the Akap4 gene causes defects in sperm flagellum and motility. Dev Biol 248(2):331–342
64. Ishikawa H, Kubo A, Tsukita S (2005) Odf2-deficient mother centrioles lack distal/subdistal appendages and the ability to generate primary cilia. Nat Cell Biol 7(5):517–524
65. Salmon NA, Reijo Pera RA, Xu EY (2006) A gene trap knockout of the abundant sperm tail protein, outer dense fiber 2, results in preimplantation lethality. Genesis 44(11):515–522
66. Zheng H et al (2007) Lack of Speml causes aberrant cytoplasm removal, sperm deformation, and male infertility. Proc Natl Acad Sci U S A 104(16):6852–6857
67. Bailey JL (2010) Factors regulating sperm capacitation. Syst Biol Reprod Med 56(5): 334–348
68. Visconti PE (2009) Understanding the molecular basis of sperm capacitation through kinase design. Proc Natl Acad Sci U S A 106(3): 667–668
69. Esposito G et al (2004) Mice deficient for soluble adenylyl cyclase are infertile because of a severe sperm-motility defect. Proc Natl Acad Sci U S A 101(9):2993–2998
70. Hess KC et al (2005) The "soluble" adenylyl cyclase in sperm mediates multiple signaling events required for fertilization. Dev Cell 9(2):249–259
71. Santi CM et al (2010) The SLO3 sperm-specific potassium channel plays a vital role in male fertility. FEBS Lett 584(5):1041–1046
72. Zeng XH et al (2011) Deletion of the Slo3 gene abolishes alkalization-activated K+ current in mouse spermatozoa. Proc Natl Acad Sci U S A 108(14):5879–5884
73. Carlson AE et al (2005) Identical phenotypes of CatSperl and CatSper2 null sperm. J Biol Chem 280(37):32238–32244
74. Carlson AE et al (2003) CatSperl required for evoked Ca2+ entry and control of flagellar function in sperm. Proc Natl Acad Sci U S A 100(25):14864–14868
75. Jin J et al (2007) Catsper3 and Catsper4 are essential for sperm hyperactivated motility and male fertility in the mouse. Biol Reprod 77(1):37–44
76. Qi H et al (2007) All four CatSper ion channel proteins are required for male fertility and sperm cell hyperactivated motility. Proc Natl Acad Sci U S A 104(4):1219–1223
77. Ren D et al (2001) A sperm ion channel required for sperm motility and male fertility. Nature 413(6856):603–609
78. Quill TA et al (2003) Hyperactivated sperm motility driven by CatSper2 is required for fertilization. Proc Natl Acad Sci U S A 100(25):14869–14874
79. Liu J et al (2007) CatSperbeta, a novel transmembrane protein in the CatSper channel complex. J Biol Chem 282(26):18945–18952
80. Wang H et al (2009) A novel, single, transmembrane protein CATSPERG is associated with CATSPERl channel protein. Biol Reprod 81(3):539–544
81. Chung JJ et al (2011) A novel gene required for male fertility and functional CATSPER channel formation in spermatozoa. Nat Commun 2:153
82. Lishko PV, Botchkina IL, Kirichok Y (2011) Progesterone activates the principal Ca2+ channel of human sperm. Nature 471(7338): 387–391
83. Strunker T et al (2011) The CatSper channel mediates progesterone-induced Ca2+ influx in human sperm. Nature 471(7338):382–386
84. Taussig LM et al (1972) Fertility in males with cystic fibrosis. N Engl J Med 287(12):586–589
85. Anguiano A et al (1992) Congenital bilateral absence of the vas deferens. A primarily genital form of cystic fibrosis. JAMA 267(13): 1794–1797
86. Claustres M (2005) Molecular pathology of the CFTR locus in male infertility. Reprod Biomed Online 10(1):14–41
87. Reynaert I et al (2000) Morphological changes in the vas deferens and expression of the cystic fibrosis transmembrane conductance regulator (CFTR) in control, deltaF508 and knock-out CFTR mice during postnatal life. Mol Reprod Dev 55(2):125–135
88. O'Hara L et al (2011) Androgen receptor expression in the caput epididymal epithelium is essential for development of the initial segment and epididymal spermatozoa transit. Endocrinology 152(2):718–729
89. Krutskikh A et al (2011) Targeted inactivation of the androgen receptor gene in murine proximal epididymis causes epithelial hypotrophy and obstructive azoospermia. Endocrinology 152(2):689–696
90. Aston KI, Carrell DT (2009) Genome-wide study of single-nucleotide polymorphisms associated with azoospermia and severe oligozoospermia. J Androl 30(6):711–725
91. Aston KI et al (2010) Evaluation of 172 candidate polymorphisms for association with

oligozoospermia or azoospermia in a large cohort of men of European descent. Hum Reprod 25(6):1383–1397
92. Summerer D (2009) Enabling technologies of genomic-scale sequence enrichment for targeted high-throughput sequencing. Genomics 94(6):363–368
93. Maxmen A (2011) Exome sequencing deciphers rare diseases. Cell 144(5):635–637
94. Meyerson M, Gabriel S, Getz G (2010) Advances in understanding cancer genomes through second-generation sequencing. Nat Rev Genet 11(10):685–696
95. Metzker ML (2010) Sequencing technologies-the next generation. Nat Rev Genet 11(1):31–46
96. Ostermeier GC et al (2002) Spermatozoal RNA profiles of normal fertile men. Lancet 360(9335):772–777
97. Yatsenko AN et al (2006) Non-invasive genetic diagnosis of male infertility using spermatozoal RNA: KLHL10 mutations in oligozoospermic patients impair homodimerization. Hum Mol Genet 15(23):3411–3419
98. Matzuk MM, Lamb DJ (2008) The biology of infertility: research advances and clinical challenges. Nat Med 14(11):1197–1213

Chapter 3

Genetics of Female Infertility Due to Anomalies of the Ovary and Mullerian Ducts

Joe Leigh Simpson

Abstract

Genetic factors are pivotal in reproductive development and subsequent reproductive processes. If disturbed, infertility can occur. In the female, genetic factors affecting the ovary and the uterus are not uncommon causes of infertility. Terminal deletions on the X long arm and X short arm and X chromosomal mosaicism have long been accepted as causes of premature ovarian failure (POF). Responsible genes on the X have not yet elucidated. Attractive candidate genes for POF also exist on autosomes, and in over a dozen genes molecular perturbations are documented in non-syndromic POF. The most common single-gene cause of POF is premutation carriers for FMR1 (fragile X syndrome). As other candidate genes and additional ethnic groups are interrogated, the proportion of POF cases due to single-gene mutation will increase. Among uterine anomalies, incomplete mullerian fusion is most common. Increased recurrence risks for first-degree relatives confirm a role for genetic factors; interrogation of candidate genes is under way.

Key words Premature ovarian failure, X-chromosome deletions, Single genes, Mutation, Candidate genes, Genetic heterogeneity, Incomplete mullerian fusion, Mullerian aplasia, Array CGH

1 Introduction

Female fertility depends on normal embryologic differentiation as well as maintenance of a healthy reproductive milieu during reproductive life-span. In this chapter we shall discuss genetic control of pivotal embryologic processes obligatory for female reproduction. Additional discussion by the author on genetics of sex differentiation is available elsewhere [1].

1.1 Embryology of Female Ovary and Reproductive Tract

Primordial germ cells originate in the endoderm of the yolk sac and migrate to the genital ridge to form the indifferent gonad. The indifferent gonad is initially indistinguishable in genetic male (46,XY) and genetic female (46,XX) embryos. Primordial germ cells migrate from the yolk sac around 28 days post-conception, reaching the gonadal ridge by 37 days, or 5 days after formation of the genital ridge. If the embryo—or more specifically the

Zev Rosenwaks and Paul M. Wassarman (eds.), *Human Fertility: Methods and Protocols*, Methods in Molecular Biology, vol. 1154, DOI 10.1007/978-1-4939-0659-8_3, © Springer Science+Business Media New York 2014

gonadal stroma—is 46,XY, indifferent gonads develop into testes, a process that begins at 43 days after conception (15 mm crown rump length—CRL). Testes become morphologically identifiable 7–8 weeks after conception (9–10 gestational or menstrual weeks). In the absence of a Y chromosome, the indifferent gonad develops into an ovary [2], beginning at 50–55 days of embryonic development. By 20 weeks of embryonic life fetal ovaries contain up to seven million germ cells, but the vast majority undergo atresia. In 45,X embryos, oocytes also differentiate, only to undergo atresia at a rate more rapid than that occurring in normal 46,XX embryos. Thus, genes on the X necessary for normal ovarian development control ovarian *maintenance, not initial differentiation.* Ductal and external genital development occurs independent of gonadal differentiation. External genitalia develop in male fashion if testosterone and dihydrotestosterone are secreted by fetal Leydig cells. Absent this, female external genitalia develop. Differentiation of mullerian structures (uterus, fallopian tubes, upper vagina) depends on functional Sertoli cells, which secrete antimullerian hormone (AMH) and prevent female internal genitalia from developing (uterus). Absent AMH and its receptor (AMHR), mullerian ducts form the uterus and fallopian tubes, and Wolffian ducts regress. This occurs in normal XX embryos as well as XY embryos (animals) that were castrated as embryos prior to testicular differentiation.

1.2 Genetic Overview of Ovarian Development

The embryologic sequence described above depends on many intact genes, operative at each step (Fig. 1). In the genital ridge, a number of genes (Emx2, Gata4, Lim1, Lhx9) have been shown to be expressed in the mouse and presumably also in humans. Murine knockouts of theses genes result in the absence of gonads and internal ducts. Perturbations of these genes have not yet been documented to affect ovarian development adversely in humans, but this would be expected.

Development of ovaries involves the expression of WNT4 and RSP01, both autosomal loci. If SRY is lacking, SOX9 remains repressed [3], thus preventing male differentiation. Continued repression of SOX9 is accomplished by FOXL2 [4], which is essential for ovarian development. Many other genes on the X chromosome and autosomes are necessary for normal ovarian development (Fig. 1). If two intact X chromosomes are not present, ovarian follicles in 45,X individuals develop but degenerate prematurely, if not by birth then at the time of expected adolescence. Genes on the second X chromosome thus are responsible for ovarian maintenance rather than primary ovarian differentiation. It follows that these genes do not undergo X inactivation.

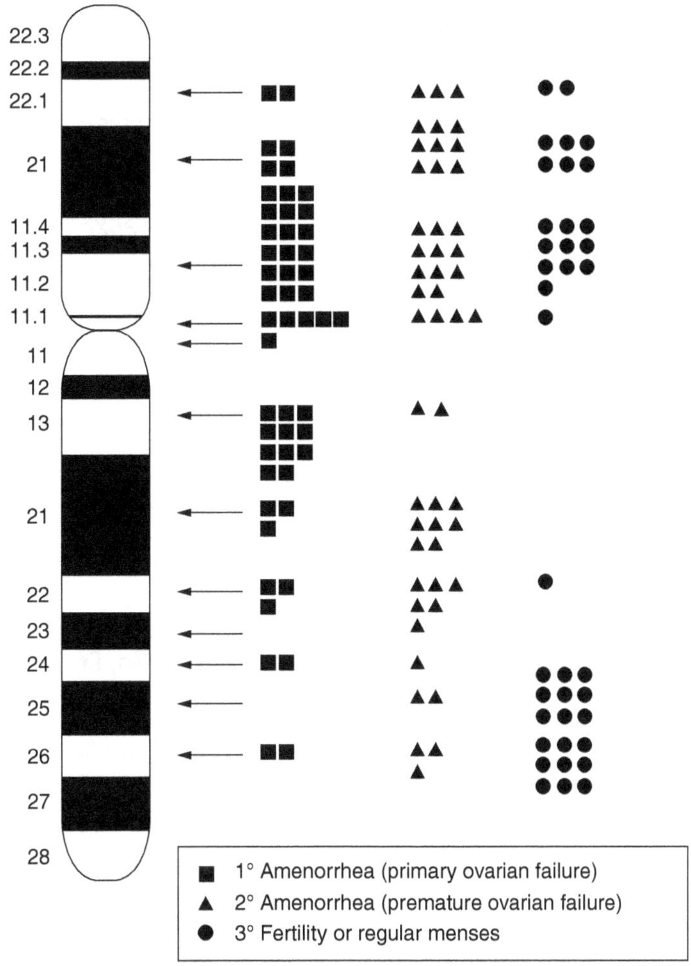

Fig. 1 Schematic diagram of the X chromosome showing ovarian function as a function of nonmosaic terminal deletions (from Simpson JL, Rajkovic A (1999) Ovarian differentiation and gonadal failure. Am J Med Genet 89:186) [13]

1.3 Genes on X Chromosome Necessary for Ovarian Development and Fertility

1.3.1 Regions on the X Short Arm (Xp) and X Long Arm (Xq)

For decades it has been known that several different regions on the X must contain pivotal genes, based on karyotype–phenotype correlations. It was not practical, however, to identify a specific gene save by serendipity (e.g., translocation involving the X). Figure 2 shows the author's 1999 tabulation of the phenotypes, based solely on terminal deletions detectable by a traditional karyotype, which has the ability to detect at best 6 Mb deletions. In the X short arm, the region of greatest importance can be deduced to be near the centromere (Xp 11), given that women with terminal deletions invariably show complete or premature ovarian failure, the latter presenting with infertility. Women having more distal terminal Xp deletions [del(X) (p21.1–p22.1.22)] menstruate more often than 45, X women, but many are still infertile or show secondary amenorrhea. Of interest, del (Xp) women may menstruate yet still be short;

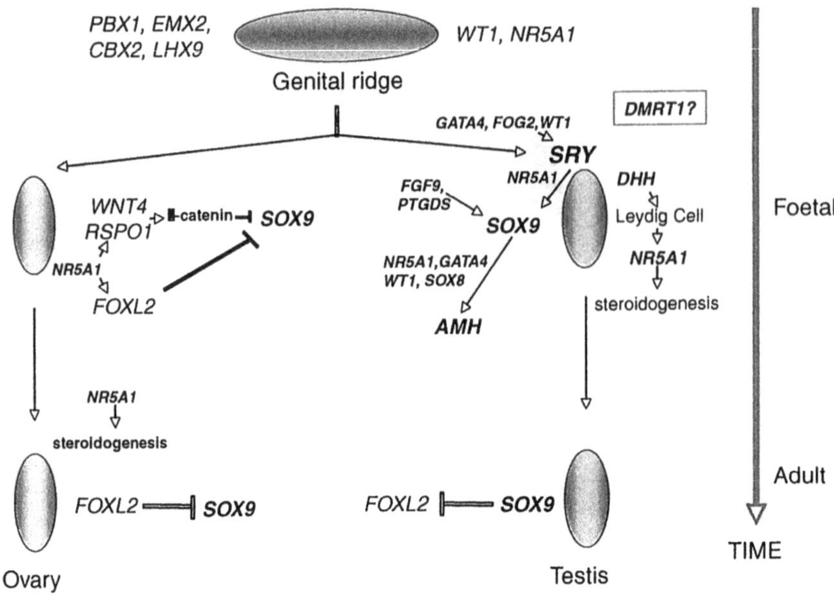

Fig. 2 Molecular and genetic events in mammalian sex determination and differentiation. McElreavey K, Bashamboo A (2011) Genetic disorders of sex differentiation. Adv Exp Med Biol 707: 91–99

thus, regions on Xp responsible for ovarian and statural determinants must differ [5–10].

In the X long arm, almost all terminal deletions originating at Xq13 are associated with primary amenorrhea, lack of breast development, and complete ovarian failure (Fig. 1) [6, 11–13]. Xq13 is thus a pivotal region for ovarian maintenance. Del(X) (q21 → 24) individuals menstruate more often than those with more proximal deletions, suggesting that in these women more proximal (Xq13) ovarian genes persist. The designation premature ovarian failure 2 (POF2) has been applied to a presumptive "locus" in the Xq13–21 region, but appellation and other gene designations applied to ovarian-pivotal genes are simplistic. In terminal Xq deletions arising at Xq25 or 26, the more common phenotype is not primary amenorrhea but premature ovarian failure (POF) [12–15]. More distal deletions arise at Xq27 or 28, and this "locus" has been labeled POF1. Distal Xq deletions are often familial [16].

1.3.2 Approaches for Identifying Ovarian Genes

Many X-linked genes have potential relevance to ovarian development, as tabulated by Lamb and Matzuk [17] and Ochalski et al. [18]. Most of these genes exert cellular functions that are not necessarily considered endocrine in nature, e.g., cell cycle, meiotic recombination, or growth factors.

Most molecular studies have involved sequencing candidate genes, but in the foreseeable future whole-gene sequencing can be expected to be the most informative. Whole-genome studies have been performed, but sample size is a problem in many studies.

Genome-wide linkage analysis has already shown an association between Xp21.3 (LOD score 3.1) and the age of natural menopause [19], and a number of single-nucleotide polymorphisms (SNPs) throughout the genome [20–23] have been associated with early menopause (age 40–45 years). A few genome-wide association studies (GWAS) of limited sample size have been performed on POF [24], while a much more definitive study is under way in Han Chinese (Chen, personal communication). DNA copy number variants (CNVs) (microdeletions or microduplications) have been sought in women with ovarian failure. DNA CNVs are too small to detect by karyotype. Usually several hundred thousand (kb) in length, CNVs may be polymorphic without clinical significance, or they may result in perturbation of genes of clinical significance, i.e., loss of function (microdeletion) or gain of function (microduplication). CNVs are detected using array CGH platforms, several commercially available. Results have not been very informative to date. In studying 97 POF patients, Knauff et al. [25] found a CNV (262 kb deletion) in Xq21.3 to be associated with POF. In this region PCHCHIIX and TGHF2LX are located. However, no deletions were found in other X regions considered on the basis of cytogenetic studies to be pivotal (the so-called POF1, POF2). Among 50 POF cases Dudding et al. [26] found only an 800 kb duplication at Xq13.3 and a duplication at Xp22.3. Aboura et al. [27] reported a statistically significant association for a 217 kb gain on Xq28, but the same gain was found in 5 of 67 controls; thus, clinical significance does not exist. No significant X deletions were found in 90 POF cases.

In contrast to the above two studies, Quilter et al. [28] found that 48 % of their 42 cases showed a CNV microduplication or deletion; however, many different X regions were involved and usually only once. Discrepancies among CNV studies are most likely explained by failure to take into account parental CNV status, a necessity to exclude paternal transmission of a polymorphic CNV that would have no clinical significance.

1.3.3 Candidate Genes on the X Short Arm

Ubiquitin-Specific Protease 9 (USP9X). This gene maps to Xp11.4 [29]. Its Drosophila orthologue is required for eye development and oogenesis. The role USP9X plays in human gonadal development is unclear, but its location in the important Xp11.4 region makes it potentially relevant.

Zinc Finger X (Zfx). Mice null for Zfx are small, less viable, less fertile, and characterized by diminished germ cell number in ovaries and testes [30]. External and internal genitalia are otherwise normal.

Bone Morphogenetic Protein 15 (BMP15). The best studied candidate gene on Xp is BMP15, a member of the transforming growth factor-beta (TGFβ) superfamily. TGF genes play pivotal roles in developmental pathways, binding and activating transmembrane serine/threonine kinase receptors. BMP is expressed during folliculogenesis and ovarian development. Human BMP15 is located on Xp.11.2, consisting of only two exons. Animal studies have long suggested that perturbations of BMP15 could be important in ovarian development. Heterozygous sheep carrying a mutation in the BMP15 gene have an increased ovulation rate, with twin and triplet births; primary ovarian failure occurs in homozygotes [31].

BMP15 knockout mutant female mice are subfertile, showing decreased ovulation rates, reduced litter size, and decreased number of litters per lifetime [32].

Supporting a role for BMP15 are reports that the BMP15–9G allele is associated with enhanced response to gonadotropin (OR 2.7, 1.3–5.7) [33]. In humans BMP was implicated in POF first by Di Pasquale and co-workers [34] who reported a heterozygous Y235C missense mutation in the second exon of the BMP15 gene in each of the two sisters having ovarian failure. The proband had streak gonads and elevated follicle-stimulating hormone (FSH) (80 MIU/ml); the younger sib had one episode of vaginal spotting but otherwise similar. The mother was homozygously normal at this allele, Y235C, transmitted from the father. The authors presented in vitro evidence for a dominant negative mechanism, presumably based on structural alterations of the mutant BMP impeding dimerization with other TGF proteins. Variants were also detected in Indian [35] and French populations [36]. However, it is unproved whether these variants are unique to POF cases or present in controls as well. Indeed, in 92 Chinese POF cases, Zhang et al. [37] found no novel SNPs.

Short Stature Homeobox (SHOX). This gene lies in the pseudoautosomal region (PAR1), a regional obligatory recombination that involves Xp22 and Yp11.3. (Pseudoautosomal regions X and Y contain homologous sequences and hence undergo synapsis during male meiosis.) In this region is SHOX, a gene involved in skeletal growth. This was initially reasoned on the basis of short stature existing in Turner syndrome (45,X) but tall stature in polysomy X (47,XXY). As would be expected, SHOX escapes X inactivation, because the genes are in the pseudoautosomal region. SHOX mutations and deletions have been associated with short stature [38, 39], Leri–Weill syndrome [40], and Langer mesomelic dysplasia [41]; duplications have also been observed. Tachdjian et al. [42] studied three POF cases known by conventional karyotype to have an Xq deletion; a del(X) q21.31 case also had an array CGH 620 kb duplication (Xp22.3 → PAR1).

Although postulated, a relationship between POF and SHOX seems unlikely. Any gene in the pseudoautosomal region is unlikely to play a major role in ovarian (or testicular) determinants because recombination would occur frequently (and obligatorily). The locus would, perhaps frequently, be lost to female offspring.

1.3.4 Candidate Genes on the X Long Arm (Xq)

XIST. The X inactivation center and XIST are located in Xq13, a region known to result in ovarian failure if a terminal deletion originates there. Loss of germ cells may or may not be the result of perturbation of XIST per se, despite disturbances of X inactivation clearly being associated with ovarian failure. The concept of a crisply delineated "critical X region" necessary for retention for normal ovarian development receives less attention than in the past, but this does not exclude the region being rich in pivotal genes.

DIAPH2 (Diaphanous). This candidate gene lies in the Xq21–24 region and is the homologue of *Drosophila melanogaster* diaphanous (dia). In Drosophila, dia is a member of a family of proteins that help establish cell polarity, govern cytokinesis, and reorganize the actin cytoskeleton. In flies, mutations cause sterility in both male and female subjects [43]. A human Xq21/autosome translocation was reported to have disrupted the last intron of DIAPH2 [44]. Like several Xq candidate genes discussed in this section, this gene escapes X inactivation.

DACH2. DACH2, which also escapes X inactivation, was identified through an X;autosome translocation identified through a patient with POF [45, 46]. DACH2 has a Drosophila homologue and is expressed in multiple tissues, but not overtly in the reproductive system. Although five heterozygous DACH2 missense mutations were found in a series of 257 Italian POF cases, controls ($N=110$) sometimes showed the same mutation. Thus, causality remains lacking. That DACH2 and POF1B (see below) are only 700 kb apart and in a gene-poor region raises the possibility of existence of a regulatory region that could be perturbed.

POF1B. The belief that "critical regions" for ovarian function existed on Xq originally led to the proposal that there existed POF1 and POF2, the former an ovarian region (gene?) on Xq21-ter and the latter on Xq13.3–21.1. Later, an X;autosome translocation led to another proposed candidate gene, called POIFB, and localized to Xq21 [47]. Lacombe et al. [48] then reported a consanguineous Lebanese family in which 46,XX sisters with POF were homozygous for a missense mutation (R329Q). Heterozygotes were unaffected, consistent with R329Q being present in the normal population. Bione et al. [49] failed to find even heterozygous perturbations in over 200 Italian POF cases; thus, POF1B is not a common explanation for POF in Italy. More recent efforts at determining the role POIFB might play in ovarian failure have focused on alteration of tight junction assembly in the actin cytoskeleton, potentially dysregulating epithelial polarity [50].

ANGIOTENSOR II (TYPE 2) (ATZ) RECEPTOR. This gene (ATZ2) is expressed in fetal tissue and perturbed in a diverse group of diseases. Because atretic granulosa cells express this gene in rodents, a plausible explanation exists for relevance to human POF. ATZ2 is also located on Xq22–23, a region of known significance (i.e., the so-called POF2). Katsuya [51] studied two families in each of which two sibs had POF; no AT2 mutations were found.

XPNPEP2. Bione and Toniolo [52] and Prueitt and colleagues [53] found disruption of XPNPEP2 in an Xq;autosome translocation involving Xq25. POF was present in the index cases. XPNPEP2 escapes X inactivation.

Familial mental retardation 1 (FMR1). About 15–20 % of women with premutation of the fragile X gene (FMR1), located on Xq24, develop POF [54]. In fragile X syndrome greater than 200 CGG repeats in exon 1 result in mental retardation in all males and 50 % of females. The normal number of CGG repeats is 27–30. With 55–200 repeats, "permutation" is said to exist because from this range expansion can occur during maternal meiosis to produce oocytes with over 200 repeats. Women with a heterozygous premutation do not show mental retardation but are at increased risk for their male offspring being affected because the number of repeats can expand during female meiosis. Females with one X chromosome who have >200 CGG repeats (heterozygotes) may be retarded but less severely and less often than males.

Schwartz and colleagues [55] found oligomenorrhea in 38 % of premutation carriers versus in 6 % controls. Allingham-Hawkins and co-workers [56] studied 1,268 controls, 50 familial POF cases, and 244 sporadic POF cases of 395 premutation carriers; 63 (16 %) underwent menopause before 40 years of age; the frequency in controls was 0.4 %. Sullivan and colleagues [57] found 12.9 % of premutation carriers (N=250; >59 repeats) to have POF versus 1.3 % (2/157) of controls. For unknown reasons, the number of CGG repeats significantly correlates with the risk of POF only within selected ranges. Thus, only a slightly increased risk of expansion is associated with 40–79 repeats. There is a higher risk with 80–99 repeats, but no further increased risk occurs after >100 repeats. This plateau is consistent with women with the full mutation (>200 CGG) not showing POF [56].

FMRI testing should be part of the work-up for POF and is formally recommended in Europe [58]. If oocyte or ovarian slice cryopreservation becomes more feasible, population screening might even be justified for fertility preservation.

1.4 Candidate Genes on Autosomes Necessary for Ovarian Development and Fertility

The importance of autosomal genes can if perturbed cause complete ovarian failure or POF. Autosomal loci have long been deduced by the existence of autosomal recessive inheritance [59] for complete ovarian failure and POF. In traditional "XX gonadal dysgenesis," streak gonads are not associated with somatic anomalies. Such women are also normal in stature [60]. "XX gonadal dysgenesis" as once defined is, however, genetically heterogeneous. A more specific diagnosis is desired. At present probably only 20 % of POF cases would yield a precise diagnosis even if all known

POF-causing genes were tested. This reflects both a dearth of identified genes and limited molecular diagnostic evaluation, i.e., failure to perform molecular studies for candidate genes.

Everything stated above concerning POF is also applicable for complete ovarian failure (primary amenorrhea) and by extension probably all ovulation disorders of embryological origin. The same genes are likely involved in all these processes. Variable expressivity within families exists for genes identified. One sibling may have bilateral streak gonads whereas another ovarian hypoplasia [59, 61–63].

1. FSH B:
   Mutations in FSH-β are rare but two are reported. Matthews and colleagues [64] described a homozygous 2 bp deletion (GT) in exon 3 at codon 61, in a woman who did not undergo thelarche or menarche. Similarly, Layman and co-workers [65] reported a compound heterozygote: in one allele there was a deletion in exon 3, codon 61, whereas in the other a missense mutation in exon 3.
2. Inactivating FSH receptor (FSHR):
   Mutations in the G-protein (FSH) receptor are not uncommon in Finland but rare elsewhere. Aittomaki and colleagues [61, 62] identified 75 women with 46,XX primary or secondary amenorrhea, based on serum FSH > 40 MIU/ml. A homozygous missense mutation in exon 7 of FSHR (C566T or Ala566Val) was found in six families [61, 62], the mutation lying in the extracellular portion of this G-protein receptor. Women heterozygous for the mutation did not show decreased fertility. The Ala566 Val mutation is uncommon outside Finland. No mutations in FSHR were found in North American women having either 46,XX hypergonadotropic hypogonadism [65] or POF [66].

   Similar findings were reported in 46,XX POF or primary amenorrhea cases from Germany [67], Brazil [68], and Mexico [69]. However, compound heterozygosity involving the mutation has been found, reported genotypes including Ile160Thr/Arg573Cys and Asp224Val/Leu602Val [70].
3. Inactivating LH receptor (LHR):
   LHR is 75 kD in length and consists of 17 exons. The gene is located on 2p near the locus for FSHR. The first 10 exons in LHR are extracellular, the 11th transmembrane, and the last 6 intracellular. Most mutations have been detected in the transmembrane domain of this G-protein receptor. Inactivating LHR mutations are more commonly reported in 46,XY individuals, causing XY sex reversal [71]. However, mutations in the LHR in 46,XX women result in the phenotype XX gonadal dysgenesis. All 46,XX cases have been recognized in sibships ascertained through their affected 46,XY siblings, who presented with Leydig cell hypoplasia and XY sex reversal.

Latronico and co-workers [72] reported a 22-year-old woman who presented with primary amenorrhea due to an LHR mutation. This woman as well as her three 46,XY sibs had a homozygous C544X mutation (X= stop codon), resulting in a truncated protein consisting of five rather than seven transmembrane domains. The 46,XX sib had breast development but only a single episode of menstrual bleeding at age 20; LH was 37 MIU/ml, and FSH was 9 MIU/ml. In another 46,XX case, Latronico and colleagues [72] recorded secondary amenorrhea; LH and FSH were 10 and 9 miU/mL, respectively. A homozygous Ala593Pro mutation was found.

Other 46,XX women with LHR mutations may show oligomenorrhea, but ovulation does not occur even though gametogenesis proceeds until the preovulatory stage. This is consistent with mouse knockout models [73].

A homozygous LHR mutation (N400S) has been reported in two Turkish sisters having the empty follicle syndrome [74].

Also of interest, activating LHR mutations have little effect in women despite activating LHR mutations causing precocious puberty in males [71].

4. Inhibin A (Inh A):
Synthesized by granulosa cells, inhibins (INHs) are heterodimeric glycoproteins that consist of an α subunit and either of the two β subunits (B_A or B_B), producing INHα or INHβA, respectively. INHs exert negative-feedback inhibition on activins and by so opposing enhance FSH secretion. The significance of dimerization is that one subunit in heterozygous mutations can lead to dominant negative effects.

Particular attention has been given to the associations between POF and one particular INHA missense mutations (or polymorphism)—G769A (Ala57Thr) [75]. Studying patients from New Zealand, Shelling and colleagues [76] found G769A in 3 of 43 POF patients (7 %) versus only 1 of 150 normal controls (0.7 %). However, doubt was cast on clinical significance because the mother of one of the three G769A individuals had the same heterozygous mutation and was clinically normal. Marozzi and co-workers [77] found G769A in 7 of 157 Italian POF individuals, 3 of 12 primary amenorrhea cases, and 0 of 36 early menopausal (40–45 years) women. Familial POF cases were relatively more likely to have G769A than sporadic cases. Dixit and colleagues [78] repeated G796A in 9 of 80 Indian POF cases; no mutations in INHβ or INBβA were found. Also studying an Indian cohort, Prakash et al. [79] found the heterozygous mutation in 3 of 30 cases of primary amenorrhea, 3 of 20 with secondary amenorrhea, and 2 of 50 controls. However, normal individuals may have the G769A transition and may be normal even if another G769A family member has POF; thus, G769A does not obligatorily confer ovarian failure in heterozygotes. Shelling and his group [80] have recently expanded the rationale for function but still have not defined mutation(s) unequivocally causing POF [75, 81–83].

5. 17A-Hydroxylase/17,20 desmolase deficiency (CYP17): This gene encodes for 17α-hydroxylase/17,20 desmolase, an enzyme pivotal for sex steroidogenesis. 46,XX individuals with a CYP17 defect have presented with primary amenorrhea or POF [84]. Thus, deficiency of this enzyme can cause 46,XX hypergonadotropic hypogonadism. The gene is located on 10q24.3, and many different mutations have been reported. Ovaries in 46,XX cases are hypoplastic, and oocytes appear incapable of exceeding 2.5 mm [85]. However, ovulation stimulation can produce oocytes capable of fertilization in vitro [86].
6. Aromatase mutations (CYP19) (46,XX) (CYP19): Conversion of androgens (Δ4-androstenedione) to estrogens (estrone) requires cytochrome P-450 aromatase (CYP19), an enzyme that is the gene product of a 40-kb gene located on chromosome 15q21.1 [87]. 46,XX aromatase deficiency may present with primary amenorrhea. Ito and co-workers [88] reported an aromatase mutation (CYP19) in a 46,XX 18-year-old Japanese woman having primary amenorrhea and cystic ovaries. The patient was a compound heterozygote, having two different point mutations in exon 10, the N-terminal exon. The mutant protein showed no activity in vitro. Conte and colleagues [89] also reported aromatase deficiency in a 46,XX woman presenting with primary amenorrhea, elevated gonadotropins, and ovarian cysts. Compound heterozygosity for two different mutations was found in exon 10. One was mutation C1303T, leading to cysteine rather than arginine; the other was G1310A, leading to tyrosine rather than cysteine.

   A phenotype different but still relevant to female infertility was reported by Mullis and co-workers [90]. Clitoral enlargement occurred at puberty, and breast development did not. Multiple ovarian follicular cysts were present. FSH was elevated; estrone and estradiol were decreased. Estrogen and progesterone therapy resulted in a growth spurt, decreased FSH, decreased androstenedione and testosterone, breast development, menarche, and decreased follicular cysts. Compound heterozygosity was found.
7. Progesterone receptor membrane component 1 (PGRMC1): This X-linked gene was interrogated in 67 POF cases, with 1 heterozygous mutation found (H165R) [91]. The change occurred in a domain necessary for nontranscriptional regulation of cytochrome P450, potentially of functional significance.
8. FOXL2/blepharophimosis–ptosis–epicanthus (BPE): FOXL2 plays a key if not the pivotal role in ovarian development (Fig. 1). Encoded on 3q21-24, FOXL2 protein must be

expressed in order to maintain SRY suppression and allow ovarian differentiation to proceed.

The initial confirmation of the importance of FOXL2 and other forkhead DNA-binding proteins in humans came by studying BPE type II syndrome. In this autosomal dominant syndrome, FOXL2 is perturbed and POF occurs [92]. Consistent with clinical features in the humans, mouse Foxl2 is expressed in eyelids and ovaries [92]. In four human families, FOXL2 mutations cosegregated with BPE and POF. Nonsense mutations included stop codons as well as a 17 bp duplication that resulted in a frameshift and, hence, truncated protein.

FOXL2 mutations are less common explanations for POF [93, 94]. In the absence of somatic features, two mutations have been found among 70 cases [93]. In 1 of 70 cases of Slovenian origin a deletion (A221–A230) removed 10 of the 14 alanines from the poly A tail [94]. In a patient of New Zealand origin a missense mutation (Tyr258Asp) was found. De Baere and colleagues [95] found no FOXL2 mutations in 30 POF patients, all lacking eyelid abnormalities. Overall, perhaps 1–2 % of isolated Caucasian POF cases have a FOXL2 mutation.

9. Newborn ovary homeobox (NOBOX):
NOBOX gene is representative of those genes that bind DNA and function as transcription factors to direct differentiation. NOBOX is oocyte specific, expressed from the primordial follicle through metaphase II. Female null mice (knockout) show ovarian failure, whereas males are normal. Although an earlier study failed to show NOBOX perturbations in 30 Japanese women [29], our group found two novel missense mutations (Arg355His and Arg360Gin) among 96 Caucasian POF cases [96]. Arg355His was present in a conserved region. Functional studies (electrophoretic mobility shift arrays, or EMSA) using Arg355His DNA showed disrupted binding of the NOBOX homeodomain to DNA. This provides the basis for postulating a dominant negative effect.

10. Growth differentiation factor 9 (GDF9):
GDF9 is a member of the TGFβ family, like BMP15 (which is also called GDF9b). GDF9 can thus form dimers with BMP15. GDF9 is an attractive candidate gene because it is expressed in oocytes. Various heterozygous mutations have been detected in some European and Asian samples [36, 82, 97] but not in others [98–99]. As noted already a deleterious heterozygous change in a gene encoding for a protein undergoing dimerization can produce a dominant negative effect. If missense mutations such as a hydrophobic amino acid replacing a hydrophilic amino acid are causative, GDF9 perturbations could account for perhaps 1–4 % of POF cases. However, like BMP15, novel

variants found only in isolated POF and never found in controls are lacking.

11. BMPR1B:
    Homozygous deletion (del 359–366) of BMPR1B, another autosomal TGFβ superfamily gene, was reported by Demirhan et al. [100] in a 16-year-old female with ovarian failure and acromesomelic chondrodystrophy. Murine knockouts for this gene are infertile [101]. Heterozygous mutations of the BMPR1B homologue in sheep can lead to increased fertility (gain of function) [102], findings similar to those shown by sheep heterozygous for BMP15 (FecX) mutations [31].
12. Factor in germline ALPHA (FIGLA):
    This 2p13.3 gene codes for a germ cell-specific basic helix–loop–helix transcription factor. It is involved in regulating zona pellucida genes. FIGLA is expressed in the embryonal ovary. In knockout models, primordial follicles are either not formed or lost soon after birth. Zhao et al. [103] studied 100 Han Chinese with POF and found 3 variants in 4 women. The missense mutation A49 was found in 2 cases: a 15–36 deletion (p.G6fsX66) that resulted in a frameshift and dysfunctional haploinsufficiency in one case and a 419–421 del (140 del N) in a fourth. Functional studies of the 140delN mutation demonstrated that FIGLA binding to the TCF3 helix–loop–helix was disrupted.
13. POU5F1:
    This transcription factor gene, located on 6p21.31, is significantly downregulated in NOBOX knockout mice, which lack ovaries. Thus, POU5F1 becomes a potential human candidate gene, potentially a downstream target of NOBOX. Wang et al. [104] sequenced 175 Chinese POF cases and found one nonsynonymous variant (Pro13Thr), a heterozygous hydrophobic-to-hydrophilic substitution.
14. PTH-responsive B1 (PTHB1):
    PTHB1 was claimed to be associated with POF in a small gene association study (24 cases; 24 controls) [105]. Any gene association or GWAS based on a sample size this small is considered to be of inadequate power. Sequencing data are awaited.
15. ADAMTS:
    Located on 5q14.1 → q15, this gene was found to be associated with POF in the discovery set of a GWAS performed on 99 Dutch POF cases and 181 controls [24]. However, the finding was not confirmed in the replication set. Given this GWAS, also underpowered like the PTHB1 alluded to previously, conclusions concerning the role this transcription factor plays in POF remain uncertain.

16. Galactosemia:
    Galactosemia is caused by deficiency of galactose 1-phosphate uridyl transferase (GALT). Kaufman and co-workers [106] reported POF in 12 of 18 galactosemic women and Waggoner and colleagues [107] in 8 of 47 (17 %) women. Pathogenesis presumably involves galactose toxicity after birth because elevated fetal levels of toxic metabolites should be cleared rapidly in utero by intact, albeit heterozygous, maternal enzymes. Consistent with this, a neonate with galactosemia showed normal ovarian histology [108].

    In a variant of galactosemia in which compound heterozygosity exists, the usually severe GALT allele (G) is present at one allele but a milder mutation (N3/4 or D/D2) homologue at its allele. These DG heterozygotes retain some enzyme activity. Despite frequently hypothesized, heterozygotes for the Duarte variants do not show POF. In fact, not all homozygotes for severe GALT are even abnormal, nor are transgenic mice in which GALT is inactivated (knockout) [109]. Badik et al. [110] also showed undiminished ovarian reserve in DG compound heterozygotes.
17. Carbohydrate-deficient glycoprotein (phosphomannomutase deficiency, PMM2):
    In type 1 carbohydrate-deficient glycoprotein (CDG) deficiency, mannose 6 phosphate cannot be converted to mannose 1 phosphate. This lipid-linked mannose-containing oligosaccharide is necessary to synthesize secretary glycoproteins. The gene is located on 16p13, and the most frequent molecular perturbation is a missense mutation [111]. Neurologic abnormalities [112] are characteristic and ovarian failure frequent. Ovaries do not show follicular activity [113, 114].
18. Autoimmune regulation/autoimmune poly-endocrinopathy-candidiasis ectodermal dystrophy (AIRE/APECED):
    The AIRE gene, located on 21q22.3, is responsible for the condition characterized by the spectrum of features listed above. In addition to these abnormalities, alopecia, vitiligo, keratopathy, malabsorption, hepatitis, and mucocutaneous candidiasis are common. Ovarian hypoplasia, usually manifested in the form of POF, exists in 55 % of APECED cases, usually in the third decade [115]. Many different AIRE perturbations have been found in this autosomal dominant disorder, a pleiotropic condition [115] showing varied expressivity. Previous reports of autosomal dominant POF associated with multiple endocrine autoimmune disorders probably have this condition. Nonsense mutations and frameshift mutations are reported. No particular mutation seems preferentially likely to cause the POF component of this pleiotropic condition.

    AIRE as a candidate gene has apparently not been sequenced in women with isolated POF, i.e., POF in the absence of autoimmune disorders.

19. Ovarian leukodystrophy (eukaryotic translation initiation factor EIF2B):
    Ovarian leukodystrophy is characterized by MRI-detectable "vanishing white matter (VWM)" that leads to variable but progressive neurological degeneration as well as ovarian failure [116, 117]. As a result of a mutation occurring in EIF2B, denatured stress-related proteins accumulate. This is of potential relevance to oogenesis, given ubiquitous oocyte degeneration. In ovarian leukodystrophy Fogli et al. [118] found variants in EIF2B2, EIF2B4, and EIF2B5. However, 0 of 93 cases with isolated POF only showed perturbations [119]. This disorder could be part of the same clinical constellation as the cerebellar ataxia disorders having ovarian failure, to be discussed below.
20. Cerebellar ataxia with XX ovarian dysgenesis:
    Ataxia and hypergonadotropic hypogonadism were first associated by Skre and colleagues [120], who in 1976 described cases in two families. In one family, a 16-year-old girl was affected, whereas in the other family three sisters were affected. In the sporadic case and in one of the three sisters, ataxia was observed soon after birth; in the two other sisters, age of onset was later during childhood. Cataracts were present in all the cases reported by Skre et al. [120].

    Hypergonadotropic hypogonadism and ataxia have since been observed on several occasions [121]. The nature of the ataxia differed among patients, for example progressive or not. Mitochondrial enzymopathy was reported by De Michele and colleagues [121], but mitochondrial studies have not otherwise been studied. Cataracts were observed only by Skre and co-workers [120], and amelogenesis only by Linssen and colleagues [122]. Neurosensory deafness reminiscent of Perrault syndrome was reported by Amor and colleagues [123]. Mental retardation is also variable [123].

    Overall, genetic heterogeneity is likely in the hypergonadotropic hypogonadism disorders showing cerebellar ataxia. A single mutant gene is unlikely to explain every single case, but not every family need be unique. No molecular studies have been conducted.
21. Symphalangism and noggin (NOG):
    NOG (17q22) is responsible for the autosomal dominant disorder proximal symphalangism (SYM1). Characteristic features include ankylosis of the proximal interphalangeal joints, carpal–tarsal fusion, brachydactyly, and deafness. Expressed in the ovary, NOG is an antagonist of bone morphogenic proteins 4 and 7 [124]. The latter are members of the TGF family of genes, discussed previously and which include BMP15 and GDF9. In one woman with this syndrome who showed POF,

a NOG mutation was found [125]. However, NOG perturbations have not been sought in isolated POF subjects.

22. Perrault syndrome:
    XX gonadal dysgenesis with neurosensory deafness constitutes Perrault syndrome [126], a long accepted autosomal recessive disorder [127–129].

    Candidate genes are most likely to merge from the connexin family because an attractive gene knockout model exists in connexin 37 [130]. Null mice for Cx37 show gonadal failure due to arrest at the antral stage of oogenesis. The connexin gene family is responsible for many forms of congenital deafness in humans.

23. Forkhead transcription 3A (FOX03A):
    Forkhead transcription genes other than FOXL2 cause ovarian follicular depletion in murine knockout models [131]. This holds true for FOX03A, which regulates G1/S transition in granulosa cells. Of 60 POF cases recruited in equal number from New Zealand and Slovenia, two showed FOX03A mutations were considered potentially significant by the authors [132]. One mutation was a single heterozygous mutation in a Slovenian woman. The non-conservative amino acid charge (Ser421Leu) seems potentially capable of inducing a conformational protein change. The other mutation was Arg506His, found in a New Zealand woman. This conservative change seems less likely to exert an untoward effect. Wang et al. [133] screened the coding regions in 114 Chinese cases and found five heterozygous nonsynonymous variants lacking in their controls. All variants were located in a highly conserved region. In the five, there was evidence of a change in protein structure.

24. FOX01A:
    FOX01A is another forkhead transcription factor gene. Watkins et al. [132] found a single conservative change (P84L) among 90 POF cases. This patient was Slovenian in origin.

25. LIM DNA-binding protein 8 (LHX8):
    LIM homeobox genes encode DNA-binding proteins. In mice Lhx8 transcripts localize in germ cells from oocytes to antral follicles. Null mice lack germ cells [134]. LIM family members contain two tandemly repeated domains that have cysteine-rich, double-zinc-finger motifs. Qin et al. [135] sequenced LHX8 in 95 Caucasian women with POF. No novel SNPs were found.

26. NANOS3:
    NANOS3 is an RNA-binding protein. In mice NANOS 3 female knockouts (KO) are infertile, but show no other phenotypic effect [136].

    Human NANOS3 consists of two exons and is expressed in germ cells. In a study of 80 Chinese and 88 American

Caucasians with POF, Qin et al. [137] found the only NANOS3 sequence variant to be a synonymous substitution already known to be present in the general population [137].

27. G-protein receptor 3 (GPR3) and G proteins:
G proteins (GP) are regulatory proteins, like FSH and LH. These hormones are ligands for specific cell surface G-protein receptors, in turn leading to intracellular signal transduction. In the mouse oocyte, the oocyte-specific G-stimulating protein-coupled receptor GPR3 is known to have a role in maintaining meiotic arrest. Female mice lacking GPR3 develop premature ovarian aging as a result of spontaneous resumption of meiosis in antral follicles, independent of the LH surge [138]. Premature oocyte attrition thus results. Located on chromosome 1, GPR3 consists of two exons. Our group interrogated GPR3 in 82 Caucasian women with POF; none (0) showed perturbations of significance [139]. One woman showed heteroduplex formation as a result of a heterozygous nucleotide substitution, C to A at position 51 (c.51C>A). However, this substitution does not alter the amino acid sequence and had already been registered in the SNP database. GPR3 mutations have thus not yet been shown to be a common explanation for POF in North American Caucasians, nor in Chinese [140].

28. KIT:
KIT is an autosomal (4q12) gene encoding the tyrosine kinase transmembrane regulator for mast/stem cell growth factor [141]. The c-kit receptor and its ligand constitute two murine loci long known to be characterized by decreased germ cells: white spotting (W) and steel (L). Thus, human KIT becomes a good candidate gene for POF. Shibanuma et al. [142] studied 40 women with unexplained POF, sequencing the coding regions. One synonymous mutation was found but this was not considered a plausible disease-causing perturbation.

29. Ring finger protein like 4 (RFPL4):
RING fingerlike protein is expressed in oocytes and, in mice, exclusively in that organ. The gene encodes an E3 ubiquitin protein ligase that helps regulate protein degradation. Human RFLP4 is located on 19q13.4 and has been shown to interact with oocyte proteins in the ubiquitin-protease degradation pathway [143]. In the context of a review, Suzumori et al. [144] mentioned that no mutations were found in their Japanese POF patients with "46,XX POF." The sample size was not stated.

30. MSH5 and DMC1:
Various pleiotropic syndromes are associated with chromosomal breakage syndromes (e.g., ataxia telangiectasia and Bloom syndrome) and have long been known to result in

ovarian failure. Genes perturbing meiosis or cell division are also logical candidates for non-syndromic POF. An example is the family of mismatch genes pivotal in repairing DNA damage. Such mutations lead to hereditary nonpolyposis colon cancer (HNPCC). Mandon-Pepin et al. [145] sought mismatch mutations in 44 POF women for DMC1, MSH4, MSH5, and SPO11. A heterozygous mutation (2547C>T) for MSH5 was found in one woman, whereas another woman showed homozygosity 3351>AC in DMC1.

31. PTEN:
    PTEN is a tumor-suppressor gene, a regulator of cell growth that could logically disturb oogenesis. Shimizu et al. [146] failed to find perturbations in 20 women with idiopathic POF.
32. Cyclin-dependent kinase inhibitor IB (CDKN1B):
    This gene, located on 12p13.1–p12, is a negative inhibitor of the cell cycle. In mice deletion of Cdkn1b, which is expressed in oocytes, results in upregulation of oogenesis and, hence, premature depletion of oocytes [17]. CDKN1B is thus a candidate gene for human POF. Ojeda et al. [147] sequenced 87 Tunisian women with POF. One nonsynonymous variant (1le119Thr) was found in a conserved region (leucine or isoleucine); the change altered hydrophobicity and could thus affect protein alignment. Control groups of Tunisian (N=137) and Colombian (N=126) women did not show the alteration. No perturbations were found in 124 Chinese POF cases [148].
33. AMH and AMH receptor type II (AMHR):
    In addition to its role in mullerian duct regression in males, AMH is an oocyte inhibitor. Murine knockout models show early depletion of primordial follicles [149]. AMH appears to play a permissive or a synergistic role in gonadal development. In humans Wang et al. failed to find plausible perturbations in 16 POF cases [150].
34. Mitochondrial genes:
    Perturbations of mitochondrial genes are good candidates for POF because the mature oocyte has the largest number of mDNA copies of any human cell. One gene related to POF has been identified.

    In progressive external ophthalmoplegia (PEO), proximal myopathy, sensory ataxia, and Parkinsonism occur. This disorder results from a mutation in the mitochondrial gene polymerase gamma. In three of the seven families studied by Luoma et al. [151], POF cosegregated with PEO. The missense mutation Y955C was found in two of the three families. This tyrosine-to-cytosine change involves a highly conserved region, making a functional effect more plausible. In the third family compound heterozygosity (N468D/A1105T) was observed in an affected woman. In a second report, Pagnamenta et al.

[152] reported individuals in three generations affected with both PEO and POF; Y955C cosegregated with PEO.

35. Germ cell failure in both sexes (XY and 46,XX):
In several sibships, both males (46,XY) and females (46,XX) have shown germ cell failure. No other organ systems were affected. Affected females show streak gonads, whereas males show germ cell aplasia (Sertoli cell-only syndrome). In two families, parents were consanguineous, and in neither were somatic anomalies observed [153, 154].

These families demonstrate that a single autosomal gene may be capable of deleteriously affecting germ cell development in both sexes. A reasonable hypothesis involves acting at a site common to early germ cell development (e.g., primordial germ cell migration or genes in the genital ridge). Several genes are expressed in the genital ridge (Fig. 2) and could be candidates. Another hypothesis is that disturbance involves germ cells migrating to the genital ridge.

In another group of families, germ cell absence occurs in both 46,XY and 46,XX sibs but coexists with distinctive patterns of somatic anomalies. Al-Awadi and co-workers [155] reported germ cell failure and an unusual form of alopecia. Scalp hair persisted in the midline, but no hair was present on sides ("manlike"). Mikati and colleagues [156] reported germ cell failure, microcephaly, short stature, mental retardation, and unusual facies (synophyrs, abnormal pinnae, micrognathia, and loss of teeth). The sibs reported by Al-Awadi and co-workers [155] were Jordanian; those reported by Mikati and colleagues [156] were Lebanese. In both families, parents were consanguineous.

1.5 Uterine Anomalies Causing Infertility

Malformations involving the internal genital ducts may cause pregnancy loss, malpresentation during pregnancy and labor, and less often infertility. In this section we discuss those uterine anomalies that may present clinically in the context of infertility.

1.5.1 Incomplete Mullerian Fusion

During embryogenesis the paired mullerian ducts fuse and canalize at the 150- to 200-mm stage, thereafter forming the upper vagina, uterus, and fallopian tubes. When the two ducts fail to fuse and canalize, incomplete mullerian fusion (IMF) exists. Various subtypes exist [63, 157]. Depending on definition (e.g., including or not arcuate uterus) IMF is not uncommon. However, clinically significant IMF is less common. Failure of fusion of mullerian ducts may result in two hemiuteri, for example, each associated with no more than one fallopian tube. If one mullerian duct fails to contribute to the definitive uterus, a rudimentary horn results. Ipsilateral renal agenesis often coexists. A rudimentary horn and several other forms of IMF may produce obstructions, leading to amenorrhea and secondarily to endometriosis.

Genetic basis of IMF is likely polygenic/multifactorial, analogous to other relatively common (1 per 1,000 incidence) birth defects restricted to a single organ system. Many familial aggregates have been reported, including multiple affected siblings as well as affected mother and daughter [158–167]. In the same kindred, affected relatives may show different forms of IMF [159]. Only one formal genetic study has been conducted, involving just 24 index cases [158], 1 of 37 (2.7 %) sisters having a clinically symptomatic uterine anomaly.

Despite paucity of data, recurrence risk of this magnitude for first-degree relatives would be consistent with predictions based on polygenic/multifactorial etiology.

Molecular sequencing studies were pursued in Han Chinese [168], for PBX1, a cofactor of the homeobox domain (HOX) (*see* Subheading 1.5.2). Pbx is expressed in murine uteri during early oogenesis knockout mice that exhibit absence of mullerian duct derivalues [169]. PBX1 was sequenced in 173 women with unicornuate ($N=55$), bicornuate ($N=76$), and septate uterus ($N=42$). No plausible disease-causing mutations were found [168].

IMF is also found in more than a dozen malformation syndromes. In Fryns syndrome, Halal syndrome, and hydrolethalus syndrome, IMF is consistently observed. IMF is less commonly observed in other disorders.

1.5.2 Hand–Foot–Genital Syndrome

Hand–foot–genital (HFG) syndrome is an autosomal dominant disorder characterized by IMF, skeletal anomalies, and urologic anomalies. In HFG syndrome skeletal anomalies exist: short first metacarpals, small distal phalanges on the thumbs, short middle fifth phalanges, and fusion of wrist bones [166, 170]. The hallux (great toe) is short because of a shortened metatarsal; the distal phalanx is small and pointed. Urinary system anomalies include urinary incontinence, a ventrally displaced urethral meatus, and malposition of the ureteral orifices in the bladder wall [166]. These urologic anomalies differ from those commonly associated with isolated IMF, which if present are typically the absence of one kidney, pelvic kidney, or duplication or absence of ureters. Vertebral anomalies in IMF also differ in type from those in HFG.

HFG is caused by perturbation of HOXA13, present on 7p14-p15. HOXA13 is one of the homeodomain genes (HOXA7–13) long known to be pivotal for internal genital duct differentiation in mammals. The first HFG family was reported [171] and later analyzed molecularly by Mortlock and Innis [172]. A HOXA13 nonsense mutation resulted in a highly conserved tryptophan being converted to a stop codon, truncating the protein by 20 amino acids. Goodman et al. [173] found a HOXA13 mutation involving a stop codon in another case and in a second family found expansion of the polyadenosine (poly A) tail. The latter suggests a dominant negative mechanism, consistent with autosomal dominant inheritance.

Goodman and colleagues [173] later studied six additional families, two previously unstudied and four previously reported. In three of the six families, nonsense mutations resulted in a truncated protein. In another family, originally described by Hennekam et al. [174], the N-terminal polyalanine tract was expanded, as found previously by Goodman et al., [173]. In a fifth family, a missense mutation was present, altering an asparagine residue in the homeodomain recognition helix that is necessary to target DNA. In 2002, Utsch and colleagues [175] reported another poly A expansion in a five-generation HFG family, supporting a dominant negative mechanism in some cases.

In conclusion, perturbation of HOXA13 clearly causes HFG. HOXA13 is clearly integral for both skeletal development and mullerian fusion, but not necessarily a common explanation for isolated IMF.

1.5.3 Transverse Vaginal Septa and Mckusick–Kaufman Syndrome

Transverse vaginal septa (TVS) are thick (2 cm) septa, usually located near the junction of the upper third and lower two-thirds of the vagina [63, 157, 176, 177], seemingly obliterating the upper genital track (cervix, uterus). The uterus appears absent on vaginal exam but can often be detected by rectal examination on imaging. Gonads (ovaries) and external genitalia (female) are normal.

Perforations may exist and are typically central in location. If no perforation exists, mucus and menstrual fluid accumulate to produce hydrocolpos or hydrometrocolpos. Conception is obviously not possible without surgical correction. If not recognized, the obstruction can lead to endometriosis, which in turn causes infertility.

In the Amish, the autosomal recessive gene responsible for TVS [178, 179] also has a pleiotropic effect causing polydactyly and cardiac defects. This disorder has a pleiotropic effect, causing polydactyly and cardiac defects [178, 180]. When present it is labelled McKusick–Kaufman syndrome (MKS). Whether all cases of TVS—Amish and non-Amish—are caused by the same MKS gene is unclear. Familial aggregates of MKS have been observed in Italian and Puerto Rican populations [181], and in these families vaginal atresia (see below) is more common than TVS.

The gene causing MKS is located on 20p12. MKS encodes a chaperonin, representative of the class of proteins that facilitate protein folding in conjunction with adenosine triphosphate hydrolysis [182]. Causation was initially deduced on the basis of H84Y/A242S compound heterozygotes cosegregating with the disorder in a large Amish pedigree [182]. Each of these two missense mutation is present in 1 per 100 Amish controls, a frequency (1 %) that when combined generates the heterozygote frequency for MKS in Amish. Neither sequence was found in 100 non-Amish controls. However, compound heterozygotes may be unaffected, suggesting that the allele frequency is higher than the 3–9 % estimated on clinical grounds alone.

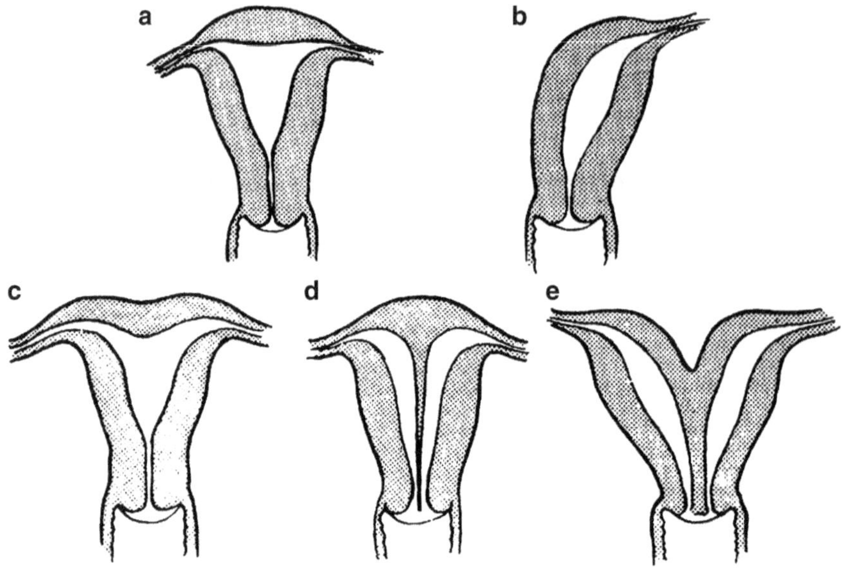

Fig. 3 Diagrammatic representation of some mullerian fusion anomalies. (**a**) Normal uterus, fallopian tubes, and vagina. (**b**) Uterus unicornis (absence of one uterine horn). (**c**) Uterus arcuatus (broadening and medial depression of a portion of the uterine fundus). (**d**) Uterus septus (persistence of a complete uterine septum). (**e**) Uterus bicornis unicollis (two hemiuteri, each leading to the same cervix) (from Simpson JL (1976) Disorders of sexual differentiation: etiology and clinical delineation. Academic, New York, NY) [63]

1.5.4 Vaginal Atresia (VA)

In this condition, the lower portion of the vagina, typically one-fifth to one-third of the total length, is replaced by 2–3 cm of fibrous tissue (Fig. 3). External genitalia are otherwise normal for women, except for absence of the hymen. It is generally believed that embryonic origin involves failure of the urogenital sinus to contribute the caudal portion of the vagina. (However, presence of VA in MKS suggests a common embryologic basis in some cases.) Superior to the obstruction is a well-differentiated upper vagina, cervix, uterine corpus, and fallopian tubes. Diagnosis requires ultrasound, MRI, or rectal examination to verify the presence of mullerian derivatives, specifically the cervix and uterus. Like TVS, conception is not possible without surgical correction. If not corrected in timely fashion, endometrics can be expected to develop and lead to secondary infertility.

No familial aggregates of *isolated* VA have been reported. However, VA is present in the context of multiple malformation syndromes [9, 157]. One syndrome was described by Winter and colleagues [183], an autosomal recessive disorder characterized by VA, renal hypoplasia or agenesis, and middle-ear anomalies (e.g., malformed incus, fixation of the malleus and incus) [183, 184]. Other syndromes in which VA has been associated include those reported by Antley–Bixler and now known to be caused by a WNT 4 mutation, Apert, Bardet Biedl, Ellis Van Creveld, Fraser (cryptophthalmos), Laurence–Moon, Pallister–Hall, and Robinow [9, 157].

1.5.5 Mullerian Aplasia (MA)

Aplasia of the mullerian ducts leads to absence of the uterine corpus, cervix, and upper (superior) vagina. Infertility is obvious and once might not have even warranted discussion in a chapter like this. However, ovaries are normal in MA and secondary sexual development is normal. ART now allows these women to have their own biologic offspring.

The vagina is formed entirely from urogenital sinus invagination. Because usually most of the vaginal length is contributed by mullerian derivatives, the vagina may be shortened to only 1–2 cm, a vaginal length that may or may not be adequate for coitus. Although by definition there is no well-differentiated uterus, bilateral remnants may persist in the form of cords. The eponym Mayer–Rokitansky–Kuster–Hauser syndrome should be considered synonymous with MA, although some authors restrict MRKH to those cases in which remnants are present.

Affected sibships are well documented [185–187]. However, discordant monozygotic twins have also been recognized [188]; thus, a single autosomal recessive gene cannot explain all cases. Carson and co-workers [189] studied 23 propositae and found no affected relatives. Van Lingen and colleagues [190] observed only one set of affected siblings among 35 cases.

Because ovaries are normal embryologically, women with MA may transit their DNA and have biologic offspring using assisted reproductive technologies (ART). Their oocytes can be stimulated, aspirated, and fertilized in vitro using the sperm from the husband or a donor and the embryo transfered into a surrogate uterus. Resulting offspring will have the genetic constitution of the MA woman and her husband. Information on inheritance of this once genetically lethal disorder can thus be derived. Surveying the US ART programs Petrozza and colleagues [191] collected 34 women with MA whose oocytes were used to generate biologic offspring in the fashion described. Of 17 female offspring, none had MA; one male child had a middle-ear defect and hearing loss. MA is thus probably multifactorial in etiology. Autosomal recessive inheritance is not excluded, but autosomal dominant is unlikely.

Molecular studies have been performed, but few have used contemporary (sequencing) methodology. Using denaturing gradient gel electrophoresis (DGGE), studies in the 1990s failed to find molecular abnormalities in various candidate genes. However, technique available at that time could exclude only large deletions. WT1 [190], PAX2 [192], AMH, and AMHR [193] were among the genes studied, and deletions so excluded. The frequency of the N314 (Duarte) allele of GALT [193, 194] was not increased in frequency. More recent and more detailed studies include those of Cheroki et al. [195], who founded no perturbations in 25 MA cases sequenced for WNT4, RAR-Gamma, and RXR-alpha [195]. Burel et al. [196] found no perturbations in HOX genes (A7–A13) nor PBX in a small sample of six cases. Ma et al. also found no PBX1 perturbations in 19 cases [168].

Philibert et al. [197] studied 28 cases for mutations in WNT4, finding one heterozygous missense mutation (L12P). No WNT4 mutations were found in three other series [198–200]. Interest has also been raised in whether certain genes might be highly methylated in MA. Sandbacka [201] did not confirm this for H19 ICR but did find aberrant methylation at 3 of 16 sites for M19.

Several groups have performed array CGH [195, 202–207], an approach potentially capable of detecting DNA CNVs that would indicate loss or gain of function of a disease-causing gene. Studies have to date involved very small sample sizes and only European populations. Still, a variety of microdeletions and microduplications have been found. These involved 1q21.1, 16p11.2, 17q12, and 22q11.2, regions of common variants in normal individuals. Familial studies, which must be performed to exclude benign polymorphisms, were not performed in most studies. Del (22q11.2) seems most likely to play a role in MA, whether de novo or familial (paternal of course) [207]. A report illustrative of problems of interpretation is that of Gervasini [208], who among 30 MA cases found duplication of Xq21 in 5 cases (two sporadic, three familial). Especially intriguing was a family of two affected MA sibs who inherited a 17 kb duplication from their father. Two other sibs having a uterus failed to inherit the microduplication. The pseudoautosomal region involved in the duplication contained SHOX, a homeobox gene that escapes X inactivation and has been noted already to be related to short stature if deficient and to Leri–Weill syndrome and Langer mesomelic dysplasia if duplicated [40, 41]. However, the array CGH findings of Gervasini et al. [208] were not confirmed in the much larger ($N = 101$) and more robust study of Sandbacka et al. [209]. This report also noted that all CNVs reported by Gervasini et al. [208] are in fact now recorded in the Database of Genomic Variants and thus considered to be without phenotypic effect. In summary, array CGH studies should, with the exception of del (22q11.2), not be considered as indicating an informative gene or region.

Although all the studies have been negative, molecular perturbations in MA have been found in a very atypical group of MA women. Biason-Lauber et al. [210] found perturbations of WNT4 in MA cases in which virilization and adrenal insufficiency existed. The relevance to isolated MA is unclear. Gervasini et al. [208] found no perturbations in 12 MA cases who also had hyperandrogenism. That WNT4 acts before AMH and is required for initial mullerian development [211] provides grounds for plausibility.

MA may be one component of malformation syndromes, discussed and tabulated elsewhere by Simpson in Simpson and Elias [9]. An example is mullerian renal cervical somite (MURCS), an acronym applied when MA coexists with renal and skeletal anomalies as well as facial clefts and cardiac anomalies [212]. MA is not uncommon in the fascio-auricular-vertebral syndrome (Goldenhar syndrome) [213], limb/pelvis/uterus hypoplasia syndrome [214, 215], thalidomide, embryopathy [216], and thrombocytopenia-absent radius (TAR) syndrome [217].

References

1. Simpson JL (2012) Disorders of the gonads, genital tract and genitalia, 6th edn. Churchill-Livingstone, New York, NY
2. Jost A (1966) Problems of fetal endocrinology: the adrenal glands. Recent Prog Horm Res 22:541–574
3. Schlessinger D, Garcia-Ortiz JE, Forabosco A, Uda M, Crisponi L, Pelosi E (2010) Determination and stability of gonadal sex. J Androl 31:16–25
4. Uhlenhaut NH, Jakob S, Anlag K, Eisenberger T, Sekido R, Kress J, Treier AC, Klugmann C, Klasen C, Holter NI, Riethmacher D, Schutz G, Cooney AJ, Lovell-Badge R, Treier M (2009) Somatic sex reprogramming of adult ovaries to testes by FOXL2 ablation. Cell 139:1130–1142
5. Fraccaro M, Maraschio P, Pasquali F, Scappaticci S (1977) Women heterozygous for deficiency of the (p21 leads to pter) region of the X chromosome are fertile. Hum Genet 39:283–292
6. Simpson JL (1986) Phenotypic-karyotypic correlations of gonadal determinants: current status and relationship to molecular studies. In: Sperling KVF (ed) Seventh international congress on human genetics. Springer, Berlin, pp 224–232
7. Simpson JL (1989) Genetic heterogeneity in XY sex reversal. Potential pitfalls in isolating the testis-determining-factor (TDF). In: Wachtel S (ed) Evolutionary mechanisms in sex determination. CRC, Baton Rouge, LA
8. Simpson JL (1988) Genetics of sex determination. In: Iizuka R, Semm K, Ohno T (eds) Human reproduction: current status, future prospect. Proceedings of VIth World Congress on Human Reproduction. Elsevier Science, Amsterdam
9. Simpson JL, Elias S (2003) Genetics in obstetrics and gynecology, 3rd edn. WB Saunders, Philadelphia
10. Simpson JL, Lebeau MM (1981) Gonadal and statural determinants on the X chromosome and their relationship to in vitro studies showing prolonged cell cycles in 45, X; 46, X, del(X)(p11); 46, X, del(X)(q13); and 46, X, del(X)(q22) fibroblasts. Am J Obstet Gynecol 141:930–940
11. Ogata T, Matsuo N (1995) Turner syndrome and female sex chromosome aberrations: deduction of the principal factors involved in the development of clinical features. Hum Genet 95:607–629
12. Simpson JL (2000) Genetic programming in ovarian development and oogenesis. In: Lobo RA, Kelsey J, Marcus R (eds) Menopause biology and pathobiology. Academic, London
13. Simpson JL, Rajkovic A (1999) Ovarian differentiation and gonadal failure. Am J Med Genet 89:186–200
14. Fitch N, de Saint Victor J, Richer CL, Pinsky L, Sitahal S (1982) Premature menopause due to a small deletion in the long arm of the X chromosome: a report of three cases and a review. Am J Obstet Gynecol 142:968–972
15. Krauss CM, Turksoy RN, Atkins L, McLaughlin C, Brown LG, Page DC (1987) Familial premature ovarian failure due to an interstitial deletion of the long arm of the X chromosome. N Engl J Med 317:125–131
16. Tharapel AT, Anderson KP, Simpson JL, Martens PR, Wilroy RS Jr, Llerena JC Jr, Schwartz CE (1993) Deletion (X)(q26.1→q28) in a proband and her mother: molecular characterization and phenotypic-karyotypic deductions. Am J Hum Genet 52:463–471
17. Matzuk MM, Lamb DJ (2008) The biology of infertility: research advances and clinical challenges. Nat Med 14:1197–1213
18. Ochalski ME, Engle N, Wakim A, Ravnan BJ, Hoffner L, Rajkovic A, Surti U (2011) Complex X chromosome rearrangement delineated by array comparative genome hybridization in a woman with premature ovarian insufficiency. Fertil Steril 95(2433):e2439–2415
19. Kok HS, van Asselt KM, van der Schouw YT, Peeters PH, Wijmenga C (2005) Genetic studies to identify genes underlying menopausal age. Hum Reprod Update 11:483–493
20. Stolk L, Zhai G, van Meurs JB, Verbiest MM, Visser JA, Estrada K, Rivadeneira F, Williams FM, Cherkas L, Deloukas P, Soranzo N, de

Keyzer JJ, Pop VJ, Lips P, Lebrun CE, van der Schouw YT, Grobbee DE, Witteman J, Hofman A, Pols HA, Laven JS, Spector TD, Uitterlinden AG (2009) Loci at chromosomes 13, 19 and 20 influence age at natural menopause. Nat Genet 41:645–647

21. Christin-Maitre S, Tachdjian G (2010) Genome-wide association study and premature ovarian failure. Ann Endocrinol (Paris) 71:218–221
22. He C, Kraft P, Chen C, Buring JE, Pare G, Hankinson SE, Chanock SJ, Ridker PM, Hunter DJ, Chasman DI (2009) Genome-wide association studies identify loci associated with age at menarche and age at natural menopause. Nat Genet 41:724–728
23. Murray A, Bennett CE, Perry JR, Weedon MN, Jacobs PA, Morris DH, Orr N, Schoemaker MJ, Jones M, Ashworth A, Swerdlow AJ (2011) Common genetic variants are significant risk factors for early menopause: results from the Breakthrough Generations Study. Hum Mol Genet 20: 186–192
24. Knauff EA, Franke L, van Es MA, van den Berg LH, van der Schouw YT, Laven JS, Lambalk CB, Hoek A, Goverde AJ, Christin-Maitre S, Hsueh AJ, Wijmenga C, Fauser BC (2009) Genome-wide association study in premature ovarian failure patients suggests ADAMTS19 as a possible candidate gene. Hum Reprod 24:2372–2378
25. Knauff EA, Blauw HM, Pearson PL, Kok K, Wijmenga C, Veldink JH, van den Berg LH, Bouchard P, Fauser BC, Franke L (2011) Copy number variants on the X chromosome in women with primary ovarian insufficiency. Fertil Steril 95(1584–1588):e1581
26. Dudding TE, Lawrence O, Winship I, Froyen G, Vandewalle J, Scott R, Shelling AN (2010) Array comparative genomic hybridization for the detection of submicroscopic copy number variations of the X chromosome in women with premature ovarian failure. Hum Reprod 25:3159–3160, author reply 3160–3151
27. Aboura A, Dupas C, Tachdjian G, Portnoi MF, Bourcigaux N, Dewailly D, Frydman R, Fauser B, Ronci-Chaix N, Donadille B, Bouchard P, Christin-Maitre S (2009) Array comparative genomic hybridization profiling analysis reveals deoxyribonucleic acid copy number variations associated with premature ovarian failure. J Clin Endocrinol Metab 94: 4540–4546
28. Quilter CR, Karcanias AC, Bagga MR, Duncan S, Murray A, Conway GS, Sargent CA, Affara NA (2010) Analysis of X chromosome genomic DNA sequence copy number variation associated with premature ovarian failure (POF). Hum Reprod 25:2139–2150
29. Jones MH, Furlong RA, Burkin H, Chalmers IJ, Brown GM, Khwaja O, Affara NA (1996) The Drosophila developmental gene fat facets has a human homologue in Xp11.4 which escapes X-inactivation and has related sequences on Yq11.2. Hum Mol Genet 5: 1695–1701
30. Luoh SW, Bain PA, Polakiewicz RD, Goodheart ML, Gardner H, Jaenisch R, Page DC (1997) Zfx mutation results in small animal size and reduced germ cell number in male and female mice. Development 124: 2275–2284
31. Hanrahan JP, Gregan SM, Mulsant P, Mullen M, Davis GH, Powell R, Galloway SM (2004) Mutations in the genes for oocyte-derived growth factors GDF9 and BMP15 are associated with both increased ovulation rate and sterility in Cambridge and Belclare sheep (Ovis aries). Biol Reprod 70:900–909
32. Dube JL, Wang P, Elvin J, Lyons KM, Celeste AJ, Matzuk MM (1998) The bone morphogenetic protein 15 gene is X-linked and expressed in oocytes. Mol Endocrinol 12:1809–1817
33. Hanevik HI, Hilmarsen HT, Skjelbred CF, Tanbo T, Kahn JA (2011) A single nucleotide polymorphism in BMP15 is associated with high response to ovarian stimulation. Reprod Biomed Online 23:97–104
34. Di Pasquale E, Beck-Peccoz P, Persani L (2004) Hypergonadotropic ovarian failure associated with an inherited mutation of human bone morphogenetic protein-15 (BMP15) gene. Am J Hum Genet 75:106–111
35. Dixit H, Rao LK, Padmalatha VV, Kanakavalli M, Deenadayal M, Gupta N, Chakrabarty B, Singh L (2006) Missense mutations in the BMP15 gene are associated with ovarian failure. Hum Genet 119:408–415
36. Laissue P, Christin-Maitre S, Touraine P, Kuttenn F, Ritvos O, Aittomaki K, Bourcigaux N, Jacquesson L, Bouchard P, Frydman R, Dewailly D, Reyss AC, Jeffery L, Bachelot A, Massin N, Fellous M, Veitia RA (2006) Mutations and sequence variants in GDF9 and BMP15 in patients with premature ovarian failure. Eur J Endocrinol 154:739–744
37. Zhang P, Shi YH, Wang LC, Chen ZJ (2007) Sequence variants in exons of the BMP-15 gene in Chinese patients with premature ovarian failure. Acta Obstet Gynecol Scand 86:585–589
38. Ellison JW, Wardak Z, Young MF, Gehron Robey P, Laig-Webster M, Chiong W (1997) PHOG, a candidate gene for involvement in

the short stature of Turner syndrome. Hum Mol Genet 6:1341–1347

39. Rao E, Weiss B, Fukami M, Rump A, Niesler B, Mertz A, Muroya K, Binder G, Kirsch S, Winkelmann M, Nordsiek G, Heinrich U, Breuning MH, Ranke MB, Rosenthal A, Ogata T, Rappold GA (1997) Pseudoautosomal deletions encompassing a novel homeobox gene cause growth failure in idiopathic short stature and Turner syndrome. Nat Genet 16:54–63
40. Belin V, Cusin V, Viot G, Girlich D, Toutain A, Moncla A, Vekemans M, Le Merrer M, Munnich A, Cormier-Daire V (1998) SHOX mutations in dyschondrosteosis (Leri-Weill syndrome). Nat Genet 19:67–69
41. Shears DJ, Vassal HJ, Goodman FR, Palmer RW, Reardon W, Superti-Furga A, Scambler PJ, Winter RM (1998) Mutation and deletion of the pseudoautosomal gene SHOX cause Leri-Weill dyschondrosteosis. Nat Genet 19:70–73
42. Tachdjian G, Aboura A, Portnoi MF, Pasquier M, Bourcigaux N, Simon T, Rousseau G, Finkel L, Benkhalifa M, Christin-Maitre S (2008) Cryptic Xp duplication including the SHOX gene in a woman with 46, X, del(X)(q21.31) and premature ovarian failure. Hum Reprod 23:222–226
43. Castrillon DH, Wasserman SA (1994) Diaphanous is required for cytokinesis in Drosophila and shares domains of similarity with the products of the limb deformity gene. Development 120:3367–3377
44. Bione S, Sala C, Manzini C, Arrigo G, Zuffardi O, Banfi S, Borsani G, Jonveaux P, Philippe C, Zuccotti M, Ballabio A, Toniolo D (1998) A human homologue of the Drosophila melanogaster diaphanous gene is disrupted in a patient with premature ovarian failure: evidence for conserved function in oogenesis and implications for human sterility. Am J Hum Genet 62:533–541
45. Mumm S, Herrera L, Waeltz PW, Scardovi A, Nagaraja R, Esposito T, Schlessinger D, Rocchi M, Forabosco A (2001) X/autosomal translocations in the Xq critical region associated with premature ovarian failure fall within and outside genes. Genomics 76: 30–36
46. Prueitt RL, Chen H, Barnes RI, Zinn AR (2002) Most X;autosome translocations associated with premature ovarian failure do not interrupt X-linked genes. Cytogenet Genome Res 97:32–38
47. Riva P, Magnani I, Fuhrmann Conti AM, Gelli D, Sala C, Toniolo D, Larizza L (1996) FISH characterization of the Xq21 breakpoint in a translocation carrier with premature ovarian failure. Clin Genet 50:267–269
48. Lacombe A, Lee H, Zahed L, Choucair M, Muller JM, Nelson SF, Salameh W, Vilain E (2006) Disruption of POF1B binding to nonmuscle actin filaments is associated with premature ovarian failure. Am J Hum Genet 79:113–119
49. Bione S, Rizzolio F, Sala C, Ricotti R, Goegan M, Manzini MC, Battaglia R, Marozzi A, Vegetti W, Dalpra L, Crosignani PG, Ginelli E, Nappi R, Bernabini S, Bruni V, Torricelli F, Zuffardi O, Toniolo D (2004) Mutation analysis of two candidate genes for premature ovarian failure, DACH2 and POF1B. Hum Reprod 19:2759–2766
50. Padovano V, Lucibello I, Alari V, Della Mina P, Crespi A, Ferrari I, Recagni M, Lattuada D, Righi M, Toniolo D, Villa A, Pietrini G (2011) The POF1B candidate gene for premature ovarian failure regulates epithelial polarity. J Cell Sci 124:3356–3368
51. Katsuya T, Horiuchi M, Minami S, Koike G, Santoro NF, Hsueh AJ, Dzau VJ (1997) Genomic organization and polymorphism of human angiotensin II type 2 receptor: no evidence for its gene mutation in two families of human premature ovarian failure syndrome. Mol Cell Endocrinol 127:221–228
52. Bione S, Toniolo D (2000) X chromosome genes and premature ovarian failure. Semin Reprod Med 18:51–57
53. Prueitt RL, Ross JL, Zinn AR (2000) Physical mapping of nine Xq translocation breakpoints and identification of XPNPEP2 as a premature ovarian failure candidate gene. Cytogenet Cell Genet 89:44–50
54. Wittenberger MD, Hagerman RJ, Sherman SL, McConkie-Rosell A, Welt CK, Rebar RW, Corrigan EC, Simpson JL, Nelson LM (2007) The FMR1 premutation and reproduction. Fertil Steril 87:456–465
55. Schwartz CE, Dean J, Howard-Peebles PN, Bugge M, Mikkelsen M, Tommerup N, Hull C, Hagerman R, Holden JJ, Stevenson RE (1994) Obstetrical and gynecological complications in fragile X carriers: a multicenter study. Am J Med Genet 51:400–402
56. Allingham-Hawkins DJ, Babul-Hirji R, Chitayat D, Holden JJ, Yang KT, Lee C, Hudson R, Gorwill H, Nolin SL, Glicksman A, Jenkins EC, Brown WT, Howard-Peebles PN, Becchi C, Cummings E, Fallon L, Seitz S, Black SH, Vianna-Morgante AM, Costa SS, Otto PA, Mingroni-Netto RC, Murray A, Webb J, Vieri F et al (1999) Fragile X premutation is a significant risk factor for premature

ovarian failure: the International Collaborative POF in Fragile X study–preliminary data. Am J Med Genet 83:322–325

57. Sullivan AK, Marcus M, Epstein MP, Allen EG, Anido AE, Paquin JJ, Yadav-Shah M, Sherman SL (2005) Association of FMR1 repeat size with ovarian dysfunction. Hum Reprod 20:402–412
58. Foresta C, Ferlin A, Gianaroli L, Dallapiccola B (2002) Guidelines for the appropriate use of genetic tests in infertile couples. Eur J Hum Genet 10:303–312
59. Simpson JL, Christakos AC, Horwith M, Silverman FS (1971) Gonadal dysgenesis in individuals with apparently normal chromosomal complements: tabulation of cases and compilation of genetic data. Birth Defects Orig Artic Ser 7:215–228
60. Simpson JL (1979) Gonadal dysgenesis and sex chromosome abnormalities. Phenotypic/karyotypic correlations. In: Vallet HL, Porter IH (eds) Genetic mechanisms of sexual development. Academic, New York, NY
61. Aittomaki K (1994) The genetics of XX gonadal dysgenesis. Am J Hum Genet 54: 844–851
62. Aittomaki K, Lucena JL, Pakarinen P, Sistonen P, Tapanainen J, Gromoll J, Kaskikari R, Sankila EM, Lehvaslaiho H, Engel AR, Nieschlag E, Huhtaniemi I, de la Chapelle A (1995) Mutation in the follicle-stimulating hormone receptor gene causes hereditary hypergonadotropic ovarian failure. Cell 82:959–968
63. Simpson JL (1976) Disorders of sexual differentiation: etiology and clinical delineation. Monograph with chapters by J.E. Jirasek, N. G. Kase, and L. Speroff. Academic, New York, NY
64. Matthews CH, Borgato S, Beck-Peccoz P, Adams M, Tone Y, Gambino G, Casagrande S, Tedeschini G, Benedetti A, Chatterjee VK (1993) Primary amenorrhoea and infertility due to a mutation in the beta-subunit of follicle-stimulating hormone. Nat Genet 5:83–86
65. Layman LC, Amde S, Cohen DP, Jin M, Xie J (1998) The Finnish follicle-stimulating hormone receptor gene mutation is rare in North American women with 46,XX ovarian failure. Fertil Steril 69:300–302
66. Liu JY, Gromoll J, Cedars MI, La Barbera AR (1998) Identification of allelic variants in the follicle-stimulating hormone receptor genes of females with or without hypergonadotropic amenorrhea. Fertil Steril 70: 326–331
67. Simoni M, Gromoll J, Nieschlag E (1997) The follicle-stimulating hormone receptor: biochemistry, molecular biology, physiology, and pathophysiology. Endocr Rev 18: 739–773
68. da Fonte Kohek MB, Batista MC, Russell AJ, Vass K, Giacaglia LR, Mendonca BB, Latronico AC (1998) No evidence of the inactivating mutation (C566T) in the follicle-stimulating hormone receptor gene in Brazilian women with premature ovarian failure. Fertil Steril 70:565–567
69. de la Chesnaye E, Canto P, Ulloa-Aguirre A, Mendez JP (2001) No evidence of mutations in the follicle-stimulating hormone receptor gene in Mexican women with 46, XX pure gonadal dysgenesis. Am J Med Genet 98:125–128
70. Touraine P, Beau I, Gougeon A, Meduri G, Desroches A, Pichard C, Detoeuf M, Paniel B, Prieur M, Zorn JR, Milgrom E, Kuttenn F, Misrahi M (1999) New natural inactivating mutations of the follicle-stimulating hormone receptor: correlations between receptor function and phenotype. Mol Endocrinol 13:1844–1854
71. Sultan C, Lumbroso S (1998) LH receptor defects. In: Kempers RD, Cohen J, Haney AF, Younger JB (eds) Fertility and reproductive medicine - Proceedings of the XVI World congress on fertility and sterility. Elsevier Science, San Francisco, CA
72. Latronico AC, Chai Y, Arnhold IJ, Liu X, Mendonca BB, Segaloff DL (1998) A homozygous microdeletion in helix 7 of the luteinizing hormone receptor associated with familial testicular and ovarian resistance is due to both decreased cell surface expression and impaired effector activation by the cell surface receptor. Mol Endocrinol 12:442–450
73. Matzuk MM, Lamb DJ (2002) Genetic dissection of mammalian fertility pathways. Nat Cell Biol 4(Suppl):s41–49
74. Yanase T, Kagimoto M, Suzuki S, Hashiba K, Simpson ER, Waterman MR (1989) Deletion of a phenylalanine in the N-terminal region of human cytochrome P-450(17 alpha) results in partial combined 17 alpha-hydroxylase/17,20-lyase deficiency. J Biol Chem 264:18076–18082
75. Woad KJ, Pearson SM, Harris SE, Gersak K, Shelling AN (2009) Investigating the association between inhibin alpha gene promoter polymorphisms and premature ovarian failure. Fertil Steril 91:62–66
76. Shelling AN, Burton KA, Chand AL, van Ee CC, France JT, Farquhar CM, Milsom SR, Love DR, Gersak K, Aittomaki K, Winship

IM (2000) Inhibin: a candidate gene for premature ovarian failure. Hum Reprod 15: 2644–2649

77. Marozzi A, Porta C, Vegetti W, Crosignani PG, Tibiletti MG, Dalpra L, Ginelli E (2002) Mutation analysis of the inhibin alpha gene in a cohort of Italian women affected by ovarian failure. Hum Reprod 17:1741–1745
78. Dixit H, Deendayal M, Singh L (2004) Mutational analysis of the mature peptide region of inhibin genes in Indian women with ovarian failure. Hum Reprod 19: 1760–1764
79. Prakash GJ, Ravi Kanth VV, Shelling AN, Rozati R, Sujatha M (2010) Mutational analysis of inhibin alpha gene revealed three novel variations in Indian women with premature ovarian failure. Fertil Steril 94:90–98
80. Shelling AN (2010) Premature ovarian failure. Reproduction 140:633–641
81. Chand AL, Ooi GT, Harrison CA, Shelling AN, Robertson DM (2007) Functional analysis of the human inhibin alpha subunit variant A257T and its potential role in premature ovarian failure. Hum Reprod 22:3241–3248
82. Chand AL, Harrison CA, Shelling AN (2010) Inhibin and premature ovarian failure. Hum Reprod Update 16:39–50
83. Dixit H, Rao LK, Padmalatha V, Kanakavalli M, Deenadayal M, Gupta N, Chakravarty B, Singh L (2005) Mutational screening of the coding region of growth differentiation factor 9 gene in Indian women with ovarian failure. Menopause 12:749–754
84. Simsek E, Ozdemir I, Lin L, Achermann JC (2005) Isolated 17,20-lyase (desmolase) deficiency in a 46,XX female presenting with delayed puberty. Fertil Steril 83:1548–1551
85. Araki S, Chikazawa K, Sekiguchi I, Yamauchi H, Motoyama M, Tamada T (1987) Arrest of follicular development in a patient with 17 alpha-hydroxylase deficiency: folliculogenesis in association with a lack of estrogen synthesis in the ovaries. Fertil Steril 47:169–172
86. Rabinovici J, Blankstein J, Goldman B, Rudak E, Dor Y, Pariente C, Geier A, Lunenfeld B, Mashiach S (1989) In vitro fertilization and primary embryonic cleavage are possible in 17 alpha-hydroxylase deficiency despite extremely low intrafollicular 17 beta-estradiol. J Clin Endocrinol Metab 68:693–697
87. Simpson ER (1998) Genetic mutations resulting in estrogen insufficiency in the male. Mol Cell Endocrinol 145:55–59
88. Ito Y, Fisher CR, Conte FA, Grumbach MM, Simpson ER (1993) Molecular basis of aromatase deficiency in an adult female with sexual infantilism and polycystic ovaries. Proc Natl Acad Sci U S A 90:11673–11677
89. Conte FA, Grumbach MM, Ito Y, Fisher CR, Simpson ER (1994) A syndrome of female pseudohermaphrodism, hypergonadotropic hypogonadism, and multicystic ovaries associated with missense mutations in the gene encoding aromatase (P450arom). J Clin Endocrinol Metab 78:1287–1292
90. Mullis PE, Yoshimura N, Kuhlmann B, Lippuner K, Jaeger P, Harada H (1997) Aromatase deficiency in a female who is compound heterozygote for two new point mutations in the P450arom gene: impact of estrogens on hypergonadotropic hypogonadism, multicystic ovaries, and bone densitometry in childhood. J Clin Endocrinol Metab 82:1739–1745
91. Mansouri MR, Schuster J, Badhai J, Stattin EL, Losel R, Wehling M, Carlsson B, Hovatta O, Karlstrom PO, Golovleva I, Toniolo D, Bione S, Peluso J, Dahl N (2008) Alterations in the expression, structure and function of progesterone receptor membrane component-1 (PGRMC1) in premature ovarian failure. Hum Mol Genet 17:3776–3783
92. Crisponi L, Deiana M, Loi A, Chiappe F, Uda M, Amati P, Bisceglia L, Zelante L, Nagaraja R, Porcu S, Ristaldi MS, Marzella R, Rocchi M, Nicolino M, Lienhardt-Roussie A, Nivelon A, Verloes A, Schlessinger D, Gasparini P, Bonneau D, Cao A, Pilia G (2001) The putative forkhead transcription factor FOXL2 is mutated in blepharophimosis/ptosis/epicanthus inversus syndrome. Nat Genet 27:159–166
93. Harris SE, Chand AL, Winship IM, Gersak K, Aittomaki K, Shelling AN (2002) Identification of novel mutations in FOXL2 associated with premature ovarian failure. Mol Hum Reprod 8:729–733
94. Gersak K, Harris SE, Smale WJ, Shelling AN (2004) A novel 30 bp deletion in the FOXL2 gene in a phenotypically normal woman with primary amenorrhoea: case report. Hum Reprod 19:2767–2770
95. De Baere E, Dixon MJ, Small KW, Jabs EW, Leroy BP, Devriendt K, Gillerot Y, Mortier G, Meire F, Van Maldergem L, Courtens W, Hjalgrim H, Huang S, Liebaers I, Van Regemorter N, Touraine P, Praphanphoj V, Verloes A, Udar N, Yellore V, Chalukya M, Yelchits S, De Paepe A, Kuttenn F, Fellous M, Veitia R, Messiaen L (2001) Spectrum of FOXL2 gene mutations in blepharophimosis-ptosis-epicanthus inversus (BPES) families demonstrates a genotype–phenotype correlation. Hum Mol Genet 10:1591–1600

96. Qin Y, Choi Y, Zhao H, Simpson JL, Chen ZJ, Rajkovic A (2007) NOBOX homeobox mutation causes premature ovarian failure. Am J Hum Genet 81:576–581
97. Castrillon DH, Miao L, Kollipara R, Horner JW, DePinho RA (2003) Suppression of ovarian follicle activation in mice by the transcription factor Foxo3a. Science 301: 215–218
98. Chand AL, Ponnampalam AP, Harris SE, Winship IM, Shelling AN (2006) Mutational analysis of BMP15 and GDF9 as candidate genes for premature ovarian failure. Fertil Steril 86:1009–1012
99. Wang B, Wen Q, Ni F, Zhou S, Wang J, Cao Y, Ma X (2010) Analyses of growth differentiation factor 9 (GDF9) and bone morphogenetic protein 15 (BMP15) mutation in Chinese women with premature ovarian failure. Clin Endocrinol 72:135–136
100. Demirhan O, Turkmen S, Schwabe GC, Soyupak S, Akgul E, Tastemir D, Karahan D, Mundlos S, Lehmann K (2005) A homozygous BMPR1B mutation causes a new subtype of acromesomelic chondrodysplasia with genital anomalies. J Med Genet 42:314–317
101. Yi SE, LaPolt PS, Yoon BS, Chen JY, Lu JK, Lyons KM (2001) The type I BMP receptor BmprIB is essential for female reproductive function. Proc Natl Acad Sci U S A 98: 7994–7999
102. Wilson T, Wu XY, Juengel JL, Ross IK, Lumsden JM, Lord EA, Dodds KG, Walling GA, McEwan JC, O'Connell AR, McNatty KP, Montgomery GW (2001) Highly prolific Booroola sheep have a mutation in the intracellular kinase domain of bone morphogenetic protein IB receptor (ALK-6) that is expressed in both oocytes and granulosa cells. Biol Reprod 64:1225–1235
103. Zhao H, Chen ZJ, Qin Y, Shi Y, Wang S, Choi Y, Simpson JL, Rajkovic A (2008) Transcription factor FIGLA is mutated in patients with premature ovarian failure. Am J Hum Genet 82:1342–1348
104. Wang J, Wang B, Song J, Suo P, Ni F, Chen B, Ma X, Cao Y (2011) New candidate gene POU5F1 associated with premature ovarian failure in Chinese patients. Reprod Biomed Online 22:312–316
105. Kang H, Lee SK, Kim MH, Song J, Bae SJ, Kim NK, Lee SH, Kwack K (2008) Parathyroid hormone-responsive B1 gene is associated with premature ovarian failure. Hum Reprod 23:1457–1465
106. Kaufman FR, Kogut MD, Donnell GN, Goebelsmann U, March C, Koch R (1981) Hypergonadotropic hypogonadism in female patients with galactosemia. N Engl J Med 304:994–998
107. Waggoner DD, Buist NR, Donnell GN (1990) Long-term prognosis in galactosaemia: results of a survey of 350 cases. J Inher Metab Dis 13:802–818
108. Levy HL, Driscoll SG, Porensky RS, Wender DF (1984) Ovarian failure in galactosemia. N Engl J Med 310:50
109. Leslie ND, Yager KL, McNamara PD, Segal S (1996) A mouse model of galactose-1-phosphate uridyl transferase deficiency. Biochem Mol Med 59:7–12
110. Badik JR, Castaneda U, Gleason TJ, Spencer JB, Epstein MP, Ficicioglu C, Fitzgerald K, Fridovich-Keil JL (2011) Ovarian function in Duarte galactosemia. Fertil Steril 96(469–473):e461
111. Bjursell C, Stibler H, Wahlstrom J, Kristiansson B, Skovby F, Stromme P, Blennow G, Martinsson T (1997) Fine mapping of the gene for carbohydrate-deficient glycoprotein syndrome, type I (CDG1): linkage disequilibrium and founder effect in Scandinavian families. Genomics 39:247–253
112. Matthijs G, Schollen E, Pardon E, Veiga-Da-Cunha M, Jaeken J, Cassiman JJ, Van Schaftingen E (1997) Mutations in PMM2, a phosphomannomutase gene on chromosome 16p13, in carbohydrate-deficient glycoprotein type I syndrome (Jaeken syndrome). Nat Genet 16:88–92
113. de Zegher F, Jaeken J (1995) Endocrinology of the carbohydrate-deficient glycoprotein syndrome type 1 from birth through adolescence. Pediatr Res 37:395–401
114. Kristiansson B, Stibler H, Wide L (1995) Gonadal function and glycoprotein hormones in the carbohydrate-deficient glycoprotein (CDG) syndrome. Acta Paediatr 84:655–659
115. Wang CY, Davoodi-Semiromi A, Huang W, Connor E, Shi JD, She JX (1998) Characterization of mutations in patients with autoimmune polyglandular syndrome type 1 (APS1). Hum Genet 103:681–685
116. Boltshauser E, Barth PG, Troost D, Martin E, Stallmach T (2002) "Vanishing white matter" and ovarian dysgenesis in an infant with cerebro-oculo-facio-skeletal phenotype. Neuropediatrics 33:57–62
117. Schiffmann R, Tedeschi G, Kinkel RP, Trapp BD, Frank JA, Kaneski CR, Brady RO, Barton NW, Nelson L, Yanovski JA (1997) Leukodystrophy in patients with ovarian dysgenesis. Ann Neurol 41:654–661

118. Fogli A, Rodriguez D, Eymard-Pierre E, Bouhour F, Labauge P, Meaney BF, Zeesman S, Kaneski CR, Schiffmann R, Boespflug-Tanguy O (2003) Ovarian failure related to eukaryotic initiation factor 2B mutations. Am J Hum Genet 72:1544–1550
119. Fogli A, Gauthier-Barichard F, Schiffmann R, Vanderhoof VH, Bakalov VK, Nelson LM, Boespflug-Tanguy O (2004) Screening for known mutations in EIF2B genes in a large panel of patients with premature ovarian failure. BMC Womens Health 4:8
120. Skre H, Bassoe HH, Berg K, Frovig AG (1976) Cerebellar ataxia and hypergonadotropic hypogonadism in two kindreds. Chance concurrence, pleiotropism or linkage? Clin Genet 9:234–244
121. De Michele G, Filla A, Striano S, Rimoldi M, Campanella G (1993) Heterogeneous findings in four cases of cerebellar ataxia associated with hypogonadism (Holmes' type ataxia). Clin Neurol Neurosurg 95:23–28
122. Linssen WH, Van den Bent MJ, Brunner HG, Poels PJ (1994) Deafness, sensory neuropathy, and ovarian dysgenesis: a new syndrome or a broader spectrum of Perrault syndrome? Am J Med Genet 51:81–82
123. Amor DJ, Delatycki MB, Gardner RJ, Storey E (2001) New variant of familial cerebellar ataxia with hypergonadotropic hypogonadism and sensorineural deafness. Am J Med Genet 99:29–33
124. Groppe J, Greenwald J, Wiater E, Rodriguez-Leon J, Economides AN, Kwiatkowski W, Affolter M, Vale WW, Belmonte JC, Choe S (2002) Structural basis of BMP signalling inhibition by the cystine knot protein Noggin. Nature 420:636–642
125. Kosaki K, Sato S, Hasegawa T, Matsuo N, Suzuki T, Ogata T (2004) Premature ovarian failure in a female with proximal symphalangism and Noggin mutation. Fertil Steril 81:1137–1139
126. Perrault M, Klotz P, Housset E (1951) Deux cas de syndrome de Turner avec surdi-mutite dans une meme fratrie. Bull Mem Soc Med Hop Paris 67:79–84
127. Josso N, de Grouchy J, Frezal J, Lamy M (1963) Le syndrome de Turner familial etude de deux families avec caryotypes XO er XX. Ann Pediatr 10:163–167
128. Nishi Y, Hamamoto K, Kajiyama M, Kawamura I (1988) The Perrault syndrome: clinical report and review. Am J Med Genet 31:623–629
129. Pallister PD, Opitz JM (1979) The Perrault syndrome: autosomal recessive ovarian dysgenesis with facultative, non-sex-limited sensorineural deafness. Am J Med Genet 4: 239–246
130. Simon AM, Goodenough DA, Li E, Paul DL (1997) Female infertility in mice lacking connexin 37. Nature 385:525–529
132. Watkins WJ, Umbers AJ, Woad KJ, Harris SE, Winship IM, Gersak K, Shelling AN (2006) Mutational screening of FOXO3A and FOXO1A in women with premature ovarian failure. Fertil Steril 86:1518–1521
133. Wang B, Mu Y, Ni F, Zhou S, Wang J, Cao Y, Ma X (2010) Analysis of FOXO3 mutation in 114 Chinese women with premature ovarian failure. Reprod Biomed Online 20:499–503
134. Pangas SA, Choi Y, Ballow DJ, Zhao Y, Westphal H, Matzuk MM, Rajkovic A (2006) Oogenesis requires germ cell-specific transcriptional regulators Sohlh1 and Lhx8. Proc Natl Acad Sci U S A 103:8090–8095
135. Qin Y, Zhao H, Kovanci E, Simpson JL, Chen ZJ, Rajkovic A (2008) Analysis of LHX8 mutation in premature ovarian failure. Fertil Steril 89:1012–1014
136. Tsuda M, Sasaoka Y, Kiso M, Abe K, Haraguchi S, Kobayashi S, Saga Y (2003) Conserved role of nanos proteins in germ cell development. Science 301:1239–1241
137. Qin Y, Zhao H, Kovanci E, Simpson JL, Chen ZJ, Rajkovic A (2007) Mutation analysis of NANOS3 in 80 Chinese and 88 Caucasian women with premature ovarian failure. Fertil Steril 88:1465–1467
138. Mehlmann LM, Saeki Y, Tanaka S, Brennan TJ, Evsikov AV, Pendola FL, Knowles BB, Eppig JJ, Jaffe LA (2004) The Gs-linked receptor GPR3 maintains meiotic arrest in mammalian oocytes. Science 306:1947–1950
139. Kovanci E, Simpson JL, Amato P, Rohozinski J, Heard MJ, Bishop CE, Carson SA (2008) Oocyte-specific G-protein-coupled receptor 3 (GPR3): no perturbations found in 82 women with premature ovarian failure (first report). Fertil Steril 90:1269–1271
140. Zhou S, Wang B, Ni F, Wang J, Cao Y, Ma X (2010) GPR3 may not be a potential candidate gene for premature ovarian failure. Reprod Biomed Online 20:53–55
141. Richards KA, Fukai K, Oiso N, Paller AS (2001) A novel KIT mutation results in piebaldism with progressive depigmentation. J Am Acad Dermatol 44:288–292
142. Shibanuma K, Tong ZB, Vanderhoof VH, Vanevski K, Nelson LM (2002) Investigation of KIT gene mutations in women with 46,XX spontaneous premature ovarian failure. BMC Womens Health 2:8
143. Suzumori N, Burns KH, Yan W, Matzuk MM (2003) RFPL4 interacts with oocyte proteins

of the ubiquitin-proteasome degradation pathway. Proc Natl Acad Sci U S A 100:550–555
144. Suzumori N, Pangas SA, Rajkovic A (2007) Candidate genes for premature ovarian failure. Curr Med Chem 14:353–357
145. Mandon-Pepin B, Touraine P, Kuttenn F, Derbois C, Rouxel A, Matsuda F, Nicolas A, Cotinot C, Fellous M (2008) Genetic investigation of four meiotic genes in women with premature ovarian failure. Eur J Endocrinol 158:107–115
146. Shimizu Y, Kimura F, Takebayashi K, Fujiwara M, Takakura K, Takahashi K (2009) Mutational analysis of the PTEN gene in women with premature ovarian failure. Acta Obstet Gynecol Scand 88:824–825
147. Ojeda D, Lakhal B, Fonseca DJ, Braham R, Landolsi H, Mateus HE, Restrepo CM, Elghezal H, Saad A, Laissue P (2011) Sequence analysis of the CDKN1B gene in patients with premature ovarian failure reveals a novel mutation potentially related to the phenotype. Fertil Steril 95(2658–2660): e2651
148. Wang B, Ni F, Li L, Wei Z, Zhu X, Wang J, Cao Y, Ma X (2010) Analysis of cyclin-dependent kinase inhibitor 1B mutation in Han Chinese women with premature ovarian failure. Reprod Biomed Online 21:212–214
149. Durlinger AL, Kramer P, Karels B, de Jong FH, Uilenbroek JT, Grootegoed JA, Themmen AP (1999) Control of primordial follicle recruitment by anti-Mullerian hormone in the mouse ovary. Endocrinology 140:5789–5796
150. Wang HQ, Takakura K, Takebayashi K, Noda Y (2002) Mutational analysis of the mullerian-inhibiting substance gene and its receptor gene in Japanese women with polycystic ovary syndrome and premature ovarian failure. Fertil Steril 78:1329–1330
151. Luoma P, Melberg A, Rinne JO, Kaukonen JA, Nupponen NN, Chalmers RM, Oldfors A, Rautakorpi I, Peltonen L, Majamaa K, Somer H, Suomalainen A (2004) Parkinsonism, premature menopause, and mitochondrial DNA polymerase gamma mutations: clinical and molecular genetic study. Lancet 364:875–882
152. Pagnamenta AT, Taanman JW, Wilson CJ, Anderson NE, Marotta R, Duncan AJ, Bitner-Glindzicz M, Taylor RW, Laskowski A, Thorburn DR, Rahman S (2006) Dominant inheritance of premature ovarian failure associated with mutant mitochondrial DNA polymerase gamma. Hum Reprod 21: 2467–2473
153. Granat M, Amar A, Mor-Yosef S, Brautbar C, Schenker JG (1983) Familial gonadal germinative failure: endocrine and human leukocyte antigen studies. Fertil Steril 40:215–219
154. Smith A, Fraser IS, Noel M (1979) Three siblings with premature gonadal failure. Fertil Steril 32:528–530
155. Al-Awadi SA, Farag TI, Teebi AS, Naguib K, el-Khalifa MY, Kelani Y, Al-Ansari A, Schimke RN (1985) Primary hypogonadism and partial alopecia in three sibs with mullerian hypoplasia in the affected females. Am J Med Genet 22:619–622
156. Mikati MA, Najjar SS, Sahli IF, Melhem RE, Mansour S, Der Kaloustian VM (1985) Microcephaly, hypergonadotropic hypogonadism, short stature, and minor anomalies: a new syndrome. Am J Med Genet 22:599–608
157. Simpson JL (1999) Genetics of the female reproductive ducts. Am J Med Genet 89: 224–239
158. Elias S, Simpson JL, Carson SA, Malinak LR, Buttram VC Jr (1984) Genetics studies in incomplete mullerian fusion. Obstet Gynecol 63:276–279
159. Ergun A, Pabuccu R, Atay V, Kucuk T, Duru NK, Gungor S (1997) Three sisters with septate uteri: another reference to bidirectional theory. Hum Reprod 12:140–142
160. Hay D (1961) Uterus unicollis and its relationship to pregnancy. J Obstet Gynaecol Br Emp 68:361–377
161. Holmes JA (1956) Congenital abnormalities of the uterus and pregnancy. Br Med J 1: 1144–1147
162. Nykiforuk NE (1938) Uterus Didelphys. Can Med Assoc J 38:175
163. Polishuk WZ, Ron MA (1974) Familial bicornuate and double uterus. Am J Obstet Gynecol 119:982–987
164. Stevenson AC, Dudgeon MY, Mc CH (1959) Observations on the results of pregnancies in women resident in Belfast. II. Abortions, hydatidiform moles and ectopic pregnancies. Ann Hum Genet 23:395–414
165. Tyler GT (1939) Didelphys, in sisters. Am J Surg 45:337–338
166. Verp MS, Simpson JL, Elias S, Carson SA, Sarto GE, Feingold M (1983) Heritable aspects of uterine anomalies. I. Three familial aggregates with Mullerian fusion anomalies. Fertil Steril 40:80–85
167. Way S (1945) The influence of minor degrees of failure of fusion of the Mullerian ducts on pregnancy and labor. J Obstet Gynaecol Br Emp 52:325–333

168. Ma J, Qin Y, Liu W, Duan H, Xia M, Chen ZJ (2011) Analysis of PBX1 mutations in 192 Chinese women with Mullerian duct abnormalities. Fertil Steril 95:2615–2617
169. Schnabel CA, Selleri L, Cleary ML (2003) Pbx1 is essential for adrenal development and urogenital differentiation. Genesis 37: 123–130
170. Donnenfeld AE, Schrager DS, Corson SL (1992) Update on a family with hand-foot-genital syndrome: hypospadias and urinary tract abnormalities in two boys from the fourth generation. Am J Med Genet 44: 482–484
171. Stern AM, Gall JC Jr, Perry BL, Stimson CW, Weitkamp LR, Poznanski AK (1970) The hand-food-uterus syndrome: a new hereditary disorder characterized by hand and foot dysplasia, dermatoglyphic abnormalities, and partial duplication of the female genital tract. J Pediatr 77:109–116
172. Mortlock DP, Innis JW (1997) Mutation of HOXA13 in hand-foot-genital syndrome. Nat Genet 15:179–180
173. Goodman FR, Bacchelli C, Brady AF, Brueton LA, Fryns JP, Mortlock DP, Innis JW, Holmes LB, Donnenfeld AE, Feingold M, Beemer FA, Hennekam RC, Scambler PJ (2000) Novel HOXA13 mutations and the phenotypic spectrum of hand-foot-genital syndrome. Am J Hum Genet 67:197–202
174. Hennekam RC (1989) Acral-genital anomalies combined with ear anomalies. Am J Med Genet 34:454–455
175. Utsch B, Becker K, Brock D, Lentze MJ, Bidlingmaier F, Ludwig M (2002) A novel stable polyalanine [poly(A)] expansion in the HOXA13 gene associated with hand-foot-genital syndrome: proper function of poly(A)-harbouring transcription factors depends on a critical repeat length? Hum Genet 110:488–494
176. Jones HW Jr, Rock JA (1983) Reparative and constructive surgery of the female genital tract. Williams & Wilkins, Baltimore
177. Lodi A (1951) Contributo clinco statistico sulle Malformazioni della vagina osservate nella Clinical Obstetrica e Gineocologica di Milano del 1906 al 1950. Ann Ostet Ginecol 73:1246–1285
178. McKusick VA (1978) The William Allan Memorial Award Lecture: Genetic nosology: three approaches. Am J Hum Genet 30: 105–122
179. McKusick VA, Bauer RL, Koop CE, Scott RB (1964) Hydrometrocolpos as a Simply Inherited Malformation. JAMA 189:813–816
180. Kaufman RL, McAlister WH, Ho CK, Hartmann AF (1974) Family studies in congenital heart disease VI. The association of severe obstructive left heart lesions, vertebral and renal anomalies; a second family. Birth Defects Orig Artic Ser 10:93–104
181. Chitayat D, Hahm SY, Marion RW, Sachs GS, Goldman D, Hutcheon RG, Weiss R, Cho S, Nitowsky HM (1987) Further delineation of the McKusick-Kaufman hydrometrocolpos-polydactyly syndrome. Am J Dis Child 141: 1133–1136
182. Stone DL, Slavotinek A, Bouffard GG, Banerjee-Basu S, Baxevanis AD, Barr M, Biesecker LG (2000) Mutation of a gene encoding a putative chaperonin causes McKusick-Kaufman syndrome. Nat Genet 25:79–82
183. Winter JS, Kohn G, Mellman WJ, Wagner S (1968) A familial syndrome of renal, genital, and middle ear anomalies. J Pediatr 72: 88–93
184. King LA, Sanchez-Ramos L, Talledo OE, Reindollar RH (1987) Syndrome of genital, renal, and middle ear anomalies: a third family and report of a pregnancy. Obstet Gynecol 69:491–493
185. Griffin JE, Edwards C, Madden JD, Harrod MJ, Wilson JD (1976) Congenital absence of the vagina. The Mayer-Rokitansky-Kuster-Hauser syndrome. Ann Intern Med 85: 224–236
186. Jones HW Jr, Mermut S (1972) Familial occurrence of congenital absence of the vagina. Am J Obstet Gynecol 114:1100–1101
187. Kupperman JH (1963) Human endocrinology, vol 1. FA Davis, Philadelphia, PA
188. Lischke JH, Curtis CH, Lamb EJ (1973) Discordance of vaginal agenesis in monozygotic twins. Obstet Gynecol 41:920–924
189. Carson SA, Simpson JL, Malinak LR, Elias S, Gerbie AB, Buttram VC Jr, Sarto GE (1983) Heritable aspects of uterine anomalies. II. Genetic analysis of Mullerian aplasia. Fertil Steril 40:86–90
190. van Lingen BL, Reindollar RH, Davis AJ, Gray MR (1998) Further evidence that the WT1 gene does not have a role in the development of the derivatives of the mullerian duct. Am J Obstet Gynecol 179:597–603
191. Petrozza JC, Gray MR, Davis AJ, Reindollar RH (1997) Congenital absence of the uterus and vagina is not commonly transmitted as a dominant genetic trait: outcomes of surrogate pregnancies. Fertil Steril 67:387–389
192. Van Lingen BL, Eccles MR, Reindollar RH, al E. (1998) Molecular genetic analysis of the

PAX2 gene in patients with gengenital absence of the uterus and vagina., Fertil Steril.

193. Zenteno JC, Carranza-Lira S, Kofman-Alfaro S (2004) Molecular analysis of the anti-Mullerian hormone, the anti-Mullerian hormone receptor, and galactose-1-phosphate uridyl transferase genes in patients with the Mayer-Rokitansky-Kuster-Hauser syndrome. Arch Gynecol Obstet 269:270–273

194. Bhagavath B, Stelling JR, Van Lingen BL et al (1998) Congenital absence of the uterus and vagina (CAUV) is not associated with the N314D allele of the galactose-1-phosphate uridyl transferase (GALT) gene. J Soc Gynecol Investig 5:140

195. Cheroki C, Krepischi-Santos AC, Rosenberg C, Jehee FS, Mingroni-Netto RC, Pavanello Filho I, Zanforlin Filho S, Kim CA, Bagnoli VR, Mendonca BB, Szuhai K, Otto PA (2006) Report of a del22q11 in a patient with Mayer-Rokitansky-Kuster-Hauser (MRKH) anomaly and exclusion of WNT-4, RAR-gamma, and RXR-alpha as major genes determining MRKH anomaly in a study of 25 affected women. Am J Med Genet A 140: 1339–1342

196. Burel A, Mouchel T, Odent S, Tiker F, Knebelmann B, Pellerin I, Guerrier D (2006) Role of HOXA7 to HOXA13 and PBX1 genes in various forms of MRKH syndrome (congenital absence of uterus and vagina). J Negat Results Biomed 5:4

197. Philibert P, Biason-Lauber A, Rouzier R, Pienkowski C, Paris F, Konrad D, Schoenle E, Sultan C (2008) Identification and functional analysis of a new WNT4 gene mutation among 28 adolescent girls with primary amenorrhea and mullerian duct abnormalities: a French collaborative study. J Clin Endocrinol Metab 93:895–900

198. Ravel C, Lorenco D, Dessolle L, Mandelbaum J, McElreavey K, Darai E, Siffroi JP (2009) Mutational analysis of the WNT gene family in women with Mayer-Rokitansky-Kuster-Hauser syndrome. Fertil Steril 91:1604–1607

199. Drash A, Sherman F, Hartmann WH, Blizzard RM (1970) A syndrome of pseudohermaphroditism, Wilms' tumor, hypertension, and degenerative renal disease. J Pediatr 76: 585–593

200. Clement-Ziza M, Khen N, Gonzales J, Cretolle-Vastel C, Picard JY, Tullio-Pelet A, Besmond C, Munnich A, Lyonnet S, Nihoul-Fekete C (2005) Exclusion of WNT4 as a major gene in Rokitansky-Kuster-Hauser anomaly. Am J Med Genet A 137:98–99

201. Sandbacka M, Bruce S, Halttunen M, Puhakka M, Lahermo P, Hannula-Jouppi K, Lipsanen-Nyman M, Kere J, Aittomaki K, Laivuori H (2011) Methylation of H19 and its imprinted control region (H19 ICR1) in Mullerian aplasia. Fertil Steril 95:2703–2706

202. Bendavid C, Pasquier L, Watrin T, Morcel K, Lucas J, Gicquel I, Dubourg C, Henry C, David V, Odent S, Leveque J, Pellerin I, Guerrier D (2007) Phenotypic variability of a 4q34→qter inherited deletion: MRKH syndrome in the daughter, cardiac defect and Fallopian tube cancer in the mother. Eur J Med Genet 50:66–72

203. Bernardini L, Gimelli S, Gervasini C, Carella M, Baban A, Frontino G, Barbano G, Divizia MT, Fedele L, Novelli A, Bena F, Lalatta F, Miozzo M, Dallapiccola B (2009) Recurrent microdeletion at 17q12 as a cause of Mayer-Rokitansky-Kuster-Hauser (MRKH) syndrome: two case reports. Orphanet J Rare Dis 4:25

204. Cheroki C, Krepischi-Santos AC, Szuhai K, Brenner V, Kim CA, Otto PA, Rosenberg C (2008) Genomic imbalances associated with mullerian aplasia. J Med Genet 45: 228–232

205. Ledig S, Schippert C, Strick R, Beckmann MW, Oppelt PG, Wieacker P (2011) Recurrent aberrations identified by array-CGH in patients with Mayer-Rokitansky-Kuster-Hauser syndrome. Fertil Steril 95:1589–1594

206. Nik-Zainal S, Strick R, Storer M, Huang N, Rad R, Willatt L, Fitzgerald T, Martin V, Sandford R, Carter NP, Janecke AR, Renner SP, Oppelt PG, Oppelt P, Schulze C, Brucker S, Hurles M, Beckmann MW, Strissel PL, Shaw-Smith C (2011) High incidence of recurrent copy number variants in patients with isolated and syndromic Mullerian aplasia. J Med Genet 48:197–204

207. Sundaram UT, McDonald-McGinn DM, Huff D, Emanuel BS, Zackai EH, Driscoll DA, Bodurtha J (2007) Primary amenorrhea and absent uterus in the 22q11.2 deletion syndrome. Am J Med Genet A 143A: 2016–2018

208. Gervasini C, Grati FR, Lalatta F, Tabano S, Gentilin B, Colapietro P, De Toffol S, Frontino G, Motta F, Maitz S, Bernardini L, Dallapiccola B, Fedele L, Larizza L, Miozzo M (2010) SHOX duplications found in some cases with type I Mayer-Rokitansky-Kuster-Hauser syndrome. Genet Med 12:634–640

209. Sandbacka M, Halttunen M, Jokimaa V, Aittomaki K, Laivuori H (2011) Evaluation

of SHOX copy number variations in patients with Mullerian aplasia. Orphanet J Rare Dis 6:53

210. Biason-Lauber A, Konrad D, Navratil F, Schoenle EJ (2004) A WNT4 mutation associated with Mullerian-duct regression and virilization in a 46,XX woman. N Engl J Med 351:792–798

211. Vainio S, Heikkila M, Kispert A, Chin N, McMahon AP (1999) Female development in mammals is regulated by Wnt-4 signalling. Nature 397:405–409

212. Duncan PA, Shapiro LR, Stangel JJ, Klein RM, Addonizio JC (1979) The MURCS association: Mullerian duct aplasia, renal aplasia, and cervicothoracic somite dysplasia. J Pediatr 95:399–402

213. Wulfsberg EA, Grigbsy TM (1990) Rokitansky sequence in association with the facio-auriculo-vertebral sequence: part of a mesodermal malformation spectrum? Am J Med Genet 37:100–102

214. Farag TI, Al-Awadi SA, Marafie MJ, Bastaki L, Al-Othman SA, Mohammed FM, Al Suliman IS, Murthy DS (1993) The newly recognised limb/pelvis-hypoplasia/aplasia syndrome: report of a Bedouin patient and review. J Med Genet 30:62–64

215. Teebi AS (1993) Limb/pelvis/uterus-hypoplasia/aplasia syndrome. J Med Genet 30:797

216. Hoffmann W, Grospietsch G, Kuhn W (1976) Thalidomide and female genital malformations. Lancet 2:794

217. Griesinger G, Dafopoulos K, Schultze-Mosgau A, Schroder A, Felberbaum R, Diedrich K (2005) Mayer-Rokitansky-Kuster-Hauser syndrome associated with thrombocytopenia-absent radius syndrome. Fertil Steril 83:452–454

Chapter 4

Gene Polymorphisms in Female Reproduction

Livio Casarini and Manuela Simoni

Abstract

This chapter presents an overview of the gene polymorphisms underlying the functions of ovarian receptors and their clinical implications in the female fecundity. A selection of genetic studies revealing significant associations between receptor polymorphisms, gene mutations, and some pathological conditions (i.e., female infertility, premature ovarian failure, polycystic ovary syndrome, endometriosis) are reviewed.

Key words Gene polymorphism, Single-nucleotide polymorphisms, Controlled ovarian stimulation, *FSHR* polymorphism, Progesterone receptor (*PGR*), FSH, LH, *MTHFR* polymorphism

1 Introduction

Gonadotropins and sex hormones play a key role in sexual development, reproductive functions, and metabolism. The action of these molecules is mediated by their receptors at gonadal level. A large number of studies attempted to evaluate the involvement of gene polymorphisms in ovarian and reproductive function. In fact, the features of the response to hormones depend on variations in the gene sequence of the receptor or the ligand and, in case of a familial disease, underlying mutations are genomic and inherited [1]. Association studies assess whether a gene variant is present more often than expected in a population to evaluate whether a medical condition or a phenotype is associated to a gene variant. Gene variants consist of insertions or deletions of one or more bases, or they may be single-base changes, known as single-nucleotide polymorphisms (SNPs) [2]. A genetic variant is considered a polymorphism when it reaches a frequency higher than 1 % in a sample population; otherwise, it is considered a gene mutation.

SNPs can occur outside a gene or within an exon, an intron, or regulatory regions, and it can be functional or silent. Functional SNPs within a gene can be the direct cause of a phenotype abnormality or may increase susceptibility to a disease. Functional SNPs in coding regions can change the protein sequence, while in

Zev Rosenwaks and Paul M. Wassarman (eds.), *Human Fertility: Methods and Protocols*, Methods in Molecular Biology, vol. 1154, DOI 10.1007/978-1-4939-0659-8_4, © Springer Science+Business Media New York 2014

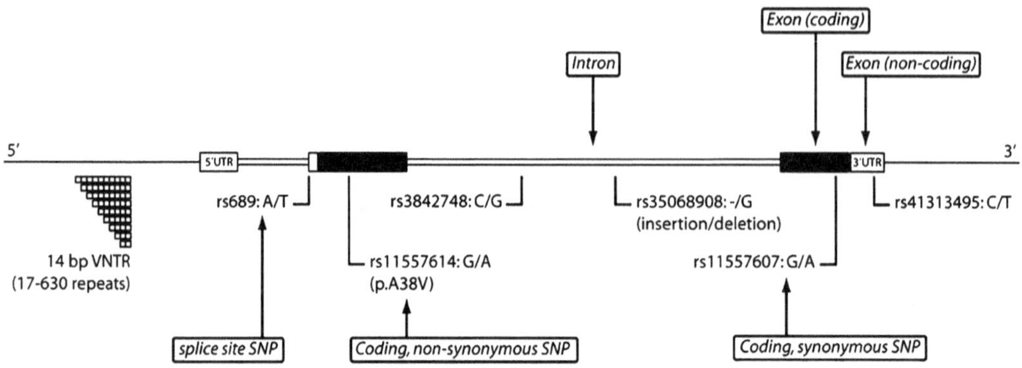

RFLP, restriction fragment length polymorphism.

Fig. 1 Graphic representation of types of genetic variants, showing insertion/deletion (ins/del) polymorphisms, both coding and noncoding SNPs, and repeat polymorphisms such as tandem repeats or VNTR. Variants are shown occurring within a gene (in this example the INS gene) but can also occur outside of genes. Other types of genetic variations that affect larger regions, such as copy number variations, are not shown. *SNP* single-nucleotide polymorphism, *VNTR* variable number of tandem repeats. Reproduced from [2] with permission from Oxford Journals

noncoding regions they may have effects on RNA transcription and processing. Silent SNPs occur in protein-coding DNA but do not change the sequence of a gene product [2] (Fig. 1).

The SNP combination in a given genome sequence determines the allelic sequences, which, in turn, are grouped in haplotype blocks. The human genome can be subdivided in haplotype blocks which are sizable regions poor in recombination events characterized by the presence of one or only a few common haplotypes [3]. Genome-wide association studies are based on haplotype blocks as marker regions, available in the HapMap online database (http://hapmap.ncbi.nlm.nih.gov/) [4, 5]. Moreover, the SNPs that occur outside a gene can be used as a tag to identify nearby functional variations within a gene. Such analyses are based on a chromosomal property called linkage disequilibrium (LD). LD refers to the observation that in a sample population, two SNPs or DNA variants that are located close to each other tend to be observed together more frequently than two variants that are located further apart [2].

In the National Center for Biotechnology Information (NCBI) database (http://www.ncbi.nlm.nih.gov/pubmed/), the SNPs are registered following a specific nomenclature together with their frequency in a specific population. For example, a well-known SNP in exon 10 of the *follicle-stimulating hormone receptor* (*FSHR*) gene is the adenine-to-guanine nucleotidic substitution at position 2039 from the transcription start site (A2039G; NCBI database acronym: rs6166). It is a non-synonymous substitution since it determines an asparagine-to-serine change at position 680 (Asn680Ser) of the amino acid chain and has a minor allele (MA) frequency of about 40 % in Caucasian population [6]. Due to its high frequency in the population, rs6166 is considered a common SNP. Moreover, it is in almost complete LD with another SNP of *FSHR* gene; the Thr307Ala is located in the exon 10 at the extracellular domain, while Asn680Ser is located in the intracellular domain [7].

Thousands of SNPs located in genes involved in sexual development or reproductive functions are registered in the NCBI database, but only some of them are the cause of a change in the cellular response resulting in modulation of hormone action or reproductive functions. Genes encoding hormones and receptors of the hypothalamus–pituitary–ovarian axis are often studied as candidate predictive factors of the ovarian response, such as the gonadotropin or the gonadotropin receptor genes [8], but enzymes such as aromatase are also prime candidates for association studies [9]. Several other markers of ovarian response have been proposed, particularly for predicting the response to stimulation with exogenous gonadotropins during assisted reproduction techniques (ART). The importance of genetic markers in controlled ovarian stimulation (COS) derives from the clinical evidence that ovarian response to exogenous gonadotropins used for ART is highly individual and variable, but only a few SNPs seem to be good candidates to be applied in clinical tests [10]. To date, Asn680Ser in the FSHR and the *PvuII* polymorphism in the *estrogen receptor 1* (*ESR1*) gene are the most studied polymorphisms which could be used as genetic markers in clinical tests. Although some studies indicate that ovarian response is modulated by some polymorphisms, the determination of an accurate and customized COS therapy is not available and the search for predictive genetic markers is ongoing [11].

2 Ovarian Response and Controlled Ovarian Stimulation

The secretion of gonadotropins by the pituitary determines the controlled progression of the ovarian cycle, and the pituitary activity is controlled via feedback by the ovarian hormones. A peak of elevated serum levels of follicle-stimulating hormone (FSH) is required to recruit a cohort of follicles at the antral stage and support their maturation, which is characterized by growth and maturation of granulosa cells [12]. Usually, only the follicle with the lowest FSH recruitment threshold can grow and becomes the Graafian follicle acquiring LH responsivity and undergoing ovulation, while the others become atretic [13]. Since the biological activity of FSH depends on the FSHR expression in the granulosa cells [14, 15], several studies have focused on the influence of this receptor on menstrual cycle dynamics [14, 16–18].

To date, it is well established that the polymorphism Asn680Ser in the *FSHR* gene is important in determining the response to FSH and is so far the unique genetic marker within the gonadotropin receptor genes. Historically, the first in vivo evidence was found in German women, with the demonstration that those homozygous for the FSHR Ser680 genotype require a higher number of FSH ampoules in ovarian hyperstimulation, compared to the homozygous Asn680 carriers [14]. Subsequent studies in ovulatory and anovulatory women confirmed this observation [19], suggesting that the FSHR Ser680 genotype is less sensitive to the FSH action in vivo, is associated with a decreased negative feedback signal to the pituitary gland, and determines longer menstrual cycle compared to the FSHR Asn680 genotype [16]. Indeed, in normo-ovulatory women, basal serum FSH levels are lower in FSHR Asn680 than Ser680 carriers during the luteo-follicular transition, confirming that Ser680 genotype is a "resistance" factor to FSH stimulation, determining a higher ovarian threshold of the gonadotropin and reflecting a different pattern of ovarian secretion during the intercycle transition phase [16, 17]. In homozygous Ser680 carriers, an earlier rise in FSH levels is observed, which provokes a duration of the menstrual cycle of about 3 days longer in connection with a decreased negative feedback of luteal secretions to the pituitary gland [16, 17].

Whether the modulation of ovarian response by the polymorphism of FSHR Asn680Ser needs to be considered during COS is still under discussion [20], and meta-analysis studies could strongly contribute to clarify the role of this polymorphism in COS. Indeed, a meta-analysis study including different European populations confirmed that the Asn680Ser is a marker associated with poor response during COS [20]. Another, extensive meta-analysis found that homozygous Ser680 women have higher basal FSH levels and require a higher dose of exogenous FSH for COS than women with genotype Asn680Asn [21]. A precise, individualized ovarian stimulation treatment designed on predictive genetic factors is however still missing, although it is well known that homozygous Ser680 women undergoing COS produce lower levels of serum estradiol compared to Asn680 carriers stimulated by an equal FSH dose [22]. In any case, whether the FSHR genotype has any effects on pregnancy rate remains unknown, since different studies have obtained contradictory results [23, 24] and analyses having the statistical power necessary to arrive to a conclusive opinion are still missing.

Another *FSHR* polymorphism, G/A located in the promoter region at position-29, could modulate ovarian response to FSH [25]. Indeed, a study found that women homozygous for the A genotype undergoing controlled ovarian hyperstimulation and IVF required the highest dose of exogenous FSH. Also, in these women the levels of estradiol concentration measured before hCG administration and the number of preovulatory follicles and of retrieved oocytes were lower compared to the other genotypes, suggesting an association between the SNP and a poor ovarian response [25]. These results are corroborated by in vitro data which demonstrate, by Western blotting and confocal microscopy, that the genotype A-29 results in reduced expression of FSHR on the surface of granulosa cells, compared with the variant G-29. Thus, poor ovarian response observed in subjects with the AA genotype at position -29 of the FSHR gene could be due to reduced receptor expression [26]. However, the association of polymorphisms in the promoter region and COS needs to be confirmed by replication in wider population groups of different ethnicity.

The luteinizing hormone (LH) receptor (LHCGR) mediates the action of two different molecules, LH and chorionic gonadotropin (CG), and it is essential for folliculogenesis, ovulation, and progression of pregnancy [27]. Some common polymorphisms located in the *LHCGR* gene (18InsLeuGly and Asn291Ser Ser312Asn) were associated with an increased activity of the receptor, and their involvement in breast cancer has been suspected [28, 29]. This model suggested that the risk of breast cancer could increase according to a modified level of estrogen exposure [30]. In fact, the frequency of LHCGR 18InsLeuGly polymorphism [28, 29] and the incidence of breast cancer [31] are higher in Northern European populations than in other populations (e.g., Asian), leading to the speculation that they may be linked [30]. However, the involvement of LHCGR polymorphisms in the modulation of the ovarian function has never been confirmed. A study has even excluded its association with ovarian hyperstimulation syndrome (OHSS) [32], and, therefore, the role of the *LHCGR* gene polymorphisms in COS response remains unclear.

The *LHB* gene encodes the β chain of LH and determines the specificity of binding to the LHCGR, while the α chain is common to FSH. LH has a key role in female and male reproduction since it triggers steroidogenesis and folliculogenesis in cooperation with FSH [33]. There are over 50 *LHB* polymorphisms recorded in the NCBI database, but those resulting in a decrease of the LH activity are a few. An LH variant (V-LH) with increased in vitro bioactivity and decreased half time in vivo has been detected in Finnish population [34, 35], in which it reaches the 28 % of frequency in homo- or heterozygosis [36]. The V-LH consists in the double-amino acid change Trp8Arg and Gly102Ser, introducing an extra glycosylation signal to the β chain [37]. The subjects homozygous for the LH polymorphism are healthy, probably due to a linkage mechanism with some SNPs in the promoter region which flattens these differences. In fact, the in vitro activity of mutant *LHB* promoter is about 40 % higher than that of the wild-type *LHB* promoter gene and the intrinsic bioactivity of V-LH is about 30 % higher than that of LH [38]. Interestingly, the detection of this

LH variant was based on its decreased immunoreactivity. Indeed, the amount of V-LH was assayed by comparison of two different assays which used an antibody that equally recognizes the V-LH and the wild-type form and another antibody against LH α/β--dimer, which does not detect the LH variant. Hence, the ratio of LH measured by the two assays is higher in wild-type subjects, and lower in homozygote carriers of the *V-LH* [36]. Trp8Arg and Gly102Ser polymorphisms appear to be involved in menstrual disorders [39, 40], while the variant 8Arg-15Thr of *LHB* seems to have a higher frequency in women undergoing in vitro fertilization (IVF) characterized by low responsiveness to treatment with recombinant FSH [41]. Instead, V-LH shows no association to pathological conditions such as endometriosis [42] or ovarian cancer [43]. However, the results obtained on the polymorphisms of *LHB* are still limited and need to be confirmed by further studies in larger sample populations and in different ethnic groups.

There are two human estrogen receptors (ER), ERα and ERβ, encoded by *ESR1* and *ESR2* genes, respectively. ERα mediates the proliferative effects of estrogen on theca cells during folliculogenesis, while ERβ leads to differentiation and antiproliferative effects on granulosa cells [44, 45]. Historically, *ESR* genes were the first candidates for studying genetic targets which could be related to the pharmacological approach to COS [46], also due to their high genetic variability [46–49]. Polymorphisms in the *ESR2* have been studied but never definitively found in association with age at menarche, menopause onset, and fertility [50]. On the contrary, the two intronic SNPs T-397C and A-351G, also called the *PvuII* and *XbaI* polymorphisms by the name of the detecting restriction enzymes [51], are likely involved in the incidence of preeclampsia [52] and abortive events [50, 53] and in the response to ovarian stimulation [47]. Taken together, these data could provide very important information to predict the COS outcome during IVF, but the molecular mechanism that underlies these features is not yet clear. The ESR1 *PvuII* polymorphism may result in the loss of a functional binding site for the *myb* family of transcription factors [54] or may be involved in epistatic mechanisms which modulate the expression of other genes of the steroidogenic cascade, such as *CYP19A1* encoding the enzyme aromatase [55], but to elucidate these molecular mechanisms further extensive in vitro studies must be done.

The vitamin B folate is involved in the regulation of the synthesis and methylation of nucleic acids; thus, it plays a key role during cell growth and proliferation including the developing ovarian follicles [11]. Indeed, a folate deficiency has been associated to increased risk of early spontaneous abortion in humans [56] and to impaired reproductive functions and early embryonic loss in the hamster [57]. The folate metabolism is important for its clinical implications since folate-deficient women undergoing COS showed lower oocyte quality and pregnancy rates and impaired ovarian function [58–62]. A molecule that plays an important role in the metabolism of folate is the enzyme methylenetetrahydrofolate reductase (MTHFR). The C677T common polymorphism in the *MTHFR* gene [63, 64] results in the amino acid change Ala222Val, which leads to a decrease of enzyme activity of about 50 % [65]. The deficiency of the enzyme activity is reflected in a defective DNA methylation and altered gene expression [66]. It is well known that a folate deficiency leads to elevated homocysteine (Hcy) serum levels, which are negatively correlated to oocyte maturity [67], in vitro embryo quality [59, 68], and pregnancy and implantation rates [69, 70]. The importance of the *MTHFR* gene polymorphism as a target was suggested by a recent publication which showed that women heterozygous for the C677T polymorphism have a higher proportion of good-quality embryo and a better chance of pregnancy than C677C homozygous women [71]. This result is consistent with the observation that the CT genotype is less present in women affected by unexplained infertility compared to control groups, in patients undergoing IVF [72]. In conclusion, the genes involved in the regulation of the folate metabolism, in particular MTHFR, could become new biomarkers to determine the individual response to COS treatment.

3 Menarche, Menopause, and Premature Ovarian Failure

The duration of the fertile age is another parameter that is plausibly affected by the presence of SNPs, since the timing of puberty and probably of menopause can be due to genetic features [73, 74]. The FSHR Ser680 polymorphism has been associated to a slightly delayed age of menarche but not with the menopausal age in Italian women [75]. An FSHR polymorphism located in the 5′ untranslated region of the gene (G/A at position -29) was found to be related with primary amenorrhea in Indian women [76]. The frequency of -29G genotype was higher in patients with primary or secondary amenorrhea than in the control group and was associated with higher FSH serum levels. However, a clear evidence of the involvement of FSHR SNP in the duration of fertile age has never been found and studies in large populations of different ethnicity have to be performed [76].

Other SNPs located in genes involved in the hypothalamic–pituitary–ovarian axis downstream to gonadotropin genes, such as the *progesterone receptor* (*PGR*) (Fig. 2), could modulate ovarian response [77] because of their direct effects on folliculogenesis, maturation and oocyte release, and endometrial implantation. For example, puberty and menstruation are complex processes that are dependent on feedback mechanisms involving the action of progesterone [78], and it is plausible that the age of menarche and

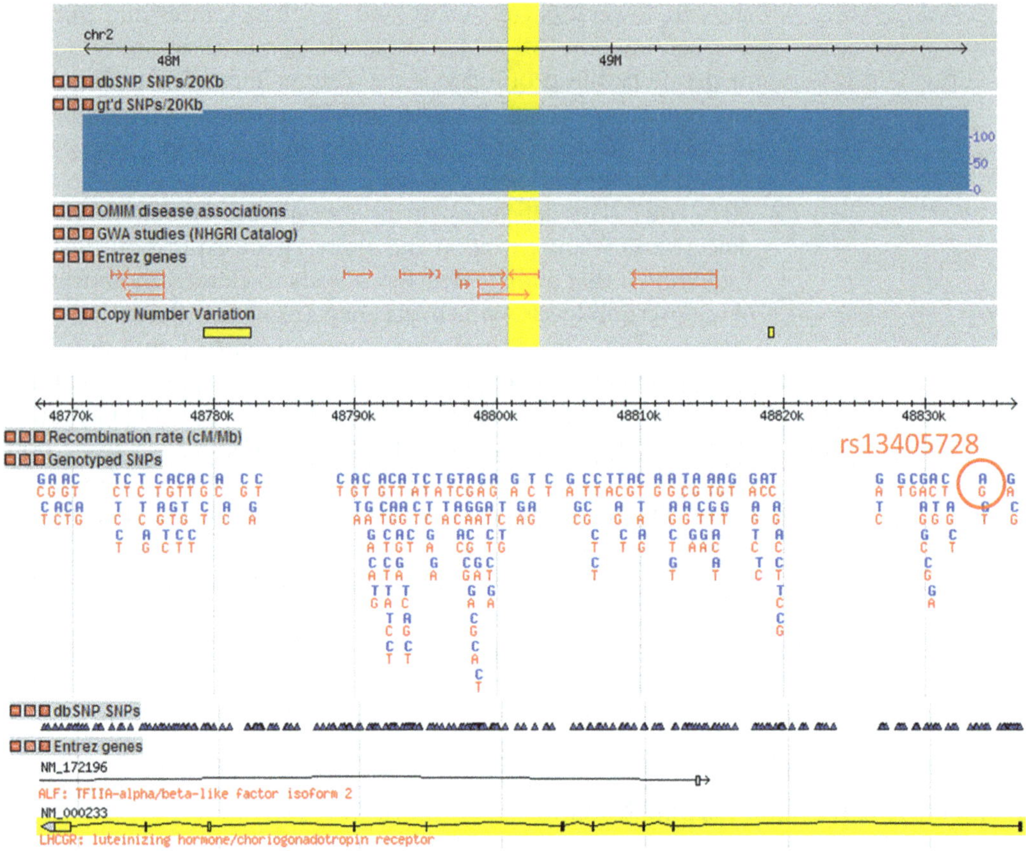

Fig. 2 The *LHCGR* and the paralogue gene *ALF1* on chromosome 2 (http://hapmap.ncbi.nlm.nih.gov). The SNP rs13405728, located in an intronic region of the *LHCGR* gene, is included in a block of linkage disequilibrium embedded in an intronic region of the gene ALF1

menstrual cycle features could be modulated by SNPs in the PGR gene [77]. A linkage research for genomic regions affecting the age of menarche found that the PGR gene is located in a region with significant logarithm of odds (LOD) scores, indicating the *non*randomness of the results of linkage analysis [79]. Indeed, the gene exists in several genetic variants. One of these variants consists in an insertion of an Alu sequence of 320 base pairs in an intron region and is named "*PROGINS*." *PROGINS* is in complete LD with the *non*-synonymous variant Val660Leu and with the synonymous variant His770His and is responsible of a decrease in the response to progesterone in vitro due to decreased mRNA stability and protein activity [80]. The Val660Leu variant was also significantly associated with spontaneous abortion and with ovarian cancer risk in vivo, but studies which assess the association between PGR polymorphisms and age at menarche are limited [81, 82].

Lastly, the important role of the enzyme aromatase in the estrogenic activity is the reason why the *CYP19A1* gene could be a candidate for the regulation of menarche. A number between 7 and 12 TTTA repeats is present in exon 4, and this polymorphism is involved in the modulation of bone metabolism in males [83]. The presence of TTTA repeats implies a bimodal distribution of alleles, with two major peaks of distribution at 7 and 11 repeats and a very low distribution of the 9 repeat allele [83]. A recent association study found that the *CYP19A1* allele characterized by 7 TTTA repeats was significantly more represented in homozygosity in a population of Greek girls characterized by early menarche than the 11 repeat allele, which, in turn, was found to be associated with late menarche [9]. This suggests that the age of menarche has a genetic component in which the *CYP19A1* gene may play an important role.

Intriguing results were obtained by genome-wide association studies. An association analysis for age at natural menopause in about 3,000 European women identified six SNPs in three loci associated with timing of ovarian aging, a risk factor for breast cancer, osteoporosis, and cardiovascular disease [84]. Surprisingly, these polymorphisms are not located inside genes involved in hypothalamic–pituitary–ovarian axis, but in *BR serine threonine kinase 1* (*BRSK1*) gene, 3′ region of the *transmembrane protein 150B* (*TMEM224*) gene, 5′ of the *suppressor of variegation 4–20 homolog 2* (*SUV420H2*), a lysine methyltransferase gene, 3′ of the hypothetical gene *LOC121793*, *rho guanine nucleotide exchange factor 7* (*ARHGEF7*) gene, and *minichromosome maintenance complex component 8* (*MCM8*) gene [84]. Moreover, the meta-analysis of genome-wide association studies has identified more than 30 new loci that influence the age at menarche and these studies are highly indicative as they involve a very wide number of samples [85–87]. At last, the SNP rs314276 within the *LIN28B* gene, a potent regulator of microRNA processing, appears to be a strong genetic determinant regulating the timing of human pubertal growth and development in boys and girls [88]. Taken together, these findings provide a basis for the associations between the age of menarche or menopause and the genotype.

4 Endometriosis

Some genetic polymorphisms in genes involved in sex steroid biosynthesis and metabolism may be reasonably associated with changes in reproductive function or increased risk of estrogen-dependent disease, e.g., *CYP19A1* gene or 17-beta-hydroxysteroid dehydrogenase type 1 (*HSD17B1* gene). Both enzymes are involved in the estradiol synthesis cascade and are essential for normal reproductive function. In addition, there are diseases, such as endometriosis, which appear to be closely related to the production and metabolism of sex steroids [89]. Endometriosis is characterized by the growth of endometrial cells outside the uterus and has a strong genetic component which justifies a large number of association studies [90, 91]. The endometriotic lesions grow and regress in relation to estrogen concentration in a dose-dependent manner, suggesting a role of polymorphisms in *CYP19A1* or *HSD17B1* genes in the pathogenesis of the disease. Specifically, the polymorphisms C1558T and Val80 in the *CYP19A1* gene, and Ser312Gly polymorphism in the *HSD17B1* gene, have been recently associated with the disease [92, 93]. However, the results obtained from association studies are often inconsistent and conflicting due to a lack of confirmation in different ethnic groups or to the limited size of the sample group [91] and the mechanism underlying the development of endometriosis at molecular level remains unknown.

5 Polycystic Ovary Syndrome

PCOS is an endocrine disorder affecting about 10 % of women during reproductive age characterized by polycystic ovaries, hyperandrogenism, abnormal menstrual cycle, and chronic anovulation. Moreover, in women who conceive successfully after treatment, an increased risk of complications during pregnancy as well as neonatal complications exists [2, 94]. Despite a large number of studies investigating the genetic features related to PCOS, the key factors involved in its pathogenesis are still unidentified.

The study of the allele frequency of FSHR polymorphisms Thr307Ala and Asn680Ser revealed that the G919/A2039 (Ala307/Asn680) haplotype is significantly linked with PCOS and could be a risk factor for the disease, although none of the FSHR polymorphisms individually analyzed was observed to be in clear association with the pathology [95] except in Korean women [96]. This finding, however, has not been independently confirmed yet. Rather, actual results suggest that FSHR polymorphisms are not clearly related to PCOS, regardless of the ethnic background, although a risk haplotype could exist. As in normo-ovulatory women, also in PCOS patients, the Ser680 allele is associated with significantly higher levels of FSH and LH and with hyperandrogenism [97]. A good approach to find candidate genes involved in PCOS might be studying the polymorphisms in genes that regulate the hypothalamic–pituitary–gonadal axis. As far as the combination of FSHR and LHCGR polymorphisms is concerned, a recent study indicated that the analyzed LHCGR SNPs are not related to PCOS and that the disease features may depend only from the FSHR SNPs [97]. On the other hand, further information may be provided by genome-wide association studies. Indeed, susceptibility locus for PCOS has been detected on chromosome 2p16.3, 2p21, and 9q33.3, particularly in connection with an SNP (rs13405728) located in an intronic region of the *LHCGR* gene [98]. This sequence is included in a block of linkage disequilibrium embedded in an intronic region of the germ cell-specific *ALF* gene sequence, a paralogue of *TFIIA-α/β-like factor* gene [99] which, moreover, was found to be altered in male infertility [100]. It is plausible that polymorphisms in the gene *ALF1* could influence female reproduction. Taken together, these findings suggest that a correlation between the region including the *LHCGR* gene and PCOS could exist, although clear associations remain to be verified.

Concerning the *LHB* gene, women with the *V-LH* were less likely to report ovarian cysts, were more likely to report infertility, and have higher early follicular phase LH concentrations compared with LH wild-type carriers [101]. In this sense, the presence of linkage mechanisms with polymorphisms in the promoter region, known to decrease the in vitro bioactivity of the LH variant [37], cannot be excluded. To assess the role of the V-LH in female fertility, its frequency was analyzed in groups of PCOS patients from Northern Europe and the United States [102]. The frequency of *V-LH* was lower in obese PCOS patients than healthy women and nonobese PCOS patients, suggesting that V-LH somehow protects obese women from developing symptomatic PCOS [102], but, again, independent studies are needed to give a solid confirmation to this observation.

6 Conclusions

The aim of association studies is to understand the relationship between common genetic variants and reproduction to achieve a customized, individual therapeutic approach. The search for biomarkers which can predict ovarian response could be a significant strategic step for achieving such a goal. To date, the studies are mainly focused on polymorphisms of the *FSHR*, *MTHFR*, and *ESR* genes, as they showed evidence in determining modulation of ovarian activity. In particular, the *FSHR* gene polymorphisms Asn680Ser and G-29A showed to modulate slightly ovarian response and are among the best candidates to be elected as markers to predict individual response to COS. Instead, the association with other ovarian features or diseases such as PCOS was only inconsistently found [103]. In any case, it is crucial to find a panel of several candidate sequences, since a single gene is probably not sufficient to constitute an effective, predictive biomarker useful for all patients [104]. For example, the combination of *FSHR–ESR1–ESR2* gene variants is statistically sufficient to predict only the 10–15 % of poor responders to treatment with recombinant FSH among the IVF patients [48]. In this sense, genome-wide association studies may provide a good way to proceed in the search for genetic markers, but some obstacles remain to be overcome. The lack of independent replication, the small sample size [105], and the choice of the sample population by ethnicity and geographic origin [106] are currently the main problems. In addition, the relationship between the cost of genetic testing and the lack of a direct benefit of the results in clinical practice [107] is the last challenge to the achievement of an individual therapeutic protocol.

References

1. Nordhoff V, Gromoll J, Simoni M (1999) Constitutively active mutations of G protein-coupled receptors: the case of the human luteinizing hormone and follicle-stimulating hormone receptors. Arch Med Res 30(6):501–509
2. Simoni M, Tempfer CB, Destenaves B, Fauser BCJM (2008) Functional genetic polymorphisms and female reproductive disorders: part I: polycystic ovary syndrome and ovarian response. Hum Reprod Update 14(5): 459–484, Ott
3. Gabriel SB, Schaffner SF, Nguyen H, Moore JM, Roy J, Blumenstiel B et al (2002) The structure of haplotype blocks in the human genome. Science 296(5576):2225–2229, Giu 21
4. Barrett JC, Fry B, Maller J, Daly MJ (2005) Haploview: analysis and visualization of LD and haplotype maps. Bioinformatics 21(2):263–265, Gen 15
5. Barrett JC (2009) Haploview: visualization and analysis of SNP genotype data. Cold Spring Harb Protoc 2009(10):pdb.ip71, Ott
6. Simoni M, Gromoll J, Höppner W, Kamischke A, Krafft T, Stähle D et al (1999) Mutational analysis of the follicle-stimulating hormone (FSH) receptor in normal and infertile men: identification and characterization of two discrete FSH receptor isoforms. J Clin Endocrinol Metab 84(2):751–755
7. Simoni M, Nieschlag E, Gromoll J (2002) Isoforms and single nucleotide polymorphisms of the FSH receptor gene: implications for human reproduction. Hum Reprod Update 8(5):413–421
8. Wunsch A, Sonntag B, Simoni M (2007) Polymorphism of the FSH receptor and ovarian response to FSH. Ann Endocrinol (Paris) 68(2–3):160–166, Giu
9. Xita N, Chatzikyriakidou A, Stavrou I, Zois C, Georgiou I, Tsatsoulis A (2010) The (TTTA)n polymorphism of aromatase (CYP19) gene is associated with age at menarche. Hum Reprod 25(12):3129–3133, Dic
10. Loutradis D, Theofanakis C, Anagnostou E, Mavrogianni D, Partsinevelos GA (2011) Genetic profile of SNP(s) and ovulation induction. Curr Pharm Biotechnol 13(3):417–425, Giu 9 [citato 2011 Giu 17]; http://www.ncbi.nlm.nih.gov/pubmed/21657995
11. Twigt JM, Hammiche F, Sinclair KD, Beckers NG, Visser JA, Lindemans J et al (2011) Preconception folic acid use modulates estradiol and follicular responses to ovarian stimulation. J Clin Endocrinol Metab 96(2): E322–E329
12. Hodgen GD (1982) The dominant ovarian follicle. Fertil Steril 38(3):281–300, Set
13. Baird DT (1987) A model for follicular selection and ovulation: lessons from superovulation. J Steroid Biochem 27(1–3):15–23
14. Perez Mayorga M, Gromoll J, Behre HM, Gassner C, Nieschlag E, Simoni M (2000) Ovarian response to follicle-stimulating hormone (FSH) stimulation depends on the FSH receptor genotype. J Clin Endocrinol Metab 85(9):3365–3369, Set
15. Zheng W, Magid M, Kramer E, Chen Y (1996) Follicle-stimulating hormone receptor is expressed in human ovarian surface epithelium and fallopian tube. Am J Pathol 148(1):47–53, Gen 1
16. Greb RR, Grieshaber K, Gromoll J, Sonntag B, Nieschlag E, Kiesel L et al (2005) A common single nucleotide polymorphism in exon 10 of the human follicle stimulating hormone receptor is a major determinant of length and hormonal dynamics of the menstrual cycle. J Clin Endocrinol Metab 90(8):4866–4872, Ago
17. Gromoll J, Simoni M (2005) Genetic complexity of FSH receptor function. Trends Endocrinol Metab 16(8):368–373, Ott
18. Kuijper EAM, Blankenstein MA, Luttikhof LJ, Roek SJM, Overbeek A, Hompes PG et al (2011) Frequency distribution of polymorphisms in the FSH receptor gene in infertility patients of different ethnicity. Reprod Biomed Online 22(Suppl 1):S60–S65
19. Loutradis D, Vlismas A, Drakakis P, Antsaklis A (2008) Pharmacogenetics in ovarian stimulation – current concepts. Ann N Y Acad Sci 1127:10–19

20. Morón FJ, Ruiz A (2010) Pharmacogenetics of controlled ovarian hyperstimulation: time to corroborate the clinical utility of FSH receptor genetic markers. Pharmacogenomics 11(11):1613–1618
21. Yao Y, Ma C, Tang H, Hu Y (2011) Influence of follicle-stimulating hormone receptor (FSHR) Ser680Asn polymorphism on ovarian function and in-vitro fertilization outcome: a meta-analysis. Mol Genet Metab 103(4):388–393
22. Behre HM, Greb RR, Mempel A, Sonntag B, Kiesel L, Kaltwasser P et al (2005) Significance of a common single nucleotide polymorphism in exon 10 of the follicle-stimulating hormone (FSH) receptor gene for the ovarian response to FSH: a pharmacogenetic approach to controlled ovarian hyperstimulation. Pharmacogenet Genomics 15(7):451–456, Lug
23. Jun JK, Yoon JS, Ku S, Choi YM, Hwang KR, Park SY et al (2006) Follicle-stimulating hormone receptor gene polymorphism and ovarian responses to controlled ovarian hyperstimulation for IVF-ET. J Hum Genet 51(8):665–670
24. Klinkert ER, te Velde ER, Weima S, van Zandvoort PM, Hanssen RGJM, Nilsson PR et al (2006) FSH receptor genotype is associated with pregnancy but not with ovarian response in IVF. Reprod Biomed Online 13(5):687–695
25. Achrekar SK, Modi DN, Desai SK, Mangoli VS, Mangoli RV, Mahale SD (2009) Poor ovarian response to gonadotrophin stimulation is associated with FSH receptor polymorphism. Reprod Biomed Online 18(4):509–515
26. Desai SS, Achrekar SK, Pathak BR, Desai SK, Mangoli VS, Mangoli RV et al (2011) Follicle-stimulating hormone receptor polymorphism (G-29A) is associated with altered level of receptor expression in Granulosa cells. J Clin Endocrinol Metab 96(9):2805–2812, Lug 13 [citato 2011 Lug 22]; http://jcem.endojournals.org/content/early/2011/07/07/jc.2011-1064.abstract
27. Ascoli M, Fanelli F, Segaloff DL (2002) The lutropin/choriogonadotropin receptor, a 2002 perspective. Endocr Rev 23(2):141–174
28. Piersma D, Berns EMJJ, Verhoef-Post M, Uitterlinden AG, Braakman I, Pols HAP et al (2006) A common polymorphism renders the luteinizing hormone receptor protein more active by improving signal peptide function and predicts adverse outcome in breast cancer patients. J Clin Endocrinol Metab 91(4):1470–1476
29. Piersma D, Verhoef-Post M, Look MP, Uitterlinden AG, Pols HAP, Berns EMJJ et al (2007) Polymorphic variations in exon 10 of the luteinizing hormone receptor: functional consequences and associations with breast cancer. Mol Cell Endocrinol 30(276(1–2)):63–70, Set
30. Powell BL, Piersma D, Kevenaar ME, van Staveren IL, Themmen APN, Iacopetta BJ et al (2003) Luteinizing hormone signaling and breast cancer: polymorphisms and age of onset. J Clin Endocrinol Metab 88(4):1653–1657
31. Key TJ, Verkasalo PK, Banks E (2001) Epidemiology of breast cancer. Lancet Oncol 2(3):133–140
32. Kerkelä E, Skottman H, Friden B, Bjuresten K, Kere J, Hovatta O (2007) Exclusion of coding-region mutations in luteinizing hormone and follicle-stimulating hormone receptor genes as the cause of ovarian hyperstimulation syndrome. Fertil Steril 87(3):603–606
33. Hillier SG, Whitelaw PF, Smyth CD (1994) Follicular oestrogen synthesis: the 'two-cell, two-gonadotrophin' model revisited. Mol Cell Endocrinol 100(1–2):51–54
34. Pettersson K, Ding YQ, Huhtaniemi I (1992) An immunologically anomalous luteinizing hormone variant in a healthy woman. J Clin Endocrinol Metab 74(1):164–171, Gen
35. Nilsson C, Jiang M, Pettersson K, Iitiä A, Mäkelä M, Simonsen H et al (1998) Determination of a common genetic variant of luteinizing hormone using DNA hybridization and immunoassays. Clin Endocrinol (Oxf) 49(3):369–376, Set
36. Haavisto AM, Pettersson K, Bergendahl M, Virkamäki A, Huhtaniemi I (1995) Occurrence and biological properties of a common genetic variant of luteinizing hormone. J Clin Endocrinol Metab 80(4):1257–1263
37. Suganuma N, Furui K, Kikkawa F, Tomoda Y, Furuhashi M (1996) Effects of the mutations (Trp8→Arg and Ile15→Thr) in human luteinizing hormone (LH) beta-subunit on LH bioactivity in vitro and in vivo. Endocrinology 137(3):831–838
38. Jiang M, Pakarinen P, Zhang FP, El-Hefnawy T, Koskimies P, Pettersson K et al (1999) A common polymorphic allele of the human luteinizing hormone beta-subunit gene: additional mutations and differential function of the promoter sequence. Hum Mol Genet 8(11):2037–2046, Ott
39. Furui K, Suganuma N, Tsukahara S, Asada Y, Kikkawa F, Tanaka M et al (1994) Identification of two point mutations in the gene coding luteinizing hormone (LH) beta-subunit, associated with immunologically anomalous LH variants. J Clin Endocrinol Metab 78(1):107–113, Gen 1
40. Ramanujam LN, Liao WX, Roy AC, Loganath A, Goh HH, Ng SC (1999) Association of

molecular variants of luteinizing hormone with menstrual disorders. Clin Endocrinol 51(2):243–246

41. Alviggi C, Clarizia R, Pettersson K, Mollo A, Humaidan P, Strina I et al (2009) Suboptimal response to GnRHa long protocol is associated with a common LH polymorphism. Reprod Biomed Online 18(1):9–14, Gen
42. Gazvani R, Pakarinen P, Fowler P, Logan S, Huhtaniemi I (2002) Lack of association of the common immunologically anomalous LH with endometriosis. Hum Reprod 17(6): 1532–1534, Giu 1
43. Akhmedkhanov A, Toniolo P, Zeleniuch-Jacquotte A, Pettersson KS, Huhtaniemi IT (2001) Luteinizing hormone, its beta-subunit variant, and epithelial ovarian cancer: the gonadotropin hypothesis revisited. Am J Epidemiol 154(1):43–49, Lug 1
44. Pelletier G, El-Alfy M (2000) Immunocytochemical localization of estrogen receptors alpha and beta in the human reproductive organs. J Clin Endocrinol Metab 85(12): 4835–4840, Dic
45. Britt KL, Findlay JK (2002) Estrogen actions in the ovary revisited. J Endocrinol 175(2): 269–276
46. Georgiou I, Konstantelli M, Syrrou M, Messinis IE, Lolis DE (1997) Oestrogen receptor gene polymorphisms and ovarian stimulation for in-vitro fertilization. Hum Reprod 12(7):1430–1433, Lug
47. Altmäe S, Haller K, Peters M, Hovatta O, Stavreus-Evers A, Karro H et al (2007) Allelic estrogen receptor 1 (ESR1) gene variants predict the outcome of ovarian stimulation in in vitro fertilization. Mol Hum Reprod 13(8):521–526, Ago
48. de Castro F, Morón FJ, Montoro L, Galán JJ, Hernández DP, Padilla ES et al (2004) Human controlled ovarian hyperstimulation outcome is a polygenic trait. Pharmacogenetics 14(5):285–293, Mag
49. Sundarrajan C, Liao W, Roy AC, Ng SC (1999) Association of oestrogen receptor gene polymorphisms with outcome of ovarian stimulation in patients undergoing IVF. Mol Hum Reprod 5(9):797–802, Set
50. Silva IV, Rezende LCD, Lanes SP, Souza LS, Madeira KP, Cerri MF et al (2010) Evaluation of PvuII and XbaI polymorphisms in the estrogen receptor alpha gene (ESR1) in relation to menstrual cycle timing and reproductive parameters in post-menopausal women. Maturitas 67(4):363–367, Dic
51. Peter I, Shearman A, Zucker D, Schmid C, Demissie S, Cupples L et al (2005) Variation in estrogen-related genes and cross-sectional and longitudinal blood pressure in the Framingham Heart Study. J Hypertens 23(12):2193–2200
52. Corbo R, Ulizzi L, Piombo L, Martinez-Labarga C, De Stefano G, Scacchi R (2007) Estrogen receptor alpha polymorphisms and fertility in populations with different reproductive patterns. Mol Hum Reprod 13(8): 537–540
53. Molvarec A, Ver A, Fekete A, Rosta K, Derzbach L, Derzsy Z et al (2007) Association between estrogen receptor [alpha] (ESR1) gene polymorphisms and severe preeclampsia. Hypertens Res 30(3):205–211
54. Herrington DM, Howard TD, Hawkins GA, Reboussin DM, Xu J, Zheng SL et al (2002) Estrogen-receptor polymorphisms and effects of estrogen replacement on high-density lipoprotein cholesterol in women with coronary disease. N Engl J Med 346(13):967–974
55. Kim S, Pyun J, Kang H, Kim J, Cha DH, Kwack K (2011) Epistasis between CYP19A1 and ESR1 polymorphisms is associated with premature ovarian failure. Fertil Steril 95(1):353–356, Gen
56. George L, Mills JL, Johansson ALV, Nordmark A, Olander B, Granath F et al (2002) Plasma folate levels and risk of spontaneous abortion. JAMA 288(15):1867–1873, Ott 16
57. Mooij PN, Wouters MG, Thomas CM, Doesburg WH, Eskes TK (1992) Disturbed reproductive performance in extreme folic acid deficient golden hamsters. Eur J Obstet Gynecol Reprod Biol 43(1):71–75, Gen 9
58. Boxmeer JC, Macklon NS, Lindemans J, Beckers NGM, Eijkemans MJC, Laven JSE et al (2009) IVF outcomes are associated with biomarkers of the homocysteine pathway in monofollicular fluid. Hum Reprod 24(5): 1059–1066, Mag
59. Ebisch IMW, Peters WHM, Thomas CMG, Wetzels AMM, Peer PGM, Steegers-Theunissen RPM (2006) Homocysteine, glutathione and related thiols affect fertility parameters in the (sub)fertile couple. Hum Reprod 21(7):1725–1733, Lug
60. Forges T, Monnier-Barbarino P, Alberto JM, Guéant-Rodriguez RM, Daval JL, Guéant JL (2007) Impact of folate and homocysteine metabolism on human reproductive health. Hum Reprod Update 13(3):225–238, Giu
61. Hecht S, Pavlik R, Lohse P, Noss U, Friese K, Thaler CJ (2009) Common 677C→T mutation of the 5,10-methylenetetrahydrofolate reductase gene affects follicular estradiol synthesis. Fertil Steril 91(1):56–61, Gen
62. Rosen MP, Shen S, McCulloch CE, Rinaudo PF, Cedars MI, Dobson AT (2007)

Methylenetetrahydrofolate reductase (MTHFR) is associated with ovarian follicular activity. Fertil Steril 88(3):632–638, Set
63. Stern LL, Mason JB, Selhub J, Choi SW (2000) Genomic DNA hypomethylation, a characteristic of most cancers, is present in peripheral leukocytes of individuals who are homozygous for the C677T polymorphism in the methylenetetrahydrofolate reductase gene. Cancer Epidemiol Biomarkers Prev 9(8):849–853, Ago
64. Friso S, Choi S, Girelli D, Mason JB, Dolnikowski GG, Bagley PJ et al (2002) A common mutation in the 5,10-methylenetetrahydrofolate reductase gene affects genomic DNA methylation through an interaction with folate status. Proc Natl Acad Sci U S A 99(8):5606–5611
65. Frosst P, Blom H, Milos R, Goyette P, Sheppard C, Matthews R et al (1995) A candidate genetic risk factor for vascular disease: a common mutation in methylenetetrahydrofolate reductase. Nat Genet 10(1):111–113, Mag
66. Ingrosso D, Cimmino A, Perna AF, Masella L, De Santo NG, De Bonis ML et al (2003) Folate treatment and unbalanced methylation and changes of allelic expression induced by hyperhomocysteinaemia in patients with uraemia. Lancet 361(9370):1693–1699, Mag 17
67. Szymański W, Kazdepka-Ziemińska A (2003) [Effect of homocysteine concentration in follicular fluid on a degree of oocyte maturity]. Ginekol Pol 74(10):1392–1396, Ott
68. Berker B, Kaya C, Aytac R, Satıroglu H (2009) Homocysteine concentrations in follicular fluid are associated with poor oocyte and embryo qualities in polycystic ovary syndrome patients undergoing assisted reproduction. Hum Reprod 24(9):2293–2302
69. Haggarty P, McCallum H, McBain H, Andrews K, Duthie S, McNeill G et al (2006) Effect of B vitamins and genetics on success of in-vitro fertilisation: prospective cohort study. Lancet 367(9521):1513–1519, Mag 6
70. Pacchiarotti A, Mohamed MA, Micara G, Linari A, Tranquilli D, Espinola SB et al (2007) The possible role of hyperhomocysteinemia on IVF outcome. J Assist Reprod Genet 24(10):459–462, Ott
71. Laanpere M, Altmäe S, Kaart T, Stavreus-Evers A, Nilsson TK, Salumets A (2011) Folate-metabolizing gene variants and pregnancy outcome of IVF. Reprod Biomed Online 22(6):603–614, Giu
72. Altmäe S, Stavreus-Evers A, Ruiz JR, Laanpere M, Syvänen T, Yngve A et al (2010) Variations in folate pathway genes are associated with unexplained female infertility. Fertil Steril 94(1):130–137, Giu
73. Meyer JM, Eaves LJ, Heath AC, Martin NG (1991) Estimating genetic influences on the age-at-menarche: a survival analysis approach. Am J Med Genet 39(2):148–154, Mag 1
74. Kaprio J, Rimpelä A, Winter T, Viken RJ, Rimpelä M, Rose RJ (1995) Common genetic influences on BMI and age at menarche. Hum Biol 67(5):739–753, Ott
75. Zerbetto I, Gromoll J, Luisi S, Reis FM, Nieschlag E, Simoni M et al (2008) Follicle-stimulating hormone receptor and DAZL gene polymorphisms do not affect the age of menopause. Fertil Steril 90(6):2264–2268, Dic
76. Achrekar SK, Modi DN, Meherji PK, Patel ZM, Mahale SD (2010) Follicle stimulating hormone receptor gene variants in women with primary and secondary amenorrhea. J Assist Reprod Genet 27(6):317–326, Giu
77. Taylor KC, Small CM, Epstein MP, Sherman SL, Tang W, Wilson MM et al (2010) Associations of progesterone receptor polymorphisms with age at menarche and menstrual cycle length. Horm Res Paediatr 74(6):421–427
78. Jabbour HN, Kelly RW, Fraser HM, Critchley HOD (2006) Endocrine regulation of menstruation. Endocr Rev 27(1):17–46
79. Guo Y, Shen H, Xiao P, Xiong D, Yang T, Guo Y et al (2006) Genome-wide linkage scan for quantitative trait loci underlying variation in age at menarche. J Clin Endocrinol Metab 91(3):1009–1014
80. Romano A, Delvoux B, Fischer D, Groothuis P (2007) The PROGINS polymorphism of the human progesterone receptor diminishes the response to progesterone. J Mol Endocrinol 38(2):331–350
81. Schweikert A, Rau T, Berkholz A, Allera A, Daufeldt S, Wildt L (2004) Association of progesterone receptor polymorphism with recurrent abortions. Eur J Obstet Gynecol Reprod Biol 113(1):67–72
82. Terry KL, De Vivo I, Titus-Ernstoff L, Sluss PM, Cramer DW (2005) Genetic variation in the progesterone receptor gene and ovarian cancer risk. Am J Epidemiol 161(5):442–451
83. Gennari L, Masi L, Merlotti D, Picariello L, Falchetti A, Tanini A et al (2004) A polymorphic CYP19 TTTA repeat influences aromatase activity and estrogen levels in elderly MEN: effects on bone metabolism. J Clin Endocrinol Metab 89(6):2803–2810, Giu 1
84. Stolk L, Zhai G, van Meurs JBJ, Verbiest MMPJ, Visser JA, Estrada K et al (2009) Loci at chromosomes 13, 19 and 20 influence age at natural menopause. Nat Genet 41(6):645–647, Giu

85. Sulem P, Gudbjartsson DF, Rafnar T, Holm H, Olafsdottir EJ, Olafsdottir GH et al (2009) Genome-wide association study identifies sequence variants on 6q21 associated with age at menarche. Nat Genet 41(6):734–738, Giu
86. Perry JRB, Stolk L, Franceschini N, Lunetta KL, Zhai G, McArdle PF et al (2009) Meta-analysis of genome-wide association data identifies two loci influencing age at menarche. Nat Genet 41(6):648–650, Giu
87. Elks CE, Perry JRB, Sulem P, Chasman DI, Franceschini N, He C et al (2010) Thirty new loci for age at menarche identified by a meta-analysis of genome-wide association studies. Nat Genet 42(12):1077–1085, Dic
88. Ong KK, Elks CE, Li S, Zhao JH, Luan J, Andersen LB et al (2009) Genetic variation in LIN28B is associated with the timing of puberty. Nat Genet 41(6):729–733, Giu
89. Kitawaki J, Kado N, Ishihara H, Koshiba H, Kitaoka Y, Honjo H (2002) Endometriosis: the pathophysiology as an estrogen-dependent disease. J Steroid Biochem Mol Biol 83(1–5):149–155, Dic
90. Falconer H, D'Hooghe T, Fried G (2007) Endometriosis and genetic polymorphisms. Obstet Gynecol Surv 62(9):616–628, Set
91. Tempfer CB, Simoni M, Destenaves B, Fauser BCJM (2009) Functional genetic polymorphisms and female reproductive disorders: part II–endometriosis. Hum Reprod Update 15(1):97–118
92. Tsuchiya M, Nakao H, Katoh T, Sasaki H, Hiroshima M, Tanaka T et al (2005) Association between endometriosis and genetic polymorphisms of the estradiol-synthesizing enzyme genes HSD17B1 and CYP19. Hum Reprod 20(4):974–978
93. Vietri MT, Cioffi M, Sessa M, Simeone S, Bontempo P, Trabucco E et al (2009) CYP17 and CYP19 gene polymorphisms in women affected with endometriosis. Fertil Steril 92(5):1532–1535
94. Norman RJ, Dewailly D, Legro RS, Hickey TE (2007) Polycystic ovary syndrome. Lancet 370(9588):685–697, Ago 25
95. Du J, Zhang W, Guo L, Zhang Z, Shi H, Wang J et al (2010) Two FSHR variants, haplotypes and meta-analysis in Chinese women with premature ovarian failure and polycystic ovary syndrome. Mol Genet Metab 100(3):292–295, Lug
96. Gu B, Park J, Baek K (2010) Genetic variations of follicle stimulating hormone receptor are associated with polycystic ovary syndrome. Int J Mol Med 26(1):107–112, Lug
97. Valkenburg O, Uitterlinden AG, Piersma D, Hofman A, Themmen APN, de Jong FH et al (2009) Genetic polymorphisms of GnRH and gonadotrophic hormone receptors affect the phenotype of polycystic ovary syndrome. Hum Reprod 24(8):2014–2022, Ago
98. Chen Z, Zhao H, He L, Shi Y, Qin Y, Shi Y et al (2011) Genome-wide association study identifies susceptibility loci for polycystic ovary syndrome on chromosome 2p16.3, 2p21 and 9q33.3. Nat Genet 43(1):55–59, Gen
99. Catena R, Argentini M, Martianov I, Parello C, Brancorsini S, Parvinen M et al (2005) Proteolytic cleavage of ALF into alpha- and beta-subunits that form homologous and heterologous complexes with somatic TFIIA and TRF2 in male germ cells. FEBS Lett 579(16):3401–3410, Giu 20
100. Huang M, Wang H, Li J, Zhou Z, Du Y, Lin M et al (2006) Involvement of ALF in human spermatogenesis and male infertility. Int J Mol Med 17(4):599–604
101. Cramer DW, Petterson KS, Barbieri RL, Huhtaniemi IT (2000) Reproductive hormones, cancers, and conditions in relation to a common genetic variant of luteinizing hormone. Hum Reprod 15(10):2103–2107, Ott
102. Tapanainen JS, Koivunen R, Fauser BCJM, Taylor AE, Clayton RN, Rajkowa M et al (1999) A new contributing factor to polycystic ovary syndrome: the genetic variant of luteinizing hormone. J Clin Endocrinol Metab 84(5):1711–1715, Mag 1
103. Lalioti MD (2011) Impact of follicle stimulating hormone receptor variants in fertility. Curr Opin Obstet Gynecol 23(3):158–167, Giu
104. Morón FJ, Galán JJ, Ruiz A (2007) Controlled ovarian hyperstimulation pharmacogenetics: a simplified model to genetically dissect estrogen-related diseases. Pharmacogenomics 8(7):775–785
105. van Disseldorp J, Franke L, Eijkemans R, Broekmans F, Macklon N, Wijmenga C et al (2011) Genome-wide analysis shows no genomic predictors of ovarian response to stimulation by exogenous FSH for IVF. Reprod BioMed Online 22(4):382–388
106. McCarthy MI, Abecasis GR, Cardon LR, Goldstein DB, Little J, Ioannidis JPA et al (2008) Genome-wide association studies for complex traits: consensus, uncertainty and challenges. Nat Rev Genet 9(5):356–369, Mag
107. Alfirevic A, Alfirevic Z, Pirmohamed M (2010) Pharmacogenetics in reproductive and perinatal medicine. Pharmacogenomics 11(1):65–79, Gen

Chapter 5

Understanding the Spermatozoon

Queenie V. Neri, Jennifer Hu, Zev Rosenwaks, and Gianpiero D. Palermo

Abstract

The former perception of the spermatozoon as a delivery device of the male genome has been expanded to include a new understanding of the cell's complex role in fertilization. Once the spermatozoon reaches the oocyte, it triggers egg activation and orchestrates the stages of pre- and post-fertilization in a preprogrammed pattern while tapping the oocyte's resources in an effort to generate a new life.

Key words ICSI, Spermatozoa, Spermatogenesis, Spermiogenesis

1 The Sperm Homunculus Revisited

The revolutionary discovery of the presence of spermatozoa in semen was made by Antonie van Leeuwenhoek (1632–1723), who in 1676 sent a letter (published in 1678) to the Royal Society of London describing and illustrating "animalcules" within the ejaculates of man and dog [1]. It became apparent that the semen was inhabited by a multitude of seemingly tadpole-like animals, the observation of which gave rise to a controversy, lasting for more than a century and a half, concerning the sperm origin, structure, and function [2]. This observation, however, did give rise to the notion that the "animalcules" played an active role in reproduction, contrary to the widely accepted idea that animals originated only from eggs.

The findings by van Leeuwenhoek over 400 years ago gave rise to certain theories as to the nature of their function including the concept of the homunculus by Hartsoeker. It is often said that in 1694, while observing human gametes through a microscope, Hartsoeker believed that he saw tiny men inside the sperm which he called homunculi or animalcules [3]. This was the beginning of the spermists' theory by Charles Bonnet (1720–1793), Lazzaro Spallanzani (1729–1799), and René Antoine Ferchault de Reaumur (1683–1757) [4, 5], who held the belief that the spermatozoon was in fact a "little man," or homunculus, that was placed inside

Zev Rosenwaks and Paul M. Wassarman (eds.), *Human Fertility: Methods and Protocols*, Methods in Molecular Biology, vol. 1154, DOI 10.1007/978-1-4939-0659-8_5, © Springer Science+Business Media New York 2014

a woman to grow into a child [6]. To these early investigators, this seemed to neatly answer the questions revolving around the mystery of conception.

With the early establishment of in vitro fertilization, the spermatozoon has paradoxically lost its identity and has been relegated to merely a mobile cell which is evaluated in the context of semen in its entirety. This attributed demoted role of the spermatozoon is the delivery of the male genome to the site of fertilization. Indeed, the evaluation of the ejaculate as a whole has been considered the most reliable assessment of the ability of a man to reproduce for the past 30 years.

2 Semen Evaluation

A semen analysis has been useful in both clinical and research settings for investigating the male's fertility status by monitoring spermatogenesis whenever the reproductive ability of the couple is in question. Although it is an imperfect tool, it remains the cornerstone of the assessment of male fertility [7].

Semen analyses must be performed according to standards set by the World Health Organization [8] in order to evaluate descriptive parameters of the ejaculate. Routine semen analysis provides highly useful information concerning the production, motility, and viability of the spermatozoa as well as the function and patency of the male genital tract. In fact, the assessment of volume and consistency of the ejaculate offers insights into the conditions of the accessory glands. Although the semen evaluation provides a useful first step in the evaluation of the infertile male, it is not a fertility test [9]. A semen analysis does not provide a clear indication of the functional integrity of the spermatozoon to undergo the due capacitation processes required to achieve egg penetration or its ability to fertilize an ovum. It is important to understand that while the results may *correlate* with fertility, the assay is not a direct measure of a patient's fertility potential [10–12].

The semen analysis is the main test requested during the evaluation of all men presenting with infertility. There may be substantial fluctuations between samples, and therefore a minimum of two properly collected samples should be examined. Ideally, the ejaculated samples should be assessed over more than one spermatogenic cycle (74 days). However, for most patients this is often not possible or desirable since this prolongs the evaluation period of their potential fertility treatment. However, if the patient's history suggests recent insult to spermatogenesis, the semen sperm assessment should be expanded over several months.

The assay measures a variety of parameters including number, characteristics of the spermatozoa, seminal plasma, its consistency and aspect, as well as presence and type of other cells. In addition, in recent years a semen analysis is often complemented with sperm functional assays that aim at providing information on the one spermatozoon capable of delivering correct complement of chromosomes to an ovum.

Spermatozoa accounts for less than 10 % of the total semen, and the remainder of the ejaculate consists of the products secreted by the accessory sex glands: the seminal vesicles (55 %), prostate gland (25 %), and bulbourethral gland (10 %). There are usually greater than 100 million spermatozoa per milliliter of semen in the ejaculate of a normal male. Although there are individual variations, men whose semen contains 20–50 million sperm cells in the entire specimen are deemed to be fertile [13]. A man with fewer than 10 million is likely to be subfertile, especially when the specimen contains many immotile cells and the motile spermatozoa are plagued by abnormal morphology. According to the new WHO criteria [8], the normal values for the fertile population are defined by the following parameters:

- Total semen volume of 1.5–5.0 ml.
- Sperm concentration of >15 million/ml.
- Motility of ≥40 %.
- ≥4 % Normal morphology.

3 Sperm Phenotyping

While quantifying the number of spermatozoa that retain progressive motility is relatively easy and somewhat intuitive, the evaluation of their shape is more complex. However, morphological assessment may provide more information on the sperm's ability to fertilize an ovum. The difficulty in assessing sperm morphology is due to the great variability in criteria utilized to evaluate its shape and size. Observations of spermatozoa recovered from the female reproductive tract, especially in postcoital endocervical mucus [14], have helped to define the characteristics of functionally competent, potentially fertilizing spermatozoa. Classification can be made from wet mounts using a phase-contrast microscope or using a variety of stains viewed with a bright-field microscope. Several stains are available, and their procedures vary in complexity [8], ranging from hematoxylin and eosin, Giemsa, Bryan–Leishman, and the Shorr stain. While the Papanicolaou modified staining procedure provides excellent nuclear and cytologic detail and has been considered for many years the best stain to observe sperm morphology, it is too time consuming and often inconsistent, leading to the increased popularity of pre-stained slides.

The now readily available pre-stained slides, such as Testsimplets® and Cell-Vu® Prestained Morphology slides, allow the observation of spermatozoa in a dynamic, extensive, and immediate manner. In addition to using spermatozoa, any cells or undefined "structures" present in semen can also be examined using these slides, and the presence of bacteria can be confirmed by their intrinsically dark DNA staining. The use of pre-stained slides, highly publicized in several atlases, has also shifted the classical concept promulgated by WHO publications of the tedious and lengthy description of individual spermatozoal defects to the new concept of evaluating the number of spermatozoa carrying a major defect and emphasizing the proportion of sperm cells presenting with a major morphological deviation [15]. Thus, this changed the focus of andrologists and embryologists to the establishment of techniques which identify spermatozoa with superior characteristics, albeit of lower numbers. This selection was accelerated by the adoption of intracytoplasmic sperm injection (ICSI), which has reframed the focus of sperm analysis to the level of the individual sperm cell.

The trend towards the selection of the individual spermatozoon has recently materialized in motile sperm organellar morphology examination (MSOME), which has been proposed to assess living male gamete phenotype sperm morphology under high magnification [16]. With this approach, screening is used to select a spermatozoon with an optimal shape for ICSI. Reports have claimed that this procedure, intracytoplasmic morphologically selected sperm injection (IMSI), yields superior clinical outcomes compared to conventional ICSI [17]. The expected beneficial impact of IMSI has been described in a series of studies where the clinical outcome of patients treated by this procedure was compared with that of couples treated by conventional ICSI [18–21].

The high-magnification morphological evaluation of viable spermatozoa is carried out using an inverted microscope equipped with interferential contrast Nomarski DIC optics. The use of maximum optical magnification (×100 lens under oil immersion), magnification selector (×1.5), and digital video-coupled magnification (×44) has led to a final video monitor magnification of around ×6,600. The nuclei are evaluated by their smoothness, symmetry, oval configuration, and homogeneity of the sperm nuclear shape, with "vacuoles" not exceeding more than 4 % of the nuclear area surface [17]. The role played by sperm nuclear vacuoles and their position within the sperm head are sill unclear. Only transmission electron microscopy (TEM) can clearly and accurately locate nuclear vacuoles [22].

The early ultrastructural studies of human sperm in the 1950s and 1960s revealed that vacuoles in the sperm nucleus are seen in the large majority of human spermatozoa regardless of the fertility of the donors. Vacuoles in human spermatozoa have been considered to be physiologic findings devoid of consequence to fertility potential [23]. Because vacuoles are normal physiological features, the proposed relationship between nuclear surface vacuoles and sperm DNA defects with consequent impaired embryo developmental competence needs to be revisited.

Other types of sperm head irregularities are surface vacuoles or indentations, craters/dents, and hollows observed on the sperm coat. In such cases, during sperm morphogenesis, the outer acrosomal membrane misforms and generates what appears to be a vacuole [24]. These vacuole-like structures disappear as the spermatozoon matures in the epididymis or at the time of the acrosome reaction [25]. In other circumstances, however, their frequency seems to increase with temperature (37 °C) and incubation time (>2 h) [26], most likely due to the plication/vacuolization of the rostral spermiolemma during capacitation. Interestingly, these vacuole-like structures or craters appear in over 90 % of spermatozoa from fertile donors with normal semen parameters [27, 28].

Higher magnification screening for sperm surface irregularities, however, did not seem to benefit the patients' clinical outcome in independent investigations [29]. This was true for patients with compromised semen parameters and for those either undergoing first or repeated ART attempts. More detailed morphological observations indicated that in human sperm heads, visible irregularities or vacuoles are almost ubiquitous and appear to be paraphysiologic findings. Analyses of both ejaculated and surgically retrieved spermatozoa also revealed the varying presence and size of vacuoles that develop during the dynamic processes of spermiogenesis and maturation. This surface irregularity did not translate to a higher incidence of DNA fragmentation or aneuploidy or to the ability of vacuolated spermatozoa to generate zygotes capable of developing to blastocysts.

4 The Sum of Its Parts

The general component of mammalian spermatozoa consists of the primary entities, the head and the flagellum, which are both enclosed by a regionally differentiated plasma membrane (Fig. 1). The most recognized organelles of the head are the nucleus and acrosome, a membrane-limited vesicle that houses several hydrolytic enzymes and is located over the apex of the nucleus. The junction of the head and tail is formed at the site of the implantation fossa at the base of the nucleus. The basal plate in this fossa receives the rounded contour of a dense capitulum that houses another important organelle, the proximal centriole. The midpiece of the tail, centrally located, is defined by an aggregated sheath of mitochondria surrounding the axoneme that terminates at the annulus. The axoneme is composed of nine jointed microtubule doublets surrounding a central pair of microtubules, as typically recognized in cilia [30].

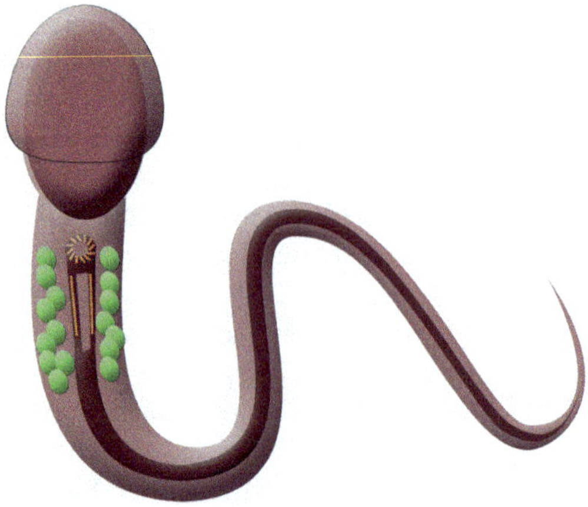

Fig. 1 Depiction of a spermatozoon. The head, composed mostly of the nucleus, is partly covered by the cap-like acrosome, an organelle containing enzymes. The tail of this sperm consists of three regions, the midpiece holds the mitochondria and centrosome, principal piece, and end piece

4.1 Acrosome

The acrosome develops over the anterior half of the spermatozoon head. It is a cap-like structure derived from the Golgi apparatus. Acrosome formation is completed during testicular maturation of germ cells during spermiogenesis.

The acrosome reaction is an important physiological event, involving fusion of the sperm plasma membrane and outer acrosomal membrane with subsequent vesiculation and release of the acrosomal content [31–33]. A functional acrosome is believed to be a prerequisite both for sperm penetration of the zona pellucida and for sperm fusion with the oolemma [34]. The acrosome reaction is a well-coordinated process [35] essential to oocyte insemination and the triggering of the first steps in embryo development [36].

In order to fertilize, mammalian sperm must first undergo capacitation in the female tract [37, 38]. Capacitation involves modifications in the sperm plasma membrane that lead to hyperactivation and permit the acrosome reaction. The acrosome reaction involves multiple fusions between the outer acrosome membrane and the overlying sperm plasma membrane, enabling the soluble contents of the acrosome to leak out through the fenestrated membranes [31], simultaneously preparing the surface over the equatorial segment for its fusogenic role [39].

ICSI bypasses the events involved in physiological sperm penetration of the oocyte and requires no specific pretreatment of sperm other than immobilization [40–42]. However, aggressive immobilization by compression of the tail prior to injection significantly improves ICSI fertilization rates [43–46]. Although the mechanism of this beneficial effect is not immediately clear, there is indirect evidence that such immobilization triggers changes in the sperm's permeability [47] and that it may expedite

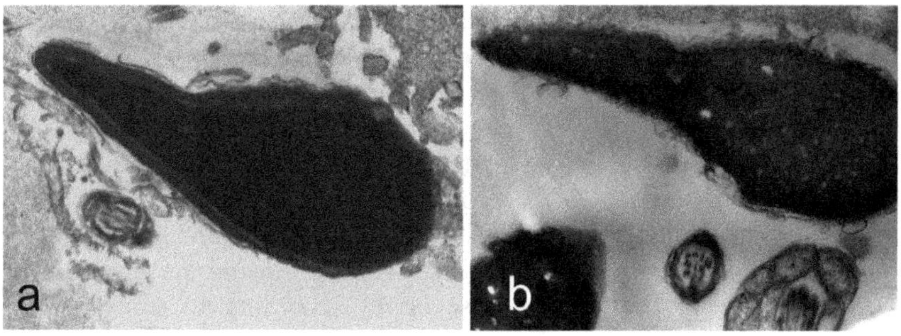

Fig. 2 (**a**) Intact human sperm head, and (**b**) after immobilization, the acrosomal membrane is lost

changes leading to sperm plasma membrane destabilization and culminate in acrosomal disruption [43, 45]. The utility of immobilization was supported by the observation that epididymal spermatozoa characterized by high lipid content in the plasma membrane [48] required more intense damage to trigger membrane destabilization. The need to aggressively immobilize a spermatozoon has also been reinforced by the knowledge that proteins present in capacitation media are required for the replacement of sperm membrane lipid components, rendering their membrane more hydrophilic, more easily to disrupt mechanically, and therefore more prone to acrosome react. In fact, the introduction of sequential media, formulated with limited glucose and protein, has resulted in complications in the execution of the ICSI procedure [49]. Because of this, re-training to induce more damage on the spermatozoon's flagellum as well as minimization of the oolemma eversion caused by withdrawal of the injection tool, presumably due to poor protein content, were needed.

The need for adequate sperm destabilization was supported by an earlier study in which spermatozoa were mechanically immobilized and inserted into the perivitelline space of mouse oocytes [50] to allow ultrathin TEM sections. This study revealed consistent alterations in the acrosomal structure including disruption of the plasma membrane, vesiculation, or even complete loss of the acrosomal cap following immobilization (Fig. 2) [50, 51]. Thus, all of the sperm assessed had undergone some membrane disorganization of the head, in contrast to the majority of control sperm. Immobilization of sperm immediately prior to the ICSI procedure is key to its consistent success [43–46, 52]. Sperm membrane permeabilization may help to expose the sperm nucleus to the ooplasm and to facilitate male pronucleus formation [53]. Removal of the sperm plasma membrane also seems to be necessary for oocyte activation. Indeed, calcium oscillations and subsequent oocyte activation appear to be induced after intracytoplasmic injection of spermatozoon-associated oocyte-activating factor [54, 55], which can exert its effect only after damage or removal of its membrane [52, 56].

4.2 Activating Factor

A soluble oocyte-activating factor has been shown to be present in the cytosolic fraction of rabbit, hamster, boar, and human spermatozoa [57–59]. In addition, it has been reported that injected spermatozoon contribute to activation of the oocyte by releasing a heat-sensitive, intracellular activating factor that is not species specific [47]. In accord with that, a cytosolic 33 kDa protein (oscillin) isolated from hamster spermatozoa induces Ca^{2+} oscillations in a pattern similar to that seen following fertilization [59] when injected into mouse eggs. According to immunohistochemical analysis, the protein in its hexamer form is located at an intracellular site in the equatorial segment of the sperm head [59], the site of sperm–oocyte fusion [60–63].

Another argument for the equatorial localization of sperm cytosolic factor (SCF) can be found in the results of experiments where an injected sperm head was able to activate and fertilize oocytes at the same rate as an intact spermatozoon [64–66]. We have purified SCF from human spermatozoa of pooled fertile donors by sequential freezing-thawing, sonication, and ultracentrifugation. The injection of 5 pl of enriched SCF elicited persistent Ca^{2+} oscillations in contrast to sham injections with culture medium that failed to induce long-lasting Ca^{2+} rises [55, 67]. Cytogenetic analysis of the resulting zygotes confirmed the involvement of the male genome. The main limitation of this study was that the SCF used is not patient specific, since it is derived from pooled specimens.

The oscillin protein isolated from hamster sperm was suggested to be the factor involved in the generation of calcium oscillations [59], and it was presented as glucosamine 6-phosphate isomerase (GPI, GenBank accession number D31766), the human homologue of hamster oscillin. The success of oscillin was short lived; in fact, we injected the recombinant product of GPI into mouse oocytes, and it failed to induce $[Ca^{2+}]_i$ oscillations. Furthermore, use of recombinant GPI, GPI mRNA, and immunodepletion did not support the role of oscillin in the initiation of calcium oscillations [68].

A sperm-specific phospholipase C isoform, PLCζ [64], triggered Ca^{2+} oscillations in the mouse indistinguishable from those at fertilization. Human PLCζ was able to elicit mouse egg activation and early embryonic development up to the blastocyst stage [69].

We postulated that the absence of this sperm-soluble factor in spermatozoa of infertile men is the plausible cause of fertilization failure in some couples even after ICSI [70]. Because fertilization failure carries such a high emotional and financial toll on our patients, the objective of this study was the identification of couples with complete fertilization failure where the male gamete was lacking the activating factor.

From September 1993 to June 2010, in 11,390 couples treated by ICSI, 2.0 % experienced absolute fertilization failure. The lack of oocyte-activating factor was confirmed in 59 couples that presented

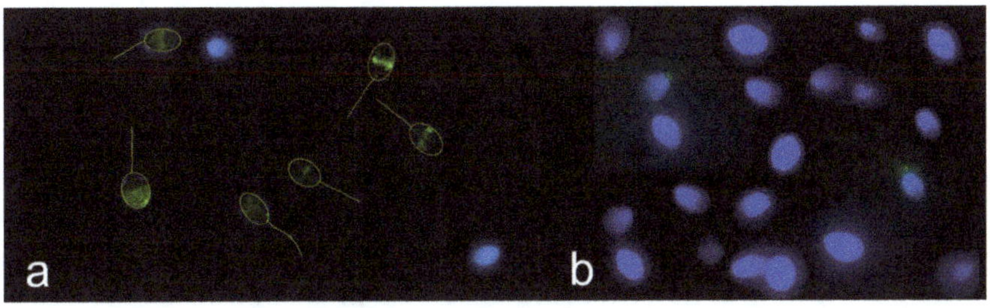

Fig. 3 Spermatozoa obtained from (**a**) fertile men positive for phospholipase C zeta (PLCζ; in *green*) and (**b**) from the study group that is lacking the soluble factor

with recurrent fertilization failure. Of those, only seven couples consented to undergo assisted oocyte activation. In all instances, the inability of the spermatozoa to induce oocyte activation was tested by injecting them into mouse oocytes [71]. In addition, all of the men included in the study had a compromised content of PLCζ in most of their spermatozoa. PLCζ in these men ranged from 0 up to 6.4 %, remarkably lower in comparison to fertile individuals at >80 % (P=0.0001) (Fig. 3).

These couples (female age 37.4 ± 4 years) had an average of 2.3 IVF cycles with no fertilization in spite of an average of 10.8 oocytes undergoing ICSI. They were treated in 11 cycles that yielded a comparable number of oocytes (11.2). In one cycle, only a single MI oocyte was retrieved that failed to mature. One man had a familial case of globozoospermia confirmed by TEM. The pretreated semen specimens had an average concentration of $58.6 \pm 40 \times 10^6$/ml with a mild motility impairment of 21.6 ± 11 % and morphology of 1.8 ± 1 %.

Prior to ICSI, spermatozoa were exposed to streptolysin O (30 min) to assist in sperm membrane permeabilization [72]. Following sperm injection, oocytes were sequentially exposed to Ca^{2+} ionophore [70, 71, 73].

Oocyte activation yielded 55.8 % (43/77) normal fertilization while 4 (5.1 %) lysed. The cleavage rate was 87.7 % with the mean number of blastomeres on day 3 being 8.3 ± 2 and a mean fragmentation rate of 10.2 % ± 3. Conceptuses were successfully transferred into the uterine cavity in all cycles with an average of 2.3 per procedure; a total of 7 (70.0 %) couples had a positive βhCG of which 4 (40.0 %) progressed to clinical pregnancies and one couple delivered a healthy baby boy.

It appeared that permeabilization of the spermatozoa helped to release the soluble-activating factor and may have contributed to sperm nuclear decondensation. This protocol demonstrates that it is possible to rescue cycles with recurrent fertilization failure after ICSI by screening for the presence of PLCζ. In fact, assisted oocyte activation and PLCζ screening provide these couples with a chance to conceive their own biological child.

4.3 DNA Packaging

The spermatozoon as a motile cell is not only capable of dynamically relocating to the appropriate site to perform its function but also distinguishes itself from other cells by its extraordinary ability to thrive and survive in hostile environments and conditions, such as acidic vaginal pH and opposing cilia motion encountered within the female genital tract. The spermatozoon's durability is a product of its fibrous sheath and the high compaction of its nucleic acid.

The packaging of DNA in mammalian cells has important implications for the biology of human infertility. Recent advances in the understanding of the structure of mammalian sperm chromatin and function have changed our perception of this tightly condensed and apparently inert chromatin. For its protection, the sperm's DNA is bundled very densely prior to sperm's transit to the oocyte. Shaping of the male gamete nucleus takes place in late spermiogenesis, as its chromatin undergoes a remarkable condensation that renders the sperm transcriptionally inert and highly resistant to digestion. Following the morphological transformation of the nucleus in the testis, as spermatozoa pass through the epididymis the chromatin is stabilized by the establishment of disulfide bonds between the thiol-rich protamines [74]. The human sperm nucleus is also composed of a DNA-condensing core and linker histones that are mostly replaced by protamines, thus changing the sperm head to a more compact and hydrodynamic shape favorable to cell motility and penetration through the egg vestments [75, 76]. However, this condensation cannot abrogate paternal genomic and chromosomal elements that are essential for the embryo to initiate and allow proper embryonic development.

The chromatin packaging of sperm cells that takes place during spermiogenesis is strikingly different from that of somatic cells [77]. It involves the replacement of somatic nucleosomal histones by a set of basic proteins, transition proteins, which are replaced by protamines in sequential steps [78] that include an increase in histone acetylation, an increase in the activity of the ubiquitin system, and a change in DNA topology resulting from the elimination of negative supercoiling [79]. It is likely that a sizeable number of single- or double-stranded DNA breaks must occur during the elongated spermatid stage to avoid supercoiling. Small basic nuclear proteins, promoting DNA condensation, may repair these strand breaks and thereby prevent the persistence of DNA damage in mature spermatozoa. Experimental evidence suggests that a large number of DNA strand breaks are indeed detected at mid-spermatogenic steps [80–82]. It has been proposed that topoisomerase II may play a role in both creating and ligating DNA nicks during spermiogenesis [83]. Upon elimination of supercoiling by the strand breaks, an efficient mechanism must be required in order to "seal" or "repair" the DNA phosphate backbone. It is possible that the DNA condensation process initiated by transition proteins and completed by protamines contributes to the repair of these DNA strand breaks. It is crucial to establish whether DNA fragmentation, which results from perturbed chromatin remodeling and condensation, can affect embryo development and fertility or if the altered complement of nuclear proteins is a major cause of impaired conception [84].

The understanding of this unique chromatin packing has important consequences for both the development of male infertility screening testing and for the understanding of sperm chromatin characteristics, which may also have implications for the outcome of ART [85–91]. It has been postulated that fertile men with normal semen parameters almost uniformly have low levels of DNA breakage, whereas infertility presents, especially with compromised semen parameters, with increased proportion of nicks and breaks in the chromatin. To complicate issues even further, up to 8 % of infertile men will have abnormal DNA integrity not corroborated by impaired semen concentration, motility, and morphology [92, 93]. A systematic observation performed in our laboratory evidenced a correlation between DNA fragmentation level measured by SCSA and TUNEL with motility [94]. It appears that the etiology of sperm DNA damage is multifactorial and may be due to intrinsic and/or external factors. Intrinsic defects that may predispose spermatozoa to DNA damage include protamine deficiency, mutations that adversely affect DNA compaction [95], or other "DNA packaging" defects. In addition, advanced male age has been related to a higher occurrence of sperm DNA damage [96–99]. Furthermore, environmental factors ranging from cigarette smoking [100, 101], genital tract inflammation, varicoceles [102], and hormone deficiencies [103] are also associated with an increased level of DNA damage, as seen in humans and animal models.

The vast majority of mammalian sperm chromatin is compacted into toroids that contain roughly 50 kb of DNA [104–106] (Fig. 4). This condensation is so complete that most of the DNA is hidden within the toroids [107]. This component of sperm DNA exists in a semicrystalline state and is resistant to nuclease digestion [108]. Mammalian protamines also contain several cysteines that are thought to confer increased stability on sperm nucleic acid by intermolecular disulfide cross-links. Sperm DNA cannot be decondensed in vitro without reducing reagents [109], and disulfide cross-links increase as sperm cells transit through the epididymis.

Sperm DNA integrity is currently assessed by destructive methods such as TUNEL, COMET, sperm chromatin dispersion (SCD) test, or a sperm chromatin structural assay (SCSA). All of these tests require fixation and destruction of the sperm being assessed [91]. The maintenance of DNA integrity is a physiological process needed for the complex packing and intertwining of the typical toroids created during spermiogenesis. Although chromatin fragmentation should be completely repaired in fully developed spermatozoa, the persistence of nicks and breaks in ejaculated

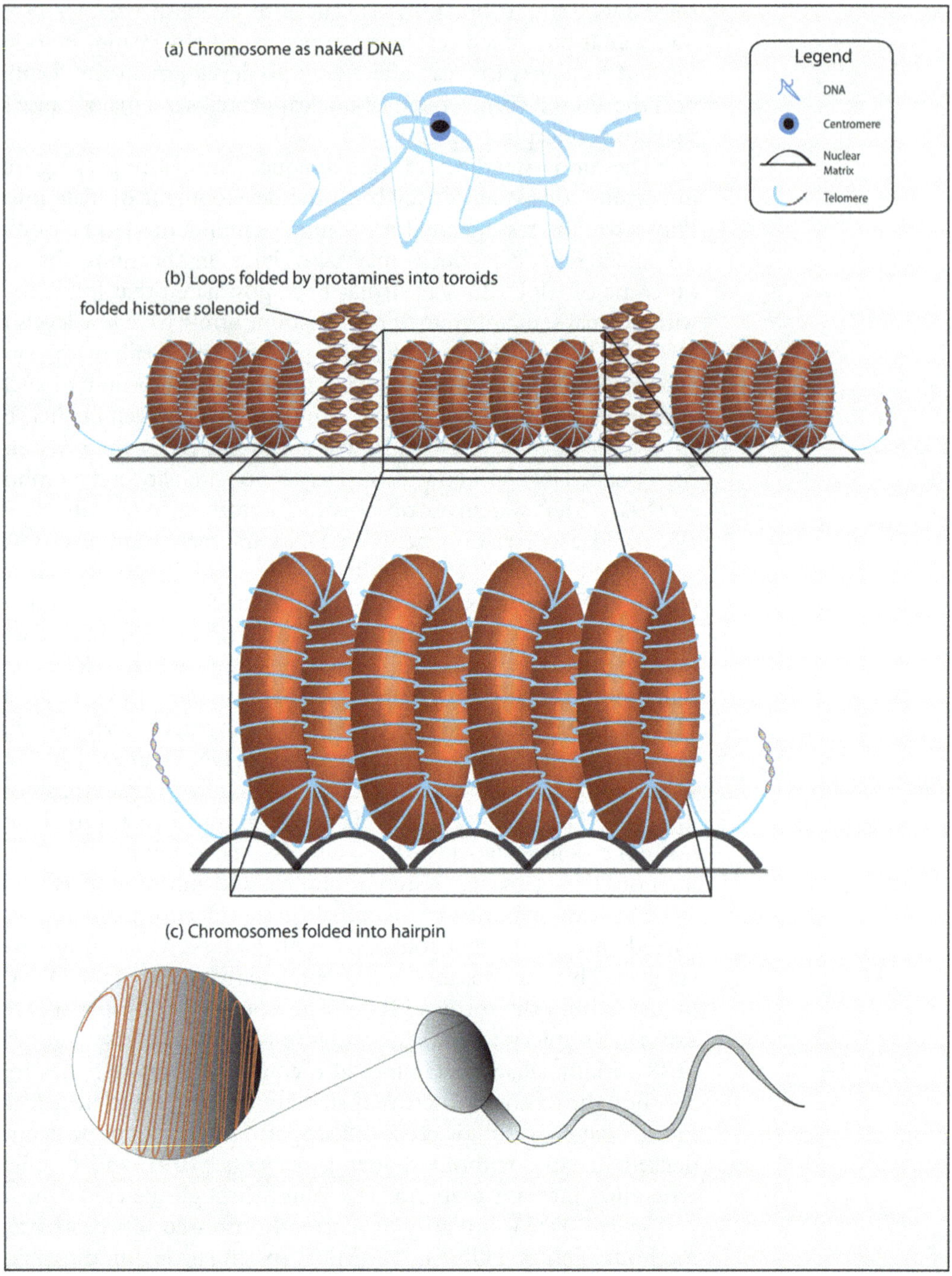

Fig. 4 (**a**) Chromosomes depicted as naked DNA. (**b**) "Donut-loop" model for sperm chromatin structure as introduced by Sotolongo et al. [108] that reveals the internal structure of the protamine–DNA fibers within the toroid (*inset*). (**c**) Chromosomes are folded into a hairpin structure in the sperm head

Table 1
Semen characteristics and DNA fragmentation level

	DFI	
No. of (%)	Normal (≤30)	Abnormal (>30)
Men	138	39
DFI range	2.1–29.7	31.0–91.4
Paternal age (M years ± SD)		
Concentration (M × 10^6/ml ± SD)	62.3 ± 41*	30.8 ± 2*
Motility (M% ± SD)	54.3 ± 11†	34.9 ± 15†
Morphology (M% ± SD)	2.9 ± 3‡	2.2 ± 2‡

*†‡Unpaired student's *t*-test, effect of DNA fragmentation on concentration, motility, and morphology, $P < 0.00001$

spermatozoa has been linked to poor embryo development and reduced implantation rates [110]. While this correlation is clear in couples attempting natural conception, artificial insemination, and in vitro fertilization, the DNA fragmentation index (DFI) is less predictive of the outcome when spermatozoa are inseminated by ICSI, where only motile spermatozoa are used. We postulate that DNA fragmentation rates measured in a particular sample do not take into consideration whether the cells are motile and therefore functionally intact [94].

DFI values obtained by SCSA carried out in 177 men were allocated according to normal (≤30) and abnormal (>30) thresholds (Table 1). Men with abnormal DFI had lower sperm concentration, motility, and morphology ($P < 0.00001$). DFI values of these patients were plotted against semen characteristics, and an inverse relationship between the declining motility and increasing DNA fragmentation was evident (Fig. 5). In fact, in men with compromised motility at an average of 19.7 ± 3 %, the DNA fragmentation rate reached over 60 %, in contrast to those with normal motility of 48.3 ± 14 % [111]. Interestingly, when these men were treated by ICSI and the outcome was compared to men with normal DFI values (≤30 %) to those with abnormal levels (>30 %), the fertilization and pregnancy rates were similar (Table 2). The unclear relationship between DNA integrity and pregnancy outcome with ICSI inseminations may be explained by the fact that only motile spermatozoa are selected for injection.

To test the resilience of sperm nuclei, mouse spermatozoa were briefly sonicated before microinjection into oocytes [66]. The sonication was used to separate the sperm heads from the tails, and only the heads were injected into oocytes. The fertilized oocytes developed into live-born pups, indicating that sonication failed to

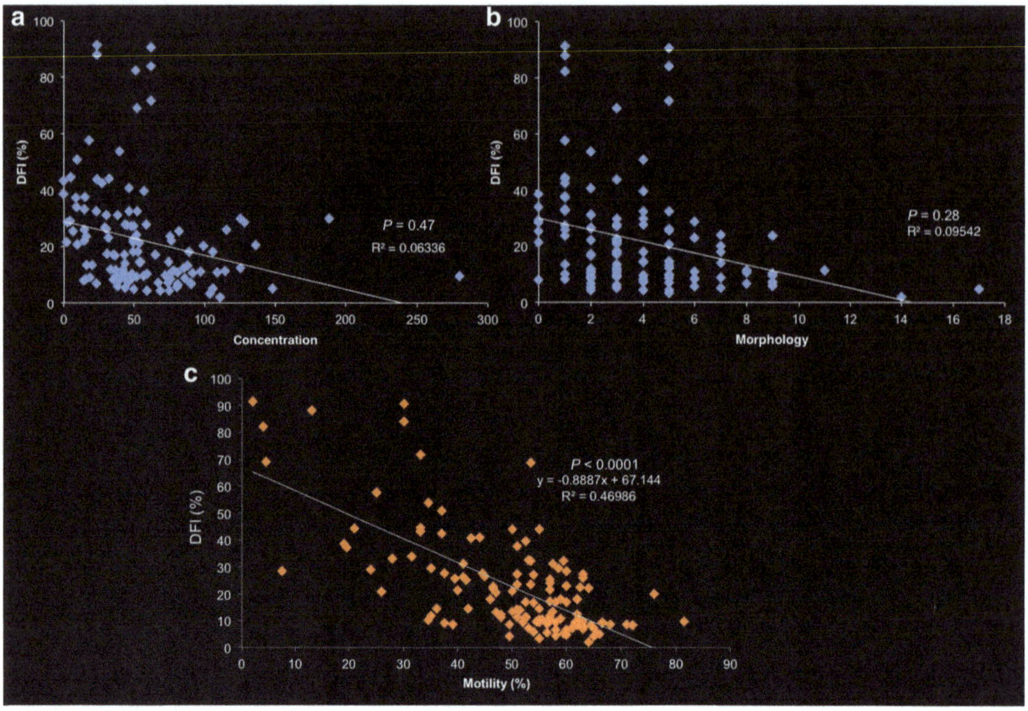

Fig. 5 DNA fragmentation index (DFI) plotted according to semen (**a**) concentration, (**b**) morphology, and (**c**) motility ($P < 0.0001$)

Table 2
Pregnancy outcome according to DNA fragmentation level

	DFI	
No. of (%)	Normal (≤30)	Abnormal (>30)
Cycles	252	80
DFI (M% ± SD)	14.5 ± 8	51.1 ± 20
Maternal age (M years ± SD)	37.9 ± 4	37.9 ± 4
Fertilization (2PN/MII injected)	1432/2011 (71.2)	445/617 (72.1)
Clinical pregnancies (+FHB)	63 (25.0)	21 (26.2)
Delivery and ongoing	58 (23.0)	18 (22.5)

Highest DFI with pregnancy was 64.9 %

meaningfully damage the sperm DNA, while sonication of somatic cells caused so many breaks in the histone-bound chromatin that the cells did not survive. The resistance of human spermatozoa to damage may explain the relative unreliability of DNA fragmentation assays in the prediction of successful embryo development and implantation. In this context it is important to note that protamines are found only in spermatozoa.

Condensation of sperm DNA into crystalline-like toroids serves a largely protective function prior to fertilization. The sperm's DNA's most important functional characteristics are then conferred by simultaneous nuclear decondensation of the sperm and oocyte. Between 2 and 15 % of mammalian sperm chromatin is bound to histones, rather than protamines [112–114]. Protamine binding also silences gene expression during spermiogenesis [115, 116]. Within 2–4 h after fertilization, protamines are completely replaced by histones, making the paternal chromatin accessible to translational processes similar to those observed in somatic cells [117, 118]. The structural organization of both histone-bound chromatin and sperm matrix attachment regions (MARs) is probably transmitted to the newly formed paternal pronucleus after fertilization; evidence suggests that both are required for proper embryogenesis. MARs consist of a proteinaceous network of the nuclear matrix that binds to chromatin at sequence-specific regions of attachment [119, 120]. This would indicate that these residual histones are not simply the result of an incomplete process of reshaping gamete's phenotype to render it more resistant to an extracorporeal journey in adverse physical and chemical conditions. These histones do not simply stand as an anchoring system among the tightly coiled toroids but also have a larger role most likely during male germ cell maturation and pre-fertilization steps. We envision the histone-labeled DNA as some sort of prompt-release, readily available DNA to support the process of spermiogenesis—for example, by orchestrating the extremely complex DNA packing during which transitional proteins followed by protamine insertion. Once sperm penetration occurs, the pre-fertilization steps that begin with the activation of the oocyte to then trigger the development of the sperm aster would ensue. In the few hours required for uncoiling of the sperm genome, replacement of protamines with histones provided by the oocyte and replication of the DNA in preparation for the first cleavage division is probably signaled by the small RNAs that order the fulfillment of these post-fertilization steps until the mammalian embryonic genome is activated.

Histones are interspersed throughout the genome, primarily at gene promoters [121, 122]. Entire gene families that are important for embryo development were preferentially associated with histones in human spermatozoa [123]. This demonstrated that histones are not randomly distributed in the sperm genome and that they are associated with specific genes. Histone-bound DNA makes up the linker regions between each protamine-bound toroid within the chromatin and possesses the highest nuclease sensitivity [108]. These histones persist during the protamine replacement following fertilization, explaining the fact that histones with specific modifications in sperm cells are also present in the paternal pronucleus. This suggests that they were never replaced [90, 124].

This understanding supports the model for histone-associated chromatin representing functional genes whose activity is important for both spermiogenesis (possibly representing residual active chromatin that persisted through chromatin condensation) [125, 126] and early fertilization [121, 123].

There is evidence that each protamine-bound toroid contains single-DNA loop domains [77, 127]. Between each protamine toroid there is a nuclease-sensitive segment of chromatin called a toroid linker and recognized as the site of attachment of DNA to the nuclear matrix. The segments that remain associated with the nuclear matrix are usually biologically active, while inactive regions segregate as independent looped domains. Thus, the nuclear matrix provides a means for organizing chromatin into discrete domains. The nuclease sensitivity confirms that these protamine linker regions are bound to histones, and this is consistent with the wide distribution of histones throughout the genome [121].

If the other two types of sperm chromatin organization are removed—protamine condensation and histone-bound nucleosomes—only the sperm matrix with associated loop domains attached and the resulting nuclei are called the sperm nuclear "halo" [128, 129]. These so-treated sperm nuclei, once injected into oocytes, formed normal pronuclei, and DNA replication proceeded [130]. However, they did not allow development to the blastocyst stage, suggesting that while the organization of DNA into loop domains by the sperm nuclear matrix is required for DNA replication, it does not support preimplantation development. From this observation we can surmise that the nuclear matrix is required for paternal pronuclear DNA replication of the one-cell embryo and that the sperm nuclear matrix may serve as checkpoint for sperm DNA integrity after fertilization.

While these considerations leave the higher order structure of the chromosome as a matter of ongoing debate [131], it appears clear that DNA is bound to histones in nucleosomes that are in various stages of condensation [132, 133]. Mammalian sperm chromosomes fall into a third category, in that they are most likely relatively homogenous in structure but are probably longer and thinner than mitotic chromosomes [134]. Mammalian sperm chromatin are folded into hairpin-like structures with the centrosomes positioned near the center of the sperm head and the telomeres of each chromosome paired and arrayed around the periphery of the sperm nucleus [135, 136] (Fig. 4). These data support the emerging view that the sperm genome provides a structural framework that includes molecular regulatory factors that are required for proper embryonic development in addition to the paternal DNA sequence.

4.4 Genetic Assessment

Chromosomal aneuploidy is the main cause of the high fetal wastage in humans. Most aneuploid pregnancies do not survive in utero, with the majority of losses occurring during the first few weeks of uterine life. Chromosome instability is a hallmark of early life, with whole-chromosome aneuploidy, mosaicism, and segmental aneuploidy being detected in 50 % [137] to 80 % [138] of very early embryos. In clinically recognized spontaneous abortions, trisomies of all chromosomes have been reported, while monosomies are rarely encountered with the exception of 45,X fetuses [139]. Aneuploid conceptions that survive constitute 0.8–1 % of all live births [140]. These babies are mostly trisomies 13, 18, and 21 and various sex chromosome aneuploidies; these represent the majority of congenital abnormalities, developmental disabilities, mental retardation, and infertility in humans. In general, autosomal trisomies (93 % of trisomy 18, 95 % of trisomy 21, and 100 % of trisomy 16) originate in the maternal line (reviewed in ref. 141), whereas sex chromosomal aneuploidies are more frequently of paternal origin (50 % of 47,XXY, 100 % of 47,XYY, and 70–80 % of 45,X) [142]. While gametic meiotic errors that lead to fetal aneuploidy occur in both the male and the female lines, the frequency of these errors is lower in spermatozoa (~9 % in sperm karyotypes) [143] than in oocytes (~20 % but as high as 60 %) [141, 144, 145].

Reliable estimates of male aneuploidy in normal men have been established by methods of analysis that allow the study of sperm chromosomes or analysis of the sperm interphase nucleus. One technique is the injection of the sperm into hamster eggs, where full complements of human sperm chromosomes can be visualized and analyzed. Although expensive, time consuming, and yielding a low number of cells, the hamster system provides a full account of aneuploidies and structural chromosome abnormalities present in each analyzed sperm cell. In another technique, fluorescent DNA probes of regions of interest are hybridized to decondensed sperm nuclei and detected by fluorescent light (fluorescent in situ hybridization, FISH). Since this approach allows analysis of a greater number of spermatozoa from the same sample, it is especially indicated for the evaluation of low-frequency anomalies. FISH's main drawbacks are that only the selected regions of interest can be visualized and scored and that any estimate of aneuploidy from this procedure refers only to the chromosomes analyzed.

With this in mind, we screened 44 patients who underwent 118 ICSI cycles by sperm FISH [146]. Fixed spermatozoa were decondensed and hybridized with three sets of probe mixtures containing locus-specific probes for chromosomes X, Y, 18, 21, 13, 15, 16, 17, and 22. Semen characteristics, shown in Table 3, were comparable to those commonly seen in our fertility practice.

After sperm scoring, men with abnormalities in ≥1.6 % of spermatozoa were considered to have a high rate of aneuploidy while those below the threshold were considered normal (controls).

Table 3
Semen parameter of men who underwent FISH analysis

Characteristics	
Men	44
Age (M years ± SD)	39.5 ± 6
Volume (M ml ± SD)	3.1 ± 1
Concentration (M × 10^6/ml ± SD)	41.4 ± 39
Motility (M% ± SD)	46.0 ± 17
Morphology (M% ± SD)	2.5 ± 2

Of those 44 men, 21 (mean age 39.1 ± 6 years) with high aneuploidy rates were treated in 56 ICSI cycles, while 23 men (mean age 39.5 ± 6 years, $n = 62$ cycles) served as controls. Autosomal disomy was the most recurrent abnormality. While compromised motility was seen only in the aneuploidy group ($P < 0.01$), that group's fertilization rate was unaffected [462/656 (70.4 %) aneuploidy vs. 372/546 (68.1 %) in the control]. The clinical pregnancy rate in the study group was 21.4 % (12/56), with a 12.5 % (7/56) delivery rate, while in the normal group it was 29.0 % (18/62) and 21 % (13/62), respectively. Interestingly, the pregnancy loss rate was 41.7 % (5/12) in the abnormal group versus 27.8 % (5/18) in the control ($P = 0.03$).

To minimize the female factor, a subanalysis was carried out in couples with a female partner ≤35 years of age. In this cohort, the sperm concentration ($P = 0.05$) and motility were compromised ($P = 0.005$), although fertilization was unaffected (65.2 % study versus 67.7 % control). The aneuploidy group's clinical pregnancy rate was 21.2 % (7/33) and its delivery rate was 15.2 % (5/33), compared to the normal cohort's 29.4 % (10/34) and 20.6 % (7/34) pregnancy and delivery rates, respectively. The pregnancy loss rate was 28.6 % (2/7) in the abnormal and 30.0 % (3/10) in the control.

Sperm chromosomal abnormalities were associated with compromised motility and sometimes with low concentration. Aneuploidy did not affect fertilization rates but negatively influenced pregnancy outcome. Interestingly, even in this series chromosomal abnormalities of the male gamete had a clear effect on embryo implantation. Performing 24 chromosome FISH on spermatozoa may increase our ability to determine the incidence of aneuploidy in a given sample (Fig. 6).

4.5 Centrosome

During fertilization, restoration of diploidy and subsequent embryonic development requires that each gamete must contain only one-half of the diploid chromosomal complement. In humans, the

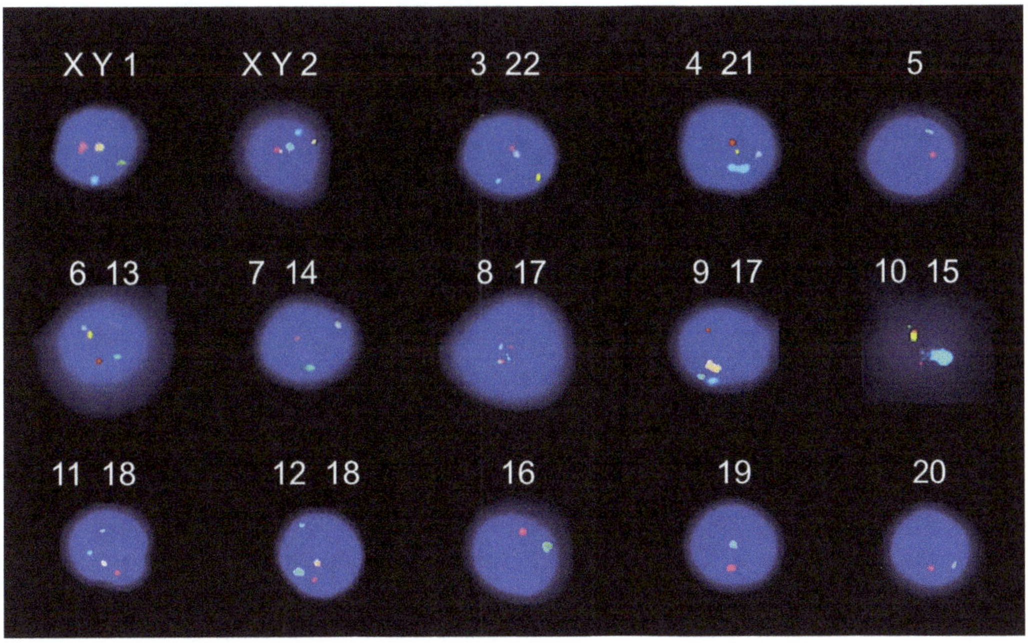

Fig. 6 Fluorescence in situ hybridization (FISH) on spermatozoa for 24 chromosomes, autosomes 1–22, and gonosomes X and Y

mature oocyte possesses all of the elements necessary for embryonic development except an active division center, which must originate from the spermatozoal centrosome. Boveri [147] first defined the term "centrosome" as a polar corpuscle containing centrioles. Later it was defined more functionally as a microtubule-organizing center (MTOC) [148]. The centrosome in somatic cells is considered to be responsible for two basic events: the nucleation of microtubules and the formulation of an efficient mitotic spindle [149].

In most cells, the centrosome consists of two morphologically distinct centrioles and the pericentriolar material (PCM). Centrioles do not seem to be present in the meiotic spindle of gametes but are present at the spindle poles during the first mitotic division in zygotes from various species [150], including humans [151]. The mature human oocyte has neither centrioles nor functional centrosomes associated with its meiotic spindle, resulting in anastral barrel-shaped structures with microtubules ending abruptly at the poles. The outer pole, however, is closely bound to the egg cortex.

In contrast to the oocyte, the human spermatozoon has two distinct centrioles. The well-defined proximal centriole, located within the connecting piece next to the basal plate of the sperm head, displays a 9+0 pattern of nine triplet microtubules surrounded by electron-dense material and flanked by nine cross-striated columns.

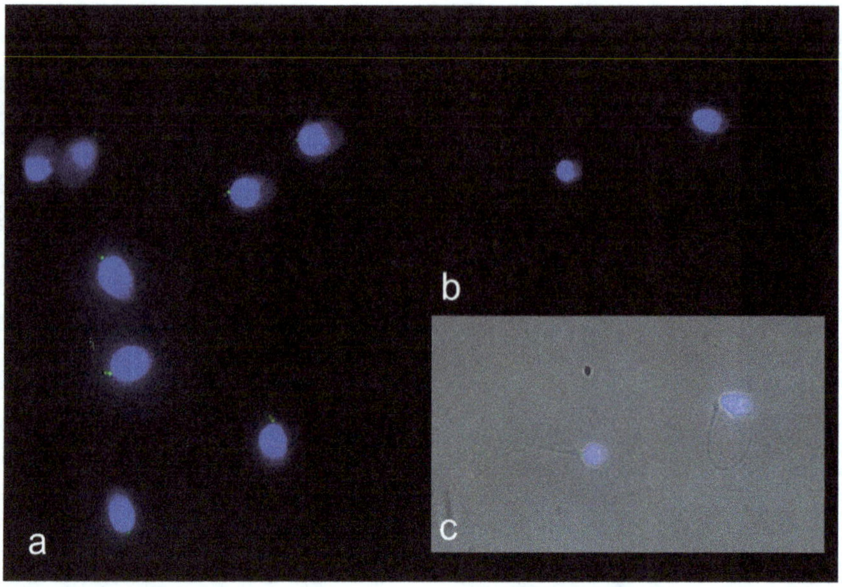

Fig. 7 An example of spermatozoa with (**a**) intact centrosomes (labeled in *green* with two signals and nuclear DNA are counterstained with Hoechst). Centrosome presence according to sperm integrity, (**b**) fluorescent labeling, and (**c**) merged with bright-field microscopy. The spermatozoon on the *left* is intact (two *green signals*) and the *right* was mechanically dissected (no signal)

The distal centriole is aligned with the axis of the flagellum perpendicular to the proximal centriole and gives rise to the sperm tail axonome during spermiogenesis [151, 152].

The absence of the sperm centrosome could be one of the causes of embryonic failure [55, 153, 154]. The utilization of biochemical and immunological techniques has now made it possible to identify proteins that are integral components of the centrosome [55, 65, 155, 156] (Fig. 7).

With the occasional exceptions (as in the mouse) [157], centriolar and centrosomal inheritance in mammals has been assumed to follow a paternal lineage, and there is now little doubt that in humans only the male gamete has an active centrosome [55, 158]. Extensive analysis by TEM has demonstrated the presence of centrioles in spermatozoa and in fertilized oocytes at syngamy, and their absence in MII oocytes confirms the paternal inheritance of the centrosome in humans [151]. Furthermore, FISH assessment of chromosome distribution has revealed that the sperm centrosome is solely responsible for organization of the first mitotic division in human embryos [158].

In keeping with this concept, we used FISH assessment to evaluate the developmental potential of embryos injected with an isolated sperm head or tail and also those with heads and separated tails [65]. The results showed that injected sperm components and dissected spermatozoa both support oocyte activation and

Table 4
Centrosomal signal after sperm dissection

	Centrosome location	
Signals	**Head**	**Tail**
2	12 (17.9)	13 (25.0)
1	8 (11.9)	9 (18.3)
⊘	47 (70.2)	38 (56.7)
Total	67	60

pronuclear formation at a rate comparable to intact spermatozoa. However, the migration and syngamy of the pronuclei are not normal, as demonstrated by the generation of abnormal embryos. The general abnormal chromosomal distribution observed within the blastomeres indicated that mechanical dissection of the spermatozoa in some way compromised centrosomal function in the zygote. It appears that with the current dissection methodology, the donation of sperm centrosomes is not yet feasible [156] (Fig. 7; Table 4). A gentler way to isolate centriole-containing tails that is more mindful of the intimate molecular structure of the head and tail junction still is needed.

5 New Insights

There is substantial interest in assessing whether RNAs (messenger RNAs [mRNAs] and small noncoding RNAs [sncRNAs]) contributed by mammalian spermatozoa may play a functional role in early embryo development [159]. However, little is known about the distribution and function of the sncRNAs within each normal human spermatozoon, estimated at ca. 24,000 [160]. The collection of sncRNAs may function in diverse processes that include gene expression, chromatin remodeling, and protection of the genome against transposition. The microRNA (miRNA) family is the best characterized of the sncRNAs. They were first identified in humans [161] and confirmed in mouse [162, 163] and porcine spermatozoa [164]. The effect of miRNAs is generally posttranscriptional mediated degradation through their interaction with the 3′ untranslated region (UTR). Some other proposed functions are developmental modifiers, as in the case of miRNAs or piwi-interacting RNA (piRNA) that protect the genome by masking repetitive and transposable elements or participate in the confrontation and consolidation at the first encounter of the male and female genomes. The role of miRNAs as epigenetic modifiers is being elucidated [165–167], and miRNA disequilibrium may be implicated in a diverse range of physiological responses which may affect spermatogenesis and influence the different presentation of azoospermia [168, 169].

Recently, the presence of small RNAs has been demonstrated in male germ cells, suggesting that they may have an essential role in spermatogenesis [170, 171]. Though assumed to be absent from the mature gamete, a restricted set of piRNAs may in fact be retained in human spermatozoa [172]. The absence of transcriptional activity in sperm has prompted the hypothesis that paternally contributed miRNAs regulate early embryonic expression, influencing offspring phenotype [169, 173]. The majority of miRNAs identified in sperm originate from promoter regions. These transcripts may bind to paternal DNA during nuclear remodeling such that they are delivered to the oocyte in association with their targeted *cis* sequences, presumably influencing their local chromatin structure. The presence of nucleosomes in these regions, some of which contain modified histones, is highly suggestive of subsequent epigenetic control in the embryo. Following fertilization, these sequences, partnered with the sperm nuclear matrix, may provide the zygote a platform for the transgenerational inheritance of paternal chromatin structure.

After fertilization, the highly methylated paternal genome is actively demethylated before establishing the new epigenetic marks necessary for early embryonic development [174]. Passive and progressive demethylation is facilitated by the exclusion of DNA methyltransferases (Dnmt1) [175] and accompanies replication to enable reprogramming and imprinting [176]. This methyltransferase is the most abundant in mammalian cells and plays a key role in maintenance of DNA methylation. These miRNAs transmitted by the male gamete immediately inhibit epigenetic marking. It has been suggested that miRNAs retained in mature spermatozoa are associated with the histone-enriched regions [123]. It is possible that miRNAs delivered by human sperm do play a functional role in the oocyte, filling the gap between sperm penetration and zygotic genome activation [105, 159].

6 Lessons Learned

The male gamete was first described in the seventeenth century as the miniaturized version of a human being that would presumably develop within the mother's womb. This early observation was dictated by the time's philosophical, religious, and cultural influences. Through the years, advanced scientific investigations have revealed the role of the spermatozoon as the paternal component in reproduction. The prominence of the oocyte as the bearer of embryo development contributed to progressively diminish the importance of the single-sperm cell. In fact, in the last 30 years the practice of andrology has been dominated by WHO guidelines [8, 177, 178] recognized worldwide as the gold standard to be used to "read" a man's ejaculate. With time, the male gamete comes to be identified with the words "semen" or "sperm," implicating the entire gamete pool, including the delivery medium.

The establishment of assisted reproductive techniques has refocused our attention towards the male gamete and has driven the development of several methods of identifying the optimal spermatozoon. Growing understanding of the spermatozoa's key developmental events, including capacitation and acrosome reaction, has guided the identification of ideal conditions for sperm culture, insemination, and cryopreservation. The development of assisted fertilization through microinjection procedures has fully restored the key role of the male gamete and has shifted the paradigm from evaluating male fertility through a multitude of cells to an analysis aimed at identifying individual sperm cells capable of inseminating an oocyte and generating a healthy conceptus. Even the recent trend of studying a spermatozoon under a 1,000-fold magnification has inspired several studies towards understanding the surface imperfections of this precious cell type.

Another scenario is unfolding for this long-time underappreciated cell through the understanding of its soluble-activating factor, which is capable of triggering the cascade of egg activation. The unveiling of the complex intertwining of the chromatin component and the presence of nucleosomes with protamine-bound DNA bring other insights. Histone-bound DNA may aid in reshaping the sperm nucleus during spermiogenesis and, through differential release of the paternal genome, guide the pre-fertilization steps. In addition, the combination of the parental genomes to generate genomically competent daughter cells is achieved through the deployment of a microtubular organelle—the centrosome—provided by the spermatozoon.

Regarding attempts to assess the sperm's genomic content in terms of its chromosomal ploidy and DNA integrity, new avenues of exploration have been opened by the recent recognition of the importance of sncRNAs.

Acknowledgements

We are very appreciative to all clinicians and scientists at the Ronald O. Perelman AND Claudia Cohen Center for Reproductive Medicine and the Urology Department. We are thankful to Justin Kocent for help in the figures and Alessia Uccelli for listing the references.

References

1. Vanleeuwenhoek A (1978) Observationes de natis e semine genitali animalculis. Philos Trans R Soc Lond 12:1040–1043
2. Cole FJ (1930) Early theories of sexual generation. The Carendon, Oxford
3. Hartsoeker N (1694) Essai de dioptriques
4. Capanna E (1999) Lazzaro spallanzani: At the roots of modern biology. J Exp Zool 285(3):178–196
5. Magner LN (2002) A history of the life sciences, 3rd edn. Marcel Dekker, New York, Revised and expanded ed
6. Hill KA (1985) Hartsoeker's homonculus: a corrective note. J Hist Behav Sci 21(2): 178–179
7. Barratt CL (2007) Semen analysis is the cornerstone of investigation for male infertility. Practitioner 251(1690):8–10, 12, 15–7
8. World Health Organization (2010) WHO laboratory manual for the examination and processing of human semen, vol 5, 5th edn. Cambridge Univ. Press, Cambridge
9. Jequier AM (2010) Semen analysis: a new manual and its application to the understanding of semen and its pathology. Asian J Androl 12(1):11–13
10. Guzick DS et al (2001) Sperm morphology, motility, and concentration in fertile and infertile men. N Engl J Med 345(19): 1388–1393
11. Sigman M, Baazeem A, Zini A (2009) Semen analysis and sperm function assays: what do they mean? Semin Reprod Med 27(2):115–123
12. Smith KD, Rodriguez-Rigau LJ, Steinberger E (1977) Relation between indices of semen analysis and pregnancy rate in infertile couples. Fertil Steril 28(12):1314–1319
13. Comhaire FH et al (1992) Objective semen analysis: has the target been reached? Hum Reprod 7(2):237–241
14. Menkveld R et al (1990) The evaluation of morphological characteristics of human spermatozoa according to stricter criteria. Hum Reprod 5(5):586–592
15. Adelman MM, Cahill EM (1989) Atlas of sperm morphology. Everbest, Hong Kong, p 123
16. Bartoov B, Berkovitz A, Eltes F (2001) Selection of spermatozoa with normal nuclei to improve the pregnancy rate with intracytoplasmic sperm injection. N Engl J Med 345(14):1067–1068
17. Bartoov B et al (2002) Real-time fine morphology of motile human sperm cells is associated with IVF-ICSI outcome. J Androl 23(1):1–8
18. Antinori M et al (2008) Intracytoplasmic morphologically selected sperm injection: a prospective randomized trial. Reprod Biomed Online 16(6):835–841
19. Bartoov B et al (2003) Pregnancy rates are higher with intracytoplasmic morphologically selected sperm injection than with conventional intracytoplasmic injection. Fertil Steril 80(6):1413–1419
20. Berkovitz A et al (2005) The morphological normalcy of the sperm nucleus and pregnancy rate of intracytoplasmic injection with morphologically selected sperm. Hum Reprod 20(1):185–190
21. Hazout A et al (2006) High-magnification ICSI overcomes paternal effect resistant to conventional ICSI. Reprod Biomed Online 12(1):19–25
22. Zamboni L (1987) The ultrastructural pathology of the spermatozoon as a cause of infertility: the role of electron microscopy in the evaluation of semen quality. Fertil Steril 48(5):711–734
23. Fawcett DW, Ito S (1958) Observations on the cytoplasmic membranes of testicular cells, examined by phase contrast and electron microscopy. J Biophys Biochem Cytol 4(2):135–142
24. Baccetti B et al (1989) Crater defect in human spermatozoa. Gamete Res 22(3):249–255
25. Kacem O et al (2010) Sperm nuclear vacuoles, as assessed by motile sperm organellar morphological examination, are mostly of acrosomal origin. Reprod Biomed Online 20(1):132–137
26. Peer S et al (2007) Is fine morphology of the human sperm nuclei affected by in vitro incubation at 37 degrees C? Fertil Steril 88(6):1589–1594
27. Tanaka A et al (2009) Are crater defects in human sperm heads physiological changes during spermiogenesis? Fertil Steril 92(3):S165
28. Watanabe S et al (2009) No relationship between chromosome aberrations and vacuole-like structures on human sperm head. Hum Reprod 24(Suppl 1):i94–i96
29. Palermo GD et al (2011) Thoughts on IMSI. In: Racowsky C et al (eds) Biennial review of infertility, vol 2. Springer, New York, p 296
30. Fawcett DW, Porter KR (1954) A study on the fine structure of ciliated epithelia. J Morphol 94:221–282
31. Barros C et al (1967) Membrane vesiculation as a feature of the mammalian acrosome reaction. J Cell Biol 34(3):C1–C5
32. Bleil JD, Wassarman PM (1983) Sperm-egg interactions in the mouse: sequence of events

and induction of the acrosome reaction by a zona pellucida glycoprotein. Dev Biol 95(2): 317–324
33. Wassarman PM (2002) Sperm receptors and fertilization in mammals. Mt Sinai J Med 69(3):148–155
34. Yanagimachi R, Bhattacharyya A (1988) Acrosome-reacted guinea pig spermatozoa become fusion competent in the presence of extracellular potassium ions. J Exp Zool 248(3):354–360
35. Bedford JM (1970) Sperm capacitation and fertilization in mammals. Biol Reprod 2(Suppl 2):128–158
36. Yanagimachi R, Noda YD (1970) Physiological changes in the postnuclear cap region of mammalian spermatozoa: a necessary preliminary to the membrane fusion between sperm and egg cells. J Ultrastruct Res 31(5–6):486–493
37. Austin CR (1951) Observations on the penetration of the sperm in the mammalian egg. Aust J Sci Res B 4(4):581–596
38. Chang MC (1951) Fertilizing capacity of spermatozoa deposited into the fallopian tubes. Nature 168(4277):697–698
39. Yanagimachi R (1994) Fertility of mammalian spermatozoa: its development and relativity. Zygote 2(4):371–372
40. Palermo G et al (1992) Pregnancies after intracytoplasmic injection of single spermatozoon into an oocyte. Lancet 340(8810):17–18
41. Palermo GD et al (1995) Intracytoplasmic sperm injection: a novel treatment for all forms of male factor infertility. Fertil Steril 63(6):1231–1240
42. Vanderzwalmen P et al (1996) Two essential steps for a successful intracytoplasmic sperm injection: injection of immobilized spermatozoa after rupture of the oolemma. Hum Reprod 11(3):540–547
43. Fishel S et al (1995) Systematic examination of immobilizing spermatozoa before intracytoplasmic sperm injection in the human. Hum Reprod 10(3):497–500
44. Gerris J et al (1995) ICSI and severe male-factor infertility: breaking the sperm tail prior to injection. Hum Reprod 10(3): 484–486
45. Palermo GD et al (1996) Aggressive sperm immobilization prior to intracytoplasmic sperm injection with immature spermatozoa improves fertilization and pregnancy rates. Hum Reprod 11(5):1023–1029
46. Van den Bergh M et al (1995) Importance of breaking a spermatozoon's tail before intracytoplasmic injection: a prospective randomized trial. Hum Reprod 10(11):2819–2820
47. Dozortsev D et al (1995) Sperm plasma membrane damage prior to intracytoplasmic sperm injection: a necessary condition for sperm nucleus decondensation. Hum Reprod 10(11):2960–2964
48. Christova Y et al (2004) Molecular diffusion in sperm plasma membranes during epididymal maturation. Mol Cell Endocrinol 216(1–2): 41–46
49. Palermo GD et al (2012) Development and current applications of assisted fertilization. Fertil Steril 97(2):248–259
50. Takeuchi T et al (2004) Does ICSI require acrosomal disruption? An ultrastructural study. Hum Reprod 19(1):114–117
51. Gomez-Torres MJ et al (2007) Sperm immobilized before intracytoplasmic sperm injection undergo ultrastructural damage and acrosomal disruption. Fertil Steril 88(3): 702–704
52. Katayama M et al (2005) Increased disruption of sperm plasma membrane at sperm immobilization promotes dissociation of perinuclear theca from sperm chromatin after intracytoplasmic sperm injection in pigs. Reproduction 130(6):907–916
53. Yanagimachi R (1998) Intracytoplasmic sperm injection experiments using the mouse as a model. Hum Reprod 13(Suppl 1):87–98
54. Dozortsev D et al (1995) Human oocyte activation following intracytoplasmic injection: the role of the sperm cell. Hum Reprod 10(2):403–407
55. Palermo GD et al (1997) Human sperm cytosolic factor triggers Ca2+ oscillations and overcomes activation failure of mammalian oocytes. Mol Hum Reprod 3(4):367–374
56. Swann K, Ozil JP (1994) Dynamics of the calcium signal that triggers mammalian egg activation. Int Rev Cytol 152:183–222
57. Stice SL, Robl JM (1990) Activation of mammalian oocytes by a factor obtained from rabbit sperm. Mol Reprod Dev 25(3):272–280
58. Swann K (1990) A cytosolic sperm factor stimulates repetitive calcium increases and mimics fertilization in hamster eggs. Development 110(4):1295–1302
59. Parrington J et al (1996) Calcium oscillations in mammalian eggs triggered by a soluble sperm protein. Nature 379(6563):364–368
60. Bedford JM, Moore HD, Franklin LE (1979) Significance of the equatorial segment of the acrosome of the spermatozoon in eutherian mammals. Exp Cell Res 119(1):119–126
61. Oko R, Sutovsky P (2009) Biogenesis of sperm perinuclear theca and its role in sperm functional competence and fertilization. J Reprod Immunol 83(1–2):2–7

62. Sutovsky P et al (2003) Interactions of sperm perinuclear theca with the oocyte: implications for oocyte activation, anti-polyspermy defense, and assisted reproduction. Microsc Res Tech 61(4):362–378
63. Sutovsky P et al (1997) The removal of the sperm perinuclear theca and its association with the bovine oocyte surface during fertilization. Dev Biol 188(1):75–84
64. Saunders CM et al (2002) PLC zeta: a sperm-specific trigger of Ca(2+) oscillations in eggs and embryo development. Development 129(15):3533–3544
65. Colombero LT et al (1999) The role of structural integrity of the fertilising spermatozoon in early human embryogenesis. Zygote 7(2): 157–163
66. Kuretake S et al (1996) Fertilization and development of mouse oocytes injected with isolated sperm heads. Biol Reprod 55(4): 789–795
67. Fissore RA, Reis MM, Palermo GD (1999) Isolation of the Ca2+ releasing component(s) of mammalian sperm extracts: the search continues. Mol Hum Reprod 5(3):189–192
68. Wolny YM et al (1999) Human glucosamine-6-phosphate isomerase, a homologue of hamster oscillin, does not appear to be involved in Ca2+ release in mammalian oocytes. Mol Reprod Dev 52(3):277–287
69. Cox LJ et al (2002) Sperm phospholipase Czeta from humans and cynomolgus monkeys triggers Ca2+ oscillations, activation and development of mouse oocytes. Reproduction 124(5):611–623
70. Neri QV et al (2010) Assessing and restoring sperm fertilizing ability. Fertil Steril 94 (4 Suppl 1):S147
71. Neri QV (2010) Tweaking human fertilization. In: Reproductive medicine. Clinical & Translation Science Center, Weill Cornell Medical College, New York. p 24
72. Sullivan EJ et al (2004) Cloned calves from chromatin remodeled in vitro. Biol Reprod 70(1):146–153
73. Heindryckx B et al (2008) Efficiency of assisted oocyte activation as a solution for failed intracytoplasmic sperm injection. Reprod Biomed Online 17(5):662–668
74. Calvin HI, Bedford JM (1971) Formation of disulphide bonds in the nucleus and accessory structures of mammalian spermatozoa during maturation in the epididymis. J Reprod Fertil Suppl 13(Suppl 13):65–75
75. Brewer L, Corzett M, Balhorn R (2002) Condensation of DNA by spermatid basic nuclear proteins. J Biol Chem 277(41): 38895–38900
76. Dadoune JP (2003) Expression of mammalian spermatozoal nucleoproteins. Microsc Res Tech 61(1):56–75
77. Ward WS (1993) Deoxyribonucleic acid loop domain tertiary structure in mammalian spermatozoa. Biol Reprod 48(6):1193–1201
78. Meistrich ML et al (1992) Highly acetylated H4 is associated with histone displacement in rat spermatids. Mol Reprod Dev 31(3): 170–181
79. Ward WS, Coffey DS (1990) Specific organization of genes in relation to the sperm nuclear matrix. Biochem Biophys Res Commun 173(1):20–25
80. Marcon L, Boissonneault G (2004) Transient DNA strand breaks during mouse and human spermiogenesis new insights in stage specificity and link to chromatin remodeling. Biol Reprod 70(4):910–918
81. McPherson S, Longo FJ (1993) Chromatin structure-function alterations during mammalian spermatogenesis: DNA nicking and repair in elongating spermatids. Eur J Histochem 37(2):109–128
82. Sakkas D et al (1995) Relationship between the presence of endogenous nicks and sperm chromatin packaging in maturing and fertilizing mouse spermatozoa. Biol Reprod 52(5): 1149–1155
83. McPherson SM, Longo FJ (1993) Nicking of rat spermatid and spermatozoa DNA: possible involvement of DNA topoisomerase II. Dev Biol 158(1):122–130
84. Boissonneault G (2002) Chromatin remodeling during spermiogenesis: a possible role for the transition proteins in DNA strand break repair. FEBS Lett 514(2–3):111–114
85. Bungum M et al (2004) The predictive value of sperm chromatin structure assay (SCSA) parameters for the outcome of intrauterine insemination IVF and ICSI. Hum Reprod 19(6):1401–1408
86. Evenson D, Jost L (2000) Sperm chromatin structure assay is useful for fertility assessment. Methods Cell Sci 22(2–3):169–189
87. Evenson DP, Larson KL, Jost LK (2002) Sperm chromatin structure assay: its clinical use for detecting sperm DNA fragmentation in male infertility and comparisons with other techniques. J Androl 23(1):25–43
88. Morris ID et al (2002) The spectrum of DNA damage in human sperm assessed by single cell gel electrophoresis (comet assay) and its relationship to fertilization and embryo development. Hum Reprod 17(4):990–998
89. Sakkas D, Manicardi GC, Bizzaro D (2003) Sperm nuclear DNA damage in the human. Adv Exp Med Biol 518:73–84

90. van der Heijden GW et al (2008) Sperm-derived histones contribute to zygotic chromatin in humans. BMC Dev Biol 8:34
91. Zini A, Sigman M (2009) Are tests of sperm DNA damage clinically useful? Pros and cons. J Androl 30(3):219–229
92. Spano M et al (2000) Sperm chromatin damage impairs human fertility. The Danish First Pregnancy Planner Study Team. Fertil Steril 73(1):43–50
93. Zini A et al (2001) Correlations between two markers of sperm DNA integrity, DNA denaturation and DNA fragmentation, in fertile and infertile men. Fertil Steril 75(4):674–677
94. Chen C et al (2011) Kinetic characteristics and DNA integrity of human spermatozoa. Hum Reprod 19(Suppl 1):i30
95. Carrell DT, Liu L (2001) Altered protamine 2 expression is uncommon in donors of known fertility, but common among men with poor fertilizing capacity, and may reflect other abnormalities of spermiogenesis. J Androl 22(4):604–610
96. Moskovtsev SI, Willis J, Mullen JB (2006) Age-related decline in sperm deoxyribonucleic acid integrity in patients evaluated for male infertility. Fertil Steril 85(2):496–499
97. Moskovtsev SI et al (2007) Sperm survival: relationship to age-related sperm DNA integrity in infertile men. Arch Androl 53(1):29–32
98. Plastira K et al (2007) The effects of age on DNA fragmentation, chromatin packaging and conventional semen parameters in spermatozoa of oligoasthenoteratozoospermic patients. J Assist Reprod Genet 24(10): 437–443
99. Singh NP, Muller CH, Berger RE (2003) Effects of age on DNA double-strand breaks and apoptosis in human sperm. Fertil Steril 80(6):1420–1430
100. Kunzle R et al (2003) Semen quality of male smokers and nonsmokers in infertile couples. Fertil Steril 79(2):287–291
101. Potts RJ et al (1999) Sperm chromatin damage associated with male smoking. Mutat Res 423(1–2):103–111
102. Saleh RA et al (2003) Evaluation of nuclear DNA damage in spermatozoa from infertile men with varicocele. Fertil Steril 80(6): 1431–1436
103. Xing W, Krishnamurthy H, Sairam MR (2003) Role of follitropin receptor signaling in nuclear protein transitions and chromatin condensation during spermatogenesis. Biochem Biophys Res Commun 312(3): 697–701
104. Brewer L et al (2003) Dynamics of protamine 1 binding to single DNA molecules. J Biol Chem 278(43):42403–42408
105. Carrell DT (2012) Epigenetics of the male gamete. Fertil Steril 97(2):267–274
106. Hud NV et al (1993) Identification of the elemental packing unit of DNA in mammalian sperm cells by atomic force microscopy. Biochem Biophys Res Commun 193(3): 1347–1354
107. Vilfan ID, Conwell CC, Hud NV (2004) Formation of native-like mammalian sperm cell chromatin with folded bull protamine. J Biol Chem 279(19):20088–20095
108. Sotolongo B, Lino E, Ward WS (2003) Ability of hamster spermatozoa to digest their own DNA. Biol Reprod 69(6):2029–2035
109. Perreault SD, Zirkin BR (1982) Sperm nuclear decondensation in mammals: role of sperm-associated proteinase in vivo. J Exp Zool 224(2):253–257
110. Tamburrino L et al (2012) Mechanisms and clinical correlates of sperm DNA damage. Asian J Androl 14(1):24–31
111. Hu JCY et al (2011) DNA fragmentation assay – a useful tool or a red herring? Fertil Steril 96(3 Suppl 1):S236
112. Bench GS et al (1996) DNA and total protamine masses in individual sperm from fertile mammalian subjects. Cytometry 23(4): 263–271
113. Hammoud S, Liu L, Carrell DT (2009) Protamine ratio and the level of histone retention in sperm selected from a density gradient preparation. Andrologia 41(2):88–94
114. Pittoggi C et al (1999) A fraction of mouse sperm chromatin is organized in nucleosomal hypersensitive domains enriched in retroposon DNA. J Cell Sci 112(Pt 20):3537–3548
115. Carrell DT, Emery BR, Hammoud S (2007) Altered protamine expression and diminished spermatogenesis: what is the link? Hum Reprod Update 13(3):313–327
116. Martins RP, Ostermeier GC, Krawetz SA (2004) Nuclear matrix interactions at the human protamine domain: a working model of potentiation. J Biol Chem 279(50): 51862–51868
117. Ajduk A, Yamauchi Y, Ward MA (2006) Sperm chromatin remodeling after intracytoplasmic sperm injection differs from that of in vitro fertilization. Biol Reprod 75(3): 442–451
118. Kopecny V, Pavlok A (1975) Incorporation of Arginine-3H into chromatin of mouse eggs shortly after sperm penetration. Histochemistry 45(4):341–345
119. Bode J et al (2000) Transcriptional augmentation: modulation of gene expression by scaffold/matrix-attached regions (S/MAR elements). Crit Rev Eukaryot Gene Expr 10(1):73–90

120. Ward WS, Coffey DS (1991) DNA packaging and organization in mammalian spermatozoa: comparison with somatic cells. Biol Reprod 44(4):569–574
121. Arpanahi A et al (2009) Endonuclease-sensitive regions of human spermatozoal chromatin are highly enriched in promoter and CTCF binding sequences. Genome Res 19(8):1338–1349
122. Wykes SM, Krawetz SA (2003) The structural organization of sperm chromatin. J Biol Chem 278(32):29471–29477
123. Hammoud SS et al (2009) Distinctive chromatin in human sperm packages genes for embryo development. Nature 460(7254): 473–478
124. van der Heijden GW et al (2006) Transmission of modified nucleosomes from the mouse male germline to the zygote and subsequent remodeling of paternal chromatin. Dev Biol 298(2):458–469
125. Martins RP, Krawetz SA (2005) Towards understanding the epigenetics of transcription by chromatin structure and the nuclear matrix. Gene Ther Mol Biol 9(B):229–246
126. Ostermeier GC et al (2005) Toward using stable spermatozoal RNAs for prognostic assessment of male factor fertility. Fertil Steril 83(6):1687–1694
127. Sotolongo B et al (2005) An endogenous nuclease in hamster, mouse, and human spermatozoa cleaves DNA into loop-sized fragments. J Androl 26(2):272–280
128. Kramer JA, Krawetz SA (1996) Nuclear matrix interactions within the sperm genome. J Biol Chem 271(20):11619–11622
129. Nadel B, De Lara J, Ward WS (1995) Structure of the rRNA genes in the hamster sperm nucleus. J Androl 16(6):517–522
130. Shaman JA, Yamauchi Y, Ward WS (2007) The sperm nuclear matrix is required for paternal DNA replication. J Cell Biochem 102(3):680–688
131. Ward WS (2010) Function of sperm chromatin structural elements in fertilization and development. Mol Hum Reprod 16(1): 30–36
132. Grewal SI, Elgin SC (2007) Transcription and RNA interference in the formation of heterochromatin. Nature 447(7143):399–406
133. Kloc A, Martienssen R (2008) RNAi, heterochromatin and the cell cycle. Trends Genet 24(10):511–517
134. Haaf T, Ward DC (1995) Higher order nuclear structure in mammalian sperm revealed by in situ hybridization and extended chromatin fibers. Exp Cell Res 219(2):604–611
135. Zalenskaya IA, Bradbury EM, Zalensky AO (2000) Chromatin structure of telomere domain in human sperm. Biochem Biophys Res Commun 279(1):213–218
136. Zalensky A, Zalenskaya I (2007) Organization of chromosomes in spermatozoa: an additional layer of epigenetic information? Biochem Soc Trans 35(Pt 3):609–611
137. Munne S et al (2004) Differences in chromosome susceptibility to aneuploidy and survival to first trimester. Reprod Biomed Online 8(1):81–90
138. Vanneste E et al (2009) Chromosome instability is common in human cleavage-stage embryos. Nat Med 15(5):577–583
139. Hassold T et al (1996) Human aneuploidy: incidence, origin, and etiology. Environ Mol Mutagen 28(3):167–175
140. Gardner RJ, Sutherland GR (2004) Chromosome abnormalities and genetic counseling, 3rd edn. Oxford University Press, New York
141. Hassold T, Hunt P (2001) To err (meiotically) is human: the genesis of human aneuploidy. Nat Rev Genet 2(4):280–291
142. Sloter E et al (2004) Effects of male age on the frequencies of germinal and heritable chromosomal abnormalities in humans and rodents. Fertil Steril 81(4):925–943
143. Templado C, Bosch M, Benet J (2005) Frequency and distribution of chromosome abnormalities in human spermatozoa. Cytogenet Genome Res 111(3–4):199–205
144. Angell RR (1991) Predivision in human oocytes at meiosis I: a mechanism for trisomy formation in man. Hum Genet 86(4):383–387
145. Jones KT (2008) Meiosis in oocytes: predisposition to aneuploidy and its increased incidence with age. Hum Reprod Update 14(2): 143–158
146. Hu JCY et al (2011) The role of sperm aneuploidy assay. Fertil Steril 96(3 Suppl 1):S24–S25
147. Boveri T (1887) Ueber die "Befruchtung der Eier von Ascaris megalocephala". Anat Anz 2:688–693
148. Pickett-Heaps J, Spurck T, Tippit D (1984) Chromosome motion and the spindle matrix. J Cell Biol 99(1 Pt 2):137s–143s
149. Bornens M et al (1987) Structural and chemical characterization of isolated centrosomes. Cell Motil Cytoskeleton 8(3):238–249
150. Le Guen P, Crozet N (1989) Microtubule and centrosome distribution during sheep fertilization. Eur J Cell Biol 48(2):239–249
151. Sathananthan AH et al (1991) Centrioles in the beginning of human development. Proc Natl Acad Sci U S A 88(11):4806–4810

152. Sathananthan AH et al (1996) The sperm centriole: its inheritance, replication and perpetuation in early human embryos. Hum Reprod 11(2):345–356
153. Moomjy M et al (1999) Sperm integrity is critical for normal mitotic division and early embryonic development. Mol Hum Reprod 5(9):836–844
154. Van Blerkom J, Davis P (1995) Evolution of the sperm aster after microinjection of isolated human sperm centrosomes into meiotically mature human oocytes. Hum Reprod 10(8): 2179–2182
155. Kimble M, Kuriyama R (1992) Functional components of microtubule-organizing centers. Int Rev Cytol 136:1–50
156. Neri QV et al (2011) Assessment of the sperm centrosome. Fertil Steril 96(3 Suppl 1): S235–S236
157. Schatten G, Simerly C, Schatten H (1991) Maternal inheritance of centrosomes in mammals? Studies on parthenogenesis and polyspermy in mice. Proc Natl Acad Sci U S A 88(15):6785–6789
158. Palermo G, Munne S, Cohen J (1994) The human zygote inherits its mitotic potential from the male gamete. Hum Reprod 9(7): 1220–1225
159. Hamatani T (2012) Human spermatozoal RNAs. Fertil Steril 97(2):275–281
160. Krawetz SA et al (2011) A survey of small RNAs in human sperm. Hum Reprod 26(12):3401–3412
161. Ostermeier GC et al (2005) A suite of novel human spermatozoal RNAs. J Androl 26(1): 70–74
162. Amanai M, Brahmajosyula M, Perry AC (2006) A restricted role for sperm-borne microRNAs in mammalian fertilization. Biol Reprod 75(6):877–884
163. Yan W et al (2008) Birth of mice after intracytoplasmic injection of single purified sperm nuclei and detection of messenger RNAs and MicroRNAs in the sperm nuclei. Biol Reprod 78(5):896–902
164. Curry E, Ellis SE, Pratt SL (2009) Detection of porcine sperm microRNAs using a heterologous microRNA microarray and reverse transcriptase polymerase chain reaction. Mol Reprod Dev 76(3):218–219
165. Khraiwesh B et al (2010) Transcriptional control of gene expression by microRNAs. Cell 140(1):111–122
166. Kim DH et al (2008) MicroRNA-directed transcriptional gene silencing in mammalian cells. Proc Natl Acad Sci U S A 105(42): 16230–16235
167. Valeri N et al (2009) Epigenetics, miRNAs, and human cancer: a new chapter in human gene regulation. Mamm Genome 20(9–10): 573–580
168. Lian J et al (2009) Altered microRNA expression in patients with non-obstructive azoospermia. Reprod Biol Endocrinol 7:13
169. Rassoulzadegan M et al (2006) RNA-mediated non-mendelian inheritance of an epigenetic change in the mouse. Nature 441(7092):469–474
170. Deng W, Lin H (2002) miwi, a murine homolog of piwi, encodes a cytoplasmic protein essential for spermatogenesis. Dev Cell 2(6):819–830
171. Kuramochi-Miyagawa S et al (2004) Mili, a mammalian member of piwi family gene, is essential for spermatogenesis. Development 131(4):839–849
172. Girard A et al (2006) A germline-specific class of small RNAs binds mammalian piwi proteins. Nature 442(7099):199–202
173. Grandjean V et al (2009) The miR-124-Sox9 paramutation: RNA-mediated epigenetic control of embryonic and adult growth. Development 136(21):3647–3655
174. Palini S et al (2011) Epigenetic regulatory mechanisms during preimplantation embryo development. Ann N Y Acad Sci 1221: 54–60
175. Carlson LL, Page AW, Bestor TH (1992) Properties and localization of DNA methyltransferase in preimplantation mouse embryos: implications for genomic imprinting. Genes Dev 6(12B):2536–2541
176. Monk M, Boubelik M, Lehnert S (1987) Temporal and regional changes in DNA methylation in the embryonic, extraembryonic and germ cell lineages during mouse embryo development. Development 99(3): 371–382
177. World Health Organization (WHO) (1980) WHO laboratory manual for the examination and processing of human semen, vol 1. Cambridge University Press, Cambridge
178. World Health Organization (WHO) (1999) WHO laboratory manual for the examination and processing of human semen, vol 4, 4th edn. Cambridge University Press, Cambridge

Chapter 6

Derivation of Human Embryonic Stem Cells (hESC)

Nikica Zaninovic, Qiansheng Zhan, and Zev Rosenwaks

Abstract

Stem cells are characterized by their absolute or relative lack of specialization their ability for self-renewal, as well as their ability to generate differentiated progeny through cellular lineages with one or more branches. The increased availability of embryonic tissue and greatly improved derivation methods have led to a large increase in the number of hESC lines.

Key words hESC derivation, Stem cells, Human embryonic stem cells, hESC culture, Feeder-free conditions

1 Stem Cells

The fertilized oocyte, zygote, and early embryo stage blastomeres have the greatest potential for stemness since they give rise to all the embryonic and extra-embryonic lineages that form the embryo. For this reason they are called *totipotent.*

Pluripotent cells are derived from the inner cell mass (ICM) of the blastocyst (BL) as they have lost their ability to contribute to trophectodermal (TE) lineages. They do, however, have the ability to divide indefinitely and give rise to all cell types arising from the three germ layers (ectoderm, mesoderm, and endoderm) under specific conditions in vitro or during pathological development. Human embryonic stem cells (hESC) and teratocarcinomas cells are examples of pluripotent cells. Key pluripotency genes (markers) include Oct-4, NANOG, and Sox2.

Multipotent cells also have self-renewing capacity but give rise to a more restricted family of cell lineages, although generally always to more than one. Hematopoietic stem cells are an example of these cell types.

Unipotent cells produce only one cell type and have the property of self-renewal, which distinguishes them from non-stem cells, e.g., skin stem cells.

Zev Rosenwaks and Paul M. Wassarman (eds.), *Human Fertility: Methods and Protocols*, Methods in Molecular Biology, vol. 1154, DOI 10.1007/978-1-4939-0659-8_6, © Springer Science+Business Media New York 2014

The distinction between these cell types is determined by experimental evidence of cellular lineages generated in vivo (chimeras or teratomas) as well as in vitro where their progeny and proliferative ability can be tested in isolation along with identification of specific characteristic molecular markers.

2 History of Embryonic Stem Cell (ESC) Research and Types of Stem Cells

2.1 Origin of the Term "Stem Cells"

The origin of the term "stem cell" can be traced back to the late nineteenth century, a time when the fertilized oocyte was referred to as a stem cell because it eventually gave rise to all cells of the organism [1].

The concept of embryonic stem cells arose from pioneering work with mouse and human teratocarcinomas. The high incidence of spontaneous testicular teratomas in mouse strain 129 (1 %) was later associated with the high efficiency of generating mouse embryonic stem cells (mESC) [2]. The pluripotent cells of teratocarcinomas, also called embryonal carcinoma (EC) stem cells, had an embryonic origin and could be obtained by transplanting mouse blastocysts to extrauterine sites such as the testes [2, 3]. These experiments suggested that it could be possible to isolate pluripotent cells directly from embryos. With advances in in vitro culture techniques, notably the use of feeder cells, EC cells were isolated. These cells grew in colonies and had large nuclei with prominent nucleoli and relatively small cytoplasmic volumes, similar to characteristics seen in ESCs [2]. In addition, these EC cells could form embryoid bodies (EB) that recapitulated the early post-implantation stages of mouse embryos [4]. The analysis of human teratocarcinomas in vitro revealed similar characteristics to those observed in the mouse [5, 6]. However, the issues of chromosomal normalcy and directed differentiation in vivo prevent the use of ECs in future transplantation studies. Once embryonic stem cells were competently isolated, interest in the use of EC cells waned, especially since these cells frequently harbored chromosomal abnormalities. Nonetheless, the initial research with EC cells was crucial in the evolution of ESC derivation and ultimately to their utility [7, 8].

2.2 Types of Stem Cells Derived from Embryos

Pioneering attempts to study early mammalian development by generating stem cell lines in vitro produced cell lines from preimplantation mammalian embryos that were poorly characterized; additionally, none of these cell lines showed the same characteristics as ESC lines [9]. The first mouse ESC (mESC) lines isolated from the ICM of developing blastocysts were described in 1981 by two independent groups using mouse embryonic fibroblasts (MEF) as feeder cells [10, 11]. These mESCs demonstrated

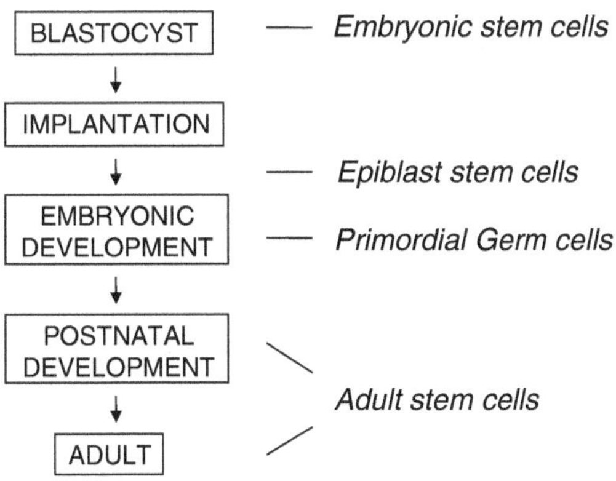

Fig. 1 Source of stem cells at different developmental stages

pluripotent characteristics in vitro and in vivo. Recently, pluripotent stem cells were isolated from the epiblast of post-implantation mouse and rat embryos (EpiSC), and as expected these cells maintained similar characteristics to ESCs, inducing similar patterns of gene expression and signaling [12, 13]. Interestingly, these EpiSC lines are distinct from mouse ESCs in their gene expression, epigenetic status, the signals controlling their differentiation, and they are more similar to human ESC cells (hESCs) [14]. It was proposed that mouse EpiSC and hESC are in a "primed" state while mouse ESC are in a naive ground state [15].

Another potential source of ESC-like stem cells is primordial germ cells. These pluripotent cells express similar pluripotency molecular markers as EC and ESC cells, as well as the capacity to differentiate into all three embryonic germ layers [16] (Fig. 1).

2.3 Human Embryonic Stem Cells

It has been demonstrated that there are significant differences between early human and early mouse development, including the cleavage rate, timing of blastulation, gene expression as well as the conditions necessary for maintaining pluripotency.

The first hESC line was derived almost two decades after the first mESC line. This delay was due to many factors including difficulties in obtaining good quality human embryos for derivation, suboptimal embryo culture conditions resulting in inefficient embryo development to the blastocyst stage, as well as the political, social, and ethical constraints inherent in the use of human embryos. The derivation of primate ESCs in 1995 contributed to the success of the first hESC derivation from a blastocyst in 1998 by the Thompson group at the University of Wisconsin [17]. The first hESC lines were derived the culturing of ICMs on MEF cells [17]. The same year, a group lead by Gerhart at John Hopkins University reported derivation of pluripotent stem cells from human primordial germ cells extracted from fetal gonadal ridges [18].

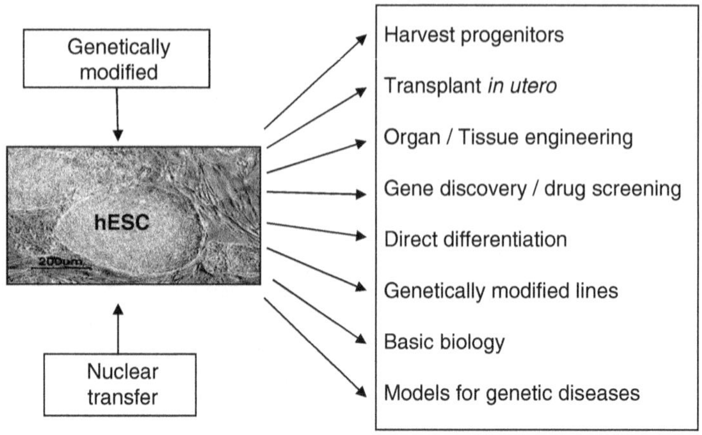

Fig. 2 Therapeutic and research potential of hESC

Soon after the first successful derivation of hESC, stem cell research became one of the fastest-growing disciplines in biomedical research. This burst of creative energy was due to the great potential of hESC in regenerative medicine, drug discovery, as well as cell replacement therapy [19–21]. Figure 2 is a schema depicting the promise of stem cell research.

2.4 Blastocyst (BL) Formation and Morphology

To better understand the underpinnings of hESC derivation, it is necessary to be well-versed in the processes involved in early human embryonic development, especially with regards to blastocyst formation. The blastocyst is characterized by the formation, differentiation, and cell allocation of different cell types, namely the ICM and trophectoderm (TE) [22]. During blastocyst development, these two morphologically distinguishable cell types are formed during embryo cavitation, where TE cells are positioned peripherally and ICM centrally within the BL cavity (Fig. 3).

The regulation of mammalian blastocyst development starts at fertilization and involves embryo polarization. This polarization is especially evident at the morula stage, at which time the outer blastomeres form TE [23]. As a result, the blastocyst is composed of the TE (identified by Cdx-2 expression) and the ICM (Cdx-2 negative), a mixed population of pluripotent cells that express Oct-4 and NANOG, and primitive endoderm cells expressing GATA6 (Fig. 3). The balance between the level of these genes influences lineage commitment in the blastocyst [24, 25]. While very little is known about their corresponding lineages in the human embryo, key pluripotent, and lineage genes are expressed [25–31]. Understanding the molecular mechanisms which define totipotency and cell lineage commitment in humans are fundamental aims of embryology and stem cell biology.

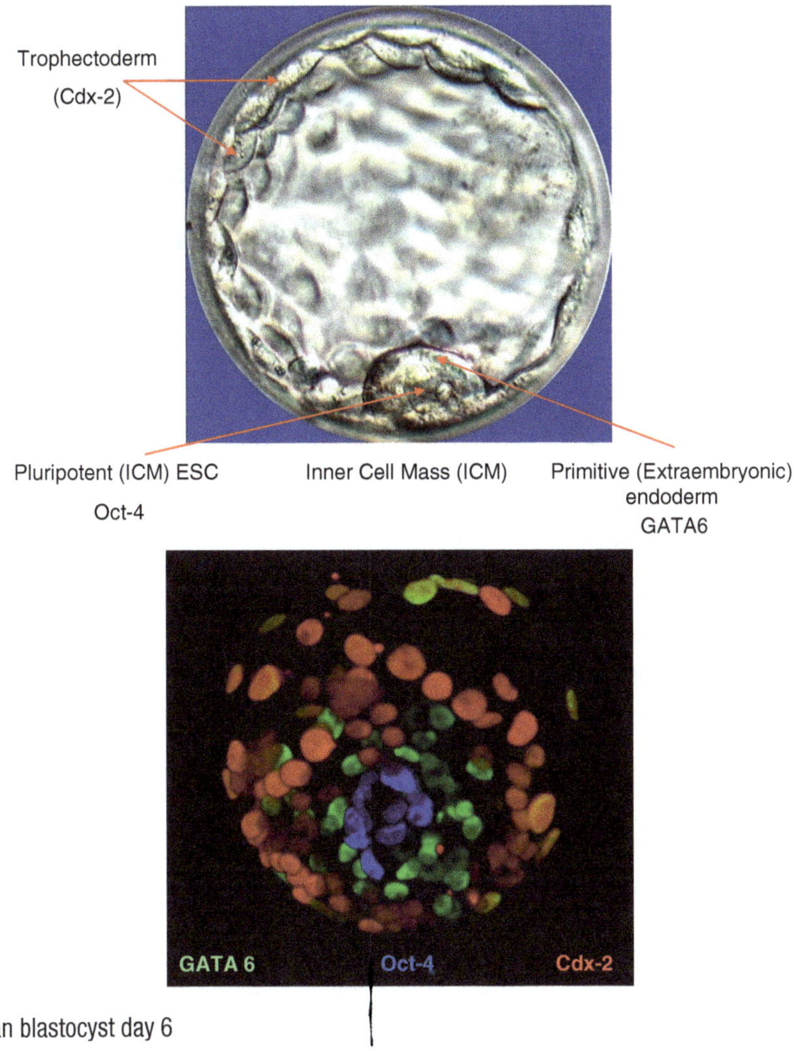

Fig. 3 Human blastocyst day 6

3 Derivation of the hESC

3.1 Derivation Methods

Currently utilized hESC derivation methods are very similar to the methods used for EC and mESC culture [7]. In recent years, the increasing number of newly derived hESC lines worldwide is indicative of the improved efficiency of derivation techniques. Derivation success rates depend on the morphological quality of the embryos, and the skill and techniques utilized by the scientists involved [32].

The standard method of hESC derivation involves derivation from the pluripotent ICM cells of the human blastocyst. hESCs can be derived after ICM isolation via immunosurgical or mechanical isolation or by whole embryo culture with subsequent isolation of the ICM outgrowth (Fig. 4) [33]. Immunosurgical ICM isolation was developed in 1975 for the study of mouse

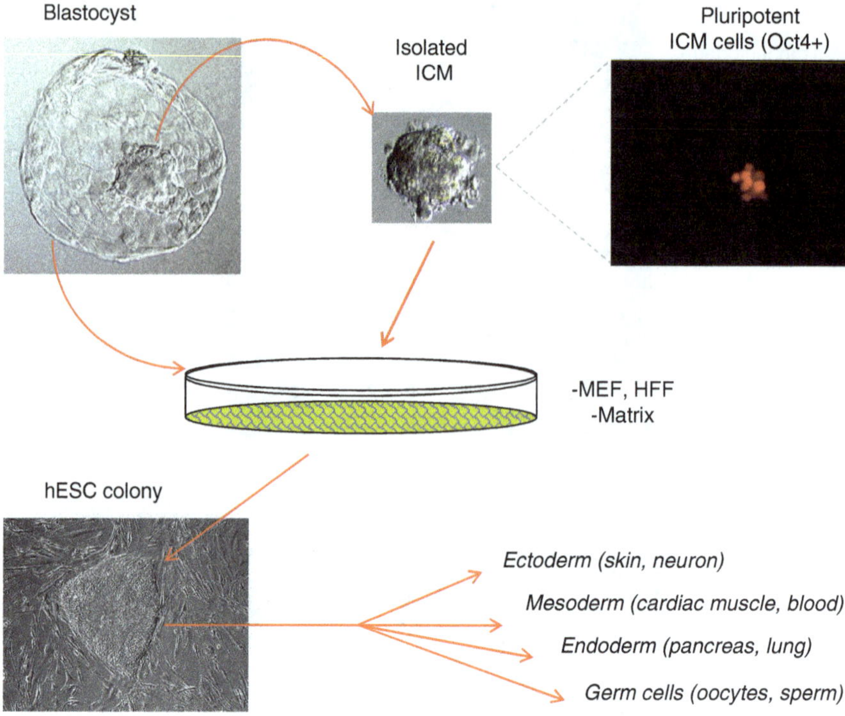

Fig. 4 Derivation of human embryonic stem cells—hESC

development [34], later also used for derivation of mESC [10, 11] and hESC [17]. It includes pre-treatment of the zona pellucida (ZP)-free blastocyst with anti-human serum followed by lysis of the trophectoderm (TE) with guinea-pig complement [34]. This method has the advantage of complete ICM isolation with direct plating on feeder cells, free of TE cells. However, the use of xenomaterials with this method reduces its usefulness.

Mechanical ICM isolation is an alternative method that utilizes needles or micromanipulators assisted by a non-contact laser [35]. This method, also known as the partial embryo culture method, includes dissection of the ICM area with subsequent culture on feeder cells [36]. This animal-free derivation method is suitable for blastocysts with prominent ICM, although the dissected ICM area will contain some TE cells [37].

The use of non-contact infrared laser for ICM isolation has been recently introduced [38, 39]. This method includes the ablation of the ZP and trophectoderm cells. In our derivation attempts, we used laser with continuous pulsations (XY clone-Staccato, Hamilton Thorne, USA) to isolate the ICM by cutting the blastocyst without ablation. With this method we are able to minimize cell damage and increase our derivation efficiency to over 50 % [40].

The easiest derivation method, whole embryo culture, involves placement of the whole ZP-free blastocyst on feeder cells.

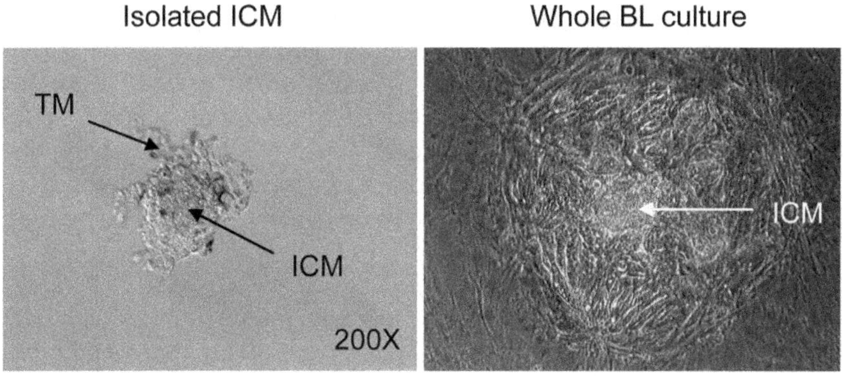

Fig. 5 hESC derivation methods

The ICM cells grow together with the TE cells as a monolayer. A significant difficulty associated with this method is the suppression of ICM transformation to ESCs by the TE outgrowth and induction of ICM differentiation [36, 41].

All derivation methods have their advantages and disadvantages; the selection of an appropriate method is influenced by the morphological quality of the blastocyst and the ICM. While all methods can be used successfully to derive hESC, it is apparent that attachment of the ICM to feeder cells or matrix is crucial (Fig. 5). However, at this time the mechanisms and factors that control ICM-ESC transformation are unclear.

3.2 Sources of hESC

The main sources of embryos for hESC derivation are either surplus embryos donated by IVF patients [17, 42] or donated embryos of poor morphologic quality (either slow developing or with suboptimal morphology) [30, 43, 44]. While ESCs are usually generated from normal diploid embryos, occasionally diploid hESCs (as analyzed by FISH) can be derived from mono-pronuclear human zygotes (1PN) [45, 46]. Parthenogenetically activated oocytes and embryos can be another source of hESC [47, 48], as they have been found to be karyotypically normal (female) and are capable of differentiating in vitro and forming teratomas with all three germ layers in vivo.

Triploid embryos formed by an extra set of male or female chromosomes can reach the BL stage [49, 50] and can be used to derive hESC lines. hESC lines derived from these human embryos have been shown to exhibit normal hESC characteristics and the ability to differentiate in vitro [51].

Human embryos are available from consenting IVF patients who donate either their fresh (immediately after transfer) or frozen (cryopreserved embryos that will not be used for transfer) embryos. Fresh embryos are more likely to be of suboptimal quality as the embryos chosen for transfer are the best embryos available.

On the other hand, while embryos may be frozen at all developmental stages, only morphologically good quality BL are frozen, thereby representing a better population than fresh research embryos. Additionally, the success of the freezing–thawing techniques will determine the number and quality of embryos available for hESC derivation [52]. Despite those variations and limitations it is clear that derivation of hESC is equally successful when using either fresh or frozen embryos [53].

3.3 Embryo Stages for Derivation of hESC

In addition to ICM isolation, mouse ESC can be derived from several pre-blastocyst stages: morula, cleavage embryos and even from isolated blastomeres, but the overall success rates are very low [54–56]. Human BL development varies in speed, expansion, number of cells, and morphology [22]. The optimal developmental rate is marked by full BL expansion with evident ICM cells that develop after 5 days post insemination or ICSI. Some blastocysts develop more slowly but can still demonstrate good morphology on day 6 or day 7. These temporal differences in development may not always reflect embryo quality, as pregnancies can be achieved and hESC can be derived from these blastocysts [22]. The highest hESC derivation rate was obtained from high-quality blastocysts on day 6 [39]. These finding were confirmed in our own derivation experiments, although the presence of ICM was prerequisite for successful derivation [40]. Interestingly, several groups have reported derivation of hESC from poor-quality embryos (unsuitable for embryo transfer) although with limited success [43, 57].

It has been reported that hESC can be derived even from blastocysts which developed as late as day 8 [58]. It is not yet clear whether blastocyst stage, number of days in culture, or BL morphology affect hESC derivation rates. Although the size and morphology of the ICM does appear to be significant, it is unknown how many pluripotent ICM cells are needed to form an embryo or hESC. Interestingly, using lineage tracing in mice, it was demonstrated that only three cells out of the total 30–50 ICM cells were required to form an embryo [59].

Human preimplantation embryos at different developmental stages can be used as sources of hESC. For example, it was shown that hESC lines were derived from morulas, indicating the pluripotent nature of these cells [60]. Based on these results and the results of mESC derivations from pre-blastocyst stage blastomeres, it was proposed that hESC lines could be derived from isolated blastomeres of cleavage stage human embryos [55, 61, 62]. This proposal could overcome political and moral obstacles associated with hESC derivation since heretofore hESC derivation required the destruction of blastocysts. In 2006, Klimanskaya et al. [63] reported derivation of hESC from single blastomeres of the 8-cell embryo. However, the overall success rate was very low (2 %) and hESC lines were obtained from disaggregated blastomeres, so the embryo was "destroyed" during the derivation process. In an effort to obtain hESC from isolated blastomeres of cleavage stage embryos, one or two blastomeres were biopsied from 8-cell embryos and cultured to form ESC. The initial efforts did not result in a hESC line [56]. However, recently the generation of a hESC line from individual human blastomeres isolated from an 8-cell embryo was reported when the single blastomere was cocultured with the parent embryo [64]. Later, single blastomere—hESC were derived in the presence of laminin and minimal xenomaterials without coculture [65]. The basis behind using laminin is to mimic the natural ICM niche, which prevents polarization of the blastomeres into TE cells. In another study, blastomeres from 4-cell embryos were used to derive two hESC lines that exhibited limited cell proliferation and chromosomal abnormalities [66]. Blastomere-derived hESC represent an interesting technical advancement and solve the inherent ethical issues we face by "destroying the embryo." It also provides the potential insight into possible differences in hESC lines derived from totipotent blastomeres versus pluripotent blastocyst ICM cells. However, it is unlikely that this technique will be soon applied clinically since it is unclear whether the biopsied embryo will in fact survive. Our own mice studies showed that successful blastomere–mESC derivation depend on cell stage and culture conditions. Two-cell stage blastomeres require a two-step protocol using blastocyst culture followed by stem cell culture. In contrast, 4–8 cell stage blastomeres benefit from the one-step protocol which placed blastomeres directly in stem cell culture [67].

3.4 Embryos from Preimplantation Genetic Screening/Diagnosis

Embryos donated after preimplantation genetic screening (PGS) analysis, including those that fertilized normally but carry chromosomal abnormalities (monosomy, trisomy, and mosaics) comprise a large group available for hESC research. The procedure involves biopsy of one or two blastomeres at the 6–8 cell stage (day 3) for analysis of chromosomal status. If the analysis reveals a chromosomal abnormality, the embryo is not transferred and can be donated for ESC derivation. Initial attempts to generate stem cell lines from such embryos achieved only limited success [68]. It has been suggested that "self-correction" of chromosomally abnormal embryos in ESC culture can occur as reported with hESC lines derived trisomic embryos [68]. One possible explanation for "self-correction" is trisomic zygote rescue, where an extra chromosome is lost during mitosis and duplication of a single chromosome may occur by uniparental disomy (UPD) [69]. Besides triploid rescue, UPD can also occur in cases with monosomy rescue, where haploid chromosomes can be duplicated [70]. Recently, there was a report of derivation of hESC lines with chromosomal abnormalities, e.g., trisomies and monosomies that originated in the embryo [71].

hESC can also be derived from embryos carrying specific genetic diseases. After preimplantation genetic diagnosis (PGD), affected embryos can be used for ESC derivation and represent an excellent in vitro model of specific genetic diseases [71–73]. hESC lines for diseases such as Huntington's Disease, Cystic Fibrosis, Fanconi Anemia, Myotonic Dystrophy, and others have been derived [71, 73].

To date, there are over 1,000 hESC lines worldwide and the majority have been characterized [74]. However, most of the published hESC research is based only on a handful of cell lines. It is unclear how many cell lines are needed to fulfill the need for diverse genetic, ethnic, HLA, and epigenetic variations for possible future tissue regeneration and transplantation. In addition, only a small number of lines are derived and cultured under xeno-free and GMP conditions (clinical grade). There are concrete guidelines and ethical standards from the European Union (http://www.hescreg.eu), the ISSCR (http://www.isscr.org), and the NIH (http://stemcells.nih.gov) for derivation of the new cell lines. In order to qualify as GMP quality cell lines, the cell lines need to be cultured in defined and control conditions by trained staff and with full documentation [75]. The NIH registry (October 2013) currently lists 234 approved hESC lines, number 211 Weill Cornell Medical College cell line (http://grants.nih.gov/stem_cells/registry).

4 hESC Culture Methods

4.1 Basic Procedures

The basic culture conditions for hESC include MEF as feeder cells to provide a matrix and a source of growth factors. The medium is supplemented with fetal bovine serum (FBS) or fetal calf serum (FCS) known to be a rich source of growth factors [17, 42]. Methods have been adapted from mouse ESC culture protocols [10, 11], and cultures require daily maintenance, including media changes, regular passaging (4–6 days), and cell cryopreservation. Passaging methods for prevention of hESC differentiation include: (1) mechanical dissociation with needles, pipette tips, or roller cutter (EZ passage tool, Invitrogen) and (2) enzymatic dissociation by collagenase IV, dispase, or xeno-free accutase and TrypLE™ Express Enzyme [76]. All protocols include dissociation of hESC colonies into cell clumps (50–100 cells) to prevent cell differentiation (Fig. 6).

Within the stem cell research community, there is a large effort to optimize conditions for derivation and culture of hESC. One future application of hESC is their therapeutic use in regenerative medicine. hESC will need to be derived and cultured in defined conditions and free of animal products (xeno-free). In addition, these hESC lines must show proliferative capacity, cell-specific differentiation and have phenotypic stability after transplantation.

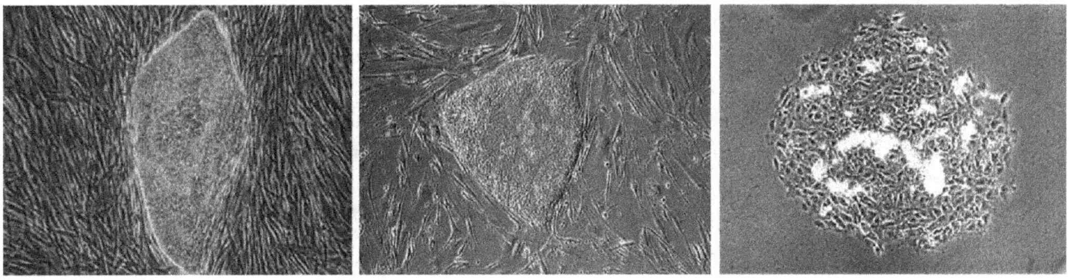

Fig. 6 hESC culture—colony morphology

The ideal culture conditions include a defined matrix, defined media components with recombinant protein supplements (if necessary), and a consistent cell passaging method [77]. Optimal conditions are required to maintain phenotypically and karyotypically stable hESC for prolonged periods of culture. To establish this well-defined culture milieu, various studies were conducted to test human feeder cells or feeder-free culture systems, as well as serum-free and animal-free culture media [78–80].

Mouse ESC can be cultured on gelatin instead of MEF feeder layers, with medium supplemented with leukemia inhibitory factor (LIF) [81, 82]. LIF acts through the activation of STAT3 to maintain pluripotency [83]. Despite its essential role in mice, LIF/STAT3 activation is not required for pluripotency in hESC [84]. Therefore, it is essential to identify the growth factors that are required to maintain the undifferentiated hESC pluripotent state. While the molecular mechanisms underlying hESC derivation and self-renewal are unclear at this time, it has been suggested that the Wnt signaling pathway and/or bFGF pathway are important [85, 86]. In addition, it was recently suggested that TGFß signaling is required for the maintenance of hESC at the undifferentiated stage [87].

Most hESC lines have been derived on MEF feeder cells [17, 42, 88]. In addition, MEF cell lines that were immortalized can be used and as they are easy to propagate [89, 90]. The growth of hESC on MEF feeder layers represents a hazard because xenobiotic pathogens may be transferred from animals to human cells [91]. Therefore, improvements in hESC cultures are necessary to optimize conditions for therapeutic application. Interestingly, no evidence of infection of hESC by MEF feeder cell-derived murine leukemia viruses has ever been reported [92].

4.2 Refinements in ESC Culture Technology

Recently, improvements in methods for culturing hESC have been introduced [33, 93, 94], as follows:

(a) *Using feeder cells of human origin instead of animal origin* [95] To avoid animal cell contamination, human feeder cells of various sources were successfully used to replace MEFs [96–99].

These include foreskin fibroblasts [95, 100], placental fibroblast feeders [101], endometrial cells [102], and adult marrow cells [97]. In addition, fibroblast-like cells derived from hESC cultures have been used for hESC derivation [99, 103–105]. Furthermore, various cells such as fetal and adult muscle and skin cells and adult fallopian tube epithelial cells were tested as well [98, 105]. From these reports we can conclude that the factors that support undifferentiated hESC growth can be found in various tissues and are neither species specific nor tissue specific.

(b) *Growing hESC in serum-free conditions using serum replacement (SR) with basic fibroblast growth factor (bFGF)* [88]
Traditionally, hESC culture medium contained FBS as a protein supplement [17, 42]. To avoid animal derived serum, human serum can be utilized [104]. Serum is a complex mixture of compounds, including some that are inhibitory to hESC growth. To optimize serum-free cultures and to overcome the problem of variability of serum lots, defined serum replacement (knockout serum replacement—KOSR) was used in combination with bFGF [88, 106, 107].

(c) *Feeder-free conditions on matrix with conditioned medium*
The feeder-free culture system was developed using Matrigel, laminin-coated or fibronectin-coated dishes with the addition of the MEF conditioned medium (MEF-CM) [108–110]. Matrigel is a matrix derived from a mouse sarcoma cell line and contains many extracellular molecules, including laminin, collagen IV, and growth factors such as FGFs [111]. The advantages of this system are that it is feeder-free and highly reproducible, but cells are still exposed to animal matrix proteins and animal-derived medium (MEF-CM). To replace Matrigel, laminin [109] or fibronectin [106] can be used to culture hESC.

(d) *Feeder-free conditions with media supplemented with only selected growth factors* [106, 112, 113]
To reduce exposure of hESC to feeder cells and animal culture media products, feeder-free systems cultured in medium with selected growth factors and without MEF-CM were developed. Matrigel was successfully used as a matrix with addition of KOSR, bFGF, Noggin, or TGF-ß1 [113, 114]. Recently, the maintenance of hESC in the undifferentiated stage in feeder-free conditions was achieved by BMP signaling antagonist Noggin and a high concentration of bFGF [112, 114, 115]. On the other hand, the addition of bFGF, lithium chloride, GABA, pipecolic acid, and TGF-ß to the defined culture medium also fosters undifferentiated hESC growth [116]. Finally, extracellular matrices from human serum can also be used to maintain hESC in culture [117].

In recent years, various xeno-free and feeder-free stem cell media have been developed [118, 119]. Some are commercially available: Nutristem (Biological Industries), mTeSR2 (Stem Cell Technology), StemPro (Invitrogen), SBX (AxCell), and VitroHES (Vitrolife).

A recent paper from the Thomson group described that simple media containing only eight xeno-free components in serum-free DMEM F12 base medium is sufficient for hESC cultures [120].

To completely eliminate animal or human feeder cells and matrices, bioactive polymers (hydrogels) have been tested for hESC culture and differentiation. These polymers can support hESC culture and represent a promising tool in ESC cell differentiation as three-dimensional (3D) culture systems [121–123].

4.3 Serum-Free and Feeder-Free Derivation of the hESC

In an attempt to derive hESC in defined conditions, including serum-free and animal-free, serum replacement in combination with human foreskin fibroblast (HFF) [124], human placental fibroblast [101], and MEF [125, 126] have been used successfully. hESC was also derived in feeder-free conditions using a mouse-derived matrix and animal-free culture medium supplemented with KOSR and bFGF [127]. In addition, hESC lines were derived on human matrix components (collagen IV, laminin, fibronectin, and vitronectin) in a defined culture medium composed of recombinant or human sources [116]. Unfortunately, two lines from this derivation expressed abnormal karyotypes: 47, XXY and trisomy 12. It is unclear if these chromosomal abnormalities are a direct consequence of the culture conditions, since abnormal copy number of chromosomes 12 and 17 have been reported in long-term hESC cultures observed after over 30 passages [128]. Recently, an hESC line was derived in complete animal-free conditions using human laminin as a matrix- and serum-free media containing purified human or recombinant proteins [129].

In order to optimize the hESC culture conditions, efforts have been made to identify components of the conditioned media needed for hESC self-renewal. Analysis of the protein composition of feeder-conditioned media of animal (MEF) and human origin reveals bFGF as a key factor in hESC self-renewal [130–132]. Indeed, the importance of bFGF for hESC self-renewal has recently been confirmed [98], and bFGF without serum components is sufficient to maintain undifferentiated hESC cells in culture [101]. In addition, an extracellular matrix and bFGF without other serum components are sufficient to derive hESC from human ICM [116].

4.4 Clonal Derivation of hESC

When hESC are derived from blastocysts, the ICM is a heterogeneous population of cells comprising of pluripotent (epiblast) cells surrounded by primitive endoderm (hypoblast) cells [31]. This heterogeneity is also reflected in the hESC cultures, where clonal isolation of hESCs reveals heterogeneity within the pluripotent stem cells [133]. In addition, clonally derived embryonic stem cells from a single mouse ICM differ in their gene expression and developmental capacity, further supporting the idea of a heterogeneous population of cells within the ICM [134]. There are no reports to date of individual human ICM cells forming ESC by clonal expansion.

hESC sublines have been clonally derived from existing lines, but at a very low efficiency of 2–4 % [88, 133, 135, 136]. This is in contrast to the mouse ESC, where clonal propagation is a standard method of mESC culture. It seems that cell–cell interaction is critical for efficient hESC propagation, since the loss of gap junctions between hESC can increase apoptosis and inhibit growth [137]. For this reason, hESC are almost always passaged in clumps of 50–100 cells. It was shown recently that clonal efficiency can be increased by up to 30 % when hESC were cultured in low oxygen (2 %) [138]. Furthermore, addition of the ROCK inhibitor Y-27632 and protein kinase C inhibitor can increase clonal efficiency through E-cadherin stabilization of cell–cell interaction [139, 140, 173].

4.5 hESC Suspension Cultures

Standard culture methods for hESC proliferation include two-dimensional cultures where colonies adhere on MEF or human feeders or matrices. These culture methods are not able to produce large numbers of cells which will be required for the potential therapeutic treatments. To generate large amounts of identical cell types at a relevant clinical scale, suspension cultures utilizing controllable bioreactors may be required. Several groups have reported using suspension hESC cultures in floating aggregates [141–143]. Media for suspension cultures varied between groups and contained different growth factors (Activin A or IL6), +/– Y-27632 inhibitor with or without components of extracellular matrix (laminin, fibronectin). Additionally, new hESC lines were derived in these conditions by using floating ICM in suspension cultures [143]. Derivation and cell proliferation appeared to be reduced compared to monolayer cultures, and the long-term genetic stability of these cells needs to be confirmed. However, it does question the dogma that hESC line derivation cultures require ICM attachments.

Several studies described improved methods to scale up hESC cultures in bioreactors [144, 145], microcarriers [146, 147], or automated cultures [148]. Future studies are necessary to demonstrate the relevance of these systems for rapid expansion and clinical application.

4.6 Genetic Integrity of hESC Lines

hESC have been successfully cultured for extended periods through numerous passages while maintaining a normal diploid karyotype [17, 42, 88]. However, hESC lines can develop abnormal karyotypes after long-term culture [149]. Long-term culture of mouse ESC can lead to diminished pluripotency and increased aneuploidy (a phenomenon which may be the major cause of failure to differentiate all tissues of the adult chimera) [150, 151]. In hESC, it was observed that karyotypic changes usually involved the addition of chromosome 12 and 17q and to a lesser extent chromosome X [128, 149, 152]. Chromosomal changes in hESC have appeared in multiple cell lines and at many different laboratories, mostly after extended passages (over 30) [149]. In addition, 16 of 30 initially diploid hESC lines showed chromosomal instability when cultured using the same conditions in the same laboratory [149]. This rate of chromosomal abnormalities in hESC is low compared to most other cell lines and surprising given the high rate of aneuploidy observed in the preimplantation human embryos from which they were derived [153].

Intriguingly, human embryonic carcinoma cells are typically aneuploid, with trisomies of chr12 and chr17q [154, 155]. Further, the pluripotency gene NANOG is located on chr12p and this region is frequently amplified in testicular germ cell tumors [156]. Overexpression of the NANOG gene, promoting self-renewal, may provide cells with an advantage in adapting to culture conditions as aneuploid hESC have a tendency to grow faster. Most chromosomal abnormalities, aneuploidies in particular, can be a reflection of the progressive adaptation of pluripotent hESC to culture conditions. It has been proposed that the chromosomal changes observed in hESC in vitro reflect in vivo tumorogenic events [149]. Specific culture conditions may contribute to chromosomal instability such as the method for passaging cells (mechanical vs. enzymatic) [157, 158]. A recent report also suggested that a feeder-free culture system is associated with more chromosomal instability then a feeder system [158]. In addition, hESC cultured in low oxygen showed reduced chromosomal abnormalities [138, 159].

4.7 Characteristics of the hESC

When grown on MEFs, hESC exhibit a high proliferative capacity, chromosomal stability in long-term culture, and the ability to differentiate into somatic cell types of all three germ layers, both in vitro and in vivo [17, 42] There are several molecular and cellular criteria by which hESC lines are assessed: self-renewal capability (indefinite proliferation), pluripotency in vitro and in vivo, differentiation into somatic and germ cells, karyotype, and good survival rates after freeze/thaw [160]. Recently, an international survey of hESC lines was established that provided a benchmark for comparison of hESC derived in different laboratories [161]. It showed that despite unique genetic backgrounds, a range of derivation techniques and various culture methods, hESC lines manifest a high degree of phenotypic similarities and molecular markers that characterize hESC identity [161].

One of the fundamental characteristics of hESC is their ability to differentiate into the cells of all three germ layers: ectoderm, mesoderm, and endoderm. This provides an opportunity to validate the cell line in vitro and in vivo (teratomas) using molecular markers of differentiated progeny, e.g., ß-tubulin III, cytokeratin, and nestin for ectoderm, brachyury and actin for mesoderm, and HNF3beta, GATA6 and GATA4 for endoderm [162]. It is necessary to perform in vivo studies of teratoma formation to evaluate the pluripotent nature of hESC. To do that, hESC are usually injected into non-obese diabetic/severe combined immunodeficient (NOD-SCID) mice that are known to form teratomas rather than reject human cells [17]. To enhance the teratoma formation, approximately 1–5 million hESC cells are injected together with Matrigel [163]. Histological analysis should reveal formation of all three germ layers.

4.8 Genetic Regulation of hESC

Oct-4 is a key pluripotency regulator in the early embryo as well as in the self-renewing ESC. Studies in mouse and human ESC indicate that Oct-4 is a component of a network of transcription factors including NANOG and Sox2 that cooperatively maintain pluripotency in ESC [164–167]. By using RNAi to suppress Oct-4, several genes that are positively (NANOG, Sox2) or negatively (Cdx-2, GATA6) regulated by Oct-4 were identified in hESC [167]. The balance between a minimal set of lineage-specific transcription factors might drive early cell-fate decisions in embryos in vivo and ESC in vitro [168]. To date, it is unclear how ICM and other embryonic cells transform into ESC, but it is evident that molecular and epigenetic changes occur during transition of these cells [169].

5 hESC Lines at Weill Medical College of Cornell University

The hESC derivation program at WCMC started in 2006 with collaboration of the Department of Reproductive Medicine (CRM) and Department of Genetic Medicine. Forty-three hESC lines were derived until recently (October 2013). Human ESC lines were derived from donated surplus embryos after IVF and PGD treatment. The majority of lines were derived in standard conditions on MEF. The first line (WCMC-1) was derived in xeno-free conditions on HFF in collaboration with the Technion, Israel. Two recent lines were derived in feeder-free conditions on Matrigel (WCMC-38, 39). Out of 43 lines, 5 lines were derived from IVF embryos and 38 from PGD embryos. Of the PGD group, 13 came from aneuploid embryos, 5 from unbalance translocation embryos, and 20 from various single gene disorders (Fig. 7). Surprisingly, lines derived from aneuploid embryos all show a normal karyotype. Other groups have reported similar results [170, 171]. Aside from the karyotypes, we further attempted to evaluate these lines for potential UPD-self-correction rescue [68]. Our results showed a high level of heterozygosity in these lines, indicating that UPD did

WCMC lines	Disease
WCMC 1	
WCMC 2	Mo 15
WCMC 3	Mo 8, 21
WCMC 4	Fragile X
WCMC 5	Fragile X
WCMC 6	
WCMC 7	
WCMC 8	BRCA2
WCMC 9	BRCA2
WCMC 10	
WCMC 11	BRCA1
WCMC 12	Tri 18
WCMC 13	Fragile X
WCMC 14	Beta-Thalassemia
WCMC 15	Hungtington, Tri13
WCMC 16	Hungtington-N/Tri13?
WCMC 17	Cystic Fibrosis
WCMC 18	UBT, [46XX, t912:21)(q24.33;q22.13)]
WCMC 19	UBT,Tri 21 Mono 12q[46XX, t912:21)(q24.33;q22.13)]
WCMC 20	UBT, Tri 21 Mono 12q[46XX, t912:21)(q24.33;q22.13)]
WCMC 21	Retinoblastoma
WCMC 22	Retinoblastoma
WCMC 23	Tri 13
WCMC 24	UBT, Mon 21[46,XX,t(7:13)(q2.2;q14.3)]
WCMC 25	XYY, Tri 18
WCMC 26	Tri 17?
WCMC 27	Hemophillia A [F8, inversion]
WCMC 28	Achondroplasia [FGFR3 G1138A] AFF
WCMC 29	MNB, 1 cell normal XY; 1 cell Abn, XX, multiple monosomies
WCMC 30	Dystonia, DYT1 AFF
WCMC 31	Dystonia, DYT1 Normal, Tri 13 XO
WCMC 32	Achondroplasia [FGFR3 G1138A] AFF
WCMC 33	Achondroplasia [FGFR3 G1138A] AFF
WCMC 34	Tri 13
WCMC 35	UBT
WCMC 36	Tri17, Tri18
WCMC 37	Fragile X syndrome, AFF
WCMC 38	XX,Mono 15 Tri 18,Tri 22; MATRIGEL
WCMC 39	XX, Tri 18? (Normal?); MATRIGEL
WCMC 40	Achondroplasia [FGFR3 G1138A] AFF
WCMC 41	Tri13
WCMC 42	Gaucher' s disease, GBA(N370S & 84GG) AFF
WCMC 43	

Fig. 7 hESC lines at WCMC of Cornell University

not occur [172]. It is unclear how aneuploid embryos generate euploid hESC lines, although a high level of embryo masaicism may explain the PGD results. In addition, it was suggested that in the initial cultures of these cell lines, where there is a mixture of aneuploid and euploid cells, normal cells have proliferative advantages over abnormal cells, suppressing abnormal cell proliferation [170].

Optimization of hESC derivation comes with experience and technical advancements. Improving ICM isolation and optimization of culture conditions are paramount in establishing a successful hESC derivation laboratory. We have found ICM isolation with laser (XY clone-Staccato, Hamilton Thorne, USA) to be technically advantageous to ICM isolation with needles. The addition of serum-free hESC medium and condition medium (hESC-CM) was another critical improvement. Clearly, the efficiency of hESC derivation is critically dependent on the preservation of a healthy ICM, the ICM isolation technique, and the hESC culture conditions [40]. These derived cells have been successfully used to generate hESC-derived endothelial cells (lentiviral transduction and clonal selection of hESC) [173, 174].

The combined use of laser ICM isolation and the use of conditioned medium allowed a hESC-line derivation rate which exceeds 50 % per blastocyst. As in every other area of medical research, experience, perseverance, and serendipity lead to advancements and success.

References

1. Ramalho-Santos M, Willenbring H (2007) On the origin of the term "stem cell". Cell Stem Cell 1:35–38
2. Martin GR, Evans MJ (1974) The morphology and growth of a pluripotent teratocarcinoma cell line and its derivatives in tissue culture. Cell 2:163–172
3. Stevens LC (1970) The development of transplantable teratocarcinomas from intratesticular grafts of pre- and postimplantation mouse embryos. Dev Biol 21:364–382
4. Martin GR, Evans MJ (1975) Differentiation of clonal lines of teratocarcinoma cells: formation of embryoid bodies in vitro. Proc Natl Acad Sci U S A 72:1441–1445
5. Andrews PW (2002) From teratocarcinomas to embryonic stem cells. Philos Trans R Soc Lond B Biol Sci 357:405–417
6. Andrews PW (1984) Retinoic acid induces neuronal differentiation of a cloned human embryonal carcinoma cell line in vitro. Dev Biol 103:285–293
7. Solter D (2006) From teratocarcinomas to embryonic stem cells and beyond: a history of embryonic stem cell research. Nat Rev Genet 7:319–327
8. Solter D (2005) What is a stem cell? Novartis Found Symp 265(3–12):92–97
9. Edwards RG (2004) In: Lanza R et al (eds) Handbook of stem cells. Elsevier Academic Press, Burlington, MA, pp 1–14
10. Evans M, Kaufman M (1981) Establishment in culture of pluripotent cells from mouse embryos. Nature 292:154–156
11. Martin G (1981) Isolation of a pluripotent cell line from early mouse embryos cultured in medium conditioned by teratocarcinoma stem cells. Proc Natl Acad Sci U S A 78:7635
12. Brons IG et al (2007) Derivation of pluripotent epiblast stem cells from mammalian embryos. Nature. doi:10.1038/05950
13. Tesar PJ et al (2007) New cell lines from mouse epiblast share defining features with human embryonic stem cells. Nature. doi:10.038/05972
14. Nichols J, Smith A (2011) The origin and identity of embryonic stem cells. Development 138:3–8
15. Nicholds J, Smith A (2009) Naïve and primed pluripotent states. Cell Stem Cell 4:487–492
16. Resnick JL et al (1992) Long-term proliferation of mouse primordial germ cells in culture. Nature 359:550–551
17. Thomson JA et al (1998) Embryonic stem cell lines derived from human blastocysts. Science 282:1145–1147
18. Shamblott MJ et al (1998) Derivation of pluripotent stem cells from cultured human primordial germ cells. Proc Natl Acad Sci U S A 95:13726–13731
19. Gerecht-Nir S, Itskovitz-Eldor J (2004) The promise of human embryonic stem cells.

Best Pract Res Clin Obstet Gynaecol 18: 843–852
20. Gerecht-Nir S, Itskovitz-Eldor J (2004) Cell therapy using human embryonic stem cells. Transpl Immunol 12:203–209
21. Golan-Mashiach M et al (2005) Design principle of gene expression used by human stem cells: implication for pluripotency. FASEB J 19:147–149
22. Veeck L, Zaninovic N (eds) (2003) An atlas of human blastocyst. Parthenon, New York
23. Edwards RG (2005) Changing genetic world of IVF, stem cells and PGD. B. Polarities and gene expression in differentiating embryo cells and stem cells. Reprod Biomed Online 1:761–776
24. Niwa H et al (2005) Interaction between Oct3/4 and Cdx2 determines trophectoderm differentiation. Cell 123:917–929
25. Heins N et al (2006) Clonal derivation and characterization of human embryonic stem cell lines. J Biotechnol 122:511–520
26. Palmieri SL et al (1994) Oct-4 transcription factor is differentially expressed in the mouse embryo during establishment of the first two extraembryonic cell lineages involved in implantation. Dev Biol 166:259–267
27. Huntriss J et al (2004) Expression of mRNAs for DNA methyltransferases and methyl-CpG-binding proteins in the human female germ line, preimplantation embryos, and embryonic stem cells. Mol Reprod Dev 67: 323–336
28. Cauffman G et al (2005) Oct-4 mRNA and protein expression during human preimplantation development. Mol Hum Reprod 11: 173–181
29. Hyslop L et al (2005) Downregulation of NANOG induces differentiation of human embryonic stem cells to extraembryonic lineages. Stem Cells 23:1035–1043
30. Zhang X et al (2006) Derivation of human embryonic stem cells from developing and arrested embryos. Stem Cells 24:2669–2676
31. Adjaye J et al (2005) Primary differentiation in the human blastocyst: comparative molecular portraits of inner cell mass and trophectoderm cells. Stem Cells 23:1514–1525
32. Stephenson EL et al (2006) Proposal for a universal minimum information convention for the reporting on the derivation of human embryonic stem cell lines. Regen Med 1: 739–750
33. Amit M, Itskovitz-Eldor J (2006) Sources, derivation, and culture of human embryonic stem cells. Semin Reprod Med 24:298–303
34. Solter D, Knowles BB (1975) Immunosurgery of mouse blastocyst. Proc Natl Acad Sci U S A 72:5099–5102
35. Moon S et al (2006) Generation, culture, and differentiation of human embryonic stem cells for therapeutic applications. Mol Ther 13:5–14
36. Kim HS et al (2005) Methods for derivation of human embryonic stem cells. Stem Cells 23:1228–1233
37. Amit M, Itskovitz-Eldor J (2002) Derivation and spontaneous differentiation of human embryonic stem cells. J Anat 200:225–232
38. Turetsky T et al (2008) Laser-assisted derivation of human embryonic stem cell lines from IVF embryos after preimplantation genetic diagnosis. Hum Reprod 23:46–53
39. Chen AE et al (2009) Optimal timing of inner cell mass isolation increases the efficiency of human embryonic stem cell derivation and allows generation of sibling cell lines. Cell Stem Cell 4:103–106
40. Zaninovic N et al (2010) Efficiency of hESC derivation: optimization of the ICM isolation and culture conditions. ASRM, Denver, Colorado
41. Heins N et al (2004) Derivation, characterization, and differentiation of human embryonic stem cells. Stem Cells 22:367–376
42. Reubinoff BE et al (2000) Embryonic stem cell lines from human blastocysts: somatic differentiation in vitro. Nat Biotechnol 18: 399–404
43. Mitalipova M et al (2003) Human embryonic stem cell lines derived from discarded embryos. Stem Cells 21:521–526
44. Chen H et al (2005) The derivation of two additional human embryonic stem cell lines from day 3 embryos with low morphological scores. Hum Reprod 20:2201–2206
45. Levron J et al (1995) Male and female genomes associated in a single pronucleus in human zygotes. Biol Reprod 52:653–657
46. Suss-Toby E et al (2004) Derivation of a diploid human embryonic stem cell line from a mononuclear zygote. Hum Reprod 19:670–675
47. Cibelli J et al (2006) Embryonic stem cells from parthenotes. Methods Enzymol 418: 117–135
48. Revazova ES et al (2007) Patient-specific stem cell lines derived from human parthenogenetic blastocysts. Cloning Stem Cells 9: 432–449
49. Plachot M (1992) Cytogenetic analysis of oocytes and embryos. Ann Acad Med Singapore 21:538–544

50. Sandalinas M et al (2001) Developmental ability of chromosomally abnormal human embryos to develop to the blastocyst stage. Hum Reprod 16:1954–1958
51. Baharvand H et al (2006) Generation of new human embryonic stem cell lines with diploid and triploid karyotypes. Dev Growth Differ 48:117–128
52. Veeck LL et al (2004) High pregnancy rates can be achieved after freezing and thawing human blastocysts. Fertil Steril 82:1418–1427
53. Sjogren A et al (2004) Human blastocysts for the development of embryonic stem cells. Reprod Biomed Online 9:326–329
54. Tesar PJ (2005) Derivation of germ-line-competent embryonic stem cell lines from preblastocyst mouse embryos. Proc Natl Acad Sci U S A 102:8239–8244
55. Chung Y et al (2006) Embryonic and extra-embryonic stem cell lines derived from single mouse blastomeres. Nature 439:216–219
56. Fong CY et al (2006) Unsuccessful derivation of human embryonic stem cell lines from pairs of human blastomeres. Reprod Biomed Online 13:295–300
57. Lerou PH et al (2008) Derivation and maintenance of human embryonic stem cells from poor-quality in vitro fertilization embryo. Nat Protoc 3:923–933
58. Stojkovic M et al (2004) Derivation of human embryonic stem cells from day-8 blastocysts recovered after three-step in vitro culture. Stem Cells 22:790–797
59. Markert CL, Petters RM (1978) Manufactured hexaparental mice show that adults are derived from three embyronic cells. Science 202:56–58
60. Strelchenko N et al (2004) Morula-derived human embryonic stem cells. Reprod Biomed Online 9:623–629
61. Solter D (2005) Politically correct human embryonic stem cells? N Engl J Med 353: 2321–2323
62. Wakayama S et al (2007) Efficient establishment of mouse embryonic stem cell lines from single blastomeres and polar bodies. Stem Cells 25:986–993
63. Klimanskaya I et al (2006) Human embryonic stem cell lines derived from single blastomeres. Nature 444(7118):481–485
64. Chung Y et al (2008) Human embryonic stem cell lines generated without embryo destruction. Cell Stem Cell. doi:10.1016/j.stem.2007.12.013
65. Ilic D et al (2009) Derivation of human embryonic stem cell lines from biopsied blastomeres on human feeders with minimal exposure to xenomaterials. Stem Cells Dev 18:1343–1350
66. Geens M et al (2009) Human embryonic stem cell lines derived from single blastomeres of two 4-cell stage embryos. Hum Reprod 24:2709–2717
67. Hao J et al (2007) Derivation of mouse embryonic stem cells (mESC) from individual and aggregated mouse blastomeres. ASRM, Washington DC
68. Munné S et al (2005) Self-correction of chromosomally abnormal embryos in culture and implications for stem cell production. Fertil Steril 84:1328–1334
69. Los F et al (1998) Uniparental disomy with and without confined placental mosaicism: a model for trisomic zygote rescue. Prenat Diagn 18:659–668
70. Shaffer LG et al (2001) American College of Medical Genetics statement of diagnostic testing for uniparental disomy. Genet Med 3:206–211
71. Verlinsky Y et al (2006) Repository of human embryonic stem cell lines and development of individual specific lines using stembrid technology. Reprod Biomed Online 13:547–550
72. Verlinsky Y et al (2005) Human embryonic stem cell lines with genetic disorders. Reprod Biomed Online 10:105–110
73. Mateizel I et al (2006) Derivation of human embryonic stem cell lines from embryos obtained after IVF and after PGD for monogenic disorders. Hum Reprod 21:503–511
74. Löser P et al (2010) Human embryonic stem cell lines and their use in international research. Stem Cells 28:240–246
75. Ährlund-Richter L et al (2009) Isolation and production of cells suitable for human therapy: challenges ahead. Cell Stem Cell 4:20–26
76. Hasegawa K et al (2010) Current technology for the derivation of pluripotent stem cell lines from human embryos. Cell Stem Cell 6:521–531
77. Hoffman LM, Carpenter MK (2005) Characterization and culture of human embryonic stem cells. Nat Biotechnol 23: 699–708
78. Crook JM et al (2007) The generation of six clinical-grade human embryonic stem cell lines. Cell Stem Cell 1:490–494
79. Ellerström C et al (2006) Derivation of a xeno-free human embryonic stem cell line. Stem Cells 24:2170–2176
80. Rajala K et al (2007) Testing of nine different xeno-free culture media for human

embryonic stem cell cultures. Hum Reprod 22:1231–1238

81. Williams RL et al (1988) Myeloid leukaemia inhibitory factor maintains the developmental potential of embryonic stem cells. Nature 336:684–687
82. Smith AG et al (1988) Inhibition of pluripotential embryonic stem cell differentiation by purified polypeptides. Nature 336:688–690
83. Niwa H et al (1998) Self-renewal of pluripotent embryonic stem cells is mediated via activation of STAT3. Genes Dev 12:2048–2060
84. Humphrey RK et al (2004) Maintenance of pluripotency in human embryonic stem cells is STAT3 independent. Stem Cells 22:522–530
85. Sato N et al (2003) Maintenance of pluripotency in human and mouse embryonic stem cells through activation of Wnt signaling by a pharmacological GSK-3-specific inhibitor. Nat Med 10(1):55–63
86. Levenstein ME et al (2006) Basic fibroblast growth factor support of human embryonic stem cell self-renewal. Stem Cells 24: 568–574
87. James D et al (2005) TGFbeta/activin/nodal signaling is necessary for the maintenance of pluripotency in human embryonic stem cells. Development 132:1273–1282
88. Amit M et al (2000) Clonally derived human embryonic stem cell lines maintain pluripotency and proliferative potential for prolonged periods of culture. Dev Biol 2000(227): 271–278
89. Park SP et al (2004) Establishment of human embryonic stem cell lines from frozen-thawed blastocysts using STO cell feeder layers. Hum Reprod 19:676–684
90. Choo A et al (2006) Immortalized feeders for the scale-up of human embryonic stem cells in feeder and feeder-free conditions. J Biotechnol 122:130–141
91. Martin MJ et al (2005) Human embryonic stem cells express an immunogenic nonhuman sialic acid. Nat Med 11:228–232
92. Amit M et al (2005) No evidence for infection of human embryonic stem cells by feeder cell-derived murine leukemia viruses. Stem Cells 23:761–771
93. Skottman H, Hovatta O (2006) Culture conditions for human embryonic stem cells. Reproduction 132:691–698
94. Mallon BS et al (2006) Toward xeno-free culture of human embryonic stem cells. Int J Biochem Cell Biol 38:1063–1075
95. Amit M et al (2003) Human feeder layers for human embryonic stem cells. Biol Reprod 68:2150–2156
96. Richards M et al (2003) Comparative evaluation of various human feeders for prolonged undifferentiated growth of human embryonic stem cells. Stem Cells 21:546–556
97. Cheng L et al (2003) Human adult marrow cells support prolonged expansion of human embryonic stem cells in culture. Stem Cells 21:131–142
98. Rosler ES et al (2004) Long-term culture of human embryonic stem cells in feeder-free conditions. Dev Dyn 229:259–274
99. Stojkovic P et al (2005) An autogeneic feeder cell system that efficiently supports growth of undifferentiated human embryonic stem cells. Stem Cells 23:306–314
100. Hovatta O et al (2003) A culture system using human foreskin fibroblasts as feeder cells allows production of human embryonic stem cells. Hum Reprod 18:1404–1409
101. Genbacev O et al (2005) Serum-free derivation of human embryonic stem cell lines on human placental fibroblast feeders. Fertil Steril 83:1517–1529
102. Lee JB et al (2005) Establishment and maintenance of human embryonic stem cell lines on human feeder cells derived from uterine endometrium under serum-free condition. Biol Reprod 72:42–49
103. Xu C et al (2004) Immortalized fibroblast-like cells derived from human embryonic stem cells support undifferentiated cell growth. Stem Cells 22:972–980
104. Richards M et al (2002) Human feeders support prolonged undifferentiated growth of human inner cell masses and embryonic stem cells. Nat Biotechnol 20:933–936
105. Wang Q et al (2005) Derivation and growing human embryonic stem cells on feeders derived from themselves. Stem Cells 23:1221–1227
106. Amit M et al (2004) Feeder layer- and serum-free culture of human embryonic stem cells. Biol Reprod 70:837–845
107. Koivisto H et al (2004) Cultures of human embryonic stem cells: serum replacement medium or serum-containing media and the effect of basic fibroblast growth factor. Reprod Biomed Online 9:330–337
108. Carpenter MK et al (2004) Properties of four human embryonic stem cell lines maintained in a feeder-free culture system. Dev Dyn 229: 243–258
109. Xu C et al (2001) Feeder-free growth of undifferentiated human embryonic stem cells. Nat Biotechnol 19:971–974
110. Carpenter M et al (2003) Characterization and differentiation of human embryonic stem cells. Cloning Stem Cells 5:79–88

111. Kleinman HK et al (1982) Isolation and characterization of type IV procollagen, laminin, and heparan sulfate proteoglycan from the EHS sarcoma. Biochemistry 21: 6188–6193
112. Xu RH et al (2005) Basic FGF and suppression of BMP signaling sustain undifferentiated proliferation of human ES cells. Nat Methods 2:185–190
113. Xu C et al (2005) Basic fibroblast growth factor supports undifferentiated human embryonic stem cell growth without conditioned medium. Stem Cells 23:315–323
114. Wang G et al (2005) Noggin and bFGF cooperate to maintain the pluripotency of human embryonic stem cells in the absence of feeder layers. Biochem Biophys Res Commun 330: 934–942
115. Li Y et al (2005) Expansion of human embryonic stem cells in defined serum-free medium devoid of animal-derived products. Biotechnol Bioeng 91:688–698
116. Ludwig TE et al (2006) Derivation of human embryonic stem cells in defined conditions. Nat Biotechnol 24:185–187
117. Stojkovic P et al (2005) Human-serum matrix supports undifferentiated growth of human embryonic stem cells. Stem Cells 23:895–902
118. Peiffer I et al (2010) Optimization of physiological xenofree molecularly defined media and matrices to maintain human embryonic stem cell pluripotency. Methods Mol Biol 584:97–108
119. Wagner KE, Vemuri MC (2010) Serum-free and feeder-free culture expansion of human embryonic stem cells. Methods Mol Biol 584:109–119
120. Chen G et al (2011) Chemically defined conditions for human iPSC derivation and culture. Nat Protoc 8:424–429
121. Yim EK, Leong KW (2005) Proliferation and differentiation of human embryonic germ cell derivatives in bioactive polymeric fibrous scaffold. J Biomater Sci Polym Ed 16:1193–1217
122. Kroupova J et al (2006) Functional polymer hydrogels for embryonic stem cell support. J Biomed Mater Res B Appl Biomater 76:315–325
123. Elisseeff J et al (2005) Advances in skeletal tissue engineering with hydrogels. Orthod Craniofac Res 8:150–161
124. Inzunza J et al (2005) Derivation of human embryonic stem cell lines in serum replacement medium using postnatal human fibroblasts as feeder cells. Stem Cells 23:544–549
125. Hong-mei P, Gui-an C (2006) Serum-free medium cultivation to improve efficacy in establishment of human embryonic stem cell lines. Hum Reprod 21:217–222
126. Chen HF et al (2007) Derivation, characterization and differentiation of human embryonic stem cells: comparing serum-containing versus serum-free media and evidence of germ cell differentiation. Hum Reprod 22: 567–577
127. Klimanskaya I et al (2005) Human embryonic stem cells derived without feeder cells. Lancet 365:1636–1641
128. Draper JS et al (2004) Recurrent gain of chromosomes 17q and 12 in cultured human embryonic stem cells. Nat Biotechnol 22: 53–54
129. Fletcher JM et al (2006) Variations in humanized and defined culture conditions supporting derivation of new human embryonic stem cell lines. Cloning Stem Cells 8:319–334
130. Prowse AB et al (2005) A proteome analysis of conditioned media from human neonatal fibroblasts used in the maintenance of human embryonic stem cells. Proteomics 5:978–989
131. Chin AC et al (2007) Identification of proteins from feeder conditioned medium that support human embryonic stem cells. J Biotechnol 130:320–328
132. Lim JW, Bodnar A (2002) Proteome analysis of conditioned medium from mouse embryonic fibroblast feeder layers which support the growth of human embryonic stem cells. Proteomics 2:1187–1203
133. Stewart MH et al (2006) Clonal isolation of hESCs reveals heterogeneity within the pluripotent stem cell compartment. Nat Methods 3:807–815
134. Lauss M et al (2005) Single inner cell masses yield embryonic stem cell lines differing in lifr expression and their developmental potential. Biochem Biophys Res Commun 331:1577–1586
135. Hewitt Z et al (2006) Fluorescence-activated single cell sorting of human embryonic stem cells. Cloning Stem Cells 8:225–234
136. Sidhu KS, Tuch BE (2006) Derivation of three clones from human embryonic stem cell lines by FACS sorting and their characterization. Stem Cells Dev 15:61–69
137. Wong RC et al (2006) Gap junctions modulate apoptosis and colony growth of human embryonic stem cells maintained in a serum-free system. Biochem Biophys Res Commun 344:181–188
138. Forsyth NR et al (2006) Physiologic oxygen enhances human embryonic stem cell clonal recovery and reduces chromosomal abnormalities. Cloning Stem Cells 8:16–23

139. Damoiseaux R et al (2009) Integrated chemical genomics reveals modifiers of survival in human embryonic stem cells. Stem Cells 27:533–542
140. Watanabe K et al (2007) A ROCK inhibitor permits survival of dissociated human embryonic stem cells. Nat Biotechnol 25:681–686
141. Zweigerdt R et al (2011) Scalable expansion of human pluripotent stem cells in suspension culture. Nat Protoc 6:689–700
142. Amit M et al (2010) Suspension culture of undifferentiated human embryonic and induced pluripotent stem cells. Stem Cell Rev Rep 6:248–259
143. Steiner D et al (2010) Derivation, propagation and controlled differentiation of human embryonic stem cells in suspension. Nat Biotechnol 28:361–364
144. Krawetz R et al (2010) Large-scale expansion of pluripotent human embryonic stem cells in stirred-suspension bioreactors. Tissue Eng Part C Methods 16:573–582
145. Singh H et al (2010) Up-scaling single cell-inoculated suspension culture of human embryonic stem cells. Stem Cell Res 4: 165–179
146. Oh SKW et al (2009) Long-term microcarrier suspension cultures of human embryonic stem cells. Stem Cell Res 2:219–230
147. Chen AKL et al (2011) Critical microcarrier properties affecting the expansion of undifferentiated human embryonic stem cells. Stem Cell Res 7:97–111
148. Thomas RJ et al (2009) Automated, scalable culture of human embryonic stem cells in feeder-free conditions. Biotechnol Bioeng 102:1636–1644
149. Baker DE et al (2007) Adaptation to culture of human embryonic stem cells and oncogenesis in vivo. Nat Biotechnol 25:207–215
150. Longo L et al (1997) The chromosome make-up of mouse embryonic stem cells is predictive of somatic and germ cell chimaerism. Transgenic Res 6:321–328
151. Liu X et al (1997) Trisomy eight in ES cells is a common potential problem in gene targeting and interferes with germ line transmission. Dev Dyn 209:85–91
152. Inzunza J et al (2004) Comparative genomic hybridization and karyotyping of human embryonic stem cells reveals the occurrence of an isodicentric X chromosome after long-term cultivation. Mol Hum Reprod 10:461–466
153. Wilton L (2002) Preimplantation genetic diagnosis for aneuploidy screening in early human embryos: a review. Prenat Diagn 22: 512–518
154. Atkin NB, Baker MC (1982) Specific chromosome change, i(12p), in testicular tumours? Lancet 2:1349
155. Skotheim RI et al (2002) New insights into testicular germ cell tumorigenesis from gene expression profiling. Cancer Res 62: 2359–2364
156. Korkola JE et al (2006) Down-regulation of stem cell genes, including those in a 200-kb gene cluster at 12p13.31, is associated with in vivo differentiation of human male germ cell tumors. Cancer Res 66:820–827
157. Mitalipova MM et al (2005) Preserving the genetic integrity of human embryonic stem cells. Nat Biotechnol 23:19–20
158. Catalina P et al (2008) Human ESCs predisposition to karyotypic instability: is a matter of culture adaptation or differential vulnerability among hESC lines due to inherent properties? Mol Cancer 7:76
159. Ezashi T et al (2005) Low O2 tensions and the prevention of differentiation of hES cells. Proc Natl Acad Sci U S A 102:4783–4788
160. Brivanlou AH et al (2003) Stem cells. Setting standards for human embryonic stem cells. Science 300:913–916
161. Adewumi O et al (2007) Characterization of human embryonic stem cell lines by the International Stem Cell Initiative. Nat Biotechnol 25:803–816
162. Spagnoli FM, Hemmati-Brivanlou A (2006) Guiding embryonic stem cells towards differentiation: lessons from molecular embryology. Curr Opin Genet Dev 16(5):469–475
163. Lawrenz B et al (2004) Highly sensitive biosafety model for stem-cell-derived grafts. Cytotherapy 6:212–222
164. Niwa HJ et al (2000) Quantitative expression of Oct-3/4 defines differentiation, dedifferentiation or self-renewal of ES cells. Nat Genet 24:372–376
165. Chambers I et al (2003) Functional expression cloning of Nanog, a pluripotency sustaining factor in embryonic stem cells. Cell 113:643–655
166. Boyer LA et al (2005) Core transcriptional regulatory circuitry in human embryonic stem cells. Cell 122:947–956
167. Babaie Y et al (2007) Analysis of Oct-4-dependent transcriptional networks regulating self-renewal and pluripotency in human embryonic stem cells. Stem Cells 25:500–510
168. Boyer LA et al (2006) Molecular control of pluripotency. Curr Opin Genet Dev 16: 455–462
169. Tang F et al (2010) Tracing the derivation of embryonic stem cells from the inner cell mass

by single-cell RNA-Seq analysis. Cell Stem Cell 6:468–478

170. Lavon N et al (2008) Derivation of euploid human embryonic stem cells from aneuploid embryos. Stem Cells 26:1874–1882
171. Biancotti JC et al (2010) Human embryonic stem cells as models for aneuploid chromosomal syndromes. Stem Cells 28: 1530–1540
172. Zaninovic N et al (2008) "Self-correction" by uniparental disomy (UPD) of chromosomally abnormal embryos does not occur in human embryonic stem cell (hESC) cultures. ESHRE 2008, Barcelona, Spain
173. James D et al (2010) Expansion and maintenance of human embryonic stem cell-derived endothelial cells by TGFbeta inhibition is Id1 dependent. Nat Biotechnol 28:161–166
174. James D et al (2011) Lentiviral transduction and clonal selection of hESCs with endothelial-specific transgenic reporters. Curr Protoc Stem Cell Biol Chapter 1:Unit1F.12

Chapter 7

The Endocrinology of the Menstrual Cycle

Robert L. Barbieri

Abstract

The ovulatory menstrual cycle is the result of the integrated action of the hypothalamus, pituitary, ovary, and endometrium. Like a metronome, the hypothalamus sets the beat for the menstrual cycle by the pulsatile release of gonadotropin-releasing hormone (GnRH). GnRH pulses occur every1–1.5 h in the follicular phase of the cycle and every 2–4 h in the luteal phase of the cycle. Pulsatile GnRH secretion stimulates the pituitary gland to secrete luteinizing hormone (LH) and follicle stimulating hormone (FSH). The pituitary gland translates the tempo set by the hypothalamus into a signal, LH and FSH secretion, that can be understood by the ovarian follicle. The ovarian follicle is composed of three key cells: theca cells, granulosa cells, and the oocyte. In the ovarian follicle, LH stimulates theca cells to produce androstenedione. In granulosa cells from small antral follicles, FSH stimulates the synthesis of aromatase (Cyp19) which catalyzes the conversion of theca-derived androstenedione to estradiol. A critical concentration of estradiol, produced from a large dominant antral follicle, causes positive feedback in the hypothalamus, likely through the kisspeptin system, resulting in an increase in GnRH secretion and an LH surge. The LH surge causes the initiation of the process of ovulation. After ovulation, the follicle is transformed into the corpus luteum, which is stimulated by LH or chorionic gonadotropin (hCG) should pregnancy occur to secrete progesterone. Progesterone prepares the endometrium for implantation of the conceptus. Estradiol stimulates the endometrium to proliferate. Estradiol and progesterone cause the endometrium to become differentiated to a secretory epithelium. During the mid-luteal phase of the cycle, when progesterone production is at its peak, the secretory endometrium is optimally prepared for the implantation of an embryo. A diagrammatic representation of the intricate interactions involved in coordinating the menstrual cycle is provided in Fig. 1.

Key words Menstrual cycle, Estradiol, Progesterone, Kisspeptin, Amphiregulin, Theca cells, Granulosa cells, Oocyte

1 Characteristics of the Human Menstrual Cycle: Cycle Length and Menstrual Flow

Treloar and colleagues reported on menstrual cycle length in 275,000 cycles reported by Caucasian women who had attended the University of Minnesota [1]. In this population, the mean age of menarche was 13 years and the mean age of menopause was 53 years. Menstrual cycle length was most regular in women between

Zev Rosenwaks and Paul M. Wassarman (eds.), *Human Fertility: Methods and Protocols*, Methods in Molecular Biology, vol. 1154, DOI 10.1007/978-1-4939-0659-8_7, © Springer Science+Business Media New York 2014

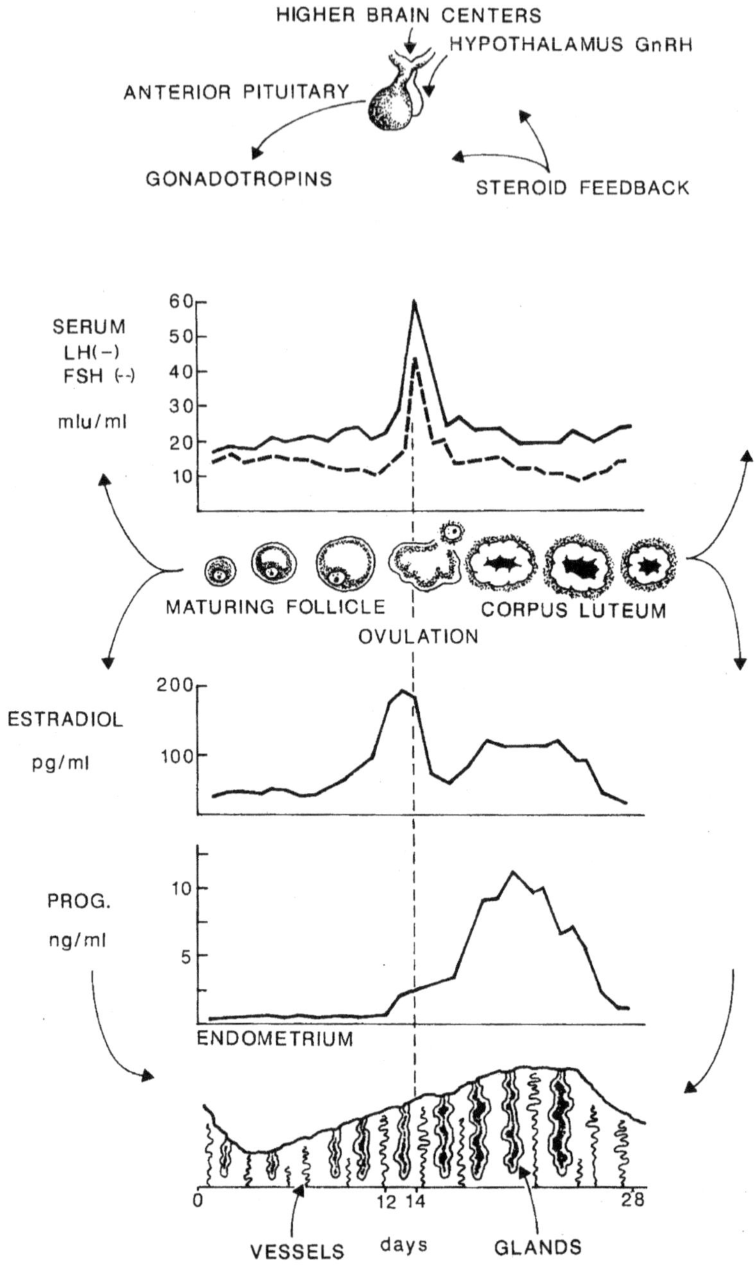

Fig. 1 Diagrammatic representation of the menstrual cycle showing the temporal relationships of the pituitary secretion of follicle stimulating hormone (FSH), luteinizing hormone (LH) with ovarian follicular estradiol production and luteal progesterone and estradiol production. The progressive response of the endometrium to the sequential changes in steroids is portrayed in the *bottom panel*

Table 1
Median and 5 % upper and lower bounds on menstrual interval in days from 275,947 menstrual cycles

Chronologic age in years	Menstrual interval in days		
	5 % Lower bound	Median	5 % Upper bound
17	22	28	40
25	23	28	37
33	22	27	34
41	22	26	32
49	15	27	>80

20 and 40 years of age, and the average menstrual cycle length was 28 days (Table 1). The median menstrual cycle length decreased from 29 days at age 20–27 years to 27 days at age 40 years. This decrease in cycle length was caused by a decrease of 2 days in the length of the follicular phase of the cycle. The length of the luteal phase remained stable. Menstrual cycle length demonstrates significant variability in the 5 years after menarche and the 5 years before menopause. In the 5 years after menarche, cycle length was reported to range from 22 to 45 days. In the 5 years before menopause cycle length ranged from 15 to 55 days. These findings have been replicated in a more contemporary study [2]. Polymenorrhea is present if the cycle length is shorter than 22 days and oligomenorrhea is present if the cycle length is longer than 35 days.

The normal duration range of menstrual bleeding is between 3 and 7 days. Menstrual bleeding lasting less than 3 days is defined as hypomenorrhea. Menstrual bleeding that is longer than 7 days is hypermenorrhea. The normal quantity of blood loss with menses is 80 ml or less. When menstrual bleeding exceed 80 ml in one cycle it is defined as menorrhagia and is correlated with depleted iron stores and an increased risk of anemia.

2 The Hypothalamus: The Metronome of Reproduction

2.1 Neuroanatomy of the Hypothalamus

The hypothalamus is bounded at its anterior border by the optic chiasm and extends caudad to the mamillary bodies. The dorsal portion of the hypothalamus constitutes the floor of the third ventricle and its lateral walls. The base of the hypothalamus, or infundibulum, contains the infundibular stalk and the hypothalamic-hyophyseal portal vascular system perfusing the pituitary gland. The hypothalamus contains approximately 10,000 gonadotropin releasing hormone (GnRH) neurons that drive the menstrual cycle by secreting GnRH in a pulsatile manner [3].

The embryonic precursors of the GnRH neurons develop in the olfactory bulb and migrate to the arcuate and preoptic nuclei. GnRH neuronal development and migration are controlled by anosmin-1, fibroblast growth factor-8 and prokineticin, and their receptors [4, 5]. Abnormal development of the olfactory bulb at an early stage of embryogenesis can result in both anosmia and amenorrhea because of the absence of the GnRH neurons. Kallman syndrome is the combination of anosmia and hypogonadotropic hypogonadism caused by the abnormal embryonic development of the olfactory and GnRH neuronal systems.

2.2 Gonadotropin Releasing Hormone (GnRH)

There are two GnRH molecules termed GnRH I and GnRH II. GnRH I was discovered and sequenced by Schally and Guillemin and plays a dominant role in reproduction [6]. GnRH II is present in most mammalian species, but its function is not clearly established in the human. GNRH II and its receptors may be critically important in checking of the LH surge (There are two GnRH receptors with GnRH I having greater affinity for the GnRH I receptor and GnRH II having greater affinity for the GnRH II receptor. This chapter focuses on GnRH I, hence referred to as GnRH.

GnRH is a decapeptide with a short half-life (5 min) in the circulation. GnRH is secreted into the pituitary portal circulation in neurosecretory bursts. Endocrine information is contained in both the pulse frequency and amplitude of the pulse. The release of GnRH in neurosecretory bursts and the short half-life of GnRH ensures that each pulse is sharp and crisp. During the follicular phase of the cycle, GnRH pulses are characterized by a high frequency, with a pulse occurring every 60–90 min and a small pulse amplitude. During the luteal phase of the cycle, GnRH pulses are characterized by low frequency, with a pulse occurring every 120–240 min and a large pulse amplitude.

The main function of the hypothalamic GnRH neurons is to receive neural signals from the brain and transform these neural signals into an endocrine output, the pulsatile release of GnRH. The arcuate nucleus is a neuroendocrine transducer that converts a series of electrical signals into an endocrine signal. GnRH is released into the pituitary portal circulation in a pulsatile manner. The frequency and amplitude of GnRH pulses are determined by multiple neuroendocrine modulators including kisspeptin, leptin, ovarian steroids, and peptides. These signals provide a link between the environment and reproductive status.

The hypothalamus monitors numerous environmental cues including body composition and body mass, nutritional status, energy expenditure (exercise), stress, and emotional state to determine the frequency and amplitude of GnRH pulses. When environmental conditions are adverse and energy expenditure chronically exceeds calorie intake, GnRH pulse frequency slows and pulse amplitude decreases resulting in low LH and FSH secretion, no growth of a dominant follicle, anovulation, and amenorrhea.

From a teleologic perspective, it is biologically inefficient to ovulate and reproduce if the environment is hostile to the nurturing of a newborn.

GnRH is the hypothalamic factor that is necessary and sufficient to drive the menstrual cycle. In classic experiments in the rhesus monkey, Knobil [7] demonstrated that if the arcuate nucleus is destroyed, resulting in the loss of GnRH secretion, the pituitary gland ceases to secrete LH and FSH, no ovarian follicles grow, and anovulatory amenorrhea ensues. The exogenous replacement of GnRH in hourly intravenous pulses restarts the reproductive system by stimulating the release of LH and FSH, thereby stimulating ovarian follicle growth, subsequent ovulation, and regular menstrual cycles.

Early investigators hypothesized that if GnRH was given at a high dose in a continuous infusion, superovulation could be achieved. Surprisingly, continuous infusion of GnRH results in the cessation of gonadotropin secretion, profound hypoestrogenism, and amenorrhea. These unexpected findings are thought to be due to downregulation and desensitization of the pituitary GnRH receptor and have been exploited by the synthesis of GnRH agonist analogues with a long half-life. These long-acting GnRH analogues acutely stimulate the release of LH and FSH (agonist effect), but when given over an interval of weeks, profoundly suppress LH and FSH, resulting in hypoestrogenism and amenorrhea. Subsequently, GnRH antagonists were synthesized which acutely suppress pituitary LH and FSH secretion. GnRH agonist and antagonist analogues are used for the treatment of breast and prostate cancer, endometriosis, and uterine leiomyoma. They are also used to prevent the LH surge in assisted reproduction cycles involving the administration of high doses of FSH.

2.3 Regulation of GnRH

The factors that control GnRH pulse frequency and amplitude are not fully characterized. Neuroendocrine factors that likely play a major role in regulating GnRH secretion include kisspeptin and leptin. Other factors that likely play an important role in regulating GnRH secretion include the gonadal steroids, estradiol, progesterone, and testosterone.

2.4 Steroid Negative and Positive Feedback on GnRH

Estradiol, progesterone, and testosterone act on both the hypothalamus and pituitary. It is generally believed that gonadal steroid feedback on LH pulse frequency is largely modulated through hypothalamic mechanisms regulating GnRH secretion. In the absence of estradiol, GnRH pulse frequency is generally increased [8] resulting in an increase in LH and FSH secretion. Estradiol replacement reduces GnRH pulse frequency. Progesterone decreases GnRH pulse frequency [9]. The combination of estradiol and progesterone markedly suppresses GnRH pulse frequency and is the cause of the anti-ovulatory contraceptive action of estrogen–progestin steroid contraceptives.

Steroid influences on LH amplitude may occur through steroid action on both the hypothalamus and/or pituitary. Acute administration of estradiol reduces GnRH pulse amplitude and also decreases LH responsiveness to GnRH resulting in pulses of LH with smaller amplitude for any given stimulus of GnRH [10]. However, when estradiol is maintained at levels >200 pg/ml for at least 2 days, LH secretion dramatically increases in response to a fixed amount of GnRH resulting in an "LH surge" [11]. In humans the ability of estradiol to stimulate an LH surge is likely mediated by events at both the hypothalamus and pituitary gland [12, 13].

2.5 Kisspeptin

Estradiol action on GnRH likely occurs "at a distance," not directly on the GnRH neuron. A developing consensus is that estradiol acts on neurons containing ER-alpha receptors that regulate the transcription of *kiss1* the gene that encodes the peptide kisspeptin. In turn, the kisspeptin releasing neuron influences the secretion of GnRH by acting through KISS1 receptors that are present on the GnRH neurons. Kisspeptin stimulates GnRH release. In humans the administration of an intravenous bolus of the C-terminal decapeptide of kisspeptin (amino acids 112–121) results in an immediate LH pulse, with an amplitude that is twice as large as LH pulses observed prior to the administration of kisspeptin. A single dose of kisspeptin stimulates LH secretion for approximately 17 min [14]. In humans loss of function mutations in genes that code for GnRH or the GnRH receptor, kisspeptin or the kisspeptin receptor are associated with hypogonadotropic hypogonadism [15, 16]. The phenotype includes low circulating levels of LH and FSH, low levels of estradiol and progesterone, absent ovulation and amenorrhea.

In the arcuate nucleus of the rat, estradiol inhibits kisspeptin secretion which in turn holds gonadotropin at basal levels. Following castration, the low levels of estradiol cause the arcuate nucleus to release more kisspeptin, thereby increasing GnRH secretion. In the rat, the anteroventral periventricular nucleus (APVP) responds to elevated estradiol levels by increasing kisspeptin secretion, and when triggered by the suprachiasmatic nucleus, the APVP releases its kisspeptin causing an LH surge and ovulation [17]. Mice with kisspeptin knockouts are capable of demonstrating an LH surge in response to chronically elevated estradiol levels, suggesting that there are additional pathways of regulation of the LH surge.

2.6 Exercise and Calorie Intake

The brain is uniquely adapted to integrate environmental signals pertaining to energy expenditure (exercise) and caloric intake in order to determine if the environment will support successful reproduction. Many non-equatorial mammals display seasonal breeding patterns which is based on the onset of anovulation when calorie intake is below energy expenditure. In the female monkey regular ovulatory cycles are observed with routine activity and a steady calorie intake at 300 kcal daily. When calorie intake is maintained at 300 kcal daily, but activity is increased to include 6 miles of additional exercise daily, anovulation and amenorrhea ensue. Increasing caloric intake to up to 600 kcal daily while maintaining 6 miles of exercise daily results in resumption of ovulation and menses. In the exercising monkey, resumption of ovulation can also be initiated without an increase in calorie intake by administering pulses of GnRH [18].

Similar findings have been observed in women. Women who exercise regularly have more anovulatory cycles than women who are sedentary [19]. In a study of sedentary women assigned to exercise plus calorie restriction or exercise plus a eucaloric diet, the exercise plus calorie restriction intervention was associated with greater declines in ovarian steroid production than exercise plus a eucaloric diet [20]. Hormones that help the brain assess the relative levels of calorie intake and energy expenditure include: leptin, insulin, thyroid hormones (thyroxine and triiodothyronine), growth hormone, insulin-like growth factor 1, cholecystokinin, glucagon-like pepetide-1, and ghrelin.

2.7 Leptin

Leptin is secreted by adipocytes and its concentration in the circulation correlates with total body adipocyte mass. Leptin acts on the brain and causes a reduction in appetite. Genetic disruption of leptin secretion or the leptin receptor results in increased food consumption and results in obesity if food is available to the animal. Leptin acts on neurons in the hypothalamus that secrete kisspeptin [21]. In very lean animals, leptin levels are low and it is believed that this results in reduced secretion of kisspeptin, in turn, causing a decrease in GnRH secretion and anovulation. In very lean animals, exogenous leptin administration is capable of increasing GnRH secretion and stimulating ovulation.

Women with hypothalamic amenorrhea (hypothalamic hypogonadism) often have low levels of leptin. Two clinical trials reported that the administration of exogenous leptin or a leptin analogue (metreleptin) to these women, resulted in the resumption of ovulatory menses in some of the subjects [22, 23]. In addition, metreleptin administration for 36 weeks increased free triiodothyronine, IGF1, and osteocalcin [23]. Leptin may exert its effect on GnRH secretion by a direct action on kisspeptin releasing neurons [24].

3 The Pituitary Gland: The Major Link Between the Brain and the Ovary

3.1 Anatomy

As noted above, the brain is uniquely adapted to integrate environmental cues and to attempt reproduction only when the environment is suitable to nurture a newborn. The pituitary gland is the major link between the brain and ovarian function. Pulsatile GnRH secretion stimulates the pituitary to secrete LH and FSH. The pituitary gland translates the tempo set by the hypothalamus into a gonadotropin signal that controls ovarian follicular activity and ovulation.

During embryonic development, the anterior lobe of the pituitary gland is derived from a pinching off of a portion of the oral cavity, Rathke's pouch. The posterior lobe of the pituitary gland consists of neural tissue with direct projections from the hypothalamus. The two lobes come into apposition to form the definitive adenohypophysis attached to the median eminence by the infundibular stalk. The adult pituitary weighs about 0.5 g and measures 10 by 12 by 6 mm. The anterior pituitary comprises 75 % of the total gland volume. The pituitary gland resides within the sphenoid bone cavity known as the sella turcica and it is separated from the cranial cavity by the dura mater, which is penetrated by the pituitary stalk. The hypothalamic-hypophysial portal system connects the brain to the anterior pituitary and transports the hypothalamic releasing hormones (GnRH, TRH, CRH, and growth hormone releasing hormone) and hypothalamic inhibitory hormones (dopamine, somatostatin) to the pituitary. The portal system is a low pressure system, with two capillary beds, one in the hypothalamus and one in the anterior pituitary.

The anterior lobe contains unique cell types that secrete gonadotropins (LH and FSH), thyrotropin, growth hormone, corticotropin, or prolactin. Gonadotropes likely differentiate in response to cellular expression of steroidogenic factor 1 (SF-1), a zinc finger nuclear receptor, and GATA-2. It is believed that one pituitary cell secretes both LH and FSH. Approximately 5–15 % of the cells in the anterior pituitary are gonadotropes. The pituitary gonadotrope receives signals from the hypothalamus in the form of pulses of GnRH. In response, the pituitary synthesizes and secretes FSH and LH in pulses that match the GnRH signal. GnRH stimulates multiple pathways in the pituitary gonadotrope, including protein kinase C, mitogen-activated protein kinase, calcium influx, and calcium–calmodulin kinase. The secretion of gonadotropins by the pituitary is also modulated by the negative feedback of steroid hormones, including estradiol progesterone and testosterone and the negative feedback of protein hormones, inhibin A and B.

3.2 Gonadotropins

Gonadotropins are glycoproteins, each composed of two subunits associated by noncovalent bonds and designated alpha and beta. The same 92 amino acid alpha subunit is present in TSH, hCG, LH, and FSH. It is the beta-subunit of each glycoprotein pituitary hormone that confers biological and immunological specificity. Gonadotropins are highly elongated molecules with intertwining of the alpha- and beta-subunit. The gonadotropins are conformationally related to the cystine knot growth factor family that includes transforming growth factor-beta, activin, and nerve growth factor. Differing carbohydrate moieties create isoforms of LH and FSH. Sulfonated forms of LH are more rapidly cleared by the liver than sialyated molecules.

Although GnRH stimulates both LH and FSH release, the two gonadotropins are differentially released throughout the menstrual cycle. Differential release is modulated by GnRH pulse frequency (LH secretion favored by moderately rapid pulse frequencies and FSH secretion by slow pulse frequencies), inhibin A and B (which are secreted from the ovary and inhibit FSH secretion), and the activins (which stimulate FSH secretion). Much of the negative feedback effects of estradiol and progesterone are mediated through changes in the hypothalamic release of GnRH rather than directly at the pituitary level. The positive feedback effects of estradiol are likely mediated at the level of both the pituitary and the hypothalamus.

LH, FSH, and hCG stimulate target cells by binding to G-protein coupled receptors. The LH receptor binds both LH and hCG. The LH receptor is expressed in thecal, granulosa, and corpus luteal cells of the ovary and Leydig cells of the testis. The FSH receptor is expressed in granulosa cells of the ovary and the Sertoli cells of the testis.

An overview of the changes in LH, FSH, total inhibin, estradiol, and progesterone throughout the menstrual cycle is provided in Fig. 2. During menses, estradiol, progesterone, and inhibin levels are low. In the absence of significant negative feedback from ovarian steroids and proteins, the hypothalamic-pituitary unit increases the secretion of GnRH, LH, and FSH, thereby stimulating the growth of a cohort of small follicles. Among the cohort of growing small follicles, one follicle will achieve more rapid growth and secrete increasing quantities of estradiol and inhibin A and B. This will result in the suppression of FSH and the cessation of growth in all but the largest, dominant follicle. A rising level of estradiol will eventually trigger an LH surge initiating the ovulatory process. Within 12 h of the initiation of the LH surge, progesterone levels start to rise due to luteinization of the follicle and estradiol levels begin to fall. During the luteal phase of the cycle, FSH levels are low, preventing the growth of a new cohort of the follicles, LH pulse frequency slows considerably and progesterone levels peak during the mid-luteal phase. If pregnancy does not occur, the corpus luteum begins to undergo luteolysis with a decrease in inhibin, estradiol, and progesterone production and a subsequent rise in pituitary secretion of LH and FSH, initiating a new round of follicle growth.

4 The Ovary

4.1 Embryonic Development

During embryonic development, primordial germ cells migrate from the yolk sac to the mesonephric ridge, where the definitive ovary develops. Germ cell development is under the influence of bone morphogenetic proteins (BMP)-4, -8b, and -2 [25].

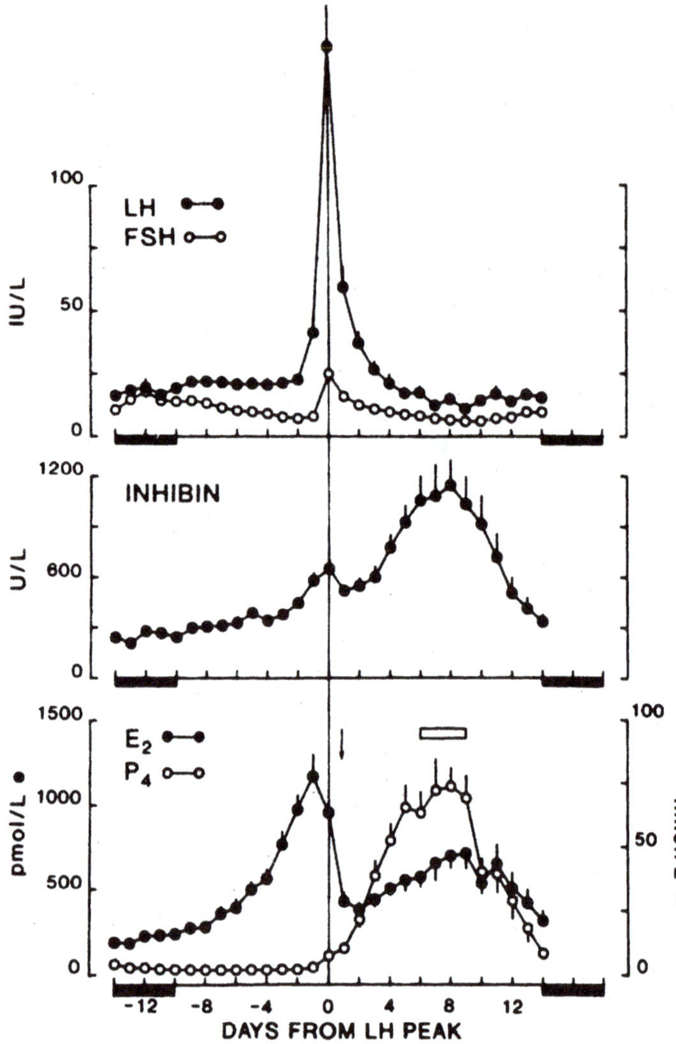

Fig. 2 Overview of the changes in luteinizing hormone (LH), follicle stimulating hormone (FSH), total inhibin, estradiol (E2), and progesterone (P4) throughout the menstrual cycle

The primitive germs cells, or oogonia, divide by mitosis and increase their numbers to several million until about 6 months of intrauterine development when mitotic division stops. Starting at 8–13 weeks of gestation, oogonia continually enter meiosis until they are all converted to oocytes which are arrested in the dictyate stage. The process is completed by 6 months after birth. Oocytes remain arrested in the meiotic prophase until stimulated to resume meiosis during the LH surge of each ovulatory cycle. The number of oocytes continually declines during life. At puberty only 300,000 oocytes remain of which only several hundred will be ovulated in a woman's lifetime.

4.2 Follicle Development

The primordial follicle represents the earliest stage of follicular development. The primordial follicle consists of the oocyte surrounded by a basal lamina and a few spindle shaped (pregranulosa) cells. The primordial follicle develops into a primary follicle when the oocyte is surrounded by cuboidal granulosa cells. In the primary follicle, cuboidal granulosa cells secrete glycoproteins that surround the oocyte in a shell, the zona pellucida. These developmental steps can occur in the absence of gonadotropin stimulation. The primary follicle becomes a secondary follicle when the granulosa cells divide and develop into multiple layers surrounding the oocyte. Secondary follicles have a band of fibroblast like cells, the theca interna, that surround the granulosa cells. During each menstrual cycle, secondary follicles are recruited into a rapid growth phase, and in ovulatory cycles, one follicle will gain dominance and go on to ovulate [26].

Both LH and FSH are necessary for the secondary follicle to fully develop into a dominant preovulatory follicle. LH stimulates the thecal cells to divide and produce androgens, which are converted to estrogens by the granulosa cells. FSH stimulates the granulosa cells to divide and to synthesize the enzymes that convert thecal androgens to estrogens, including the aromatase enzyme (Cyp19). At the beginning of the menstrual cycle (day 1 of the cycle is the first day of menstrual bleeding), only a few small antral follicles (2–6 mm in diameter) are present in each ovary. One antral follicle will gain dominance and start to grow rapidly, achieving a preovulatory diameter of approximately 20–25 mm. The granulosa cell complement of a large follicle is in the range of 50 million cells, a reflection of the rapid cell division occurring during this phase of follicle growth. Advanced in vitro systems permit the maturation of preantral follicles to antral follicles with an associated mature oocyte [27]. These experimental systems will advance our understanding of the factors involved in follicle growth and maturation.

4.3 The Two Cell Theory and Steroidogenesis

The follicle has three main components: theca and granulosa cells and the oocyte. Theca cells express LH receptors and secrete androgens, and a small amount of progesterone, in response to LH stimulation. Granulosa cells from small follicles express FSH receptors and respond to FSH stimulation by synthesizing aromatase, the enzyme which converts thecal androstenedione to estrone. Granulosa cells from large follicles also have LH receptors, in preparation for responding to the LH surge [28].

There is a delicate balance between LH and FSH stimulation of the thecal and granulosa cell compartments. In polycystic ovary syndrome, excessive pituitary secretion of LH (evidenced by an elevated ratio of LH to FSH) results in excess stimulation of androstenedione production from the theca and insufficient conversion of the androstenedione to estrone. Consequently the follicular microenvironment is characterized by excessive androstenedione and low estradiol concentration in the follicular fluid. These androgen dominant follicles contain a suboptimal number of granulosa cells for their size and an oocyte that has difficulty resuming meiosis.

Follicle growth is directly dependent on FSH stimulation. In women with amenorrhea due to hypothalamic hypogonadism (insufficient calorie intake, excessive exercise, excessive stress), the pituitary secretion of FSH is very low. In the absence of FSH stimulation, the small secondary follicles do not grow and no dominant follicle develops. In the absence of a dominant follicle, estrogen secretion is insufficient to trigger an LH surge, thereby resulting in anovulation and amenorrhea.

In the ovary, the key steroidogenic pathways involve converting cholesterol to progesterone, androgens, and estrogens. The cyclopentanophenanthrene-ring structure is the basic carbon skeleton for all steroid hormones. There are three physiologically important gonadal steroid nuclei: estrane (18 carbon atoms), androstane (19 carbons atoms), and pregnane (21 carbon atoms). Outside of pregnancy, the quantitatively important sources of steroid production are the ovaries, the testes, and the adrenals. Each of these glands makes a pattern of steroid hormones unique to the gland. The ovary secretes large quantities of estradiol and progesterone. The testes secretes testosterone. The adrenal secretes cortisol and aldosterone. In addition, each gland produces some common core steroids, such as progesterone, 17-hydroxyprogesterone, and androstenedione (Fig. 3).

Cholesterol is the parent steroid from which all gonadal steroids, glucocorticoids, and mineralocorticoids are derived. Circulating lipoproteins provide significant quantities of cholesterol for metabolism to gonadal steroids. Once the lipoprotein packet is metabolized within the gonadal cell, cholesterol is transported to the mitochondira where it is converted to pregnenolone by the cholesterol cleavage enzyme. Pregnenolone is then metabolized to the other core steroids (progesterone, 17-hdroxyprogesterone, and androstenedione) by cyp17 and 3beta hydroxysteroid dehydrogenase. In turn, progesterone can either be secreted or, in the adrenal, metabolized to aldosterone. 17-Hydroxyprogesterone can be secreted or metabolized in the adrenal to cortisol. Androstenedione can be secreted or metabolized to testosterone or estradiol (Fig. 3).

4.4 The Oocyte

The oocyte is maintained in prophase of meiosis I until the LH surge signals the oocyte to resume meiosis. The factors that cause and maintain the arrest of meiosis are not well characterized. cAMP may play an important role in the arrest of meiosis and G-protein coupled receptors likely maintain intracellular cAMP concentrations in the oocyte. Oocytes in mature antral follicles resume meiotic maturation following exposure to the LH surge. One potential mechanism for this effect is that LH induces the

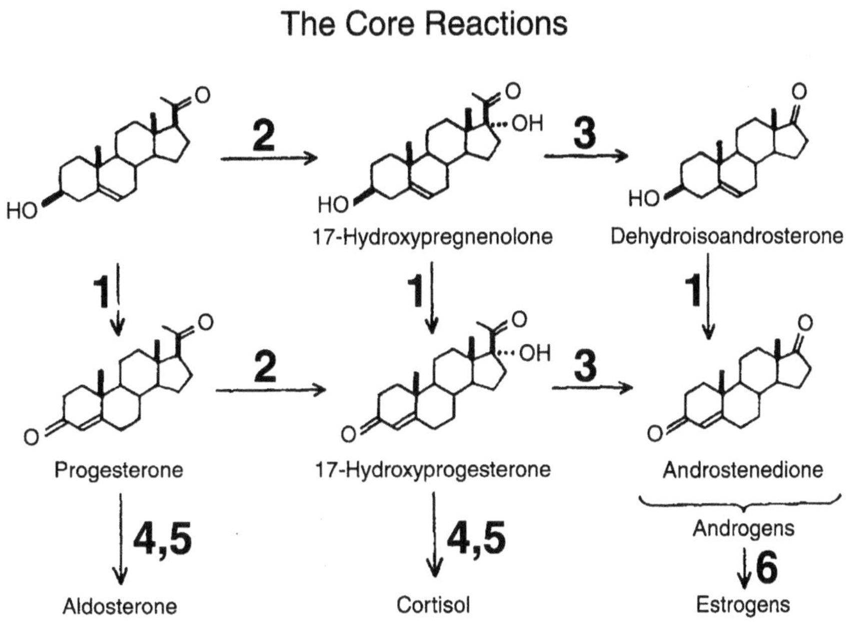

Fig. 3 The core reactions in steroidogenesis. The ovary, testis, and the adrenal produce six common core steroids. The core delta-5 steroids are pregnenolone, 17-alpha hydroxypregnenolone, and dehydroepiandrosterone. The core delta-4 steroids are progesterone, 17-alpha hydroxyprogesterone, and androstenedione. The core steroids are important precursors for the production of aldosterone, cortisol, testosterone, and estradiol. Enzyme1 is 3-beta hydroxysteroid dehydrogenase isomerase. Enzyme 2 is 17-alpha hydroxylase CYP17. Enzyme 3 is the 17, 20-layse. Enzyme 4 is 21-hydroxyase. Enzyme 5 is 11-beta hydroxylase. Enzyme 6 is aromatase

secretion of EGF-like growth factors, including amphiregulin, that stimulate the oocyte to resume meiosis (see below section on amphiregulin). The resumption of meiosis by the oocyte can be observed with the light microscope and involves the sequential loss of the nuclear envelope (germinal vesicle breakdown) followed by the extrusion of one half of the chromosomes in the first polar body. The oocyte proceeds to metaphase of meiosis II and becomes arrested in metaphase II before it is physically released from the ovary. Completion of meiosis occurs following fertilization.

The oocyte secretes proteins, including growth-differentiation factor-9 (GDF-9) and bone-morphogenic protein-15 (BMP-15), that influence granulosa cell function. Growth differentiation factor-9 (GDF-9) is a member of the TGF-beta superfamily and is highly expressed in oocytes. In rodent models, GDF-9 stimulates granulosa cell differentiation and the secretion of multiple granulosa-derived glycoproteins that form the cumulus. GDF-9 suppresses granulosa cell LH receptor expression which reduces the risk of premature luteinization of the cumulus cells. BMP-15 is a member of the TGF-beta superfamily and produced in high concentrations by the oocyte. It is structurally related to GDF-9. In sheep a point mutation in the receptor for BMP-15 is associated with an increased ovulation rate [29, 30].

4.5 Follicle Microenvironment

An important feature of the ovarian follicle is that granulosa cells and oocytes are separated from the systemic circulation by a basement membrane between the theca cells and the granulosa cells. All proteins entering the follicle must pass through the basement membrane to reach the inner follicle. This feature permits two adjacent follicles to develop unique intrafollicular microenvironments. Many studies suggest that follicles that have a mircroenvironment with high concentrations of FSH and estradiol are the follicles that are most likely to grow rapidly, gain dominance, and ovulate. This is not surprising because FSH and estradiol are potent stimulators of granulosa cell mitosis. Follicles that contain low concentrations of FSH and estradiol grow poorly and eventually undergo atresia. The follicle destined to ovulate can be identified by the following characteristics: it has an optimal number of granulosa cells for its size and it contains large amounts of estradiol and low amounts of the androgens, androstenedione, and testosterone. The follicles destined for atresia have low numbers of granulosa cells for their size and more androgens than estrogens in the follicular fluid [31].

The development of the dominant follicle creates an environment in which the oocyte matures and is ready to resume meiosis. Sufficient estradiol is secreted from its granulosa cells to trigger the LH surge which will cause the follicle to ovulate when it is ripe. The estradiol produced by the same follicle induces endometrial proliferation which prepares the uterus for embryo implantation. It has been demonstrated that if the preovulatory follicle is surgically destroyed, the circulating concentration of estradiol drops quickly, the FSH level rises and it takes about 14 days for another follicle to develop to the same stage [32].

Proteomic studies of human follicular fluid indicate the presence of approximately 400 unique proteins. Computational analysis suggests that these proteins are involved in the following pathways: (1) steroidogenesis, (2) cell to cell signaling, (3) antioxidative homeostasis, (4) interleukin signaling, (5) liver X receptor/retinoid X receptor activation, (6) insulin-like growth factor signaling, and (7) lipid metabolism [33]. With regards to the insulin-like growth factor signaling system, the proteomic studies demonstrated the presence of IGF-2 and the IGF-binding proteins 2, 3, 5, and 7.

4.6 Selection and Dominance

At the beginning of the follicular phase of the cycle there are approximately five antral follicles (range 0–15 follicles) measuring approximately 4 mm in diameter (range 2–9 mm in diameter) present in each ovary [34]. By day 12 of the cycle one follicle of approximately 18–20 mm in diameter dominates one ovary and is the follicle destined to ovulate. The process by which one small antral follicle is selected from amongst its peers to develop into the large preovulatory follicle is called selection. The mechanisms subserving selection are not full characterized, but the status of the follicle microenvironment is crucial to the process of selection.

On day 4 of the follicular phase no single follicle has achieved dominance. By day 8 of the cycle a dominant follicle has emerged. Once a dominant follicle is established, no new large antral follicles will grow until the dominant follicle undergoes ovulation, and the resulting corpus luteum undergoes atresia (if pregnancy is not established). The dominant follicle prevents the growth of other new large follicles by secreting inhibin-B inhibin-A and estradiol, which markedly suppress FSH secretion, thereby blocking the contemporaneous growth of a new large follicle. The processes of selection and dominance help ensure that for primates, on average, only one follicle ovulates each cycle. From a teleological perspective, the tremendous energy demands of a primate pregnancy, birth, and feeding of a newborn make it preferable to make singleton pregnancy the norm.

During follicular development the surface area of the follicle doubles approximately 19 times [35]. The increase in follicle size is due both to an increase in granulosa and theca cell number and a marked increase in follicle fluid. The aquaporins (AQPs) are a family of cell membrane proteins that allow water to pass through cell membranes at a rate approximately 100 times greater than the passage of water by diffusion. The expression of AQPs increase in granulosa and theca cells as the follicle grows, with a marked increase in AQP2 and AQP3 during the early ovulatory phase and an increase in AQP1 in the late ovulatory phase [36]. The increase in AQP expression likely mediates the fluid increase in the follicle as it gains dominance and prepares for ovulation.

4.7 Ovulation

The timing of ovulation is determined by the dominant ovarian follicle. When the dominant ovarian follicle produces enough estradiol to sustain levels in the range of 200–300 pg/ml for 48 h the hypothalamus responds by increasing kisspeptin and GnRH secretion resulting in a surge in pituitary secretion of LH and FSH. The gonadotropin surge is characterized by an increase in LH pulse frequency and amplitude resulting in a marked increase in serum LH. Before the LH surge, estradiol concentration rises rapidly, with a doubling time of approximately 60 h, concomitant with the growth of the follicle. The mean duration of the LH surge is approximately 48 h with an ascending limb of 14 h, a plateau of 14 h, and descending limb of approximately 20 h. The gonadotropin surge is characterized by smaller increases in FSH, with a similar time course.

The LH surge stimulates four events in the ovary that result in ovulation: (1) an increase in intrafollicular proteolytic enzymes (e.g., plasmin) which destroy the basement membrane of the follicle and allows follicular rupture, (2) luteinization of the granulosa and theca cells which results in a marked increase in progesterone production, (3) resumption of oocyte meiosis in preparation for fertilization, and (4) the growth of blood vessels into the follicle which prepares it to become a corpus luteum. The beginning of the LH surge precedes ovulation by about 36–44 h.

4.8 Amphiregulin

Substantial evidence from multiple species indicates that the LH surge stimulates cAMP signaling pathways and the expression of EGF-like growth factors, including amphiregulin, epiregulin, and betacellulin. In turn, amphiregulin activates EGF receptors within the follicle which trigger oocyte nuclear maturation and prepares the oocyte for fertilization [37]. In humans, amphiregulin is the dominant EGF-like molecule. Amphiregulin is not present in human follicle before the LH surge, but levels increase markedly after the onset of the LH surge. Amphiregulin appears to stimulate cumulus expansion and oocyte maturation [38]. Human germinal vesicle oocytes obtained from small follicles and exposed to amphiregulin or epiregulin in vitro are much more likely to progress to metaphase II than control oocytes. Supplementation of in vitro maturation medium with amphiregulin or epiregulin may be an effective intervention to enhance oocyte competence [39].

4.9 The Corpus Luteum

The main purpose of the corpus luteum is to secrete progesterone to prepare the endometrium for embryo implantation. Corpus luteum secreted progesterone maintains a developing pregnancy until the placenta becomes the main source of progesterone, at 7–8 menstrual weeks of gestation [40]. During the luteal phase, pituitary LH stimulates the corpus luteum to make progesterone. During early pregnancy hCG secreted from the conceptus stimulates the corpus luteum to make progesterone and support the growth of the conceptus within the endometrium [41].

The corpus luteum is derived from both theca and granulosa cells. In the corpus luteum the small cells are thought to be derived from the theca and the large cells from the granulosa cells. Both cell types produce progesterone. Theca-derived luteal cells produce androgen precursors that are aromatized to estradiol by the granulosa-derived luteal cells [42].

The luteal cells obtain cholesterol for progesterone synthesis from circulating lipoprotein cholesterol through endocytosis. Cholesterol is transported to the inner membrane of the mitochondria and converted to pregnenolone by the P450scc, cholesterol side chain cleavage enzyme. The rate limiting step in progesterone production by the corpus luteum appears to be the translocation of cholesterol from the outer mitochondrial membrane to the inner membrane where the P450scc is located [43]. Steroidogenic acute regulatory protein (StAR) is essential to the efficient translocation of cholesterol to the inner mitochondrial membrane. StAR synthesis is controlled by LH, progesterone, and three transcription regulatory factors: steroidogenic-factor 1, GATA-4, and CCAAT/enhancer-binding protein beta. Prior to the LH surge, StAR is present in very low levels in the human granulosa cell, consistent with the low level of progesterone production. In the periovulatory interval, the theca cells contain high concentrations of StAR, suggesting that the increase in progesterone secretion observed at the time of the LH surge is derived from the theca cells [44]. In the non-fertile cycle, the corpus luteum undergoes luteolysis, a process of regression that results in a decrease in progesterone and estradiol secretion. An early molecular event in luteolysis is a decrease in intracellular StAR protein concentration [45].

4.10 Ovarian Regulatory Proteins

Estradiol secreted by the ovary suppresses pituitary gland secretion of FSH. The ovarian granulosa cells also secrete potent protein inhibitors of FSH secretion, inhibin A, and inhibin B. Inhibins are dimeric proteins that contain an alpha-subunit with a disulfide linkage to either a beta-A subunit (inhibin A) or a beta-B subunit (inhibin B). In the follicular phase of the menstrual cycle, the small developing follicles predominantly secrete inhibin B. In the preovulatory and luteal phases of the menstrual cycle, the granulosa cells of the large dominant follicle and the luteinized granulosa cells in the corpus luteum predominantly secrete inhibin A. The feedback of the inhibins on pituitary FSH secretion is critical to the development of a single dominant follicle. Once a dominant follicle is selected, it secretes sufficient quantities of inhibin A and inhibin B to suppress FSH and prevent the growth of other follicles, during that cycle. Of note, in the pituitary gland there is an intraglandular paracrine system whereby the combination of two subunits of inhibin as homodimers (beta-A/beta-A) activin A or beta-B/beta-B (activin B) are capable of stimulating FSH secretion. Follistatin, a protein present in the pituitary and follicular fluid binds the activins, inactivating them, and therefore can act to suppress FSH secretion.

4.11 Anti-Mullerian Hormone

Anti-mullerian hormone (AMH) is a dimeric glycoprotein member of the TGF-beta superfamily. The classical role of AMH is to induce the degeneration of the mullerian ducts during male sexual differentiation. In addition, AMH is produced by granulosa cells of small follicles and acts in the ovary through two receptors, type II which is specific for AMH and type I which is also binds other members of the BMP family. As follicles grow, the antral fluid concentration of AMH decreases, while the concentration of inhibin-B increases [46]. In mice, loss of AMH results in a reduction in the FSH-responsiveness of growing follicles and an increased rate of follicle atresia. In the human loss of all follicles is associated with very low circulating AMH levels. In girls with Turner syndrome, AMH levels are below normal at all ages, from birth to adulthood [47]. AMH levels are relatively stable throughout the menstrual cycle. Unlike FSH, which must be measured during menses (cycle day 2, 3, or 4) to predict the likelihood of fertility treatment success, AMH is a useful predictor of ovarian reserve when measured at any time in the menstrual cycle [48].

In clinical practice, the ovarian follicle pool can be assessed by measuring cycle day 3 FSH or inhibin B, AMH or by counting the number of secondary follicles by transvaginal ultrasound. Of these methods for assessing the ovarian follicle pool, measurement of AMH and ultrasound determination of antral follicle count appear to be superior to the measurement of cycle day 3 FSH or inhibin B [49]. In addition both AMH and antral follicle count are good predictors of the risk of developing ovarian hyperstimulation in response to FSH treatment for infertility [50]. Based on this observation it has been proposed that AMH levels could be used to assign women to optimal regimens of controlled ovarian hyperstimulation based on the AMH assessment of their oocyte pool size [51]. For example, women with normal AMH concentrations might be best treated with moderate doses of gonadotropin, while women with elevated AMH concentrations, suggesting a large ovarian follicle pool, might be best treated with a lower gonadotropin dose to prevent excess ovarian response to FSH [52]. In addition, markers of the size of the oocyte pool (AMH, antral follicle count, age of female) are predictive of the magnitude of the ovarian response to fertility treatments involving ovarian stimulation [53].

4.12 Ovarian Follicle Depletion

An immutable feature of human ovarian biology is a relentless decline in the number of ovarian follicles with aging. Statistical models that have been used to describe the decline include: power, differential, biphasic, Gompertz, and exponential. The power and differential equation models have been reported to best model the rate of follicle loss observed with aging assessed by counting follicles in the ovarian surgical specimens [54]. The basic observation governing follicle loss over time is that a relatively fixed *percentage* of follicles are lost during every time interval. This exponential loss of ovarian follicles is similar to the rate of decay in radioactive materials, over each time period, a fixed percentage of the radioactive atoms decay. Some investigators have reported the presence of ovarian follicle stem cells in murine model systems that could lead to a regeneration of the ovarian follicle pool. Other investigators doubt that ovarian stem cells play an important physiological role in regenerating oocytes during adult life [55].

5 The Endometrium

5.1 Anatomy

The reproductive purpose of the primate menstrual cycle is to generate a single oocyte for fertilization and to simultaneously prepare the endometrium for implantation of an embryo.

The endometrium has classically been divided into two main layers, the upper functionalis layer and the lower basalis layer. The functionalis layer is the layer in which endometrial proliferation, secretory changes, implantation, and menstrual sloughing occur. The functionalis layer consists of a compact zone adjacent to the lumen and a spongy zone abutting the basalis. The basalis lies between the spongy zone and the myometrium and can repopulate all the cells in the functional zone following menstrual sloughing. The basal zone likely contains stem cells [56]. The endometrium is responsive to steroid hormones and local protein factors.

5.2 Endocrinology of the Endometrium

Estradiol stimulates endometrial epithelial cell proliferation, gland growth, and vascularization of the glands. Estradiol increases the endometrial synthesis of its own intracellular receptor, augmenting its effect, and the production of progesterone receptors, preparing the endometrium to respond to the luteal production of progesterone. Estrogen likely achieves many of its effects on the endometrium by stimulating the epithelium and stroma to secrete protein products that create cross-talk between glands and stroma. After ovulation, the endometrium responds to the rapid rise in progesterone with secretory changes in the luminal cells, further gland development, decidualization of the stromal cells, and the development of spiral vessels. If no pregnancy occurs, the decrease of estradiol and progesterone that follows atrophy of the corpus luteum results in the collapse of the endometrial glands, constriction of the blood vessels, and the sloughing of the endometrium.

Protein factors that regulate endometrial growth include WNT7A, Frizzled (FZD) receptors, and Dishevelled (Dsh) [57, 58]. Estradiol stimulates WNT7A production in the luminal epithelium. WNT7A diffuses in a gradient towards the basal layer and stimulates cell proliferation. The sequence of events that subserves WNT7A control of endometrial proliferation includes WNT7A binding to the FZD receptor which phosphorylates the protein Dsh, which turns off glycogen synthase kinase-beta, thereby reducing the degradation of beta-catenin and increasing intracellular beta-catenin concentrations. Beta-catenin is a transcription factor that stimulates cell proliferation [59].

During the luteal phase of the cycle, endometrial cell proliferation is reduced by progesterone stimulation of Dickkopf-1 (DKK-1) which binds to LRP6, blocking the FZD receptor, thereby blocking WNT7A action [60]. This reduces beta-catenin levels and reduces cell proliferation. Knockdown of the progesterone receptor prevents the induction of DKK-1 [61].

Progesterone stabilizes the endometrium and prevents unscheduled bleeding by influencing the production of key proteins. Progesterone blocks stromal cell production of matrix metalloproteinase (MMP) 1, 3, and 9. MMP 1, 3, and 9 degrade extravascular and stromal matrix. By blocking production of the

MMPs, progesterone stabilizes the stromal and vascular supporting matrix. Progesterone also stimulates stromal cell production of tissue factor (TF), a cell surface protein that participates in the extrinsic pathway of coagulation through the binding of activated Factor VII [62]. By stimulating TF production, progesterone helps reduce the risk of unscheduled bleeding from the endometrium. Progesterone also stimulates stroma cell production of plasminogen activator inhibitor 1 (PAI-1). PAI-1 blocks fibrinolysis, thereby stabilizing clots [63].

In anovulatory menstrual cycles, the absence of a progesterone stimulus results in excessive production of MMP 1, 3, and 9, decreased production of TF and PAI-1, thereby leading to unpredictable menstrual bleeding in both timing and amount. In ovulatory cycles with excessively high estradiol and abnormally low progesterone secretion (commonly observed in perimenopausal women), a similar pattern of dysfunctional protein secretion and bleeding may occur. Women with heavy menstrual bleeding, even in ovulatory cycles, have been demonstrated to have elevated endometrial concentrations of tissue plasminogen activator and plasmin, which reduce the stability of the clots. Tranexamic acid is an inhibitor of plasminogen and the plasminogen activator–plasmin complex. Administration of tranexamic acid to women with heavy menstrual bleeding results in the stabilization of fibrin and reduces the amount of menstrual bleeding [64, 65].

5.3 Classical Dating of the Endometrium

Classical dating of the endometrium involves histological observation of changes that occur on 8 dimensions in the functional layer of the endometrium [66]. The endometrium of the lower uterine segment is not as hormonally responsive as the epithelium of the uterine fundus, and it should not be used for histological dating. Three dimensions: gland mitoses, stromal mitoses, and pseudostratification of nuclei are prominent in the follicular phase of the cycle. Three dimensions: basal glandular vacuolation, secretory changes, and stromal edema are prominent in the luteal phase of the cycle. Two dimensions: pseudodecidual reaction and leukocyte infiltration are prominent in the peri-menses phase of the cycle.

In a 28 menstrual day cycle, ovulation typically occurs on cycle day 14. The first day of the menstrual cycle is considered to be the first day of bleeding. Menses typically lasts 4–7 days. The proliferative phase of the menstrual cycle begins at the termination of menses and extends to ovulation at cycle day 14. The postovulatory or secretory phase extends from cycle day 14 to cycle day 28 and the onset of menstrual bleeding.

5.4 Proliferative Phase

Following menses, under the stimulation of estrogen, the endometrium gradually regrows. The early proliferative phase extends from cycle day 4–7. The middle proliferative phase from cycle 8–11 and the late proliferative phase from cycle day 11–14.

In the early proliferative phase, the endometrium is less than 3 mm thick. The glands are short, narrow, tubular, and straight and lined by low columnar cells that have round nuclei near the base. A few mitoses are present in the glands. The endoplasmic reticulum and Golgi apparatus are not well developed. The stroma is compact and has few mitoses. In the middle proliferative phase, the glands are longer and have a slightly curved effect. There is early pseudostratification of the nuclei that appear superimposed in layers. There are numerous mitoses in the glands. In the late proliferative phase, the glands are tortuous as a result of active growth. The luminal cells increase in height and become pseudostratified. There are numerous mitoses and pseudostratification of the nuclei. The stoma is dense and has numerous mitoses.

5.5 Secretory Phase

Ovulation heralds the beginning of the secretory phase. In the secretory phase the endometrium that has been primed by exposure to estradiol differentiates under the influence of progesterone. The first 7 days of the secretory phase are characterized by the following sequential changes: the appearance of basal vacuoles and continuing mitoses in the glands and stroma (for up to 3 days after ovulation), an increase in glandular tortuosity with prominent subnuclear vacuoles, initial accumulation of glycogen-rich vacuoles at the base of the luminal cells and the onset of prominent acidophilic secretions in the gland lumen. In the last 7 days of the secretory phase stomal changes become prominent including the following sequential changes: maximal stromal edema, development of highly coiled spiral arteries, condensation of the stroma around the spiral arterioles, lymphocytic infiltration of the stroma, and focal necrosis of the stoma around the spiral arterioloes. NK lymphocyte cells encircle arterioles and develop cell to cell connections with stromal cells [67].

5.6 Implantation

A detailed discussion of implantation is beyond the scope of this review. Implantation requires the coordinated development of a blastocyst stage embryo and a prepared luteal-phase endometrium [68]. In the human the implantation window is from days 19–24 of an idealized 28-day menstrual cycle. Implantation involves the interaction of microvilli on the surface of the hatched syncytiotrophoblasts and microprotrusions on the apical surface of endometrial glands called pinopodes. Shortly thereafter, the syncytiotrophoblasts penetrate the endometrial epithelium and the endometrium then regrows to cover the site of penetration.

The autocrine, paracrine, and endocrine mechanisms involved in implantation vary significantly among species. Progesterone is essential for implantation and pregnancy maintenance in most mammals. In the murine model, leukemia inhibitory factor (LIF), an interleukin 6-like cytokine, plays an important role in implantation. In LIF deficient mice embryos do not implant in the endometrium [69].

In the mouse, treatment with an LIF-antagonist blocks embryo implantation [70]. Similarly, interleukin-11 (IL11) appears to play a key role in mouse implantation. In female mice lacking the receptor for interleukin-11 the endometrium is not receptive for embryo implantation [71]. In humans LIF is primarily expressed in the mid-luteal phase at the apical pole of endometrial glandular cells [72]. In both the human and mouse, endometrial expression of CD98, a cell surface ligand that is induced by LIF, epidermal growth factor, estradiol, and hCG appears to play a role in regulating endometrial receptivity [73]. CD98 may associate with galectins, which are capable of binding carbohydrate moieties.

Perpetuation of a species requires efficient and successful reproduction. In primates, successful reproduction is dependent on an ovulatory menstrual cycle. In turn an ovulatory menstrual cycle requires the intricate integration of the hypothalamus, pituitary, ovary, and endometrium. The hypothalamus is the conductor of the menstrual cycle, setting the beat through the pulsatile release of GnRH. The pituitary gland translates the beat set by the hypothalamus into a signal, LH and FSH secretion, that can be understood by the ovarian follicle. The ovarian follicle is composed of three key cells: theca cells, granulosa cells, and the oocyte. In the ovarian follicle, LH stimulates theca cells to produce androstenedione. In the granulosa cells from small antral follicles, FSH stimulates the cells to aromatize theca-derived androstenedione to estradiol. A critical quantity of estradiol, produced from a large dominant antral follicle, causes positive feedback in the hypothalamus, likely through the kisspeptin system, resulting in an increase in GnRH secretion and an LH surge. The LH surge causes the initiation of the process of ovulation. After ovulation, the follicle is transformed into the corpus luteum, which, stimulated by LH, secretes progesterone. Progesterone prepares the endometrium for implantation of the conceptus. During the mid-luteal phase of the cycle, when progesterone production is at its peak, the secretory endometrium is optimally prepared for the implantation of an embryo. In turn, implantation of an embryo and successful pregnancy permits perpetuation of the species.

References

1. Treloar AE, Boynton RE, Borchelt BG et al (1967) Variation of the human menstrual cycle through reproductive life. Int J Fertil 12: 77–126
2. Betsey EM, Pinole AP (1997) Menstrual bleeding patterns in untreated women. Contraception 55:57–65
3. Rance NE, Young WS, McMullen NT (1994) Topography of neurons expressing luteinizing hormone releasing hormone gene transcripts in the human hypothalamus and basal forebrain. J Comp Neurol 339:573–586
4. Crowley WF, Pitteloud N, Seminara S (2008) New genes controlling human reproduction and how you find them. Trans Am Clin Climatol Assoc 119:29–38
5. Martin C, Balasubramanian R, Dwyer AA et al (2011) The role of prokineticin 2 pathway in

human reproduction: evidence from the study of human and murine gene mutations. Endocr Rev 32:225–246

6. Schally AV, Arimura A, Bowers CY et al (1968) Hypothalamic neurohormones regulating anterior pituitary function. Recent Prog Horm Res 24:497–588
7. Knobil E (1980) The neuroendocrine control of the menstrual cycle. Recent Prog Horm Res 35:53–88
8. Kesner J, Wilson R, Kaufman J et al (1987) Unexpected responses of the hypothalamic gonadotropin-releasing hormone 'pulse generator' to physiological estradiol inputs in the absence of the ovary. Proc Natl Acad Sci 84: 8745–8749
9. Soules MR, Steiner R, Clifton D et al (1984) Progesterone modulation of pulsatile luteinizing hormone secretion in normal women. J Clin Endocrinol Metab 58:378–383
10. Nakai Y, Plant TM, Hess D et al (1978) On the sites of the negative and positive feedback actions of estradiol in the control of gonadotropin secretion in the rhesus monkey. Endocrinology 102:1008–1014
11. Karsch FJ, Weick R, Butler W et al (1973) Induced LH surges in the rhesus monkey: strength-duration characteristics of the estrogen stimulus. Endocrinology 92:1740–1747
12. Levine J, Norman RL, Gliessman P et al (1985) In vivo gonadotropin-releasing hormone release and serum luteinizing hormone measurements in ovariectomized, estrogen-treated rhesus macaques. Endocrinology 117:711–721
13. Maeda K, Ohkura S, Uenoyama Y et al (2010) Neurobiological mechanisms underlying GnRH pulse generation by the hypothalamus. Brain Res 1364:103–115
14. Chan YM, Butler JP, Pinnell NE et al (2011) Kisspeptin resets the hypothalamic GnRH clock in men. J Clin Endocrinol Metab 96: E908–E915
15. Mimri R, Lebenthal Y, Lazar L et al (2011) A novel loss of function mutation in GPR54/ KISS1R leads to hypogonadotropic hypogonadism in a highly consanguineous family. J Clin Endocrinol Metab 96:E536–E545
16. Pitteloud N, Durrani S, Raivio T et al (2010) Complex genetics in idiopathic hypogonadotropic hypogonadism. Front Horm Res 39: 142–153
17. Williams WP, Jarjisian SG, Mikkelsen JD et al (2011) Circadian control of kisspeptin and a gated GnRH response mediate the preovulatory luteinizing hormone surge. Endocrinology 152:595–606
18. Williams N, Caston-Balderrama AL, Helmreich D et al (2001) Longitudinal changes in reproductive hormones and menstrual cyclicity in cynomolgus monkeys during strenuous exercise training: abrupt transition to exercise-induced amenorrhea. Endocrinology 142: 2381–2389
19. DeSouz MJ, Toombs RJ, Scheid JL et al (2010) High prevalence of subtle and severe menstrual disturbances in exercising women: confirmation using daily hormone measures. Hum Reprod 25:491–503
20. Williams NI, Reed JL, Leidy HJ et al (2010) Estrogen and progesterone exposure is reduced in response to energy deficiency in women aged 25 to 40 years. Hum Reprod 25: 2328–2339
21. Bjorbaek C, Kahn BB (2004) Leptin signaling in the central nervous system and the periphery. Recent Prog Horm Res 59:305–331
22. Welt CK, Chan JL, Bullen J et al (2004) Recombinant human leptin in women with hypothalamic amenorrhea. N Engl J Med 351: 987–997
23. Chou SH, Chamberland JP, Liu X et al (2011) Leptin is an effective treatment for hypothalamic amenorrhea. Proc Natl Acad Sci U S A 108(16):6585–6590
24. Castellano JM, Bentsen AH, Mikkelsen JD et al (2010) Kisspeptins: bridging energy homeostasis and reproduction. Brain Res 1364:129–138
25. Oktem O, Urman B (2010) Understanding follicle growth in vivo. Hum Reprod 12: 2944–2954
26. Dorrington JH, Armstrong DT (1979) Effects of FSH on gonadal functions. Recent Prog Horm Res 35:301–342
27. Xu M, Fazleabas AT, Shikanov A et al (2011) In vitro oocyte maturation and preantral follicle culture from the luteal-phase baboon ovary produce mature oocytes. Biol Reprod 84: 689–697
28. Young JM, McNeilly AS (2010) Theca: the forgotten cell of the ovarian follicle. Reproduction 140:489–504
29. Galloway SM, Gregan SM, Wilson T et al (2002) Bmp15 mutations and ovarian function. Mol Cell Endocrinol 191:15–18
30. Chu MX, Liu ZH, Jiao CL et al (2007) Mutations in BMPR-IB and BMP-15 genes associated with litter size in small tailed Han sheep. J Anim Sci 85:598–603
31. McNatty KP, Smith DM, Makris A et al (1979) The microenvironment of the human antral follicle. J Clin Endocrinol Metab 49:851–860
32. DiZerega GS, Hodgen GD (1981) Folliculogenesis in the primate ovarian cycle. Endocr Rev 2:27–49
33. Yoo SW, Savchev S, Sergott L et al (2011) A large network of interconnected signaling pathways

in human ovarian follicles is supported by the gene expression activity of the granulosa cells. Reprod Sci 18:476–484

34. Haadsma ML, Groen H, Fidler V et al (2008) The predictive value of ovarian reserve tests for spontaneous pregnancy in subfertile ovulatory women. Hum Reprod 23:1800–1807
35. Rodgers RJ, Lavranos TC, van Wezel IL et al (1999) Development of the ovarian follicular epithelium. Mol Cell Endocrinol 151: 171–179
36. Thoroddsen A, Dahm-Kahler P, Lind AK et al (2011) The water permeability channels aquaporins 1 to 3 are differentially expressed in granulosa and theca cells of the preovulatory follicle during precise stages of human ovulation. J Clin Endocrinol Metab 96:1021–1028
37. Conti M, Hsieh M, Park JY et al (2006) Role of epidermal growth factor network in ovarian follicles. Mol Endocrinol 20:715–723
38. Zamah AM, Hsieh M, Chen J et al (2010) Human oocyte maturation is dependent on LH-stimulated accumulation of the epidermal growth factor-like growth factor, amphiregulin. Hum Reprod 25:2569–2578
39. Ben-Ami I, Komsky A, Bern O et al (2011) In vitro maturation of human germinal vesicle stage oocytes: role of epidermal growth factor-like growth factors in the culture medium. Hum Reprod 26:76–81
40. Csapo AI, Pulkkinen MO, Wiest WG (1973) Effects of luteectomy and progesterone replacement therapy in early pregnant patients. Am J Obstet Gynecol 115:759–765
41. Jarvela IY, Ruokonen A, Tekay A (2008) Effect of rising hCG levels on the human corpus luteum during early pregnancy. Hum Reprod 23:2775–2781
42. Kohen P, Castro O, Palomina A et al (2003) The steroidogenic response and corpus luteum expression of the steroidogenic acute regulatory protein after human chorionic administration at different times in the human luteal phase. J Clin Endocrinol Metab 88:3421–3430
43. Strauss JF, Kallen CB, Christenson LK et al (1999) The steroidogenic acute regulatory protein (StAR) a window into the complexities of intracellular cholesterol trafficking. Recent Prog Horm Res 54:369–394
44. Kiriakidou M, McAllister JM, Sugawara T et al (1996) Expression of steroidogenic acute regulatory protein (StAR) in the human ovary. J Clin Endocrinol Metab 81:4122–4128
45. Stocco C, Telleria C, Gibori G (2007) The molecular control of corpus luteum formation, function and regression. Endocr Rev 28: 117–149
46. Andersen CY, Schmidt KT, Kristensen SG et al (2010) Concentrations of AMH and inhibin-B in relation to follicular diameter in normal human small antral follicles. Human Reprod 25:1282–1287
47. Hagen CP, Aksglaede L, Sorensen K et al (2010) Serum levels of anti-Mullerian hormone as a marker of ovarian function in 926 health females from birth to adulthood and in 172 Turner syndrome patients. J Clin Endocrinol Metab 95:5003–5010
48. Ledger WL (2010) Clinical utility of measurement of anti-mullerian hormone in reproductive endocrinology. J Clin Endocrinol Metab 95:5144–5154
49. Hansen KR, Hodnett GM, Knowlton N et al (2011) Correlation of ovarian reserve tests with histologically determined primordial follicle number. Fertil Steril 95:170–175
50. Broer SL, Dolleman M, Opmeer BC et al (2011) AMH and AFC as predictors of excessive response in controlled ovarian hyperstimulation: a meta-analysis. Hum Reprod Update 17:46–54
51. Genro VK, Grynberg M, Scheffer JB et al (2011) Serum anti-Mullerian hormone levels are negatively related to follicular output rate (FORT) in normo-cycling women undergoing controlled ovarian hyperstimulation. Hum Reprod 26:671–677
52. Nelson SM, Yates RW, Lyall H et al (2009) Anti-mullerian hormone-based approach to controlled ovarian stimulation for assisted conception. Hum Reprod 24:867–875
53. Al-Azemi M, Killick ST, Duffy S et al (2011) Multi-maker assessment of ovarian reserve predicts oocyte yield after ovulation induction. Hum Reprod 26:414–422
54. Coxworth JE, Hawkes K (2010) Ovarian follicle loss in humans and mice: lessons from statistical model comparison. Hum Reprod 25: 1796–1805
55. Begum S, Papaioannou VE, Gosden RG (2008) The oocyte population is not renewed in transplanted adult ovaries. Hum Reprod 23:2326–2330
56. Gargett CE, Masuda H (2010) Adult stem cells in the endometrium. Mol Hum Reprod 16: 818–834
57. Sonderegger S, Pollheimer J, Knofler M (2010) Wnt signaling in implantation, decidualization and placental differentiation – review. Placenta 31:839–847
58. Tulac S, Nayak NR, Kao LC et al (2003) Identification, characterization and regulation of canonical Wnt signaling pathway in human endometrium. J Clin Endocrinol Metab 88: 3860–3866

59. van Amerongen R, Nusse R (2009) Towards an integrated view of Wnt signaling in development. Development 136:3205–3214
60. Tulac S, Overgaard MT, Hamilton AE et al (2006) Dickkopf-1, an inhibitor of Wnt signaling is regulated by progesterone in human endometrial stromal cells. J Clin Endocrinol Metab 91:1453–1461
61. Cloke B, Huhtinen K, Fusi L et al (2008) The androgen and progesterone receptors regulate distinct gene networks and cellular functions in decidualizing endometrium. Endocrinology 149:4462–4474
62. Liu X, Nie J, Guo SW (2011) Elevated immunoreactivity to tissue factor and its association with dysmenorrhea severity and the amount of menses in adenomyosis. Hum Reprod 26: 337–345
63. Nordengren J, Pilka R, Noskova V et al (2004) Differential localization and expression of urokinase plasminogen activator (uPA), its receptor (uPAR) and its inhibitor PAI-1 mRNA and protein in endometrial tissue during the menstrual cycle. Mol Hum Reprod 10:655–663
64. Dunn CJ, Goa KL (1999) Tranexamic acid: a review of its use in surgery and other indications. Drugs 57:1005–1032
65. Jensen JT, Parke S, Mellinger U et al (2011) Effective treatment of heavy menstrual bleeding with estradiol valerate and dienogest. Obstet Gynecol 117:777–787
66. Noyes RW, Hertig AT, Rock J (1950) Dating the endometrial biopsy. Fertil Steril 1:3–11
67. King A (2000) Uterine leukocytes and decidualization. Hum Reprod Update 6:28–36
68. Norwitz ER, Schust DJ, Fisher SJ (2001) Implantation and the survival of early pregnancy. N Engl J Med 345:1400–1408
69. Paiva P, Menkhorst E, Salamonsen L et al (2009) Leukemia inhibitory factor and interleukin-11: critical regulators in the establishment of pregnancy. Cytokine Growth Factor Rev 20:319–328
70. Menkhorst E, Zhang JG, Sims NA et al (2011) Vaginally administered PEGylated LIF antagonist blocked embryo implantation and eliminated non-target effects on bone in mice. PLoS One 6:e19665
71. Robb L, Li R, Hartley L, Nandurkar HH et al (1998) Infertility in female mice lacking the receptor for interleukin 11 is due to a defective uterine response to implantation. Nat Med 4:303–308
72. Cullinan EB, Abbondanzo SJ, Anderson PS et al (1996) Leukemia inhibitory factor and LIF receptor expression in human endometrium suggests a potential autocrine/paracrine function in regulating embryo implantation. Proc Natl Acad Sci 93:3115–3120
73. Dominquez F, Simon C, Quinonero A et al (2010) Human endometrial CD98 is essential for blastocyst adhesion. PLoS One 5:e13380

Chapter 8

Assisted Reproductive Techniques

Jack Yu Jen Huang and Zev Rosenwaks

Abstract

Assisted reproductive technologies (ART) encompass fertility treatments, which involve manipulations of both oocyte and sperm in vitro. This chapter provides a brief overview of ART, including indications for treatment, ovarian reserve testing, selection of controlled ovarian hyperstimulation (COH) protocols, laboratory techniques of ART including in vitro fertilization (IVF), and intracytoplasmic sperm injection (ICSI), embryo transfer techniques, and luteal phase support. This chapter also discusses potential complications of ART, namely ovarian hyperstimulation syndrome (OHSS) and multiple gestations, and the perinatal outcomes of ART.

Key words Assisted reproductive technology (ART), Ovarian reserve, Follicle stimulating hormone (FSH), Anti-Müllerian hormone (AMH), Controlled ovarian hyperstimulation (COH), In vitro fertilization (IVF), Intracytoplasmic sperm injection (ICSI), Embryo transfer, Ovarian hyperstimulation syndrome (OHSS), Luteal phase support

1 Introduction

Since the first reports of pregnancies and live births following in vitro fertilization (IVF) by Steptoe and Edwards [1, 2], there have been tremendous improvements in the field of assisted reproductive technologies (ART). Today, several million babies have been born using ART [3] and IVF or "test-tube baby" has become a household term.

ART encompasses fertility treatments which require manipulations of both oocyte and sperm in vitro [4]. The objective of this chapter is to provide an overview of ART, including indications for treatment, ovarian reserve testing, and selection of controlled ovarian hyperstimulation (COH) protocols, laboratory aspects of ART, embryo transfer (ET) techniques, and luteal phase support. We also discuss the ART protocols utilized at Weill Cornell. Lastly, complications, namely ovarian hyperstimulation syndrome (OHSS) and multiple gestations, and the perinatal outcomes of ART are discussed. Detailed discussion of the specific ART techniques are provided in subsequent chapters.

Zev Rosenwaks and Paul M. Wassarman (eds.), *Human Fertility: Methods and Protocols*, Methods in Molecular Biology, vol. 1154, DOI 10.1007/978-1-4939-0659-8_8,

The most commonly performed ART procedure is IVF. IVF involves a sequence of events starting with COH with exogenous administration of gonadotropins to stimulate the development of ovarian follicles, followed by transvaginal ultrasound (US)-guided retrieval of oocytes, fertilization of oocytes with sperm in vitro, culture of the resultant embryos, and transfer of embryos to the recipient. An important innovation in ART is assisted fertilization by intracytoplasmic sperm injection (ICSI), which involves the injection of a single sperm into the cytoplasm of a mature oocyte. Other modalities of ART include embryo assisted hatching (AH), autologous endometrial coculture (AECC), preimplantation genetic diagnosis (PGD) or screening (PGS), cryopreservation of gametes, embryos, and ovarian tissue, frozen-thawed embryo transfer (FET), the use of donor gametes and gestational carriers. Prior to the advent of ART, other less often utilized procedures include laparoscopic tubal transfer of gametes (gamete intrafallopian transfer; GIFT), zygotes (zygote intrafallopian transfer; ZIFT), and embryos (tubal embryo transfer; TET). Due to their invasiveness and the necessity to utilize general anesthesia during these procedures, they have become almost obsolete. They are only utilized when transcervical embryo transfer is technically difficult to perform.

2 Indications

IVF was first reported as a treatment option for patients with severe tubal disease [1, 2]. With improvement in the efficacy of IVF and the introduction of ICSI, the indications for IVF have expanded to include infertility due to severe male factor, severe endometriosis, ovulatory dysfunction, diminished ovarian reserve, and unexplained infertility, especially where conventional treatments have failed (Fig. 1). IVF is also the best treatment option for couples with multifactorial infertility problems.

2.1 Tubal Factor

Tubal factor infertility accounts for 30 % of cases of female infertility and 14 % percent of diagnoses among couples who undergo ART treatments in the USA [5]. The etiologies of tubal obstruction can be intrinsic (ascending salpingitis and salpingitis isthmica nodosa) or extrinsic (surgical sterilization, endometriosis, and peritonitis). Tubal damage is often caused by Chlamydia trachomatis, gonorrhea, and multi-bacterial infections.

Prior to the introduction of IVF, reconstructive tubal surgery had been the only treatment option for patients with tubal obstruction. At present, IVF is the treatment of choice for women over the age of 35 years with significant tubal disease and those with other coexisting infertility problems [6, 7]. IVF should also be offered to patients who remain infertile 1 year following tubal surgery.

2.1.1 Hydrosalpinx

The presence of communicating hydrosalpinx is associated with a 50 % reduction in pregnancy rate in patients undergoing IVF

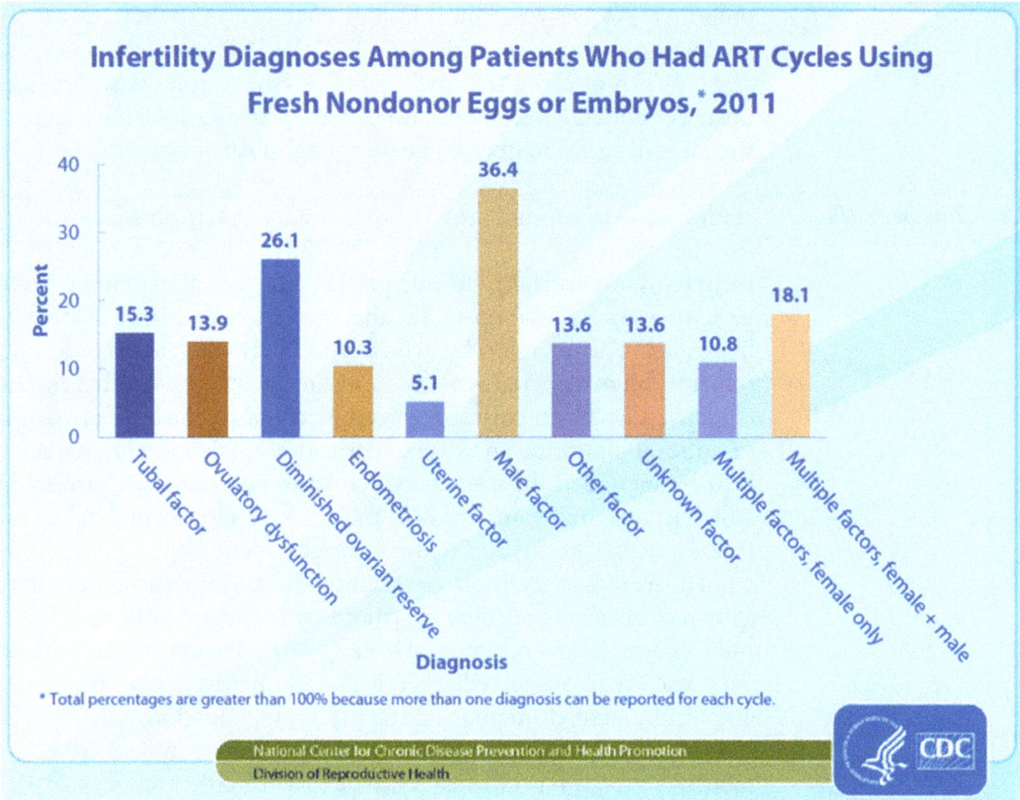

Fig. 1 CDC/SART 2011 report: diagnoses among couples who had ART cycles using fresh nondonor eggs or embryos

treatment [8, 9]. Hydrosalpingeal fluid has been shown to be embryotoxic and may adversely affect embryo implantation [10]. For women with hydrosalpinx, laparoscopic salpingectomy resulted in a twofold increase in ongoing pregnancy rates [8, 11] and, therefore, should be considered prior to undergoing IVF treatment [9, 12]. Alternatively, laparoscopic or hysteroscopic tubal occlusion can be performed to improve IVF pregnancy rates [9, 13–16]. Other less well-studied surgical options are salpingostomy [17] and drainage at the time of oocyte retrieval [18]. The latter approach may lead to infectious complications.

2.2 Male Factor

Abnormal semen parameters may be a contributing factor in up to 40 % of infertile couples and represents approximately 36 % of infertility diagnoses among couples who undergo ART treatments [5, 19]. After the advent of ICSI, the proportion of male factor infertility cases presenting for ART has been increasing. For patients with mild male factor infertility, whose semen parameters are not improved despite medical or surgical treatments, timed intrauterine insemination (IUI) may be offered [20]. Parameters associated with successful IUI treatment include greater than ten million total motile sperms and 14 % normal morphology based on strict Kruger criteria [21, 22]. Total sperm count of less than one million and 4 % normal morphology have been shown to be associated with poor success with IUI [21, 23]; IVF is indicated in these cases. IVF is also the treatment of choice in couples with previous unsuccessful IUI treatment and other coexisting infertility factors, such as advanced maternal age and tubal obstruction.

2.3 Endometriosis

Endometriosis is found in 9–50 % of patients undergoing laparoscopy for the evaluation of infertility [24, 25]. Clinical manifestations include dysmenorrhea, chronic pelvic pain, and dyspareunia. Some patients may be asymptomatic and present only with a history of infertility. The manner by which endometriosis causes infertility remains enigmatic, particularly in instances where no tubal disease or pelvic distortion exist. Proposed mechanisms include distortion of adnexal anatomy, an adverse peritoneal environment characterized by increased inflammatory cytokines and oxidative stress [3], which in turn may interfere with follicular development [26], ovum pick up, fertilization, and embryo development.

Laparoscopic excision or ablation of endometriosis has been shown to alleviate infertility in symptomatic patients with minimal or mild (stage I-II) endometriosis [27, 28]. Asymptomatic patients with known or suspected stage I–II endometriosis may be treated empirically with clomiphene citrate (CC) or gonadotropin and IUI [29]. Patients with known or suspected moderate and severe endometriosis (stage III–IV) may be treated with either surgery or IVF [30]. There is no prospective randomized controlled trial (RCT) comparing the efficacy of the two treatment modalities. Surgical treatment is preferred in patients who are symptomatic and those with endometriomas greater than 4 cm [31]. IVF is indicated if there are coexisting causes of infertility, such as tubal obstruction, advanced maternal age, and abnormal semen parameters or when conventional treatment is unsuccessful [31]. A 3-month course of gonadotropin releasing hormone (GnRH) agonist administered before starting IVF has been shown to improve the ongoing pregnancy rate [32].

2.4 Ovulatory Dysfunction

Ovulatory dysfunction is the most common etiology of female infertility, accounting for 25 % of cases [5]. Most of these patients present with either oligomenorrhea or amenorrhea. The most common etiology of anovulation is polycystic ovary syndrome (PCOS), characterized most frequently by the triad of polycystic ovaries (PCO), oligo- or amenorrhea, and clinical and biochemical signs of androgen excess [33]. Ovulatory dysfunction may also be caused by endocrinopathies, such as thyroid disorders and hyperprolactinemia. Thyroid stimulating hormone (TSH) should be screened and thyroxin replacement should be administered if hypothyroidism is the underlying etiology. Hyperprolactinemia alone or when associated with elevated TSH, as a result of primary hypothyroidism, may cause anovulation and should be corrected with bromocriptine, cabergoline after exclusion of pituitary macroadenomas, or thyroid replacement, respectively.

Patients with WHO group 1 (hypogonadotropic-hypogonadism) ovulatory disorder respond to exogenous gonadotropins or to pulsatile GnRH infusion [34]. Patients with normogonadotropic-normogonadal ovulatory disorders (WHO group 2), including those with PCOS, can be successfully treated with ovulation induction (OI) combined with timed intercourse or IUI. OI is usually successful following treatment with CC [35], exogenous gonadotropins, aromatase inhibitors (letrozole, anastrazole) [36], metformin (in women with insulin resistance) [37], or selective estrogen receptor modulators (tamoxifen) [38].

Patients with PCO seen on US, even in the absence of clinical features of PCOD, have an increased risk of over-responding to gonadotropins and of developing OHSS and high order multiple gestations [39, 40]. In patients who exhibit high responsiveness to OI with gonadotropins, conversion to IVF represents an effective yet safe alternative to proceeding with IUI or cycle cancellation [41]. IVF is also indicated in patients who do not conceive following conventional OI treatments [42] and couples with other coexisting infertility factors.

2.5 Unexplained Infertility

Unexplained infertility is defined as the absence of an identifiable cause of infertility despite a thorough investigation demonstrating tubal patency, normal semen parameters, ovulation, normal ovarian reserve, and a normal endometrial cavity. The incidence of unexplained infertility ranges from 10 to 30 % [43].

Treatment options include expectant management, IUI, empiric treatment with CC, CC combined with IUI, gonadotropins/IUI, and IVF. IVF is the most effective treatment option for couples with unexplained infertility, resulting in the highest per cycle pregnancy rate in the shortest time interval [44]. In one reported study, the pregnancy rates following IVF, gonadotropin and IUI, CC and IUI, and expectant management were in the ranges of 20–30 %, 10–15 %, 7–9 %, and 1–3 %, respectively [44].

Empirical treatment algorithms for couples with unexplained infertility typically involves three cycles of CC/IUI, followed by three cycles of gonadotropin/IUI, and by IVF if the patients remain unsuccessful. In a recent prospective RCT involving women with unexplained infertility between the ages of 21 and 39 years, patients were randomized to either an accelerated treatment algorithm (IVF following three unsuccessful CC/IUI treatment cycles) or the conventional treatment algorithm (IVF following three unsuccessful CC/IUI and three unsuccessful gonadotropin/IUI treatment cycles). The time to pregnancy was significantly shorter in the accelerated arm compared with the conventional arm (hazard ratio 1.25; 95 % CI, 1.00–1.56) [45]. The accelerated treatment algorithm was also more cost-effective, compared to the conventional treatment group. Age appears to be the single most important determining factor of success. Patients over 40 years of age should proceed with IVF following three unsuccessful gonadotropin/IUI treatment cycles [46].

Recent experience suggests that IVF should be utilized sooner than later in women older than 37 years (personal communication).

2.6 Preimplantation Genetic Diagnosis and Screening

Biopsy can be performed on the polar bodies of oocytes, blastomere(s) of day 3 embryos or trophectoderm of day 5 blastocysts. The cell(s) can then be analyzed for single gene defects (known as PGD) or be screened for aneuploidy (known as PGS) [47]. PGD allows the detection of significant genetic diseases, such as cystic fibrosis or sickle cell anemia, before embryo transfer and conception [48, 49]. The embryos can also be screened for aneuploidy using fluorescent in situ hybridization (FISH), microarray, and polymerase chain reaction (PCR) technologies [50]. PGS may be beneficial in the evaluation of patients with advanced maternal age, history of recurrent abortions, especially if a balanced translocation has been identified in one of the parents, or when there is a history of repeated implantation failure despite good embryo morphology [47].

2.7 Fertility Preservation

In recent years, there is greater awareness among reproductive endocrinologists, oncologists, and patients of the impact of cancer treatment on fertility [51]. Women at risk of premature ovarian failure due to gonadotoxic chemotherapy or radiation treatment should be offered the possibility of fertility preservation. The only strategy of female fertility preservation recognized by the American Society of Clinical Oncology (ASCO) and the American Society of Reproductive Medicine (ASRM) is COH followed by retrieval and cryopreservation of oocytes, or COH followed by IVF using sperm from a male partner or a donor, and cryopreservation of the resultant embryos [52, 53]. According to ASRM, oocyte vitrification (rapid freezing) is no longer considered experimental and represents an attractive fertility preservation option for women without a male partner [54]. Other fertility preservation procedures include in vitro maturation (IVM) and cryopreservation of ovarian tissue [55, 56].

3 Ovarian Reserve Testing

Predicting patient response to COH represents a significant clinical challenge. A sensitive predictive marker of ovarian reserve could be helpful in designing optimal COH protocols for anticipated high and low responders, for determining the starting dose of gonadotropins, and in efforts to avoid adverse events such as OHSS and cycle cancellation. However, a single sensitive marker of ovarian reserve has yet to be developed, albeit anti-Müllerian hormone (AMH) measurements appear to show promise.

3.1 Age and Ovarian Reserve

Ovarian reserve refers to the genetically predetermined pool of female germ cells or primordial follicles. The pool of ovarian germ cells peaks at 16–20 weeks of fetal gestation, containing

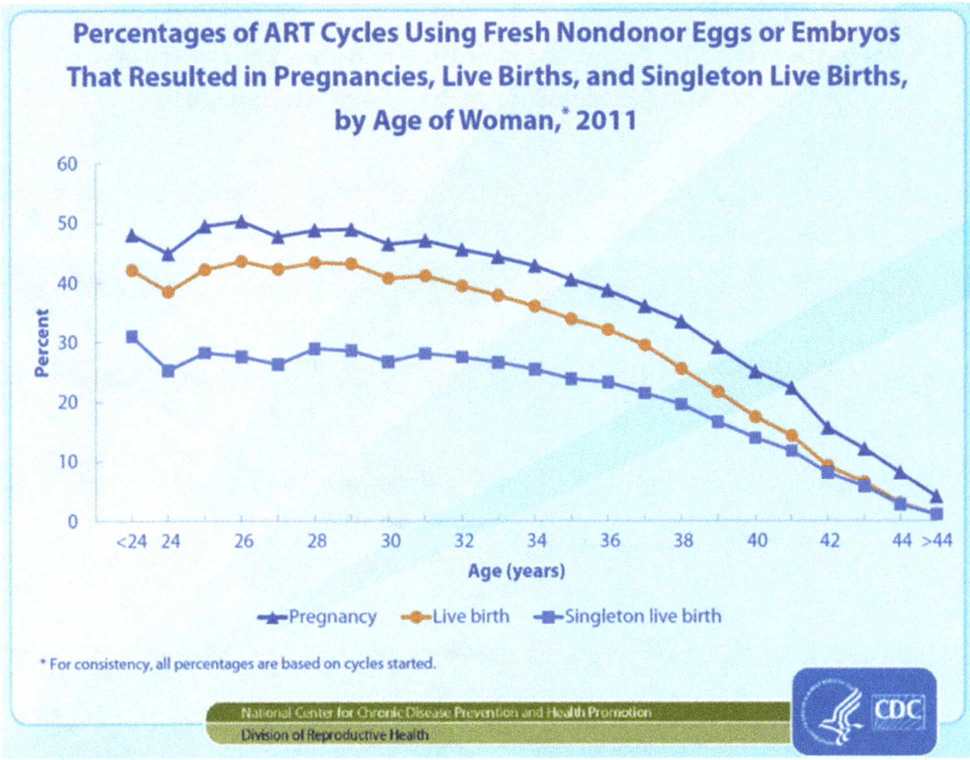

Fig. 2 CDC/SART 2011 report: pregnancy and live birth rates by age of woman following ART cycles using fresh nondonor oocytes or embryos

approximately 6–7 million oogonia. From this point onward, depletion in the germ cells occurs in a bi-exponential fashion [57]. At birth, the ovary contains 1–2 million primordial follicles decreasing to 300,000–500,000 at puberty. The decline in the primordial follicles remains constant until the age of 37. From this point onward, there is a dramatic increase in follicular atresia, leading to menopause 10–15 years later [58]. During the reproductive period, only 400–500 oocytes ovulate. The fate of the majority of primordial follicles is atresia, presumably via apoptosis [59].

Given the relationship between advanced maternal age and the decline in fertility [58], it is not surprising that age is the most important determining factor of success in women undergoing IVF (Fig. 2). Although ART may overcome infertility in younger women, it does not reverse the age-dependent decline in fertility [60]. In fact, among patients undergoing IVF for various indications, diminished ovarian reserve appears to be associated with the poorest prognosis. According to the 2011 Centers for Disease Control (CDC) and Society for Assisted Reproductive Technology (SART) report, the percentages of IVF cycles that resulted in live births were comparable by diagnoses with the exception of diminished ovarian reserve: tubal factor 29 %, ovulatory dysfunction

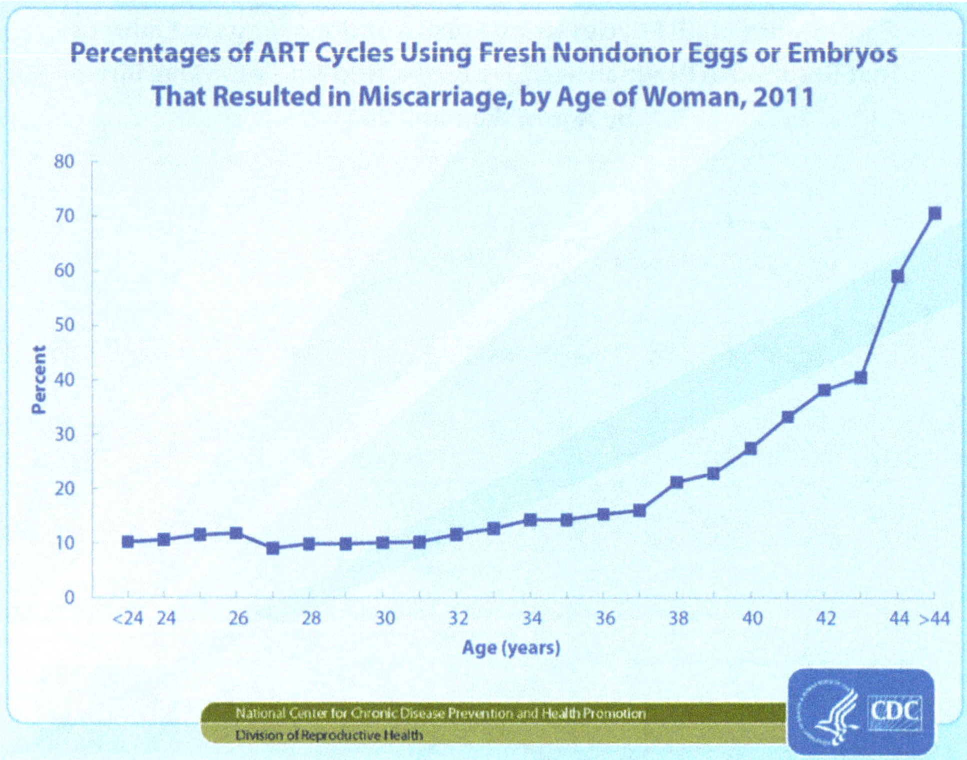

Fig. 3 CDC/SART 2011 report: miscarriage rates following ART treatments using fresh nondonor eggs or embryos

37 %, male factor 33 %, endometriosis 30 %, unexplained cause 32 %, and diminished ovarian reserve 17 % [5].

Advanced maternal age is associated with a decline in the number of oocytes retrieved, embryos available for transfer and embryo quality [61–63], ultimately resulting in lower implantation [63], pregnancy, and live birth rates [5]. Among patients of 30 years of age, the pregnancy and live birth rates per cycle stated were 48 and 42 %, respectively [5]. The corresponding rates decreased to 43 and 36 %, respectively in patients who were 35 years of age [5]. Among patients of 40 years old age, the corresponding rates significantly decreased to 25 and 17 % [5]. In fact, the cumulative live birth rates after six cycles of ART was 86 % in patients younger than 35 years of age; the corresponding rate was only 42 % in patients over the age 40 years [60].

In addition to the decline in fertility, the incidence of spontaneous miscarriages also increases with advanced maternal age (Fig. 3) [64]. Aging is associated with a decline in the both the pool of primordial follicles and the quality of oocytes. The increased miscarriage rate is attributed to a higher prevalence of aneuploidy in aging oocytes as a result of meiotic nondisjunction [65]. In a review of 288 patients over the age of 45 years who underwent IVF treatment at our institution, 20 % of patients did not start because of an elevated FSH or ovarian cyst and 30 % of cycles were cancelled due to poor response to COH. The pregnancy rate per transfer was 21 %. Of these, 85 % resulted in miscarriage. The overall delivery rate was 3 % per retrieval. Only five patients had live births; all were age 45 years [66].

3.2 Ovarian Reserve Testing

Assessment of ovarian reserve can be divided into serum endocrine markers, ultrasonographic markers, and dynamic evaluations. Numerous ovarian reserve markers have been proposed and evaluated. Endocrine parameters include basal follicle-stimulating hormone (FSH), estradiol (E_2), inhibin B, and, more recently, anti-Müllerian hormone (AMH) levels. Ultrasonographic assessment of the ovaries, including antral follicle count (AFC), ovarian volume, ovarian blood flow, has also been extensively evaluated. Less commonly used dynamic evaluations of ovarian reserve include CC challenge testing (CCCT) [67], exogenous FSH ovarian test (EFORT) [68], and gonadotropin releasing hormone (GnRH) agonist stimulation test (GAST) [69, 70].

3.2.1 Hormonal Markers

FSH

Depletion of primordial follicles with aging results in a decline in inhibin B production by their associated granulosa cells. The lack of inhibin B negative feedback in turn results in an increase in pituitary FSH production [71]. It has been shown that serum FSH level begins to rise one or two decades before menopause [72]. To date, there is no absolute consesus as to the cut off which defines what constitutes an abnormal early follicular phase FSH level on day 2 or 3 of menstrual cycle. Reported abnormal FSH levels range from 12 to 15 mIU/L [73, 74]. Indeed, elevated serum FSH level during the early follicular phase of the menstrual cycle has been associated with poor IVF treatment outcomes, including lower peak E_2, higher cycle cancellation rate, decreased number of oocytes retrieved, and reductions in fertilization rate, number of embryos available for transfer, and pregnancy rates [73, 74].

The impact of elevated FSH on the reproductive outcomes of young women is less clear. A study from our institution found that women under the age of 40 with elevated serum FSH levels had lower oocyte yield compared to those with normal FSH levels. However, the implantation and pregnancy rates were unaffected. In women over the age of 40 years, elevated FSH levels were associated with decreased implantation and clinical pregnancy rates. Other studies have found reductions in implantation rates among young women with elevated FSH [74–76]. A single elevated FSH level may not be sufficiently accurate to predict outcomes since there are intercycle fluctuations in FSH levels. However, one study reported that no pregnancies occurred in patients with a history of three or more elevated FSH levels regardless of age [77].

Estradiol

Abnormal basal E_2 levels on cycle day 2 or 3 have been defined as greater than 75 pg/ml [78, 79]. Premature luteal FSH elevation can result in early follicular recruitment, manifested by elevated basal E_2 level. Patients with an elevated basal E_2 level were noted to have fewer oocytes retrieved, lower pregnancy rates, and a higher cancellation rates compared with those with normal basal E_2 levels [78, 79]. One study found that no pregnancies occurred in patients with E_2 greater than 75 pg/mL [79]. Similar to basal FSH levels, there is intercycle and intracycle variability in E_2 levels. The E_2 values may also vary depending on the types of quantitative assays. In fact, a recent meta-analysis found basal E_2 to have a very low predictive accuracy for poor response and non-pregnancy [80].

Anti-Müllerian Hormone

AMH, a member of the transforming growth factor-β superfamily, is produced by granulosa cells surrounding preantral and early antral follicles [81]. The advantage of AMH compared to the other serum ovarian reserve markers is that AMH can be measured at anytime of the menstrual cycle. Serum AMH levels have been shown to be cycle independent and consistent throughout menstrual cycles [82, 83]. Serum AMH levels have also been shown to positively correlate with antral follicle counts (AFC) and the number of oocytes retrieved [84, 85], making it a reasonable predictor of ovarian response and OHSS risk [40, 86]. Lee et al. found AMH and serum E_2 on day of hCG as the two most reliable predictors of OHSS [86]. Using a cut-off value for AMH of greater than 3.36 ng/ml, the sensitivity was 90.5 % and specificity was 81.3 %.

There is no defined AMH cut-off level for predicting diminished ovarian response [87]. Using a cut-off of AMH level ≤1.26 ng/ml, Gnoth et al. reported a sensitivity of 97 and 41 % specificity in detecting poor response [88]. Another study found a cut-off of AMH <0.1 ng/ml to be associated with 76 % sensitivity and 88 % specificity in predicting poor response and 22 % sensitivity and 89 % specificity in predicting non-pregnancy [89].

Inhibin B

Inhibin is a heterodimeric glycoprotein of the TGF-β superfamily, consisting of alpha and beta subunits, that is secreted by the granulosa cells. Inhibin B rises in the early follicular phase and then progressively decreases until after the midcycle LH surge [90]. In women with diminished ovarian reserve, inhibin secretion is low in the early follicular phase.

Day 3 levels of inhibin B have been proposed as markers of ovarian reserve. Inhibin B concentrations of less than 45 pg/ml were shown to be associated with lower responses to COH, fewer oocytes retrieved, higher cancellation rates, and reduced clinical pregnancy rates [91]. However, a recent meta-analysis found that the accuracy of serum inhibin B as a predictor of poor response and non-pregnancy to be modest at a very low threshold level [80].

3.2.2 Ultrasonographic Markers

Antral Follicle Count

The number of 2–10 mm diameter follicles seen on transvaginal US on the cycle day 2 or 3 has been shown to be a good predictor of ART success [92]. AFC was found to be superior to chronological age, total ovarian volume, and basal FSH, E_2, and inhibin B levels in predicting poor ovarian response to gonadotropins [93]. More recently, a meta-analysis showed that AFC was as good as AMH in predicting response to ovarian stimulation for IVF [84]. In experienced hands, AFC is an excellent if not the best predictor of ovarian response (Personal Communications)

Ovarian Volume and Blood Flow

The volume of the ovaries diminishes with advancing maternal age corresponding with the age-related diminishing pool of primordial follicles [94]. An ovarian volume less than 3 cm^3 has been associated with higher basal FSH level, total gonadotropin requirement, and cycle cancellation rate and fewer oocytes retrieved [95]. A meta-analysis found ovarian volume to be inferior to AFC in predicting poor response [96]. Ovarian and uterine blood flow have also been investigated as potential predictors of success [97]. Ovarian stromal peak systolic velocity has been reported as an important predictor of ovarian response in patients with a normal basal serum FSH level [98].

3.2.3 Dynamic Tests

Clomiphene Citrate Challenge Test (CCCT)

CCCT involves measurement of baseline FSH level on cycle day 3, followed by administration of 100 mg CC on cycle day 5–9, and FSH measurement on cycle day 10 [67]. Women with low ovarian reserve exhibit diminished pituitary inhibin B and E_2 negative feedback after the CC challenge manifested by abnormally elevated pituitary secretion of FSH [91]. An elevated FSH value greater than two standard deviations above the mean is defined to be abnormal and is indicative of diminished ovarian reserve [67]. Abnormal CCCT has been associated with decreased IVF success [67, 99, 100]. The cost-effectiveness of using CCCT in the evaluation of infertile patients has been questioned. A meta-analysis showed that CCCT offered no additional prognostic value [101, 102]. In fact, basal FSH and the CCCT are similar in predicting clinical pregnancy in women undergoing infertility treatment.

Exogenous FSH Ovarian Reserve Testing (EFORT)

EFORT involves measurement of basal serum FSH and E_2 levels on cycle day 3. Following administration of 300 IU FSH, serum E_2 level is repeated 24 h later [68]. Ninety percent of the women whose EFORT parameters were considered to be normal (basal FSH <11 mIU/ml and ΔE_2 >30 pg/ml) had adequate response to ovarian stimulation. Eighty-one percent of women in whom both parameters were considered to be abnormal (basal FSH >11 mIU/ml and ΔE_2 <30 pg/ml) had poor response to ovarian stimulation.

GnRH Stimulation Test

Administration of GnRH agonists initially evokes a surge in pituitary gonadotropins and is followed by desensitization gonadotropin receptors and down-regulation of gonadotrope function. The surge in FSH results in an increased production of E_2 by the granulosa cells. The GnRH stimulation test (GAST) evaluates changes in serum E_2 levels on cycle day 2 and 3 following administration of 0.1 mg triptorelin [69]. Changes in serum E_2 levels less than

80–180 pmol/l have been shown to be associated with poor response to COH [69, 103, 104]. GAST is dependent on the pituitary production of gonadotropins and the ovarian response to the subsequent increase in pituitary secretions of FSH and LH. A meta-analysis found that the accuracy of GAST is comparable to that of AFC in predicting poor response [80].

4 Evaluation of Couples Undergoing IVF

4.1 General Health

The general health of all couples undergoing IVF should be evaluated prior to treatment. Evaluation should include a thorough review of the medical and surgical history, as well as outcomes of previous pregnancies and fertility treatments, including COH-IUI and IVF. A complete physical examination should be performed with particular attention paid to body mass index (BMI) and evidence of hirsutism, thyroid, and prolactin dysfunction. A thorough pelvic exam should be performed with attention paid to any cervical, uterine, or ovarian abnormal findings.

In addition, all patients should undergo baseline laboratory evaluation, including complete blood counts, blood type and screen, thyroid function tests, prolactin, and rubella and varicella immune status. Additional tests may be required for patients with existing medical conditions, such as hypertension, diabetes mellitus, cardiac disease, and any inherited or acquired thrombophilias. These patients should receive preconception evaluation by internal medicine subspecialists and obtain medical clearance prior to starting ART treatment.

4.2 Ovarian Reserve Testing

Patient should have updated ovarian reserve testing, such as a basal serum FSH, AMH, E_2 levels, or AFC, as it is particularly helpful in determining the appropriate gonadotropin starting dosage and predicting the response to COH treatment.

4.3 Evaluation of the Uterine Cavity and Adnexa

Assessment of the uterus and the adnexa should include a bimanual examination and pelvic US. Previous hysterosalpingogram (HSG) studies if available should be reviewed for the presence of uterine filling defects and hydrosalpinges. Additional evaluation of the uterine cavity by a mid-cycle pelvic ultrasound or saline infusion sonogram (SIS) may be warranted prior to starting IVF to rule out submucosal myomas or polyps. Submucosal myomas, adversely impact implantation, lowering clinical pregnancy, and live birth rates by as much as 70 % and by increasing miscarriage rates threefold [105–107].

Patients with abnormal radiological findings on HSG or SIS should be evaluated with a hysteroscopy. Hysteroscopy is the gold standard for the investigation of the uterine cavity, permitting direct visualization of the cavity and resection of any submucosal myomas, polyps, or adhesions. The cost-effectiveness of routine diagnostic hysteroscopy prior to IVF is controversial. In one retrospective review, endometrial pathology was found in 23 % of infertile patients who underwent diagnostic hysteroscopies [108]. Endometrial polyps and submucosal myomas accounted for 8 and 4 % of the pathological findings, respectively. Although some ART programs include office hysteroscopy as part of the evaluation prior to IVF, clear benefits to this approach have not been shown in large prospective RCTs. The ESHRE guidelines concluded that hysteroscopy is only indicated to confirm and treat suspected intrauterine pathology [109].

4.4 Trial Transfer

A trial transfer should be performed in order to determine the length and contour of the uterine cavity. Trial transfers also identify patients with potentially difficult embryo transfers (ET) due to cervical stenosis or an acute cervical-uterine angle. Information obtained from the trial transfer aids physicians in selecting the appropriate ET catheter and determining the need for US guidance during ET. Performing trial transfers has been shown to improve IVF success [110].

Trial transfers are usually performed during the mid-cycle, on menstrual cycle days 10–12, prior to the start of IVF treatment. Alternatively, trial transfers can be performed at the time of an SIS, hysteroscopy, or during oocyte retrievals. Performing the trial transfer at the time of oocyte retrieval does not appear to have a deleterious effect on the endometrium. In fact, the timing of trial transfers does not appear to affect implantation or pregnancy rates [111]. At our institution, a semi-rigid catheter (Cook, Ob/Gyn, Spencer, IN, USA) is used to ascertain the depth and position of the uterine cavity. The depth of the cavity is determined by measuring the distance from the top of the fundal cavity to the external cervical os. Since the uterine cavity may be enlarged or distorted following a pregnancy, the trial transfer is usually repeated following a live birth [112].

Patients who are noted to have cervical stenosis should be counselled about the potential negative impact of a difficult transfer and be offered cervical dilatation. The ideal time to perform cervical dilation is before the start of the IVF cycle as an elective procedure or during the oocyte retrieval procedure [113, 114]. Alternatively, placement of laminaria stents or a Malecot catheter may be offered for patients with a history of difficult ET or refractory cervical stenosis [115, 116].

4.5 Evaluation of the Male Partner

The evaluation of the male partner is addressed in more details in the subsequent chapter. A recent semen analysis should be available to ensure that there is adequate sperm concentration for IVF insemination and to facilitate the decision whether there is a need for assisted fertilization by ICSI.

4.6 Screening of Sexually Transmissible Diseases

The patient and partner should be screened for sexually transmitted diseases, including hepatitis B and C (hepatitis B surface antigen and hepatitis C antibody), HIV 1/2 (ELISA), syphilis (RPR or VDRL), Chlamydia, and gonorrhea (nucleic acid amplification tests).

5 Controlled Ovarian Hyperstimulation (COH)

The success of IVF treatment can be optimized by adopting an individualized, patient-directed approach to COH. Key components involve selection of an appropriate COH protocol and dosage of gonadotropins, close monitoring of follicular growth and serum E_2 levels, adjustment of gonadotropin dosage to avoid hyper-response, and individualized timing of human chorionic gonadotropin (hCG) injection. This approach to COH monitoring improves oocyte and embryo quality, pregnancy and implantation rates, and most importantly, minimizes the risk of complications, especially OHSS.

5.1 Selection of an Appropriate COH Protocol

The first IVF live birth was achieved following retrieval of a single oocyte from an unstimulated, natural menstrual cycle [2]. However, natural cycle IVF is inefficient as it is associated with high cancellation rates due to the risk of spontaneous ovulation and low implantation and pregnancy rates following transfer of a single embryo [117]. With the introduction of gonadotropins, natural cycle IVF has been largely abandoned in favor of the more efficient approach with COH using gonadotropins [118, 119].

The basic principle of contemporary COH protocols involves pituitary suppression with either a GnRH agonist or antagonist and ovarian stimulation with exogenous gonadotropins [120]. The use of human menopausal gonadotropin (hMG) and FSH promotes the development of multiple follicles, resulting in increased oocyte yield, number of embryos available for selection and transfer, and improved pregnancy rates.

An important step in choosing an appropriate COH protocol is to be able to anticipate patient response, i.e., will she be a normal, high or low responder to gonadotropins?

5.1.1 Normal Responders

Normal responders are characterized by favorable prognostic factors including young maternal age (<35 years), normal BMI, adequate ovarian reserve (AFC between 6 and 10), normal AMH level, normal basal FSH and E_2 levels (FSH <10 mIU/ml, E_2 <75 pg/ml), short duration of subfertility, a history of previous live birth, or previous successful IVF treatment [121].

Mid-Luteal GnRH Agonist Down Regulation or "Long" Protocol

The most commonly used COH protocol has been the mid-luteal, GnRH agonist down regulation or "long" protocol (Fig. 4) [122]. Daily injection of a GnRH agonist, most commonly 1.0 mg

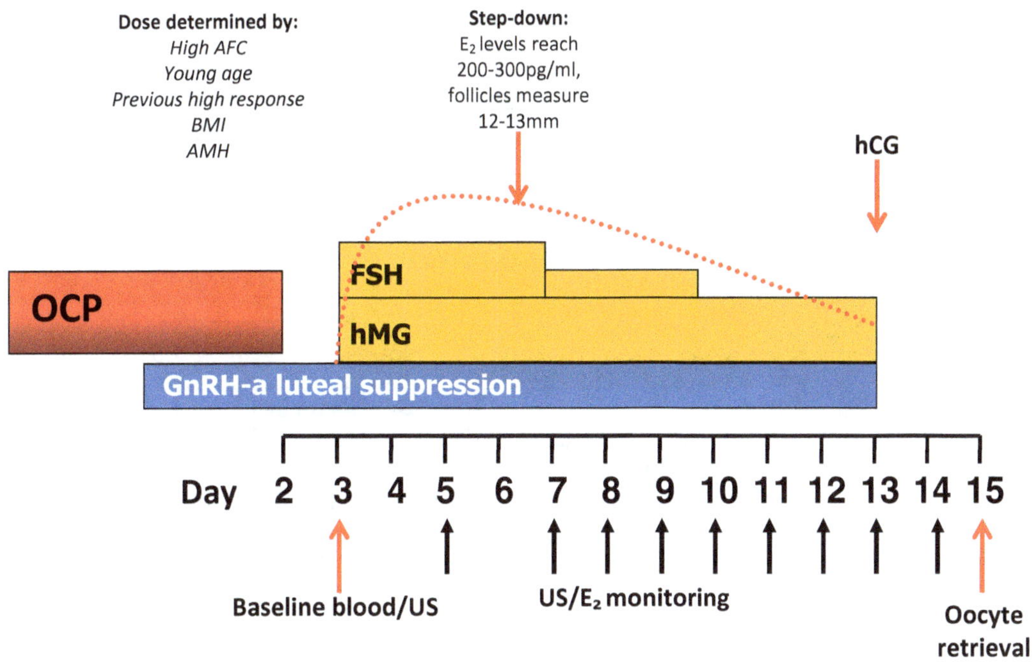

Fig. 4 GnRH agonist "long" protocol

leuprolide acetate (LA), is initiated in the mid-luteal phase of the preceding cycle. Patients present on day 3 of the subsequent menstrual cycle for baseline serum and ultrasound (US) assessment. Upon determining adequate pituitary suppression, appropriately suppressed E_2 levels, and a thin endometrial stripe by US, daily exogenous gonadotropin stimulation is initiated. Daily LA injection is often reduced to 0.5 mg.

Administration of GnRH agonists has several disadvantages. These include side effects of hot flushes and headaches and development of ovarian cysts. GnRH agonist COH protocols may also cause excessive pituitary suppression, which may require longer gonadotropin stimulation and increased exogenous gonadotropin doses.

GnRH Antagonist "Short" Protocol

More recently, the introduction of GnRH antagonists for ART has simplified COH protocols. GnRH antagonist protocols are also commonly known as "short" protocols (Fig. 5). GnRH antagonists compete with endogenous GnRH binding to the pituitary GnRH receptors, thereby rapidly inhibiting secretion of gonadotropins [123, 124]. Proposed advantages of GnRH antagonists include a more physiological follicular recruitment process, absence of the suppressive effects of GnRH agonists, and reduction in the duration of COH and gonadotropin requirement. Moreover, the use of GnRH antagonists improves patient compliance, reduces the physical and psychological treatment burden, and has been shown to lead to a lower patient dropout rate [125].

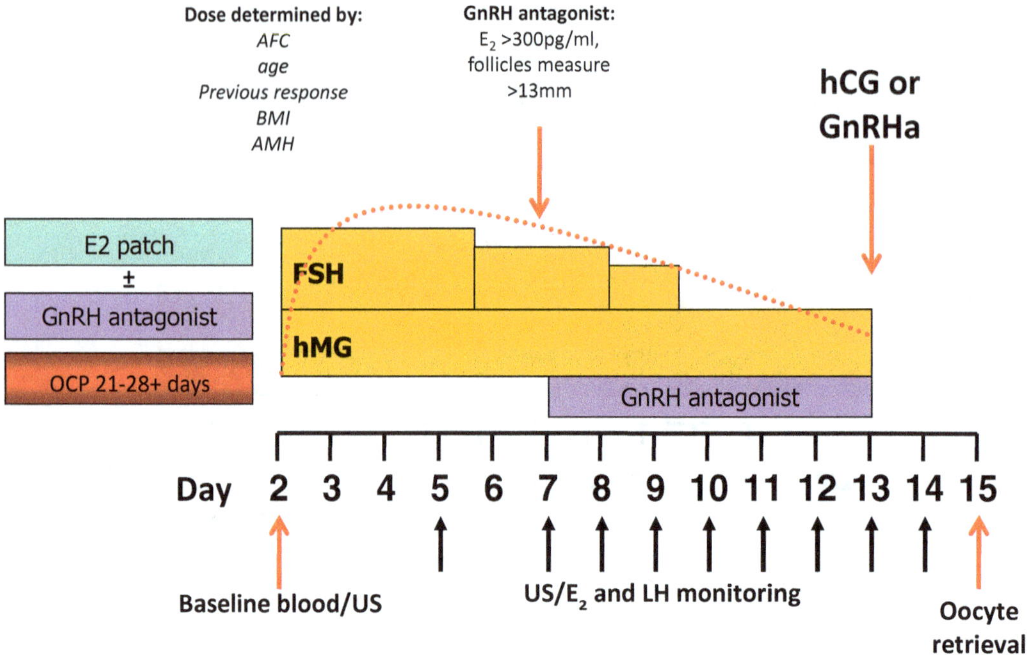

Fig. 5 Ovarian stimulation with GnRH antagonists

GnRH antagonists can be administered either in single doses or in multiple doses. In the single-dose GnRH antagonist COH protocol, a 3 mg dose of the GnRH antagonist, cetrorelix, is typically administered on cycle day 7 of the COH cycle. This dosage has been shown to effectively prevent premature LH surges [126]. If hCG injection is not administered 4 days following the single dose of cetrorelix, a low-dose GnRH antagonist, 0.25 mg cetrorelix or ganirelix, should be given daily until the day of hCG. Approximately 10 % of single-dose GnRH antagonist COH cycles require additional daily GnRH antagonist administration [127].

In the multiple-dose protocol, GnRH antagonist administration can be fixed or flexible. In the fixed protocol, administration of 0.25 mg, ganirelix or cetrorelix is typically started on cycle day 6 and is continued daily until the day of hCG. In the flexible protocol, daily GnRH antagonist is administered upon reaching a predetermined designated E2 threshold or a follicular diameter is reached. We administer daily ganirelix when serum E_2 levels reach 300 pg/ml or when the leading follicle attains a diameter of 13 mm or greater. A recent meta-analysis found no difference in pregnancy rates between flexible and fixed GnRH antagonist protocols [128]. The flexible GnRH antagonist protocols require fewer ampoules of gonadotropins than the fixed protocol.

5.1.2 High Responders

The management of patients who are at risk of developing an exaggerated response to COH represents a formidable challenge. An important consideration is the prevention of OHSS. Known risk factors for OHSS include young age, lean body weight, and a past history of OHSS. Women with PCO seen on US, irrespective of having other clinical features of PCOS, are at greater risk of developing OHSS. OHSS incidence has been reported to be as high as 30 % in patients with PCOS. In patients undergoing COH treatment, high gonadotropin doses, high absolute levels (greater than 3,000 pg/ml) or rapidly rising E_2 levels also represent risk factors for the development of OHSS.

Strategies for the prevention of OHSS include identifying patients at risk, individualization of COH protocols, and judicious use of gonadotropins. Our general philosophy has been to utilize moderate stimulation protocols aiming at retrieving 5–15 oocytes, while maintaining peak E_2 levels at less than 2,500 pg/ml on the day of the ovulatory trigger.

Oral Contraceptive Pill-GnRH Agonist Dual Suppression Protocol

A COH protocol that is applicable to patients at risk of developing OHSS is the oral contraceptive pill (OCP)-GnRH agonist dual suppression protocol [129]. Patients are pretreated with OCP for 28 days. LA 1 mg is started on day 21, overlapping the OCP for 7 days. Once adequate suppression has been documented, low dose of hMG or FSH is initiated in a combination tailored to the individual patient; LA is reduced to 0.5 mg daily.

In a retrospective analysis of the complications among 973 donor egg cycles using the OCP-GnRH agonist dual suppression protocol at CRMI, there was only one case of moderate OHSS requiring admission to hospital for observation, intravenous hydration, and prophylactic subcutaneous heparin [130]. No paracentesis or other interventions were required. An additional benefit of OCP-GnRH-agonist dual suppression is a reduction in incidence of functional ovarian cysts that may develop because of the initial stimulatory effect of the GnRH agonist.

GnRH Antagonist Protocol/ GnRH Agonist Ovulatory Trigger

GnRH antagonist protocols have traditionally been used for the treatment of poor responders. Recently, this protocol has been extended to include younger women at risk of developing OHSS. In the setting of high responders, the use of GnRH antagonist has been shown to decrease the incidence of OHSS compared to GnRH agonist-based COH protocols [131, 132]. In fact, patients treated with GnRH antagonist COH protocols were less likely to require interventions to prevent OHSS, such as coasting and cycle cancellation [131, 132]. GnRH antagonist cycles also permit the use of GnRH agonists to trigger an endogenous LH ovulatory surge, which further reduces the incidence of moderate and severe OHSS [133]. Our first line protocol for patients at risk of OHSS and for all oocyte donors is the OCP-GnRH antagonist/ gonadotropins protocol.

Compared to the long GnRH agonist protocols, GnRH antagonist cycles were initially thought to be associated with lower clinical, ongoing pregnancy, and live-birth rates [131, 132]. A recent meta-analysis did not confirm these early observations [134]. The observed reduction in pregnancy rates was thought to be due, in part to the use of GnRH agonist for the ovulatory trigger, which had been shown to result in premature luteolysis of the corpora lutea and poor luteal steroid production [135, 136]. In the oocyte donor population, COH stimulation with GnRH agonist and antagonist protocols resulted in comparable ongoing pregnancy rates in the donor oocyte recipients [137].

5.1.3 Poor Responders

Diminished ovarian reserve has become increasingly common among patients undergoing in IVF with the reported prevalence ranging from 10 to 25 % [5, 138]. The variations in the prevalence rate can be attributed to a lack of a universally accepted definition of poor responders. Most definitions of poor response are based on arbitrary levels of peak serum E_2 or oocyte yield. The term poor responder was first proposed by Garcia et al. in 1983 and was defined as peak E_2 concentration less than 300 pg/ml following COH with 150 IU of hMG [139]. Others defined poor response as peak serum E_2 concentration less than 500 pg/ml [140, 141] or E_2 less than 100 pg/ml following 5 days of COH [142]. Poor response has also been defined by advanced age (>40 years) [143], elevated basal FSH level (over 10–15 mIU/L) [143–146], previous cancelled cycle, abnormal clomiphene challenge test (CCCT) [67], prolonged duration of COH [147], increased daily and total gonadotropin ampoules used (>44) [146, 148], and the harvest of less than 3–5 oocytes [149].

Recently, ESHRE defined poor ovarian response as having at least two of the following three features: (1) Advanced maternal age (≥40 years) or any other risk factor for diminished ovarian reserve; (2) previous history of poor ovarian response (fewer than 3 oocytes retrieved with a conventional COH protocol); (3) an abnormal ovarian reserve test (AFC <5–7 follicles or AMH <0.5–1.1 ng/ml) [150]. It should be noted, however, that there are women who appear to respond adequately in spite of having the above characteristics (personal communication).

The management of patients with diminished ovarian reserve remains a major challenge. IVF treatment among poor responders is associated with poor clinical outcomes, including high no-start rates due to abnormally elevated baseline serum FSH level, high cycle cancellation rate due to poor response, and low fecundity rates. The optimal COH protocol for poor responders remains elusive. The ultimate goal of COH protocols for poor responders is to prevent premature follicle selection and to optimize uniform follicular development.

Luteal Estradiol Patch/ GnRH Antagonist Protocol

It has been shown that older women have premature elevation of FSH during the antecedent luteal phase of the menstrual cycle [72]. Among this patient population, ovarian antral follicles may respond to the prematurely rising levels in FSH, resulting in premature dominant follicle selection and accelerated development of one or two dominant follicles [151–153].

Administration of either oral estradiol (E_2) tablets or GnRH antagonist injections in the preceding luteal phase has been shown to effectively inhibit the premature rise in FSH; thus preventing premature follicular recruitment and allowing for a greater uniformity of antral follicle development [154, 155]. We reported a novel IVF protocol for poor responders involving luteal administration of both transdermal E_2 patches and GnRH antagonists followed by a flexible GnRH antagonist COH protocol using high-dose gonadotropins [156]. Patients applied 0.1 mg E_2 patch on alternate day, starting 10 days following LH surge. One day after starting the first E_2 patch, 0.25 mg GnRH antagonist is also administered for 3 consecutive days. Stimulation with gonadotropins is begun on day 2 of their ensuing menstrual cycle after assessment of baseline serum FSH, LH, and E_2 levels and US examination. Compared to prior unsuccessful IVF treatment cycles, the luteal E_2 patch/GnRH antagonist COH protocols resulted in improved clinical outcomes, including a lower cancellation rate and higher oocyte retrieved and embryos transferred [156]. We have utilized E_2 patches alone for a number of years and have shown it to be as effective as administration of E_2 patches with a GnRH antagonist. A recent study has confirmed this observation [157].

Co-Flare and Micro-Flare Protocols

Other commonly used treatment strategies include the use of GnRH agonist flare (co-flare) or microdose GnRH agonist (micro-flare) COH protocols in combination with high dose of gonadotropins [158–160].

The co-flare COH protocol takes advantage of the initial stimulation of GnRH receptors and consequent "flare up" secretion of endogenous gonadotropins which enhances the effect of exogenously administered gonadotropins. The co-flare COH protocol involves administration of 1 mg LA from cycle day 2 to cycle day 4 and 0.5 mg LA from cycle day 5 onwards [161]. High-dose gonadotropin (300–450 IU) stimulation is usually begun on cycle day 3. The flare response appears to be a good predictor of clinical outcomes. Doubling of serum E_2 level from cycle day 2 to 3 was associated with a lower cancellation rate and higher peak serum E_2, oocyte yield and live birth rates [161]. Theoretical concerns about the co-flare protocol relate to the enhanced ovarian androgen production, corpus luteum rescue, and the luteinization effect, all of which may adversely affect oocyte quality and pregnancy rates.

The OCP-microdose flare protocol is another commonly applied COH strategy for poor responders [149, 159]. OCP is administered for 14–21 days. Twice daily microdose Lupron (MDL) (40 μg) is started on the third day following the last dose of OCP. High-dose gonadotropins (300–450 IU) are started on the third day of MDL. The proposed advantage of MDL, which is equivalent to 1/50th the concentration of commonly utilized LA, is the ability to stimulate endogenous FSH release without increasing ovarian androgen production, corpus luteum rescue, and the luteinization effect.

Modification of the co-flare and MDL protocols by the addition of estrogen patches and GnRH antagonists in the antecedent luteal phase have also been proposed and may improve the ovarian response to COH [162].

Clomid and Letrozole Protocols

An attractive COH protocol for young patients with a history of poor embryo quality or poor response is a combination of either CC or letrozole with gonadotropins [163–165]. Administration of CC and letrozole leads to a decrease in E_2 negative feedback at the level of the pituitary. The resulting increase in endogenous gonadotropin secretion may enhance the ovarian response [166]. Other advantages include reduced hMG requirements and enhanced luteal phase progesterone (P4) secretions. The CC/hMG and letrozole/hMG protocols entail starting a 5-day course of 100 mg CC or 2.5–5 mg letrozole on cycle day 2. Stimulation with gonadotropins is started on cycle day 4.

Modified Natural Cycle Protocol

In recent years, especially since the introduction of GnRH antagonists, there is a resurging interest in modified natural cycle IVF as a treatment option for poor responders [167]. Modified natural cycle IVF involves utilizing ganirelix 0.25 mg and 150 IU hMG when the leading follicle reached 13 mm. Compared to natural cycle IVF, modified natural IVF cycles for IVF were more likely to result in ET when compared to natural IVF cycles [168, 169]. In the setting of poor responders, particularly those who develop less than three follicles despite aggressive high-dose gonadotropin protocols, modified natural-cycle IVF may represents a cost-effective alternative [170, 171].

5.2 Determination of Gonadotropin Dosage

The dose of gonadotropins should be individualized based on patient characteristics, including age, BMI, previous response to COH, and ovarian reserve markers, namely basal AFC, FSH, E_2, and AMH levels [84]. In normal responders, the usual starting gonadotropin dose ranges between 150 and 225 IU of FSH either alone or in combination with LH (either recombinant LH or urinary LH). Patients who are at risk of developing OHSS should be started with relatively low dose of gonadotropins (100–112.5 IU per day as 75 IU per day does not appear to be adequate for most women). In poor responders, the starting doses of gonadotropins are usually 300 IU FSH alone or in combination with 150–300 IU of hMG.

In a RCT comparing starting dose of 300, 450, and 600 IU FSH along with the MDL flare protocols, there were no differences in cancellation and pregnancy rates in poor responders, suggesting that 300 IU of FSH is a sufficient dose for most women [172]. However, the flare FSH response may partly explain the lack of differences observed.

5.3 Monitoring

5.3.1 Baseline Blood Works and Ultrasound

Baseline hormonal assays and pelvic US are performed on menstrual cycle day 2 for patients treated with the GnRH antagonist short protocol and on menstrual cycle day 3 for women who are down regulated with the long GnRH agonist protocol. Patients pretreated with OCP or E_2 patches are instructed to present on cycle day 2 of the withdrawal bleed, while women who are on the OCP/MDL protocol present for baseline blood works and US on the third day following the last dose of OCP.

Baseline hormonal evaluations include FSH, LH, E_2, beta hCG, and P4 determinations. E_2 is measured by radio immune assay (RIA) as it is more accurate and sensitive for quantitative assessment of E_2 levels below 100 pg/ml. Newly automated ELISA (DPC) assays accurately assess E_2 at concentrations over 100 pg/ml. Normal baseline parameters include FSH of <13 mIU/ml, RIA E_2 <less than 75 pg/ml, and P4 <1 ng/ml. Pelvic US is performed to evaluate endometrial thickness e (ES), AFC quantification and to ensure that no ovarian cysts or pathology exists.

Management options for patients with elevated baseline FSH and E_2 levels include cancellation of cycle or treatment with OCP for 2–4 weeks. Luteal priming with estrogen patches alone or E2 patches and GnRH antagonists should be considered in subsequent IVF cycles.

The management of an ovarian cyst is based on its size, appearance, and E_2 level. The presence of functional ovarian cysts, especially when associated with high E_2 level, have been shown to adversely affect IVF outcomes. Management options include cycle cancellation, administration of GnRH antagonists for 3–5 days or until resolution of the cyst, or treatment with OCP for 2–4 weeks. The finding of a nonfunctional, simple cyst with E_2 level of <75 pg/ml generally does not warrant delaying controlled ovarian stimulation.

We have recently shown that a thickened baseline ES, defined as being greater than 5 mm, on day 3 following luteal GnRH-agonist down regulation is associated with reduced implantation, clinical pregnancy, and live birth rates. Compared to cycles that started COH on day 3 (in the presence of thickened ES), cycles that were delayed for 3–4 days on LA had significantly higher clinical pregnancy (49 % vs. 37 %) and live birth (39 % vs. 22 %) rates [173]. The optimal management of thickened ES on cycle day 2 of GnRH antagonist protocols remains to be determined. Management options include delaying the starts with administration of GnRH antagonists for 2–3 days or simply repeating the evaluation in the following 2 days to ensure that the endometrium has shed.

5.3.2 Monitoring of Follicle Growth and Serum Estradiol Levels

Patient E_2 response to COH is assessed following 2–3 days of gonadotropin stimulation to allow for delicate adjustment. Monitoring of daily LH concentrations is only necessary in patients undergoing GnRH antagonists cycles, whereas monitoring of LH is unnecessary in patients undergoing long GnRH agonist protocols.

In the setting of high E_2 (over 300 pg/ml), the gonadotropin dose is reduced. E_2 level is repeated and US is performed the following day. In the setting of low E_2 (RIA value under 100 pg/ml), serum FSH level is verified to ensure gonadotropin injection is done properly. The gonadotropin dose usually remains unchanged for the first 4–5 days of stimulation in order to allow the gonadotropin threshold to reach a steady state. Repeat serum E_2 level and US are usually performed on cycle day 5 or 6 when the leading follicles reach 13 mm in diameter, monitoring of serum E_2 and US is performed daily.

In most ART centers, the follicle diameter is assessed using two-dimensional US. There is no standard approach to measuring follicular diameters [174]. Some ART programs use the single largest diameter, whereas others calculate the mean of two, three, or four diameters, measured in one or two planes as a surrogate of the true follicular size. Sonography-based automated volume calculation (SonoAVC) is a promising new technology that provides rapid and accurate automated measurements of the mean follicular diameter and a volume-based diameter for each follicle in the ovaries [175]. There appears to be an excellent correlation between SonoAVC-calculated follicular volume and true follicular volume [176–178].

5.4 Adjustment of Gonadotropin Dose

Once a gonadotropin threshold is established and E_2 levels reach 250–300 pg/ml while at the same time several follicles of 11–12 mm in average diameter are observed by US, the dose of gonadotropin can be reduced in a step-down manner. On the other had for high responders, a step-up approach for titrating gonadotropin dosage may be considered. In the step-up up protocol, patients are started on very low doses of gonadotropins (75–112.5 IU). The gonadotropin dose is gradually increased until the follicles reach 12 mm in diameter at which point one can begin to reduce the dosage in a step-down fashion.

For most patients, gonadotropin dose is reduced by 25–50 % during the course of stimulation. This step-down protocol appears to pharmacologically mirror the FSH dynamics of the natural cycle [179], allowing the larger follicles to continue to develop in a relatively low gonadotropin milieu. This step-down approach also improves synchronization of follicular maturation [180]. For high responders, an important advantage of the step-down protocol is the reduction in incidence of OHSS. The step-down protocol is also applicable to poor responders. In an RCT comparing step-down versus high fixed dose (450 IU FSH) regimens in GnRH agonist COH protocol, no difference was observed in clinical outcomes [181]. The step-down protocol conferred the additional benefits of reduced gonadotropin dosage and cost of medications.

5.4.1 Coasting

In the presence of high or rapidly rising serum E_2 level (2,500–3,000 pg/ml) and multiple immature follicles (greater than 20), a commonly applied strategy to prevent OHSS is coasting [182]. This strategy involves withholding gonadotropins while the GnRH agonist and antagonist are continued. By withholding the FSH stimulatory effect on the granulosa cells, coasting reduces the growth of FSH-dependent small and intermediate size follicles without affecting the growth of the larger dominant follicles. Coasting is presumably most effective when the leading follicles reach 15 mm in diameter before gonadotropins are withheld. When coasting is initiated before follicles reach 15 mm diameter, an abrupt arrest in follicular growth associated with rapid declines in the plasma estradiol level and oocyte quality is often observed.

A disadvantage of coasting is the risk of cycle cancellation. We usually delay hCG administration until the E_2 level decreases to less than 3,000 pg/ml [182]. However, consideration should be given to cancel the cycle if the E_2 concentration drops by more than 30 % or if the patient is coasted for four or more days. Oocyte quality is often poor under these circumstances. It has also been shown that prolonged coasting resulted in significantly lowered implantation and pregnancy rates [183].

5.5 Timing of hCG Injection

The timing of hCG injection should be individualized based on several factors, including follicle diameter, E_2 levels, prior cycle response and embryo quality, and the particular COH protocol. In normal responders, we typically administer hCG when at least two follicles measuring over 17 mm in average diameter are observed on US. Ideally, the E_2 level should be greater than 400 pg/ml for 3 or 4 days.

When there is a history of retrieval of predominantly immature oocytes, consideration should be given to administering hCG at greater follicular diameters (i.e., 19–20 mm). Similarly, when patients utilize CC or letrozole COH protocols, hCG injection is usually given when the leading follicles reach 19–21 mm in average diameters. Conversely, in patients with a history of poor embryo quality despite retrieval of mostly mature oocytes, consideration should be given to administer hCG earlier than the prior IVF cycle or at smaller follicular diameters. hCG administration should also be considered when there is a plateau or drop in E_2 once the leading follicle has reached 17–18 mm on US.

5.5.1 hCG Trigger in High Responders

In efforts to reduce the risk of OHSS in high responders, we have followed a flexible approach to hCG dosage administration tailored to patient response. For patients who are at risk of developing OHSS, the dose of hCG for triggering ovulation should be reduced to 3,300–5,000 IU. It has been our policy to administer 5,000 IU when E_2 is between 1,500 and 2,500 pg/ml. hCG dose is reduced to 4,000 IU if the E_2 is between 2,500 and 3,000 pg/ml. The decision to administer hCG is individualized in patients with E_2 of 3,000 pg/ml. hCG is withheld if the E_2 significantly exceeds 3,000 pg/ml if the patient has more than 12 follicles and complains of bloating discomfort. Alternatively, if the patient has six or fewer follicles and is asymptomatic, a reduced dose of 3,300 IU hCG can be administered.

5.5.2 GnRH Agonist Trigger

An important advantage of GnRH antagonist COH protocols is the possibility of using GnRH agonists to induce the LH surge. A single injection of 1–2 mg LA is usually adequate to trigger an LH surge. The mechanisms by which a GnRH agonist trigger prevents OHSS is related to the shorter half-life of the elicited endogenous LH, subsequent pituitary suppression, and withdrawal of LH support for the corpora lutea, all of which results in early luteolysis [135, 136]. LH remains in the circulation for less than 24 h and has very little luteotrophic effect [184]. In contrast, serum hCG levels remain 7–10 days following intramuscular injection [185].

There are drawbacks to using GnRH agonists to trigger ovulation. The GnRH agonist may not be effective in patients with overly suppressed LH secretion or those with hypothalamic amenorrhea. Moreover, the use of GnRH agonists to trigger ovulation appears to be associated with low clinical and ongoing pregnancy rates and an increase in pregnancy loss. The above observations are believed to be related to the associated luteal dysfunction and impaired steroid secretion following administration of GnRH agonists [135, 136].

Several investigators explored using hCG in doses of 1,000–2,500 IU to supplement the GnRH-induced ovulatory trigger. In an effort to reduce the pregnancy loss rate associated with the agonist trigger, Humaidan et al. examined the effects of administering 1,500 IU hCG either 12 or 35 h after the GnRH agonist and found that the live birth rates were comparable to those observed following hCG trigger [186]. Shapiro et al. used a combination of GnRH agonist with hCG, using hCG doses between 1,000 and 2,500 IU [187]. The two groups had similar ongoing pregnancy rates. No case of OHSS was observed. Thus, a "dual trigger" using a GnRH agonist and 1,500 IU hCG may be effective in promoting final oocyte maturation without increasing the risk of OHSS. Based on the above findings, we often implement the GnRH agonist-1,500 U hCG dual trigger for patients who are at risk of OHSS

5.5.3 Monitoring Following hCG Injection

It has been our practice to measure hCG/LH and E_2 levels on the morning following the ovulatory trigger hCG injection. Serum hCG and LH levels serve as quality assurance measures to ensure hCG and GnRH agonist have been administered properly. In the rare event that the serum hCG or LH levels are undetectable, a repeat hCG dose is given and serum E_2 and hCG levels are repeated on the following day. If the post-hCG E_2 level drops by over 30 %, consideration should be given to cycle cancellation. A significant fall in E_2 levels may be a harbinger of poor follicular integrity and may be associated with poor oocyte and embryo quality.

Patients with E_2 levels >3,000 pg/ml on the day of hCG or E_2 levels >4,000 pg/ml on the day after hCG require close monitoring for signs and symptoms of impending OHSS. Patients are evaluated on the third and, if necessary, fifth day following oocyte retrieval. The evaluation includes a complete physical examination, measurement of waist circumference, weight, pelvic US for ovarian size and presence of ascites, blood evaluation of hematocrit, liver transaminases, and kidney function tests. In women with symptoms indicative of clinical progression of OHSS, all embryos are cryopreserved, thus avoiding pregnancy associated OHSS.

6 Oocyte Retrieval

In the early days of IVF, oocyte pick-up (OPU) was performed laparoscopically under general anesthesia [1, 2]. Due to the invasiveness of the procedure, this approached has been abandoned. Today, transvaginal US-guided follicle aspiration has become the gold standard for OPU [188–190]. OPU is performed as an outpatient procedure under conscious sedation, or local, epidural, spinal, and general anesthesia. Conscious sedation is the most commonly used method of analgesia and anesthesia during OPU [191] and is performed in 95 % cases in the USA [192]. In certain European countries, such as Germany, OPU is more commonly performed under general anesthesia. A recent meta-analysis found no single anesthetic method to be superior for pain relief or IVF outcome [193].

OPU is scheduled for 34–36 h following hCG administration. In patients with a history of retrieval of mainly immature oocytes, it may be beneficial to schedule the OPU 36–39 h post-hCG injection in order to allow a greater yield of mature oocytes.

OPU is performed with the patient being placed in the dorsal lithotomy position. The vagina and perineum are prepped with povidone–iodine or hexachlorophene solution to minimize the risk of iatrogenic pelvic infection, followed by copious irrigation with sterile saline solution. Although the incidence of infection following OPU is low (0.4 %) [130, 194], intravenous administration of prophylactic antibiotic should be considered for patients with a history of pelvic inflammatory disease, severe endometriosis, and pelvic adhesions following ruptured appendicitis or multiple prior pelvic surgical procedures. We administer a course of oral tetracycline following OPU. All oocyte donors receive 2 g of intravenous cefoxitin prior to OPU.

A high frequency vaginal US transducer (5–7 MHz) is used to visualize the ovaries. OPU is accomplished through US-guided aspiration with either a single- or double-lumen, 16 or 17-gauge disposable needle. Double-lumen needles, composed of one channel to withdraw follicular fluid and another to instill isotonic saline into the follicle, enable simultaneous or intermittent flushing and aspiration of ovarian follicles. A consistent vacuum pressure of 80–100 mmHg assists in the collection of follicular fluid, which is immediately transported to the embryology lab for evaluation.

Follicular aspiration followed by flushing has been suggested in order to increase the number of oocytes recovered [195]. Recent meta-analysis showed that flushing was not associated with improved clinical or ongoing pregnancy rates or an increase in oocyte yield [196]. In fact, flushing of follicles resulted in increased operative time and opiate analgesic requirement. Nonetheless, occasionally follicular flushing may be beneficial in patients with few follicles, such as in natural or minimal stimulation cycles and those undergoing IVM treatment [197].

7 In Vitro Fertilization

In the embryology lab, the follicular aspirates are transferred to a HEPES-buffered media, in order to maintain physiologic pH. On a heated (37 °C) microscope stage, the cumulus–oocyte complexes (COC) are identified in the follicular aspirates and cleared of any blood clot or debris. Evaluation of oocyte maturity is based on the morphological appearance of the COC. Immature oocytes, germinal vesicle (GV) or metaphase I, have densely packed cumulus and coronal layers. Mature, metaphase II (MII) oocytes have highly dispersed cumulous cells and a radiating coronal layer. The COCs are transferred to culture media, maintained at 37 °C in an atmosphere of 5 % CO_2.

A semen specimen is usually obtained by masturbation immediately prior or following OPU. Sperm preparation is performed by either the "swim up" or the density-gradient method to isolate a high concentration of motile sperm. The isolated motile sperm is incubated in media containing high concentrations of protein for 30 min to 4 h in order to achieve capacitation.

The decision to fertilize the oocyte by insemination or ICSI is based on the available post-wash total motile sperm. Conventional IVF is performed 4–6 h after OPU. Each mature oocyte is incubated with 50–100,000 sperm/ml for 12–18 h at 37 °C in 5 %

CO_2 and 98 % relative humidity. Fertilization is confirmed by the presence of 2 pronuclei (PN) and the extrusion of the second polar body 18 h post-IVF. The fertilization rate achieved by IVF is in the range of 50–70 %. Oocytes of intermediate maturity, metaphase I (MI), may undergo IVM in culture medium for 12–24 h prior to fertilization, although pregnancy rates of embryos derived from in vitro matured oocytes are suboptimal.

It is crucial to identify oocytes with 3 PN due to their inability to divide normally as they appear identical to 2 PN embryos on the day of transfer. Polypoid embryos are the result of either polyspermic fertilization or fertilization of a diploid oocyte (digyny). Retention of the second polar body during the second meiotic division is the most common cause of digynic triploid embryos but may less frequently result from retention of the first polar body during the first meiosis division. The estimated incidence of polypoidy is in the range of 5–10 %. The incidence is much higher in immature oocyte (up to 30 %) [198]. Transfer of triploid embryos may lead to implantation failure or miscarriage [199].

8 Intracytoplasmic Sperm Injection (ICSI)

ICSI circumvents the need for the sperm to penetrate the zona pellucida (ZP) and may overcome impairment in spermatozoa–oocyte interaction. In 1992, Palermo et al. reported the first series of live births following ICSI for couples with severely impaired sperm characteristics, in whom IVF and subzonal insemination had failed [200]. Today, ICSI has become an established procedure and an integral part of ART, enabling males previously considered infertile to father children [201]. During the past 10 years, the number of fresh IVF cycles performed with ICSI more than doubled, from 28,090 in 1999 to 67,328 in 2008. Today, ICSI is performed in 64 % of nondonor, fresh IVF cycles in the USA [5].

8.1 Indications

1. The most common indication for ICSI is severely impaired sperm characteristics, including severe oligospermia (fewer than five million motile sperm/ml), severe asthenospermia (less than 5 % progressive motility), and severe teratospermia (less than 4 % normal morphology based on strict Kruger criteria) [202]. In these settings, poor fertilization rates have been observed using conventional IVF [202]. The use of ICSI has enabled the fertilization rates to reach 70–75 % [200, 203] rates, which are comparable to the fertilization rates achieved by IVF in patients without severe male factor infertility.
2. ICSI is indicated for patients with a history of prior or repeated fertilization failure with standard IVF.
3. ICSI is always performed whenever surgically retrieved spermatozoa are recovered either by microsurgical sperm

aspiration (MESA) [204] or testicular sperm extraction (TESE) [204, 205]. The advent of IVF-ICSI has also enabled patients with Klinefelter syndrome to have biological offspring [206]. There have been sporadic reports of pregnancies following surgically retrieved spermatid injection in the setting of non-obstructive azoospermia [207].

4. ICSI is commonly applied to achieve fertilization when spermatozoa have been previously cryopreserved, as cryopreservation may result in deterioration of all semen parameters [208]. In a review of 118 couples who underwent 169 IVF cycles at our institution using semen samples that were cryopreserved before cancer treatment, fertilization with ICSI resulted in significantly higher fertilization and live birth rates compared to IVF [209].
5. The presence of antisperm antibodies may affect sperm motility and impair fertilization [210]. ICSI has been highly effective in increasing fertilization rates in patients with exceedingly high sperm antibody titers (greater than 80 %) as detected by mixed antiglobulin reaction (MAR) or direct IgG and IgA immunobead testing [205].
6. ICSI is routinely performed to fertilize in vitro matured and cryopreserved-thawed oocytes. IVM and cryopreservation have been shown to cause hardening of the ZP which may impair sperm penetration [211].
7. ICSI is indicated for patients undergoing IVF and PGD for single gene defects. Conventional insemination technique may result in extra spermatozoa attaching to the ZP which may interfere with the PCR analysis.

8.2 The ICSI Procedure

ICSI is performed on mature MII oocytes. The cumulus cells are enzymatically removed using hyaluronidase to allow confirmation of maturity. Removal of the cumulus cells also enhances the visualization of the oocyte during the ICSI procedure. When available, a motile, morphologically "normal" spermatozoan is selected for injection. A holding pipette gently maintains the oocyte position with the polar body oriented to the 12 o'clock position. The injection pipette containing the spermatozoon is introduced sequentially through the ZP and the oolema at the 3 o'clock position until a break in the oolema is observed. The sperm is released and the pipette slowly removed. ICSI is commonly performed at a 90° angle from the first polar body in order to avoid the meiotic spindle apparatus, which is usually located adjacent to the first polar body.

The meiotic spindle may not always be predicted by the location of the first polar body [212]. There is a theoretical concern that the meiotic spindle apparatus may be damaged by the microinjection needle. Based on the birefrigent property of the meiotic spindles, computer-assisted polarization microscopy systems may be used to monitor the meiotic spindle position during ICSI [213]. The presence of a normal birefrigent meiotic spindle apparatus at the time of ICSI has been associated with improved oocyte and embryonic developmental capacity [214, 215].

When triploid fertilization is found following ICSI, these represent digynic embryos. As with dispermic fertilization following conventional IVF, these embryos should not be transferred.

9 Embryo Culture

Following fertilization by IVF or ICSI, the zygotes are transferred to the embryo culture media. A variety of culture media are available commercially. They can be broadly categorized into monoculture or sequential culture systems. The monoculture media were developed to support zygote development to the blastocyst stage without changing the culture media. Traditional monoculture embryo media were formulated to mimic human tubal fluid (HTF). At present, most commercially available embryo culture systems involve the use of sequential culture media. The formulations of the sequential embryo culture system are based on the concept of different nutrient and energy requirements of cleavage stage and blastocyst stage embryos. Pre-compacting embryos require pyruvate and nonessential amino acids as nutrient sources while the post-compacting embryos prefer glucose and essential amino acids as energy sources [216–218]. The sequential culture system involves two different culture media. The first medium provides optimal support for the development of cleavage stage embryos. These embryos are subsequently transferred to a second culture medium to optimize extended culture to the blastocyst stage.

The efficiency of in vitro culture system is not only dependent on its chemical formulation, it is also critically affected by pH, gas concentration, and temperature. Other factors affecting embryo culture system efficiency include volume of the culture media droplet, the type of protein supplement, and the presence of reactive oxygen species [219, 220].

Although the first IVF pregnancy was achieved following the transfer of blastocysts, worldwide [2], the majority of embryo transfers are performed on day 3 cleavage stage embryos. According to the 2008 SART registry, only 35 % of all transfers occurred on day 5 [5]. The main concern has been the efficacy of the in vitro culture system and the reduced developmental potential of in vitro cultured embryos. Recent advances in cell culture media have led to a shift in ART practice from transferring day 3 cleavage ET to day 5 blastocyst stage embryos. Extended culture to the blastocyst stage allows for selection of embryos with the highest implantation and developmental potential, therefore reducing the number of embryos transferred and decreasing the risk for multiple births.

Compared to day 3 transfer of cleavage stage embryos, blastocyst transfer on day 5 or 6 was associated with higher implantation (Day 3: 12–20 % vs. Day 5/6: 30–60 %) and live-birth rates (Day 3: 29.4 % vs. Day 5/6: 36.0 %) [221]. In fact, in patients under the age of 40, elective single blastocyst transfer significantly reduces the incidence of twins without reducing the overall pregnancy rate [221].

The beneficial effects of blastocyst transfer appear to be reduced in patients with multiple prior unsuccessful IVF cycles. In patients with three or more previous unsuccessful day 3 transfer cycles who were randomized to day 3 versus day 5 transfers, the implantation rate (21 % vs. 6 %) and clinical pregnancy rate (22 % vs. 13 %) were higher with blastocyst transfer, but the live birth rates were comparable between the two groups (10 % vs. 13 % blastocyst) [221]

The decision to proceed with extended embryo culture is based on several good prognostic factors, including young age, good ovarian reserve, prior successful IVF cycles, and the development of six or more, high quality, cleavage stage embryos on day 3 post-retrieval.

There are drawbacks to extended embryo culture, including fewer surplus embryos available for cryopreservation, the possibility of arrested embryo development, and failure to transfer any embryo [222]. These drawbacks are less likely to occur in patients with good prognostic factors. A recent meta-analysis found the cancellation rates to be comparable between day 3 and day 5–6 transfers in patient with good prognosis [221]. Blastocyst culture has also been associated with a higher incidence of monozygotic twins, altered sex ratio in favor of males [223], and possibly increased incidence of imprinting disorders [224, 225].

10 Additional Laboratory Procedures

10.1 Assisted Hatching

The ZP serves as an important barrier to polyspermic fertilization and protects the developing embryo from the surrounding female genital tract during transport. Upon entry to the uterine cavity, the blastocyst secretes an activator of zona lysins, which dissolves the ZP [226, 227], allowing the trophoblastic cells to interact directly with endometrial cells at the site of implantation.

The goal of AH is to improve implantation by creating an artificial breach in the ZP. AH is performed of day 3 embryos in order to permit adequate blastomere adherence just before compaction. A variety of techniques have been employed for AH. These include mechanical partial zona drilling or dissection (PZD) [228], thinning of the ZP using Tyrode's acid [229], Piezo vibro-accustic manipulators [230], and lasers [231]. Cohen et al. first reported the beneficial effects of AH using PZD [232] and, subsequently, Tyrode's acid solution in improving the implantation rates in patients over the age of 38 and those with elevated basal FSH levels [228].

PZD is performed by tangential piercing of the ZP from the 11 to 1 o'clock position. This segment is then rubbed with the holding pipette until a slit is made in the zona. With the acid Tyrode's technique, a micropipette containing Tyrode's acid is positioned close to the zona and the solution is gently delivered. In zona thinning, the zona is attenuated without creating an actual gap in the zona. Laser AH using both contact and non-contact delivery systems are the most commonly used AH techniques today.

The available evidence does not support universal application of AH [233]. A recent Cochrane meta-analysis of 28 RCTs found an increase in pregnancy rate but not in live birth rate following treatment with AH [234]. AH may have a beneficial effect in selected population of patients, namely those with advanced maternal age, diminished ovarian reserve, and repeated implantation failures following IVF. The use of AH has also been advocated in certain conditions, including thickened ZP and in vitro conditions which cause hardening of the ZP, such as cryopreservation and IVM [235, 236].

Potential complications related to AH include damage to the blastomere during micromanipulation and possible compromise to the embryo's developmental potential. AH may also increase the incidence of monozygotic twinning [237, 238].

As originally reported, patients who are scheduled to undergo embryo AH are treated with a course of oral methylprednisolone following oocyte retrieval. It has been shown that methylprednisolone improved implantation of PZD treated embryos possibly through an immunosuppressive effect [239]. However, the exact effect of methylprednisolone on the interaction between immune cells and the micromanipulated embryos is not well elucidated.

10.2 Autologous Endometrial Coculture (AECC)

The main objective of AECC is to improve embryo developmental potential. AECC intends to mimic the uterine environment by co-incubation of the embryos with endometrial cells in vitro. An endometrial biopsy is performed using a Pipelle endometrial suction curette 5–12 days after ovulation in a preceding cycle [240]. The stromal and glandular cells are isolated and cryopreserved. The cells are thawed on the day of hCG administration. Zygotes are placed on this reconstituted layer of endometrial cells and are cultured until post-retrieval day 3. Proposed mechanisms of action of AECC include the release of cytokines (GM-CSF, LIF, and the IL-1 system) from the stromal and epithelial cells of the endometrium [241–243], secretion of embryotrophic factors [244], and detoxification of the culture medium by removal of heavy metal cations, free radicals, or metabolic inhibitors [245, 246].

Patients with a history of poor embryo quality and/or repeated implantation failure may benefit from AECC [240, 247–250]. We have shown that AECC improved embryo quality and pregnancy outcomes in patients with repeated implantation failure [240, 247, 251]. A recent meta-analysis confirmed the beneficial effects of AECC in improving embryo quality and increasing implantation, clinical, and ongoing pregnancy rates [252].

11 Embryo Transfer

The ET procedure is considered to be a most critical step in the IVF process (See Schattman chapter). ET is usually performed on day 3 for replacement of cleavage stage embryo or on day 5 or day 6 for replacement of blastocysts. The main objective is to transfer the embryo(s) to the appropriate location within the uterine cavity in an atraumatic manner in an effort to maximize the chance of implantation. Variables affecting the outcomes of ET have been studied extensively. In particular, pregnancy rates after ET have been shown to be influenced by the choice of ET catheter, performance of trial transfer, use of US guidance, and most recently the supplementation of the ET media [253].

11.1 Embryo Transfer Catheter

Several catheters are commercially available and can be broadly categorized as either "soft" (Cook–Cook, Ob/Gyn, Inc., Bloomington, IN; Wallace–Marlow Technologies, Willoughby, OH) or "firm" (Tomcat and Tefcat–Kendell Health Care, Hampshire, MA; TDT and Frydman–Laboratoire CCD, Paris, France). Two meta-analyses found that soft catheters are associated with higher clinical pregnancy rates compared to firm catheters [254, 255]. There appears to be little difference in clinical efficacy between the Cook and Wallace soft catheters [255, 256]. One study noted catheter insertion failure to be more common with the Wallace catheter than with the Cook catheter [256]. For patients with a stenotic cervical os or a distorted cervical-uterine junction, firm catheters may facilitate the ET but may be associated with more bleeding, trauma, and stimulation of uterine contractions.

11.2 Patient Preparation

The use of an appropriate speculum is important in order to achieve good visualization of the cervical os and to straighten the cervical-utero angle. The cervical-utero angle can also be straightened by performing the ET under a full bladder [257] or applying active traction by placing a suture or tenaculum on the cervix [258].

Following application of a bivalve speculum, the cervix is cleansed with saline or culture media to decrease bacterial contamination. Excess mucus on the cervix should be atraumatically removed. Mucus in the cervical canal can be aspirated using a syringe. Excess cervical mucus may cause ascending bacterial infection of the uterine cavity and has been shown to adversely affect the clinical and ongoing pregnancy and implantation rates [259]. The mucus can also obstruct the tip of the transfer catheter or entrap the embryo, thus resulting in misplacement or embryo loss(s).

11.3 Trial Transfer

We routinely perform a trial transfer immediately prior to the actual ET utilizing a soft ET catheter. This trial transfer is performed in order to confirm the depth and direction of the uterine cavity and to determine any potential difficulty that may otherwise occur during the actual transfer. Pre-cycle trial transfers may predict difficult ET but may not accurately predict the length of the uterine cavity at the time of the actual ET [260]. It has been shown that the uterine position may change between the time of the trial transfer and that of the actual ET [261, 262]. A retrospective study from our institution found that 19 % of patients had a discrepancy of ≥1.5 cm and 30 % had a difference of ≥1 cm in uterine cavity length between measurement at the time of the trial transfer and the actual ET [112]. This was more likely to occur in patients with a retroverted uterus.

11.4 Ultrasound Guidance

Since first reported by Strickler et al. in 1986 [263], the use of US guidance for ET has become more prevalent as it facilitates ET and reduces the incidence of difficult transfers [264]. A recent meta-analysis of 17 RCTs comparing US-guided ET versus ET by clinical touch concluded that US-guided ET improved the clinical and ongoing pregnancy rates [265]. US guidance enables the clinician to visualize the catheter depth at the time of transfer, allowing proper placement of the embryo(s) and minimizing trauma to the endometrial lining. US guidance is particularly helpful in patients with uterine myomas and those with previous caesarean section scar defects in which the ET catheter can be misplaced. Disadvantages of US-guided ET include patient discomfort from both maintaining a full bladder and abdominal compression with the US probe. US-guided Et also requires a second operator and may increase procedure time.

11.5 Embryo Transfer Techniques

For patients with a stenotic cervical canal or a distorted cervical-uterine angle, adjusting the outer sheath of the catheter to approximate the angle of the cervix may facilitate the navigation of the catheter through the cervical canal [266]. Alternatively, a malleable stylet can be used to pass the outer sheath of the soft catheter through the cervical canal to the level of the internal os. The inner catheter loaded with embryos is then introduced through the outer sheath. This technique is known as "embryo afterloading."

Once the length of the endometrial cavity has been determined and the appropriate transfer catheter has been selected, the embryologist is instructed to mould the outer sheath of the catheter to approximate the angle of the cervix. The transfer catheter is then loaded with the embryo(s) and handed to the clinician. The transfer catheter is inserted through the cervical canal into the uterine cavity. Attention should be paid to ensure atraumatic passage through the cervical canal and to avoid contacting the uterine fundus or causing trauma to the endometrium. Bleeding or disruption of the endometrium leads to prostaglandin release and may induce uterine contractions. Blood or cervical mucus at the tip of the ET catheter also increases the risk of retained embryo(s) [267].

The ideal placement of the embryos is 1–2 cm below the fundus of the uterine cavity. Most commonly, ET is performed at 1–2 cm lower than the distance measured at trial transfer using a soft non-tapered Wallace ET catheter. Placement of the embryos close to the fundal wall may increase the risk of ectopic pregnancy [268, 269]. Conversely, placement of the embryos close to the internal cervical os may increase the incidence of cervical pregnancy [270].

Following placement of the embryo(s), the catheter is slowly withdrawn from the uterus. Some physicians prefer to wait 30 s before withdrawing the catheter, while others remove the catheter immediately following transfer. The 30 s delay does not appear to improve clinical pregnancy rates [271]. Pressure on the plunger of the syringe should be maintained until the catheter is completely withdrawn from the cavity in order to minimize negative pressure within the ET catheter. After the ET, the catheter is returned to the embryologist. Under the microscope, the catheter is flushed and carefully inspected for possible retained embryos. Pregnancy rates do not appear to be compromised when embryos are retained and are immediately retransferred into the uterine cavity [267].

11.6 Embryo Transfer Media

The embryo(s) are typically loaded onto tip of the transfer catheter in 20 μl of commercially available ET media, which are most commonly composed of synthetic HTF and macromolecules, such as albumin or serum substitute supplement. The composition of the medium surrounding the embryo(s) at the time of the ET may have an important role in embryo implantation. Recent studies have focused on supplementation of transfer medium with hyaluronan (HA), a glycosaminoglycan found in human fluid secretions and extracellular matrix of the reproductive tract [272]. The levels of HA have been shown to increase dramatically on the day of implantation [273], supporting its role in embryo implantation. Supplementation of ET medium with HA may improve clinical outcomes following ET [274–277]. In a RCT involving 1,282 patients, enrichment of transfer medium with HA significantly improved the clinical pregnancy (54.6 % vs. 48.5 %, OR 1.28) and implantation (32 % vs. 25 %, OR 1.43) rates. Two other studies of a smaller sample size failed to find any improvement in clinical pregnancy rates in patients using HA-enriched medium for ET [278, 279]. The beneficial effect of HA-enriched medium may be more evident in selected groups of patients, namely those with tubal factor infertility and those with recurrent (>4) unsuccessful IVF cycles [276, 278].

11.7 Bed Rest

In many ART programs, patients are advised to have bed rest for 30 min following ET. The rationale behind this recommendation is that decreased physical activity promotes embryo retention within the uterine cavity. A recent meta-analysis found insufficient evidence to support the practice of bed rest following ET in improving pregnancy outcomes [280].

11.8 Operator Experience

Although variables affecting the outcomes of ET have been studied extensively, the efficacy of ET appears to be most dependent on operator experience and skill [281]. Pregnancy rates have been shown to be improved with increased operator experience. A recent RCT comparing Cook and Frydman catheters involving three experienced physicians showed that variations in pregnancy rates between the two ET catheters were mostly operator dependent [281].

12 Luteal Support

Normal physiologic function of the corpus luteum requires the presence of adequate granulosa and theca cell components, which in turn require adequate pulsatile LH support during the luteal phase [282]. Estrogen secretion from the granulosa cells promotes proliferation of endometrial cells in the basal layer and upregulation of P4 receptors [283]. P4 secretion from the theca cells induces endometrial glandular secretion and decidualization of the stromal layer [284]. Adequate concentrations of both estrogen and P4 are essential for optimal endometrial maturation before embryo implantation. Withdrawal of LH secretion at the end of the menstrual cycle results in regression of the corpus luteum or luteolysis. In the event of embryo implantation, hCG secretion by the syncytiotrophoblast cells maintains the steroidal production of corpus luteum.

The importance of P4 in supporting early pregnancy was first demonstrated by Csapo et al. in a series of experiments [285]. Removal of the corpus luteum, or luteectomy, prior to 7 weeks gestation resulted in pregnancy loss [285]. Luteectomy performed after 7 weeks resulted in a transient decrease in the serum P4 level but the pregnancies persisted. In a subsequent study, Csapo et al. demonstrated that pregnancy can be maintained by exogenous P4 supplementation in luteectomized patients [285]. These studies demonstrated that P4 secretion from the corpus luteum is essential for maintenance of pregnancy up to 7 weeks gestation. The estimated onset of shift from luteal to placental steroidogenesis, the luteoplacental shift, occurs between 7 and 9 weeks gestation. Further experiments utilizing the egg donation model suggested even an earlier shift at 6–7 weeks [286].

12.1 Possible Etiologies of Luteal Phase Defect in IVF Cycles

Although COH and oocyte retrieval results in the development of multiple corpora lutea, in some women there is inadequate steroidal production from the corpora lutea following IVF treatment, resulting in luteal phase defects (LPD) [287]. Although several etiologies of LFD have been proposed, the exact mechanism of the LPD remains debatable. One proposed etiology in IVF cycles attributes the defect to the removal of large quantities of granulosa cells during the oocyte retrieval resulting in impaired P4 synthesis by the corpora lutea [288, 289]. Another proposed etiology relates to the supraphysiologic levels of steroids secreted by the corpora lutea following COH. Fauser and Devroey postulated that the supraphysiologic level of steroids inhibits LH secretion via a long-loop negative feedback inhibition at the level of the hypothalamic-pituitary axis [290]. hCG administration for final oocyte maturation may also exert a similar suppressive effect on LH production via a short-loop feedback mechanism [291]. The most widely supported cause of the LPD relates to the delayed recovery in gonadotrope functions following prolonged pituitary down-regulation with GnRH agonist [292]. Recovery of gonadotrope functions may require 2–3 weeks [292, 293]. Although the duration of pituitary suppression is shorter after treatment with GnRH antagonists, GnRH antagonist treatment also predisposes to LPD. Luteal serum LH and P4 levels and the duration of the luteal phase were shown to be diminished in patients treated with GnRH antagonist COH protocols [294, 295].

Based on the current understanding of the LPD in patients undergoing COH and IVF, hormonal luteal phase support (LPS) has become a standard practice in ART. LPS is given in various forms of P4, estrogen, hCG, or in combination. The initiation of LPS varies among different ART programs, ranging from the day of HCG, the day of oocyte retrieval, and the day following retrieval to the day of ET. A RCT evaluated the different time of onset of LPS and found no effect on the clinical outcomes in GnRH agonist down-regulated COH cycles [296]. Typically, we initiate LPS on the day following retrieval. The optimal dose, duration, or type of treatment also remains to be determined. Nonetheless, two meta-analyses confirmed that treatment outcome improved with all forms of LPS when compared to COH-IVF cycles where no luteal support was given [297, 298].

12.2 Progesterone Support

A variety of P4 formulations are currently available, including oral, vaginal, rectal, and intramuscular (IM) forms [299]. IM injection of P4 in oil has been widely used for LPS. A typical dose is 50 mg daily, which achieves serum P4 concentration at or above the physiological range. The main drawbacks of IM P4 injections are the commonly associated side effects, including pain at the injection site, rash, inflammatory reactions, cellulitis, and abscess formation.

P4 support can also be administered in the form of vaginal tablet inserts (Endometrin 100 mg, Ferring Pharmaceuticals, Inc., Parsippany, NJ), suppositories of micronized P4 (Prometrium, 100 mg, Solvay Pharmaceuticals Inc., Marietta, GA) and, most recently, gel (Crinone 8 %, Wyeth Laboratories, St. Davids, PA). Common daily dosages range from 200 to 600 mg. The clinical efficacy of vaginal P4 gel and other forms of vaginal P4 appear to be comparable [300]. Compared to IM P4, vaginal P4 administration achieves a lower serum concentration, but the local endometrial concentration appears to be greater [301]. Vaginal P4 supplementation is favored by many clinicians as it is associated with improved patient satisfaction and less discomfort.

Orally administered P4 is rapidly metabolized in the hepatobiliary system and has been shown to be the least effective method of LPS. Compared to IM or vaginal P4, LPS in the form of oral P4 was associated with reduced implantation rates [302, 303].

Comparison between the efficacy of IM and vaginal P4 for LPS led to conflicting results. Two RCTs comparing IM and vaginal P4-favored IM P4. Patients treated with IM P4 had higher pregnancy, implantation, and live birth rates [304, 305]. A 2008 Cochrane meta-analysis found IM P4 to be associated with higher ongoing pregnancy and live birth rates than the vaginal route [306]. In recent years, there is increasing evidence in the literature to suggest that vaginal and IM P4 have comparable efficacy. An open-label, multicenter RCT involving 1,184 women found vaginal P4 gel and IM P4 to have comparable clinical and ongoing pregnancy rates [307]. A 2009 meta-analysis found no significant difference between IM and vaginal P4 forms in terms of clinical and ongoing pregnancy rates [308].

12.3 Estrogen Supplementation

Optimal endometrial development and maturation require adequate concentrations of both estrogen and P4. Estrogen secretion from the granulosa cells promotes proliferation of endometrial cells in the basal layer and upregulation of P4 receptors [283]. It has been shown that COH following GnRH agonist down regulation lead to a decline in mid-luteal phase serum E2 and P4 levels [309]. Moreover, low luteal E2 levels have been associated with poor IVF outcomes [310].

Based on the above findings, several RCTs evaluated the clinical benefit of supplementing luteal phase P4 with E2, in the forms of either vaginal suppositories or patches [311–319]. Three meta-analyses concluded that the addition of E_2 to P4 for LPS did not improve outcomes in both GnRH agonist and antagonist IVF cycles [320–322].

Luteal E2 supplementation appears to be crucial in patients who received GnRH agonists for the ovulatory trigger. Luteal function is compromised following the GnRH agonist trigger, resulting in premature luteolysis and poor steroidal production by the corpus luteum [135, 136]. Thus, it is critical to provide these patients with luteal E2 in addition to P4 support. This can be achieved by administration of daily E_2 and P4 suppositories starting on the day following oocyte retrieval. We prefer to support the luteal phase with alternating day application of 0.1 mg E_2 patch and 50 mg intramuscular P4. Patients are monitored on a weekly basis in order to maintain the serum P4 above 20 ng/ml and serum E_2 level above 200 pg/ml.

12.4 hCG Support

hCG administration can be utilized to improve corpora lutea function by stimulating production of E_2 and P_4 [323]. For LPS, hCG is usually administered in several small doses of hCG (1,500–2,500 IU) 5–10 days following oocyte retrieval [324, 325]. The clinical efficacy of hCG administration for LPS appears to be comparable to P4 [298]. However, its use can significantly increase the risk of OHSS [298].

12.5 When to Stop LPS?

There is no universal consensus on the optimal duration for LPS. A survey of 21 ART programs showed great variations in LPS protocols, ranging from stopping LPS on the day of positive hCG to continuation for up to 12 weeks of pregnancy [326]. Withdrawal of vaginal P4 at the time of a positive pregnancy test did not appear to influence the miscarriage rate [327, 328] but was associated with increased bleeding episodes [328]. A recent RCT compared stopping LPS on the day of the first viability US scan versus continuing LPS for three additional weeks following the first US scan (up to 9 weeks gestational age) [326]. No differences in miscarriage rates or vaginal bleeding were observed between the two groups.

13 Complications of ART

Complications and risks of ART can be related to the use of gonadotropin stimulation or the actual retrieval procedures. Common side effects of gonadotropins include bloating, weight gain, nausea, mood swings, and breast tenderness [130]. The incidence of serious and minor complications experienced by women undergoing COH and OPU are generally rare [130, 329, 330]. Adnexal torsion is a rare complication related to COH. In a case series of 2,495 IVF cycles, there were two cases of adnexal torsion [330]. The incidence may be higher among patients with OHSS. Complications related to oocyte retrieval include bleeding, infection, ovarian abscess, ureterovaginal fistula, and pseudo-aneurysm of the iliac artery [130, 329, 330]. In a review 587 donors who underwent 973 cycles of COH and 886 OPUs at our institution, there were two cases of ovarian torsions (0.2 %), two significant infections (0.2 %), and one case of ruptured ovarian cyst necessitating hospitalization (0.1 %) [130]. Recently, one report suggested that woman undergoing COH for IVF have an increased prevalence of thromboembolic disease compared to the general population [331].

ART is associated with three notable complications: OHSS, multiple pregnancy, and ectopic pregnancy.

13.1 OHSS

A complication that occurs almost exclusively with the use of gonadotropin stimulation is OHSS [39]. The overall incidence of OHSS is estimated to be in the range of 0.4–10 % [332]. Approximately 0.0005–4 % of IVF patients develop severe OHSS [39, 333].

OHSS is an iatrogenic medical complication of COH with gonadotropins arising because of the exposure of multiple ovarian follicles to hCG. The fundamental physiologic changes observed in OHSS relate to vascular hyperpermeability secondary to the elaboration of hyperpermeability factors from the overstimulated, enlarged ovaries. The resulting transudation of protein rich fluid from the intravascular to the extravascular compartment can lead to hemoconcentration, electrolyte imbalance, and hepatic and renal dysfunctions. OHSS can also be associated with life-threatening complications in severe cases, including pleural or pericardial effusion, adult respiratory distress syndrome, thromboembolic events, myocardial or cerebral infarctions, and even death [75, 334–337].

13.2 Multiple Gestations

High-order pregnancy remains the most significant complication of ART as it is associated with increased maternal and fetal morbidity and mortality [338]. ART is associated with a tenfold increase in the incidence of multiple-fetal gestations. According to the 2011 CDC/SART report, the incidence of twin gestations was 29 % and the incidence of triplets and higher-order gestations was 3.5 % [4]. The increased risk of multiple gestations is related to the number of embryos transferred. Transferring two more embryos not only increases the likelihood of live births but also increases the risk of multiple pregnancy [4]. In patients who had a single embryo transferred, 21 % of transfers resulted in live births. Of these live births, 2.4 % were twins and 0.1 % were triplets or more. In patients who had two embryos transferred, the live birth rate per transfer increased to 43 %. However, 33 % were twins and 1 % was triplets or more. Among patients who have three embryos transferred, the live birth rate was 35 % per transfer. Of these live births, 29 % were twins and 4 % were triplets or more.

In an effort the reduce the incidence of multiple pregnancy, the American Society of Reproductive Medicine (ASRM) and Society for Assisted Reproductive Technologies (SART) have released guidelines to guide ART programs and patients in determining the appropriate number of embryos to transfer (Table 1) [339]. These recommendations are based on the patient's age and

Table 1
American Society for Reproductive Medicine and Society for Assisted Reproductive Technology guidelines on the number of embryos transferred

	Age			
	<35 years	35–37 years	38–40 years	41–42 years
Cleavage stage embryos				
Favorable	1–2	2	3	5
All others	2	3	4	5
Blastocysts				
Favorable	1	2	2	3
All others	2	2	3	3

Favorable prognosis includes: first IVF cycle, previous successful IVF cycle, good embryo quality, surplus embryos available for cryopreservation

the presence of favorable characteristics, including first cycle of IVF, previous successful IVF outcome, good-quality embryos, and surplus embryos for cryopreservation. In patients under the age of 35 years with favorable characteristics, consideration should be given to transferring of a single embryo. Although the risk of multiple pregnancies is significantly reduced, this risk is not completely eliminated. Blastocyst transfer has been associated with an increased incidence of monozygotic twins [340, 341].

13.3 Ectopic Pregnancy

ART is associated with an increased incidence of ectopic pregnancy. The reported incidence of ectopic pregnancy after IVF and ET ranges from 2 to 11 % [330, 342], compared to that observed in the general population of 1.2–1.4 % [343]. ART also increases the risk of heterotopic pregnancies, a condition where implantation can occur simultaneous at multiple sites (e.g., uterus or fallopian tubes, cervix or abdomen). In the general population, the occurrence of heterotopic pregnancy is extremely rare; the incidence is estimated to be 1 in 30,000 [344]. The incidence of heterotopic pregnancy has been reported to be as high as 1 % in patients undergoing IVF [345, 346]. The higher incidence of ectopic and heterotopic pregnancies may be related to tubal disease or tubal dysfunction in patients undergoing ART.

14 Perinatal Outcomes from ART

Today, more than five million babies have been born using ART worldwide [347]. Numerous studies have evaluated the obstetrical and perinatal outcomes following IVF and ICSI [348–351].

It has been noted that women with infertility, regardless of whether they have undergone ART treatment, have increased risks of adverse obstetrical and perinatal outcomes. After adjusting for age and parity, subfertile women were found have a twofold increase in the risks of pre-eclampsia and placental abruption and a threefold increase in developing placenta previa [352]. They were also more likely to require induction of labor, Caesarean section (twofold), instrumental delivery (twofold), and deliver low birth weight and preterm infants [352].

After controlling for maternal age, parity, and multiple gestations, singleton pregnancies conceived following IVF were found to have a higher likelihood of obstetrical complications, including gestational diabetes (twofold) [349], gestational hypertension (1.6-fold) [349], and placenta previa (threefold) [349]. IVF singleton pregnancies were also more likely to require induction of labor (1.6-fold) and Caesarean section (twofold) [349].

Singleton pregnancies conceived from IVF were also noted to be associated with increased perinatal morbidity and mortality (1.8-fold) [353], including preterm delivery (twofold) [353, 354], low birth weight (1.8-fold) [349, 354], very low birth weight (threefold) [349, 353, 354], small for gestational age(1.6-fold) [349, 353], and requirement for admission to neonatal intensive units (NICU) [349, 353].

Similarly, singleton pregnancies conceived following IVF-ICSI were found to be associated with increased risk of gestational diabetes (1.3-fold), gestational hypertension (6.4-fold), abruptio placenta (1.8-fold), and placenta previa (1.8-fold) [350]. Pregnancies were also more likely to result in Caesarean section, preterm delivery, and delivery of low birth weight and very low birth weight infants [350].

The association between ART treatment and congenital malformations is controversial. Most studies are limited by their small sample size, varying definitions of congenital abnormality, inadequate control of possible confounding variables, and a lack of appropriate comparison data. The reported prevalence of major defects varied from 3.0 to 9.0 % [355]. Most studies did not demonstrate an increased risk of major birth defects in children conceived with ART [356–359]. Comparing IVF to IVF with ICSI, there was no difference in the prevalence of major birth defects [355]. More recently, data from Australian birth registry showed a twofold increased risk of major congenital abnormalities in both singletons and twins following IVF (9.0 %) and IVF-ICSI (8.6 %) compared to spontaneous conceptions (4.2 %) [360]. A prospective cohort study from Germany also found increased prevalence of major malformations in children conceived with IVF-ICSI(8.7 %) compared to a population-based control cohort (6.1 %) (relative risk, 1.44 [1.25–1.65]) [350]. When considering organ-specific major malformations, the use of ICSI has been associated with an increased risk of hypospadias [356, 361]. The use of ART has also been associated with an increased incidence of imprinting disorders in the offspring, namely Beckwith–Wiedemann syndrome [224, 225, 362, 363] and Angelman's syndrome [364, 365].

Long-term follow-up studies of children conceived following IVF and ICSI appeared reassuring [359, 366–368]. No differences were observed in terms of psychomotor, cognitive, intellectual, or psychological development between IVF, IVF-ICSI, and spontaneously conceived children [366–368]. However, children conceived following IVF-ICSI were more likely to require surgical interventions, physiotherapy, and dietary therapy than naturally conceived children at 5 years of age [369, 370].

15 Future Directions

The field of ART is rapidly evolving. Two areas of ART in particular that have undergone dramatic improvements in recent years are preimplantation genetic diagnosis (PGD) and noninvasive methods of embryo selection. Traditionally, aneuploidy screening by PGS involves blastomere biopsy of day 3 embryos and chromosome analysis using FISH [371]. The two main drawbacks of this approach have been the high incidence of chromosomal mosaicism in day 3 embryos and the limited sensitivity of FISH since not all 23 sets of chromosomes are evaluated [372, 373]. There is a move towards performing trophectoderm biopsy on day 5 blastocyst and application of more accurate molecular diagnostic technologies, including single nucleotide polymorphism (SNP) and comparative genomic hybridization (CGH) microarray technologies [374–376]. A major limitation of this approach is the time required to complete the analysis (usually 48 h). Therefore, the biopsied embryos need to be cryopreserved and transferred in a subsequent FET cycle. Recently, the use of quantitative PCR for 24 chromosome aneuploidy has reduced the processing time to 4 h, making same day ET a feasible option [377].

Another area of intense research interest is noninvasive methods of embryo selection using transcriptomic, proteomic, and metabolomic technologies. Selection of embryos with optimal implantation potential is of great importance to the field of ART. Current embryo selection is based on morphological appearance and possesses limited sensitivity in predicting implantation and pregnancy. There is a pressing need to identify biomarkers of embryo developmental potential, which will enhance embryo selection, improve the efficacy of single ET, and reduce the risk of multiple gestations. Transcriptomics involves quantitative analysis of messenger RNA. An example is transcriptomic profiling of cumulus cells using microarray to identify potential biomarkers of oocyte quality and embryo development [378]. Recent advances in proteomic analysis using mass spectrometry, high performance liquid chromatography, and protein microarray have enabled identification of amino acids and proteins within limited amounts of biological fluids [379]. Analysis of spent culture media revealed significant correlations between amino acid and protein turnover with embryonic developmental potential and clinical pregnancy and live birth rates [380, 381]. Metabolomic analysis involves the use of spectrophotometric assays, such as Raman and near-infrared (NIR), to detect cellular metabolites in the surrounding environment (also known as secretome or exometabolome) [382]. Metabolomic profiling of spent embryo culture media using Raman and NIR are two promising approaches in predicting the reproductive potential of embryos [383–385]. Noninvasive imaging of human embryos has also been evaluated as potential predictor of embryonic development. Using time-lapse image analysis and gene expression profiling, imaging parameters were shown to predict blastocyst formation with 93 % sensitivity and specificity [386]. Further studies are needed to help identify the most viable embryo.

References

1. Steptoe PC, Edwards RG (1976) Reimplantation of a human embryo with subsequent tubal pregnancy. Lancet 307(7965): 880–882
2. Steptoe PC, Edwards RG (1978) Birth after reimplantation of a human embryo. Lancet 312(8085):366
3. International Committee for Monitoring Assisted Reproductive Technology, de Mouzon J, Lancaster P, Nygren KG, Sullivan E, Zegers-Hochschild F et al (2009) World collaborative report on assisted reproductive technology, 2002. Hum Reprod 24(9):2310–2320
4. Centers for disease control and prevention (2010) What is assisted reproductive technology? http://www.cdc.gov/art/
5. Centers for disease control and prevention (2013) 2011 Assisted reproductive technology success rates: national summary and fertility clinic reports. http://www.cdc.gov/art/ART2011/index.htm
6. Benadiva CA, Kligman I, Davis O, Rosenwaks Z (1995) In vitro fertilization versus tubal surgery: is pelvic reconstructive surgery obsolete? Fertil Steril 64(6):1051–1061
7. Practice Committee of American Society for Reproductive Medicine (2008) The role of tubal reconstructive surgery in the era of assisted reproductive technologies. Fertil Steril 90(5 Suppl 1):S250–S253
8. Strandell A, Lindhard A, Waldenstrom U, Thorburn J, Janson PO, Hamberger L (1999) Hydrosalpinx and IVF outcome: a prospective, randomized multicentre trial in Scandinavia on salpingectomy prior to IVF. Hum Reprod 14(11):2762–2769
9. Johnson NP, van Voorst S, Sowter MC, Strandell A, Mol BW (2010) Surgical treatment for tubal disease in women due to undergo in vitro fertilisation. Cochrane Database Syst Rev 20(2):CD002125
10. Strandell A, Lindhard A (2002) Why does hydrosalpinx reduce fertility?: the importance of hydrosalpinx fluid. Hum Reprod 17(5):1141–1145
11. Déchaud H, Daurès JP, Arnal F, Humeau C, Hédon B (1998) Does previous salpingectomy improve implantation and pregnancy rates in patients with severe tubal factor infertility who are undergoing in vitro fertilization? A pilot prospective randomized study. Fertil Steril 69(6):1020–1025
12. The Practice Committee of the American Society for Reproductive Medicine (2006) Salpingectomy for hydrosalpinx prior to in vitro fertilization. Fertil Steril 86(5 Suppl 1) S200–S201
13. Surrey ES, Schoolcraft WB (2001) Laparoscopic management of hydrosalpinges before in vitro fertilization-embryo transfer: salpingectomy versus proximal tubal occlusion. Fertil Steril 75(3):612–617
14. Rosenfield RB, Stones RE, Coates A, Matteri RK, Hesla JS (2005) Proximal occlusion of

hydrosalpinx by hysteroscopic placement of microinsert before in vitro fertilization-embryo transfer. Fertil Steril 83(5):1547

15. Darwish AM, El Saman AM (2007) Is there a role for hysteroscopic tubal occlusion of functionless hydrosalpinges prior to IVF/ICSI in modern practice? Acta Obstet Gynecol Scand 86(12):1484–1489
16. Mijatovic V, Veersema S, Emanuel MH, Schats R, Hompes PGA (2010) Essure hysteroscopic tubal occlusion device for the treatment of hydrosalpinx prior to in vitro fertilization-embryo transfer in patients with a contraindication for laparoscopy. Fertil Steril 93(4):1338–1342
17. Bayrak A, Harp D, Saadat P, Mor E, Paulson R (2006) Recurrence of hydrosalpinges after cuff neosalpingostomy in a poor prognosis population. J Assist Reprod Genet 23(6): 285–288
18. Sowter M, Akande V, Williams J, Hull M (1997) Is the outcome of in-vitro fertilization and embryo transfer treatment improved by spontaneous or surgical drainage of a hydrosalpinx? Hum Reprod 12(10):2147–2150
19. Thonneau P, Marchand S, Tallec A, Ferial ML, Ducot B, Lansac J et al (1991) Incidence and main causes of infertility in a resident population (1,850,000) of three French regions (1988–1989). Hum Reprod 6(6): 811–816
20. Lee RKK, Hou JW, Ho HY, Hwu YM, Lin MH, Tsai YC et al (2002) Sperm morphology analysis using strict criteria as a prognostic factor in intrauterine insemination. Int J Androl 25(5):277–280
21. Van Voorhis BJ, Barnett M, Sparks AET, Syrop CH, Rosenthal G, Dawson J (2001) Effect of the total motile sperm count on the efficacy and cost-effectiveness of intrauterine insemination and in vitro fertilization. Fertil Steril 75(4):661–668
22. Miller DC, Hollenbeck BK, Smith GD, Randolph JF, Christman GM, Smith YR et al (2002) Processed total motile sperm count correlates with pregnancy outcome after intrauterine insemination. Urology 60(3): 497–501
23. Campana A, Sakkas D, Stalberg A, Bianchi PG, Comte I, Pache T et al (1996) Intrauterine insemination: evaluation of the results according to the woman's age, sperm quality, total sperm count per insemination and life table analysis. Hum Reprod 11(4):732–736
24. Missmer SA, Hankinson SE, Spiegelman D, Barbieri RL, Marshall LM, Hunter DJ (2004) Incidence of laparoscopically confirmed endometriosis by demographic, anthropometric, and lifestyle factors. Am J Epidemiol 160(8):784–796
25. Strathy JH, Molgaard CA, Coulam CB, Melton L Jr (1982) Endometriosis and infertility: a laparoscopic study of endometriosis among fertile and infertile women. Fertil Steril 38(6):667–672
26. Toya M, Saito H, Ohta N, Saito T, Kaneko T, Hiroi M (2000) Moderate and severe endometriosis is associated with alterations in the cell cycle of granulosa cells in patients undergoing in vitro fertilization and embryo transfer. Fertil Steril 73(2):344–350
27. Marcoux S, Maheux R, Bérubé S (1997) Laparoscopic surgery in infertile women with minimal or mild endometriosis. N Engl J Med 337(4):217–222
28. Jacobson TZ, Duffy JM, Barlow D, Farquhar C, Koninckx PR, Olive D (2010) Laparoscopic surgery for subfertility associated with endometriosis. Cochrane Database Syst Rev 1:CD001398
29. Tummon IS, Asher LJ, Martin JSB, Tulandi T (1997) Randomized controlled trial of superovulation and insemination for infertility associated with minimal or mild endometriosis. Fertil Steril 68(1):8–12
30. Adamson GD, Pasta DJ (1994) Surgical treatment of endometriosis-associated infertility: meta-analysis compared with survival analysis. Am J Obstet Gynecol 171(6):1488–1504
31. Kennedy S, Bergqvist A, Chapron C, D'Hooghe T, Dunselman G, Greb R et al (2005) ESHRE guideline for the diagnosis and treatment of endometriosis. Hum Reprod 20(10):2698–2704
32. Surrey ES, Silverberg KM, Surrey MW, Schoolcraft WB (2002) Effect of prolonged gonadotropin-releasing hormone agonist therapy on the outcome of in vitro fertilization-embryo transfer in patients with endometriosis. Fertil Steril 78(4):699–704
33. The Rotterdam ESHRE/ASRM-sponsored PCOS consensus workshop group (2004) Revised 2003 consensus on diagnostic criteria and long-term health risks related to polycystic ovary syndrome (PCOS). Hum Reprod 19(1):41–47
34. Martin K, Hall J, Adams J, Crowley W Jr (1993) Comparison of exogenous gonadotropins and pulsatile gonadotropin-releasing hormone for induction of ovulation in hypogonadotropic amenorrhea. J Clin Endocrinol Metab 77(1):125–129
35. Greenblatt RB (1961) Chemical induction of ovulation. Fertil Steril 12:402–404

36. Holzer H, Casper R, Tulandi T (2006) A new era in ovulation induction. Fertil Steril 85(2):277–284
37. Legro RS, Barnhart HX, Schlaff WD, Carr BR, Diamond MP, Carson SA et al (2007) Clomiphene, metformin, or both for infertility in the polycystic ovary syndrome. N Engl J Med 356(6):551–566
38. Messinis IE, Nillius SJ (1982) Comparison between tamoxifen and clomiphene for induction of ovulation. Acta Obstet Gynecol Scand 61(4):377–379
39. MacDougall MJ, Tan SL, Jacobs HS (1992) In vitro fertilization and the ovarian hyperstimulation syndrome. Hum Reprod 7(5): 597–600
40. Agrawal R, Tan SL, Wild S, Sladkevicius P, Engmann L, Payne N et al (1999) Serum vascular endothelial growth factor concentrations in in vitro fertilization cycles predict the risk of ovarian hyperstimulation syndrome. Fertil Steril 71(2):287–293
41. Nisker J, Tummon I, Daniel S, Kaplan B, Yuzpe A (1994) Conversion of cycles involving ovarian hyperstimulation with intrauterine insemination to in-vitro fertilization. Hum Reprod 9(3):406–408
42. Ashkenazi J, Feldberg D, Dicker D, Yeshaya A, Ayalon D, Goldman JA (1989) IVF-ET in women with refractory polycystic ovarian disease. Eur J Obstet Gynecol Reprod Biol 30(2):157–161
43. Templeton AA, Penney GC (1982) The incidence, characteristics, and prognosis of patients whose infertility is unexplained. Fertil Steril 37(2):175–182
44. Guzick DS, Grefenstette I, Baffone K, Berga SL, Krasnow JS, Stovall DW et al (1994) Infertility: Infertility evaluation in fertile women: a model for assessing the efficacy of infertility testing. Hum Reprod 9(12): 2306–2310
45. Reindollar RH, Regan MM, Neumann PJ, Levine B-S, Thornton KL, Alper MM et al (2010) A randomized clinical trial to evaluate optimal treatment for unexplained infertility: the fast track and standard treatment (FASTT) trial. Fertil Steril 94(3):888–899
46. Tsafrir A, Simon A, Margalioth EJ, Laufer N (2009) What should be the first-line treatment for unexplained infertility in women over 40 years of age: ovulation induction and IUI, or IVF? Reprod Biomed Online 19(Suppl 4):47–56
47. Harton G, Braude P, Lashwood A, Schmutzler A, Traeger-Synodinos J, Wilton L et al (2011) ESHRE PGD consortium best practice guidelines for organization of a PGD centre for PGD/preimplantation genetic screening. Hum Reprod 26(1):14–24
48. Xu K, Shi ZM, Veeck LL, Hughes MR, Rosenwaks Z (1999) First unaffected pregnancy using preimplantation genetic diagnosis for sickle cell anemia. JAMA 281(18): 1701–1706
49. Simpson JL (2010) Preimplantation genetic diagnosis at 20 years. Prenat Diagn 30(7): 682–695
50. Harper JC, Harton G (2010) The use of arrays in preimplantation genetic diagnosis and screening. Fertil Steril 94(4):1173–1177
51. President's Cancer Panel 2003–2004 Annual Report (2004) Living beyond cancer: finding a new balance. US Department of Health and Human Services, p 1–87
52. Lee SJ, Schover LR, Partridge AH, Patrizio P, Wallace WH, Hagerty K et al (2006) American Society of Clinical Oncology recommendations on fertility preservation in cancer patients. J Clin Oncol 24(18):2917–2931
53. Ethics Committee of the American Society for Reproductive Medicine (2005) Fertility preservation and reproduction in cancer patients. Fertil Steril 83(6):1622–1628
54. American Society for Reproductive Medicine (2013) Mature oocyte cryopreservation: a guideline. Fertil Steril 99:37
55. Tulandi T, Huang JYJ, Tan SL (2008) Preservation of female fertility: an essential progress. Obstet Gynaecol 112(5): 1160–1172
56. Cobo A, Diaz C (2011) Clinical application of oocyte vitrification: a systematic review and meta-analysis of randomized controlled trials. Fertil Steril 96(2):277–285
57. Gougeon A, Ecochard R, Thalabard JC (1994) Age-related changes of the population of human ovarian follicles: increase in the disappearance rate of non-growing and early-growing follicles in aging women. Biol Reprod 50(3):653–663
58. Faddy MJ, Gosden RG, Gougeon A, Richardson SJ, Nelson JF (1992) Accelerated disappearance of ovarian follicles in mid-life: implications for forecasting menopause. Hum Reprod 7(10):1342–1346
59. Vaskivuo TE, Anttonen M, Herva R, Billig H, Dorland M, te Velde ER et al (2001) Survival of human ovarian follicles from fetal to adult life: apoptosis, apoptosis-related proteins, and transcription factor GATA-4. J Clin Endocrinol Metab 86(7):3421–3429
60. Malizia BA, Hacker MR, Penzias AS (2009) Cumulative live-irth rates after in vitro fertilization. N Engl J Med 360(3):236–243

61. Ziebe S, Loft A, Petersen JH, Andersen AG, Lindenberg S, Petersen K et al (2001) Embryo quality and developmental potential is compromised by age. Acta Obstet Gynecol Scand 80(2):169–174
62. Hull MG, Fleming CF, Hughes AO, McDermott A (1996) The age-related decline in female fecundity: a quantitative controlled study of implanting capacity and survival of individual embryos after in vitro fertilization. Fertil Steril 65(4):783–790
63. Spandorfer SD, Chung PH, Kligman I, Liu HC, Davis OK, Rozenwaks Z (2000) An analysis of the effect of Age on implantation rates. J Assist Reprod Genet 17(6):303–306
64. Liu H-C, Rosenwaks Z (1991) Early pregnancy wastage in IVF (in vitro fertilization) patients. J Assist Reprod Genet 8(2):65–72
65. Spandorfer SD, Davis OK, Barmat LI, Chung PH, Rosenwaks Z (2004) Relationship between maternal age and aneuploidy in in vitro fertilization pregnancy loss. Fertil Steril 81(5):1265–1269
66. Spandorfer SD, Bendikson K, Dragisic K, Schattman G, Davis OK, Rosenwaks Z (2007) Outcome of in vitro fertilization in women 45 years and older who use autologous oocytes. Fertil Steril 87(1):74–76
67. Navot D, Rosenwaks Z, Margalioth EJ (1987) Prognostic assessment of female fecundity. Lancet 2(8560):645–647
68. Fanchin R, de Ziegler D, Olivennes F, Taieb J, Dzik A, Frydman R (1994) Endocrinology: exogenous follicle stimulating hormone ovarian reserve test (EFORT): a simple and reliable screening test for detecting 'poor responders' in in-vitro fertilization. Hum Reprod 9(9):1607–1611
69. Winslow KL, Toner JP, Brzyski RG, Oehninger SC, Acosta AA, Muasher SJ (1991) The gonadotropin-releasing hormone agonist stimulation test: a sensitive predictor of performance in the flare-up in vitro fertilization cycle. Fertil Steril 56(4):711–717
70. Muasher SJ, Oehninger S, Simonetti S, Matta J, Ellis LM, Liu HC et al (1988) The value of basal and/or stimulated serum gonadotropin levels in prediction of stimulation response and in vitro fertilization outcome. Fertil Steril 50(2):298–307
71. Burger H, Cahir N, Robertson D, Groome N, Dudley E, Green A et al (1998) Serum Inhibins A and B fall differentially as FSH rises in perimenopausal women. Clin Endonol 48(6):809–813
72. Klein N, Illingworth P, Groome N, McNeilly A, Battaglia D, Soules M (1996) Decreased inhibin B secretion is associated with the monotropic FSH rise in older, ovulatory women: a study of serum and follicular fluid levels of dimeric inhibin A and B in spontaneous menstrual cycles. J Clin Endocrinol Metab 81(7):2742–2745
73. Cameron TT, O'Shea FC, Rolland JM, Hughes EG, Kretser DMD, Healy DL (1988) Occult ovarian failure: a syndrome of infertility, regular menses, and elevated follicle-stimulating hormone concentrations. J Clin Endocrinol Metab 67(6):1190–1194
74. Toner JP, Philput CB, Jones GS, Muasher S (1991) Basal follicle-stimulating hormone level is a better predictor of in vitro fertilization performance than age. Fertil Steril 55(4):784–791
75. Akande VA, Fleming CF, Hunt LP, Keay SD, Jenkins JM (2002) Biological versus chronological ageing of oocytes, distinguishable by raised FSH levels in relation to the success of IVF treatment. Hum Reprod 17(8):2003–2008
76. El-Toukhy T, Khalaf Y, Hart R, Taylor A, Braude P (2002) Young age does not protect against the adverse effects of reduced ovarian reserve: an eight year study. Hum Reprod 17(6):1519–1524
77. Roberts JE, Spandorfer S, Fasouliotis SJ, Kashyap S, Rosenwaks Z (2005) Taking a basal follicle-stimulating hormone history is essential before initiating in vitro fertilization. Fertil Steril 83(1):37–41
78. Smotrich DB, Widra EA, Gindoff PR, Levy MJ, Hall JL, Stillman RJ (1995) Prognostic value of day 3 estradiol on in vitro fertilization outcome. Fertil Steril 64(6):1136–1140
79. Licciardi FL, Liu HC, Rosenwaks Z (1995) Day 3 estradiol serum concentrations as prognosticators of ovarian stimulation response and pregnancy outcome in patients undergoing in vitro fertilization. Fertil Steril 64(5):991–994
80. Broekmans FJ, Kwee J, Hendriks DJ, Mol BW, Lambalk CB (2006) A systematic review of tests predicting ovarian reserve and IVF outcome. Hum Reprod Update 12(6):685–718
81. Visser JA, Themmen APN (2005) Anti-Müllerian hormone and folliculogenesis. Mol Cell Endocrinol 234(1–2):81–86
82. Fanchin R, Taieb J, Lozano DHM, Ducot B, Frydman R, Bouyer J (2005) High reproducibility of serum anti-Mullerian hormone measurements suggests a multi-staged follicular secretion and strengthens its role in the assessment of ovarian follicular status. Hum Reprod 20(4):923–927

83. La Marca A, Giulini S, Tirelli A, Bertucci E, Marsella T, Xella S et al (2007) Anti-Mullerian hormone measurement on any day of the menstrual cycle strongly predicts ovarian response in assisted reproductive technology. Hum Reprod 22(3):766–771
84. Broer SL, Mol BW, Hendriks D, Broekmans FJ (2009) The role of antimullerian hormone in prediction of outcome after IVF: comparison with the antral follicle count. Fertil Steril 91(3):705–714
85. Nelson SM, Yates RW, Lyall H, Jamieson M, Traynor I, Gaudoin M et al (2009) Anti-Mullerian hormone-based approach to controlled ovarian stimulation for assisted conception. Hum Reprod 24(4):867–875
86. Lee T-H, Liu C-H, Huang C-C, Wu Y-L, Shih Y-T, Ho H-N et al (2008) Serum anti-mullerian hormone and estradiol levels as predictors of ovarian hyperstimulation syndrome in assisted reproduction technology cycles. Hum Reprod 23(1):160–167
87. La Marca A, Sighinolfi G, Radi D, Argento C, Baraldi E, Artenisio AC et al (2010) Anti-Müllerian hormone (AMH) as a predictive marker in assisted reproductive technology (ART). Hum Reprod Update 16(2): 113–130
88. Gnoth C, Schuring AN, Friol K, Tigges J, Mallmann P, Godehardt E (2008) Relevance of anti-Mullerian hormone measurement in a routine IVF program. Hum Reprod 23(6): 1359–1365
89. Muttukrishna S, Suharjono H, McGarrigle H, Sathanandan M (2004) Inhibin B and anti-Mullerian hormone: markers of ovarian response in IVF/ICSI patients? BJOG 111(11):1248–1253
90. Groome NP, Illingworth PJ, O'Brien M, Cooke I, Ganesan TS, Baird DT et al (1994) Detection of dimeric inhibin throughout the human menstrual cycle by two-site enzyme immunoassay. Clin Endocrinol 40(6):717–723
91. Hofmann GE, Danforth DR, Seifer DB (1998) Inhibin-B: the physiologic basis of the clomiphene citrate challenge test for ovarian reserve screening. Fertil Steril 69(3):474–477
92. Chang M-Y, Chiang C-H, Hsieh T-T, Soong Y-K, Hsu K-H (1998) Use of the antral follicle count to predict the outcome of assisted reproductive technologies. Fertil Steril 69(3):505–510
93. Bancsi LF, Broekmans FJ, Eijkemans MJ, de Jong FH, Habbema JD, te Velde ER (2002) Predictors of poor ovarian response in in vitro fertilization: a prospective study comparing basal markers of ovarian reserve. Fertil Steril 77(2):328–336
94. Wallace WH, Kelsey TW (2004) Ovarian reserve and reproductive age may be determined from measurement of ovarian volume by transvaginal sonography. Hum Reprod 19(7):1612–1617
95. Lass A, Skull J, McVeigh E, Margara R, Winston RM (1997) Measurement of ovarian volume by transvaginal sonography before ovulation induction with human menopausal gonadotrophin for in-vitro fertilization can predict poor response. Hum Reprod 12(2): 294–297
96. Hendriks DJ, Kwee J, Mol BW, te Velde ER, Broekmans FJ (2007) Ultrasonography as a tool for the prediction of outcome in IVF patients: a comparative meta-analysis of ovarian volume and antral follicle count. Fertil Steril 87(4):764–775
97. Engmann L, Sladkevicius P, Agrawal R, Bekir J, Campbell S, Tan SL (1999) The pattern of changes in ovarian stromal and uterine artery blood flow velocities during in vitro fertilization treatment and its relationship with outcome of the cycle. Ultrasound Obstet Gynecol 13(1):26–33
98. Engmann L, Sladkevicius P, Agrawal R, Bekir JS, Campbell S, Tan SL (1999) Value of ovarian stromal blood flow velocity measurement after pituitary suppression in the prediction of ovarian responsiveness and outcome of in vitro fertilization treatment. Fertil Steril 71(1):22–29
99. Csemiczky G, Harlin J, Fried G (2002) Predictive power of clomiphene citrate challenge test for failure of in vitro fertilization treatment. Acta Obstet Gynecol Scand 81(10):954–961
100. Yanushpolsky EH, Hurwitz S, Tikh E, Racowsky C (2003) Predictive usefulness of cycle day 10 follicle-stimulating hormone level in a clomiphene citrate challenge test for in vitro fertilization outcome in women younger than 40 years of age. Fertil Steril 80(1):111–115
101. Jain T, Soules MR, Collins JA (2004) Comparison of basal follicle-stimulating hormone versus the clomiphene citrate challenge test for ovarian reserve screening. Fertil Steril 82(1):180–185
102. Hendriks DJ, Mol B-WJ, Bancsi LFJMM, te Velde ER, Broekmans FJM (2006) The clomiphene citrate challenge test for the prediction of poor ovarian response and nonpregnancy in patients undergoing in vitro fertilization: a systematic review. Fertil Steril 86(4):807–818

103. Hendriks DJ, Broekmans FJ, Bancsi LFJMM, Looman CWN, de Jong FH, te Velde ER (2005) Single and repeated GnRH agonist stimulation tests compared with basal markers of ovarian reserve in the prediction of outcome in IVF. J Assist Reprod Genet 22(2): 65–74
104. Ranieri DM, Quinn F, Makhlouf A, Khadum I, Ghutmi W, McGarrigle H et al (1998) Simultaneous evaluation of basal follicle-stimulating hormone and 17[beta]-estradiol response to gonadotropin-releasing hormone analogue stimulation: an improved predictor of ovarian reserve. Fertil Steril 70(2): 227–233
105. Medicine PCotASfR (2008) Myomas and reproductive function. Fertil Steril 90(Suppl 5):125
106. Klatsky PC, Tran ND, Caughey AB, Fujimoto VY (2008) Fibroids and reproductive outcomes: a systematic literature review from conception to delivery. Am J Obstet Gynecol 198(4):357–366
107. Pritts EA, Parker WH, Olive DL (2009) Fibroids and infertility: an updated systematic review of the evidence. Fertil Steril 91(4): 1215–1223
108. Karayalcin R, Ozcan S, Moraloglu O, Ozyer S, Mollamahmutoglu L, Batloglu S (2010) Results of 2500 office-based diagnostic hysteroscopies before IVF. Reprod Biomed Online 20(5):689–693
109. The ESHRE Capri Workshop Group (2000) Optimal use of infertility diagnostic tests and treatments. Hum Reprod 15(3):723–732
110. Mansour R, Aboulghar M, Serour G (1990) Dummy embryo transfer: a technique that minimizes the problems of embryo transfer and improves the pregnancy rate in human in vitro fertilization. Fertil Steril 54(4): 678–681
111. Katariya KO, Bates GW, Robinson RD, Arthur NJ, Propst AM (2007) Does the timing of mock embryo transfer affect in vitro fertilization implantation and pregnancy rates? Fertil Steril 88(5):1462–1464
112. Shamonki MI, Schattman GL, Spandorfer SD, Chung PH, Rosenwaks Z (2005) Ultrasound-guided trial transfer may be beneficial in preparation for an IVF cycle. Hum Reprod 20(10):2844–2849
113. Groutz A, Lessing JB, Wolf Y, Yovel I, Azem F, Amit A (1997) Cervical dilatation during ovum pick-up in patients with cervical stenosis: effect on pregnancy outcome in an in vitro fertilization-embryo transfer program. Fertil Steril 67(5):909–911
114. Abusheikha N, Lass A, Akagbosu F, Brinsden P (1999) How useful is cervical dilatation in patients with cervical stenosis who are participating in an in vitro fertilization-embryo transfer program? the bourn hall experience. Fertil Steril 72(4):610–612
115. Glatstein IZ, Pang SC, McShane PM (1997) Successful pregnancies with the use of laminaria tents before embryo transfer for refractory cervical stenosis. Fertil Steril 67(6): 1172–1174
116. Yanushpolsky EH, Ginsburg ES, Fox JH, Stewart EA (2000) Transcervical placement of a Malecot catheter after hysteroscopic evaluation provides for easier entry into the endometrial cavity for women with histories of difficult intrauterine inseminations and/or embryo transfers: a prospective case series. Fertil Steril 73(2):402–405
117. Feldman B, Seidman DS, Levron J, Bider D, Shulman A, Shine S et al (2001) In vitro fertilization following natural cycles in poor responders. Gynecol Endocrinol 15(5):328–334
118. Wortham JWJ, Veeck LL, Witmyer J, Sandow BA, Jones HWJ (1983) Vital initiation of pregnancy (VIP) using human menopausal gonadotropin and human chorionic gonadotropin ovulation induction: phase II–1981. Fertil Steril 40(2):170–177
119. Laufer N, DeCherney AH, Haseltine FP, Polan ML, Mezer HC, Dlugi AM et al (1983) The use of high-dose human menopausal gonadotropin in an in vitro fertilization program. Fertil Steril 40(6):734–741
120. Macklon NS, Stouffer RL, Giudice LC, Fauser BCJM (2006) The science behind 25 years of ovarian stimulation for in vitro fertilization. Endocr Rev 27(2):170–207
121. van Loendersloot LL, van Wely M, Limpens J, Bossuyt PMM, Repping S, van der Veen F (2010) Predictive factors in in vitro fertilization (IVF): a systematic review and meta-analysis. Hum Reprod Update 16(6): 577–589
122. Barbieri RL, Hornstein MD (1999) Assisted reproduction-in vitro fertilization success is improved by ovarian stimulation with exogenous gonadotropins and pituitary suppression with gonadotropin-releasing hormone analogues. Endocr Rev 20(3):249–252
123. Klingmüller D, Schepke M, Enzweiler C, Bidlingmaier F (1993) Hormonal responses to the new potent GnRH antagonist Cetrorelix. Acta Endocrinol (Copenh) 128(1):15–18
124. Tarlatzis BC, Fauser BC, Kolibianakis EM, Diedrich K, Devroey P (2006) GnRH antag-

onists in ovarian stimulation for IVF. Hum Reprod Update 12(4):333–340

125. Devroey P, Aboulghar M, Garcia-Velasco J, Griesinger G, Humaidan P, Kolibianakis E et al (2009) Improving the patient's experience of IVF/ICSI: a proposal for an ovarian stimulation protocol with GnRH antagonist co-treatment. Hum Reprod 24(4):764–774
126. Olivennes F, Alvarez S, Bouchard P, Fanchin R, Salat-Baroux J, Frydman R (1998) The use of a GnRH antagonist (Cetrorelix) in a single dose protocol in IVF-embryo transfer: a dose finding study of 3 versus 2 mg. Hum Reprod 13(9):2411–2414
127. Olivennes F, Belaisch-Allart J, Emperaire J-C, Dechaud H, Alvarez S, Moreau L et al (2000) Prospective, randomized, controlled study of in vitro fertilization-embryo transfer with a single dose of a luteinizing hormone-releasing hormone (LH-RH) antagonist (cetrorelix) or a depot formula of an LH-RH agonist (triptorelin). Fertil Steril 73(2):314–320
128. Al-Inany H, Aboulghar MA, Mansour RT, Serour GI (2005) Optimizing GnRH antagonist administration: meta-analysis of fixed versus flexible protocol. Reprod Biomed Online 10(5):567–570
129. Damario MA, Barmat L, Liu HC, Davis OK, Rosenwaks Z (1997) Dual suppression with oral contraceptives and gonadotrophin releasing-hormone agonists improves in-vitro fertilization outcome in high responder patients. Hum Reprod 12(11):2359–2365
130. Maxwell KN, Cholst IN, Rosenwaks Z (2008) The incidence of both serious and minor complications in young women undergoing oocyte donation. Fertil Steril 90(6):2165–2171
131. Al-Inany HG, Abou-Setta AM, Aboulghar M (2006) Gonadotrophin-releasing hormone antagonists for assisted conception. Cochrane Database Syst Rev 3:CD001750
132. Al-Inany HG, Abou-Setta AM, Aboulghar M (2007) Gonadotrophin-releasing hormone antagonists for assisted conception: a Cochrane review. Reprod Biomed Online 14(5):640–649
133. Youssef MA, Van der Veen F, Al-Inany HG, Griesinger G, Mochtar MH, Aboulfoutouh I et al (2011) Gonadotropin-releasing hormone agonist versus HCG for oocyte triggering in antagonist assisted reproductive technology cycles. Cochrane Database Syst Rev 19(1):CD008046
134. Al-Inany HG, Youssef MA, Aboulghar M, Broekmans F, Sterrenburg M, Smit J et al (2011) Gonadotrophin-releasing hormone antagonists for assisted reproductive technology. Cochrane Database Syst Rev 5: CD001750
135. Emperaire JC, Parneix I, Ruffie A (2004) Luteal phase defects following agonist-triggered ovulation: a patient-dependent response. Reprod Biomed Online 9(1): 22–27
136. Humaidan P, Ejdrup Bredkjær H, Bungum L, Bungum M, Grøndahl ML, Westergaard L et al (2005) GnRH agonist (buserelin) or hCG for ovulation induction in GnRH antagonist IVF/ICSI cycles: a prospective randomized study. Hum Reprod 20(5):1213–1220
137. Bodri D, Sunkara SK, Coomarasamy A (2011) Gonadotropin-releasing hormone agonists versus antagonists for controlled ovarian hyperstimulation in oocyte donors: a systematic review and meta-analysis. Fertil Steril 95(1):164–169
138. Keay SD, Liversedge NH, Mathur RS, Jenkins JM (1997) Assisted conception following poor ovarian response to gonadotrophin stimulation. Br J Obstet Gynaecol 104(5): 521–527
139. Garcia JE, Jones GS, Acosta AA, Wright GJ (1983) Human menopausal gonadotropin/ human chorionic gonadotropin follicular maturation for oocyte aspiration: phase I, 1981. Fertil Steril 39(2):167–173
140. Brzyski RG, Muasher SJ, Droesch K, Simonetti S, Jones GS, Rosenwaks Z (1988) Follicular atresia associated with concurrent initiation of gonadotropin-releasing hormone agonist and follicle-stimulating hormone for oocyte recruitment. Fertil Steril 50(6): 917–921
141. Raga F, Bonilla-Musoles F, Casañ EM, Bonilla F (1999) Recombinant follicle stimulating hormone stimulation in poor responders with normal basal concentrations of follicle stimulating hormone and oestradiol: improved reproductive outcome. Hum Reprod 14(6):1431–1434
142. Schoolcraft W, Schlenker T, Gee M, Stevens J, Wagley L (1997) Improved controlled ovarian hyperstimulation in poor responder in vitro fertilization patients with a microdose follicle-stimulating hormone flare, growth hormone protocol. Fertil Steril 67(1):93–97
143. Karande V, Morris R, Rinehart J, Miller C, Rao R, Gleicher N (1997) Limited success using the "flare" protocol in poor responders in cycles with low basal follicle-stimulating hormone levels during in vitro fertilization. Fertil Steril 67(5):900–903
144. Droesch K, Muasher SJ, Brzyski RG, Jones GS, Simonetti S, Liu HC et al (1989) Value

of suppression with a gonadotropin-releasing hormone agonist prior to gonadotropin stimulation for in vitro fertilization. Fertil Steril 51(2):292–297

145. Feldberg D, Farhi J, Ashkenazi J, Dicker D, Shalev J, Ben-Rafael Z (1994) Minidose gonadotropin-releasing hormone agonist is the treatment of choice in poor responders with high follicle-stimulating hormone levels. Fertil Steril 62(2):343–346

146. Faber BM, Mayer J, Cox B, Jones D, Toner JP, Oehninger S et al (1998) Cessation of gonadotropin-releasing hormone agonist therapy combined with high-dose gonadotropin stimulation yields favorable pregnancy results in low responders. Fertil Steril 69(5):826–830

147. Toth TL, Awwad JT, Veeck LL, Jones HWJ, Muasher SJ (1996) Suppression and flare regimens of gonadotropin-releasing hormone agonist. Use in women with different basal gonadotropin values in an in vitro fertilization program. J Reprod Med 41(5):321–326

148. Shaker AG, Fleming R, Jamieson ME, Yates RW, Coutts JR (1992) Absence of effect of adjuvant growth hormone therapy on follicular responses to exogenous gonadotropins in women: normal and poor responders. Fertil Steril 58(5):919–923

149. Surrey ES, Bower J, Hill DM, Ramsey J, Surrey MW (1998) Clinical and endocrine effects of a microdose GnRH agonist flare regimen administered to poor responders who are undergoing in vitro fertilization. Fertil Steril 69(3):419–424

150. Ferraretti AP, La Marca A, Fauser BCJM, Tarlatzis B, Nargund G, Gianaroli L et al (2011) ESHRE consensus on the definition of 'poor response' to ovarian stimulation for in vitro fertilization: the Bologna criteria. Hum Reprod 26(7):1616–1624

151. Santoro N, Isaac B, Neal-Perry G, Adel T, Weingart L, Nussbaum A et al (2003) Impaired folliculogenesis and ovulation in older reproductive aged women. J Clin Endocrinol Metab 88(11):5502–5509

152. Klein N, Battaglia D, Fujimoto V, Davis G, Bremner W, Soules M (1996) Reproductive aging: accelerated ovarian follicular development associated with a monotropic follicle-stimulating hormone rise in normal older women. J Clin Endocrinol Metab 81(3):1038–1045

153. Klein NA, Harper AJ, Houmard BS, Sluss PM, Soules MR (2002) Is the short follicular phase in older women secondary to advanced or accelerated dominant follicle development? J Clin Endocrinol Metab 87(12):5746–5750

154. Fanchin R, Cunha-Filho JS, Schonäuer LM, Kadoch IJ, Cohen-Bacri P, Frydman R (2003) Coordination of early antral follicles by luteal estradiol administration provides a basis for alternative controlled ovarian hyperstimulation regimens. Fertil Steril 79(2):316–321

155. Fanchin R, Castelo Branco A, Kadoch IJ, Hosny G, Bagirova M, Frydman R (2004) Premenstrual administration of gonadotropin-releasing hormone antagonist coordinates early antral follicle sizes and sets up the basis for an innovative concept of controlled ovarian hyperstimulation. Fertil Steril 81(6):1554–1559

156. Dragisic KG, Davis OK, Fasouliotis SJ, Rosenwaks Z (2005) Use of a luteal estradiol patch and a gonadotropin-releasing hormone antagonist suppression protocol before gonadotropin stimulation for in vitro fertilization in poor responders. Fertil Steril 84(4):1023–1026

157. Elassar A, Mann JS, Engmann L, Nulsen J, Benadiva C (2011) Luteal phase estradiol versus luteal phase estradiol and antagonist protocol for controlled ovarian stimulation before in vitro fertilization in poor responders. Fertil Steril 95(1):324–326

158. Garcia J, Padilla S, Bargati J, Baramki T (1990) Follicular phase gonadotropin-releasing hormone agonist and human gonadotropins: a better alternative for in vitro fertilization. Fertil Steril 53:302–305

159. Scott R, Navot D (1994) Enhancement of ovarian responsiveness with microdoses of gonadotropin-releasing hormone agonist during ovulation induction for in vitro fertilization. Fertil Steril 61(5):880–885

160. Manzi D, Thorton K, Scott L, Nulsen J (1994) The value of increasing the dose of human menopausal gonadotropins in women who initially demonstrate a poor response. Fertil Steril 62:251–256

161. Spandorfer S, Navarro J, Kump LM, Liu H-C, Davis OK, Rosenwaks Z (2001) "Co-flare" stimulation in the poor responder patient: predictive value of the flare response. J Assist Reprod Genetics 18(12):629–633

162. Huang JYJ, Singer T, Williams Z, Abdallah R, Davis O, Rosenwaks Z (2010) Combining luteal estradiol patch and GnRH antagonist suppression with co-flare and microdose flare stimulation for IVF in poor responders. Fertil Steril 94(Suppl 1):S164–S165

163. Benadiva CA, Davis O, Kligman I, Liu HC, Rosenwaks Z (1995) Clomiphene citrate and hMG: an alternative stimulation protocol for selected failed in vitro fertilization patients. J Assist Reprod Genet 12(1):8–12

164. D'Amato G, Caroppo E, Pasquadibisceglie A, Carone D, Vitti A, Vizziello GM (2004) A novel protocol of ovulation induction with delayed gonadotropin-releasing hormone antagonist administration combined with high-dose recombinant follicle-stimulating hormone and clomiphene citrate for poor responders and women over 35 years. Fertil Steril 81(6):1572–1577
165. Schoolcraft WB, Surrey ES, Minjarez DA, Stevens JM, Gardner DK (2008) Management of poor responders: can outcomes be improved with a novel gonadotropin-releasing hormone antagonist/letrozole protocol? Fertil Steril 89(1):151–156
166. Mitwally MFM, Casper RF (2002) Aromatase inhibition improves ovarian response to follicle-stimulating hormone in poor responders. Fertil Steril 77(4):776–780
167. Pelinck MJ, Vogel NEA, Hoek A, Arts EGJM, Simons AHM, Heineman MJ (2005) Minimal stimulation IVF with late follicular phase administration of the GnRH antagonist cetrorelix and concomitant substitution with recombinant FSH: a pilot study. Hum Reprod 20(3):642–648
168. Huang JYJ, Singer T, Abdallah R, Rauch E, Rosenwaks Z, Spandorfer S (2010) Comparison of natural cycles versus modified natural cycles in poor responders undergoing IVF. Fertil Steril 94(4 Suppl 1):S248–S249
169. Kolibianakis E, Zikopoulos K, Camus M, Tournaye H, Van Steirteghem A, Devroey P (2004) Modified natural cycle for IVF does not offer a realistic chance of parenthood in poor responders with high day 3 FSH levels, as a last resort prior to oocyte donation. Hum Reprod 19(11):2545–2549
170. Morgia F, Sbracia M, Schimberni M, Giallonardo A, Piscitelli C, Giannini P et al (2004) A controlled trial of natural cycle versus microdose gonadotropin-releasing hormone analog flare cycles in poor responders undergoing in vitro fertilization. Fertil Steril 81(6):1542–1547
171. Elizur SE, Aslan D, Shulman A, Weisz B, Bider D, Dor J (2005) Modified natural cycle using GnRH antagonist can be an optional treatment in poor responders undergoing IVF. J Assist Reprod Genet 22(2):75–79
172. Berkkanoglu M, Ozgur K (2010) What is the optimum maximal gonadotropin dosage used in microdose flare-up cycles in poor responders? Fertil Steril 94(2):662–665
173. Huang JYJ, Singer T, Abdallah R, Williams Z, Davis O, Rosenwaks Z (2010) A thickened endometrial stripe on day 3 of leuprolide acetate down-regulated IVF cycles is associated with reduced clinical pregnancy and live birth rates. Fertil Steril 94(4 Suppl 1):S51–S52
174. Forman RG, Robinson J, Yudkin P, Egan D, Reynolds K, Barlow DH (1991) What is the true follicular diameter: an assessment of the reproducibility of transvaginal ultrasound monitoring in stimulated cycles. Fertil Steril 56(5):989–992
175. Ata B, Tulandi T (2011) Ultrasound automated volume calculation in reproduction and in pregnancy. Fertil Steril 95(7):2163–2170
176. Salama S, Arbo E, Lamazou F, Levailllant JM, Frydman R, Fanchin R (2010) Reproducibility and reliability of automated volumetric measurement of single preovulatory follicles using SonoAVC. Fertil Steril 93(6):2069–2073
177. Lamazou F, Arbo E, Salama S, Grynberg M, Frydman R, Fanchin R (2010) Reliability of automated volumetric measurement of multiple growing follicles in controlled ovarian hyperstimulation. Fertil Steril 94(6):2172–2176
178. Raine-Fenning N, Jayaprakasan K, Chamberlain S, Devlin L, Priddle H, Johnson I (2009) Automated measurements of follicle diameter: a chance to standardize? Fertil Steril 91(4 Suppl 1):1469–1472
179. van Santbrink EJ, Hop WC, van Dessel TJ, de Jong FH, Fauser BC (1995) Decremental follicle-stimulating hormone and dominant follicle development during the normal menstrual cycle. Fertil Steril 64(1):37–43
180. Fauser BC, Donderwinkel P, Schoot DC (1993) The step-down principle in gonadotropin treatment and the role of GnRH analogues. Baillieres Clin Obstet Gynaecol 7(2):309–330
181. Cedrin-Durnerin I, Bständig B, Hervé F, Wolf J-P, Uzan M, Hugues J-N (2000) A comparative study of high fixed-dose and decremental-dose regimens of gonadotropins in a minidose gonadotropin-releasing hormone agonist flare protocol for poor responders. Fertil Steril 73(5):1055–1056
182. Abdallah R, Kligman I, Davis O, Rosenwaks Z (2010) Withholding gonadotropins until human chorionic gonadotropin administration. Semin Reprod Med 28(06):486,92
183. Nardo LG, Cheema P, Gelbaya TA, Horne G, Fitzgerald CT, Pease EHE et al (2006) The optimal length of 'coasting protocol' in women at risk of ovarian hyperstimulation syndrome undergoing in vitro fertilization. Hum Fertil 9(3):175–180
184. Kol S (2004) Luteolysis induced by a gonadotropin-releasing hormone agonist is the key to prevention of ovarian hyperstimulation syndrome. Fertil Steril 81(1):1–5

185. Damewood MD, Shen W, Zacur HA, Schlaff WD, Rock JA, Wallach EE (1989) Disappearance of exogenously administered human chorionic gonadotropin. Fertil Steril 52(3):398–400
186. Humaidan P, Ejdrup Bredkjær H, Westergaard LG, Yding Andersen C (2010) 1,500 IU human chorionic gonadotropin administered at oocyte retrieval rescues the luteal phase when gonadotropin-releasing hormone agonist is used for ovulation induction: a prospective, randomized, controlled study. Fertil Steril 93(3):847–854
187. Shapiro BS, Daneshmand ST, Garner FC, Aguirre M, Hudson C (2011) Comparison of "triggers" using leuprolide acetate alone or in combination with low-dose human chorionic gonadotropin. Fertil Steril 95(8):2715–2717
188. Lenz S, Lauritsen JG (1982) Ultrasonically guided percutaneous aspiration of human follicles under local anesthesia: a new method of collecting oocytes for in vitro fertilization. Fertil Steril 38(6):673–677
189. Schulman JD, Dorfmann A, Jones S, Joyce B, Hanser J (1985) Outpatient in vitro fertilization using transvaginal oocyte retrieval and local anesthesia. N Engl J Med 312(25):1639
190. Flood JT, Muasher SJ, Simonetti S, Kreiner D, Acosta AA, Rosenwaks Z (1989) Comparison between laparoscopically and ultrasonographically guided transvaginal follicular aspiration methods in an in vitro fertilization program in the same patients using the same stimulation protocol. J In Vitro Fert Embryo Transf 6(3):180–185
191. Trout SW, Vallerand AH, Kemmann E (1998) Conscious sedation for in vitro fertilization. Fertil Steril 69(5):799–808
192. Ditkoff EC, Plumb J, Selick A, Sauer MV (1997) Anesthesia practices in the United States common to in vitro fertilisation (IVF) centers. J Assist Reprod Genet 14:145–147
193. Kwan I, Bhattacharya S, Knox F, McNeil A (2006) Conscious sedation and analgesia for oocyte retrieval during IVF procedures: a Cochrane review. Hum Reprod 21(7):1672–1679
194. Weinreb EB, Cholst IN, Ledger WJ, Danis RB, Rosenwaks Z (2010) Should all oocyte donors receive prophylactic antibiotics for retrieval? Fertil Steril 94(7):2935–2937
195. El Hussein E, Balen AH, Tan S-L (1992) A prospective study comparing the outcome of oocytes retrieved in the aspirate with those retrieved in the flush during transvaginal ultrasound directed oocyte recovery for in-vitro fertilization. BJOG 99(10):841–844
196. Wongtra-Ngan S, Vutyavanich T, Brown J (2010) Follicular flushing during oocyte retrieval in assisted reproductive techniques. Cochrane Database Syst Rev 9:CD004634
197. Hill MJ, Levens ED (2010) Is there a benefit in follicular flushing in assisted reproductive technology? Curr Opin Obstet Gynecol 22(3):208–212
198. Van Der Ven HH, Al-Hasani S, Diedrich K, Hamerich U, Lehmann F, Krebs D (1985) Polyspermy in in vitro fertilization of human oocytes: frequency and possible causes. Ann N Y Acad Sci 442(1):88–95
199. Kang HJ, Rosenwaks Z (2008) Triploidy: the breakdown of monogamy between sperm and egg. Int J Dev Biol 52(5–6):449–454
200. Palermo G, Joris H, Devroey P, Van Steirteghem AC (1992) Pregnancies after intracytoplasmic injection of single spermatozoon into an oocyte. Lancet 340(8810):17–18
201. The Practice Committee of the American Society for Reproductive Medicine (2008) Intracytoplasmic sperm injection (ICSI). Fertil Steril 90(5 Suppl 1):S187
202. Vawda AI, Gunby J, Younglai EV (1996) Andrology: semen parameters as predictors of in-vitro fertilization: the importance of strict criteria sperm morphology. Hum Reprod 11(7):1445–1450
203. Lanzendorf SE, Maloney MK, Veeck LL, Slusser J, Hodgen GD, Rosenwaks Z (1988) A preclinical evaluation of pronuclear formation by microinjection of human spermatozoa into human oocytes. Fertil Steril 49(5):835–842
204. Devroey P, Liu J, Nagy Z, Tournaye H, Silber SJ, Van Steirteghem AC (1994) Normal fertilization of human oocytes after testicular sperm extraction and intracytoplasmic sperm injection. Fertil Steril 62(3):639–641
205. Devroey P, Liu J, Nagy Z, Goossens A, Tournaye H, Camus M et al (1995) Pregnancies after testicular sperm extraction and intracytoplasmic sperm injection in non-obstructive azoospermia. Hum Reprod 10(6):1457–1460
206. Denschlag D, Tempfer C, Kunze M, Wolff G, Keck C (2004) Assisted reproductive techniques in patients with Klinefelter syndrome: a critical review. Fertil Steril 82(4):775–779
207. Mansour RT, Fahmy IM, Taha AK, Tawab NA, Serour GI, Aboulghar MA (2003) Intracytoplasmic spermatid injection can result in the delivery of normal offspring. J Androl 24(5):757–764

208. Marcus-Braun N, Braun G, Potashnik G, Har-vardi I (2004) Effect of cryopreservation on quality and fertilization capacity of human sperm. Eur J Obstet Gynecol Reprod Biol 116(1):63–66
209. Hourvitz A, Goldschlag DE, Davis OK, Gosden LV, Palermo GD, Rosenwaks Z (2008) Intracytoplasmic sperm injection (ICSI) using cryopreserved sperm from men with malignant neoplasm yields high pregnancy rates. Fertil Steril 90(3):557–563
210. Witkin SS, Viti D, David SS, Stangel J, Rosenwaks Z (1992) Relation between antisperm antibodies and the rate of fertilization of human oocytes in vitro. J Assist Reprod Genet 9(1):9–13
211. Chian R-C, Huang JYJ, Gilbert L, Son W-Y, Holzer H, Cui SJ et al (2009) Obstetric outcomes following vitrification of in vitro and in vivo matured oocytes. Fertil Steril 91(6):2391–2398
212. Hardarson T, Lundin K, Hamberger L (2000) The position of the metaphase II spindle cannot be predicted by the location of the first polar body in the human oocyte. Hum Reprod 15(6):1372–1376
213. Wang W-H, Meng L, Hackett RJ, Odenbourg R, Keefe DL (2001) The spindle observation and its relationship with fertilization after intracytoplasmic sperm injection in living human oocytes. Fertil Steril 75(2):348–353
214. Rienzi L, Ubaldi F, Martinez F, Iacobelli M, Minasi MG, Ferrero S et al (2003) Relationship between meiotic spindle location with regard to the polar body position and oocyte developmental potential after ICSI. Hum Reprod 18(6):1289–1293
215. Moon JH, Hyun CS, Lee SW, Son WY, Yoon SH, Lim JH (2003) Visualization of the metaphase II meiotic spindle in living human oocytes using the Polscope enables the prediction of embryonic developmental competence after ICSI. Hum Reprod 18(4):817–820
216. Gardner DK, Lane M, Calderon I, Leeton J (1996) Environment of the preimplantation human embryo in vivo: metabolite analysis of oviduct and uterine fluids and metabolism of cumulus cells. Fertil Steril 65(2):349–353
217. Gardner DK (1998) Changes in requirements and utilization of nutrients during mammalian preimplantation embryo development and their significance in embryo culture. Theriogenology 49(1):83–102
218. Lane M, Gardner DK (2007) Embryo culture medium: which is the best? Best Pract Res Clin Obstet Gynaecol 21(1):83–100
219. Blake D, Svalander P, Jin M, Silversand C, Hamberger L (2002) Protein supplementation of human IVF culture media. J Assist Reprod Genet 19(3):137–143
220. Agarwal A, Gupta S, Sharma R (2005) Oxidative stress and its implications in female infertility: a clinician's perspective. Reprod Biomed Online 11(5):641–650
221. Blake DA, Farquhar CM, Johnson N, Proctor M (2007) Cleavage stage versus blastocyst stage embryo transfer in assisted conception. Cochrane Database Syst Rev 4:CD002118
222. Levitas E, Lunenfeld E, Har-Vardi I, Albotiano S, Sonin Y, Hackmon-Ram R et al (2004) Blastocyst-stage embryo transfer in patients who failed to conceive in three or more day 2–3 embryo transfer cycles: a prospective, randomized study. Fertil Steril 81(3):567–571
223. Ménézo YJR, Chouteau J, Torelló MJ, Girard A, Veiga A (1999) Birth weight and sex ratio after transfer at the blastocyst stage in humans. Fertil Steril 72(2):221–224
224. DeBaun MR, Niemitz EL, Feinberg AP (2003) Association of in vitro fertilization with Beckwith-Wiedemann syndrome and epigenetic alterations of LIT1 and H19. Am J Hum Genet 72(1):156–160
225. Maher ER, Brueton LA, Bowdin SC, Luharia A, Cooper W, Cole TR et al (2003) Beckwith-Wiedemann syndrome and assisted reproduction technology (ART). J Med Genet 40(1):62–64
226. Schiewe MC, Hazeleger NL, Sclimenti C, Balmaceda JP (1995) Physiological characterization of blastocyst hatching mechanisms by use of a mouse antihatching model. Fertil Steril 62(2):288–294
227. O'Sullivan CM, Ungarian JLR, Singh K, Liu S, Hance J, Rancourt DE (2004) Uterine secretion of ISP1 & 2 tryptases is regulated by progesterone and estrogen during pregnancy and the endometrial cycle. Mol Reprod Dev 69(3):252–259
228. Cohen J, Elsner C, Kort H, Malter H, Massey J, Mayer MP et al (1990) Impairment of the hatching process following IVF in the human and improvement of implantation by assisting hatching using micromanipulation. Hum Reprod 5(1):7–13
229. Cohen J, Alikani M, Trowbridge J, Rosenwaks Z (1992) Implantation enhancement by selective assisted hatching using zona drilling of human embryos with poor prognosis. Hum Reprod 7(5):685–691
230. Nakayama T, Fujiwara H, Yamada S, Tastumi K, Honda T, Fujii S (1999) Clinical application of a new assisted hatching method using

a piezo-micromanipulator for morphologically low-quality embryos in poor-prognosis infertile patients. Fertil Steril 71(6): 1014–1018

231. Cohen J (1991) Assisted hatching of human embryos. J In Vitro Fert Embryo Transf 8(4):179–190

232. Cohen J, Malter H, Fehilly C, Wright G, Elsner C, Kort H et al (1988) Implantation of embryos after partial opening of oocyte zona pellucida to facilitate sperm penetration. Lancet 2(8603):162

233. The Practice Committee of the American Society for Reproductive Medicine (2006) The role of assisted hatching in in vitro fertilization: a review of the literature. A committee opinion. Fertil Steril 86(5 Suppl 1): S124–S126

234. Das S, Blake D, Farquhar C, Seif MM (2009) Assisted hatching on assisted conception (IVF and ICSI). Cochrane Database Syst Rev 15(2):CD001894

235. Ng EHY, Naveed F, Lau EYL, Yeung WSB, Chan CCW, Tang OS et al (2005) A randomized double-blind controlled study of the efficacy of laser-assisted hatching on implantation and pregnancy rates of frozen–thawed embryo transfer at the cleavage stage. Hum Reprod 20(4):979–985

236. Al-Nuaim LA, Jenkins JM (2002) Assisted hatching in assisted reproduction. BJOG 109(8):856–862

237. Hershlag A, Paine T, Cooper GW, Scholl GM, Rawlinson K, Kvapil G (1999) Monozygotic twinning associated with mechanical assisted hatching. Fertil Steril 71(1):144–146

238. Schieve LA, Meikle SF, Peterson HB, Jeng G, Burnett NM, Wilcox LS (2000) Does assisted hatching pose a risk for monozygotic twinning in pregnancies conceived through in vitro fertilization? Fertil Steril 74(2):288–294

239. Cohen J, Malter H, Wright G, Kort H, Massey J, Mitchell D (1989) Partial zona dissection of human oocytes when failure of zona pellucida penetration is anticipated. Hum Reprod 4(4):435–442

240. Spandorfer SD, Barmat LI, Navarro J, Liu H-C, Veeck L, Rosenwaks Z (2002) Importance of the biospy date in autologous endometrial cocultures for patients with multiple implantation failures. Fertil Steril 77(6): 1209–1213

241. Spandorfer SD, Barmat LI, Liu HC, Mele C, Veeck L, Rosenwaks Z (1998) Granulocyte macrophage-colony stimulating factor production by autologous endometrial co-culture is associated with outcome for in vitro fertilization patients with a history of multiple implantation failures. Am J Reprod Immunol 40(5):377–381

242. Spandorfer SD, Clarke R, Bovis L, Liu H-C, Neuer A, Witkin SS et al (2000) Interleukin-1 levels in the supernatant of conditioned media of embryos grown in autologous endometrial coculture: correlation with embryonic development and outcome for patients with a history of multiple implantation failures after IVF. Am J Reprod Immunol 43(1):6–11

243. Spandorfer SD, Navarro J, Levy D, Black AR, Liu HC, Veeck L et al (2001) Autologous endometrial coculture in patients with in vitro-fertilization (IVF) failure: correlations of outcome with leukemia inhibiting factor production. Am J Reprod Immunol 46: 375–380

244. Barmat LI, Worrilow KC, Paynton BV (1997) Growth factor expression by human oviduct and buffalo rat liver coculture cells. Fertil Steril 67(4):775–779

245. Loutradis D, John D, Kiessling AA (1987) Hypoxanthine causes a 2-cell block in random-bred mouse embryos. Biol Reprod 37(2):311–316

246. Fukui Y, McGowan LT, James RW, Pugh PA, Tervit HR (1991) Factors affecting the in-vitro development to blastocysts of bovine oocytes matured and fertilized in vitro. J Reprod Fertil 92(1):125–131

247. Barmat LI, Liu H-C, Spandorfer SD, Kowalik A, Mele C, Xu K et al (1999) Autologous endometrial co-culture in patients with repeated failures of implantation after in vitro fertilization-embryo transfer. J Assist Reprod Genet 16(3):121–127

248. Spandorfer SD, Barmat L, Navarro J, Burmeister L, Veeck L, Clarke R et al (2002) Autologous endometrial coculture in patients with a previous history of poor quality embryos. J Assist Reprod Genet 19(7):309–312

249. Spandorfer SD, Pascal P, Parks J, Clark R, Veeck L, Davis OK et al (2004) Autologous endometrial coculture in patients with IVF failure: outcome of the first 1,030 cases. J Reprod Med 49(6):463–467

250. Spandorfer S, Soslow R, Clark R, Fasouliotis S, Davis O, Rosenwaks Z (2006) Histologic characteristics of the endometrium predicts success when utilizing autologous endometrial coculture in patients with IVF failure. J Assist Reprod Genet 23(4):185–189

251. Barmat LI, Liu H-C, Spandorfer SD, Xu K, Veeck L, Damario MA et al (1998) Human preembryo development on autologous endometrial coculture versus conventional medium. Fertil Steril 70(6):1109–1113

252. Kattal N, Cohen J, Barmat LI (2008) Role of coculture in human in vitro fertilization: a meta-analysis. Fertil Steril 90(4):1069–1076
253. Derks RS, Farquhar C, Mol BW, Buckingham K, Heineman MJ (2009) Techniques for preparation prior to embryo transfer. Cochrane Database Syst Rev 4:CD007682
254. Abou-Setta AM, Al-Inany HG, Mansour RT, Serour GI, Aboulghar MA (2005) Soft versus firm embryo transfer catheters for assisted reproduction: a systematic review and meta-analysis. Hum Reprod 20(11):3114–3121
255. Buckett WM (2006) A review and meta-analysis of prospective trials comparing different catheters used for embryo transfer. Fertil Steril 85(3):728–734
256. Saldeen P, Abou-Setta AM, Bergh T, Sundström P, Holte J (2008) A prospective randomized controlled trial comparing two embryo transfer catheters in an ART program. Fertil Steril 90(3):599–603
257. Lewin A, Schenker J, Avrech O, Shapira S, Safran A, Friedler S (1997) The role of uterine straightening by passive bladder distension before embryo transfer in IVF cycles. J Assist Reprod Genet 14(1):32–34
258. Abou-Setta AM (2007) Effect of passive uterine straightening during embryo transfer: a systematic review and meta-analysis. Acta Obstet Gynecol Scand 86(5):516–522
259. Fanchin R, Harmas A, Benaoudia F, Lundkvist U, Olivennes F, Frydman R (1998) Microbial flora of the cervix assessed at the time of embryo transfer adversely affects in vitro fertilization outcome. Fertil Steril 70(5):866–870
260. Miller K, Frattarelli J (2007) The pre-cycle blind mock embryo transfer is an inaccurate predictor of anticipated embryo transfer depth. J Assist Reprod Genet 24(2):77–82
261. Henne MB, Milki AA (2004) Uterine position at real embryo transfer compared with mock embryo transfer. Hum Reprod 19(3): 570–572
262. Yang W-J, Lee RK-K, Su J-T, Lin M-H, Hwu Y-M (2007) Uterine position change between mock and real embryo transfers. Taiwan J Obstet Gynecol 46(2):162–165
263. Strickler RC, Christianson C, Crane JP, Curato A, Knight AB, Yang V (1985) Ultrasound guidance for human embryo transfer. Fertil Steril 43(1):54–61
264. Matorras R, Urquijo E, Mendoza R, Corcóstegui B, Expósito A, Rodríguez-Escudero FJ (2002) Ultrasound-guided embryo transfer improves pregnancy rates and increases the frequency of easy transfers. Hum Reprod 17(7):1762–1766
265. Brown J, Buckingham K, Abou-Setta AM, Buckett W (2010) Ultrasound versus 'clinical touch' for catheter guidance during embryo transfer in women. Cochrane Database Syst Rev 20(1):CD006107
266. Sallam HN, Agameya AF, Rahman AF, Ezzeldin F, Sallam AN (2002) Ultrasound measurement of the uterocervical angle before embryo transfer: a prospective controlled study. Hum Reprod 17(7): 1767–1772
267. Nabi A, Awonuga A, Birch H, Barlow S, Stewart B (1997) Multiple attempts at embryo transfer: does this affect in-vitro fertilization treatment outcome? Hum Reprod 12(6):1188–1190
268. Yovich JL, Turner SR, Murphy AJ (1985) Embryo transfer technique as a cause of ectopic pregnancies in in vitro fertilization. Fertil Steril 44(3):318–321
269. Waterstone J, Curson R, Parsons J (1991) Embryo transfer to low uterine cavity. Lancet 337(8754):1413
270. Bennett S, Waterstone J, Parsons J, Creighton S (1993) Two cases of cervical pregnancy following in vitro fertilization and embryo transfer to the lower uterine cavity. J Assist Reprod Genet 10(1):100–103
271. Martínez F, Coroleu B, Parriego M, Carreras O, Belil I, Parera N et al (2001) Ultrasound-guided embryo transfer: immediate withdrawal of the catheter versus a 30 second wait. Hum Reprod 16(5):871–874
272. Salamonsen L, Shuster S, Stern R (2001) Distribution of hyaluronan in human endometrium across the menstrual cycle. Cell Tissue Res 306(2):335–340
273. Carson DD, Dutt A, Tang J-P (1987) Glycoconjugate synthesis during early pregnancy: hyaluronate synthesis and function. Dev Biol 120(1):228–235
274. Simon A, Safran A, Revel A, Aizenman E, Reubinoff B, Porat-Katz A et al (2003) Hyaluronic acid can successfully replace albumin as the sole macromolecule in a human embryo transfer medium. Fertil Steril 79(6):1434–1438
275. Urman B, Yakin K, Ata B, Isiklar A, Balaban B (2008) Effect of hyaluronan-enriched transfer medium on implantation and pregnancy rates after day 3 and day 5 embryo transfers: a prospective randomized study. Fertil Steril 90(3):604–612
276. Friedler S, Schachter M, Strassburger D, Esther K, El Ron R, Raziel A (2007) A randomized clinical trial comparing recombinant hyaluronan/recombinant albumin versus human tubal fluid for cleavage stage embryo

transfer in patients with multiple IVF-embryo transfer failure. Hum Reprod 22(9): 2444–2448

277. Bontekoe S, Blake D, Heineman MJ, Williams EC, Johnson N (2010) Adherence compounds in embryo transfer media for assisted reproductive technologies. Cochrane Database Syst Rev 7:CD007421

278. Valojerdi M, Karimian L, Yazdi P, Gilani MA, Madani T, Baghestani A (2006) Efficacy of a human embryo transfer medium: a prospective, randomized clinical trial study. J Assist Reprod Genet 23(5):207–212

279. Loutradi KE, Prassas I, Bili E, Sanopoulou T, Bontis I, Tarlatzis BC (2007) Evaluation of a transfer medium containing high concentration of hyaluronan in human in vitro fertilization. Fertil Steril 87(1):48–52

280. Li B, Zhou H, Li W (2011) Bed rest after embryo transfer. Eur J Obstet Gynecol Reprod Biol 155(2):125–128

281. Yao Z, Vansteelandt S, Van der Elst J, Coetsier T, Dhont M, De Sutter P (2009) The efficacy of the embryo transfer catheter in IVF and ICSI is operator-dependent: a randomized clinical trial. Hum Reprod 24(4):880–887

282. Jones GS (1991) Luteal phase defect: a review of pathophysiology. Curr Opin Obstet Gynecol 3(5):641–648

283. Milgrom E, Thi L, Atger M, Baulieu EE (1973) Mechanisms regulating the concentration and the conformation of progesterone receptor(s) in the uterus. J Biol Chem 248(18):6366–6374

284. Clarke CL, Sutherland RL (1990) Progestin regulation of cellular proliferation. Endocr Rev 11(2):266–301

285. Csapo AI, Pulkkinen MO, Ruttner B, Sauvage JP, Wiest WG (1972) The significance of the human corpus luteum in pregnancy maintenance. I. Preliminary studies. Am J Obstet Gynecol 112(8):1061–1067

286. Rosenwaks Z, Navot D, Veeck L, Liu HC, Steingold K, Kreiner D et al (1988) Oocyte donation. The Norfolk program. Ann N Y Acad Sci 541:728–741

287. Smitz J, Devroey P, Camus M, Deschacht J, Khan I, Staessen C et al (1988) The luteal phase and early pregnancy after combined GnRH-agonist/HMG treatment for superovulation in IVF or GIFT. Hum Reprod 3(5):585–590

288. Garcia J, Jones C, Acosta A, Wright G (1981) Corpus luteum function after follicle aspiration for oocyte retrieval. Fertil Steril 36: 565–572

289. Frydman R, Testart J, Giacomini P, Imbert MC, Martin E, Nahoul K (1982) Hormonal and histological study of the luteal phase in women following aspiration of the preovulatory follicle. Fertil Steril 38:312–317

290. Fauser BCJM, Devroey P (2003) Reproductive biology and IVF: ovarian stimulation and luteal phase consequences. Trends Endocrinol Metab 14(5):236–242

291. Miyake A, Aono T, Kinugasa T, Tanizawa O, Kurachi K (1979) Suppression of serum levels of luteinzing hormone by short- and long-loop negative feedback in ovariectomized women. J Endocrinol 80(3):353–356

292. Broekmans FJ, Bernardus RE, Berkhout G, Schoemaker J (1992) Pituitary and ovarian suppression after early follicular and mid-luteal administration of a LHRH agonist in a depot formulation: decapeptyl CR. Gynecol Endocrinol 6(3):153–161

293. Smitz J, Erard P, Camus M, Devroey P, Tournaye H, Wisanto A et al (1992) Pituitary gonadotrophin secretory capacity during the luteal phase in superovulation using GnRH-agonists and HMG in a desensitization or flare-up protocol. Hum Reprod 7(9): 1225–1229

294. Beckers NGM, Macklon NS, Eijkemans MJ, Ludwig M, Felberbaum RE, Diedrich K et al (2003) Nonsupplemented luteal phase characteristics after the administration of recombinant human chorionic gonadotropin, recombinant luteinizing hormone, or gonadotropin-releasing hormone (GnRH) agonist to induce final oocyte maturation in in vitro fertilization patients after ovarian stimulation with recombinant follicle-stimulating hormone and GnRH antagonist cotreatment. J Clin Endocrinol Metab 88(9): 4186–4192

295. Tavaniotou A, Devroey P (2006) Luteal hormonal profile of oocyte donors stimulated with a GnRH antagonist compared with natural cycles. Reprod Biomed Online 13(3): 326–330

296. Mochtar MH, Van Wely M, Van der Veen F (2006) Timing luteal phase support in GnRH agonist down-regulated IVF/embryo transfer cycles. Hum Reprod 21(4):905–908

297. Pritts EA, Atwood AK (2002) Luteal phase support in infertility treatment: a meta-analysis of the randomized trials. Hum Reprod 17(9):2287–2299

298. Daya S, Gunby J (2004) Luteal phase support in assisted reproduction cycles. Cochrane Database Syst Rev 3:CD004830

299. Tavaniotou A, Smitz J, Bourgain C, Devroey P (2000) Comparison between different routes of progesterone administration as luteal phase support in infertility treatments. Hum Reprod Update 6(2):139–148

300. Polyzos NP, Messini CI, Papanikolaou EG, Mauri D, Tzioras S, Badawy A et al (2010) Vaginal progesterone gel for luteal phase support in IVF/ICSI cycles: a meta-analysis. Fertil Steril 94(6):2083–2087
301. Miles RA, Paulson RJ, Lobo RA, Press MF, Dahmoush L, Sauer M (1994) Pharmacokinetics and endometrial tissue levels of progesterone after administration by intramuscular and vaginal routes: a comparative study. Fertil Steril 62(3):485–490
302. Friedler S, Raziel A, Schachter M, Strassburger D, Bukovsky I, Ron-El R (1999) Luteal support with micronized progesterone following in-vitro fertilization using a down-regulation protocol with gonadotrophin-releasing hormone agonist: a comparative study between vaginal and oral administration. Hum Reprod 14(8):1944–1948
303. Licciardi FL, Kwiatkowski A, Noyes NL, Berkeley AS, Krey LL, Grifo JA (1999) Oral versus intramuscular progesterone for in vitro fertilization: a prospective randomized study. Fertil Steril 71(4):614–618
304. Perino M, Brigandì FG, Abate FG, Costabile L, Balzano E, Abate A (1997) Intramuscular versus vaginal progesterone in assisted reproduction: a comparative study. Clin Exp Obstet Gynecol 24(4):228–231
305. Propst AM, Hill JA, Ginsburg ES, Hurwitz S, Politch J, Yanushpolsky EH (2001) A randomized study comparing Crinone 8 % and intramuscular progesterone supplementation in in vitro fertilization-embryo transfer cycles. Fertil Steril 76(6):1144–1149
306. Daya S, Gunby JL (2008) Luteal phase support in assisted reproduction cycles. Cochrane Database Syst Rev 3:CD004830
307. Levine H (2000) Luteal support in IVF using the novel vaginal progesterone gel crinone 8 %: results of an open-label trial in 1184 women from 16 U.S. centers. Fertil Steril 74(4):836–837
308. Zarutskie PW, Phillips JA (2009) A meta-analysis of the route of administration of luteal phase support in assisted reproductive technology: vaginal versus intramuscular progesterone. Fertil Steril 92(1):163–169
309. Hutchinson-Williams KA, Lunenfeld B, Diamond MP, Lavy G, Boyers SP, DeCherney AH (1989) Human chorionic gonadotropin, estradiol, and progesterone profiles in conception and nonconception cycles in an in vitro fertilization program. Fertil Steril 52(3):441–445
310. Liu HC, Davis O, Berkeley A, Graf M, Rosenwaks Z (1991) Late luteal estradiol patterns are a better prognosticator of pregnancy outcome than serial beta-human chorionic gonadotropin concentrations. Fertil Steril 56(3):421–426
311. Smitz J, Bourgain C, Van Waesberghe L, Camus M, Devroey P, Van Steirteghem AC (1993) A prospective randomized study on oestradiol valerate supplementation in addition to intravaginal micronized progesterone in buserelin and HMG induced superovulation. Hum Reprod 8(1):40–45
312. Lewin A, Benshushan A, Mezker E, Yanai N, Schenker JG, Goshen R (1994) The role of estrogen support during the luteal phase of in vitro fertilization-embryo transplant cycles: a comparative study between progesterone alone and estrogen and progesterone support. Fertil Steril 62(1):121–125
313. Farhi J, Weissman A, Steinfeld Z, Shorer M, Nahum H, Levran D (2000) Estradiol supplementation during the luteal phase may improve the pregnancy rate in patients undergoing in vitro fertilization-embryo transfer cycles. Fertil Steril 73(4):761–766
314. Gorkemli H, Ak D, Akyurek C, Aktan M, Duman S (2004) Comparison of pregnancy outcomes of progesterone or progesterone + estradiol for luteal phase support in ICSI-ET cycles. Gynecol Obstet Invest 58(3):140–144
315. Lukaszuk K, Liss J, Lukaszuk M, Maj B (2005) Optimization of estradiol supplementation during the luteal phase improves the pregnancy rate in women undergoing in vitro fertilization-embryo transfer cycles. Fertil Steril 83(5):1372–1376
316. Fatemi HM, Kolibianakis EM, Camus M, Tournaye H, Donoso P, Papanikolaou E et al (2006) Addition of estradiol to progesterone for luteal supplementation in patients stimulated with GnRH antagonist/rFSH for IVF: a randomized controlled trial. Hum Reprod 21(10):2628–2632
317. Ceyhan ST, Basaran M, Kemal Duru N, Yilmaz A, Göktolga Ü, Baser I (2008) Use of luteal estrogen supplementation in normal responder patients treated with fixed multidose GnRH antagonist: a prospective randomized controlled study. Fertil Steril 89(6):1827–1830
318. Engmann L, DiLuigi A, Schmidt D, Benadiva C, Maier D, Nulsen J (2008) The effect of luteal phase vaginal estradiol supplementation on the success of in vitro fertilization treatment: a prospective randomized study. Fertil Steril 89(3):554–561
319. Serna J, Cholquevilque JL, Cela V, Martínez-Salazar J, Requena A, Garcia-Velasco JA (2008) Estradiol supplementation during the

luteal phase of IVF-ICSI patients: a randomized, controlled trial. Fertil Steril 90(6): 2190–2195

320. Gelbaya TA, Kyrgiou M, Tsoumpou I, Nardo LG (2008) The use of estradiol for luteal phase support in in vitro fertilization/intracytoplasmic sperm injection cycles: a systematic review and meta-analysis. Fertil Steril 90(6):2116–2125
321. Kolibianakis EM, Venetis CA, Papanikolaou EG, Diedrich K, Tarlatzis BC, Griesinger G (2008) Estrogen addition to progesterone for luteal phase support in cycles stimulated with GnRH analogues and gonadotrophins for IVF: a systematic review and meta-analysis. Hum Reprod 23(6):1346–1354
322. Jee BC, Suh CS, Kim SH, Kim YB, Moon SY (2010) Effects of estradiol supplementation during the luteal phase of in vitro fertilization cycles: a meta-analysis. Fertil Steril 93(2): 428–436
323. Hutchinson-Williams KA, DeCherney AH, Lavy G, Diamond MP, Naftolin F, Lunenfeld B (1990) Luteal rescue in in vitro fertilization-embryo transfer. Fertil Steril 53(3):495–501
324. Herman A, Raziel A, Strassburger D, Soffer Y, Bukovsky I, Ron-El R (1996) The benefits of mid-luteal addition of human chorionic gonadotrophin in in-vitro fertilization using a down-regulation protocol and luteal support with progesterone. Hum Reprod 11(7): 1552–1557
325. Peñarrubia J, Balasch J, Fábregues F, Creus M, Casamitjana R, Ballescá JL et al (1998) Human chorionic gonadotrophin luteal support overcomes luteal phase inadequacy after gonadotrophin-releasing hormone agonist-induced ovulation in gonadotrophin-stimulated cycles. Hum Reprod 13(12): 3315–3318
326. Aboulghar MA, Amin YM, Al-Inany HG, Aboulghar MM, Mourad LM, Serour GI et al (2008) Prospective randomized study comparing luteal phase support for ICSI patients up to the first ultrasound compared with an additional three weeks. Hum Reprod 23(4): 857–862
327. Tryde Schmidt KL, Ziebe S, Popovic B, Lindhard A, Loft A, Andersen AN (2001) Progesterone supplementation during early gestation after in vitro fertilization has no effect on the delivery rate. Fertil Steril 75(2): 337–341
328. Nyboe Andersen A, Popovic-Todorovic B, Schmidt KT, Loft A, Lindhard A, Højgaard A et al (2002) Progesterone supplementation during early gestations after IVF or ICSI has no effect on the delivery rates: a randomized controlled trial. Hum Reprod 17(2):357–361
329. Aragona C, Mohamed MA, Espinola MSB, Linari A, Pecorini F, Micara G et al (2011) Clinical complications after transvaginal oocyte retrieval in 7,098 IVF cycles. Fertil Steril 95(1):293–294
330. Roest J, Mous H, Zeilmaker G, Verhoeff A (1996) The incidence of major clinical complications in a Dutch transport IVF programme. Hum Reprod Update 2(4): 345–353
331. Henriksson P, Westerlund E, Wallén H, Brandt L, Hovatta O, Ekbom A (2013) Incidence of pulmonary and venous thromboembolism in pregnancies after in vitro fertilisation: cross sectional study. BMJ 346:e8632
332. MacDougall MJ, Tan SL, Balen A, Jacobs HS (1993) A controlled study comparing patients with and without polycystic ovaries undergoing in-vitro fertilization. Hum Reprod 8(2): 233–237
333. Chen D, Burmeister L, Goldschlag D, Rosenwaks Z (2003) Ovarian hyperstimulation syndrome: strategies for prevention. Reprod Biomed Online 7(1):43–49
334. Golan A, Ron-El R, Herman A, Soffer Y, Wainraub Z, Caspi E (1989) Ovarian hyperstimulation syndrome: an update review. Obstet Gynecol Surv 44:430–440
335. Navot D, Bergh PA, Laufer N (1992) Ovarian hyperstimulation syndrome in novel reproductive technologies: prevention and treatment. Fertil Steril 58(2):249–261
336. Cluroe AD, Synek BJ (1995) A fatal case of ovarian hyperstimulation syndrome with cerebral infarction. Pathology 27(4):344–346
337. Akdemir R, Uyan C, Emiroglu Y (2002) Acute myocardial infarction secondary thrombosis associated with ovarial hyperstimulation syndrome. Int J Cardiol 83(2):187–189
338. Salihu HM, Aliyu MH, Rouse DJ, Kirby RS, Alexander GR (2003) Potentially preventable excess mortality among higher-order multiples. Obstet Gynaecol 102(4):679–684
339. The Practice Committee of the American Society for Reproductive Medicine, The Practice Committee of the Society for Assisted Reproductive Technology (2009) Guidelines on number of embryos transferred. Fertil Steril 92(5):1518–1519
340. Milki AA, Jun SH, Hinckley MD, Behr B, Giudice LC, Westphal LM (2003) Incidence of monozygotic twinning with blastocyst transfer compared to cleavage-stage transfer. Fertil Steril 79(3):503–506

341. Knopman J, Krey LC, Lee J, Fino ME, Novetsky Akiva P, Noyes N (2010) Monozygotic twinning: an eight-year experience at a large IVF center. Fertil Steril 94(2):502–510
342. Azem F, Yaron Y, Botchan A, Amit A, Yovel I, David MP et al (1993) Ectopic pregnancy after in vitro fertilization-embryo transfer (IVF-ET): the possible role of the ET technique. J Assist Reprod Genet 10(4):302–304
343. Chow WH, Daling JR, Cates W Jr, Greenberg RS (1987) Epidemiology of ectopic pregnancy. Epidemiol Rev 9(1):70–94
344. Reece EA, Petrie RH, Sirmans MF, Finster M, Todd WD (1983) Combined intrauterine and extrauterine gestations: a review. Am J Obstet Gynecol 146(3):323–330
345. Marcus SF, Macnamee M, Brinsden P (1995) Pregnancy: heterotopic pregnancies after in-vitro fertilization and embryo transfer. Hum Reprod 10(5):1232–1236
346. Rizk B, Tan SL, Morcos S, Riddle A, Brinsden P, Mason BA et al (1991) Heterotopic pregnancies after in vitro fertilization and embryo transfer. Am J Obstet Gynecol 164(1 Pt 1):161–164
347. ESHRE Guidelines & Legal. ART fact sheet (2010). http://www.eshre.eu/ESHRE/English/Guidelines-Legal/ART-fact-sheet/page.aspx/1061
348. Ochsenkühn R, Strowitzki T, Gurtner M, Strauss A, Schulze A, Hepp H et al (2003) Pregnancy complications, obstetric risks, and neonatal outcome in singleton and twin pregnancies after GIFT and IVF. Arch Gynecol Obstet 268(4):256–261
349. Jackson RA, Gibson KA, Wu YW, Croughan MS (2004) Perinatal outcomes in singletons following in vitro fertilization: a meta-analysis. Obstet Gynecol 103(2):551–563
350. Katalinic A, Rösch C, Ludwig M (2004) Pregnancy course and outcome after intracytoplasmic sperm injection: a controlled, prospective cohort study. Fertil Steril 81(6): 1604–1616
351. Allen VM, Wilson RD, Cheung A, Genetics Committee of the Society of Obstetricians and Gynaecologists of Canada (SOGC), Reproductive Endocrinology Infertility Committee of the Society of Obstetricians and Gynaecologists of Canada (SOGC) (2006) Pregnancy outcomes after assisted reproductive technology. J Obstet Gynaecol Can 28(3):220–250
352. Thomson F, Shanbhag S, Templeton A, Bhattacharya S (2005) Obstetric outcome in women with subfertility. BJOG 112(5): 632–637
353. Helmerhorst FM, Perquin DAM, Donker D, Keirse MJNC (2004) Perinatal outcome of singletons and twins after assisted conception: a systematic review of controlled studies. BMJ 328(7434):261
354. Schieve LA, Ferre C, Peterson HB (2004) Perinatal outcome among singleton infants conceived through assisted reproductive technology in the United States. Obstet Gynaecol 103(6):1144–1153
355. Lie RT, Lyngstadass A, Orstavik KH, Bakketeig LS, Jacobsen G, Tanbo T (2005) Birth defects in children conceived by ICSI compared with children conceived by other IVF-methods; a meta-analysis. Int J Epidemiol 34(3):696–701
356. Wennerholm U-B, Bergh C, Hamberger L, Lundin K, Nilsson L, Wikland M et al (2000) Incidence of congenital malformations in children born after ICSI. Hum Reprod 15(4):944–948
357. Bonduelle M, Wilikens A, Buysse A, Van Assche E, Wisanto A, Devroey P et al (1996) Prospective follow-up study of 877 children born after intracytoplasmic sperm injection (ICSI), with ejaculated epididymal and testicular spermatozoa and after replacement of cryopreserved embryos obtained after ICSI. Hum Reprod 11(Suppl 4):131–155
358. Van Steirteghem A (1998) Outcome of assisted reproductive technology. N Engl J Med 338(3):194–195
359. Knoester M, Helmerhorst FM, Vandenbroucke JP, van der Westerlaken LAJ, Walther FJ, Veen S (2008) Perinatal outcome, health, growth, and medical care utilization of 5- to 8-year-old intracytoplasmic sperm injection singletons. Fertil Steril 89(5): 1133–1146
360. Hansen M, Kurinczuk JJ, Bower C, Webb S (2002) The risk of major birth defects after intracytoplasmic sperm injection and in vitro fertilization. N Engl J Med 346(10): 725–730
361. Ericson A, Kallen B (2001) Congenital malformations in infants born after IVF: a population-based study. Hum Reprod 16(3):504–509
362. Sutcliffe AG, Peters CJ, Bowdin S, Temple K, Reardon W, Wilson L et al (2006) Assisted reproductive therapies and imprinting disorders–a preliminary British survey. Hum Reprod 21(4):1009–1011
363. Gicquel C, Gaston V, Mandelbaum J, Siffroi J-P, Flahault A, Le Bouc Y (2003) In vitro fertilization may increase the risk of Beckwith-Wiedemann Syndrome related to the abnormal imprinting of the KCNQ1OT gene. Am J Hum Genet 72(5):1338–1341

364. Cox GF, Bürger J, Lip V, Mau UA, Sperling K, Wu B-L et al (2002) Intracytoplasmic sperm injection may increase the risk of imprinting defects. Am J Hum Genet 71(1): 162–164
365. Ørstavik KH, Eiklid K, van der Hagen CB, Spetalen S, Kierulf K, Skjeldal O et al (2003) Another case of imprinting defect in a girl with Angelman syndrome who was conceived by intracytoplasmic sperm injection. Am J Hum Genet 72(1):218–219
366. Place I, Englert Y (2003) A prospective longitudinal study of the physical, psychomotor, and intellectual development of singleton children up to 5 years who were conceived by intracytoplasmic sperm injection compared with children conceived spontaneously and by in vitro fertilization. Fertil Steril 80(6): 1388–1397
367. Bonduelle M, Ponjaert I, Steirteghem AV, Derde M-P, Devroey P, Liebaers I (2003) Developmental outcome at 2 years of age for children born after ICSI compared with children born after IVF. Hum Reprod 18(2):342–350
368. Ponjaert-Kristoffersen I, Tjus T, Nekkebroeck J, Squires J, Verte D, Heimann M et al (2004) Psychological follow-up study of 5-year-old ICSI children. Hum Reprod 19(12): 2791–2797
369. Bonduelle M, Wennerholm U-B, Loft A, Tarlatzis BC, Peters C, Henriet S et al (2005) A multi-centre cohort study of the physical health of 5-year-old children conceived after intracytoplasmic sperm injection, in vitro fertilization and natural conception. Hum Reprod 20(2):413–419
370. Bonduelle M, Bergh C, Niklasson A, Palermo GD, Wennerholm U-B (2004) Medical follow-up study of 5-year-old ICSI children. Reprod BioMed Online 9(1):91–101
371. Munne S, Weier HU, Stein J, Grifo J, Cohen J (1993) A fast and efficient method for simultaneous X and Y in situ hybridization of human blastomeres. J Assist Reprod Genet 10(1):82–90
372. Baart EB, van den Berg I, Martini E, Eussen HJ, Fauser BCJM, Opstal DV (2007) FISH analysis of 15 chromosomes in human day 4 and 5 preimplantation embryos: the added value of extended aneuploidy detection. Prenat Diagn 27(1):55–63
373. Wilton L, Thornhill A, Traeger-Synodinos J, Sermon KD, Harper JC (2009) The causes of misdiagnosis and adverse outcomes in PGD. Hum Reprod 24(5):1221–1228
374. Treff NR, Su J, Kasabwala N, Tao X, Miller KA, Scott RT Jr (2010) Robust embryo identification using first polar body single nucleotide polymorphism microarray-based DNA fingerprinting. Fertil Steril 93(7): 2453–2455
375. Schoolcraft WB, Fragouli E, Stevens J, Munne S, Katz-Jaffe MG, Wells D (2010) Clinical application of comprehensive chromosomal screening at the blastocyst stage. Fertil Steril 94(5):1700–1706
376. Harper J, SenGupta S (2011) Preimplantation genetic diagnosis: state of the ART. Hum Genet 2011:1–12
377. Scott RT Jr, Tao X, Taylor D, Ferry KM, Treff NR (2010) A prospective randomized controlled trial demonstrating significantly increased clinical pregnancy rates following 24 chromosome aneuploidy screening: biopsy and analysis on day 5 with fresh transfer. Fertil Steril 94(4 Suppl 1):S2
378. Assou S, Haouzi D, De Vos J, Hamamah S (2010) Human cumulus cells as biomarkers for embryo and pregnancy outcomes. Mol Hum Reprod 16(8):531–538
379. Katz-Jaffe MG, McReynolds S, Gardner DK, Schoolcraft WB (2009) The role of proteomics in defining the human embryonic secretome. Mol Hum Reprod 15(5): 271–277
380. Katz-Jaffe MG, Schoolcraft WB, Gardner DK (2006) Analysis of protein expression (secretome) by human and mouse preimplantation embryos. Fertil Steril 86(3):678–685
381. Brison DR, Houghton FD, Falconer D, Roberts SA, Hawkhead J, Humpherson PG et al (2004) Identification of viable embryos in IVF by non-invasive measurement of amino acid turnover. Hum Reprod 19(10): 2319–2324
382. Nagy ZP, Sakkas D, Behr B (2008) Symposium: innovative techniques in human embryo viability assessment. Non-invasive assessment of embryo viability by metabolomic profiling of culture media ('metabolomics'). Reprod Biomed Online 17(4):502–507
383. Seli E, Sakkas D, Scott R, Kwok SC, Rosendahl SM, Burns DH (2007) Noninvasive metabolomic profiling of embryo culture media using Raman and near-infrared spectroscopy correlates with reproductive potential of embryos in women undergoing in vitro fertilization. Fertil Steril 88(5): 1350–1357
384. Scott R, Seli E, Miller K, Sakkas D, Scott K, Burns DH (2008) Noninvasive metabolomic

profiling of human embryo culture media using Raman spectroscopy predicts embryonic reproductive potential: a prospective blinded pilot study. Fertil Steril 90(1):77–83

385. Seli E, Vergouw CG, Morita H, Botros L, Roos P, Lambalk CB et al (2010) Noninvasive metabolomic profiling as an adjunct to morphology for noninvasive embryo assessment in women undergoing single embryo transfer. Fertil Steril 94(2):535–542

386. Wong CC, Loewke KE, Bossert NL, Behr B, De Jonge CJ, Baer TM et al (2010) Non-invasive imaging of human embryos before embryonic genome activation predicts development to the blastocyst stage. Nat Biotechnol 28(10):1115–1121

Chapter 9

Novel Markers of Male Infertility

Michael Funaro and Darius A. Paduch

Abstract

Diagnostic tests should detect disease, have prognostic value, and aid in clinical decision making. Nowhere else in laboratory medicine does one have to interpret a subject's results within the dynamic of a couple as in reproductive medicine. Abnormal markers of male reproduction do not necessarily mean sterility, but instead indicate problems with spermatogenesis, sperm maturation, transport through epididymis and ejaculatory duct, or abnormal ejaculatory function. Decades of research suggest that one test will never fit all scenarios and a battery of assays evaluating different aspects of male reproduction will likely have the best prognostic value. There is a strong need for standardization and harmonization of evolving assays to establish their clinical relevance. Next-generation genome sequencing and the discovery of small noncoding RNAs in sperm already are changing the field and permit further insight into the biology of male reproduction as well as offer new diagnostic tests.

Key words Sperm, Acrosome, Male infertility, microRNA, Genetics

1 Background

Infertility affects approximately 15 % of couples trying to conceive; its prevalence differs between countries and socioeconomic and ethnic groups [1]. In the United States, it is thought that one in seven couples will encounter infertility [2, 3]. Male factor infertility is generally thought to be directly accountable for 30 % of all infertility cases and partially accountable in another 20 % [4]. Male infertility is attributed to numerous developmental, infectious, vascular, environmental, and genetic factors as well as personal habits [5], though a significant number of cases remain idiopathic in origin. Conception and delivery of a healthy child depend on "normal" sperm being able to advance through the female reproductive tract, fertilize the egg, and provide genetic material to enable optimal embryo development and pregnancy. In this case, a man lacks sperm even if testicular sperm extraction

Zev Rosenwaks and Paul M. Wassarman (eds.), *Human Fertility: Methods and Protocols*, Methods in Molecular Biology, vol. 1154, DOI 10.1007/978-1-4939-0659-8_9, © Springer Science+Business Media New York 2014

is attempted. Such a scenario occurs in men with bilateral anorchia or AZFa and AZFb classic deletions.

In all other situations, it is most prudent to assume that both male and female factors contribute to a couple's infertility, but to different extents. The arbitrary degree to which the male contributes to the overall infertility depends on the depth and extent of the evaluation of both the male and female patient. It is critical to remember that an abnormality in any "biomarker" of male reproductive potential is only a probable cause, and not an absolute indication of infertility; reproduction involves two individuals. The immunological and biochemical ecosystem in the female tract can modify and impede the abilities of sperm to perform its normal functions. One of the best examples for this is the development of anti-sperm antibodies in a female partner. Hence, although the same man might be fertile with different partners, he may be deemed infertile with his current one.

The quest for reliable biomarkers in male fertility evaluation stems largely from the challenges associated with understanding and evaluating spermatogenic function in the testis as well as the poor accuracy and specificity of semen analysis itself. Traditional evaluation methods perform suboptimally as predictors of success in natural conception, intrauterine insemination, and in vitro fertilization.

Testicular biopsy was previously the cornerstone in the evaluation of many types of infertility [6]. Testicular biopsy is fundamentally problematic in that it is an invasive procedure which requires anesthesia and can pose undue risk for the health of the testis; it typically provides a very small sample of tissue; and the associated histology is often unable to reveal the true cause of infertility [6]. Collectively these disadvantages render the analysis of sperm and sperm-derived biomarkers a superior first-line approach for determining the causes of male infertility.

The health and characteristics of sperm and their developmental precursors have been proposed and subsequently employed as suitable surrogates in the evaluation of reproductive fitness [7]. Evaluation of spermatozoa is commonly assessed with respect to numerous physiological parameters using what are now widely developed and standardized semen analysis techniques [8]. Aside from physical parameters, there are also numerous assays and other methodologies developed and under development to determine the quality of the genetic components and competency of sperm in the context of achieving fertilization, both through spontaneous pregnancy and assisted reproductive techniques (ART). The recognition and development of these makers are ongoing, and the application of this evaluation in predicting and treating clinical outcomes is the subject of ongoing debate.

One cannot forget that cutoff values used in differentiating between normal and abnormal results will differ depending on the context and outcome measures. For example, what may be "normal" for intracytoplasmic sperm injection (ICSI) with respect to sperm count is often inadequate for intrauterine insemination (IUI). Thus, any abnormality in male biomarkers of fertility should be further evaluated and thoroughly interpreted in the context of the patient, his health, and the planned course of therapeutic action if one is concerned about his health in addition to his reproductive function and potential.

2 Semen Analysis

The analysis of sperm on a microscopic level has taken place for centuries. A medical student working with Antoni van Leeuwenhoek first reported seeing "animalcules" in seminal fluid in 1677 [9]. Semen analyses serve as a diagnostic cornerstone in male fertility. On the most basic level, it measures the function of the testicles for sperm production and accesses the fluid volume contributed by accessory glands [10]. The World Health Organization (WHO) has standardized semen analysis in a protocol which is widely followed in andrology clinics [11].

2.1 Assessment of Motility

Recent changes have been made to the WHO guidelines for the evaluation of sperm motility. Sperm are now categorized as progressively motile, nonprogressively motile, and immotile. Progressive motility constitutes spermatozoa moving actively, linearly, or in large circles, while nonprogressive motility constitutes lesser motility with reduced progression or limited circles [12]. It is generally recognized that the percentage of progressively motile sperm is associated with pregnancy rates [13, 14]. While this association exists, overall motility remains a poor predictor of fertility.

2.2 Assessment of Concentration and Volume

Semen analysis typically seeks the concentration of sperm in the ejaculate and the overall volume of ejaculate. Semen volume should be greater than 1.5 ml, with deficient volume resulting from incomplete sample collection, low testosterone, retrograde ejaculation, or ejaculatory duct obstruction [15]. The relevancy of sperm count and progressive motility in natural conception is still debatable, though these parameters are major determinants of IUI [16]. Oligozoospermia refers to sperm concentration below the WHO reference value (15 million/ml). Azoospermia refers to complete absence of sperm in the ejaculate. Asthenozoospermia is the state of having total and/or forward progressive motility below WHO lower reference values (40 and 32 %, respectively) [16].

3 Assessment of Morphology

Spermatozoa are evaluated against morphology criteria established by the WHO based on reported data about morphology and reproductive outcomes. The correlation between the percentage of normal morphology and fertilization rates in vivo [17] and in vitro [18] is typically regarded as sufficient justification for in vitro fertilization. Key criteria included in an assessment of sperm morphology is an oval-shaped head with smooth contour, head length of 4–5 μm, and sperm width of 2.5–3 μm. Length-to-width ratio is also considered, as is the definition of the acrosomal region, comprising 40–70 % of the area of the head. The midsection and tail are also evaluated; the midsection should be slender and in proportion to the head, while the tail should be uniform, uncoiled, and approximately 45 μm in length [19]. Per the WHO protocol, most borderline and marginal forms are seen to fall outside of the relatively narrow range of normal variation and so the criteria are strict in this regard.

Particular areas of concern in abnormal spermatozoa include vacuolated heads (>20 % of head area occupied by unstained vacuolar areas), heads with reduced acrosomal area (<40 %), double heads, and other gross deformities. Common defects of the midsection include bent neck, asymmetrical mounting of mid-piece and head, and gross irregularities. Abnormal tails are commonly short, multiple, bent, of irregular girth, or otherwise defective [12].

4 Limitations of Semen Analysis

Though it has become a cornerstone and starting point for the evaluation of the infertile male, a basic semen analysis is generally regarded as insufficient in determining the fertility status of an individual male [20]. The analysis is arguably a poor indicator of sperm function with regard to fertility, and its prognostic value is limited and the subject of ongoing debate [21, 22]. Semen parameters have tremendous amount of inherent biological heterogeneity and vary by country, region, and individual and can even vary widely in consecutive samples delivered by an individual. These tests provide little to no insight into the location, cause, or time frame of the origins of abnormalities in ejaculated sperm. The inherent weaknesses and limitations of a standard semen analysis have undoubtedly spurred the demand for a series of informative diagnostic biomarkers in sperm for the evaluation of male infertility.

5 Sperm Function Testing

A number of biomarkers are able to give better insight into the function of sperm—more specifically, they are better able to evaluate whether or not sperm are fit and likely to succeed in fertilization. Ideal sperm markers, and ultimately sperm function tests, would identify (1) diagnosis of a specific sperm dysfunction, (2) prediction of fertilization or pregnancy rates, and lastly (3) indication of therapies suitable for the particular dysfunction [12, 23]. Thus far, attention has focused closely on zona pellucida binding and reactions involving the sperm acrosome, an organelle located in the anterior portion of spermatozoa. While the acrosome has received considerable attention, there are many markers in further development. Numerous genetic abnormalities as well as DNA integrity, and the degree to which these markers affect morphology and perhaps more importantly reproductive fitness and sperm function, have been the subject of in-depth investigation.

6 Sperm Acrosome

Studies have shown a strong correlation between sperm morphology and the likelihood of success in sperm binding to the zona pellucida [24, 25]. In the process of sperm binding to the oocyte, the contents of the acrosome are exposed. Included within the acrosome are enzymes, notably acrosin, which digest the zona pellucida and permit sperm penetration and fusion of sperm and oocyte. Recent evidence points to lock-and-key-type mechanisms whereby chemical elements of the zona pellucida bind to species-restricted counterpart molecules on the sperm. This ensures species-restricted fertilization [26].

In cases where the sperm and egg make contact but fertilization remains unsuccessful, careful analysis of the acrosome reaction and an understanding of the defects in the acrosome suggest potential therapeutic methods. It has been demonstrated that the morphology of the acrosome has significant implications both for the success of the binding to the zona pellucida and subsequent passage through the zona pellucida in order to achieve successful fertilization [24, 27]. The ability of the acrosome reaction to take place is crucial in guiding treatment. In the absence of the acrosome reaction, zona penetration cannot occur. In cases where the acrosome reaction does not take place in the presence of the partner's zona pellucida or a donor zona pellucida, ICSI is suggested. Studies have attempted to identify specific markers for deficient or defective acrosomes, alleviating the need to evaluate binding between spermatozoa and zona pellucida [28, 29]. The majority of studies have looked at acrosin and acrosin-related proteins in order to identify either changes in acrosin itself, and immediately related proteins, or the quantity of acrosin available to help to mediate the acrosome reaction with the zona pellucida [30, 31]. Additional studies have also looked at recombinant versions of glycoproteins of the zona pellucida, ZP1, ZP2, and ZP3 [32, 33]. None of the acrosome assays have gained widespread clinical acceptance, and their performance highly depends on investigator's own experience with the particular assay.

7 Sperm DNA Integrity

During spermatogenesis and spermiogenesis, the DNA in sperm becomes highly compacted and histones are mostly replaced by protamines. This complex, enzymatically controlled process leads to a high level of organization of DNA and protein into a structure called chromatin. Many authors use the terms DNA integrity and chromatin integrity interchangeably, but most assays measure only specific characteristics of chromatin.

Sperm DNA integrity is recognized as a significant parameter of sperm quality and is believed to serve as a meaningful marker in predicting the outcome of ART procedures and ultimately male infertility. Studies have demonstrated an association between sperm chromatin integrity and male fertility potential [34]. In addition, there have been reports of a distinct, inverse relationship between the percentage of sperm with abnormal chromatin and male fertility, especially when the proportion of aberrant cells are greater than 40 % as measured by the sperm chromatin structure assay (SCSA) [35]. Sperm chromatin damage is thought to specifically put couples at risk for recurrent spontaneous abortions and poor embryo development [36, 37]. However, we are just beginning to understand the role of sperm chromatin integrity in male fertility. DNA strand breaks, epigenetic alterations, chromosome microdeletions, unrepaired mismatch errors, and aneuploidy or a combination of these factors contribute to compromised fertility; thus, it is unlikely that one test will fit all patients [38]. Beyond the immediate nature of the DNA damage, it is also not apparent how DNA integrity assays can be effectively used in clinical practice [39].

In the use of IVF and other techniques for assisted fertilization, it is important to consider the biological mechanisms that are not active and how their absence may necessitate the use of quality control or sorting to achieve optimal results with these techniques. Biological sperm selection mechanisms end up rejecting all but very few spermatozoa that are released with ejaculation [40]. The existence of large numbers of spermatozoa may be seen as a compensatory mechanism, given the tendency for errors and defects in sperm [40]. While the competitive nature of sperm selection is largely removed, simply trying to recreate this may not be of the greatest relevance to achieving optimal IVF outcomes and preventing problems with embryo development. It is presently unclear precisely how the events in the female reproductive tract preceding fertilization screen for the quality of sperm DNA quality [41]. However, there is strong evidence to suggest that chromatin damage plays a crucial role here [42].

Available methods for detecting sperm chromatin damage include SCSA, TUNEL, Halosperm, and Comet assays [43]. While these assays are able to test for DNA integrity, they require permanent fixation of the sperm. Permanent fixation of the sperm renders them no longer suitable for clinical use; therefore, these assays are consumptive [41]. In these consumptive assays there is no opportunity to select between sperm with varying degrees of chromatin damage for immediate use with IVF, related therapies, or future preservation. However our group has recently published a novel method of early detection of sperm DNA damage which also allows for nondestructive selection of normal sperm using fluorescence-activated cell sorting.

7.1 TUNEL Assay

The TUNEL assay was initially developed for use with somatic cells and was subsequently applied to sperm cells [44]. TUNEL assay transfers a labeled nucleotide to the 3′-hydroxyl group of damaged DNA strands. This process occurs with the use of the enzyme deoxynucleotidyl transferase [45]. Following application of the fluorescence-labeled nucleotide, the investigator then monitors the fluorescence intensity of each sperm. Typically designation for any particular sperm is whether it is "undamaged" (no fluorescence) or "damaged" (with fluorescence) (Fig. 1). This process is carried out either by a technician using a fluorescence light microscope or, alternatively, through the use of a flow cytometer [46]. The laboratory then typically reports the number of sperm which are TUNEL positive.

The TUNEL assay has been shown to be effective in predicting the likelihood of success in ART procedures like IVF. In general, higher rates of TUNEL-positive sperm have been associated with greater failure rate of IVF and higher incidence of adverse

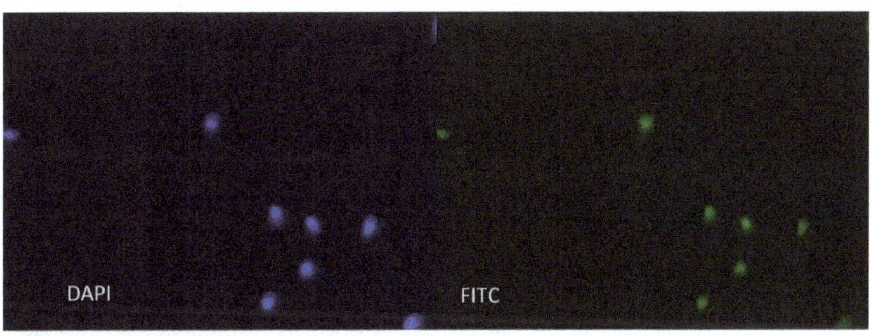

Fig. 1 Undamaged (*left*) and damaged (*right*, with fluorescence) sperm

outcomes in pregnancy [47]. As such, many institutions implement a cutoff point, which, when exceeded, prompts them to recommend against IVF or other ART procedures. The percentage of TUNEL-positive sperm is generally believed to increase with age in the general population. However, there is variation and a host of other environmental and individual factors influence the percentage of TUNEL-positive sperm.

The TUNEL assay generally achieves good results, and many clinicians who regularly employ the TUNEL assay report good clinical utility. However, there are drawbacks and aspects of the assay which must be regarded carefully. There is no commonly employed and "off-the-shelf" TUNEL assay for sperm, and so most institutions have developed their own protocols using general-purpose TUNEL testing kits from vendors. In addition, institutions choose between the use of flow cytometry or microscopic observation over several fields to quantify fluorescence and generate a result. In the latter case, it is essential that the readings be done by well-trained laboratory personnel; a single technician is ideal, as this ensures consistent results. Given differences in protocols between institutions, results and TUNEL percent cutoffs for clinical outcomes are often institution specific. There are ongoing calls for standardization between protocols and reported percentages, which would lead to increased flexibility in comparing results between institutions and also development of practice guidelines. Standardization and development of sperm assay-specific TUNEL kits will likely lead to an increase in the prevalence of use.

In our lab, we have begun to look at changes in the percent of TUNEL-positive sperm in cohorts of men presenting for infertility. While it is widely held that the proportion of TUNEL-positive sperm is elevated in infertile men, little is known about the precise relationship between age and TUNEL results. Sperm DNA damage has been shown to increase with age in the general population, but less attention has been paid to age-related trends expressly in infertile men [48]. Our study is beginning to show that there is an age-related trend in the percent of TUNEL-positive sperm amongst infertile men—older men who are infertile have a greater frequency of DNA damage in their sperm versus younger men who may still be infertile and exceed normal levels of sperm DNA damage (Table 1). There is some promise that with additional studies TUNEL may be used to improve reproductive outcomes in those with abnormally high rates of DNA fragmentation in their sperm as they age.

7.2 COMET Assay

The comet (single-cell gel electrophoresis) assay evaluates DNA fragmentation with the use of a microscope, on a cell-by-cell basis. In summary, the assay consists of mixing sperm with liquified agarose gel and then subjecting them to electrophoresis. The cells are first lysed. Fragmented DNA is smaller in size and thus exits the sperm head, while the intact DNA remains in place, as travel distance is size dependent [49]. Upon electrophoresis, the fragments

Table 1
Percent of TUNEL-positive sperm increases with age

Age	Number of subjects	Mean % TUNEL (+) sperm	Std error	Lower 95 %	Upper 95 %	Percent of subjects with TUNEL >10 %
21–30	35	6.58	1.36	3.9	9.2	17 %*
31–40	179	8.05	0.60	6.9	9.2	28 %*
41–50	84	8.05	0.88	6.3	9.8	29 %*
51–80	19	14.54	1.84	10.9	18.2	47 %*

All assays were performed by the same technologist. Statistically significant differences were observed, ANOVA $p<0.001$. Men > 51 (*) had statistically significant higher mean % of TUNEL (+) sperm than any other younger group of men

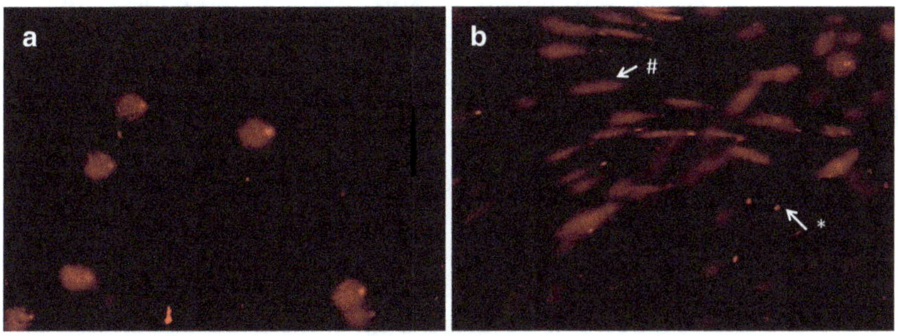

Fig. 2 COMET assay. Change in electrophoresis conditions will modify the shape of the DNA dispersion (**a**—low voltage, **b**—high voltage). #—sperm with complete dispersion of DNA (abnormal), *—sperm with minimal dispersion of DNA (normal) (own data)

splay out in the shape of a comet; the name of the assay derives from this (Fig. 2) [44]. Much like TUNEL, it has been shown that elevated numbers of sperm demonstrating DNA damage through the comet assay correlate with reduced rates of fertilization with IVF and ICSI and also poorer outcome and more frequent early spontaneous abortion when sperm are used in ART techniques [50, 51]. The process of evaluation is like some TUNEL protocols in that it relies on microscopic observation, typically of a few hundred sperm. It also lacks a consistent protocol, and so different institutions achieve different results. Those institutions that use the comet assay frequently report good results and find that it is valuable in guiding subsequent steps in infertility treatment.

7.3 Sperm Chromatin Structure Assay

Variants of this assay have been available since the late 1970s, and the assay is now widely available as the SCSA assay (sperm chromatin structure assay) (SCSA Diagnostics, Brookings, SD). The assay relies on the properties of acridine orange (AO), which has metachromatic properties and shifts from green to red color when it associates with single-stranded DNA or RNA. The results are then reported in three ways: as αt, which represents the red/

red + green fluorescence, total green florescence, and another calculated parameter, COMPαt. COMPαt represents the number of sperm outside of the normal population and more recently has been representative of a DNA fragmentation index (DFI) [49]. AO has poor penetration into native-state condensed DNA. Therefore, SCSA protocols typically use high temperature or low pH to denature DNA and allow more thorough penetration of DNA. SCSA testing requires the use of flow cytometry and other expensive equipment and the use of a reference for successful calculation of COMPαt. The protocol has changed slightly since its inception but is largely conserved between institutions. There is some range of cutoff values in guiding treatment, which vary by institution, but typically DFI values above 30 % are associated with infertility and failure of ICSI and IVF [35, 52]. Results of SCSA correlate well with TUNEL and other assays [53, 54].

8 Apoptosis Versus Compaction Error

Breaks in DNA occur naturally during the process of sperm chromatin compaction. However, work in our laboratory has shown that breaks in DNA may be induced by the activation of the mitochondrial pathway of apoptosis using nitroprusside sodium. The decrease in mitochondrial membrane potentials precedes the breaks in DNA and early apoptotic changes, thus showing that inhibition of mitochondrial pathways of sperm DNA damage may potentially benefit men with high levels of DNA breakage (Fig. 3).

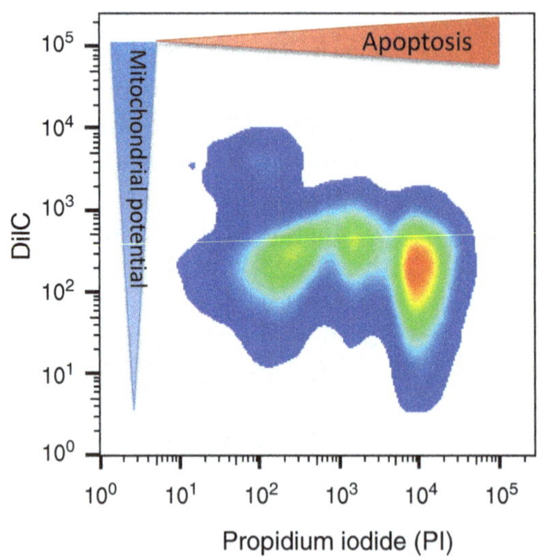

Fig. 3 Flow cytometry analysis shows that mitochondrial potential decreases (measured using DiIC) with progression of apoptosis (PI). Change in mitochondrial potential is an early event in sperm apoptosis (own data)

8.1 Sperm Protamines and Histones

Successful and fertile sperm with the potential to fertilize require the proper generation of mature testicular spermatids as a prerequisite [55]. The progenitor cells for spermatids are spermatocytes—each spermatocyte splits into two haploid spermatids. As a spermatocyte, somatic histones that are present in the cells are exchanged for testis-specific spermatocyte cells. In round spermatids, both of these histone elements are replaced with a protamine that causes a halt in transcription following chromatin condensation [56]. Following this, there is an uncoupling of the processes of transcription, translation, and protein synthesis in the development of the spermatids.

Recent research has focused on the ratio between histones and protamines. It has been demonstrated that an abnormal ratio of histones to protamines is associated with male infertility [57]. Ratios of protamines and histones can be detected using Western blots, but because DNA and proteins are very tightly packed, it is difficult to destabilize chromatin and release proteins from DNA. Harsh conditions can change epitope antigenicity as well as lead to variability in extraction efficiency.

8.2 Sperm RNA Profiling

The recent discovery of a significant amount of RNA in spermatozoa contradicted the previously held belief that the paternal contribution was limited to one copy of the genome [58]. Furthermore, detection of RNA in sperm has raised the intriguing question of its possible role in embryonic development [59]. Krawetz showed that human sperm carry mRNA as well as small noncoding RNAs (sncRNAs). In human sperm from ejaculate, the following sncRNAs can be detected: microRNAs (miRNAs) (7 %), Piwi-interacting piRNAs (17 %), repeat-associated small RNAs (65 %), and quiescent RNAs (11 %) of sncRNAs [60]. That such a complex population of male-derived sncRNAs is available for delivery upon fertilization suggests their critical role as modifiers in early post-fertilization and makes them an exciting target in the evaluation of male infertility.

miRNAs and small RNAs (sRNAs) in general are considered key regulators of posttranslational mRNA modifications that affect the expression of multiple genes at the same time. Small RNAs have been identified within the testis and sperm. Achieving a better understanding of the role of sncRNA in male reproduction is an exciting and very promising area of research that will bring major breakthroughs in our understanding of human reproduction. Comparative analysis of our own sequencing miRNA data (human testis) showed that 70 % of human miRNAs are highly conserved between species, thus allowing for relatively easy study of mechanism of action of candidate miRNA in animal models.

The first evidence of the role of miRNA in male infertility was derived from the Dicer knockout (KO) mouse model. Unlike females, *Dicer*$^{-/-}$ male mice are infertile [61]. Although many miRNAs (mir-125b, let-7 family, mir-26a, mir-30c) are ubiquitously and highly expressed in most tissues, other miRNAs have high tissue specificity. Two clusters, mir-17/92 and mir-106b/25, are involved in spermatogonial differentiation [62]. Human germ cell tumors, including testicular cancer, have been linked to abnormal expression of mir-372, 373, and let-7c [63]. Discovery of sperm-specific miRNAs and experimental verification that they are involved in embryo development further underscore the importance of miRNA in male infertility [60].

Although 58 distinctive miRNAs expressed in testis and specific to germ cells have been listed in a recent review by McIver et al., most of the reported miRNAs were either identified using deep sequencing data from whole mouse and porcine testis or derived from rodent spermatogonial stem cells [64, 65]. Human data on the expression of miRNAs in a well-characterized population of fertile and infertile men is lacking, and the few manuscripts published used either commercially available human RNA contaminated by epididymis or a limited set of 2–3 subjects. PIWI proteins (P-element-induced wimpy testis in Drosophila) are responsible for maintaining incomplete differentiation of stem cells and stability of cell division rates in germ line cells [66]. PIWI regulates germ cell differentiation through interacting with classic miRNA [67].

Recently, a novel class of longer-than-average miRNA (26-31) known as PIWI-interacting RNAs (piRNAs) has been identified in spermatogenic cells in the testes of mammals [68]. piRNA originates from repetitive regions of the genome such as retrotransposons or heterochromatin. PIWI and their associated piRNA are believed to form an endogenous system of silencing the expression of genetic elements like retrotransposons and thus prevent gene products from such regions from interfering with normal spermatogenesis [66].

miRNAs are extremely stable and easy to detect using body fluids or tissues. Work in our laboratory indicates that miRNAs not only can be an excellent marker of abnormal sperm function, but also dramatically improve our insight into the role sperm plays after fertilization. This will hopefully lead to much better discriminatory criteria for clinical tests for male infertility than are currently available. Using automatic tissue profiling, we discovered that miRNA fingerprinting is highly specific for selection of histologically diverse patterns in human testis, thus experimentally proving the concept that novel marker profiling will ultimately lead to paradigm shift in the evaluation of infertile male (Fig. 4). Expression of miRNA is highly similar between men with different genetic backgrounds as long as they have similar histopathology (Fig. 5).

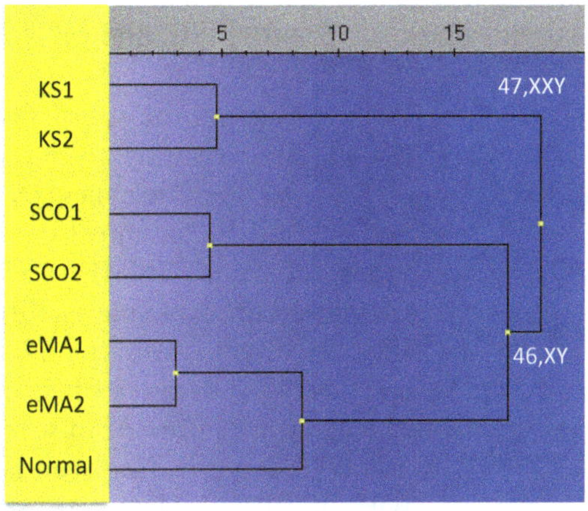

Fig. 4 Dendrogram based on a set of 22 human miRNAs allows for automatic and operator-independent classification of human testicular histology into Sertoli cell-only syndrome (SCO), early maturation arrest (eMA), and normal testicular histology. In addition first branch differentiates between Klinefelter's (KS) (47,XXY) and non-Klinefelter's (46,XY) miRNA expression profile within testis (own data)

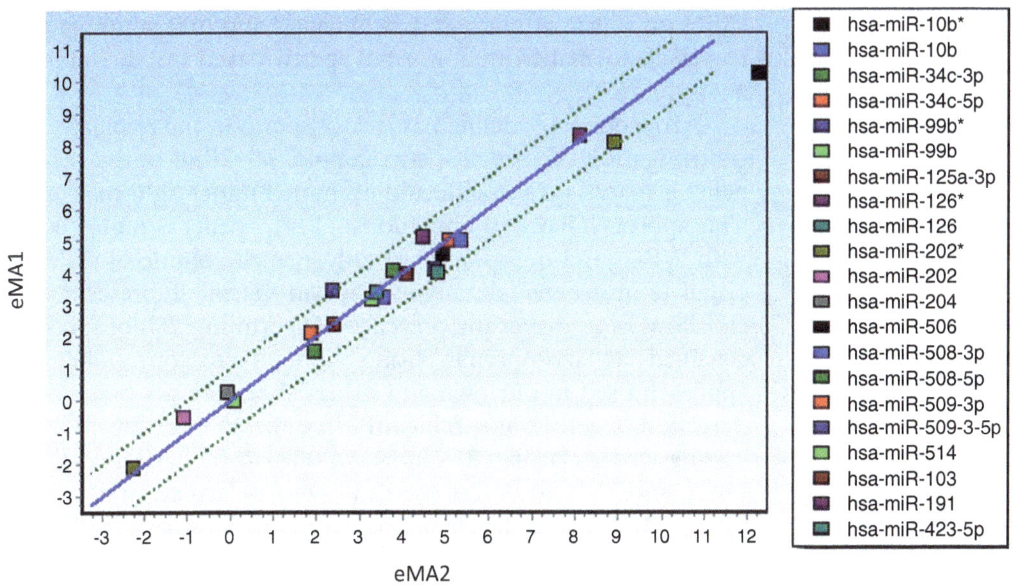

Fig. 5 Expression profile of miRNA in testis is highly similar in two subjects with the same histopathological defect (early maturation arrest—eMA) and different genetic background. RR = 0.92, $p < 0.004$, *green lines* depict onefold difference (own data)

9 Chromosomal Markers, Sperm FISH, and Molecular Detection of Sperm

A number of chromosomal abnormalities account for approximately 5 % of infertility in the male population and about 15 % of infertility among azoospermic males [69]. The majority of these abnormalities and aberrations fall into karyotype anomalies and Y chromosome microdeletions (YCMD) [55]. The most common abnormality of the Y chromosome is Klinefelter's syndrome (KS), which is marked by the genotype 47,XXY [70]. KS exists on a spectrum with varying degrees of mosaicism—expression of additional X chromosomes present in some but not all cells. Surgical retrieval of haploid sperm has been successful in 30–70 % of individuals with KS [71]. However, studies have indicated the elevated potential for genetically imbalanced haploid spermatozoa, particularly in non-mosaic-type KS patients, which increases the potential for KS to occur in offspring conceived via ICSI [72]. It is ultimately believed that irregularities in both sex and autosomal chromosomes can manifest in offspring of KS patients resulting from ICSI [73]. While the use of fluorescent in situ hybridization (FISH) on sperm and interphase cells has dramatically increased the ability to detect chromosomal abnormalities in sperm and reproductive tissues, at this point indications for FISH on sperm are not well established and no technique exists to select normal from abnormal sperm based on chromosomal aneuploidy.

Azoospermia is defined as lack of sperm in the ejaculate. After centrifugation of a small sperm sample, 10–20 μl of the resulting pellet is placed on the slide and examined under light microscopy. This approach has two limitations: (1) the pellet is highly cellular with WBCs and debris and (2) only a small volume of pellet and ejaculate in general is examined. Sperm-specific fluorescent markers allow us to detect the presence of sperm in 72 % of specimens classified as azoospermic using light microscopy. The other approach used in our own and other laboratories is based on the detection of germ cell-specific mRNA or proteins extracted by processing whole ejaculated volume. In our lab we have used DDX4, DAZ, and VCY to detect sperm in whole ejaculate and testicular tissue. DDX4 has 100 % sperm detection sensitivity and 100 % negative predictive value even in specimens with azoospermia measured by light microscopy. Such tests are better suited to detect small number of sperm in ejaculate; this fact has tremendous implications in our management of male infertility cases. If sperm are found in the ejaculates using novel molecular techniques, it is then very likely that therapy aimed at improving spermatogenesis may help and that, if needed, testicular biopsy and microsurgical testicular sperm extraction will be successful.

9.1 Genetic Markers

In the past, a common approach used to study genetic causes of infertility was to screen single genes for single-nucleotide polymorphism (SNP), leading to loss or gain of function of protein [74]. However, despite over two decades of intense research by multiple groups, including our own, such a labor-intensive and time-consuming approach has failed to result in the development of clinically useful tools to diagnose or treat infertility in men [75]. Male genetic testing for infertility is limited to cytogenetics to exclude Klinefelter's syndrome (47,XXY) and other numerical and structural chromosomal abnormalities and screening for YCMD [76]. Genome-wide association studies (GWAS) have been proposed to identify genetic causes of male infertility [77]. However, the main caveat of this approach stems from the phenotypic heterogeneity of male infertility, which makes interpretation of identified clusters of SNPs very difficult. Genome-wide association studies using isolated germ cells in humans to decrease phenotypic heterogeneity are not feasible because of the high number of testicular biopsies from normal and infertile men needed for statistical interpretation for such a potential project.

Thus, a more focused approach would be to target specific switches of multiple genes specifically expressed in testis using material derived from human testis. This approach would lead to faster translation of results of animal research and development of clinically meaningful applications. One cannot forget that the results of genetic screening performed typically using DNA extracted from peripheral blood may not necessarily represent genetic aberrations in ejaculated sperm. Recent data indicate that certain mutations that lead to severe phenotypes in offspring give spermatogonial survival advantage within testis and promote clonal expansion of sperm with particular genetic defects. Thus, the future genetic analysis of the infertile male may shift from peripheral blood to sperm itself. This approach has become possible since next-generation sequencing technology has led to a dramatic drop in the price of deep sequencing. To be clinically useful, sequencing data will require novel algorithms to summarize the thousands of identified SNPs and assign the observed patterns in the binary class of normal and abnormal results that clinicians and infertile couples can utilize in their decision making.

References

1. Cao XW et al. (2011) A review of WHO laboratory manual for the examination and processing of human semen, (5th edn). Zhonghua Nan Ke Xue 17(12):1059–1063
2. Chandra A et al (2005) Fertility, family planning, and reproductive health of U.S. women: data from the 2002 National Survey of Family Growth. Vital Health Stat 23(25):1–160
3. Mosher WD (1985) Reproductive impairments in the United States, 1965–1982. Demography 22(3):415–430
4. Nangia AK, Likosky DS, Wang D (2010) Distribution of male infertility specialists in relation to the male population and assisted reproductive technology centers in the United States. Fertil Steril 94(2):599–609

5. Skakkebaek NE, Giwercman A, de Kretser D (1994) Pathogenesis and management of male infertility. Lancet 343(8911):1473–1479
6. Dohle GR, Elzanaty S, van Casteren NJ (2012) Testicular biopsy: clinical practice and interpretation. Asian J Androl 14(1):88–93
7. Ostermeier GC et al (2005) Toward using stable spermatozoal RNAs for prognostic assessment of male factor fertility. Fertil Steril 83(6):1687–1694
8. Anton E, Krawetz SA (2012) Spermatozoa as biomarkers for the assessment of human male infertility and genotoxicity. Syst Biol Reprod Med 58(1):41–50
9. Howards SS (1997) Antoine van Leeuwenhoek and the discovery of sperm. Fertil Steril 67(1):16–17
10. De Jonge C (2012) Semen analysis: looking for an upgrade in class. Fertil Steril 97(2): 260–266
11. Lu JC, Huang YF, Lu NQ (2010) WHO laboratory manual for the examination and processing of human semen: its applicability to andrology laboratories in China. Zhonghua Nan Ke Xue 16(10):867–871
12. Franken DR, Oehninger S (2012) Semen analysis and sperm function testing. Asian J Androl 14(1):6–13
13. Zinaman MJ et al (2000) Semen quality and human fertility: a prospective study with healthy couples. J Androl 21(1):145–153
14. Larson KL et al (2000) Sperm chromatin structure assay parameters as predictors of failed pregnancy following assisted reproductive techniques. Hum Reprod 15(8): 1717–1722
15. Stahl PJ, Stember DS, Goldstein M (2012) Contemporary management of male infertility. Annu Rev Med 63:525–540
16. Duran HE et al (2002) Intrauterine insemination: a systematic review on determinants of success. Hum Reprod Update 8(4):373–384
17. Kruger TF et al (1986) Sperm morphologic features as a prognostic factor in in vitro fertilization. Fertil Steril 46(6):1118–1123
18. Van Waart J et al (2001) Predictive value of normal sperm morphology in intrauterine insemination (IUI): a structured literature review. Hum Reprod Update 7(5):495–500
19. Menkveld R et al (1990) The evaluation of morphological characteristics of human spermatozoa according to stricter criteria. Hum Reprod 5(5):586–592
20. Brugh VM 3rd, Lipshultz LI (2004) Male factor infertility: evaluation and management. Med Clin North Am 88(2):367–385
21. Chia SE, Tay SK, Lim ST (1998) What constitutes a normal seminal analysis? Semen parameters of 243 fertile men. Hum Reprod 13(12):3394–3398
22. Guzick DS et al (2001) Sperm morphology, motility, and concentration in fertile and infertile men. N Engl J Med 345(19):1388–1393
23. Muller CH (2000) Rationale, interpretation, validation, and uses of sperm function tests. J Androl 21(1):10–30
24. Menkveld R et al (1991) Detection of sperm antibodies on unwashed spermatozoa with the immunobead test: a comparison of results with the routine method and seminal plasma TAT titers and SCMC test. Am J Reprod Immunol 25(2):88–91
25. Gupta SK, Bhandari B (2011) Acrosome reaction: relevance of zona pellucida glycoproteins. Asian J Androl 13(1):97–105
26. Litscher ES, Williams Z, Wassarman PM (2009) Zona pellucida glycoprotein ZP3 and fertilization in mammals. Mol Reprod Dev 76(10):933–941
27. Ikawa M et al (2010) Fertilization: a sperm's journey to and interaction with the oocyte. J Clin Invest 120(4):984–994
28. Kumar V et al (2009) Proteomic analysis of heparin-binding proteins from human seminal plasma: a step towards identification of molecular markers of male fertility. J Biosci 34(6):899–908
29. Perdrix A et al (2011) Assessment of acrosome and nuclear abnormalities in human spermatozoa with large vacuoles. Hum Reprod 26(1):47–58
30. Chaudhury K et al (2005) Acrosin activity as a potential marker for sperm membrane characteristics in unexplained male infertility. Fertil Steril 83(1):104–109
31. Mitchell V et al (2004) Proacrosin, an acrosomal marker for the detection of spermatogenic cells in ejaculates from azoospermic men. Gynecol Obstet Fertil 32(9):779–784
32. Furlong LI, Harris JD, Vazquez-Levin MH (2005) Binding of recombinant human proacrosin/acrosin to zona pellucida (ZP) glycoproteins. I. Studies with recombinant human ZPA, ZPB, and ZPC. Fertil Steril 83(6):1780–1790
33. Veaute C et al (2010) Anti-human proacrosin antibody inhibits the zona pellucida (ZP)-induced acrosome reaction of ZP-bound spermatozoa. Fertil Steril 93(7):2456–2459
34. Zini A et al (2001) Correlations between two markers of sperm DNA integrity, DNA denaturation and DNA fragmentation, in fertile and infertile men. Fertil Steril 75(4):674–677

35. Spano M et al (2000) Sperm chromatin damage impairs human fertility. The Danish First Pregnancy Planner Study Team. Fertil Steril 73(1):43–50
36. Kumar K et al (2012) Predictive value of DNA integrity analysis in idiopathic recurrent pregnancy loss following spontaneous conception. J Assist Reprod Genet 29(9):861–867
37. Brahem S et al (2011) Semen parameters and sperm DNA fragmentation as causes of recurrent pregnancy loss. Urology 78(4):792–796
38. Tarozzi N et al (2007) Clinical relevance of sperm DNA damage in assisted reproduction. Reprod Biomed Online 14(6):746–757
39. Sharma RK, Said T, Agarwal A (2004) Sperm DNA damage and its clinical relevance in assessing reproductive outcome. Asian J Androl 6(2):139–148
40. Manning JT, Chamberlain AT (1994) Sib competition and sperm competitiveness: an answer to "why so many sperms?" and the recombination/sperm number correlation. Proc Biol Sci 256(1346):177–182
41. Henkel R (2012) Sperm preparation: state-of-the-art–physiological aspects and application of advanced sperm preparation methods. Asian J Androl 14(2):260–269
42. Holt WV, Van Look KJ (2004) Concepts in sperm heterogeneity, sperm selection and sperm competition as biological foundations for laboratory tests of semen quality. Reproduction 127(5):527–535
43. Dugum M, Sandlow JI, Brannigan RE (2011) Sperm DNA damage evaluation techniques. J Androl 32(3):207–209
44. Evenson DP, Wixon R (2006) Clinical aspects of sperm DNA fragmentation detection and male infertility. Theriogenology 65(5): 979–991
45. Sharma R, Masaki J, Agarwal A (2013) Sperm DNA fragmentation analysis using the TUNEL assay. Methods Mol Biol 927:121–136
46. Perreault SD et al (2000) Evaluation of aneuploidy and DNA damage in human spermatozoa: applications in field studies. Andrologia 32(4–5):247–254
47. Henkel R et al (2003) DNA fragmentation of spermatozoa and assisted reproduction technology. Reprod Biomed Online 7(4):477–484
48. Templado C et al (2011) Advanced age increases chromosome structural abnormalities in human spermatozoa. Eur J Hum Genet 19(2):145–151
49. Schlegel PN, Paduch DA (2005) Yet another test of sperm chromatin structure. Fertil Steril 84(4):854–859
50. Morris ID et al (2002) The spectrum of DNA damage in human sperm assessed by single cell gel electrophoresis (Comet assay) and its relationship to fertilization and embryo development. Hum Reprod 17(4):990–998
51. Schmid TE et al (2003) Genetic damage in oligozoospermic patients detected by fluorescence in-situ hybridization, inverse restriction site mutation assay, sperm chromatin structure assay and the Comet assay. Hum Reprod 18(7):1474–1480
52. Spano M et al (1998) The applicability of the flow cytometric sperm chromatin structure assay in epidemiological studies. Asclepios. Hum Reprod 13(9):2495–2505
53. Darzynkiewicz Z, Li X, Gong J (1994) Assays of cell viability: discrimination of cells dying by apoptosis. Methods Cell Biol 41:15–38
54. Potts RJ et al (1999) Sperm chromatin damage associated with male smoking. Mutat Res 423(1–2):103–111
55. Steger K, Cavalcanti MC, Schuppe HC (2011) Prognostic markers for competent human spermatozoa: fertilizing capacity and contribution to the embryo. Int J Androl 34(6 Pt 1):513–527
56. Steger K (1999) Transcriptional and translational regulation of gene expression in haploid spermatids. Anat Embryol (Berl) 199(6): 471–487
57. Zhang X, San Gabriel M, Zini A (2006) Sperm nuclear histone to protamine ratio in fertile and infertile men: evidence of heterogeneous subpopulations of spermatozoa in the ejaculate. J Androl 27(3):414–420
58. Kramer JA, Krawetz SA (1997) RNA in spermatozoa: implications for the alternative haploid genome. Mol Hum Reprod 3(6): 473–478
59. Krawetz SA (2005) Paternal contribution: new insights and future challenges. Nat Rev Genet 6(8):633–642
60. Krawetz SA et al (2011) A survey of small RNAs in human sperm. Hum Reprod 26(12):3401–3412
61. Hayashi K et al (2008) MicroRNA biogenesis is required for mouse primordial germ cell development and spermatogenesis. PLoS One 3(3):e1738
62. Tong MH et al (2012) Two miRNA Clusters, Mir-17-92 (Mirc1) and Mir-106b-25 (Mirc3), are involved in the regulation of spermatogonial differentiation in mice. Biol Reprod 86(3):72
63. Gillis AJ et al (2007) High-throughput microRNAome analysis in human germ cell tumours. J Pathol 213(3):319–328

64. McIver SC et al (2012) miRNA and mammalian male germ cells. Hum Reprod Update 18(1): 44–59
65. He Z et al (2009) Small RNA molecules in the regulation of spermatogenesis. Reproduction 137(6):901–911
66. Saito K et al (2006) Specific association of Piwi with rasiRNAs derived from retrotransposon and heterochromatic regions in the Drosophila genome. Genes Dev 20(16):2214–2222
67. Megosh HB et al (2006) The role of PIWI and the miRNA machinery in Drosophila germline determination. Curr Biol 16(19):1884–1894
68. Kim VN (2006) Small RNAs just got bigger: Piwi-interacting RNAs (piRNAs) in mammalian testes. Genes Dev 20(15):1993–1997
69. Ferlin A et al (2007) Male infertility: role of genetic background. Reprod Biomed Online 14(6):734–745
70. Mehta A, Paduch DA (2012) Klinefelter syndrome: an argument for early aggressive hormonal and fertility management. Fertil Steril 98(2):274–283
71. Schiff JD et al (2005) Success of testicular sperm extraction [corrected] and intracytoplasmic sperm injection in men with Klinefelter syndrome. J Clin Endocrinol Metab 90(11):6263–6267
72. Morel F et al (2003) An increased incidence of autosomal aneuploidies in spermatozoa from a patient with Klinefelter's syndrome. Fertil Steril 79(Suppl 3):1644–1646
73. Lanfranco F et al (2004) Klinefelter's syndrome. Lancet 364(9430):273–283
74. Paduch DA, Mielnik A, Schlegel PN (2005) Novel mutations in testis-specific ubiquitin protease 26 gene may cause male infertility and hypogonadism. Reprod Biomed Online 10(6):747–754
75. Massart A et al (2012) Genetic causes of spermatogenic failure. Asian J Androl 14(1):40–48
76. Stahl PJ et al (2010) A decade of experience emphasizes that testing for Y microdeletions is essential in American men with azoospermia and severe oligozoospermia. Fertil Steril 94(5): 1753–1756
77. Carrell DT, Aston KI (2011) The search for SNPs, CNVs, and epigenetic variants associated with the complex disease of male infertility. Syst Biol Reprod Med 57(1–2):17–26

Chapter 10

Luteal Phase Support in ART Treatments

Yuval Or, Edi Vaisbuch, and Zeev Shoham

Abstract

In a normal spontaneous menstrual cycle, the luteal phase is characterized by the production and secretion of estradiol (E) and progesterone (P) from the corpus luteum (CL) in an episodic manner. The steroidogenesis of the CL is dependent on continued tonic luteinizing hormone (LH) secretion (Fritz and Speroff, Clinical gynecologic endocrinology and infertility, 8th edn. Wolters Kluwer, Lippincott Williams & Wilkins, Philadelphia, 2011). The dependence of the CL was further supported by the prompt luteolysis that followed the administration of GnRH analogues or withdrawal of GnRH when ovulation has been induced by the administration of pulsatile GnRH (Hutchison and Zeleznik, Endocrinology 115:1780–1786, 1984; Fraser et al., Hum Reprod 12:430–435, 1997). Progesterone concentrations normally rise sharply after ovulation, reaching a peak approximately 8 days after the LH surge. Since the secretion of E and P during the luteal phase is episodic and correlates closely with LH pulses, relatively low mid-luteal progesterone levels can be found in the course of a totally normal luteal phase (Fritz and Speroff, Clinical gynecologic endocrinology and infertility, 8th edn. Wolters Kluwer, Lippincott Williams & Wilkins, Philadelphia, 2011).

Key words Luteal phase, Progesterone support, Estrogen supplementation, hCG support, Gonadotropin support, LH, Luteinizing hormone

1 Introduction

In the normal cycle, the luteal phase lasts between 11 and 17 days with an average of 14 days. The luteal phase cannot be extended indefinitely even with LH exposure, indicating that the demise of the CL is due to an active luteolytic mechanism which is unknown [1]. The survival of the CL can be prolonged by the stimulus of the rapidly increasing hormone secreted from the implanted blastocyst, the human chorionic gonadotropin (HCG). The function of the CL is crucial during the first 7–9 weeks of pregnancy [1]. Edwards and Steptoe [2] in their very first publication of human IVF success suggested that the luteal phase of stimulated IVF cycles is abnormal. In the mid-1980s, retrospective studies suggested that anovulatory patients undergoing fertility treatment with clomiphene citrate or

Zev Rosenwaks and Paul M. Wassarman (eds.), *Human Fertility: Methods and Protocols*, Methods in Molecular Biology, vol. 1154, DOI 10.1007/978-1-4939-0659-8_10, © Springer Science+Business Media New York 2014

gonadotropins without IVF also have increased delivery rate after administration of progesterone for luteal support [3, 4].

Over the years, several theories have been raised to explain the etiology of luteal phase defects in stimulated IVF cycles including (1) granulose cell removal during ovum pickup; (2) hCG-induced suppression of LH; and (3) GnRH analogue-induced pituitary suppression. Kerin et al. [5] showed that carefully performed aspiration of the follicles during IVF treatment did not lead to impaired steroid function of the subsequent corpus luteum. In IVF treatment hCG is administered to achieve resumption of meiosis in the meiotic arrested oocyte. Because of the similarity of hCG to the LH molecule, it was suggested that a negative short-loop feedback mechanism might exist that controls LH secretion. According to this hypothesis, hCG may negatively affect pituitary LH secretion by reducing hypothalamic GnRH secretion. Tavaniotou and Devroey [6] demonstrated that hCG injection does not downregulate LH secretion in the luteal phase of the normal, unstimulated cycle in normoovulatory women. GnRH agonists are widely used in order to achieve pituitary suppression in IVF treatment. Smitz et al. [7] in their research conclude that in IVF treatments, gonadotropin function remains impaired until the end of the luteal phase after desensitization and flare-up GnRH agonist and gonadotropin stimulation protocols and by that effect causes and magnifies luteal phase defects. Albano et al. [8] showed that even in antagonist protocol cycles and despite the rapid recovery of pituitary function, corpus luteum function seems to be impaired in cycles that are stimulated with hMG and the GnRH antagonist. Fauser and Devroey [9] concluded that the luteal phase of stimulated IVF cycles is abnormal due to the high follicular phase estrogen levels attained with hyperstimulation. The occurrence of an abnormal luteal phase in IVF cycles, with characteristic features of raised progesterone levels, along with a significantly reduced luteal phase length is well established. Slowing down of the GnRH pulse generator along with diminished LH pulse amplitude is responsible for the demise of the corpus luteum. Luteolysis can be induced by the administration of GnRH agonist or antagonist [9]. During IVF treatment ovarian stimulation produces multiple corpora lutea which secrete estrogen and progesterone at levels that are much higher than secreted by the normal physiological corpus luteum. The duration of the secretion of estrogen and progesterone in the stimulated ovary is shorter, and the decline is more abrupt [10]. The high levels of estrogen and progesterone leading to diminished LH secretion and the abrupt decline of estrogen and progesterone are all responsible for the abnormal luteal phase in IVF treatments; thus, it is universally accepted that luteal phase support (LPS) is required in order to achieve better implantation and pregnancy rates. Indeed, since the first report on successful LPS almost three decades ago [11], agents such as hCG, progesterone, GnRH agonist, and estrogen alone or in combination administered at different dosages and through different routes of administration for varying durations have been studied. Yet, the question of which treatment provides the best luteal support is still open.

2 Progesterone Support

The first method described to support luteal phase defect was administration of progesterone supplementation. Progesterone is central to the endometrial preparation before implantation occurs as well as to pregnancy support. In 2004, Daya and Gunby [12] under the Cochrane collaboration published a systematic review of luteal phase support in assisted reproduction cycles including 59 studies and concluded that when P was administered for luteal support, the odds of clinical pregnancy significantly increased by 34 %, compared to no treatment. In 2011, Van der Linden and colleagues [13] updated this Cochrane review and found that the odds ratio (OR) for clinical pregnancy when administering P for luteal support was 1.83 (95 % CI 1.29–2.61) comparing to placebo or no treatment. The OR for live birth rate was 2.95 (95 % CI 1.02–8.56). The authors concluded that there is a significant effect of progesterone administration for luteal support [13].

Progesterone for luteal support in IVF can be administered by several routes including oral, intramuscular, transdermal, sublingual, nasal, vaginal, or rectal. The oral route of micronized progesterone failed to cause any secretory transformation of the endometrium in patients with premature ovarian failure as shown by Bourgain et al. [14]. The first passage of progesterone through the liver after oral ingestion results in a massive metabolism and leads to too low systemic levels of P necessary to achieve adequate support to the endometrium. Women who took oral P supplement had lower implantation and pregnancy rates and higher miscarriage rates compared to those who took either vaginal or intramuscular P [10]. A meta-analysis performed by Zarutskie and Phillips [15] showed that the vaginal and the intramuscular routes are equally effective in supporting the endometrium after IVF-ET cycles. In the Cochrane review the reduction in clinical pregnancy rate with the oral route, compared to the intramuscular or the vaginal routes, did not reach statistical significance [13].

The intramuscular administration of progesterone in oil delivers the hormone to the target organ in a highly effective manner and without the first-pass metabolism in the liver. However, this route might cause several side effects including pain, inflammatory reactions, abscess formation at the injection sites, and allergic responses. A rare complication of eosinophilic pneumonia has been reported as a serious complication of intramuscular P [16]. The usual dose is 25–100 mg daily. Due to the inconvenience of daily intramuscular injections and the side effects of the drug many women and physicians avoid the intramuscular route.

Another option for administering progesterone intramuscularly is the use of the synthetic progestin, 17 α-hydroxyprogesterone caproate. Pregnancy and miscarriage rates were no different with this approach [17]. However, theoretical concerns about teratogenicity when compared to natural progesterone have resulted in its limited use [10].

The vaginal route offers several advantages: it is painless, convenient, and accepted by most patients; it does not require special equipment or personnel; and it rarely produces allergic reaction. Several formulations for the delivery of progesterone by the vaginal route exist as follows: (1) vaginal administration of oral micronized progesterone (Utrogestan)—effective in the dosage of 200–300 mg × three times a day; (2) vaginal tablets of progesterone (Endometrin) in the dosage of 100 twice a day; and (3) vaginal gel of progesterone 8 % (Crinone) in the dosage of 1.125 g once a day.

Over the years, direct comparisons between the intramuscular and the vaginal routes demonstrated conflicting results. Some randomized clinical trials demonstrated that daily intramuscular progesterone in oil injections is superior to daily vaginal micronized progesterone, while others failed to show a significant difference between these two routes. In their meta-analysis, Pritts and Atwood [18] found that both clinical pregnancy and delivery rates improved when intramuscular progesterone was used with a combined relative risk of 1.33 (95 % CI 1.02–1.75) and 2.06 (1.48–2.88), respectively. However, a later meta-analyses by Zarutskie and Phillips [15] found a comparable effect between the vaginal and intramuscular routes of progesterone administration on the endpoints of clinical pregnancy (OR = 0.91, 95 % CI 0.74–1.13) and ongoing pregnancy (OR = 0.94, 95 % CI 0.71–1.26).

Similarly, Devroey and his team in a series of studies demonstrated that the vaginal route was at least as good as the intramuscular one. These findings were established after comparing endometrial histological findings and pregnancy and miscarriage rates [14, 19, 20]. Surprisingly, these clinical outcomes were observed even though serum progesterone levels were abnormally low [21], suggesting the direct delivery of the hormone from the vagina to the target organ—the endometrium. This postulated mechanism has been termed the "first uterine pass effect" [22]. The Cochrane review did not find any significant differences when comparing the IM route to vaginal or rectal route [13].

In a nonpregnant cycle, when using progesterone by the vaginal route the onset of menses is not delayed beyond the normal schedule, while when the intramuscular route is used menses may be delayed as long as the progesterone therapy is continued. However, there is no evidence that this bleeding leads to either lower pregnancy rates or higher miscarriage rates [23, 24].

The progesterone-containing ring is another mode of vaginal progesterone luteal support. Continuous release of progesterone in the dose of 10–20 nmol/L/day was demonstrated to be as effective as vaginal or intramuscular treatment in IVF and egg donor cycles [25].

Another debated issue relates to the time of onset and discontinuation of luteal support. Regarding the onset of supplementation, there was no significant difference in the live birth rates in in vitro fertilization-embryo transfer cycles when progesterone was started on the day of hCG administration, the day after hCG, the day of ovum retrieval, or the day of day-3 embryo transfer [26]. Regarding discontinuation, some investigators limited supplementation to the day of a positive β-hCG while others continued treatment up to 12 weeks of pregnancy. Nyboe Andersen and colleagues [27] reported in a prospective study that prolongation of progesterone supplementation for 3 more weeks beyond the day of a positive pregnancy test had no influence on miscarriage rates and concluded that progesterone supplementation can safely be withdrawn at the time of a positive hCG test. Aboulghar et al. [28] concluded from their randomized study that prolongation of progesterone supplementation beyond the day of fetal heart ascertainment had no influence on the miscarriage rate. Despite the lack of evidence to support continuation of LPS beyond the day a clinical pregnancy is established, many believe that progesterone supplementation should be continued up to the 10th week of gestation. Among the explanations for such a practice is the increased risk for vaginal bleeding following the discontinuation of progesterone [28] and the associated emotional stress this engenders.

3 Estrogen Supplementation

Both estrogen and progesterone are secreted from the corpus luteum simultaneously. Although estradiol does not mediate luteinization directly, it is probably required for P receptor replenishment. Yet, conventional luteal support in IVF treatments does not include estrogen therapy, even though both estrogen and progesterone fall prematurely in most cases after ovarian stimulation [14]. In randomized control studies comparing luteal support of progesterone alone with support of both progesterone and estradiol valerate in a regimen of GnRH antagonist and recombinant FSH IVF treatment, no statistical difference was found in endocrine profiles and in ongoing pregnancy rates [29, 30]. In a systematic review and meta-analysis, there were no significant differences between progesterone alone or combined progesterone and estrogen regimens in IVF treatments [31]. In a recent review by the Cochrane collaboration, Van der Linden et al. [13] concluded that in IVF GnRHa regimens there were no significant differences in clinical or ongoing pregnancy, miscarriage, or live birth rates when progesterone combined with estrogen was compared to progesterone alone for luteal phase support. Of note, in the subgroup of patients that used transdermal estrogen support there was a significant effect on clinical pregnancy rates in favor of estrogen supplementation [13]. However, Var et al., in a recent randomized controlled trial, showed that the additional administration of estrogen to progesterone for luteal phase support significantly improved pregnancy and implantation rates and decreased miscarriage rates in comparison to luteal support with progesterone only [32].

4 Human Chorionic Gonadotropin Support

When ovulation occurs, the remaining granulosa cells in the follicle acquire LH receptors and rapidly undergo luteinization under the influence of LH. Both LH and HCG can activate these LH receptors. In IVF regimens using GnRH agonists in order to prevent premature LH surge, the downregulation of the pituitary and the blockage of LH secretion remain for at least 10 days after the administration of GnRHa has been stopped. Since in normal spontaneous cycles the survival of the CL can be prolonged by the stimulus of rapidly increasing hCG secreted from the implanted blastocyst, administering hCG to IVF patients in order to preserve CL activity appears reasonable.

Ovarian hyperstimulation syndrome (OHSS) is an iatrogenic complication of ovulation induction with exogenous gonadotropin treatments. hCG has a crucial role in the pathophysiology and the development of OHSS. Administration of hCG for luteal support increases the risk of developing OHSS and increasing its severity [1].

Luteal support by hCG is administered intramuscularly or subcutaneously two to four times during the luteal phase in a dose ranging from 1,000 to 5,000 IU for each injection. According to the Cochrane review of Daya and Gunby [12] from 2004, when hCG was compared to placebo or no treatment in GnRHa IVF regimens, there was an increase of 50 % in the odds of clinical pregnancy, but it did not reach statistical significance. The miscarriage rate was significantly reduced, and the ongoing pregnancy rate was significantly increased. Nonetheless, the odd of OHSS was 20-fold higher with hCG use. For cycles without GnRHa there was no significant difference in any outcome [12]. In the updated Cochrane review, the authors concluded that when hCG was compared to placebo or no treatment, there was a significant effect on the ongoing pregnancy rate [OR 1.75 (95 % CI 1.09–2.81)] but no effect on live birth, clinical pregnancy, and miscarriage rates. There was a significant effect in favor of placebo regarding OHSS of OR 0.28 (95 % CI 0.14–0.54) [13].

When comparing progesterone to hCG for luteal support, the authors found no significant differences in the clinical or the ongoing pregnancy, miscarriage, or multiple pregnancy rates. Still, there was an increased likelihood for OHSS following the use of hCG alone compared to progesterone [13]. When comparing progesterone alone to progesterone + hCG, no significant differences were found in the clinical or the ongoing pregnancy, miscarriage, or multiple pregnancy rates, However the OR of OHSS was 0.45 (95 % CI 0.26–0.79) [13].

5 Gonadotropin-Releasing Hormone Agonists for Luteal Support

Special consideration should be given to patients who undergo IVF treatment using GnRH antagonist regimen. In this regimen, GnRH agonists can be used instead of hCG for oocyte maturation and for luteal support and thereby reduce the occurrence of OHSS. Humaidan et al. [33] on behalf of the "Copenhagen GnRH Agonist Triggering Workshop Group" reviewed this narrative based on expert presentations, group discussions, literature search including randomized controlled trials, and authors' knowledge. There is an important difference between the half-life of the LH in a natural cycle (~60 min), after administrating hCG (>24 h) and after administrating the GnRH agonist. Therefore the duration of the LH surge after GnRH agonist administration was shorter (20 h) when compared to the natural LH surge (48 h). These differences necessitate special approaches of luteal support. A meta-analysis performed by Griesinger et al. [34] in 2006 showed that in comparison to hCG, GnRH agonist administration is associated with a significantly reduced likelihood of achieving a clinical pregnancy (0.21, 0.05–0.84; $P = 0.03$). The odds of first-trimester pregnancy loss is increased after GnRH agonist triggering; however, the confidence interval crosses unity (11.51, 0.95–138.98; $P = 0.05$). The occurrence of OHSS was significantly reduced [34]. This reduction of pregnancy rate was attributed to the effect of luteal phase deficiency and not to the oocyte quality [35]. In these studies, the luteal support was the standard support with vaginal P and oral estrogen. It was speculated that the luteal endometrial environment after GnRH analogs differs from that observed after hCG triggering. Therefore, conventional support with progesterone and estrogen might be insufficient [33]. A recent meta-analysis showed that implemented luteal LH activity supplementation or intensive luteal support with progesterone and estrogen resulted in delivery rates comparable to those with hCG triggering, with only 6 % different. OHSS was completely eliminated [33].

Intensive luteal support included intramuscular progesterone 50 mg per day from the day of ovum pick up until the 10th week of gestation and estradiol transdermal patches of 0.1 mg every other day [36].

In the Cochrane review comparison between progesterone versus progesterone + GnRH revealed a significant live birth rate in favor of GnRH + progesterone—OR of 2.44 (95 % CI 1.62–3.67) and significant pregnancy rate—OR 1.36 (95 % 1.11–1.66) [13].

6 Conclusions and Recommendations

From the beginning of ART it has been clear that the luteal phase is insufficient and luteal support should be administered in order to improve the pregnancy and live birth rates.

The following agents have been used for luteal support in ART patients: progesterone, estrogen, hCG, LH, and GnRH analog. All can be administered as a single agent or combined with each other's regimens.

After performing this literature review we believe that progesterone is the essential agent for luteal support and should be administered to patients undergoing ART treatments. The vaginal route offers the same benefits as the intramuscular route but is more convenient and acceptable to patients. In IVF-ET cycles there were no significant differences in live birth rates when progesterone was started on the day of hCG administration, the day after hCG, the day of ovum pick up, or the day of embryo transfer; therefore, it can be tailored according to the patient- and the physician-preferred protocol. Although it was found that there were no significant differences in pregnancy and live birth rates when progesterone treatment was stopped on the day of positive βhCG test or 3 weeks later, many believe that progesterone supplementation should be continued up to the 10th week of gestation.

Since there were no significant differences in clinical or ongoing pregnancy, miscarriage, or live birth rates when estrogen was added to progesterone compared to progesterone alone, it is not necessary to administer estrogen for luteal support.

Although hCG administration for luteal support results in a better pregnancy and live birth rates compared to placebo, it can result in severe OHSS. Since there were no significant differences in the clinical or the ongoing pregnancy, miscarriage, or multiple pregnancy rates when comparing progesterone to hCG for luteal support while the risk of OHSS in the progesterone treatment is minimal, it is questionable whether the use of HCG for luteal support is appropriate.

In GnRH antagonist protocols, especially when a GnRH agonist trigger is utilized, special attention should be given to choosing an appropriate support regimen. Luteal LH activity supplementation or intensive luteal support with progesterone and estrogen should be considered.

References

1. Fritz MA, Speroff L (2011) Clinical gynecologic endocrinology and infertility, 8th edn. Wolters Kluwer, Lippincott Williams & Wilkins, Philadelphia
2. Edwards RG, Steptoe PC, Purdy JM (1980) Establishing full-term human pregnancies using cleaving embryos grown in vitro. Br J Obstet Gynaecol 87(9):737–756
3. Check JH et al (1987) The efficacy of progesterone in achieving successful pregnancy: I. Prophylactic use during luteal phase in anovulatory women. Int J Fertil 32(2):135–138
4. Check JH, Chase JS, Nowroozi K, Nazari A (1989) Ovulation induction and pregnancies in women with ovarian failure (reversing menopause). In: Parthenon publishing, recent developments in fertility & sterility, proceedings of the XIII World Congress on fertility and sterility, Marrakesh, October 1989, vol 6. pp 43–47
5. Kerin JF et al (1981) Human luteal phase function following oocyte aspiration from the immediately preovular graafian follicle of spontaneous ovular cycles. Br J Obstet Gynaecol 88(10):1021–1028
6. Tavaniotou A, Devroey P (2003) Effect of human chorionic gonadotropin on luteal luteinizing hormone concentrations in natural cycles. Fertil Steril 80(3):654–655
7. Smitz J et al (1992) Pituitary gonadotrophin secretory capacity during the luteal phase in superovulation using GnRH-agonists and HMG in a desensitization or flare-up protocol. Hum Reprod 7(9):1225–1229
8. Albano C et al (1998) The luteal phase of nonsupplemented cycles after ovarian superovulation with human menopausal gonadotropin and the gonadotropin-releasing hormone antagonist Cetrorelix. Fertil Steril 70(2): 357–359
9. Fauser BC, Devroey P (2003) Reproductive biology and IVF: ovarian stimulation and luteal phase consequences. Trends Endocrinol Metab 14(5):236–242
10. Gardner DK, Weissman A, Howles CM, Shoham Z (eds) (2009) Textbook of assisted reproductive techniques laboratory and clinical perspectives, 3rd edn. Martin Dunitz, London
11. Leeton J, Trounson A, Jessup D (1985) Support of the luteal phase in in vitro fertilization programs: results of a controlled trial with intramuscular Proluton. J In Vitro Fert Embryo Transf 2(3):166–169
12. Daya S, Gunby J (2004) Luteal phase support in assisted reproduction cycles. Cochrane Database Syst Rev (3):CD004830
13. van der Linden M et al (2011) Luteal phase support for assisted reproduction cycles. Cochrane Database Syst Rev (10):CD009154
14. Bourgain C et al (1990) Effects of natural progesterone on the morphology of the endometrium in patients with primary ovarian failure. Hum Reprod 5(5):537–543
15. Zarutskie PW, Phillips JA (2009) A meta-analysis of the route of administration of luteal phase support in assisted reproductive technology: vaginal versus intramuscular progesterone. Fertil Steril 92(1):163–169
16. Bouckaert Y et al (2004) Acute eosinophilic pneumonia associated with intramuscular administration of progesterone as luteal phase support after IVF: case report. Hum Reprod 19(8):1806–1810
17. Costabile L et al (2001) A prospective randomized study comparing intramuscular progesterone and 17alpha-hydroxyprogesterone caproate in patients undergoing in vitro fertilization-embryo transfer cycles. Fertil Steril 76(2):394–396
18. Pritts EA, Atwood AK (2002) Luteal phase support in infertility treatment: a meta-analysis of the randomized trials. Hum Reprod 17(9):2287–2299
19. Devroey P et al (1989) Progesterone administration in patients with absent ovaries. Int J Fertil 34(3):188–193
20. Smitz J et al (1992) A prospective randomized comparison of intramuscular or intravaginal natural progesterone as a luteal phase and early pregnancy supplement. Hum Reprod 7(2):168–175
21. Miles RA et al (1994) Pharmacokinetics and endometrial tissue levels of progesterone after administration by intramuscular and vaginal routes: a comparative study. Fertil Steril 62(3): 485–490
22. De ZD (1995) Hormonal control of endometrial receptivity. Hum Reprod 10(1):4–7
23. Roman E et al (2000) Analysis of the bleeding pattern in assisted reproduction cycles with luteal phase supplementation using vaginal micronized progesterone. Hum Reprod 15(7): 1435–1439
24. Gurbuz B et al (2003) Bleeding patterns in women using intramuscular progesterone for luteal support in in-vitro fertilisation cycles. J Obstet Gynaecol 23(3):267–270
25. Zegers-Hochschild F et al (2000) Prospective randomized trial to evaluate the efficacy of a vaginal ring releasing progesterone for IVF and oocyte donation. Hum Reprod 15(10):2093–2097

26. Mochtar MH, Van Wely M, Van der Veen F (2006) Timing luteal phase support in GnRH agonist down-regulated IVF/embryo transfer cycles. Hum Reprod 21(4):905–908
27. Nyboe Andersen A et al (2002) Progesterone supplementation during early gestations after IVF or ICSI has no effect on the delivery rates: a randomized controlled trial. Hum Reprod 17(2):357–361
28. Aboulghar MA et al (2008) Prospective randomized study comparing luteal phase support for ICSI patients up to the first ultrasound compared with an additional three weeks. Hum Reprod 23(4):857–862
29. Fatemi HM et al (2006) Addition of estradiol to progesterone for luteal supplementation in patients stimulated with GnRH antagonist/rFSH for IVF: a randomized controlled trial. Hum Reprod 21(10):2628–2632
30. Fatemi HM et al (2007) The luteal phase of recombinant follicle-stimulating hormone/gonadotropin-releasing hormone antagonist in vitro fertilization cycles during supplementation with progesterone or progesterone and estradiol. Fertil Steril 87(3):504–508
31. Kolibianakis EM et al (2008) Estrogen addition to progesterone for luteal phase support in cycles stimulated with GnRH analogues and gonadotrophins for IVF: a systematic review and meta-analysis. Hum Reprod 23(6):1346–1354
32. Var T et al (2011) A comparison of the effects of three different luteal phase support protocols on in vitro fertilization outcomes: a randomized clinical trial. Fertil Steril 95(3):985–989
33. Humaidan P, Kol S, Papanikolaou EG, on behalf of the "the Copenhagen GnRH Agonist Triggering Workshop Group" (2011) GnRH agonist for triggering of final oocyte maturation: time for a change of practice? Hum Reprod Update 17(4):510–524
34. Griesinger G et al (2006) GnRH agonist for triggering final oocyte maturation in the GnRH antagonist ovarian hyperstimulation protocol: a systematic review and meta-analysis. Hum Reprod Update 12(2):159–168
35. Eldar-Geva T et al (2007) Similar outcome for cryopreserved embryo transfer following GnRH-antagonist/GnRH-agonist, GnRH-antagonist/HCG or long protocol ovarian stimulation. Reprod Biomed Online 14(2):148–154
36. Engmann L et al (2008) The use of gonadotropin-releasing hormone (GnRH) agonist to induce oocyte maturation after cotreatment with GnRH antagonist in high-risk patients undergoing in vitro fertilization prevents the risk of ovarian hyperstimulation syndrome: a prospective randomized controlled study. Fertil Steril 89(1):84–91
37. Hutchison JS, Zeleznik AJ (1984) The rhesus monkey corpus luteum is dependent on pituitary gonadotropin secretion throughout the luteal phase of the menstrual cycle. Endocrinology 115(5):1780–1786
38. Fraser HM et al (1997) Initiation of high dose gonadotrophin-releasing hormone antagonist treatment during the late follicular phase in the macaque abolishes luteal function irrespective of effects upon the luteinizing hormone surge. Hum Reprod 12(3):430–435

Chapter 11

General Principles of Cryopreservation

Roger G. Gosden

Abstract

Cryopreservation quickly became a cornerstone technology in assisted reproduction because the banking of gametes, embryos, and gonadal tissues has increased the effectiveness of assisted reproductive technology cycles for infertility treatment as well as fertility preservation for patients at risk of premature sterilization. Cryopreservation protocols, both slow/equilibrium cooling and vitrification methods, have evolved empirically and still depend heavily on operator skill, but further automation promises to improve reproducibility and uniformity of results.

Key words Cryopreservation, Fertility, Oocyte, Ovary, Vitrification

1 Role in Reproductive Medicine

Cryopreservation provides a service technology for long-term storage of cells and tissues in many fields of biology and medicine, but perhaps never more importantly than in reproductive medicine. It quickly found a role at the center of assisted reproductive technology (ART) treatment; it is also a key technology in fertility preservation for patients, both male and female, at risk of sterilization during cancer treatment or from other causes. Cryopreservation can even be claimed to have been born, if not conceived, in reproductive medicine, because it was in this field that the original breakthrough took place. The first practical applications were for semen, with enormous ramifications for industry and medicine, starting in the early 1950s with artificial insemination for cattle followed by semen banking for men with cancer or as "fertility insurance" before undergoing vasectomy. There were few if any other applications in reproductive medicine until the 1980s when clinical IVF started to revolutionize the treatment of infertility, heralding an urgent need to cryopreserve new cell types.

Cryotechnology for embryos and eggs was not available for the first IVF pregnancies, but neither was it required because natural menstrual cycles were used. Since only one oocyte was available it

Zev Rosenwaks and Paul M. Wassarman (eds.), *Human Fertility: Methods and Protocols*, Methods in Molecular Biology, vol. 1154, DOI 10.1007/978-1-4939-0659-8_11, © Springer Science+Business Media New York 2014

was inseminated immediately, and the embryo transferred fresh. But arguments for generating additional oocytes using ovarian stimulation quickly became compelling and soon became standard practice [1]. When up to a dozen or even more oocytes are available a proportionate number of embryos can be generated for transfer, and hence there is a higher chance of pregnancy. Moreover, a controlled ovarian cycle enables more precise timing of oocyte collection. At first there was a temptation, not to say tendency, to transfer all the morphologically normal embryos together, but this raises the hazard of multiple pregnancy which is often regarded as the most serious drawback of IVF treatment. In the early days practitioners and their patients therefore were on the horns of a dilemma: whether to cautiously transfer only one or a few embryos and discard the rest or to transfer all good-quality embryos, accepting the risk and perhaps offering "selective reduction" if multiple fetal hearts were detected by ultrasound. In either case, it was an acute ethical dilemma, and the wastage of potentially healthy conceptuses was perverse for patients struggling to overcome infertility.

First reports of successful embryo cryopreservation were greeted, on the one hand, with enthusiasm because they addressed the vexing problem of multiple pregnancy by saving spare embryos for subsequent freeze–thaw cycles and, on the other, caution because of the then unknown effects of freezing on the health of children-to-be. While the field quickly embraced cryopreservation, lingering concerns remained while there was only a paucity of clinical data. Animal models, principally mice, had already indicated that embryos suffer no long-term harm from cryopreservation [2] and, besides, preimplantation stages precede the critical period starting with gastrulation when environmental factors can cause teratogenic effects. To date, more than half a million babies have been born worldwide from frozen embryos and, as far as can be determined from surveys and individual clinics, there is no excess incidence of birth defects, and as time goes by there is growing assurance of their long-term health [3].

The technology can now be regarded as "mature," albeit still imperfect. Not all embryos survive after thawing; success rates vary between centers and even between laboratory personnel and patients using the same protocols. Evidently green fingers still count, and clinical outcomes will continue to vary until technology eliminates the operator factor and there is better understanding of the character and causes of cryoinjury.

2 Cryoinjury: All About Ice

Mammals never evolved mechanisms for tolerating freeze injury, unlike some non-homeothermic animals that can avoid frostbite and even tolerate core freezing in winter. To survive low temperatures cells need to become sufficiently dehydrated, which normally requires the action of cryoprotective agents (CPAs). When temperatures fall below 0 °C aqueous fluids do not freeze immediately but supercool, an unstable state that ends abruptly with the growth of ice crystals. In practice the degree of supercooling is curtailed by artificially nucleating ice, which occurs in the extracellular space before starting in the tiny volume inside cells. Cryoinjury is not so much caused by the physical impalement of cells by ice crystals as the toxic effects of concentrated salts, damaged proteins, and osmotic stress.

Cellular injury and its avoidance during freezing are illustrated by a simple experiment with red blood cells which we gave to students at Edinburgh University some years ago. These cells are useful models because hemolysis, signifying the loss of semipermeable membrane properties, is a "red flag," the fraction of cells affected being indicated by light absorption at the λ_{max} for hemoglobin in the fluid. When a suspension of cells in isotonic saline is centrifuged after freezing and thawing, virtually all the hemoglobin is released into the supernatant solution, whereas a comparable tube of unfrozen blood cells produces a red pellet under a clear solution, indicating that they are still intact. Few other cell types have the advantage of being freely suspended in medium, and none have such a conspicuous intracellular marker, although lactate dehydrogenase is readily measured. Now if glycerol at, say, 1.5 mol/L is added to a fresh suspension of the same cells in saline a completely different result is obtained after freezing, thawing, and centrifuging compared to the first tube. The cells remain intact, resembling the unfrozen control specimen by forming a red pellet with little if any hemolysis. Glycerol avoids cellular injury normally caused by freezing, but this is getting ahead of the story of cryoprotective agents.

3 Discovery of Cryopreservation

In the 1930s, B.J. Luyet, an American immigrant from Switzerland, formulated principles for preserving cells at low temperatures [4]. His experiments were only partially successful, but he deserves credit for advancing the field from a virtual state of speculation to a proper scientific footing. His aim was to vitrify colloids and cells, a process in which a glassy, amorphous solid is formed instead of ice crystals. Vitrification requires rapid cooling and a high concentration of glass-forming solute(s) to generate sufficient viscosity that water molecules fail to organize into a crystalline structure. While vitrification is currently gaining momentum and could become the standard procedure for many cell types, it was not the technology that originally enabled cryopreservation to be applied practically. That breakthrough came rather unexpectedly and from a different source.

In the late 1940s, Sir Alan Parkes, the cryobiologist Audrey Smith, and their graduate student, Chris Polge, tested the ability of fructose solutions to protect chicken, rabbit, and human spermatozoa from freeze injury. The motility of spermatozoa offered researchers the convenience of a natural viability indicator which they could observe simply by phase contrast microscopy. Their choice was also propitious because sperm cryopreservation was needed for the cattle breeding industry. According to the story, their first experimental trials had limited success, but one day almost all of the cells recovered swimming motion after thawing from −79 °C. When they repeated the experiment with fresh solutions of fructose very few cells survived. Such a tantalizing result is the stuff of nightmares for experimental scientists, but fortunately they soon realized that some of the stock bottles in the refrigerator had lost their labels, and two of them had been switched between wrong bottles. The effective solution in their experiments contained not fructose, as they had presumed, but glycerol [5].

Over the six decades since their discovery, glycerol has proved to be very successful for cryopreserving a range of cell types, although it is not effective for every cell. For example, the relative impermeability of cell membranes renders it ineffective for oocytes. To protect cells from ice formation a CPA must penetrate them to a high concentration, implying that the molecule should be small, highly water soluble, and of course "nontoxic." But even a natural metabolite, like glycerol, can be toxic when exposed to cells at the super-physiological concentrations required for cryopreservation; hence, CPAs are usually loaded and unloaded close to the freezing point when metabolic activity is very low. Polge et al. [5] also tested 1,2-propanediol and ethylene glycol, both of which are effective CPAs later finding important applications in low-temperature preservation. Dimethylsulfoxide (DMSO) was introduced soon afterwards, and this small handful of CPAs still accounts for most routine practice, although other compounds are added to create a cocktail producing superior results. Secondary CPAs, generally sugars, serve for enhancing dehydration, whereas the primary CPAs have multiple effects, not all of which are well understood but include dehydration, replacement of cell water, mitigation of the so-called solution effects of concentrated salts, and stabilizing protein structure.

If a cryomicroscope had been available to Polge and his colleagues, they would have been able to observe the process of freezing at the cellular level, notably the creeping growth of ice crystals in the extracellular space and their absence inside cells. This process is fundamentally different to vitrification technology in which the entire specimen forms a glassy solid devoid of ice. In fact using their method the protoplasm vitrifies when it becomes sufficiently dehydrated and viscous with CPAs, proteins, and other molecules while being surrounded by frozen lakes of extracellular fluid. To achieve this state, cooling must be gradual with concentrations of the primary CPA in the molar range. Thus, this method became known as slow cooling or equilibrium cooling, the principles being explained by the pioneering work of Lovelock [6], Mazur [7], and others.

The optimal rate of cooling is cell specific: too slowly puts the cells at jeopardy from the toxic effects of exposure to concentrated CPAs and too rapidly exposes them to the risk of intracellular freezing from inadequate dehydration. When cells are loaded with CPAs they transiently shrink by osmosis because their membranes are more permeable to water than to the solute. When ice nucleates in the extracellular space the overall concentration of solutes rises, causing a secondary and lasting dehydration of the cells [7]. Consequently, those with low membrane permeability require very slow rates of cooling of the order 0.1–0.5 °C/min during the critical stages of dehydration until the temperature falls below about −35 °C. Afterwards the specimen can be cooled faster or even plunged into liquid nitrogen. This is a time-consuming protocol, sometimes taking several hours.

Vitrification technology is rapidly gaining popularity for a variety of reasons, first among them being the avoidance of ice formation. In addition, it is rapid, inexpensive, and (deceptively) easy, but there are, of course, some drawbacks. Since the concentration of CPAs for vitrification is much higher (of the order 5–6 mol/L) toxicity is a serious problem, somewhat mitigated by using more than one CPA at proportionately reduced concentrations, step gradients to reduce osmotic stress CPAs, and loading and unloading at temperatures close to 0 °C. Each step must be carried out rapidly (in seconds) before plunging into liquid nitrogen to reach the glass transition temperature at rates of ~20,000 °C/min or even more rapidly [8]. The faster the cooling rate the less concentrated the CPA solution needs to be to achieve vitrification. Vitrification technology is therefore distinctive compared to the standard slow freezing method, but they share the same CPA compounds and a requirement for rapid rewarming to avoid crystallization.

Such high rates of cooling are achieved by mounting specimens on fine tools in a minimum of fluid, sometimes merely a film <0.1 μl. Commonly used tools, such as cryoloops, open-pulled straws, cryotops, and cryoleafs, can expose cells to the risk of contamination in non-sterile liquid nitrogen, but pressures from regulatory bodies, such as the Federal Drugs Administration, are encouraging manufacturers to add protective sleeves and barriers, although these reduce the rates of cooling. Vitrification may not be the answer for every cell type and presents special problems for preserving bulky tissues and whole organs, but it is nonetheless an important advance for cryopreservation in reproductive biology and medicine.

4 State of the Art

Although embryo freezing quickly became a standard practice that no IVF clinic could operate without, oocytes have proved more difficult to cryopreserve and the history of this technology has been far from smooth. After initial reports in the mid-1980s of viable pregnancies from frozen oocytes [9, 10], the practice almost disappeared for a decade, and only relatively recently has it restarted and begun to shake off the label of "experimental." It was as if a voluntary moratorium was cast for years over oocyte banking.

There were several reasons for caution. At first there were large differences between the post-thawing survival rates of embryos and oocytes, the latter often being <50 % [11]. This gap was not simply a consequence of the oocyte being a single cell compared with many at cleavage and blastocyst stages because pronucleate eggs also performed well during cryopreservation [12]. Subsequently it narrowed with the introduction of higher concentrations of sucrose to increase cellular dehydration [13], although this progress did not address the problems of low fertilization and pregnancy rates or, even more sinister, the possibility of higher risks of aneuploidy.

Fertilization rates improved when ICSI became available for bypassing the zona pellucida, which was thought to be hardened against sperm entry by cortical granule exocytosis during the cryopreservation process [14]. Today, ICSI is used almost universally for cryopreserving oocytes based on the rationale that since there is no clear disadvantage with ICSI nothing can be gained by taking a chance with standard insemination.

Concerns about aneuploidy have haunted the field for longer. Animal studies revealed that microtubules in the metaphase II spindle become depolymerized by cooling, even to room temperatures [15]. Hence by implication the chromatids could drift free from the spindle equator, the pairs failing afterwards to segregate faithfully during anaphase at the poles. Using the PolScope, the spindle can be observed in living cells to reassemble after rewarming, the process appearing to be less effective in human and cow oocytes than in mice [16], although not all studies have found evidence of cytogenetic abnormalities in thawed human oocytes [17].

Nevertheless, oocyte cryopreservation resumed in the later 1990s, healthy babies being reported after improved survival and fertilization rates attributed to ICSI and replacement of DMSO with propanediol [18, 19]. Propanediol had already become the standard CPA for embryos [20]. Reassuring data are accumulating to confirm the safety of the slow freezing technique, at least in the most experienced laboratories. Oocyte freezing is now an alternative to embryo banking in IVF cycles and for fertility preservation, even as insurance against aging and natural menopause.

Increasingly, however, it is vitrification technology that is being used for oocytes, as well as for cleavage-stage embryos and blastocysts, based on excellent post-vitrification survival and pregnancy rates in the centers reporting large numbers of cases [21]. There is an ongoing debate about the pros and cons of the two methods, but a final conclusion still seems remote, randomized clinical trials being virtually impractical since there is no universal protocol for slow freezing and a profusion for vitrification. Time will tell whether the much-vaunted results of vitrification in the "super-clinics" can be reproduced on the smaller scale of most American clinics. In the long run, operator skill and experience are likely to count for less than at present because advances in technology will provide a far more controlled process than pairs of hands, especially for vitrification.

Vitrification may also have new applications for sperm banking, despite the success of conventional methods. Since the advent of ICSI, men with severe oligospermia can be helped to become genetic parents, creating the new challenge of preserving very small numbers of these minute cells. Since vitrification technology excels with tiny samples it might address this problem [22]. It is also advocated for gonadal tissues [23], although permeation of high concentrations of CPAs and ultra-rapid cooling are serious obstacles to progress with bulky specimens. To date, all cryopreserved ovarian tissues leading to live births after transplantation were prepared by slow freezing methods since the first case report [24]. Testicular tissue has occasionally been cryopreserved in the same way for prepubertal children at risk of sterilization from cancer or the effects of treatment, although to my knowledge there are no cases of retransplantation to date. While the track record of slow freezing for tissues is encouraging, it is certainly not ideal because freezing of the extracellular fluid is likely to disrupt tissue architecture, likely causing more acute damage in the testis than the ovary which is a more plastic organ.

There is a confusing range of options for patients needing to preserve their reproductive cells and tissues, some of which are still experimental. A young woman with excellent chances of surviving a formerly dire disease but who is likely to be sterilized by treatment could opt to cryopreserve her embryos if a male partner is available, her oocytes, or her ovarian tissue, using either slow freezing or vitrification methods. In the distant future she might even be offered banking of an intact ovary or germ cells deriving from cultured stem cells. Males have a narrower range of choices, although most have the advantage of semen banking which does not require any invasive procedure. In each case, however, they are making a serious investment to protect their reproductive futures by cryotechnology. The preservation of some cell types is already routine, and technologies for the rest are making progress, giving hope all round.

References

1. Wortham JW Jr, Veeck LL, Witmyer J et al (1983) Vital initiation of pregnancy (VIP) using human menopausal gonadotropin and human chorionic gonadotropin ovulation induction: phase I—1981. Fertil Steril 39: 785–792
2. Glenister PH, Whittingham DG, Lyon MF (1984) Further studies on the effect of radiation during the storage of frozen 8-cell mouse embryos at -196 degrees C. J Reprod Fertil 70:229–234
3. Aytoz A, Van den Abbeel E, Bonduelle M et al (1999) Obstetric outcome of pregnancies after the transfer of cryopreserved and fresh embryos obtained by conventional in-vitro fertilization and intracytoplasmic sperm injection. Hum Reprod 14:2619–2624
4. Luyet BJ (1937) The vitrification of organic colloids and of protoplasm. Biodynamica 1:1–14
5. Polge C, Smith AU, Parkes AS (1949) Revival of spermatozoa after vitrification and dehydration at low temperatures. Nature 164:666
6. Lovelock JE (1953) The haemolysis of human red blood-cells by freezing and thawing. Biochim Biophys Acta 10:414–426
7. Mazur P (1963) Kinetics of water loss from cells at subzero temperatures and the likelihood of intracellular freezing. J Gen Physiol 47:347–369
8. Rall WF, Fahy GM (1985) Ice-free cryopreservation of mouse embryos at -196°C by vitrification. Nature 313:573–575
9. Chen C (1986) Pregnancy after human oocyte cryopreservation. Lancet 19:884–886
10. Van Uem JF, Siebzehnrübl ER, Schuh B et al (1987) Birth after cryopreservation of unfertilized oocytes. Lancet 28:752–753
11. Gook DA, Edgar DH (2007) Human oocyte cryopreservation. Hum Reprod Update 13: 591–605
12. Veeck LL, Amundson CH, Brothman LJ et al (1993) Significantly enhanced pregnancy rates per cycle through cryopreservation and thaw of pronuclear stage oocytes. Fertil Steril 59: 1202–1207
13. Fabbri R, Porcu E, Marsella T et al (2001) Human oocyte cryopreservation: new perspectives regarding oocyte survival. Hum Reprod 16:411–416
14. Johnson MH, Pickering SJ, George MA (1988) The influence of cooling on the properties of the zona pellucida of the mouse oocyte. Hum Reprod 3:383–387
15. Pickering SJ, Braude PR, Johnson MH et al (1990) Transient cooling to room temperature can cause irreversible disruption of the meiotic spindle in the human oocyte. Fertil Steril 54:102–108
16. Aman RR, Parks JE (1994) Effects of cooling and rewarming on the meiotic spindle and chromosomes of in vitro-matured bovine oocytes. Biol Reprod 50:103–110
17. Cobo A, Rubio C, Gerli S et al (2001) Use of fluorescence in situ hybridization to assess the chromosomal status of embryos obtained from cryopreserved oocytes. Fertil Steril 75:354–360
18. Porcu E, Fabbri R, Seracchioli R et al (1997) Birth of a healthy female after intracytoplasmic sperm injection of cryopreserved human oocytes. Fertil Steril 68:724–726
19. Borini A, Bianchi V (2011) Oocyte cryopreservation—slow freezing. In: Donnez J, Kim SS (eds) Principles and practice of fertility preservation. Cambridge University Press, Cambridge
20. Lassalle B, Testart J, Renard JP (1985) Human embryo features that influence the success of cryopreservation with the use of 1,2-propanediol. Fertil Steril 44:645–651
21. Kuwayama M (2007) Highly efficient vitrification for cryopreservation of human oocytes and embryos: the Cryotop method. Theriogenology 67:73–80
22. Isachenko E, Isachenko V, Katkov II et al (2003) Vitrification of human spermatozoa without cryoprotectants: review of problems and practical success. Reprod Biomed Online 6:191–200
23. Silber S, Kagawa N, Kuwayama M et al (2010) Duration of fertility after fresh and frozen ovary transplantation. Fertil Steril 94:2191–2196
24. Donnez J, Dolmans MM, Demylle D et al (2004) Live birth after orthotopic transplantation of cryopreserved ovarian tissue. Lancet 364:1405–1410

Part II

Methodology

Chapter 12

In Vitro Maturation of Immature Human Oocytes for Clinical Application

Ri-Cheng Chian and Yun-Xia Cao

Abstract

Meiotic progression in the oocyte is defined as oocyte maturation from reinitiation of the first meiotic division from the germinal vesicle (GV) stage to the metaphase-II (M-II) stage (Fig. 1), (Cha and Chian, Hum Reprod Update 4:103–120, 1998). Priming with FSH or HCG prior to immature oocyte retrieval improves oocyte maturation and pregnancy rates. The size of follicles may be an important feature for IVM treatment. The combination of natural-cycle IVF with immature oocyte retrieval followed by IVM is an attractive treatment for young women with all types of infertility without recourse to ovarian stimulation with an acceptable pregnancy rate.

Key words Oocytes, IVM, IVF, FSH, HCG - Pregnancy

1 Introduction

Pincus and Enzmann [2] first reported that the mammalian oocytes can be matured spontaneously to M-II stage in vitro when the immature oocytes were removed from their antral follicles. They reported that the immature human oocyte can be matured in vitro following removal from follicles cultured for 12 h as was observed with rabbit oocytes [3]. In fact, it is difficult to mature immature human oocytes from GV stage to M-II stage in vitro following culture for only 12 h. The time course for human oocyte maturation in vitro was corrected by Edwards [4], when he reported that immature human oocytes require at least 34 h in culture in order to become mature and that in fact most immature human oocytes required more than 36 h of culture in vitro to achieve maturity (Fig. 1).

Zev Rosenwaks and Paul M. Wassarman (eds.), *Human Fertility: Methods and Protocols*, Methods in Molecular Biology, vol. 1154, DOI 10.1007/978-1-4939-0659-8_12,

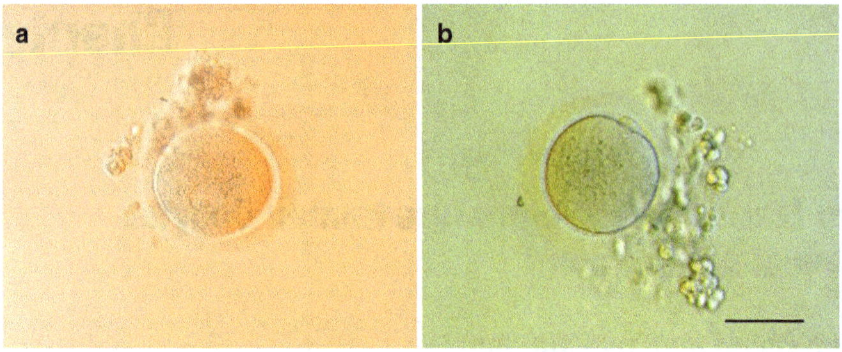

Fig. 1 The human GV-stage and M-II-stage oocytes. (**a**) The intact GV-stage oocyte is observed clearly in the ooplasm, and (**b**) the first polar body of M-II-stage oocyte was extruded into perivitelline space (PVS) under microscope (scale bar: 60 μm)

Interestingly, the first human embryos generated in vitro were from in vitro-matured oocytes, not from in vivo-matured oocytes [5, 6]. Although great attempts were made in the early days with immature human oocytes for in vitro maturation (IVM), followed by in vitro fertilization (IVF) [7], the first live birth from IVF was the result of an in vivo-matured oocyte retrieved in the natural cycle [8], not from in vitro-matured oocytes. Although the immature oocytes were obtained from the ovaries of patients undergoing surgery, due to ethical concerns embryos produced from IVM oocytes were not transferred. Moreover, early efforts to collect either immature or mature human oocytes for IVF were challenging.

With breakthroughs and developments in laparoscopy and endocrinology, it became possible to retrieve mature oocytes from the leading or the dominant follicle before ovulation occurred [9, 10]. The mature oocytes were fertilized in vitro, and the resulting embryos were used for transfer. Finally, this development resulted in the world's first IVF baby [8]. Nevertheless, in the early days, IVF technology utilizing natural-cycle retrievals was not very efficient. This was due in part to the fact that only one or two follicles developed and that often the oocyte was not retrieved successfully and no embryo was available for transfer. Moreover, the need to frequently monitor for the LH surge and the unpredictability of retrieval timing made their efforts quite inconvenient. Thus, natural-cycle IVF soon transitioned to ovarian stimulation in order to produce more oocytes and improve IVF efficiency. Initially, clomiphene citrate (CC), or CC combined with gonadotropins, was utilized, later to be supplanted by human menopause gonadotropins alone [11–13]. Subsequently other stimulation protocols were developed including gonadotropins combined with GnRH agonists and antagonists in "short" or "long" protocols, all with the aim of increasing the efficiency of multiple oocyte recovery. Although the optimal number of oocytes has not been clearly defined, all stimulation protocols aim to ideally retrieve 10–15 mature oocytes with the ultimate goal of increasing the number of embryos available for transfer and increasing the chance for achieving a pregnancy.

However, some women are extremely sensitive to stimulation with exogenous gonadotropins and are at increased risk of developing ovarian hyperstimulation syndrome (OHSS), a life-threatening condition [14]. Moreover, repeated ovarian stimulation may cause untoward severe side effects and may lead to other complications including ovarian torsion [15–17]. Therefore, IVM of immature human oocytes became increasingly attractive to many scientists and clinicians. IVM of immature oocytes is a reproductive technology that enables the immature oocytes to mature in vitro in order to fertilize those oocytes to generate viable embryos and live births. The most beneficial point for IVM of immature human oocytes is the avoidance of side effects from ovarian stimulation as well as reduction of the cost of gonadotropins and the associated intensive monitoring necessary during ovarian stimulation. Indeed, over the past three decades scientists and clinicians have continued their quest for the ideal "mild" stimulation protocol or even avoiding stimulation while achieving high success rates.

One of these approaches has been IVF technology. The first live birth from IVM oocytes when immature oocytes derived from ovarian follicles during surgery were donated to another couple [18]. The first live birth from a woman using her own oocytes following immature oocyte retrieval and IVM was from a woman with polycystic ovary syndrome (PCOS) [19]. The immature oocytes were retrieved by using the transvaginal ultrasound-guided oocyte retrieval technique, which has become the routine method for oocyte collection. The technology available to retrieve immature oocytes from women with PCOS has led to the development of the IVM technology and the basic research necessary to apply it to clinical practice. In this chapter, we focus on discussing human IVM technology and its efficiency for clinical application as well as the further development potential.

2 Methodology of IVM

The clinical application of IVM oocytes has come a long way since the first live birth in the early 1990s. The first pregnancy in a woman with anovulatory infertility following IVM of immature oocytes and IVF was reported by Trounson et al. [19]. Another pregnancy was reported in a group of patients with PCOS treated with IVM combined with intracytoplasmic sperm injection (ICSI) and assisted hatching (AH) [20]. Subsequent studies indicated that although immature oocytes recovered from untreated patients with PCOS can be matured, fertilized, and developed in vitro, the implantation rate of these cleaved embryos was disappointingly low [20–22]. However, recent data indicate that with an alternative IVM treatment in these patients, using priming with follicle-stimulating hormone (FSH) or HCG before immature oocyte retrieval, the clinical pregnancy and implantation rates can be significantly improved [23]. There are several thousand IVM babies born, based on many groups' contributions. There is no single unique IVM technology which is applied worldwide. However, IVM technology can be classified based on the hormone treatment utilized before immature oocyte retrieval: (1) priming with FSH and (2) priming with HCG.

2.1 Priming with FSH

As an alternative approach, a truncated course of ovarian stimulation with FSH before immature oocyte retrieval has been applied, indicating that FSH pretreatment promotes efficient recovery of immature oocytes and maturation in vitro [24]. It has been reported that the immature oocytes from stimulated cycles from normal cycling woman without HCG can be matured and fertilized in vitro and can result in live births [25]. However, another report indicated that FSH priming with a fixed dose (150 IU/day) for 3 days from day 3 of the menstrual cycle does not increase the number of oocytes obtained per aspiration and does not improve oocyte maturation, cleavage rates, or embryo development in women with normal menstrual cycles [26]. Furthermore, Suikkari et al. [27] reported that although low-dose FSH priming started in the luteal phase improves the efficiency of immature oocyte recovery and maturation and fertilization rates, the average number of immature oocytes collected and the rates of oocyte maturation and fertilization are not different in women with regular menstrual cycles compared to women with irregular cycles and PCOS. Nevertheless, it has been reported that priming with rFSH during the follicular phase before harvesting of immature oocytes from patients with PCOS improves oocyte maturation potential and implantation rates, resulting in significantly higher pregnancy (29 %, 7/24) and implantation (21 %, 8/37) rates [28]. More recently, Mikkelsen et al. [29] also reported that there were no differences in the rates of oocyte maturation, fertilization, cleavage, or implantation after either 2 or 3 days of coasting between FSH priming and aspiration of immature human oocytes for IVM in normal menstrual cycling women who were primed with 150 IU FSH/day for 3 days starting on day 3. It seems that optimizing IVM treatment for different groups of patients is important and that after individualized IVM treatment the pregnancy and implantation rates per embryo transfer can reach 23 and 14 %, respectively [30].

Interestingly, it has been reported that FSH priming with 75 IU/day for 6 days in combination with HCG priming for 36 h before immature oocyte retrieval has no additional benefit in women with PCOS [31]. Although there are conflicting reports on the benefits of using FSH priming in women with regular menstrual ovaries or irregular menstrual cycles of PCOS, theoretically the use of FSH priming at the beginning of follicular or luteal phases may enhance follicular development and maturational competence of immature oocytes in vivo. However, further research is required to confirm the beneficial effects of priming with FSH before immature oocyte retrieval in both women with regular menstrual cycles or in PCOS patients with irregular menstrual cycles.

2.2 Priming with HCG

A few GV-stage oocytes may be retrieved from the stimulated cycles even 36 h after HCG administration. These immature oocytes are capable of undergoing IVM and then normal fertilization and development. Although successful pregnancies have been established using these in vitro-matured oocytes [32–35], the pregnancy rate has been unacceptably low. It has been reported that morphological and molecular differences exist between immature oocytes collected from stimulated cycles and those collected during Caesarean section [36]. In addition, it has been found that the time course of germinal vesicle breakdown (GVBD) and oocyte maturation is different in these oocytes, although the final rates of oocyte maturation are not different in the groups [1]. It appears that the oocytes retrieved from follicles in women undergoing ovarian stimulation respond to HCG that may promote the initiation of oocyte maturation in vivo. It has been demonstrated that the time course of oocyte maturation in vitro is hastened and the rate of oocyte maturation is increased by priming with 10,000 IU HCG 36 h before retrieval of immature oocytes from women with PCOS [37, 38] (Fig. 2). Therefore, it is possible that pregnancy rate may potentially be improved by priming with HCG prior to immature oocyte retrieval [39]. This hypothesis was confirmed by other reports [31, 40–42].

Lin et al. [31] reported a 36.4 % clinical pregnancy rate in 33 cycles primed with HCG before immature oocyte retrieval in women with PCOS, suggesting that there is a beneficial effect of HCG priming on IVM treatment. Interestingly, it has been reported that the time course and maturation rates are different in GV-stage oocytes depending on the morphology of their associated cumulus cells after HCG priming [43]. Therefore, it seems that with HCG priming may not only promote oocyte-initiated maturation to metaphase I from relatively larger sized follicles (>10 mm in diameter) but also enhance the acquisition of maturational and development competence in vivo of GV-stage oocytes from small follicles. However, the exact mechanism of how HCG promotes these changes on small follicle-derived GV-stage oocytes is largely unclear. Further research is required to understand the functional role of HCG priming before immature oocyte retrieval on the small size of follicles in women with PCOS.

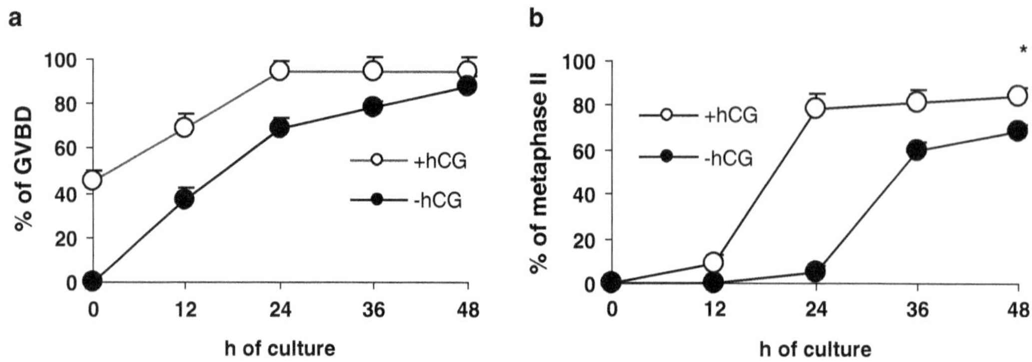

Fig. 2 The time course of human oocyte maturation during culture in vitro. (**a**) There were 46.2 % ± 5.2 of GVBD oocytes at the time of retrieval in HCG-primed group. The final percentages of GVBD were not different between the oocytes retrieved from the HCG-primed group. (**b**) There were 78.2 % ± 7.1 of mature oocytes at 24 h after culture in HCG-primed group. The final percentages of oocyte maturation were significantly different between the oocytes retrieved from the HCG-primed group (*dash-filled circle dash*) and the non-hCG-primed (*dash open circle dash*) group (total of 183 oocytes). *Asterisk* indicates significant difference between two groups at the time point (at least $P < 0.05$) (Chian et al. [38])

2.3 Immature Oocyte Retrieval

Transvaginal ultrasound-guided follicular aspiration has now become the preferred procedure of choice for immature oocyte retrieval in IVM treatment cycles. The same principles applied to IVF oocyte retrieval are also valid for IVM patients. A smaller gauge needle (19 or 20G) is preferable. This causes less pain and less damage to the smaller follicles, thereby allowing for a high efficiency of collection.

Because the intrafollicular pressure is already higher in small follicles, the aspiration vacuum pressure is reduced to 85 mmHg, which is approximately half the conventional IVF aspiration pressure. A higher aspiration pressure provokes an increase in the number of denuded oocytes. Good ultrasonographic visualization is the key point for successful immature oocyte retrieval. The follicular sizes vary, and certain follicles may be difficult to aspirate or, even if they are aspirated, no oocytes may be recovered, especially from the very-small-size follicles (<4 mm in diameter). Aspirates are collected in 10 ml culture tubes containing approximately 2 ml of heparinized warming medium (usually containing 2 units/ml of heparin). It is possible to use 0.9 % saline containing 2 units/ml heparin.

There are two ways to look for and collect immature oocytes from follicular aspirates. (1) Dish search: The follicular aspirates are poured directly into a petri dish and examined for immature oocytes under a stereomicroscope. (2) Cell strainer: The follicular aspirates are filled through a cell strainer (70 μm nylon). After filtering, the collected aspirates can be rinsed with prewarmed flushing medium and transferred to a petri dish to search for immature oocytes under a stereomicroscope. All handling procedures should be conducted on warm stages or plates at 37 °C.

2.4 In Vitro Maturation of Immature Oocytes

Human oocytes acquire a series of competences during follicular development (oocyte growth and maturation) that play critical roles at fertilization and subsequent early embryonic development. Oocyte maturation in vitro is profoundly affected by culture conditions. Although numerous data have been accumulated from animal studies, the current rationale for choosing a specific medium for IVM of immature human oocytes appears to stem largely from adapting methods developed from culturing other cell types.

Different energy substrates and nutrients can greatly influence oocyte meiotic and cytoplasmic maturation [44, 45]. Glucose, pyruvate, and lactate are the main substrates for energy metabolism in somatic cells and oocytes. Glutamine can also serve as an energy substrate to improve in vitro nuclear maturation of hamster and rabbit oocytes [46, 47]. Oocyte utilization of pyruvate is closely dependent upon cumulus cells that can convert glucose or lactate into pyruvate to be used by oocytes [48]. Pyruvate or oxaloacetate, but not glucose, lactate, or phosphoenolpyruvate, supports the maturation of denuded mouse oocytes through meiosis to M-II stage [49]. It has been confirmed that mitochondrial oxidative metabolism is much more important than anaerobic glucose metabolism for energy production in the mammalian oocytes [50]. Synthesis of pyruvate from glucose in the cumulus cells provides additional evidence that these cells are able to influence the nutritional environment of the maturing oocytes [51]. It has been shown that sodium pyruvate in non-serum maturation medium supports and promotes nuclear maturation of bovine cumulus-denuded oocytes [52]. However, it has been reported that pyruvate alone is insufficient for oocyte cytoplasmic maturation [53]. Nevertheless, it has been indicated that the expression pathway of glycolytic metabolism reflects the presence of different mechanisms involved in gene expression/regulation at the transcriptional and translational level and their accumulation during human oocyte maturation [54]. In addition, it has been suggested that metabolism of glucose through the Embden–Meyerhof pathway is important during bovine oocyte maturation in vitro [55]. There is no direct information about human oocytes.

Spontaneous mouse oocyte maturation in vitro, in either the presence or the absence of meiotic inhibitor, is associated with a decrease in oocyte cAMP levels [56]. In mice, glucose treatment of cumulus–oocyte complexes (COCs) produced elevated cAMP levels, which were associated with a decreased incidence of GVBD in hypoxanthine-supplemented medium [57]. Pyruvate directly affects nuclear maturation in mouse oocytes [58]. Although it has been reported that glucose may have an inhibitive effect on cumulus-free human oocyte maturation during culture in vitro [59], the results from our laboratory indicate that oocyte maturation medium with glucose is beneficial to bovine and human oocyte nuclear and cytoplasmic maturation in vitro in proper concentrations [45, 60].

Essential ad/or nonessential amino acids are commonly added to serum-supplemented or serum-free culture media for mammalian embryo development in vitro. In many species, it has been known that the addition of amino acids to the culture medium is beneficial for embryonic development. Apart from amino acid use for protein synthesis, they play important roles as osmolytes [48], intracellular buffers [61], heavy metal chelators and energy sources [62], as well as precursors for versatile physiological regulators, such as nitric oxide and polyamines [63]. It has also been shown that the culture medium with amino acids affects glucose metabolism in mouse blastocysts in vitro [64]. Although it has been shown that amino acids support rabbit [47], hamster [46], porcine [65], and bovine [44] oocyte maturation, amino acid requirements for human oocyte maturation in culture are not fully understood. The data from our laboratory indicate that essential amino acids supplemented to a simple chemically defined medium is absolutely required for bovine oocyte cytoplasmic maturation to support subsequent embryonic development and nonessential amino acids with essential amino acids have a synergic effect on oocyte cytoplasmic maturation [66, 67].

It has been reported that the addition of water-soluble vitamins, particularly inositol, to the embryo culture medium enhances the hatching of rabbit and hamster blastocysts [68, 69]. Vitamins affect glucose metabolism in mouse [64] and sheep embryos [70]. However, there is a paucity of information about the effects of vitamins in culture medium on the maturational and developmental competence of immature oocytes. The results from this laboratory demonstrate that the presence of vitamins in oocyte maturation medium is important for subsequent bovine embryonic development [71]. Based on animal model studies, we have designed a new IVM medium and shown that this IVM medium is beneficial for nuclear and cytoplasmic maturation of immature human oocytes derived from both stimulated IVF and unstimulated cycles [60].

Complex culture media supplemented with various sera, gonadotropins (FSH and LH), and estradiol have been most widely used in research or clinical application of human oocytes IVM [72]. In vitro studies using growth factors have shown that meiotic resumption in COCs can be induced by adding those growth factors into culture media. Supplementation of insulin into IVM medium seems to stimulate oocyte maturation and morphologic development of mouse and bovine blastocysts [73, 74].

The immature COCs are cultured in the IVM medium in the incubator and allowed to begin the maturation process for 24–48 h. Twenty-four hours after maturation in culture, all of the COCs are stripped for identification of oocyte maturity. COCs will be denuded using a finely drawn glass pipette following 1 min of exposure to a commercially available hyaluronidase solution. The mature oocytes are then subjected to insemination by either

IVF or ICSI after stripping. The remaining immature oocytes (GV and M-I) will continue to mature in culture for another 24 h. Forty-eight hours after oocyte retrieval (or oocyte maturation in culture), the remaining stripped oocytes are reexamined and if any have matured (M-II) at this point, they will be inseminated immediately by either IVF or ICSI.

2.5 In Vitro Fertilization of In Vitro-Matured Oocytes

ICSI is recommended for the insemination of in vitro-matured oocytes because we believe that this method offers a greater chance of successful fertilization over conventional IVF inseminations. ICSI is a common procedure for this reason. Although it is preferable to prepare sperm freshly before ICSI, it does not appear problematic to use sperm prepared on the day of or 1 day after egg collection for oocytes matured 48 h after egg retrieval. Commercially available ICSI medium and polyvinylpyrrolidone (PVP) solution can be used to prepare the ICSI dish. It is also appropriate to use the oocyte-washing medium for preparation of the ICSI droplets because the pH of the oocyte-washing medium is quite stable at room temperature and atmosphere. However, it is important to note that the ICSI dish should be prepared at least 1 h before ICSI and kept at 37 °C in the incubator or on a warm stage or plate for equilibration. After ICSI, the individual oocyte is transferred into a droplet (20 μl) of embryo culture medium in a Petri dish for culture in the incubator.

2.6 Preparation of the Endometrium

Preparing the endometrial lining requires using progesterone. Two hundred milligrams (200 mg) of intravaginal progesterone (Prometrium) is administered to the patient three times daily or progesterone injections are given intramuscularly with instruction starting from the day after egg collection and continuing until the pregnancy test. On the day of embryo transfer (ET), endometrial thickness should be measured by transvaginal ultrasound scan. At this point, the endometrial thickness should be at least ≥7.0 mm. If the endometrial thickness is <7.0 mm, the embryo should be cryopreserved and transferred in a subsequent cycle.

2.7 Embryo Transfer

It seems that most ET in IVM treatment can be done on day 2 or day 3 after ICSI because no extra benefit is derived by culturing the embryos to the blastocyst stage if the available number of embryos is small. In general, ET should be performed on day 2 after ICSI if the number of embryos obtained is ≤3; ET should be performed on day 3 after ICSI if the number of embryos obtained is ≥4. ET with blastocysts should only be considered if a total of more than four good-quality four-cell-stage embryos are achieved on day 2 of embryo assessment after ICSI.

The scoring of cleavage-stage embryos for transfer is crucial for pregnancy potential. Since the oocytes may not be matured and inseminated at the same time following maturation in culture, the developmental stages of embryos may be variable in the same patient. Therefore, before ET, all embryos for each patient should be pooled and selected for transfer. The final outcome of pregnancy may depend to a great extent on the experience of the embryologist. The cleavage speed of embryos and the morphological marker of each cleaved blastomere are usually used for scoring the quality of embryo generated from in vitro-matured oocytes. It is recommended that a maximum of two embryos be transferred into the uterus, based on the quality of embryos or if blastocysts are obtained the number of embryos for transfer should be only one or two. It does not appear in the context of IVM cycles that transferring a greater number of poor-quality embryos increases pregnancy and implantation rates.

One of the final key contributory factors to a successful pregnancy is embryo transfer. Careful attention must be paid to both the scientific and clinical aspects of this event. A trial or a mock transfer prior to the actual transfer provides very useful information to assess the curve of the cervical canal and ascertain the position of the uterus to avert foreseeable problems during the actual transfer. It is important that as much mucus as possible is removed from the cervix with a sterile cotton bud before ET. An abdominal ultrasound-guided ET may be recommended in selected cases in order to confirm that the embryos with the fluid contents of the catheter are in the uterus.

3 Clinical Outcome

For last 10 years, significant progress has been made to improve pregnancy and implantation rates from in vitro-matured human oocytes. IVM technology can now be offered as a successful option to infertile women with PCOS. As mentioned above, it has been demonstrated that immature oocyte priming with FSH or HCG prior to immature oocyte retrieval improves oocyte maturation rate and embryo quality as well as pregnancy rates in women infertile with PCOS. The size of follicles may be important for the subsequent embryonic development, but the developmental competence of oocytes derived from small antral follicles seems not to be adversely affected by the presence of a dominant follicle. We have collected information on approximately 1,500 IVM babies born worldwide (Table 1). It seems that there are no differences between IVM babies and naturally conceived babies in terms of birth weight and birth defects. Currently it has been estimated that more than 5,000 healthy infants have been born following immature oocyte retrieval and IVM. In general, the clinical pregnancy and implantation rates have reached 35–40 % and 10–15 %, respectively, in infertile women with PCOS [75].

Table 1
IVM babies born data collected from different countries and area (until the end of 2010)

Country and area	Number of IVM babies born
Australia	19
Brazil	15
Bulgaria	11
Canada	132
China	292
Columbia	7
Denmark	34
Finland	52
France	50
Greece	1
Israel	3
Italy	157
Japan	69
Jordan	1
Norway	4
South Korea	455
Slovenia	7
Sweden	22
Taiwan	20
Turkey	8
UK	8
USA	6
Vietnam	59
Total	*1,432*

Data of IVM babies born were collected from following IVF clinics or centers or organizations: (1) Fertility Specialists WA, Bethesda Hospital, 25 Queenslea Drive, Claremont WA 6010, Australia; (2) Nilo Frantz Human Reproduction and Research Center, Nilo Pecanha Avenue 1221, 10th Floor, Porto Alegre, RS, Brazil; (3) Sofia Center of Reproductive Medicine and IVF, Sofia, Bulgaria; (4) McGill Reproductive Center, McGill University Health Center (MUHC), 687 Pine Avenue W, Montreal, Quebec, Canada, H3A 1A1; (5) Medical Center for Human Reproduction, Department of Obstetrics and Gynecology, Peking University Third Hospital, Beijing, 100191, China; (6) Center for Reproductive Medicine, The First Affiliated Hospital with Nanjing Medical University, Nanjing, 210029, China; (7) Center for Reproductive Medicine,

(continued)

Table 1 (continued)

The First Affiliated Hospital of Anhui Medical University, Hefei, 230022, China; (8) Reproductive Medicine Center, The First Affiliated Hospital of Wenzhou Medical College, Wenzhou, 325000, China; (9) Reproductive Medical Center, The First Affiliated Hospital of Sun Yat-sen University, Guangzhou, 510700, China; (10) The Women's Clinic and IVF Center, The Hong Kong Sanatorium and Hospital, China; (11) CECOLFES, Bogota, Colombia; (12) The Fertility Clinics, Herlev University Hospital, DK-2730, Herlev, Denmark; (13) Infertility Clinic, The Family Federation of Finland, Helsinki, Finland; (14) Service de Gynecologie-Obstetrique et Medecine de la Reproduction, Hospital Antoine Beclere, Clamart, France; (15) Laboratoire de Biologie de la Reproduction-IFREARES, 20 Route de Revel, 31400-Toulouse, France; (16) IAKENTRO, 4 Ag, Vasiliou St, Thessaloniki, Greece; (17) IVF Unit, Assaf Harofeh Medical Center, Tel-Aviv University, Zerifin, Israel; (18) Biogenesi Reproductive Medicine Centre, Istituti Clinici Zucchi, V Zucchi, 24-Monza, Italy; (19) IVF Namba Clinic and IVF Osaka Clinic, Osaka, Japan; (20) Yoshida Ladies' Clinic, Sendai, Japan; (21) Kyono ART Clinic, Sendai, Japan; (22) ART and Genetic Department, AL-Khalidi Medical Center, Jordan; (23) Maria Fertility Hospital, Seoul, South Korea; (24) Bioingeniør, Fertilitetssenteret ved Aleris Sykehus, Fredrik Stangs gt. 11–13, O264 Oslo, Norway; (25) Department of Reproductive Medicine and Gynecological Endocrinology, University Clinical Centre Maribor, Slovenia; (26) Fertility Unit, Karolinska Institutet, Department of Clinical Science, Technology and Intervention, Karolinska University Hospital, Novum, SE 14186 Stockholm, Sweden; (27) IVF Unit, Department of Obstetrics and Gynecology, Shin Kong Wu Ho-Su Memorial Hospital, Taipei, Taiwan; (28) IVF Unit, Department of Obstetrics and Gynecology, Dokuz Eylul University, Izmir, Turkey; (29) Gurgan Clinic Women Health, Infertility and IVF Center, Cankaya Caddesi, No. 20/3, Ankara, Turkey; (30) Oxford Fertility Unit, Level 4, Women's Centre, John Radcliffe Hospital, Oxford, OX3 9DU, UK; (31) Delaware Valley Institute of Fertility and Genetics, Marlton, NJ 08053, USA; (32) HCM Society for Reproductive Medicine (HOSREM), 84T/8 Tran Dinh Xu Street, District 1, Ho Chi Minh City, Vietnam

4 Development of IVM Technology: Natural-Cycle IVF/M

In women, although only a single follicle usually grows to the preovulatory stage and releases its oocytes for potential fertilization, there are many small follicles that also develop during the same follicular phase of the menstrual cycle. It seems that approximately 20 antral follicles are recruited and only one is selected through to the preovulatory stages of development during each menstrual cycle [76]. Recently, based on daily transvaginal ultrasonography, it has been documented that there are two or three waves of ovarian follicles which develop in women during the menstrual cycle challenging the traditional theory that a single cohort of antral follicles grows during the follicular phase of the menstrual cycle [77, 78]. Animal model studies support these findings and suggest that oocyte quality and early embryonic developmental competence of immature oocyte following maturation in vitro are not detrimentally affected by the presence of the dominant follicle in the ovaries [79, 80].

It has been demonstrated in humans that atresia does not occur in the non-dominant follicles even after the dominant follicle is selected during folliculogenesis, because immature oocytes retrieved from non-dominant follicles have been successfully matured in vitro and fertilized and have resulted in several pregnancies and healthy live births [23]. Therefore, one very attractive possibility for enhancement of the success of natural cycle of IVF treatment is its combination with immature oocyte retrieval and IVM. If the mature oocyte from the dominant follicle and immature oocytes from small follicles are both collected, the chances of a pregnancy should be greatly increased as we manage to mature these immature oocytes and produce several viable embryos. However, it seems very important to prevent ovulation from the dominant follicle due to a natural LH surge when the patient is treated with natural-cycle IVF combined with IVM. Our experiences indicate that 10,000 IU HCG can be administered 36 h before oocyte retrieval when the size of the dominant follicle reached 10–14 mm in diameter. Most oocytes collected from the dominant follicles are at M-II stage. A pilot study suggested that the clinical pregnancy rate can reach approximately 45–50 % per ET following natural cycle in IVF combined with IVM in a selected group of patients [81]. The detailed protocol for natural-cycle IVF/M is described in Fig. 3.

Literature reports for pregnancy rate per embryo transfer in natural-cycle IVF vary between 0 and 30 % [82–84]. A number of problems arise in natural-cycle IVF treatment alone, including an increased risk of retrieval failure during oocyte collection leading to cancellation of treatment cycle. This disadvantage seems to be overcome with natural-cycle IVF and IVM (IVF/M). It has been shown that more than a half of infertile women who came to infertility clinic for IVF treatment can be treated with natural-cycle IVF/M or IVM alone when these treatments have been chosen primarily and that natural-cycle IVF/M is an efficient treatment, especially for women under the age of 35 years. Therefore, natural-cycle IVF/M is clearly a more efficient treatment than natural-cycle IVF alone.

IVM treatment is offered as a successful option to infertile women with PCOS because there are many antral follicles in the ovaries in this group of patients. PCOS is the most widespread endocrine disorder among women of reproductive age as well as the most common cause of anovulatory infertility and has been shown to exist in approximately 10 % of the general population [85]. Lim et al. [86] further confirmed that 15.4 % of infertile women have PCOS, and these women can be treated with IVM alone with reported pregnancy and implantation rates that can reach 41.3 and 16.6 %, respectively. This study demonstrated that natural-cycle IVF/M together with IVM alone can be used to treat more than half of infertile patients with acceptable pregnancy rate. Therefore, IVM technology may supplant traditional IVF in this particular group of infertile patients. Further optimization of IVM techniques may lead to more popular use of this technology in all infertile patients.

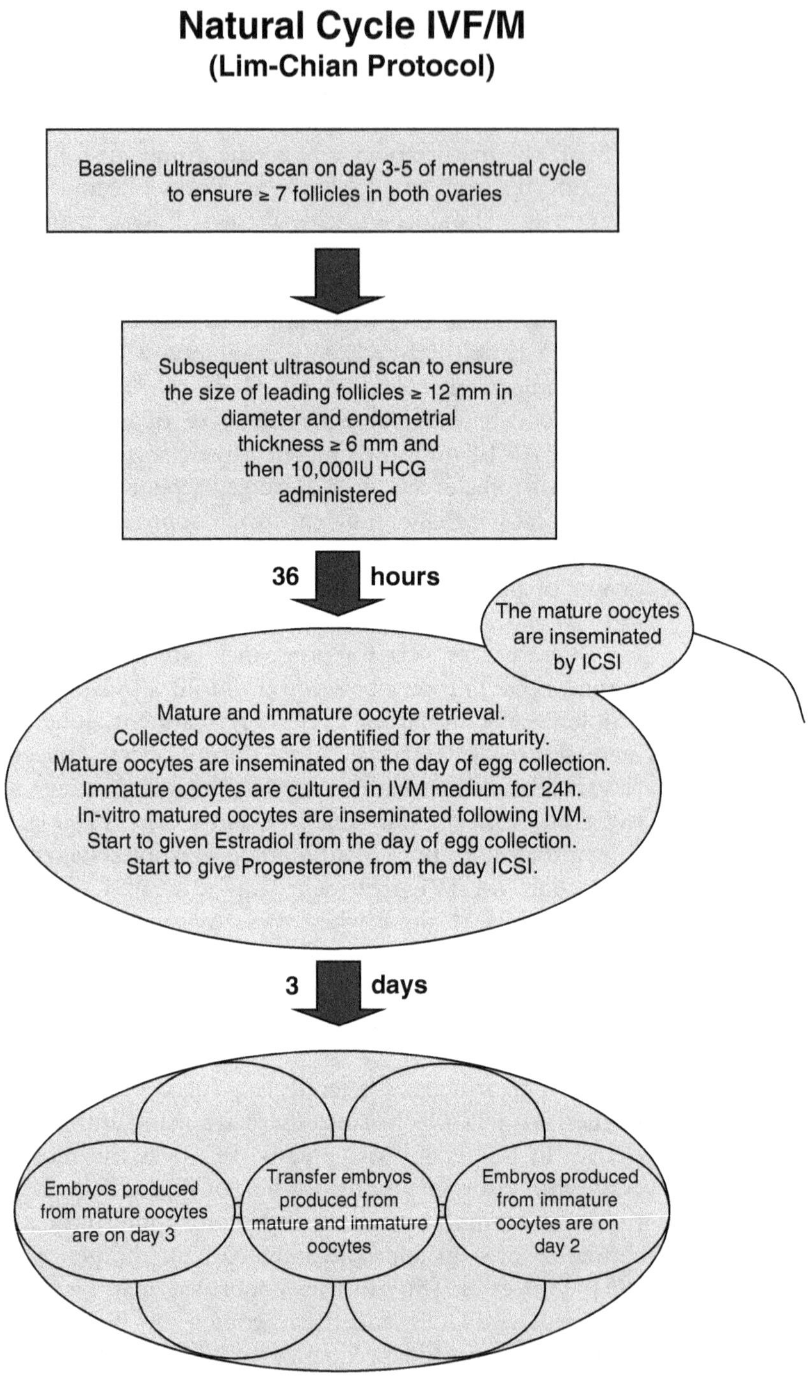

Fig. 3 Detailed protocol for natural-cycle IVF combined with in vitro maturation (Lim-Chian protocol) (Lim et al. [86])

5 Conclusions

Priming with FSH or HCG prior to immature oocyte retrieval improves oocyte maturation and pregnancy rates especially in women with PCOS. The size of follicles may be an important feature for the subsequent embryonic development, but the developmental competence of oocytes derived from the small antral follicles seems not to be adversely affected by the presence of a dominant follicle. Approximately 5,000 healthy infants have been born following immature oocyte retrieval and IVM [87]. In general, the clinical pregnancy and implantation rates per ET have reached 35–40 % respectively, in women with PCOS. Therefore, as an option, IVM treatment can be offered to infertile women with PCOS. The combination of natural-cycle IVF with immature oocyte retrieval followed by IVM is an attractive treatment for young women with all types of infertility without recourse to ovarian stimulation with an acceptable pregnancy rate.

References

1. Cha KY, Chian RC (1998) Maturation in vitro of immature human oocytes for clinical use. Hum Reprod Update 4:103–120
2. Pincus G, Enzmann EV (1935) The comparative behaviours of mammalian eggs in vitro and in vivo. II. The activation of tubal eggs in the rabbit. J Exp Med 62:665–675
3. Pincus G, Saunders B (1939) The comparative behaviours of mammalian eggs in vitro and in vivo. VI. The maturation of human ovarian ova. Anat Rec 75:537
4. Edwards RG (1965) Maturation in vitro of human ovarian oocytes. Lancet 2(286): 926–929
5. Rock J, Menkin MF (1994) In vitro fertilization and cleavage of human ovarian eggs. Science 100:105–107
6. Menkin MF, Rock J (1948) In vitro fertilization and cleavage of human ovarian eggs. Am J Obstet Gynecol 55(3):440–452
7. Edwards RG, Bavister BD, Steptoe PC (1969) Early stages of fertilization in vitro of human oocytes matured in vitro. Nature 221(5181): 632–635
8. Edwards RG, Steptoe PC (1978) Birth after the reimplantation of a human embryo. Lancet 312:366
9. Steptoe PC, Edwards RG (1970) Laparoscopic recovery of preovulatory human oocytes after priming of ovaries with gonadotrophins. Lancet 1(7649):683–689
10. Edwards RG, Steptoe PC, Purdy JM (1970) Fertilization and cleavage in vitro of preovulator human oocytes. Nature 227(5265):1307–1309
11. Lopata A et al (1978) In vitro fertilization of preovulatory oocytes and embryo transfer in infertile patients treated with clomiphene and human chorionic gonadotropin. Fertil Steril 30(1):27–35
12. Johnston I et al (1981) In vitro fertilization: the challenge of the eighties. Fertil Steril 36(6):699–706
13. Johnes HW Jr, Jones GS, Andrews MC et al (1982) The program for in vitro fertilization at Norfolk. Fertil Steril 38:14–21
14. Beerendonk CC, van Dop PA, Braat DD et al (1998) Ovarian hyperstimulation syndrome: fact and fallacies. Obstet Gynecol Surv 53:439–449
15. Tarlatzis BC et al (1995) Ovarian stimulation and ovarian tumours: a critical reappraisal. Hum Reprod Update 1(3):284–301
16. Duckitt K, Templeton AA (1998) Cancer in women with infertility. Curr Opin Obstet Gynecol 10(3):199–203
17. Smits G, Olatunbosun O, Delbaere A et al (2003) Ovarian hyperstimulation syndrome due to a mutation in the follicle-stimulating hormone receptor. N Engl J Med 349:760–766
18. Cha KY, Koo JJ, Ko JJ et al (1991) Pregnancy after in vitro fertilization of human follicular oocytes collected from nonstimulated cycle,

their culture in vitro and their transfer in a donor oocyte program. Fertil Steril 55: 109–113
19. Trounson A, Wood C, Kausche A (1994) In vitro maturation and the fertilization and developmental competence of oocytes recovered from untreated polycystic ovarian patients. Fertil Steril 62(2):353–362
20. Barnes FL et al (1995) Blastocyst development and birth after in-vitro maturation of human primary oocytes, intracytoplasmic sperm injection and assisted hatching. Hum Reprod 10(12):3243–3247
21. Barnes FL et al (1996) Production of embryos from in vitro-matured primary human oocytes. Fertil Steril 65(6):1151–1156
22. Cha KY, Han SY, Chung HM et al (2000) Pregnancies and deliveries after in vitro maturation culture followed by in vitro fertilization and embryo transfer without stimulation in women with polycystic ovary syndrome. Fertil Steril 73:978–983
23. Chian RC, Buckett WM, Abdul Jalil AK et al (2004) Natural-cycle in vitro fertilization combined with in vitro maturation of immature oocytes is a potential approach in infertility treatment. Fertil Steril 82:1675–1678
24. Wynn P, Picton HM, Krapez JA et al (1998) Pretreatment with follicle stimulating hormone promotes the number of human oocytes reaching metaphase II by in vitro maturation. Hum Reprod 13:3132–3138
25. Liu J, Katz E, Garcia JE et al (1997) Successful in vitro maturation of human oocytes not exposed to human chorionic gonadotropin during ovulation induction, resulting in pregnancy. Fertil Steril 67:566–568
26. Mikkelsen AL, Smith SD, Lindenberg S (1999) In-vitro maturation of human oocytes from regularly menstruating women may be successful without follicle stimulating hormone priming. Hum Reprod 14:1847–1851
27. Suikkari AM, Tulppala M, Tuuri T et al (2000) Luteal phase start of low-dose FSH of follicles results in an efficient recovery, maturation and fertilization of immature human oocytes. Hum Reprod 15:747–751
28. Mikkelsen AL, Lindenberg S (2001) Benefit of FSH priming of women with PCOS to the in vitro maturation procedure and the outcome: a randomized prospective study. Reproduction 122(4):587–592
29. Mikkelsen AL, Høst E, Blaabjerg J et al (2003) Time interval between FSH priming and aspiration of immature human oocytes for in-vitro maturation: a prospective randomized study. Reprod Biomed Online 16:116–120
30. Hreinsson J, Rosenlund B, Fridén B et al (2003) Recombinant LH is equally effective as recombinant hCG in promoting oocyte maturation in a clinical in-vitro maturation programme: a randomized study. Hum Reprod 18:2131–2136
31. Lin YH et al (2003) Combination of FSH priming and hCG priming for in-vitro maturation of human oocytes. Hum Reprod 18(8):1632–1636
32. Veeck LL et al (1983) Maturation and fertilization of morphologically immature human oocytes in a program of in vitro fertilization. Fertil Steril 39(5):594–602
33. Check ML, Brittingham D, Check JH et al (2001) Pregnancy following transfer of cryopreserved-thawed embryos that had been a result of fertilization of all in vitro matured metaphase or germinal stage oocytes. Clin Exp Obstet Gynecol 28:69–70
34. Nagy ZP et al (1996) Pregnancy and birth after intracytoplasmic sperm injection of in vitro matured germinal-vesicle stage oocytes: case report. Fertil Steril 65(5):1047–1050
35. Edirisinghe WR et al (1997) Birth from cryopreserved embryos following in-vitro maturation of oocytes and intracytoplasmic sperm injection. Hum Reprod 12(5):1056–1058
36. Chian RC, Park SE, Park EH et al (1997) Molecular and structural characteristics between immature human oocytes retrieved from stimulated and unstimulated ovaries. In: Leung PCK, Gomel V (eds) Vitro fertilization and assisted reproduction. Monduzzi Editore, Bologna
37. Chian RC et al (1999) Pregnancies resulting from in vitro matured oocytes retrieved from patients with polycystic ovary syndrome after priming with human chorionic gonadotropin. Fertil Steril 72(4):639–642
38. Chian RC et al (2000) Prospective randomized study of human chorionic gonadotrophin priming before immature oocyte retrieval from unstimulated women with polycystic ovarian syndrome. Hum Reprod 15(1):165–170
39. Chian RC et al (1999) Priming with human chorionic gonadotropin before retrieval of immature oocytes in women with infertility due to the polycystic ovary syndrome. N Engl J Med 341(21):1624–1626
40. Hwang JL et al (2002) Oocyte donation using immature oocytes from a normal ovulatory woman. Acta Obstet Gynecol Scand 81(3): 274–275
41. Nagele F et al (2002) Successful pregnancy resulting from in-vitro matured oocytes retrieved at laparoscopic surgery in a patient

with polycystic ovary syndrome. Hum Reprod 17(2):373–374

42. Son WY et al (2002) Blastocyst development and pregnancies after IVF of mature oocytes retrieved from unstimulated patients with PCOS after in-vivo HCG priming. Hum Reprod 17(1):134–136
43. Yang S et al (2001) Expression of luteinizing hormone receptor (LH-R), follicle stimulating hormone receptor (FSH-R) and epidermal growth factor receptor (EGF-R) in cumulus cells of the oocytes collected from PCOS patients in hCG-priming IVM/F-ET program. Fertil Steril 76(3):S38–S39
44. Rose-Hellekant TA, Libersky-Williamson EA, Bavister BD (1998) Energy substrates and amino acids provided during in vitro maturation of bovine oocytes alter acquisition of developmental competence. Zygote 6(4):285–294
45. Chung JT, Tan SL, Chian RC (2002) Effect of glucose on bovine oocyte maturation and subsequent fertilization and early embryonic development in vitro. Biol Reprod 66:176
46. Gwatkin RB, Haidri AA (1973) Requirements for the maturation of hamster oocytes in vitro. Exp Cell Res 76(1):1–7
47. Bae IH, Foote RH (1975) Utilization of glutamine for energy and protein synthesis by cultured rabbit follicular oocytes. Exp Cell Res 90(2):432–436
48. Biggers JD, Lawitts JA, Lechene CP (1993) The protective action of betaine on the deleterious effects of NaCl on preimplantation mouse embryos in vitro. Mol Reprod Dev 34(4): 380–390
49. Biggers JD, Whittingham DG, Donahue RP (1967) The pattern of energy metabolism in the mouse oocyte and zygote. Proc Natl Acad Sci USA 58(2):560–567
50. Gandolfi F et al (1998) Comparative analysis of calf and cow oocytes during in vitro maturation. Mol Reprod Dev 49(2):168–175
51. Leese HJ, Barton AM (1984) Pyruvate and glucose uptake by mouse ova and preimplantational embryos. J Reprod Fertil 72:9–13
52. Geshi M, Takenouchi N, Yamauchi N et al (2000) Effects of sodium pyruvate in nonserum maturation medium on maturation, fertilization, and subsequent development of bovine oocytes with or without cumulus cells. Biol Reprod 63:1730–1734
53. Zheng P et al (2001) Maturation of rhesus monkey oocytes in chemically defined culture media and their functional assessment by IVF and embryo development. Hum Reprod 16(2):300–305
54. el Mouatassim S et al (1999) Glucose metabolism during the final stage of human oocyte maturation: genetic expression of hexokinase, glucose phosphate isomerase and phosphofructokinase. Zygote 7(1):45–50
55. Krisher RL, Bavister BD (1999) Enhanced glycolysis after maturation of bovine oocytes in vitro is associated with increased developmental competence. Mol Reprod Dev 53(1): 19–26
56. Downs SM et al (1989) Maintenance of meiotic arrest in mouse oocytes by purines: modulation of cAMP levels and cAMP phosphodiesterase activity. Gamete Res 23(3):323–334
57. Downs SM (1995) The influence of glucose, cumulus cells, and metabolic coupling on ATP levels and meiotic control in the isolated mouse oocyte. Dev Biol 167(2):502–512
58. Kim H, Schuetz AW (1991) Regulation of nuclear membrane assembly and maintenance during in vitro maturation of mouse oocytes: role of pyruvate and protein synthesis. Cell Tissue Res 265(1):105–112
59. Cekleniak NA et al (2001) A novel system for in vitro maturation of human oocytes. Fertil Steril 75(6):1185–1193
60. Chian RC, Tan SL (2002) Maturational and developmental competence of cumulus-free immature human oocytes derived from stimulated and intracytoplasmic sperm injection cycles. Reprod Biomed Online 5(2):125–132
61. Edwards LJ, Williams DA, Gardner DK (1998) Intracellular pH of the mouse preimplantation embryo: amino acids act as buffers of intracellular pH. Hum Reprod 13(12):3441–3448
62. Bavister BD (1995) Culture of preimplantation embryos: facts and artifacts. Hum Reprod Update 1(2):91–148
63. Wu G, Morris SM Jr (1998) Arginine metabolism: nitric oxide and beyond. Biochem J 336 (Pt 1):1–17
64. Lane M, Gardner DK (1998) Amino acids and vitamins prevent culture-induced metabolic perturbations and associated loss of viability of mouse blastocysts. Hum Reprod 13(4): 991–997
65. Ka HH et al (1997) Amino acids in maturation medium and presence of cumulus cells at fertilization promote male pronuclear formation in porcine oocytes matured and penetrated in vitro. Biol Reprod 57(6):1478–1483
66. Rezaei N, Abdul Jalil AK, Chung JT et al (2003) Role of essential and non-essential amino acids contained in maturation medium on bovine oocyte maturation and subsequent fertilization and early embryonic development in vitro. Theriogenology 59:497
67. Rezaei N, Chian R (2005) Effects of essential and non-essential amino acids on in-vitro maturation, fertilization and development of

immature bovine oocytes. Iran J Reprod Med 3:36–41

68. Fahy MM, Kane MT (1992) Inositol stimulates DNA and protein synthesis, and expansion by rabbit blastocysts in vitro. Hum Reprod 7(4):550–552
69. Kane MT, Bavister BD (1988) Vitamins requirements for development of eight-cell hamster embryos to hatching blastocysts in vitro. Biol Reprod 39:1137–1143
70. Gardner DK et al (1994) Enhanced rates of cleavage and development for sheep zygotes cultured to the blastocyst stage in vitro in the absence of serum and somatic cells: amino acids, vitamins, and culturing embryos in groups stimulate development. Biol Reprod 50(2):390–400
71. Abdul Jalil AK, Rezaei N, Chung JT et al (2002) Effect of vitamins during oocyte maturation on subsequent embryonic development in vitro. In: 48th annual meeting CFAS2002. p TP-24
72. Trounson A, Anderiesz C, Jones GM et al (1998) Oocyte maturation. Hum Reprod 13(Suppl 3):52–62
73. Ocana-Quero JM, Pinedo-Merlin M, Ortega-Mariscal E et al (1998) Influence of human and bovine insulin on in vitro maturation, fertilization and cleavage rates of bovine oocytes. Arch Zootec 47:85–93
74. Reed ML, Estrada JL, Illera MJ et al (1993) Effects of epidermal growth factor, insulin-like growth factor-1, and dialyzed porcine follicular fluid on porcine oocyte maturation in vitro. J Exp Zool 266:74–78
75. Chian RC, Lim JH, Tan SL (2004) State of the art in in-vitro oocyte maturation. Curr Opin Obstet Gynecol 16(3):211–219
76. Hillier SG (1994) Current concepts of the role of FSH and LH in folliculogenesis. Hum Reprod 9:188–191
77. Baerwald AR, Adams GP, Pierson RA (2003) A new model for ovarian follicular development during the human menstrual cycle. Fertil Steril 80(1):116–122
78. Baerwald AR, Adams GP, Pierson RA (2003) Characterization of ovarian follicular wave dynamics in women. Biol Reprod 69(3): 1023–1031
79. Smith LC et al (1996) Oocyte quality in small antral follicles in the presence or absence of a large dominant follicle in cattle. J Reprod Fertil 106(2):193–199
80. Chian RC, Chung JT, Downey BR et al (2002) Maturational and developmental competence of immature oocytes retrieved from bovine ovaries at different phases of folliculogenesis. Reprod Biomed Online 4:129–134
81. Lim JH, Yang SH, Chian RC (2007) New alternative to infertility treatment for women without ovarian stimulation. Reprod Biomed Online 14(5):547–549
82. MacDougall MJ et al (1994) Comparison of natural with clomiphene citrate-stimulated cycles in in vitro fertilization: a prospective, randomized trial. Fertil Steril 61(6):1052–1057
83. Thornton MH, Francis MM, Paulson RJ (1998) Immature oocyte retrieval: lessons from unstimulated IVF cycles. Fertil Steril 70(4): 647–650
84. Janssens RM et al (2000) In-vitro fertilization in a spontaneous cycle: easy, cheap and realistic. Hum Reprod 15(2):314–318
85. Carmina E, Lobo RA (1999) Polycystic ovary syndrome (PCOS): arguably the most common endocrinopathy is associated with significant morbidity in women. J Clin Endocrinol Metab 84(6):1897–1899
86. Lim JH et al (2009) Selection of patients for natural cycle in vitro fertilization combined with in vitro maturation of immature oocytes. Fertil Steril 91(4):1050–1055
87. Chian RC, Uzelac PS, Nargund G (2014) In vitro maturation of human immature oocytes for fertility preservation. Fertil Steril 99(5):1173–1181

Chapter 13

GnRH Antagonist-Based Protocols for In Vitro Fertilization

David Reichman and Zev Rosenwaks

Abstract

In this chapter we elaborate on the intricacies of employing GnRH antagonists for prevention of the endogenous luteinizing hormone surge during IVF. Several practical aspects of implementing a GnRH antagonist-based stimulation protocol are described; selection of the correct dose, choosing when to start the antagonist, programming of cycle starts, selection of the appropriate gonadotropins, and triggering of final oocyte maturation are elucidated.

Key words GnRH antagonist, IVF, Short protocol

1 Introduction

GnRH antagonists offer several advantages over agonists for preventing premature LH surges during controlled ovarian stimulation for IVF. Whereas GnRH agonists are associated with an initial gonadotropin-stimulatory flare effect, GnRH antagonist down-regulation is immediate, thus allowing for a shorter period of suppression, fewer hypo-estrogenic side effects, and a reduced duration of gonadotropin stimulation [1–3]. Most importantly, compared to GnRH agonist protocols, GnRH antagonist protocols are associated with a lower incidence of ovarian hyperstimulation syndrome (OHSS); their utility has the added option of triggering final oocyte maturation with a GnRH agonist (GnRHa) for patients at risk of developing OHSS [4–6]. While GnRH antagonist analogues have long been available, initial use was first limited by anaphylactoid reactions, and later by reports that third-generation preparations were associated with lower pregnancy rates [7–9]. As experience has increased with GnRH antagonists, so too have pregnancy rates improved; the most recent Cochrane review revealed no difference in success rates between agonist- and antagonist-based protocols [10]. Given this equivalent efficacy and

Zev Rosenwaks and Paul M. Wassarman (eds.), *Human Fertility: Methods and Protocols*, Methods in Molecular Biology, vol. 1154, DOI 10.1007/978-1-4939-0659-8_13, © Springer Science+Business Media New York 2014

the advantages listed above, GnRH antagonists have now become the preferred method of ovarian stimulation for an increasing number of reproductive specialists.

GnRH receptor antagonists were first synthesized in the early 1970s [11, 12]. These first-generation agents had favorable side effect profiles, but insufficient potency for clinical use [13]. Second-generation agents, such as NalArg and detirelix, had improved potency and water solubility, but were characterized by histamine-mediated systemic side effects which greatly limited their use [8, 14]. Schally and colleagues synthesized Cetrorelix in the late 1980s, one of the two third-generation antagonists currently commercially available [15]. While several third-generation antagonists were synthesized, two ultimately were cultivated for clinical use. Cetrorelix and Ganirelix (Merck Serono, Darmstadt, Germany) are both available as 0.25mg subcutaneous injections for daily dosing; Cetrorelix was previously offered as a 3mg "single dose" formulation, but has been removed from the U.S. market and is being phased out of non-U.S. territories. In the pages that follow, practical methods for the integration of GnRH antagonists for down-regulation in IVF clinical practice will be outlined; where applicable, controversy regarding various techniques will be highlighted, and the existing literature synthesized.

2 Outline

1. GnRH antagonist dose selection
2. Fixed versus flexible initiation
3. Programming stimulation start with luteal pretreatment
4. Gonadotropin stimulation
5. Triggering of oocyte maturation
6. GnRH agonist trigger

2.1 GnRH Antagonist Dose Selection

The preferred GnRH antagonist dose should ideally prevent the endogenous LH surge while at the same time retain sufficient basal levels of LH to support steroidogenesis. Several studies in the late 1990s sought to establish the minimal effective dose for third-generation daily GnRH antagonists to suppress ovulation and maximize clinical outcomes. In one of the first dose finding studies, Devroey's group demonstrated no premature LH surges in 60 patients exposed to either 0.5 mg or 0.25 mg of Cetrorelix with a fixed protocol starting on day 6; Two patients receiving 0.1 mg, however, experienced a premature rise in LH, and further enrollment was curtailed [16]. A phase-II multicenter randomized controlled trial from ESHRE's Ganirelix Dose-Finding Study

Group followed shortly thereafter, and examined doses of 0.0625 mg, 0.125 mg, 0.25 mg, 0.5 mg, 1.0 mg, and 2.0 mg starting in a fixed protocol from day 7, with rFSH employed for stimulation [17]. Whereas the lowest doses were associated with the highest estradiol responses, they were associated with a 16 % (0.0625 mg) and 9 % (0.125 mg) risk of a premature LH rise of ≥10 mIU/mL. Conversely, higher doses were associated with more profound suppression (LH<1 mIU/mL), decreased implantation, and higher rates of miscarriage. The authors concluded that a daily GnRH antagonist dose of 0.25 mg was associated with the most favorable clinical pregnancy rate while reasonably minimizing the risk of premature LH rise ≥10 mIU/mL (1.4 %). It is this daily dose of GnRH antagonist that has been most widely adopted across the field.

There are occasional patients in clinical practice who will "break through" on the 0.25 mg dose, manifested by a rise in LH level as well as an increase in progesterone. For these patients, the dose can be increased to 0.5 mg daily in the next cycle to ensure effective suppression of the GnRH-induced LH surge. It should be noted that a recent study implicated premature elevations in LH (≥10 mIU/mL) in the absence of a rise in progesterone with poorer clinical outcomes, albeit in a small study population [18]. The mechanism whereby this rise may jeopardize clinical outcomes remains to be elucidated. Of note, 91.7 % of the patients experiencing an LH rise in this study had not yet been prescribed a GnRH antagonist, suggesting that the criteria for initiating the antagonists' use may have been significantly flawed.

As an alternative to daily administration, a 3 mg "single-dose" subcutaneous injection of Cetrorelix acetate (Cetrotide; Merck Serono, Darmstadt, Germany) is available in non-U.S. territories, which can be supplemented with daily 0.25 mg injections if further suppression is required after 3 days (it should be noted, however, that this formulation is being phased out of production) [19–21]. In a multicenter randomized controlled trial based in the USA, single dose (3 mg) GnRH antagonist was compared to daily dose administration (0.25 mg) in 185 patients undergoing IVF, with initiation of the antagonist when the lead follicle was ≥14 mm; no cycles were characterized by LH rise ≥10 mIU/mL, median injections were 1 versus 4 for the 3 mg versus 0.25 mg dosages, and pregnancy rates (43.2 % single-dose vs. 46.0 % multi-dose) were comparable for the two groups despite lower serum estradiols in those patients receiving the longer acting formulation [22]. Live-birth rates were not reported. A subsequent Taiwanese study performed a randomized controlled trial of single dose Cetrorelix (starting day 7) versus daily GnRH antagonist administration (starting day 5) in 40 patients, and found that patients receiving the single dose preparation exhibited lower levels of estradiol and fewer zygotes, but no statistically significant differences in implantation (22.2 multi-dose [MD] vs. 11.1 % single-dose [SD]) or clinical pregnancy (50 % MD vs. 25 % SD) [23]. This study was methodologically limited and underpowered to statistically evaluate differences in pregnancy, and should thus be interpreted with caution.

2.2 Fixed Versus Flexible Protocols

There are two primary approaches for initiating of GnRH antagonists during ovarian stimulation. In a fixed protocol, the GnRH antagonist is administered on a set day of stimulation, which can be as early as day 1 or as late as day 8, but most routinely is either on day 5 or 6. Conversely, flexible protocols withhold antagonist start until either the lead follicle has crossed a threshold diameter (typically 13–15 mm depending on the center) or when the serum estradiol has surpassed a threshold level (~300 pg/mL in our practice) (Fig. 1).

Several randomized control trials have compared fixed versus flexible GnRH initiation and its effects on outcomes, but all have been underpowered to analyze differences in clinical pregnancy rates or live-births. Ludwig et al. first compared initiation of the antagonist on day 6 of stimulation ($n = 20$) versus initiation when the lead follicle was 14 mm ($n = 20$ single dose, $n = 20$ multi-dose) [24]. Although uniform criteria for the timing of hCG were not used, less vials of Cetrotide were required with the flexible

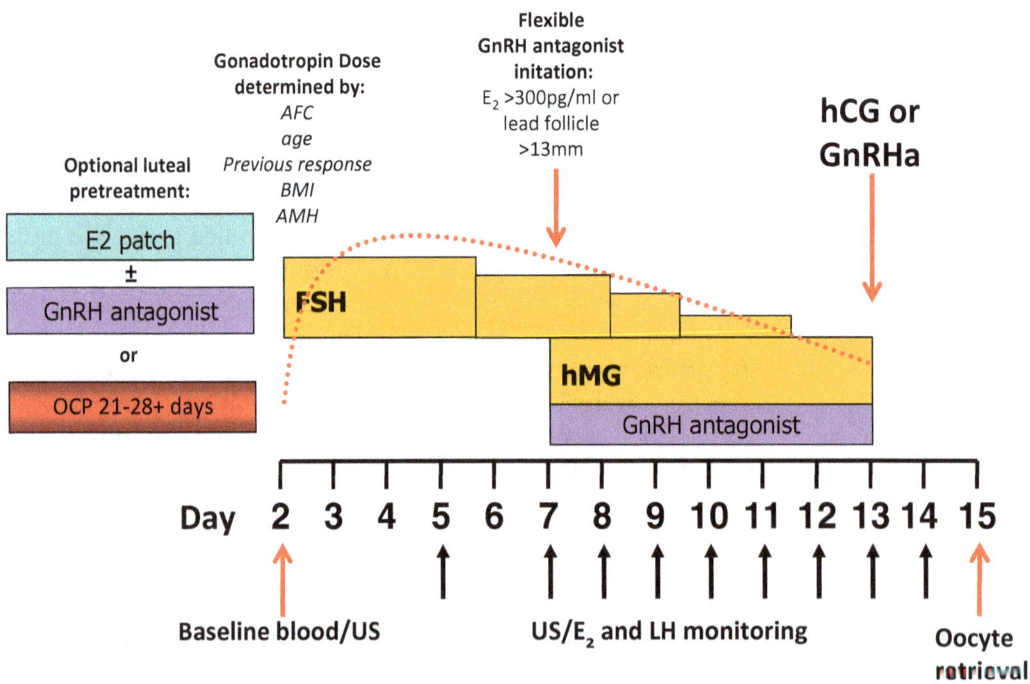

Fig. 1 Ovarian stimulation with GNRH antagonist down-regulation: flexible start

approach, and less gonadotropins were consumed during stimulation. Fixed initiation on day 6 was associated with fewer oocytes retrieved (6 vs. 11); however, pregnancy rates were not examined given the limited number of patients enrolled. The following year, Kolibianakis et al. published their pseudo-randomized (by date of birth) trial of fixed (day 6) versus flexible (lead follicle ≥15 mm) GnRH antagonist administration in 111 patients, which revealed no overall differences in implantation or pregnancy rates between the two groups [25]. Four patients in the flexible protocol experienced premature rise in LH, as opposed to only 1 patient in the fixed arm.

Shortly thereafter, Escudero et al. reported their comparison of a fixed (day 6) versus flexible (≥14 mm lead follicle) protocol in 109 patients ≤35 years old [26]. Serum E2 levels were higher for patients on the flexible protocol, and these patients required fewer GnRH-antagonist injections. No other differences, however, were noted between groups. While the authors noted that poor responder patients were more likely to be cancelled on the fixed protocol for lack of response, no premature LH surges occurred on either protocol. Lastly, the Dutch Ganirelix Study Group published their multicenter randomized control trial comparing fixed (day 6) versus flexible (≥15 mm lead follicle) GnRH antagonist use in 205 IVF patients [27]. Their study was powered to detect a difference of 2 oocytes retrieved between groups as opposed to any difference in pregnancy outcomes. Premature luteinization was encountered in 1 % of fixed protocol cycles versus 2 % of flexible protocols. No differences were detected in number of oocytes retrieved, embryos available for transfer, or ongoing pregnancy rates.

The four aforementioned RCTs were examined in a subsequent meta-analysis in 2005 [28]. The flexible protocol reduced the number of antagonist ampules used (OR –1.2, 95 % CI –1.26 to –1.15) and the amount of gonadotropins needed for stimulation (OR 95.5 IU, 95 % CI 74.8–116.1). No differences were noted, however, in pregnancy rate per woman randomized (OR 0.7, 95 % CI 0.45–1.1).

Although none of the above-noted four trials reported a significant increase in premature LH surges with the flexible protocol, it is likely that initiation of the antagonist in flexible protocols should occur earlier than the 15 mm threshold employed in these studies. Citing the high rate of surge prior to the initiation of the antagonist in the published literature, Kolibianakis et al. modified the flexible start criteria to a mean follicular diameter of ≥12 mm or serum E2 ≥150 pg/mL [29]. In a randomized controlled trial of 146 IVF patients, they demonstrated that this flexible protocol required an average of 2 additional ampules of the antagonist, while resulting in equal ongoing clinical pregnancy rates.

In our own practice, which is characterized by a high number of poor responders, we employ a flexible protocol wherein the GnRH antagonist is administered when the lead follicle mean diameter is ≥13 mm or the serum estradiol exceeds 300 pg/mL, whichever occurs first. We have found this practice effectively prevents premature rises in LH while at the same time maximizes response and decreases unnecessary consumption of GnRH antagonist doses.

2.3 Programming Cycle Starts: Luteal Pretreatment Strategies

There are several physiologic and logistic arguments potentially favoring the programming of cycle starts with luteal pretreatment in GnRH antagonist cycles. The hormonal suppression of follicle-stimulating hormone (FSH) in the late luteal phase theoretically ensures that a more uniform cohort of follicles will be responsive to gonadotropin stimulation, and potentially maximizes the number of oocytes retrieved in patients with diminished ovarian reserve [30, 31]; moreover, luteal pretreatment allows for additional flexibility for programs needing to control the number of patients starting stimulation on any given day, thereby ensuring that patient load stays within manageable and safe parameters. Such pretreatment can be performed either with oral contraceptive pills, estradiol alone, GnRH antagonists, or a combination of the latter two options.

When oral contraceptive pills are employed, patients are typically instructed to begin OCPs within the first 5 days of menses, and continue use for between 14 and 28 days. While pretreatment with OCPs can aid in scheduling cycle starts both for patients and for centers, their use has been called into question based on recent data that have suggested a potential decrease in pregnancy rates when they are employed [32]. Several randomized controlled trials have been performed to evaluate the effect of oral contraceptive pill pretreatment on IVF outcomes [33–37]. A meta-analysis revealed a reduced chance of ongoing pregnancy (RR 0.80, 95 % CI 0.66–0.97, with an absolute difference rate of 5 %), increased time required for stimulation (RR 1.33, 95 % CI 0.61–2.05), as well as increased gonadotropin consumption (RR 360.3, 95 % CI 157.5–563.1) when OCPs are used [32]. The authors concluded that for every 20 patients pretreated with oral contraceptives, one would not conceive as a result of OCP use. The mechanism for this supposed reduction in pregnancies remains unclear, but it has been posited that lower levels of luteinizing hormone following oral contraceptive treatment could play a role [32]. Of note, all of the studies described above used rFSH in isolation without additional LH supplementation. In contrast, a large retrospective study examining pregnancy rates in 604 patients pretreated with OCPs versus 1,195 patients who started stimulation on day 2 of a natural cycle found no difference in implantation or clinical pregnancy rates [38].

Luteal estradiol treatment is another alternative for pretreatment prior to stimulation. Estradiol can be administered either orally or as a transdermal patch; patients place the first 0.1 mg estradiol patch 9–10 days following the previous cycle's LH surge, and change the patch every 48 h until menses occurs, and baseline ultrasonography and blood work are performed thereafter. Similar to oral contraceptive pill use, estradiol pretreatment does increase the length of stimulation and gonadotropins used, but improves homogeneity of the follicular cohort and increases the number of follicles reaching coordinated maturity [39]. Alternatively, a single 3 mg dose of Cetrotide on day 11 following the LH surge has also been suggested [40], as well as a pretreatment regimen combining these two modalities [31]. Such protocols have demonstrated clinical utility in the treatment of poor responder patients [31, 41, 42].

No prospective study to date has been adequately powered to establish whether pretreatment with luteal estradiol in antagonist cycles improves pregnancy rates when compared to cycles where no estradiol pretreatment is utilized. The flexible use of luteal estradiol has, however, been used successfully in IVF programs to avoid weekend retrievals. Blockeel et al. randomized 86 patients to either conventional day 2 start with no pretreatment versus pretreatment with estradiol valerate (4 mg daily) starting from cycle day 25 for 6–10 days, with the intent to avoid weekend retrievals [43]. Using this protocol, only 2.7 % of patients underwent weekend retrieval, as opposed to 20.5 % of controls, with similar clinical pregnancy rates between groups (38.1 % vs. 38.6 %).

In our clinical practice, we find that starting stimulation on day 2 without pretreatment is convenient, effective, and the simplest logistic approach for the majority of patients. We reserve estradiol pretreatment with or without concurrent GnRH antagonist administration for those poor responders who are likely to have significant luteal elevations in FSH, or in those patients who tend to develop a dominant follicle despite normal ovarian reserve. Oral contraceptive pill pretreatment is useful for coordinating the starts of those patients who are coming from afar, albeit with the caveat that the literature suggests that such treatments may slightly diminish the likelihood of success; we have not observed this decrease in our own experience, albeit we routinely supplement FSH with human menopausal gonadotropins upon administration of GnRH antagonists.

2.4 Gonadotropin Stimulation

Gonadotropin stimulation is most commonly started on day 2 of the menstrual cycle in an antagonist-based protocol. While the nuances of selecting the gonadotropin starting dose are beyond the scope of this manuscript, several studies have sought to standardize approaches to dosing [44–46]. A balance exists between maximizing the number of oocytes retrieved while minimizing the risk of ovarian hyperstimulation syndrome; striking this balance is paramount for the safety of the patient. As an alternative to starting stimulation on day 2, day 3 starts for programs seeking to avoid weekend retrievals have recently been evaluated, and revealed no difference in the mean number of oocytes retrieved or the rate of ongoing pregnancies [47]. The authors thus proposed starting on day 2 from Friday to Tuesday, and starting on day 3 starts from Wednesday to Thursday, an approach which minimized retrieval over weekends. While an interesting logistic approach for programs working less than a 7 day work-week, the results of this trial should be carefully interpreted, as this was a retrospective analysis of a previously performed RCT which was not designed to examine day of start, nor powered for analyzing differences in pregnancy rates for this intervention [48]. Until additional data is available, we would prefer to administer a 0.25 mg dose of GnRH antagonist for patients requiring 1- or 2-day delay in cycle start.

It is not uncommon to see a plateau in serum estradiol the day following the first GnRH antagonist injection. Two studies to date have evaluated the practice of increasing the gonadotropin dose by 75 international units concurrent with the addition of the GnRH antagonist [49, 50]. Those studies found no improvement in IVF cycle parameters or success rates as compared to controls who maintained their original dose. Although several programs routinely increase the dose of gonadotropins when the antagonist is added, the literature and our experience do not support a need for this practice [51].

LH supplementation's benefit to ovarian stimulation for IVF remains controversial [52, 53]. Given that initiation of the GnRH antagonist is associated with a drop in mean serum luteinizing hormone levels, and given that LH plays a significant role in ovarian hormone production, it remains our practice to supplement with a daily dose of human menopausal gonadotropin concurrent with the initiation of the antagonist. For instance, a patient on 150 IU of rFSH is converted to either 75 IU of hMG plus 75 IU of rFSH, or simply changed to 150 IU of hMG. A recent Cochrane review supports this practice, showing a subtle but statistically significant reduction in live-birth rates (OR 0.85, 95 % CI 0.74–0.99) when hMG is not used for stimulation [54]. Only one of the studies included in this meta-analysis, however, employed a GnRH antagonist-based protocol for down-regulation [55]. We do not recommend recombinant luteinizing hormone over hMG, given the expense of this medication and the lack of available data demonstrating superiority over pure FSH preparations alone [56–61]. It should be noted, however, that the addition of low dose hCG to pure or recombinant FSH has shown promise as an exogenous source of LH, and can be considered as an alternative to highly purified hMG.

2.5 Triggering of Oocyte Maturation

The timing of hCG administration for the triggering of final oocyte maturation remains variable among IVF centers, with some programs triggering at a lead average follicle diameter of 17 mm and others waiting for 20 mm or greater. To date, only a small body of literature exists regarding the optimal timing of hCG in antagonist cycles. In 2004, Kolibianakis et al. published the results of a randomized controlled trial examining trigger timing in 413 IVF patients [62]. Patients were randomized to administration of hCG as soon as ≥3 follicles ≥17 mm were present on ultrasound, versus 2 days after these criteria were met. A 10.6 % reduction in ongoing pregnancy (35.6 % vs. 25.0 %, $p=0.027$) was observed when patients were stimulated for an additional 2 days; no difference in number of retrieved oocytes, number of zygotes, or embryo quality was noted (oocyte maturity was not specifically evaluated). A follow-up study by the same group applied the above dichotomous triggering criteria in ten donors, who subsequently underwent endometrial biopsy on the day of oocyte retrieval; prolongation of the follicular phase via delayed triggering resulted in endometrial advancement (median = 3 days) in all samples, with no secretory changes observed with earlier triggering [63].

Subsequently, a retrospective analysis compared "ideal" triggering at 17 mm lead follicular size ($n=798$) versus delay ($n=472$) or advancement ($n=372$) of retrieval by 1 day to avoid retrievals occurring on a Saturday or Sunday. No difference in implantation or live-birth rates was observed between these three groups [64]. A follow-up randomized controlled trial by Devroey's group examined triggering at an earlier time-point (≥3 follicles ≥16 mm) versus 1 day later, and demonstrated a reduction in the number of mature oocytes retrieved (6.1 vs. 9.2, $p=.009$) with no difference in live-birth rates (34.6 % vs. 40.7 %) with earlier triggering [65]. Lastly, a retrospective analysis of the Engage randomized control trial compared trigger when ≥3 follicles reached ≥17 mm ($n=524$) versus a 1 day delay ($n=201$), and revealed no difference in ongoing pregnancy rates (37.8 % vs. 41.8 %) for trigger at 17 mm versus 1 day later, respectively [47]. Of note, this trial was not designed to specifically evaluate the timing of the ovulatory trigger.

Given our own clinical experience, we feel strongly that the timing of hCG administration is crucial for IVF success. We find that an individualized approach which tailors the timing of hCG to the patient's previous IVF history and yield of mature oocytes in past cycles maximizes clinical outcomes, and reduces the retrieval of oocytes which are post-mature. Our general approach is to trigger ovulation when ≥2 follicles have reached ≥17 mm on average diameter, and to change these criteria in future cycles either with earlier or later time-points depending on oocyte maturity and embryo quality observed in the previous cycles.

2.6 Use of GnRH Agonist Trigger

A distinct advantage of GnRH antagonist-based cycles is the ability to administer a GnRH agonist for the triggering of final oocyte maturation. The sustained luteotropic effect of exogenously administered human chorionic gonadotropin is known to exacerbate the risk of ovarian hyperstimulation syndrome; in contrast, use of a GnRH agonist engenders a comparatively brief release of endogenous luteinizing hormone and FSH from the pituitary [66]. Our protocol is to administer a 2 mg subcutaneous dose of leuprolide acetate together with a reduced dose of hCG (1,500 IU) in those rare cycles where estradiol levels exceed 3,000 pg/mL; while specific criteria to gauge the success of this trigger prior to retrieval have yet to be conclusively determined, an LH level of >15 mIU/mL the morning after trigger generally predicts the retrieval of mature oocytes [67–69].

Although it should not be viewed as being completely reliable or a fail-safe approach, the use of a GnRH agonist trigger is associated with a marked reduction in the risk of OHSS [70–75]. Engmann et al. randomized 65 high-responder patients to either conventional hCG triggering in a GnRH agonist down-regulated cycle versus GnRH agonist trigger with aggressive luteal support in a GnRH antagonist cycle, and reported a 31.0 % versus 0 % risk of OHSS, respectively. Humaidan et al. subsequently randomized 302 IVF/ICSI patients undergoing antagonist down-regulated stimulation to either conventional trigger (10,000 IU hCG) versus GnRH agonist triggering (0.5 mg buserelin with 1,500 IU hCG for luteal support), and reported no cases of OHSS with the latter intervention (as compared with 2 % after conventional trigger) [75]. Given the impressive reduction of OHSS associated with GnRH agonist triggering, it is not surprising that it has been readily adopted by many programs as the principal modality for triggering oocyte maturation in donors [76–79].

The premature luteolysis associated with GnRH agonist triggers prevented its early widespread adoption. Whereas the endogenous LH surge in the natural cycle lasts approximately 48 h, GnRH agonist triggering evokes a shortened surge lasting only 24–36 h, leading to premature luteolysis [66, 80, 81]. Indeed, implantation rates are markedly reduced if estrogen is not sufficiently supplemented until conventional luteal-placental shift occurs at approximately 7 weeks gestation [82]. Several different approaches have been developed to overcome the luteolytic effects of the GnRH agonist trigger [83–85]. Approaches include aggressive support with both estrogen and progesterone until at least 7 weeks gestation, supplementation at the time of trigger with low dose (1,000–1,500 IU) hCG, or else segmental treatment using a freeze-thaw approach [68, 86–88]. The lowering of ongoing pregnancy rates noted in the initial studies of GnRH agonist trigger has now been mitigated when such interventions are employed to appropriately address the luteal phase [89]. In general, our preference is for a hybrid approach with 1,500 hCG co-administered at the time of trigger, together with aggressive estradiol and progesterone luteal support commencing CD 15 and continuing until luteal placental shift is noted and adequate estrogen and progesterone is being produced by the fetal-placental unit.

3 Conclusion

GnRH antagonist-based IVF cycles have become a mainstay of modern assisted reproductive technology, eclipsing GnRH agonist down-regulation as the principal approach for pituitary suppression during ovarian stimulation. As an increasing number of fertility programs adopt this framework for stimulation, it is our hope that fewer cases of ovarian hyperstimulation syndrome will be encountered, patients will better tolerate stimulation, and pregnancy rates will continue to be maximized or optimized.

References

1. Huirne JA, Lambalk CB (2001) Gonadotropin-releasing-hormone-receptor antagonists. Lancet 358:1793–1803
2. Borm G, Mannaerts B (2000) Treatment with the gonadotrophin-releasing hormone antagonist ganirelix in women undergoing ovarian stimulation with recombinant follicle stimulating hormone is effective, safe and convenient: results of a controlled, randomized, multicentre trial. The European Orgalutran Study Group. Hum Reprod 15:1490–1498
3. Matikainen T, Ding YQ, Vergara M, Huhtaniemi I, Couzinet B, Schaison G (1992) Differing responses of plasma bioactive and immunoreactive follicle-stimulating hormone and luteinizing hormone to gonadotropin-releasing hormone antagonist and agonist treatments in postmenopausal women. J Clin Endocrinol Metab 75:820–825
4. Ludwig M, Felberbaum RE, Devroey P, Albano C, Riethmuller-Winzen H, Schuler A et al (2000) Significant reduction of the incidence of ovarian hyperstimulation syndrome (OHSS) by using the LHRH antagonist Cetrorelix (Cetrotide) in controlled ovarian stimulation for assisted reproduction. Arch Gynecol Obstet 264:29–32
5. Felberbaum RE, Albano C, Ludwig M, Riethmuller-Winzen H, Grigat M, Devroey P et al (2000) Ovarian stimulation for assisted reproduction with HMG and concomitant midcycle administration of the GnRH antagonist cetrorelix according to the multiple dose protocol: a prospective uncontrolled phase III study. Hum Reprod 15:1015–1020
6. Itskovitz-Eldor J, Kol S, Mannaerts B (2000) Use of a single bolus of GnRH agonist triptorelin to trigger ovulation after GnRH antagonist ganirelix treatment in women undergoing ovarian stimulation for assisted reproduction, with special reference to the prevention of ovarian hyperstimulation syndrome: preliminary report: short communication. Hum Reprod 15:1965–1968
7. Phillips A, Hahn DW, McGuire JL, Ritchie D, Capetola RJ, Bowers C et al (1988) Evaluation of the anaphylactoid activity of a new LHRH antagonist. Life Sci 43:883–888
8. Hahn DW, McGuire JL, Vale WW, Rivier J (1985) Reproductive/endocrine and anaphylactoid properties of an LHRH-antagonist, ORF 18260 [Ac-DNAL1(2), 4FDPhe2, D-Trp3, D-Arg6]-GnRH. Life Sci 37: 505–514
9. Al-Inany HG, Abou-Setta AM, Aboulghar M (2006) Gonadotrophin-releasing hormone antagonists for assisted conception. Cochrane Database Syst Rev (3):CD001750
10. Al-Inany HG, Youssef MA, Aboulghar M, Broekmans F, Sterrenburg M, Smit J et al (2011) Gonadotrophin-releasing hormone antagonists for assisted reproductive technol-

ogy. Cochrane Database Syst Rev (5): CD001750

11. De la Cruz A, Coy DH, Vilchez-Martinez JA, Arimura A, Schelly AV (1976) Blockage of ovulation in rats by inhibitory analogs of luteinizing hormones-releasing hormone. Science 191:195–197
12. Vilchez-Martinez JA, Schally AV, Coy DH, Coy EJ, Debeljuk L, Arimura A (1974) In vitro inhibition of LH release by a synthetic antagonist of LH-releasing hormone (LH-RH). Endocrinology 95:213–218
13. Cetel NS, Rivier J, Vale W, Yen SS (1983) The dynamics of gonadotropin inhibition in women induced by an antagonistic analog of gonadotropin-releasing hormone. J Clin Endocrinol Metab 57:62–65
14. Schmidt F, Sundaram K, Thau RB, Bardin CW (1984) [Ac-D-NAL(2)1,4FD-Phe2, D-Trp3, D-Arg6]-LHRH, a potent antagonist of LHRH, produces transient edema and behavioral changes in rats. Contraception 29:283–289
15. Bajusz S, Csernus VJ, Janaky T, Bokser L, Fekete M, Schally AV (1988) New antagonists of LHRH. II. Inhibition and potentiation of LHRH by closely related analogues. Int J Pept Protein Res 32:425–435
16. Albano C, Smitz J, Camus M, Riethmuller-Winzen H, Van Steirteghem A, Devroey P (1997) Comparison of different doses of gonadotropin-releasing hormone antagonist Cetrorelix during controlled ovarian hyperstimulation. Fertil Steril 67:917–922
17. (1998) A double-blind, randomized, dose-finding study to assess the efficacy of the gonadotrophin-releasing hormone antagonist ganirelix (Org 37462) to prevent premature luteinizing hormone surges in women undergoing ovarian stimulation with recombinant follicle stimulating hormone (Puregon). The ganirelix dose-finding study group. Hum Reprod 13:3023–3031
18. Dovey S, McIntyre K, Jacobson D, Catov J, Wakim A (2011) Is a premature rise in luteinizing hormone in the absence of increased progesterone levels detrimental to pregnancy outcome in GnRH antagonist in vitro fertilization cycles. Fertil Steril 96:585–589
19. Olivennes F, Alvarez S, Bouchard P, Fanchin R, Salat-Baroux J, Frydman R (1998) The use of a GnRH antagonist (Cetrorelix) in a single dose protocol in IVF-embryo transfer: a dose finding study of 3 versus 2 mg. Hum Reprod 13:2411–2414
20. Olivennes F, Diedrich K, Frydman R, Felberbaum RE, Howles CM (2003) Safety and efficacy of a 3 mg dose of the GnRH antagonist cetrorelix in preventing premature LH surges: report of two large multicentre, multinational, phase IIIb clinical experiences. Reprod Biomed Online 6:432–438
21. Olivennes F, Belaisch-Allart J, Emperaire JC, Dechaud H, Alvarez S, Moreau L et al (2000) Prospective, randomized, controlled study of in vitro fertilization-embryo transfer with a single dose of a luteinizing hormone-releasing hormone (LH-RH) antagonist (cetrorelix) or a depot formula of an LH-RH agonist (triptorelin). Fertil Steril 73:314–320
22. Wilcox J, Potter D, Moore M, Ferrande L, Kelly E (2005) Prospective, randomized trial comparing cetrorelix acetate and ganirelix acetate in a programmed, flexible protocol for premature luteinizing hormone surge prevention in assisted reproductive technologies. Fertil Steril 84:108–117
23. Lee TH, Wu MY, Chen HF, Chen MJ, Ho HN, Yang YS (2005) Ovarian response and follicular development for single-dose and multiple-dose protocols for gonadotropin-releasing hormone antagonist administration. Fertil Steril 83:1700–1707
24. Ludwig M, Katalinic A, Banz C, Schroder AK, Loning M, Weiss JM et al (2002) Tailoring the GnRH antagonist cetrorelix acetate to individual patients' needs in ovarian stimulation for IVF: results of a prospective, randomized study. Hum Reprod 17:2842–2845
25. Kolibianakis EM, Albano C, Kahn J, Camus M, Tournaye H, Van Steirteghem AC et al (2003) Exposure to high levels of luteinizing hormone and estradiol in the early follicular phase of gonadotropin-releasing hormone antagonist cycles is associated with a reduced chance of pregnancy. Fertil Steril 79:873–880
26. Escudero E, Bosch E, Crespo J, Simon C, Remohi J, Pellicer A (2004) Comparison of two different starting multiple dose gonadotropin-releasing hormone antagonist protocols in a selected group of in vitro fertilization-embryo transfer patients. Fertil Steril 81:562–566
27. Mochtar MH (2004) The effect of an individualized GnRH antagonist protocol on folliculogenesis in IVF/ICSI. Hum Reprod 19:1713–1718
28. Al-Inany H, Aboulghar MA, Mansour RT, Serour GI (2005) Optimizing GnRH antagonist administration: meta-analysis of fixed versus flexible protocol. Reprod Biomed Online 10:567–570
29. Kolibianakis EM, Venetis CA, Kalogeropoulou L, Papanikolaou E, Tarlatzis BC (2011) Fixed versus flexible gonadotropin-releasing hormone antagonist administration in in vitro fertilization: a randomized controlled trial. Fertil Steril 95:558–562

30. de Ziegler D, Jaaskelainen AS, Brioschi PA, Fanchin R, Bulletti C (1998) Synchronization of endogenous and exogenous FSH stimuli in controlled ovarian hyperstimulation (COH). Hum Reprod 13:561–564
31. Dragisic KG, Davis OK, Fasouliotis SJ, Rosenwaks Z (2005) Use of a luteal estradiol patch and a gonadotropin-releasing hormone antagonist suppression protocol before gonadotropin stimulation for in vitro fertilization in poor responders. Fertil Steril 84:1023–1026
32. Griesinger G, Kolibianakis EM, Venetis C, Diedrich K, Tarlatzis B (2010) Oral contraceptive pretreatment significantly reduces ongoing pregnancy likelihood in gonadotropin-releasing hormone antagonist cycles: an updated meta-analysis. Fertil Steril 94:2382–2384
33. Kim CH, Jeon GH, Cheon YP, Jeon I, Kim SH, Chae HD et al (2009) Comparison of GnRH antagonist protocol with or without oral contraceptive pill pretreatment and GnRH agonist low-dose long protocol in low responders undergoing IVF/intracytoplasmic sperm injection. Fertil Steril 92:1758–1760
34. Cedrin-Durnerin I, Bstandig B, Parneix I, Bied-Damon V, Avril C, Decanter C et al (2007) Effects of oral contraceptive, synthetic progestogen or natural estrogen pre-treatments on the hormonal profile and the antral follicle cohort before GnRH antagonist protocol. Hum Reprod 22:109–116
35. Kolibianakis EM, Papanikolaou EG, Camus M, Tournaye H, Van Steirteghem AC, Devroey P (2006) Effect of oral contraceptive pill pretreatment on ongoing pregnancy rates in patients stimulated with GnRH antagonists and recombinant FSH for IVF. A randomized controlled trial. Hum Reprod 21:352–357
36. Huirne JA, van Loenen AC, Donnez J, Pirard C, Homburg R, Schats R et al (2006) Effect of an oral contraceptive pill on follicular development in IVF/ICSI patients receiving a GnRH antagonist: a randomized study. Reprod Biomed Online 13:235–245
37. Rombauts L, Healy D, Norman RJ (2006) A comparative randomized trial to assess the impact of oral contraceptive pretreatment on follicular growth and hormone profiles in GnRH antagonist-treated patients. Hum Reprod 21:95–103
38. Pinkas H, Sapir O, Avrech OM, Ben-Haroush A, Ashkenzi J, Fisch B et al (2008) The effect of oral contraceptive pill for cycle scheduling prior to GnRH-antagonist protocol on IVF cycle parameters and pregnancy outcome. J Assist Reprod Genet 25:29–33
39. Fanchin R, Salomon L, Castelo-Branco A, Olivennes F, Frydman N, Frydman R (2003) Luteal estradiol pre-treatment coordinates follicular growth during controlled ovarian hyperstimulation with GnRH antagonists. Hum Reprod 18:2698–2703
40. Fanchin R, Castelo Branco A, Kadoch IJ, Hosny G, Bagirova M, Frydman R (2004) Premenstrual administration of gonadotropin-releasing hormone antagonist coordinates early antral follicle sizes and sets up the basis for an innovative concept of controlled ovarian hyperstimulation. Fertil Steril 81:1554–1559
41. Weitzman VN, Engmann L, DiLuigi A, Maier D, Nulsen J, Benadiva C (2009) Comparison of luteal estradiol patch and gonadotropin-releasing hormone antagonist suppression protocol before gonadotropin stimulation versus microdose gonadotropin-releasing hormone agonist protocol for patients with a history of poor in vitro fertilization outcomes. Fertil Steril 92:226–230
42. Hill MJ, McWilliams GD, Miller KA, Scott RT Jr, Frattarelli JL (2009) A luteal estradiol protocol for anticipated poor-responder patients may improve delivery rates. Fertil Steril 91:739–743
43. Blockeel C, Engels S, De Vos M, Haentjens P, Polyzos NP, Stoop D et al (2012) Oestradiol valerate pretreatment in GnRH-antagonist cycles: a randomized controlled trial. Reprod Biomed Online 24:272–280
44. Popovic-Todorovic B, Loft A, Lindhard A, Bangsboll S, Andersson AM, Andersen AN (2003) A prospective study of predictive factors of ovarian response in 'standard' IVF/ICSI patients treated with recombinant FSH. A suggestion for a recombinant FSH dosage normogram. Hum Reprod 18:781–787
45. Olivennes F, Howles CM, Borini A, Germond M, Trew G, Wikland M et al (2009) Individualizing FSH dose for assisted reproduction using a novel algorithm: the CONSORT study. Reprod Biomed Online 18:195–204
46. Yovich J, Stanger J, Hinchliffe P (2012) Targeted gonadotrophin stimulation using the PIVEt algorithm markedly reduces the risk of OHSS. Reprod Biomed Online 24:281–292
47. Levy MJ, Ledger W, Kolibianakis EM, Ijzerman-Boon PC, Gordon K (2013) Is it possible to reduce the incidence of weekend oocyte retrievals in GnRH antagonist protocols? Reprod Biomed Online 26:50–58
48. Devroey P, Boostanfar R, Koper NP, Mannaerts BM, Ijzerman-Boon PC, Fauser BC (2009) A double-blind, non-inferiority RCT comparing

corifollitropin alfa and recombinant FSH during the first seven days of ovarian stimulation using a GnRH antagonist protocol. Hum Reprod 24:3063–3072
49. Aboulghar MA, Mansour RT, Serour GI, Al-Inany HG, Amin YM, Aboulghar MM (2004) Increasing the dose of human menopausal gonadotrophins on day of GnRH antagonist administration: randomized controlled trial. Reprod Biomed Online 8:524–527
50. Propst AM, Bates GW, Robinson RD, Arthur NJ, Martin JE, Neal GS (2006) A randomized controlled trial of increasing recombinant follicle-stimulating hormone after initiating a gonadotropin-releasing hormone antagonist for in vitro fertilization-embryo transfer. Fertil Steril 86:58–63
51. Shapiro DB, Mitchell-Leef D, Carter M, Nagy ZP (2005) Ganirelix acetate use in normal- and poor-prognosis patients and the impact of estradiol patterns. Fertil Steril 83:666–670
52. Martin-Johnston M, Beltsos AN, Grotjan HE, Uhler ML (2007) Adding human menopausal gonadotrophin to antagonist protocols—is there a benefit? Reprod Biomed Online 15:161–168
53. Al-Inany HG, Abou-Setta AM, Aboulghar MA, Mansour RT, Serour GI (2009) Highly purified hMG achieves better pregnancy rates in IVF cycles but not ICSI cycles compared with recombinant FSH: a meta-analysis. Gynecol Endocrinol 25:372–378
54. van Wely M, Kwan I, Burt AL, Thomas J, Vail A, Van der Veen F et al (2011) Recombinant versus urinary gonadotrophin for ovarian stimulation in assisted reproductive technology cycles. Cochrane Database Syst Rev (2): CD005354
55. Bosch E, Vidal C, Labarta E, Simon C, Remohi J, Pellicer A (2008) Highly purified hMG versus recombinant FSH in ovarian hyperstimulation with GnRH antagonists—a randomized study. Hum Reprod 23:2346–2351
56. Kolibianakis EM, Kalogeropoulou L, Griesinger G, Papanikolaou EG, Papadimas J, Bontis J et al (2007) Among patients treated with FSH and GnRH analogues for in vitro fertilization, is the addition of recombinant LH associated with the probability of live birth? A systematic review and meta-analysis. Hum Reprod Update 13:445–452
57. Cedrin-Durnerin I, Grange-Dujardin D, Laffy A, Parneix I, Massin N, Galey J et al (2004) Recombinant human LH supplementation during GnRH antagonist administration in IVF/ICSI cycles: a prospective randomized study. Hum Reprod 19:1979–1984
58. Griesinger G, Schultze-Mosgau A, Dafopoulos K, Schroeder A, Schroer A, von Otte S et al (2005) Recombinant luteinizing hormone supplementation to recombinant follicle-stimulating hormone induced ovarian hyperstimulation in the GnRH-antagonist multiple-dose protocol. Hum Reprod 20: 1200–1206
59. Levi-Setti PE, Cavagna M, Bulletti C (2006) Recombinant gonadotrophins associated with GnRH antagonist (cetrorelix) in ovarian stimulation for ICSI: comparison of r-FSH alone and in combination with r-LH. Eur J Obstet Gynecol Reprod Biol 126:212–216
60. Sauer MV, Thornton MH 2nd, Schoolcraft W, Frishman GN (2004) Comparative efficacy and safety of cetrorelix with or without midcycle recombinant LH and leuprolide acetate for inhibition of premature LH surges in assisted reproduction. Reprod Biomed Online 9:487–493
61. Mochtar MH, Van der V, Ziech M, van Wely M (2007) Recombinant luteinizing hormone (rLH) for controlled ovarian hyperstimulation in assisted reproductive cycles. Cochrane Database Syst Rev (2):CD005070
62. Kolibianakis EM, Albano C, Camus M, Tournaye H, Van Steirteghem AC, Devroey P (2004) Prolongation of the follicular phase in in vitro fertilization results in a lower ongoing pregnancy rate in cycles stimulated with recombinant follicle-stimulating hormone and gonadotropin-releasing hormone antagonists. Fertil Steril 82:102–107
63. Kolibianakis EM, Bourgain C, Papanikolaou EG, Camus M, Tournaye H, Van Steirteghem AC et al (2005) Prolongation of follicular phase by delaying hCG administration results in a higher incidence of endometrial advancement on the day of oocyte retrieval in GnRH antagonist cycles. Hum Reprod 20: 2453–2456
64. Tremellen KP, Lane M (2010) Avoidance of weekend oocyte retrievals during GnRH antagonist treatment by simple advancement or delay of hCG administration does not adversely affect IVF live birth outcomes. Hum Reprod 25:1219–1224
65. Kyrou D, Kolibianakis EM, Fatemi HM, Tarlatzis BC, Tournaye H, Devroey P (2011) Is earlier administration of human chorionic gonadotropin (hCG) associated with the probability of pregnancy in cycles stimulated with recombinant follicle-stimulating hormone and gonadotropin-releasing hormone (GnRH) antagonists? A prospective randomized trial. Fertil Steril 96:1112–1115

66. Itskovitz J, Boldes R, Levron J, Erlik Y, Kahana L, Brandes JM (1991) Induction of preovulatory luteinizing hormone surge and prevention of ovarian hyperstimulation syndrome by gonadotropin-releasing hormone agonist. Fertil Steril 56:213–220
67. Kummer NE, Feinn RS, Griffin DW, Nulsen JC, Benadiva CA, Engmann LL (2013) Predicting successful induction of oocyte maturation after gonadotropin-releasing hormone agonist (GnRHa) trigger. Hum Reprod 28:152–159
68. Griffin D, Benadiva C, Kummer N, Budinetz T, Nulsen J, Engmann L (2012) Dual trigger of oocyte maturation with gonadotropin-releasing hormone agonist and low-dose human chorionic gonadotropin to optimize live birth rates in high responders. Fertil Steril 97:1316–1320
69. Chen SL, Ye DS, Chen X, Yang XH, Zheng HY, Tang Y et al (2012) Circulating luteinizing hormone level after triggering oocyte maturation with GnRH agonist may predict oocyte yield in flexible GnRH antagonist protocol. Hum Reprod 27:1351–1356
70. Lewit N, Kol S, Manor D, Itskovitz-Eldor J (1996) Comparison of gonadotrophin-releasing hormone analogues and human chorionic gonadotrophin for the induction of ovulation and prevention of ovarian hyperstimulation syndrome: a case-control study. Hum Reprod 11:1399–1402
71. Imoedemhe DA, Chan RC, Sigue AB, Pacpaco EL, Olazo AB (1991) A new approach to the management of patients at risk of ovarian hyperstimulation in an in-vitro fertilization programme. Hum Reprod 6:1088–1091
72. Kol S (2004) Luteolysis induced by a gonadotropin-releasing hormone agonist is the key to prevention of ovarian hyperstimulation syndrome. Fertil Steril 81:1–5
73. Babayof R, Margalioth EJ, Huleihel M, Amash A, Zylber-Haran E, Gal M et al (2006) Serum inhibin A, VEGF and TNFalpha levels after triggering oocyte maturation with GnRH agonist compared with HCG in women with polycystic ovaries undergoing IVF treatment: a prospective randomized trial. Hum Reprod 21:1260–1265
74. Engmann L, DiLuigi A, Schmidt D, Nulsen J, Maier D, Benadiva C (2008) The use of gonadotropin-releasing hormone (GnRH) agonist to induce oocyte maturation after cotreatment with GnRH antagonist in high-risk patients undergoing in vitro fertilization prevents the risk of ovarian hyperstimulation syndrome: a prospective randomized controlled study. Fertil Steril 89:84–91
75. Humaidan P, Ejdrup Bredkjaer H, Westergaard LG, Yding Andersen C (2010) 1,500 IU human chorionic gonadotropin administered at oocyte retrieval rescues the luteal phase when gonadotropin-releasing hormone agonist is used for ovulation induction: a prospective, randomized, controlled study. Fertil Steril 93:847–854
76. Acevedo B, Gomez-Palomares JL, Ricciarelli E, Hernandez ER (2006) Triggering ovulation with gonadotropin-releasing hormone agonists does not compromise embryo implantation rates. Fertil Steril 86:1682–1687
77. Galindo A, Bodri D, Guillen JJ, Colodron M, Vernaeve V, Coll O (2009) Triggering with HCG or GnRH agonist in GnRH antagonist treated oocyte donation cycles: a randomised clinical trial. Gynecol Endocrinol 25:60–66
78. Melo M, Busso CE, Bellver J, Alama P, Garrido N, Meseguer M et al (2009) GnRH agonist versus recombinant HCG in an oocyte donation programme: a randomized, prospective, controlled, assessor-blind study. Reprod Biomed Online 19:486–492
79. Sismanoglu A, Tekin HI, Erden HF, Ciray NH, Ulug U, Bahceci M (2009) Ovulation triggering with GnRH agonist vs. hCG in the same egg donor population undergoing donor oocyte cycles with GnRH antagonist: a prospective randomized cross-over trial. J Assist Reprod Genet 26:251–256
80. Hoff JD, Quigley ME, Yen SS (1983) Hormonal dynamics at midcycle: a reevaluation. J Clin Endocrinol Metab 57:792–796
81. Gonen Y, Balakier H, Powell W, Casper RF (1990) Use of gonadotropin-releasing hormone agonist to trigger follicular maturation for in vitro fertilization. J Clin Endocrinol Metab 71:918–922
82. Humaidan P, Bredkjaer HE, Bungum L, Bungum M, Grondahl ML, Westergaard L et al (2005) GnRH agonist (buserelin) or hCG for ovulation induction in GnRH antagonist IVF/ICSI cycles: a prospective randomized study. Hum Reprod 20:1213–1220
83. Engmann L, Benadiva C (2012) Agonist trigger: what is the best approach? Agonist trigger with aggressive luteal support. Fertil Steril 97:531–533
84. Garcia-Velasco JA (2012) Agonist trigger: what is the best approach? Agonist trigger with vitrification of oocytes or embryos. Fertil Steril 97:527–528

85. Humaidan P (2012) Agonist trigger: what is the best approach? Agonist trigger and low dose hCG. Fertil Steril 97:529–530
86. Benadiva C, Engmann L (2012) Intensive luteal phase support after GnRH agonist trigger: it does help. Reprod Biomed Online 25:329–330
87. Devroey P, Polyzos NP, Blockeel C (2011) An OHSS-Free Clinic by segmentation of IVF treatment. Hum Reprod 26:2593–2597
88. Griesinger G, Schultz L, Bauer T, Broessner A, Frambach T, Kissler S (2011) Ovarian hyperstimulation syndrome prevention by gonadotropin-releasing hormone agonist triggering of final oocyte maturation in a gonadotropin-releasing hormone antagonist protocol in combination with a "freeze-all" strategy: a prospective multicentric study. Fertil Steril 95:2029–2033, 2033.e1
89. Humaidan P, Kol S, Papanikolaou EG (2011) GnRH agonist for triggering of final oocyte maturation: time for a change of practice? Hum Reprod Update 17:510–524

Chapter 14

Ovarian Stimulation for IVF: Mild Approaches

O. Hamdine, F.J. Broekmans, and B.C.J.M. Fauser

Abstract

In contrast to current approaches, the aim of mild stimulation is to develop safer and more patient-friendly protocols in which the risks of the treatment as a whole are minimized. Mild stimulation is defined as the method when exogenous gonadotropins are administered at lower doses, and/or for a shorter duration in GnRH antagonist co-treated cycles, or when oral compounds (antiestrogens, aromatase inhibitors) are used for ovarian stimulation for IVF, with the aim of limiting the number of oocytes obtained to fewer than eight. In this chapter we discuss the relevant physiology of follicle development, the development of milder stimulation protocols, the implications of mild stimulation, the current state of affairs, and future developments.

Key words Follicle, Mild ovarian stimulation, Clomiphene citrate, GnRH antagonist, GnRH agonist, Aromatase inhibitors, hCG/LH and GnRH antagonist co-treatment, rFSH/GnRH antagonist co-treatment, IVF, In vitro fertilization

1 Introduction

Ovarian stimulation has been an integral part of assisted reproductive treatment (ART) for the past 30 years. It has been applied with the aim of increasing the number of oocytes to compensate for inefficiencies in laboratory procedures, enabling the selection of one or more embryos for fresh transfer and the cryopreservation of surplus embryos [1].

Currently, a "long" gonadotropin-releasing hormone (GnRH) agonist protocol, either initiated during the luteal phase of the preceding cycle or after OC pretreatment, combined with high doses of exogenous follicle-stimulating hormone (FSH) remains the most frequently used stimulation protocol for IVF treatment [2]. The aim of this protocol is to generate between 8 and 15 oocytes [3]. This conventional approach is complex, expensive, time consuming and may give rise to considerable patient discomfort and risks for complications such as ovarian hyperstimulation syndrome (OHSS) [2]. Other negative effects are high emotional stress levels and substantial dropout rates [4].

Zev Rosenwaks and Paul M. Wassarman (eds.), *Human Fertility: Methods and Protocols*, Methods in Molecular Biology, vol. 1154, DOI 10.1007/978-1-4939-0659-8_14, © Springer Science+Business Media New York 2014

Another major complication related to IVF treatment is the risk of multiple pregnancies. Such pregnancies bear an increased risk for late miscarriage, preeclampsia, growth retardation, gestational diabetes, and preterm delivery. Subsequently, the risk of suffering major cerebral damage or development of long-term health sequelae in the offspring is substantially increased compared to singleton pregnancies. The occurrence of multiple pregnancies after IVF is directly related to the number of embryos transferred. One or more embryos are usually selected and transferred to compensate for the low implantation potential of human embryos [1]. Much like embryos after natural conception, a high rate of aneuploidy is encountered in embryos arising from IVF. This may provide an explanation for the failure to obtain high implantation rates in IVF [5]. Uncertainties remain regarding long-term health risks although studies published to date seem reassuring. So far, no causative association has been shown between ovarian stimulation with exogenous gonadotropins and increased risk of benign or malignant ovarian disease [2, 6–8].

Improved laboratory performance has reduced the need for large quantities of oocytes [6, 9]. Supportive evidence regarding a potentially negative effect of supraphysiological estradiol (E_2) levels on endometrial receptivity [10, 11], corpus luteum function [12, 13], oocyte and embryo quality [14–16] indicates that more modest responses to ovarian stimulation might have a beneficial effect upon the implantation potential of the embryo [4, 17].

2 Physiology of Follicle Development

Folliculogenesis commences when primordial follicles enter the growing pool of primary follicles. The initial growth of primordial follicles occurs continuously and presumably in a random fashion. This process is referred to as "primary" recruitment [18]. The size of the follicle pool is determined during fetal life. The germ cell number reaches its maximum of approximately seven million by 20 weeks of gestation. From this point on, the germ cell number falls dramatically [19]. Approximately 1,000 follicles per month are recruited from the resting pool at a fixed rate, accelerating beyond the age of 35, although a more gradual acceleration in this wastage process may also be plausible. The majority of these primordial follicles undergo atresia through apoptosis before reaching the antral stages (Fig. 1) [18].

Complete follicular development takes several months. Follicle-stimulating hormone (FSH) receptors become detectable on granulosa cells during early preantral follicle development. Early follicle development prior to the antral stages is considered to be independent of FSH. Only antral follicles may be considered as FSH responsive, and will constitute the continuous cohort from which

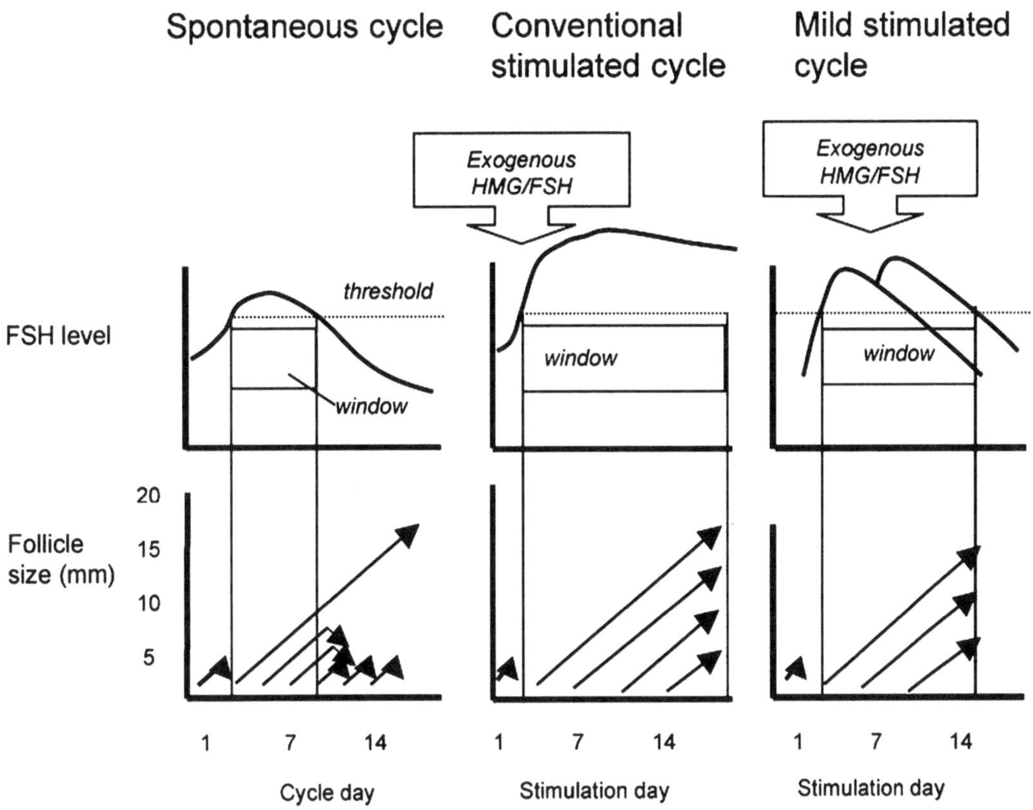

Fig. 1 Schematic representation of the FSH threshold and window concept in relation to monofollicular selection in a spontaneous cycle, multifollicular development in conventional ovarian hyperstimulation, and the interference with decreasing FSH levels by administrating exogenous FSH, resulting in multifollicular growth. (From: Macklon et al. Endocrine Reviews 2006)

they may grow into dominance under the influence of endogenous FSH in the follicular phase of the menstrual cycle. The number of follicles with augmented aromatase activity and subsequent E_2 biosynthesis is determined by the duration and magnitude of FSH stimulation [20]. High FSH levels during the luteo-follicular transition rescue a limited number of antral follicles from apoptosis due to the survival action of FSH. The number of follicles available for this cyclic recruitment is estimated to be about 10 per ovary, with a large age-dependent variation [18].

Although each growing follicle may initially have an equal potential to reach full maturation, only those antral follicles that are at a more advanced developmental stage during this intercycle rise in FSH gain gonadotropin dependence and continue to grow. These levels surpass the so-called threshold for ovarian stimulation. The threshold should be surpassed to ensure ongoing preovulatory follicle development [20]. This process is referred to as "secondary" recruitment [18].

In the subsequent follicular phase, FSH levels plateau and gradually decrease during the mid to late follicular phase due to the negative feedback of inhibin B and E_2. Decreasing follicular phase FSH levels effectively restrict the time where FSH levels remain above the threshold, also referred to as the "FSH window." It is crucial for the selection of a single dominant follicle [21–23]. Only the most mature follicle continues to grow due to increased sensitivity for FSH and enhanced LH receptor expression. As a result, the remaining antral follicles are deprived of adequate FSH stimulation and will therefore cease to mature and undergo atresia [18]. The FSH window concept emphasizes the importance of the duration of elevated FSH levels rather than the magnitude of FSH elevation [20].

Ovarian stimulation makes use of the concept that disruption of decrementing FSH levels leads to the development of multiple dominant follicles [4]. By using high doses of exogenous gonadotropins, FSH levels can be increased above the FSH threshold for an extended period of time. The process of single dominant follicle selection is overruled and all growing follicles reaching the stage of gonadotropin dependence will continue to grow allowing multifollicular development to occur [22].

3 Developing the Concept of Mild Stimulation

The first successful IVF treatment was performed in a natural cycle [24]. However, pregnancy rates are generally very low when natural cycles are used for IVF [25], caused by frequent inability to obtain the single oocyte and the inherent low implantation potential of human embryos in general, even if the natural selected oocyte is the source for this embryo [26].

3.1 GnRH Agonists

The introduction of GnRH agonists to prevent a premature LH rise, premature oocyte maturation, and luteinization had a very favorable impact on outcomes in IVF. GnRH agonists have high affinity for the GnRH receptor, and continuous occupation leads to desensitization due to the clustering and internalization of pituitary GnRH receptors [2]. The initial administration of GnRH agonists is associated with increased FSH and LH secretion, also referred to as the "flare" effect. Prolonged administration results in downregulation of pituitary GnRH receptors, eventually leading to the suppression of FSH and LH secretion [25]. Due to this "flare" effect, the use of a GnRH agonist long suppression protocol for ovarian stimulation is time consuming. Undesirable effects inherent to the use of GnRH agonists are the incidental formation of ovarian cysts due to the "flare" effect, complaints of estrogen deprivation and the need for increased amounts of exogenous gonadotropins due to ongoing suppression of endogenous gonadotropins [27].

Ovarian stimulation can only commence when pituitary downregulation has occurred which usually takes a period of 2 weeks. GnRH agonists are administered either in a long or short protocol [2, 3]. Pituitary recovery after cessation of GnRH agonist administration takes around 14 days, necessitating the application of luteal support in order to ensure sufficient progesterone exposure [28, 29]. Extensive evidence for the supremacy of the long suppression protocol has led to its widespread use in IVF [2].

3.2 GnRH Antagonists

In 1999 GnRH antagonists were introduced into IVF clinical practice to prevent a premature LH rise [30]. GnRH antagonists compete directly with endogenous GnRH by occupying the GnRH receptor and cause a rapid and immediate, reversible suppression of gonadotropin secretion [31]. The immediate suppression and recovery of pituitary function renders GnRH antagonists particularly suitable for short-term use in IVF [2, 27, 30]. In contrast to GnRH agonists, GnRH antagonists are not associated with an initial stimulation of gonadotropins. Therefore, cyst formation and the effects of hypo-estrogenemia are avoided [4, 30].

The initiation of FSH administration in a GnRH antagonist regimen is cycle phase dependent and is usually started on cycle day 2 or 3. GnRH antagonists may be administered at any time during the early or mid-follicular phase of a menstrual cycle to prevent a premature LH rise, although a fixed time moment may lead to a more favorable outcome [2, 32]. Several studies have been performed to determine the minimal effective dose and treatment schedule in IVF patients. Three general approaches have emerged. A single large dose can be administered in the late follicular phase on stimulation day 8 or 9. In the multiple-dose protocol, daily small doses are given from stimulation day 6 onward. Alternatively, in the flexible protocol, daily small doses could be initiated depending on the size of the dominant follicle or the estradiol level. The GnRH antagonist is continued until the day that human chorionic gonadotropin (hCG) to trigger final oocyte maturation is given [33–36].

The duration of treatment is shortened by 1–2 days and slightly fewer follicles are seen at the time of hCG injection compared with a GnRH agonist. Therefore, the number of retrieved oocytes tends to be lower [35]. Initial studies have shown significantly lower pregnancy rates following GnRH antagonist treatment in comparison to GnRH agonists [30, 37, 38]. However, two recent meta-analyses could not identify a significant difference in the live birth rate between the two types of analogues used for pituitary suppression [32, 39]. A possible explanation for the difference in cycle outcome could be the higher LH, E_2 and progesterone levels which are encountered during the early follicular phase in GnRH antagonists. In fixed day 6 GnRH antagonist cycles, ongoing pregnancy rates are significantly lower in patients with elevated progesterone levels at initiation of stimulation [40]. Moreover, there is a possible association between elevated LH levels 2 days after commencing a GnRH antagonist in a fixed day 6 protocol and lower ongoing pregnancy rates [41]. Earlier expression of progesterone receptors in the follicular phase and down-regulation of estrogen receptors by the exposure of supraphysiological steroid hormone levels may lead to earlier closure of the implantation window [42, 43]. Indeed, once endometrium is primed by estradiol, the duration of progesterone exposure plays an important role in endometrium receptivity [10, 35].

Treatment with GnRH antagonists was associated with a lower risk of complications such as OHSS [32, 39] and a lower burden of treatment [44]. The introduction of GnRH antagonist has allowed for the development of more patient-friendly protocols.

3.3 Natural Cycle IVF

Natural cycle IVF has been largely replaced by ovarian stimulation to improve success rates per started cycle. It is carried out in a spontaneous menstrual cycle without administration of any medication at any time during the cycle, aiming to retrieve a single oocyte at the lowest possible cost [3].

Natural cycle IVF is physically and emotionally less demanding, with extremely low risks for multiple pregnancies and no risks for OHSS. It is cheaper than IVF with controlled ovarian hyperstimulation (IVF-COH) but also less effective. The results are hampered by high cancellation rates due to premature LH surges, premature ovulation and reduced chances for successful oocyte retrieval. Consequently, ongoing pregnancy rates are unacceptably low (7.2 % per started cycle) [45]. However, to improve effectiveness, natural cycle IVF could be offered as a series of treatment cycles. After four cycles, the cumulative probability of pregnancy is 46 % with an associated live birth rate of 32 % [46]. Although this is comparable to a single IVF-COH cycle, the increased number of oocyte retrievals and the associated risks of infection and bleeding should be taken into account.

It has been suggested that patients with a poor response after IVF-COH might benefit from this approach [47]. However, there is no superior protocol that significantly improves success rates in poor responders [48].

3.4 Modified Natural Cycle IVF

The outcomes of natural cycle IVF can be improved by using a GnRH antagonist to prevent LH surges and premature ovulations. The aim of modified natural cycle IVF (MNC-IVF) is to collect a naturally single oocyte with a reduced chance of cycle cancellation. GnRH antagonists are commenced in the late follicular phase, after follicular dominance has developed. The addition of exogenous gonadotropins substitutes for the drop in FSH and E_2 levels, thereby supporting the ongoing growth of a dominant follicle (add-back therapy) [49, 50]. MNC-IVF has the same advantages as natural cycle IVF. Despite these advantages, low efficacy has restricted its widespread use [51, 52].

A large prospective cohort study analyzed the cumulative pregnancy rates after three cycles of MNC-IVF. The overall ongoing pregnancy rate per started cycle was 8.3 % and 20.8 % after three cycles [53]. After nine cycles the cumulative pregnancy rate increased up to 40.6 %. The pregnancy rate per cycle remained constant in higher cycle numbers, possibly due to selective dropout of poor prognosis patients. However, dropout rates were rather high (47.8 %), suggesting a substantial physical and emotional burden for patients [50]. Sequential treatment with MNC-IVF followed by IVF-COH did not compromise overall success rates, while twin pregnancy rates remained low. After sequential treatment the cumulative ongoing pregnancy rates and live birth rates were 51.5 % and 50.0 % [52].

Although MNC-IVF is a safe and patient-friendly treatment option, when choosing either MNC-IVF or IVF-COH, the advantages, disadvantages and the expected pregnancy rates of both treatments should be taken into consideration.

3.5 Clomiphene Citrate

Clomiphene citrate (CC) is an oral antiestrogen which stimulates ovarian function by increasing pituitary gonadotropin secretion due to blockage of E_2 steroid feedback. Steroid negative feedback remains intact, thereby making it a relatively safe product. Advantages are the oral administration, low costs, and widespread availability [2].

Before the introduction of GnRH agonists, combined CC/gonadotropin regimens were considered the standard of care. Multiple follicle development can be achieved by the addition of exogenous gonadotropins. Due to the synergistic effect of these compounds the amount of gonadotropins required is reduced, thereby shortening the duration of stimulation. In addition, gonadotropins may counteract the undesired antiestrogenic effects of CC on the endometrium, which have been held responsible for the relatively low embryo implantation rates. Compared to CC alone, this combination might improve pregnancy rates [2, 54, 55].

However, various CC protocols for IVF treatment exist, but the optimal protocol is still unknown. Compared to natural cycle IVF, ovarian stimulation with CC resulted in a higher number of oocyte retrievals and higher pregnancy rates [56, 57]. On the other hand, better performance was observed following a GnRH agonist long protocol [58]. Furthermore, the addition of recombinant LH to a CC/gonadotropin protocol led to comparable cancellation and pregnancy rates as conventional stimulation [59]. To decrease the incidence of LH surges, GnRH antagonist can be added to CC. Although some studies report similar pregnancy rates following a standard GnRH agonist protocol and CC/gonadotropin/GnRH antagonist treatment [60–62], others report significantly lower pregnancy rates [63].

In conclusion, it remains unclear whether the addition of GnRH antagonists or recombinant LH improves the outcome of CC/gonadotropin protocols. Due to the high diversity of the studies published, conclusions cannot be drawn regarding the possible benefits of CC in ovarian stimulation for IVF. Whether CC has a place in mild ovarian stimulation treatment remains to be seen.

3.6 Aromatase Inhibitors

For more than 20 years, aromatase inhibitors have been used primarily for the treatment of advanced breast cancer in postmenopausal women. Aromatase inhibitors block the conversion of androgens to estrogens. Estrogen-negative feedback at the pituitary–hypothalamic axis is reduced, leading to increased gonadotropin secretion and stimulation of ovarian follicular growth and development [64, 65].

Administration of aromatase inhibitors in the early follicular phase reduces the need for exogenous gonadotropins and the burden of treatment on the patient. Aromatase inhibitors have several advantages which make them attractive as an agent for use in mild stimulation protocols; the oral route of administration, the absence of ovarian hyperstimulation, fewer side effects and a shorter half-life than CC. Finally, they do not cause estrogen receptor depletion, thereby having no adverse effects on estrogen target tissues [65]. However, a major concern is the possible teratogenicity demonstrated in animal studies [66]. In today's gynecological practice, the use of aromatase inhibitors is still "off-label" although letrozole is frequently used in fertility preservation in patients with estrogen-dependent cancers, particularly breast cancer [64].

Studies concerning the use of aromatase inhibitors in IVF treatment are limited and small sized (Table 1). Initially, early studies mainly concerned poor ovarian responders. A tendency towards higher clinical pregnancy rates was observed both in good prognosis patients [67] and in poor responders [68] following a letrozole/FSH/GnRH antagonist protocol. Compared to a GnRH agonist regimen, similar pregnancy rates were obtained in poor responders [69], whereas others showed no improvement in IVF outcome [70]. Nevertheless, the numbers are too small to draw any conclusions. Larger randomized trials are needed to determine the benefit of aromatase inhibitors in IVF treatment.

3.7 Late Follicular Phase rFSH/GnRH Antagonist Co-treatment

As mentioned previously, decreasing FSH levels and the subsequent closure of the FSH window are essential for the selection of a single dominant follicle [20, 22]. This mechanism can be disrupted by administrating low doses of exogenous gonadotropins during the mid-follicular phase. The decrease in FSH

Table 1
Characteristics of randomized trials involving the use of aromatase inhibitors for controlled ovarian hyperstimulation

Study	Inclusion criteria	Study protocol	Control stimulation protocol	Main outcome
Goswami et al. (2004) [69]	> 35 years, failure in 1–3 IVF attempts due to poor ovarian response following long GnRH agonist	CD 3–7: 2.5 mg/day Letrozole CD 3 and 8: 75 IU/day FSH (n = 13)	Long GnRH agonist protocol, after 2 weeks 300 IU/day FSH, adjustable dose (n = 25)	Ongoing pregnancy rate per cycle 23 % versus 24 % (NS)
Garcia-Velasco et al. (2005)	No age limit, 1 previous cancelled IVF attempt due to poor response	OAC pretreatment. SD 1–4: 225 IU/day FSH and 150 IU/day hMG. Dose adjustment from SD 5 SD 1–5: 2.5 mg/day letrozole GnRH antagonist from leading follicle 14 mm (n = 71)	OAC pretreatment. SD 1–4: 225 IU/day FSH and 150 IU/day hMG. Dose adjustment from SD 5 GnRH antagonist from leading follicle 14 mm (n = 76)	Pregnancy rate per cycle 22.4 % versus 15.2 % (NS)
Verpoest et al. (2006) [67]	Normo-ovulatory patients, <39 years, normal BMI, regular indication for IVF or ICSI	CD 2–6: 2.5 mg/day letrozole CD 2 onwards: 150 IU/day FSH fixed dose CD 6: fixed start GnRH antagonist (n = 10)	CD 2 onwards: 150 IU/day FSH fixed dose CD 6: fixed start GnRH antagonist (n = 10)	Ongoing pregnancy rate per cycle 50 % versus 20 %
Ozmen et al. (2009) (abstract) [68]	Poor responders, regular indication for ICSI (n = 70)	CD3: 5 mg/day letrozole and 450 IU/day FSH fixed dose Flexible GnRH antagonist	CD3: 450 IU/day FSH fixed dose Flexible GnRH antagonist	Clinical pregnancy rate per embryo transfer 25.8 % versus 20 % (NS)
Davar et al. (2010) [70]	No age limit, 1 previous cancelled IVF attempt due to poor response	OAC pretreatment. CD 3: 300–450 IU/day FSH or hMG CD 3–7: 5 mg/day letrozole GnRH antagonist from leading follicle 14 mm (n = 45)	OAC pretreatment. GnRH agonist flare protocol with 300–450 IU/day FSH	Clinical pregnancy rate 4.4 % versus 12.2 % (NS) Clinical pregnancy rate per embryo transfer 5.3 % versus 14.3 % (NS)

The number of included cycles is equal to the number of included patients unless stated otherwise. Outcomes were significantly different unless stated otherwise

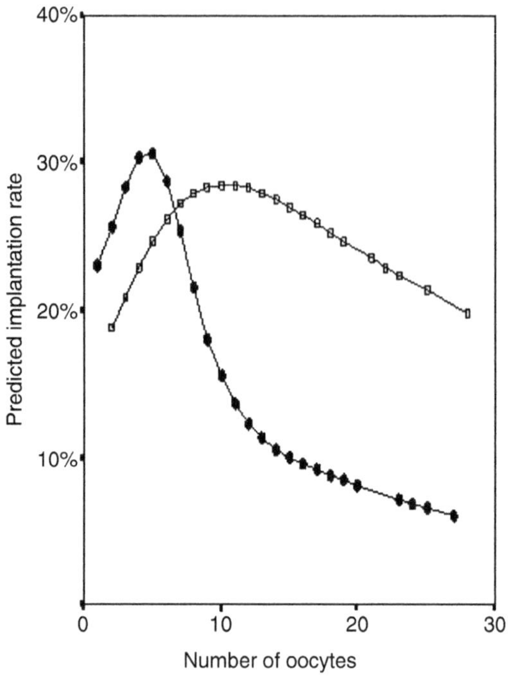

Fig. 2 Implantation rate according to the number of oocytes retrieved following mild (*black dots*) or conventional ovarian stimulation (*open squares*). (From: Verberg et al. Human Reproduction Update 2009)

levels is prevented, resulting in widening of the FSH window, and thereby rescuing the recruited follicles from atresia and stimulating them to dominance (Fig. 2) [20, 71]. The clinical availability of GnRH antagonists has enabled this concept to be introduced into IVF practice [72].

Table 2 represents the characteristics of the randomized studies involving this mild "late start" approach. A pilot study showed that multiple follicle development could be induced when ovarian stimulation was commenced on cycle day 5. A fixed daily dose of 150 IU recombinant FSH resulted in ongoing growth of a limited number of dominant follicles and a sufficient number of oocytes retrieved as compared with 100 IU/day [73]. However, commencing treatment on cycle day 7 is associated with a lower percentage of women presenting with multiple dominant follicle development [72, 74].

The use of a mild stimulation protocol starting on cycle day 5 was investigated in a randomized controlled trial. One hundred and forty-two normo-ovulatory patients were allocated to a long GnRH agonist protocol or to one of two GnRH antagonist protocols either commencing recombinant FSH on cycle day 2 or on cycle day 5. GnRH antagonist co-treatment was initiated on the day the leading follicle reached 14 mm in diameter. Overall, no differences were found among the three groups comparing pregnancy rate per started cycle.

Table 2
Characteristics of randomized controlled trials involving the efficacy mild "late start" ovarian stimulation

Study	Inclusion criteria	Study protocol	Control stimulation protocol	Main outcome
De Jong et al. (2000) [73]	Normo-ovulatory patients, <38 years, regular indication for IVF	Ovarian stimulation with 100 IU/day FSH from CD 5 onward. GnRH antagonist from CD 8 or when ≥ 1 follicle 13 mm. No luteal support (n = 8)	Ovarian stimulation with 150 IU/day FSH from CD 5 onward. GnRH antagonist from CD 8 or when ≥ 1 follicle 13 mm. No luteal support (n = 7)	Multiple follicle development 63 % versus 100 %. Ongoing pregnancy rate 25 % versus 14 % (NS)
Hohmann et al. (2003) [75]	Normo-ovulatory patients, <38 years, normal BMI, regular indication for IVF or IVF/ICSI	Ovarian stimulation with 150 IU/day FSH (fixed doses) from CD 5 onward. GnRH antagonist from leading follicle 14 mm (n = 45)	A. Long GnRH agonist protocol, fixed doses 150 IU/day FSH after 2 weeks (n = 49) B. From CD 2 fixed doses 150 IU/day FSH, GnRH antagonist from leading follicle 14 mm (n = 48)	Ongoing pregnancy rate 16 % versus 18 % (group A) versus 17 % (group B) (NS)
Heijnen et al. (2007) [44]	Normo-ovulatory patients, <38 years, normal BMI, regular indication for IVF or ICSI	Ovarian stimulation with 150 IU/day FSH (fixed doses) from CD 5 onward. GnRH antagonist from leading follicle 14 mm. Single embryo transfer (n = 205, 444 cycles)	Long GnRH agonist protocol, fixed doses 150 IU/day FSH after 2 weeks. Double embryo transfer (n = 199, 325 cycles)	Ongoing pregnancy rate per year of treatment 47 % versus 51 % (NS) 1 year cumulative proportion of pregnancies leading to term live birth: 43.2 % versus 44.6 %
Baart et al. (2007) [14]	Normo-ovulatory patients, <38 years, normal BMI, regular indication for IVF, sperm count > 5 million/ml. First cycles	Ovarian stimulation with 150 IU/d FSH (fixed doses) from CD 5 onward. GnRH antagonist from leading follicle 14 mm (n = 55)	Long GnRH agonist protocol, fixed doses 225 IU/d FSH after 2 weeks (n = 40)	Lower number of chromosomal abnormal embryos following mild ovarian stimulation

The number of included cycles is equal to the number of included patients unless stated otherwise. Outcomes were significantly different unless stated otherwise

Mild stimulation resulted in a shorter stimulation and 27 % reduction in exogenous FSH. A high cancellation rate was compensated by improved embryo quality concomitant with a higher chance of undergoing embryo transfer [75].

In a large non-inferiority trial in 404 patients, a mild approach combined with single embryo transfer, resulted in a similar cumulative proportion of pregnancies leading to term live births in 1 year compared to profound conventional stimulation with double embryo transfer, while reducing multiple pregnancy rates and overall costs [44].

Both studies show a lower number of oocytes after mild stimulation. While a low response during ovarian stimulation is normally associated with ovarian aging and poor oocyte quality, in mild stimulation this may represent a normal response. In fact, it represents a physiological response to the subtle interference with single dominant follicle selection. A certain degree of oocyte selection occurs because only the most mature follicles are stimulated, leading to high quality oocytes and embryos and higher implantation rates (Fig. 3) [9, 75].

Taken together, the finding that a large number of oocytes is not required for a successful IVF program, combined with a shorter stimulation and acceptable pregnancy rates, a mild approach represents an important step towards more patient friendly IVF treatment.

3.8 Late Follicular Phase hCG/LH and GnRH Antagonist Co-treatment

An alternative mild stimulation approach is late follicular phase replacement of FSH by LH. Developing follicles reach a critical diameter of 10 mm around the mid-follicular phase. At this stage granulosa cells express LH receptors leading to an increased responsiveness to LH. This action of LH is independent of FSH [20].

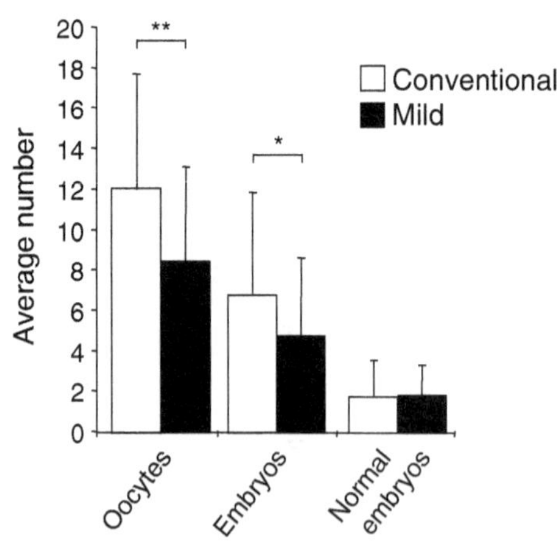

Fig. 3 The average number of oocytes retrieved, embryos, and chromosomally normal embryos following conventional treatment and mild stimulation. (From Baart et al. Human Reproduction 2007)

Sullivan et al. showed convincingly that late follicular phase follicle development can be sustained by just LH [76]. Hence, FSH can be replaced by LH (or hCG) for the stimulation of more mature follicles. Several studies have been conducted to assess this concept in the context of ovarian stimulation for IVF. It has been demonstrated that after priming with recombinant FSH, multiple follicle development could be sustained with low doses of hCG [77]. Furthermore, while cycle outcome was comparable to traditional COH regimens, complete replacement of FSH with low-dose hCG was associated with a reduced number of preovulatory follicles, higher fertilization rates, and significantly reduced gonadotropin consumption [78]. Additionally, in a more recent RCT, substitution of FSH by low-dose hCG led to a reduced gonadotropin consumption, whereas oocyte yield and pregnancy rates remained comparable to the standard GnRH antagonist treatment [79]. A slightly different approach was used in a large RCT. Ovarian stimulation was continued until either two codominant follicles reached 13–14 mm or the patient reached stimulation day 6, after which the FSH dose was lowered to 75 IU/day and combined with 200 IU of hCG. Compared to a standard GnRH antagonist and a standard long GnRH agonist protocol, this novel regime resulted in a significant reduction in FSH needed and therefore minimized costs. However, no difference was found in the number of (mature) oocytes retrieved, implantation rates, ongoing pregnancy rates and incidence of OHSS [80]. Similar results were obtained by Koichi et al. [81]. None of these studies reported premature luteinization in the hCG/LH protocol.

Late follicular phase administration of low-dose hCG with GnRH antagonist co-treatment can be safely applied in a selected group of patients. Decreased gonadotropin consumption minimizes treatment costs, whereas pregnancy rates remain comparable to standard protocols. However, a reduced incidence of OHSS could not be established, despite the reported reduction in the number of small preovulatory follicles. Additional studies are needed to investigate whether this mild stimulation approach is suitable in the majority of patients.

4 Implications of Mild Stimulation Protocols

4.1 Embryo Quality

It has become evident that ovarian stimulation may have detrimental effects on embryo quality, as assessed by morphology as well as the chromosomal constitution of the embryos [14, 82, 83]. This could be explained by disruption of the process of natural selection during ovarian hyperstimulation. Moreover, exposure of growing follicles to the potentially negative effects of ovarian stimulation could affect embryo quality [2, 4].

Little information is available with regard to possible consequences of interfering with natural selection. In vitro studies in rodents have shown detrimental effects on embryo development after exposing oocytes to high doses exogenous gonadotropins and concurrent high estradiol levels [84, 85]. Both oocyte maturation and the completion of meiosis may be disturbed, leading to chromosomal aneuploid oocytes and embryos [86]. The same negative impact is encountered in human embryos. High E_2 levels are deleterious to embryo adhesion in vitro, mainly because of a direct toxic effect on the embryo that may occur at the cleavage stage [16]. Additionally, high rates of chromosomal aneuploidy and mosaicism have been reported in human IVF embryos [14, 82, 83, 87].

Mild ovarian stimulation does not result in decreased pregnancy rates which might be explained by the reduced interference with natural follicle selection caused by this approach. In a prospective randomized trial, embryo aneuploidy rates were compared following a GnRH agonist long protocol and a mild stimulation protocol, by using preimplantation genetic screening. The mild stimulation protocol was associated with a higher proportion of chromosomally normal embryos (Fig. 4). Another interesting observation was that following mild stimulation, a low oocyte yield was associated with an even further decrease in the proportion of aneuploid embryos, suggesting that conventional stimulation impairs oocyte and embryo quality [14]. Indeed, a low number of oocytes retrieved is associated with optimal implantation rates retrieved and may therefore represent an appropriate response to mild stimulation. On the other hand, a similar response is indicative of ovarian aging when occurring after conventional ovarian stimulation [88].

However, retrieval of a lower number of oocytes might reduce the number of supernumerary embryos for cryopreservation, which could be a potential disadvantage of mild stimulation. The observation that mild stimulation could still result in a high oocyte yield and concurring higher proportions of abnormal embryos in some patients, underscores the need of further development of minimal stimulation approaches [14].

4.2 Endometrial Receptivity and Luteal Phase Consequences

Ovarian stimulation with gonadotropins affects luteal phase function and alters endometrial receptivity. Ovarian steroids induce specific endometrial changes which have a negative effect on the implantation potential of embryos. There is evidence that implantation of human embryos does not occur if endometrial advancement in the early luteal phase is more than 3 days [42]. This phenomenon is seen in GnRH agonist as well as in GnRH antagonist cycles.

The use of gonadotropins and GnRH analogues always leads to a deficient luteal phase. The main cause is related to the multifollicular development achieved during ovarian stimulation.

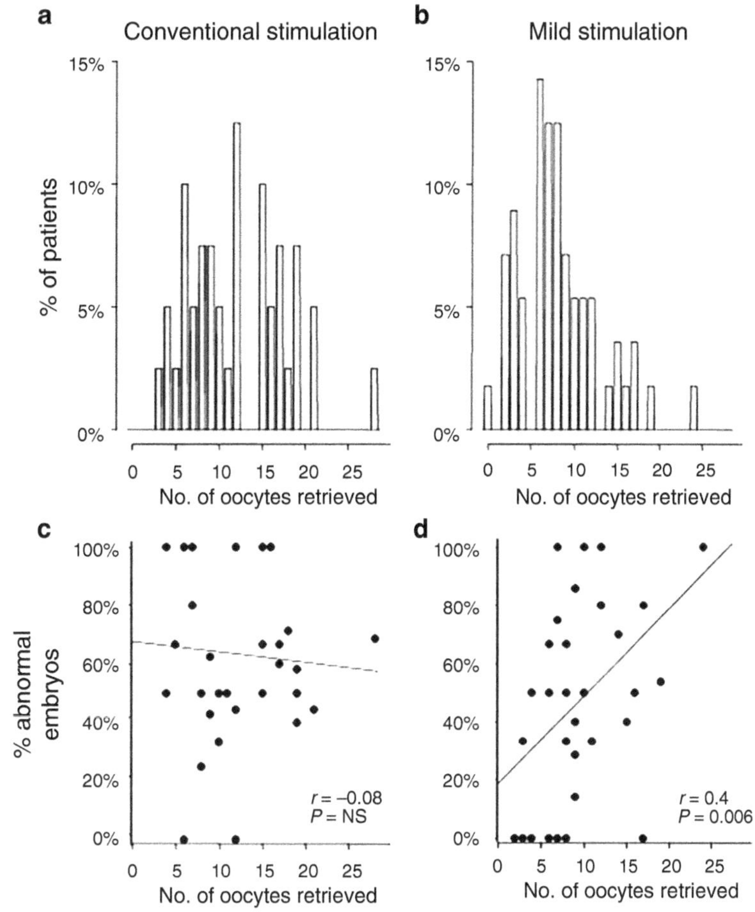

Fig. 4 Distribution of the number of oocytes retrieved per patient and the relationship between number of oocytes and the percentage of abnormal embryos generated following conventional treatment and mild stimulation. (From Baart et al. Human Reproduction 2007)

It has been suggested that an insufficient luteal phase is mostly related to supraphysiological steroid levels in the late follicular phase and early luteal phase, which directly inhibit pituitary secretion of LH [12, 13, 89, 90]. Luteal support should therefore be provided after ovarian stimulation combined with GnRH analogues [12]. Luteal phase length can be supported by administrating hCG or by supplementation with estrogens and/or progesterone [10].

Reducing the amount of steroids present could result in increased endometrial receptivity. Indeed, an improvement in implantation rates and pregnancy rates has been demonstrated when using a gonadotropin step-down regimen in high responders [91]. Mild stimulation approaches could therefore improve implantation rates as they aim at a more physiological response. Further studies are needed to clarify this point.

4.3 Psychological Burden

Ovarian stimulation is a demanding and stressful treatment for patients. Psychological stress is the main reason why couples discontinue IVF treatment before they have received all reimbursed treatment cycles [92]. The chance of dropout seems to be significantly increased when male subfertility is the treatment indication and when embryo transfer has not been performed [93]. A dropout rate of 54 % was observed in 450 couples who did not achieve childbirth, mainly because of psychological stress and a poor prognosis [94]. Moreover, an increase of both anxiety and depression was seen after unsuccessful treatment [95]. Apart from the psychological impact of treatment failure, IVF treatment itself, with its daily injections, possible side-effects, ultrasounds and invasive procedures, might be a cause of psychological distress [96], which has a direct negative effect on the chance of conceiving [97, 98]. Furthermore, multiple pregnancy may cause more emotional distress in patients, whereas less stress is experienced in singleton pregnancies [99].

Mild ovarian stimulation provides a shorter and more patient-friendly treatment with fewer complications and possibly a decrease in anxiety and treatment-related stress [100]. The psychological implications of mild stimulation combined with single embryo transfer during a first IVF cycle, were assessed in a large randomized trial. Women undergoing conventional stimulation experienced more physical discomfort before the start of ovarian stimulation. However, during subsequent treatment stages, no differences were found in physical discomfort between the two groups, although the mild stimulation group suffered for shorter period of time. At oocyte retrieval the mild stimulation group experienced more negative affect, possibly due to the attitude of clinicians towards the mild approach. Overall, mild stimulation with SET did not lead to more psychological complaints than conventional IVF [96]. After multiple cycles of conventional treatment, women showed more often clinically relevant symptoms of depression after treatment failure as compared to women who underwent multiple cycles of mild stimulation. This suggests that treatment burden becomes more severe with every cycle during conventional treatment as compared to mild stimulation [101, 102].

Taken together, mild stimulation constitutes a more patient friendly treatment regimen and the decrease of psychological distress and dropout rates can compensate for the slightly lower pregnancy rate per cycle observed with the mild approach [93].

4.4 Health Economics

Conventional stimulation with the transfer of two embryos leads to a high probability of multiple pregnancies, resulting in high costs due to intensive antenatal surveillance, increased chances for complications, perinatal and postpartum care. The high rate of multiple pregnancies and subsequent high costs in IVF can be greatly reduced by a single embryo transfer policy. This would lead to a considerable increase in health and indirect cost benefits [1, 103, 104].

Mild stimulation protocols are shorter than conventional protocols with decreased chances for complications such as OHSS; hence the per cycle cost of this treatment will be lower [30, 75]. Although natural cycle IVF is more cost-effective than stimulated cycles [46], this is not the case in MNC-IVF [51] and CC stimulated cycles [105]. However, the administration of exogenous gonadotropins in the mid-follicular phase has been shown to reduce costs [44, 75]. Since this approach reduces dropout rates in IVF, the overall number of cycles per patient may increase, resulting in similar live birth rate per treatment period compared with conventional treatment with double embryo transfer [106, 107]. Despite the increased number of cycles, the overall costs per term live birth are lower following mild stimulation. The reduction in costs is mainly attributable to a decrease in multiple pregnancies and preterm births. Indeed, conventional treatment resulted in unacceptably higher costs per extra term live birth [44, 108]. In conclusion, from an economic perspective a mild approach is more advantageous per term live birth, despite an increased number of IVF cycles per year.

4.5 Current State of Affairs and Future Developments

A better understanding of ovarian physiology, advances in ultrasound technology, laboratory technology, and the availability of GnRH antagonists have led to the development of novel, milder stimulation approaches. In recent years, IVF is increasingly viewed in terms of long-term health outcomes for women and their offspring and in terms of cost-effectiveness. Nevertheless, the acceptance of milder stimulation approaches has been rather slow.

Currently, the focus on pregnancy rates per started cycle encourages maximal stimulation and transfer of multiple embryos. Adopting the endpoint live birth per started treatment would encourage patient-friendly treatment regimens and single embryo transfer. It would be a major step towards reducing the burden of IVF [2, 44].

There is a general resistance in clinics to explore the use of GnRH antagonist and milder strategies due to overall success rates of standard GnRH agonist treatment [6]. Standard treatment may seem to have an advantage over mild stimulation because more embryos will be obtained, resulting in more embryos available for cryopreservation. However, this advantage might be counterbalanced by obtaining relatively more good quality embryos after mild stimulation [14]. Moreover, a modest number of oocytes following mild stimulation is associated with the optimal chance of achieving pregnancy [9]. Accepting a lower oocyte yield opens the possibility of using milder regimens such as clomiphene citrate or aromatase inhibitors.

Although pregnancy rates per cycle tend to be lower in mild stimulation protocols with single embryo transfer, the overall delivery rate at term during a 1-year treatment period was similar to conventional IVF [44]. Improved performance of embryo cryopreservation programs may provide further opportunities in mild IVF cycles.

The use of mild stimulation approaches reduces the cost per cycle, however, it remains too expensive to improve access to IVF treatment in developing countries [6]. Furthermore, as long as IVF is not readily available to, or affordable for, the general population, a preference towards conventional treatment strategies with higher pregnancy rates and more complications will remain [4].

Recently, more thought is being given to individualized treatment programs. Several studies have shown that the use of certain patient characteristics, such as body weight, age and ovarian reserve parameters, may reduce cancellation rates [88, 109, 110]. Additionally, the development of long-acting gonadotropin preparations has led to a reduction in the number of total injections and to more patient convenience [111].

Nowadays, more women use IVF for unconventional reasons such as preimplantation genetic diagnosis (PGD) and embryo or oocyte cryopreservation in women with cancer. This group might benefit from a mild approach because it allows more rapid completion of the stimulation cycle compared with a GnRH agonist long protocol [6]. Both PCOS patients, who are at high risk for developing OHSS, and poor responders, who fail to respond to any type of ovarian stimulation, might also benefit from a mild approach [4].

In conclusion, mild stimulation is a patient-friendly approach with reduced chances for complications. However, more research is needed to develop individualized treatment regimens leading to a milder ovarian response. Secondly, the effectiveness of mild stimulation needs to be tested in women older than 38 years of age and in poor responders. Further improvement of embryo quality, embryo selection and cryopreservation of supernumerary embryos will result in higher overall cumulative pregnancy rates per cycle when applying a strict single embryo transfer policy. Additionally, there is a need to develop cheaper stimulation protocols to improve general access to IVF treatment. Finally, a shift of focus is needed towards milder stimulation strategies to reduce the burden of IVF for patients, their offspring and society.

References

1. Fauser BC, Devroey P, Macklon NS (2005) Multiple birth resulting from ovarian stimulation for subfertility treatment. Lancet 365(9473):1807–1816
2. Macklon NS, Stouffer RL, Giudice LC, Fauser BC (2006) The science behind 25 years of ovarian stimulation for in vitro fertilization. Endocr Rev 27(2):170–207
3. Nargund G, Fauser BC, Macklon NS, Ombelet W, Nygren K, Frydman R (2007) The ISMAAR proposal on terminology for ovarian stimulation for IVF. Hum Reprod 22(11):2801–2804
4. Verberg MF, Macklon NS, Nargund G, Frydman R, Devroey P, Broekmans FJ et al (2009) Mild ovarian stimulation for IVF. Hum Reprod Update 15(1):13–29
5. Boue A, Boue J, Gropp A (1985) Cytogenetics of pregnancy wastage. Adv Hum Genet 14:1–57
6. Fauser BC, Nargund G, Andersen AN, Norman R, Tarlatzis B, Boivin J et al (2010) Mild ovarian stimulation for IVF: 10 years later. Hum Reprod 25(11):2678–2684
7. Klip H, Burger CW, Kenemans P, van Leeuwen FE (2000) Cancer risk associated with subfertility and ovulation induction: a review. Cancer Causes Control 11(4):319–344
8. Klip H, van Leeuwen FE, Schats R, Burger CW (2003) Risk of benign gynaecological diseases and hormonal disorders according to responsiveness to ovarian stimulation in IVF: a follow-up study of 8714 women. Hum Reprod 18(9):1951–1958
9. Verberg MF, Eijkemans MJ, Macklon NS, Heijnen EM, Baart EB, Hohmann FP et al (2009) The clinical significance of the retrieval of a low number of oocytes following mild ovarian stimulation for IVF: a meta-analysis. Hum Reprod Update 15(1):5–12
10. Devroey P, Bourgain C, Macklon NS, Fauser BC (2004) Reproductive biology and IVF: ovarian stimulation and endometrial receptivity. Trends Endocrinol Metab 15(2):84–90
11. Macklon NS, Fauser BC (2000) Impact of ovarian hyperstimulation on the luteal phase. J Reprod Fertil Suppl 55:101–108
12. Beckers NG, Macklon NS, Eijkemans MJ, Ludwig M, Felberbaum RE, Diedrich K et al (2003) Nonsupplemented luteal phase characteristics after the administration of recombinant human chorionic gonadotropin, recombinant luteinizing hormone, or gonadotropin-releasing hormone (GnRH) agonist to induce final oocyte maturation in in vitro fertilization patients after ovarian stimulation with recombinant follicle-stimulating hormone and GnRH antagonist cotreatment. J Clin Endocrinol Metab 88(9):4186–4192
13. Fauser BC, Devroey P (2003) Reproductive biology and IVF: ovarian stimulation and luteal phase consequences. Trends Endocrinol Metab 14(5):236–242
14. Baart EB, Martini E, Eijkemans MJ, Van OD, Beckers NG, Verhoeff A et al (2007) Milder ovarian stimulation for in-vitro fertilization reduces aneuploidy in the human preimplantation embryo: a randomized controlled trial. Hum Reprod 22(4):980–988
15. Kok JD, Looman CW, Weima SM, Te Velde ER (2006) A high number of oocytes obtained after ovarian hyperstimulation for in vitro fertilization or intracytoplasmic sperm injection is not associated with decreased pregnancy outcome. Fertil Steril 85(4): 918–924
16. Valbuena D, Martin J, de Pablo JL, Remohi J, Pellicer A, Simon C (2001) Increasing levels of estradiol are deleterious to embryonic implantation because they directly affect the embryo. Fertil Steril 76(5):962–968
17. Sunkara SK, Rittenberg V, Raine-Fenning N, Bhattacharya S, Zamora J, Coomarasamy A (2011) Association between the number of eggs and live birth in IVF treatment: an analysis of 400 135 treatment cycles. Hum Reprod 26(7):1768–1774
18. McGee EA, Hsueh AJ (2000) Initial and cyclic recruitment of ovarian follicles. Endocr Rev 21(2):200–214
19. Baker TG (1963) A quantitative and cytological study of germ cells in human ovaries. Proc R Soc Lond B Biol Sci 158:417–433
20. Fauser BC, Van Heusden AM (1997) Manipulation of human ovarian function: physiological concepts and clinical consequences. Endocr Rev 18(1):71–106
21. Scheele F, Schoemaker J (1996) The role of follicle-stimulating hormone in the selection of follicles in human ovaries: a survey of the literature and a proposed model. Gynecol Endocrinol 10(1):55–66
22. Schipper I, de Jong FH, Fauser BC (1998) Lack of correlation between maximum early follicular phase serum follicle stimulating hormone concentrations and menstrual cycle characteristics in women under the age of 35 years. Hum Reprod 13(6):1442–1448

23. van Santbrink EJ, Hop WC, van Dessel TJ, de Jong FH, Fauser BC (1995) Decremental follicle-stimulating hormone and dominant follicle development during the normal menstrual cycle. Fertil Steril 64(1):37–43
24. Edwards RG, Steptoe PC, Purdy JM (1980) Establishing full-term human pregnancies using cleaving embryos grown in vitro. Br J Obstet Gynaecol 87(9):737–756
25. Barbieri RL, Hornstein MD (1999) Assisted reproduction-in vitro fertilization success is improved by ovarian stimulation with exogenous gonadotropins and pituitary suppression with gonadotropin-releasing hormone analogues. Endocr Rev 20(3):249–252
26. Evers JL (2002) Female subfertility. Lancet 360(9327):151–159
27. Fauser BC, Devroey P (2005) Why is the clinical acceptance of gonadotropin-releasing hormone antagonist cotreatment during ovarian hyperstimulation for in vitro fertilization so slow? Fertil Steril 83(6):1607–1611
28. Broekmans FJ, Hompes PG, Lambalk CB, Schoute E, Broeders A, Schoemaker J (1996) Short term pituitary desensitization: effects of different doses of the gonadotrophin-releasing hormone agonist triptorelin. Hum Reprod 11(1):55–60
29. Donderwinkel PF, Schoot DC, Pache TD, de Jong FH, Hop WC, Fauser BC (1993) Luteal function following ovulation induction in polycystic ovary syndrome patients using exogenous gonadotrophins in combination with a gonadotrophin-releasing hormone agonist. Hum Reprod 8(12):2027–2032
30. Tarlatzis BC, Fauser BC, Kolibianakis EM, Diedrich K, Rombauts L, Devroey P (2006) GnRH antagonists in ovarian stimulation for IVF. Hum Reprod Update 12(4):333–340
31. Klingmuller D, Schepke M, Enzweiler C, Bidlingmaier F (1993) Hormonal responses to the new potent GnRH antagonist Cetrorelix. Acta Endocrinol (Copenh) 128(1):15–18
32. Al-Inany HG, Youssef MA, Aboulghar M, Broekmans F, Sterrenburg M, Smit J et al (2011) Gonadotrophin-releasing hormone antagonists for assisted reproductive technology. Cochrane Database Syst Rev (5):CD001750
33. Albano C, Smitz J, Camus M, Riethmuller-Winzen H, Van SA, Devroey P (1997) Comparison of different doses of gonadotropin-releasing hormone antagonist Cetrorelix during controlled ovarian hyperstimulation. Fertil Steril 67(5):917–922
34. Ganirelix Dose Finding Study Group (1998) A double-blind, randomized, dose-finding study to assess the efficacy of the gonadotrophin-releasing hormone antagonist ganirelix (Org 37462) to prevent premature luteinizing hormone surges in women undergoing ovarian stimulation with recombinant follicle stimulating hormone (Puregon). The ganirelix dose-finding study group. Hum Reprod 13(11):3023–3031
35. Huirne JA, Homburg R, Lambalk CB (2007) Are GnRH antagonists comparable to agonists for use in IVF? Hum Reprod 22(11):2805–2813
36. Olivennes F, Alvarez S, Bouchard P, Fanchin R, Salat-Baroux J, Frydman R (1998) The use of a GnRH antagonist (Cetrorelix) in a single dose protocol in IVF-embryo transfer: a dose finding study of 3 versus 2 mg. Hum Reprod 13(9):2411–2414
37. Al-Inany H, Aboulghar M (2002) GnRH antagonist in assisted reproduction: a Cochrane review. Hum Reprod 17(4):874–885
38. Al-Inany HG, Abou-Setta AM, Aboulghar M (2006) Gonadotrophin-releasing hormone antagonists for assisted conception. Cochrane Database Syst Rev (3):CD001750
39. Kolibianakis EM, Collins J, Tarlatzis BC, Devroey P, Diedrich K, Griesinger G (2006) Among patients treated for IVF with gonadotrophins and GnRH analogues, is the probability of live birth dependent on the type of analogue used? A systematic review and meta-analysis. Hum Reprod Update 12(6):651–671
40. Kolibianakis EM, Zikopoulos K, Smitz J, Camus M, Tournaye H, Van Steirteghem AC et al (2004) Elevated progesterone at initiation of stimulation is associated with a lower ongoing pregnancy rate after IVF using GnRH antagonists. Hum Reprod 19(7):1525–1529
41. Kolibianakis EM, Zikopoulos K, Schiettecatte J, Smitz J, Tournaye H, Camus M et al (2004) Profound LH suppression after GnRH antagonist administration is associated with a significantly higher ongoing pregnancy rate in IVF. Hum Reprod 19(11):2490–2496
42. Kolibianakis E, Bourgain C, Albano C, Osmanagaoglu K, Smitz J, Van SA et al (2002) Effect of ovarian stimulation with recombinant follicle-stimulating hormone, gonadotropin releasing hormone antagonists, and human chorionic gonadotropin on endometrial maturation on the day of oocyte pick-up. Fertil Steril 78(5):1025–1029
43. Papanikolaou EG, Bourgain C, Kolibianakis E, Tournaye H, Devroey P (2005) Steroid receptor expression in late follicular phase endometrium in GnRH antagonist IVF cycles is already altered, indicating initiation of early luteal phase

transformation in the absence of secretory changes. Hum Reprod 20(6):1541–1547

44. Heijnen EM, Eijkemans MJ, De Klerk C, Polinder S, Beckers NG, Klinkert ER et al (2007) A mild treatment strategy for in-vitro fertilisation: a randomised non-inferiority trial. Lancet 369(9563):743–749
45. Pelinck MJ, Hoek A, Simons AH, Heineman MJ (2002) Efficacy of natural cycle IVF: a review of the literature. Hum Reprod Update 8(2):129–139
46. Nargund G, Waterstone J, Bland J, Philips Z, Parsons J, Campbell S (2001) Cumulative conception and live birth rates in natural (unstimulated) IVF cycles. Hum Reprod 16(2):259–262
47. Morgia F, Sbracia M, Schimberni M, Giallonardo A, Piscitelli C, Giannini P et al (2004) A controlled trial of natural cycle versus microdose gonadotropin-releasing hormone analog flare cycles in poor responders undergoing in vitro fertilization. Fertil Steril 81(6):1542–1547
48. Kyrou D, Kolibianakis EM, Venetis CA, Papanikolaou EG, Bontis J, Tarlatzis BC (2009) How to improve the probability of pregnancy in poor responders undergoing in vitro fertilization: a systematic review and meta-analysis. Fertil Steril 91(3):749–766
49. Nargund G, Frydman R (2007) Towards a more physiological approach to IVF. Reprod Biomed Online 14(5):550–552
50. Pelinck MJ, Vogel NE, Arts EG, Simons AH, Heineman MJ, Hoek A (2007) Cumulative pregnancy rates after a maximum of nine cycles of modified natural cycle IVF and analysis of patient drop-out: a cohort study. Hum Reprod 22(9):2463–2470
51. Pelinck MJ, Vogel NE, Hoek A, Arts EG, Simons AH, Heineman MJ (2005) Minimal stimulation IVF with late follicular phase administration of the GnRH antagonist cetrorelix and concomitant substitution with recombinant FSH: a pilot study. Hum Reprod 20(3):642–648
52. Pelinck MJ, Knol HM, Vogel NE, Arts EG, Simons AH, Heineman MJ et al (2008) Cumulative pregnancy rates after sequential treatment with modified natural cycle IVF followed by IVF with controlled ovarian stimulation. Hum Reprod 23(8):1808–1814
53. Pelinck MJ, Vogel NE, Hoek A, Simons AH, Arts EG, Mochtar MH et al (2006) Cumulative pregnancy rates after three cycles of minimal stimulation IVF and results according to subfertility diagnosis: a multicentre cohort study. Hum Reprod 21(9):2375–2383
54. Nelson LM, Hershlag A, Kurl RS, Hall JL, Stillman RJ (1990) Clomiphene citrate directly impairs endometrial receptivity in the mouse. Fertil Steril 53(4):727–731
55. Ronen J, Bosschieter J, Wiswedel K, Hendriks S, Levin M (1988) Ovulation induction for in vitro fertilisation using clomiphene citrate and low-dose human menopausal gonadotrophin. Int J Fertil 33(2):120–122
56. Ingerslev HJ, Hojgaard A, Hindkjaer J, Kesmodel U (2001) A randomized study comparing IVF in the unstimulated cycle with IVF following clomiphene citrate. Hum Reprod 16(4):696–702
57. MacDougall MJ, Tan SL, Hall V, Balen A, Mason BA, Jacobs HS (1994) Comparison of natural with clomiphene citrate-stimulated cycles in in vitro fertilization: a prospective, randomized trial. Fertil Steril 61(6): 1052–1057
58. Dhont M, Onghena A, Coetsier T, De SP (1995) Prospective randomized study of clomiphene citrate and gonadotrophins versus goserelin and gonadotrophins for follicular stimulation in assisted reproduction. Hum Reprod 10(4):791–796
59. Weigert M, Krischker U, Pohl M, Poschalko G, Kindermann C, Feichtinger W (2002) Comparison of stimulation with clomiphene citrate in combination with recombinant follicle-stimulating hormone and recombinant luteinizing hormone to stimulation with a gonadotropin-releasing hormone agonist protocol: a prospective, randomized study. Fertil Steril 78(1):34–39
60. Fiedler K, Ludwig M (2003) Use of clomiphene citrate in in vitro fertilization (IVF) and IVF/intracytoplasmic sperm injection cycles. Fertil Steril 80(6):1521–1523
61. Lin YH, Hwang JL, Seow KM, Huang LW, Hsieh BC, Tzeng CR (2006) Comparison of outcome of clomiphene citrate/human menopausal gonadotropin/cetrorelix protocol and buserelin long protocol—a randomized study. Gynecol Endocrinol 22(6):297–302
62. Williams SC, Gibbons WE, Muasher SJ, Oehninger S (2002) Minimal ovarian hyperstimulation for in vitro fertilization using sequential clomiphene citrate and gonadotropin with or without the addition of a gonadotropin-releasing hormone antagonist. Fertil Steril 78(5):1068–1072
63. Mansour R, Aboulghar M, Serour GI, Al-Inany HG, Fahmy I, Amin Y (2003) The use of clomiphene citrate/human menopausal gonadotrophins in conjunction with GnRH antagonist in an IVF/ICSI program is not a

cost effective protocol. Acta Obstet Gynecol Scand 82(1):48–52

64. Lee VC, Ledger W (2011) Aromatase inhibitors for ovulation induction and ovarian stimulation. Clin Endocrinol (Oxf) 74(5): 537–546
65. Mitwally MF, Casper RF (2001) Use of an aromatase inhibitor for induction of ovulation in patients with an inadequate response to clomiphene citrate. Fertil Steril 75(2):305–309
66. Tiboni GM, Marotta F, Rossi C, Giampietro F (2008) Effects of the aromatase inhibitor letrozole on in utero development in rats. Hum Reprod 23(8):1719–1723
67. Verpoest WM, Kolibianakis E, Papanikolaou E, Smitz J, Van SA, Devroey P (2006) Aromatase inhibitors in ovarian stimulation for IVF/ICSI: a pilot study. Reprod Biomed Online 13(2):166–172
68. Ozmen B, Sonmezer M, Atabekoglu CS, Olmus H (2009) Use of aromatase inhibitors in poor-responder patients receiving GnRH antagonist protocols. Reprod Biomed Online 19(4):478–485
69. Goswami SK, Das T, Chattopadhyay R, Sawhney V, Kumar J, Chaudhury K et al (2004) A randomized single-blind controlled trial of letrozole as a low-cost IVF protocol in women with poor ovarian response: a preliminary report. Hum Reprod 19(9):2031–2035
70. Davar R, Oskouian H, Ahmadi S, Firouzabadi RD (2010) GnRH antagonist/letrozole versus microdose GnRH agonist flare protocol in poor responders undergoing in vitro fertilization. Taiwan J Obstet Gynecol 49(3): 297–301
71. Fauser BC, Donderwinkel P, Schoot DC (1993) The step-down principle in gonadotrophin treatment and the role of GnRH analogues. Baillieres Clin Obstet Gynaecol 7(2):309–330
72. Macklon NS, Fauser BC (2003) Mild stimulation in in vitro fertilization. Ann N Y Acad Sci 997:105–111
73. de Jong D, Macklon NS, Fauser BC (2000) A pilot study involving minimal ovarian stimulation for in vitro fertilization: extending the "follicle-stimulating hormone window" combined with the gonadotropin-releasing hormone antagonist cetrorelix. Fertil Steril 73(5):1051–1054
74. Hohmann FP, Laven JS, de Jong FH, Eijkemans MJ, Fauser BC (2001) Low-dose exogenous FSH initiated during the early, mid or late follicular phase can induce multiple dominant follicle development. Hum Reprod 16(5):846–854
75. Hohmann FP, Macklon NS, Fauser BC (2003) A randomized comparison of two ovarian stimulation protocols with gonadotropin-releasing hormone (GnRH) antagonist cotreatment for in vitro fertilization commencing recombinant follicle-stimulating hormone on cycle day 2 or 5 with the standard long GnRH agonist protocol. J Clin Endocrinol Metab 88(1):166–173
76. Sullivan MW, Stewart-Akers A, Krasnow JS, Berga SL, Zeleznik AJ (1999) Ovarian responses in women to recombinant follicle-stimulating hormone and luteinizing hormone (LH): a role for LH in the final stages of follicular maturation. J Clin Endocrinol Metab 84(1):228–232
77. Filicori M, Cognigni GE, Tabarelli C, Pocognoli P, Taraborrelli S, Spettoli D et al (2002) Stimulation and growth of antral ovarian follicles by selective LH activity administration in women. J Clin Endocrinol Metab 87(3):1156–1161
78. Filicori M, Cognigni GE, Gamberini E, Parmegiani L, Troilo E, Roset B (2005) Efficacy of low-dose human chorionic gonadotropin alone to complete controlled ovarian stimulation. Fertil Steril 84(2):394–401
79. Blockeel C, De VM, Verpoest W, Stoop D, Haentjens P, Devroey P (2009) Can 200 IU of hCG replace recombinant FSH in the late follicular phase in a GnRH-antagonist cycle? A pilot study. Hum Reprod 24(11):2910–2916
80. Serafini P, Yadid I, Motta EL, Alegretti JR, Fioravanti J, Coslovsky M (2006) Ovarian stimulation with daily late follicular phase administration of low-dose human chorionic gonadotropin for in vitro fertilization: a prospective, randomized trial. Fertil Steril 86(4):830–838
81. Koichi K, Yukiko N, Shima K, Sachiko S (2006) Efficacy of low-dose human chorionic gonadotropin (hCG) in a GnRH antagonist protocol. J Assist Reprod Genet 23(5):223–228
82. Katz-Jaffe MG, Trounson AO, Cram DS (2005) Chromosome 21 mosaic human preimplantation embryos predominantly arise from diploid conceptions. Fertil Steril 84(3):634–643
83. Munne S, Magli C, Adler A, Wright G, de Boer K, Mortimer D et al (1997) Treatment-related chromosome abnormalities in human embryos. Hum Reprod 12(4):780–784
84. Ertzeid G, Storeng R (1992) Adverse effects of gonadotrophin treatment on pre- and post-

implantation development in mice. J Reprod Fertil 96(2):649–655

85. Van der Auwera I, D'Hooghe T (2001) Superovulation of female mice delays embryonic and fetal development. Hum Reprod 16(6):1237–1243
86. Hodges CA, Ilagan A, Jennings D, Keri R, Nilson J, Hunt PA (2002) Experimental evidence that changes in oocyte growth influence meiotic chromosome segregation. Hum Reprod 17(5):1171–1180
87. Baart EB, Martini E, van den Berg I, Macklon NS, Galjaard RJ, Fauser BC et al (2006) Preimplantation genetic screening reveals a high incidence of aneuploidy and mosaicism in embryos from young women undergoing IVF. Hum Reprod 21(1):223–233
88. Verberg MF, Eijkemans MJ, Macklon NS, Heijnen EM, Fauser BC, Broekmans FJ (2007) Predictors of low response to mild ovarian stimulation initiated on cycle day 5 for IVF. Hum Reprod 22(7):1919–1924
89. Pellicer A, Valbuena D, Cano F, Remohi J, Simon C (1996) Lower implantation rates in high responders: evidence for an altered endocrine milieu during the preimplantation period. Fertil Steril 65(6):1190–1195
90. Simon C, Cano F, Valbuena D, Remohi J, Pellicer A (1995) Clinical evidence for a detrimental effect on uterine receptivity of high serum oestradiol concentrations in high and normal responder patients. Hum Reprod 10(9):2432–2437
91. Simon C, Garcia Velasco JJ, Valbuena D, Peinado JA, Moreno C, Remohi J et al (1998) Increasing uterine receptivity by decreasing estradiol levels during the preimplantation period in high responders with the use of a follicle-stimulating hormone step-down regimen. Fertil Steril 70(2):234–239
92. Cousineau TM, Domar AD (2007) Psychological impact of infertility. Best Pract Res Clin Obstet Gynaecol 21(2):293–308
93. Verberg MF, Eijkemans MJ, Heijnen EM, Broekmans FJ, de Klerk C, Fauser BC et al (2008) Why do couples drop-out from IVF treatment? A prospective cohort study. Hum Reprod 23(9):2050–2055
94. Olivius C, Friden B, Borg G, Bergh C (2004) Why do couples discontinue in vitro fertilization treatment? A cohort study. Fertil Steril 81(2):258–261
95. Verhaak CM, Smeenk JM, van Minnen A, Kremer JA, Kraaimaat FW (2005) A longitudinal, prospective study on emotional adjustment before, during and after consecutive fertility treatment cycles. Hum Reprod 20(8):2253–2260
96. de Klerk C, Heijnen EM, Macklon NS, Duivenvoorden HJ, Fauser BC, Passchier J et al (2006) The psychological impact of mild ovarian stimulation combined with single embryo transfer compared with conventional IVF. Hum Reprod 21(3):721–727
97. Smeenk JM, Verhaak CM, Eugster A, van Minnen A, Zielhuis GA, Braat DD (2001) The effect of anxiety and depression on the outcome of in-vitro fertilization. Hum Reprod 16(7):1420–1423
98. Smeenk JM, Verhaak CM, Vingerhoets AJ, Sweep CG, Merkus JM, Willemsen SJ et al (2005) Stress and outcome success in IVF: the role of self-reports and endocrine variables. Hum Reprod 20(4):991–996
99. Glazebrook C, Sheard C, Cox S, Oates M, Ndukwe G (2004) Parenting stress in first-time mothers of twins and triplets conceived after in vitro fertilization. Fertil Steril 81(3):505–511
100. Fauser BC, Devroey P, Yen SS, Gosden R, Crowley WF Jr, Baird DT et al (1999) Minimal ovarian stimulation for IVF: appraisal of potential benefits and drawbacks. Hum Reprod 14(11):2681–2686
101. de Klerk C, Macklon NS, Heijnen EM, Eijkemans MJ, Fauser BC, Passchier J et al (2007) The psychological impact of IVF failure after two or more cycles of IVF with a mild versus standard treatment strategy. Hum Reprod 22(9):2554–2558
102. Hojgaard A, Ingerslev HJ, Dinesen J (2001) Friendly IVF: patient opinions. Hum Reprod 16(7):1391–1396
103. Ledger WL, Anumba D, Marlow N, Thomas CM, Wilson EC (2006) The costs to the NHS of multiple births after IVF treatment in the UK. BJOG 113(1):21–25
104. Lukassen HG, Schonbeck Y, Adang EM, Braat DD, Zielhuis GA, Kremer JA (2004) Cost analysis of singleton versus twin pregnancies after in vitro fertilization. Fertil Steril 81(5):1240–1246
105. Kovacs P, Matyas S, Bernard A, Kaali SG (2004) Comparison of clinical outcome and costs with CC + gonadotropins and gnrha + gonadotropins during Ivf/ICSI cycles. J Assist Reprod Genet 21(6):197–202
106. Eijkemans MJ, Heijnen EM, de Klerk C, Habbema JD, Fauser BC (2006) Comparison of different treatment strategies in IVF with cumulative live birth over a given period of time as the primary end-point: methodological con-

siderations on a randomized controlled non-inferiority trial. Hum Reprod 21(2):344–351

107. Heijnen EM, Macklon NS, Fauser BC (2004) What is the most relevant standard of success in assisted reproduction? The next step to improving outcomes of IVF: consider the whole treatment. Hum Reprod 19(9):1936–1938

108. Polinder S, Heijnen EM, Macklon NS, Habbema JD, Fauser BJ, Eijkemans MJ (2008) Cost-effectiveness of a mild compared with a standard strategy for IVF: a randomized comparison using cumulative term live birth as the primary endpoint. Hum Reprod 23(2):316–323

109. Freiesleben NL, Lossl K, Bogstad J, Bredkjaer HE, Toft B, Loft A et al (2008) Predictors of ovarian response in intrauterine insemination patients and development of a dosage nomogram. Reprod Biomed Online 17(5):632–641

110. Olivennes F, Howles CM, Borini A, Germond M, Trew G, Wikland M et al (2009) Individualizing FSH dose for assisted reproduction using a novel algorithm: the CONSORT study. Reprod Biomed Online 18(2):195–204

111. Fauser BC, Mannaerts BM, Devroey P, Leader A, Boime I, Baird DT (2009) Advances in recombinant DNA technology: corifollitropin alfa, a hybrid molecule with sustained follicle-stimulating activity and reduced injection frequency. Hum Reprod Update 15(3):309–321

Chapter 15

IVF Stimulation: Protocols for Poor Responders

Owen K. Davis

Abstract

One of the most vexing challenges in the practice of reproductive medicine is the management of the "poor responder," specifically the patient manifesting an inadequate follicular response to ovarian stimulation. Poor response predicts a reduction in the number of mature oocytes retrieved, with the consequences of fewer embryos available for selection and transfer, reduced pregnancy rates, and a markedly decreased likelihood of residual embryos for cryopreservation. This chapter reviews the definition and prediction of poor response and discusses strategies that have been developed and incorporated into the reproductive endocrinologist's clinical armamentarium in an effort to optimize outcomes for these women. It should be stated at the outset that no single approach is successful for all patients, and that there is currently no firm clinical consensus regarding the relative efficacy of the different stimulation protocols. This arises in part due to inconsistency regarding the definition of what constitutes a poor response and a paucity of well-designed, randomized controlled trials.

Key words IVF, Poor responders, GnRH-agonist flare protocols, Diminished ovarian reserve, Estrogen priming, Clomiphene citrate, Aromatase inhibitors, Luteal phase synchronization

1 Introduction

Although the first human birth following assisted reproduction resulted from the retrieval of a single oocyte in a natural menstrual cycle [1], adjunctive controlled ovarian hyperstimulation (COH) rapidly became incorporated into routine clinical practice. The efficiency of IVF (in vitro fertilization) is significantly enhanced through the retrieval of multiple oocytes and the generation of multiple embryos. This permits the transfer of more than one embryo in cases where appropriate (e.g., advanced maternal age), but also allows for more refined embryo selection (whether through morphologic assessment, pre-implantation genetic screening, etc.), thus optimizing *per embryo* implantation rates. With the increasingly prevalent performance of elective double and, ideally, single embryo transfer, selection of the embryo(s) with the greatest implantation potential is of ascending importance. The availability of supernumerary oocytes and/or embryos

Zev Rosenwaks and Paul M. Wassarman (eds.), *Human Fertility: Methods and Protocols*, Methods in Molecular Biology, vol. 1154, DOI 10.1007/978-1-4939-0659-8_15, © Springer Science+Business Media New York 2014

for cryopreservation is a further benefit of multiple-oocyte retrieval, and augments the potential cumulative pregnancy rate following a single IVF cycle. In cases of fertility preservation, where oocyte/embryo cryopreservation prior to gonadotoxic therapy is the immediate goal of an IVF cycle, the recruitment and recovery of multiple oocytes will clearly heighten the chances for a patient's future reproductive success.

2 Definition of the Poor Responder

A "poor response" to COH has been variably defined through the specification of individual or combined thresholds for cycle day-specific or peak estradiol attained, numbers of follicles, numbers of oocytes retrieved, daily or total gonadotropin dosage requirements, and cycle cancellation. Some studies have included subjects as "poor responders" based on baseline demographic or screening criteria including age and basal FSH level rather than limiting inclusion to those patients specifically manifesting a poor historic clinical response. The heterogeneity of definitions has in no small measure contributed to the lack of clarity in the literature as a whole. For example, some studies have defined a "poor response" as the attainment of fewer than six [2] or four follicles [3]. Other studies define "poor response" as the retrieval of fewer than three [4], four [5] or five [6] oocytes. Further, some studies target a peak estradiol threshold of 500 pg/ml [7] or a day six estradiol level of <100 pg/ml during stimulation [8]. In short, it is clear that the ART literature would benefit from the adoption of uniform criteria for defining a poor response. It is to be expected that the reported efficacy of various proposed stimulation protocols for the "poor responder" will be highly dependent on how stringently the condition is defined, and this should be borne in mind when interpreting published series or trials. Further, if a poor response is defined by low oocyte/embryo yield in a single ART cycle, particularly in a patient without additional predictors of diminished ovarian reserve (*see* below), future response even on the same protocol could be more robust due to inter-cycle variability alone. Definitional inconsistency aside, it is nonetheless imperative that the practitioner be conversant with the various treatment options for these challenging patients and that further well-designed studies be undertaken.

3 Risk Factors for Poor Response

The dominant risk factor for poor response to COH is chronological maternal age, due in large measure to the aging demographic of women undergoing treatment at many ART centers.

Indeed, a poor response to ovarian stimulation appears to be one of the earliest discernable manifestations of reproductive aging [9]. It should be emphasized, however, that a woman's age alone cannot predict her response to COH, and that other markers and predictors of ovarian reserve should be incorporated into clinical decision-making.

A number of other acquired, environmental, and iatrogenic factors have been associated with diminished ovarian response. Cigarette smoking has been implicated more broadly in earlier onset of menopause, but also specifically in a diminished response to COH [10]. Women with moderate or severe endometriosis will commonly present with ovarian endometriomas, which are associated with reduced oocyte yield at retrieval [11]. It has further been suggested that surgical resection of endometriomas [12], non-endometriotic ovarian cysts [13], and ovarian drilling for polycystic ovary syndrome [14] may also lead to compromised ovarian reserve and oocyte yield.

Antineoplastic chemotherapy and radiation therapy can result in irreversible gonadotoxic effects, and the magnitude of this impact is both dose- and agent-dependent, and directly related to the patient's age at the time of diagnosis and treatment. As an increasing number of female cancer patients present either for fertility preservation or treatment, this must be taken into consideration in devising stimulation protocols; here, ovarian reserve testing is imperative. In these cases, the risk of a poor ovarian response is increased [15]. Less commonly, identifiable genetic factors may be responsible for diminished ovarian reserve and a potentially reduced response to ovarian stimulation: such would include women with Turner's mosaicism, and women harboring a clinically significant expansion of CGG repeat sequences in the FMR-1 gene (fragile X mutation).

4 Prediction of Poor Response: Ovarian Reserve Testing

Several risk factors for poor response to ovarian stimulation have been cited in the foregoing discussion, with advanced chronological age being perhaps the most common. However, no single risk factor can absolutely predict how a given patient will respond to COH, thus necessitating a more accurate assessment of "ovarian reserve" (most commonly taken to be the residual, quantitative endowment of her follicles/oocytes). Put simply, age alone will neither predict good ovarian reserve in a younger women nor poor ovarian reserve in an older woman. A number of tests have been proposed in an effort to characterize a patient's ovarian reserve, thus predicting her likely response to COH and permitting an appropriate prospective choice of an optimal stimulation protocol. Perhaps the best single predictor of the magnitude of response to ovarian stimulation is the patient's past response [16]. The "treatment naïve" patient therefore poses a somewhat greater challenge regarding choice of initial protocol. A variety of approaches have been described and incorporated into practice, including basal and dynamic hormonal testing, and ovarian imaging. Dynamic tests (e.g., the "clomiphene citrate challenge test") will not be reviewed in this chapter, as they entail the administration of pharmacologic agents and have not been demonstrated to be clinically superior to appropriately selected and interpreted basal hormone levels and ultrasonographic ovarian morphometrics.

4.1 Basal Hormonal Testing

Although an association between reproductive aging and subtle elevations in serum follicle-stimulating hormone (FSH) levels had been previously noted [17], the first report of an association between basal FSH levels and IVF outcomes was published in 1988 [18]. In most currently employed assays, the upper limit of normal for FSH on day 2–4 of the follicular phase is 10–12 mIU/ml, with levels of >15 mIU/ml connoting significant elevations. It is generally advised that each center define its own cutoff for normal basal FSH. Serum FSH levels demonstrate significant intercycle variability, and this appears to be most pronounced in women with diminished ovarian reserve. The highest FSH value is generally the most predictive; a normal value does not establish normal ovarian reserve. Measurement of basal FSH should optimally incorporate simultaneous assessment of the basal serum estradiol level, as estradiol elevations can suppress the FSH, thus confounding interpretation. It has been suggested that an elevated day 3 estradiol level (e.g., above 60–75 pg/ml), in and of itself, predicts a significant reduction in ovarian response to COH [19]. While elevations of basal FSH prognosticate a lower response to COH in terms of oocyte yield and higher cancellation rates in both young and older women, it should be emphasized that younger patients will still manifest significantly higher implantation and pregnancy rates than their older counterparts, largely due to lower rates of embryonic aneuploidy [20]. As with most tests of ovarian reserve, basal FSH and estradiol levels are of limited value in prediction of pregnancy, and treatment should not be withheld solely on the basis of abnormal testing [21]; rather, these tests should be utilized to better counsel the patient in terms of her likely response and to permit optimal selection of the stimulation protocol.

Basal day 3 inhibin B has also been suggested to predict ART outcomes, with low values (e.g., <45 ng/ml) correlating with reduced response and pregnancy rates [22]. Over time, however, measurement of inhibin B has not been demonstrated to confer a predictive advantage over basal FSH/estradiol levels, and has a limited ability to predict ovarian response and pregnancy rates following the assisted reproductive technologies.

Anti-Mullerian hormone (AMH) has emerged as one of the more useful single markers of ovarian reserve. A member of the TGFB superfamily, AMH is a heterodimer secreted by the granulosa compartment of developing follicles through the early antral stage, independent of gonadotropin stimulation. AMH is relatively constant, manifesting very modest intra-cycle [23] and inter-cycle variability [24]. Although an international standard is not yet established, normal ovarian reserve is generally inferred if the AMH level exceeds approximately 0.90–1.0 ng/ml using a sensitive ELISA. There is a growing literature suggesting that the serum AMH level is a better predictor of ovarian response during ART than other available biochemical markers of ovarian reserve, including basal FSH/estradiol and inhibin B [25]. Here again, it is likely that AMH will prove a better index of a woman's ovarian follicular endowment and hence response to stimulation than to the occurrence or non-occurrence of pregnancy, per se, due to the lack of a strict correspondence between oocyte quality and quantity.

4.2 Sonographic Ovarian Imaging

A correlation between aging and diminished ovarian volume, as measured by ultrasonography, was appreciated as early as 1987 [26]. Mean ovarian volume has since proven a relatively poor predictor of ovarian response [27] to COH, however, and is seldom employed as an isolated metric in current clinical practice. In contrast, assessment of the basal antral follicle count (AFC) has proven to be a useful and relatively simple biophysical modality. AFC assessment is typically performed in the early follicular phase, utilizing a high-frequency transvaginal ultrasound transducer, and represents the total number of follicles <10 mm, bilaterally. Although a variety of cutoffs have been suggested, adequate ovarian reserve is generally predicted in the setting of an AFC in the range of eight to ten follicles. The AFC manifests minimal inter-cycle variability, and appears to correlate with and have similar predictive value to the AMH level [28]. The accuracy of the AFC will be expected to vary with the quality of the ovarian images obtained, which can be affected by such variables as ovarian position and obesity.

Taken together, the AFC and AMH levels have become particularly useful tools in the evaluation of ovarian reserve, and should be interpreted in the context of a given patient's age and prior response to COH. Taken together, these clinical data facilitate appropriate counseling of patients and the choice of an appropriately tailored stimulation protocol. It should be emphasized that no single test or combination of tests can predict success or failure with certainty, and should therefore not be used to exclude the properly informed patient from the option of treatment with autologous gametes.

5 Stimulation Strategies for Poor Responders

A number of strategies have been suggested and employed in an effort to enhance the response to stimulation in women with DOR. The true absolute and relative efficacy of these various approaches is difficult to interpret due to several confounding factors, including most prominently: heterogeneous patient populations (e.g., variable definitions and criteria for "poor response"), the incorporation of several different ancillary interventions into proposed protocols which clouds the relative import of each component strategy and the retrospective design of the majority of these studies, which cannot permit exclusion of the likely significant impact of random inter-cycle variability and "regression to the mean."

The putative physiologic bases for the described strategies typically incorporate one or more of the following principles: decreased pituitary suppression, exposure of the ovaries to increased concentrations of exogenous gonadotropins, enhanced secretion of endogenous gonadotropins, and improved synchronization of follicular development in the preceding luteal phase. The remainder of this review will examine the more prevalent of these clinical approaches.

5.1 Reduction or Elimination of GnRH Agonist

The use of a GnRH agonist (GnRHa) for down-regulation, or the "long" GnRHa protocol, typically entails mid-luteal initiation of a GnRHa (e.g., day 21) with resulting pituitary suppression in the ensuing follicular phase at the time stimulation commences. This approach effectively prevents the occurrence of untimely endogenous LH-surges and has been a standard and effective approach in ART over the past two decades. The commercial availability of GnRHa's (e.g., leuprolide acetate) preceded that of the GnRH-antagonists; the latter avoid the biphasic effects (stimulation followed by down-regulation of endogenous gonadotropins) of the agonists, but the earlier generations of antagonists were limited by untoward reactions due to histamine release. Nonetheless, despite the current availability and increasingly dominant use of GnRH-antagonists in ART, the "long" GnRHa protocols still persist and are a main-stay of ovarian stimulation at many centers.

It has been theorized that excessive suppression of the pituitary will increase the required dosage of gonadotropins and reduce the response and success rate in low response patients. Indeed, a number of studies have shown that halving the standard dose of GnRHa results in less profound suppression of endogenous gonadotropins than, and a similar rate of premature LH-surges to the standard dosage, with similar clinical outcomes in normal response patients [29]. The strategy of lowering the agonist dosage in the treatment of low responders has been studied retrospectively, and has been suggested to reduce the incidence of cycle cancellation and to increase the numbers of oocytes and embryos obtained, with a trend toward improved pregnancy rates [30]. It has also been suggested that cessation of GnRHa administration with the commencement of stimulation ("stop" protocol), following mid-luteal down-regulation, may improve outcomes for low responders [31]. The available published data suggest that it is reasonable to employ a low-dose or a "stop" regimen when stimulating the poor responder with a long GnRHa protocol, but fall short of a definitive conclusion of improved outcomes.

5.2 High-Dose Gonadotropins

Increasing the daily dosage of administered gonadotropins from a standard dose of 150–300 IU/day up to 450–600 IU/day is an intuitively appealing approach to the stimulation of the poor responder. High-dose stimulation is, in fact, widely employed in this specific context. However, the efficacy of this strategy is unproven; several published studies have demonstrated lack of benefit. In one RCT, for example, increasing the initial dose of FSH from 150 to 300 IU/day in women with low antral follicle counts (<5) improved neither the median number of recovered oocytes nor the ongoing pregnancy rate [32]. Similarly, in another randomized trial, doubling the gonadotropin dosage in the course of the treatment of poor responders failed to improve the outcome in ART cycles [33].

This failure of high-dose stimulation to materially improve outcomes in low responders appears paradoxical. One possible explanation could be desensitization of FSH receptors during prolonged gonadotropin exposure [34]. It is also possible that high-dose stimulation in the early follicular phase is unable to overcome the more profound impact of early follicular recruitment in the preceding luteal phase, a phenomenon that may play a particularly significant role in women with DOR, who often manifest elevated "ambient" FSH levels. This early follicular recruitment may in part account for the shortening of the follicular phase seen in the peri-menopause.

Moreover, concerns have been raised regarding a possible negative impact of high doses of gonadotropins on oocyte/embryo quality. Some published reports have suggested increased chromosomal abnormality rates in murine oocytes exposed to high levels of gonadotropins [35]. One RCT described a higher incidence of embryonic aneuploidy following human IVF with conventional versus "mild" gonadotropin stimulation [36]. Nonetheless, absenting the development of demonstrably effective strategies in this difficult patient population, it is likely that the use of high-dose stimulation regimens will remain a common practice.

5.3 Short Agonist/Flare Protocols

GnRH-agonist "flare" protocols have been proposed as a means of clinically exploiting the initial stimulatory effect of these agents, i.e. the initial augmentation or "flare" of endogenous gonadotropin secretion.

Traditional "short" agonist protocols entail initiation of full dose agonist (e.g., leuprolide acetate 0.5–1 mg) in the early follicular phase (day 2 or 3), with co-administration of gonadotropins 1 or 2 days after the start of the agonist. Concern that the conventional short protocol might result in premature elevation of LH and, consequently, serum androgens and progesterone [37] led to the development of the "microdose flare" regimen. Here, the dose of agonist is markedly reduced; leuprolide acetate is generally initiated at a dosage of 40 μg twice daily throughout the follicular phase of stimulation. Microdose agonist protocols usually follow pretreatment with an oral contraceptive (OCP) with the intent of preventing corpus luteum rescue from the previous cycle. High-dose gonadotropin administration (e.g., 300–600 IU/day) commences following 2 days of the microdose agonist alone, and is continued in either a fixed or step-down regimen until the day of hCG trigger. An early study suggested an improved follicular response in a group of 34 women who had historically manifested a low response on a conventional long agonist protocol [38]. Another small trial in which patients served as their own historical controls similarly demonstrated improved IVF outcomes; all patients had previously been cancelled on a long agonist protocol [39]. In this report, a microdose flare approach was employed along with adjunctive growth hormone (GH). The cancellation rate in this cohort was12.5 %, with an ongoing pregnancy rate of 43.8 %.

Since these initial promising reports, the published data examining short agonist protocols have been decidedly mixed, with a number of RCT's failing to demonstrate statistical superiority of the "flare" approach to either the long agonist protocol [40] or to GnRH-antagonist protocols [41]. Here again, the literature should be evaluated with caution given the heterogeneity in the studied patient populations. Inclusion criteria for defining poor responders vary among these studies with respect to patient age, previous treatment history, basal FSH and quantitative response with respect to follicles, oocytes and peak estradiol attained. It can be stated unequivocally that the likelihood of a clinically meaningful improvement in outcome will be strongly correlated with the severity of ovarian depletion.

5.4 Luteal Phase Synchronization: "Estrogen Priming"

One concern in the low responder patient population is the possibility of early follicular recruitment in the preceding luteal phase, which can lead to the ultimate development of one or a few dominant follicle(s) during stimulation for IVF. This tendency may be exacerbated by elevated endogenous FSH levels in women with DOR. In theory, if the lead follicles are essentially predetermined by cycle day 2 or 3, the subsequent stimulation protocol may be insufficient to recruit additional less advanced follicles from the gonadotropin sensitive pool, irrespective of gonadotropin dosage.

Specifically, enhanced synchronization of the early antral follicles could have the potential to improve the ensuing response in ART cycles. Fanchin and colleagues demonstrated that luteal phase administration of estradiol resulted in a reduced discrepancy in antral follicle sizes in the early follicular phase [42]. This same group subsequently reported that luteal phase administration of a GnRH-antagonist could similarly coordinate the sizes of early antral follicles [43].

The Cornell group described an "estrogen-priming" protocol in 2005 [44] which incorporated both luteal phase estradiol and antagonist suppression in the treatment of low responders; this regimen entails administration of micronized transdermal estradiol via 0.1 mg skin patches, commencing on the 10th day following the endogenous LH-surge, with replacement of the patch on alternate days until the start of menses, at which point the final patch is left in place. As initially described, this protocol also employed luteal GnRH-antagonist (ganirelix acetate, 0.25 mg) co-administration for 3 consecutive days, commencing on the 11th day following the LH-surge. High-dose gonadotropin stimulation is then undertaken on cycle day 2, with step-down as appropriate and the initiation of GnRH-antagonist in a flexible protocol. This preliminary report demonstrated a significant improvement in the number of normally fertilized oocytes obtained with a trend toward a reduced cancellation rate and increased numbers of total and mature oocytes. The ongoing pregnancy rate in this challenging cohort of patients was 26.2 %. This protocol was subsequently compared to the OCP/microflare approach in a retrospective cohort study of a younger group of low responders (mean age: 32 years) but with elevated mean basal FSH levels (12.2 and 12.9 mIU/ml respectively); the estrogen-priming protocol compared favorably with respect to both peak estradiol response and mean number of embryos transferred [45]. In this recent study, the ongoing pregnancy rate was 48 % per completed cycle in the estrogen-priming group versus 33 % in the OCP/microflare group, although this difference was not statistically significant. Other groups have examined the impact of luteal estradiol priming in poor responders and described improvements in both embryo yield and quality [46].

5.5 Adjunctive Treatments of the Low Response Patient

In addition to the strategies outlined in the foregoing discussion, a number of adjunctive agents have been employed in an effort to enhance ovarian responsiveness to COH in the poor response patient including clomiphene citrate, aromatase inhibitors, and growth hormone (GH). Here again, the literature is mixed, but these agents deserve at least brief mention.

The addition of clomiphene citrate to gonadotropins in ART protocols has a long history, dating back to the advent of IVF. The increased risk of premature LH-surges in clomiphene-based protocols has been largely mitigated by the introduction of the GnRH-antagonists. Clomiphene stimulates the secretion of endogenous gonadotropins and may be co-administered in either a sequential or simultaneous manner with exogenous gonadotropins. Although at least one RCT comparing combined clomiphene and high-dose gonadotropins to a long GnRH-agonist protocol demonstrated a significant improvement in the cancellation rate and numbers of oocytes and good quality embryos obtained [47], neither this nor a number of case–control studies have demonstrated a consistent improvement in pregnancy rates.

Aromatase inhibitors (AI), like clomiphene, stimulate follicular development through a central mechanism; here, the inhibition of the conversion of androgen substrate to estradiol diminishes the negative feedback effects of estradiol at the level of the hypothalamus, with augmentation of endogenous gonadotropin release. The possibility that a combination of adjunctive AI with gonadotropin stimulation could improve the ovarian response in poor responders was initially proposed in 2002 [48]. In the setting of ART, the AI (typically letrozole at a daily oral dosage of 2.5–5 mg/day) is administered for 5 days starting at the beginning of the follicular phase and overlapping with gonadotropins after the first 2 days of the AI alone. An antagonist is employed to prevent an untimely LH surge. At least one RCT demonstrated an improved ongoing pregnancy rate in poor responders when compared head-to-head with an OCP/flare protocol [49]. Further study will be required to clearly establish the utility of adjunctive AI in the stimulation of poor responders.

GH appears to exert both direct and indirect effects on ovarian function, including follicular recruitment and development, oocyte maturation and steroidogenesis [50, 51]. In women with a relative GH deficiency, GH may augment ovarian responsiveness to gonadotropin stimulation [52]. Adjunctive administration of GH in COH protocols for both normal response and low response patients has been studied, with varied results vis-à-vis the magnitude of response and pregnancy rates; one meta-analysis suggested a marginally significant improvement in pregnancy rates following the use of GH for IVF [53]. Large RCT's will be required to define the possible benefits of GH co-treatment in this context.

6 Conclusions

Poor responders remain a major challenge in assisted reproduction. Poor response to COH is largely a result of diminished ovarian reserve, which is frequently but by no means exclusively a function of advanced chronologic age. When a poor response is predicted by prior treatment history and/or hormonal testing and sonographic assessment of the ovaries, implementation of an appropriate stimulation protocol is therefore a particularly integral part of the patient's overall treatment plan. A number of treatment strategies have been described in the literature and reviewed in this chapter. The lack of uniform criteria for categorizing patients as "poor responders" and the paucity of large RCT's has hampered efforts to attain a clinical consensus as to which protocol(s) hold the greatest promise for these patients. It is also clear that response to COH per se cannot predict either the implantation rate or likelihood of a successful pregnancy; it is important to recognize that the per embryo implantation rate is largely commensurate with the patient's chronologic age and oocyte quality, and this must be factored into clinical decisions regarding treatment strategies with autologous oocytes versus the alternative of oocyte donation. It is to be hoped that future well-designed trials will permit optimization of outcomes for this significant subset of the overall ART patient population.

References

1. Steptoe PC, Edwards RG (1978) Birth after reimplantation of a human embryo. Lancet 2:366
2. Saldeen P, Llen K, Sundstrom P (2007) The probability of successful IVF outcome after poor ovarian response. Acta Obstet Gynecol Scand 86:457–461
3. Morgia F, Sbracia M, Schimberni M et al (2004) A controlled trial of natural cycle versus microdose gonadotropin-releasing hormone analog flare cycles in poor responders undergoing in vitro fertilization. Fertil Steril 81(6):1542–1547
4. Marci R, Oaserta D, Dolo V, Tatone C, Pavan A, Moscarini M (2005) GnRH antagonist in IVF poor-responder patients: results of a randomized trial. Reprod Biomed Online 11(2):189–193
5. Klinkert E, Broekmans F, Looman C, te Velde E (2004) A poor response in the first in vitro fertilization cycle is not necessarily related to a poor prognosis in subsequent cycles. Fertil Steril 81(5):1247–1253
6. Frankfurter D, Dayal M, Dubey A, Peak D, Gindoff P (2007) Novel follicular-phase gonadotropin-releasing hormone antagonist stimulation protocol for in vitro fertilization in the poor responder. Fertil Steril 88(5): 1442–1445
7. Baka S, Makrakis E, Tzanakaki D et al (2006) Cancellation of a first cycle is not predictive of a subsequent failure. Ann N Y Acad Sci 1092: 418–425
8. Khalaf Y, El-Toukhy T, Taylor A, Braude P (2002) Increasing the gonadotrophin dose in the course of an in vitro fertilization cycle does not rectify an initial poor response. Eur J Obstet Gynecol Reprod Biol 103:146–149
9. de Boer E, Tonkelaar I, te Velde E, Burgur C, van Klip H, Leeuwen F (2003) Increased risk of early menopausal transition and natural menopause after poor response at first IVF treatment. Hum Reprod 18(7):1544–1552
10. El-Nemr A, Al-Shawaf T, Sabatini L, Wilson C, Lower AM, Grudzinskas JG (1998) Effect of smoking on ovarian reserve and ovarian stimulation in in-vitro fertilization and embryo transfer. Hum Reprod 13(8):2192–2198
11. Suzuki T, Izumi S, Matsubayashi H, Awaji H, Yoshikata K, Makino T (2005) Impact of ovarian endometrioma on oocytes and pregnancy outcome in in vitro fertilization. Fertil Steril 83(4):908–913
12. Esinler I, Bozdag G, Aybar F, Bayar YH (2006) Outcome of in vitro fertilization/intracytoplasmic sperm injection after laparoscopic cystectomy for endometriomas. Fertil Steril 85(6):1730–1735
13. Somigliana E, Ragni G, Infantino M, Benedetti F, Arnoldi M, Crosignani PG (2006) Does laparoscopic removal of nonendometriotic benign ovarian cysts affect ovarian reserve? Acta Obstet Gynecol Scand 85(1):74–77
14. Kandil M, Selim M (2005) Hormonal and sonographic assessment of ovarian reserve before and after laparoscopic ovarian drilling in

polycystic ovary syndrome. BJOG 112(10):1427–1430

15. Ginsburg E, Yanushpolsky E, Jackson K (2001) In vitro fertilization for cancer patients and survivors. Fertil Steril 75(10):705–710
16. Kligman I, Rosenwaks Z (2001) Differentiating clinical profiles: predicting good responders, poor responders, and hyper-responders. Fertil Steril 76(6):1185–1190
17. Sherman BM, Korenmann SG (1975) Hormonal characteristics of the human menstrual cycle throughout reproductive life. J Clin Invest 55(4):699–706
18. Muasher SJ, Oehinger S, Simonetti S et al (1988) The value of basal and/or stimulated serum gonadotropin levels in prediction of stimulation response and in vitro fertilization outcome. Fertil Steril 50(2):298–307
19. Licciardi F, Liu H, Rosenwaks Z (1995) Day 3 estradiol serum concentrations as prognosticators of ovarian stimulation response and pregnancy outcome in patients undergoing in vitro fertilization. Fertil Steril 64(5):991–994
20. Van Rooij I, Bancsi L, Broekmans F, Looman C, Habbema J, te Velde E (2003) Women older than 40 years of age and those with elevated follicle-stimulating hormone levels differ in poor response rate and embryo quality in in vitro fertilization. Fertil Steril 79(3):482–488
21. Lambalk C (2003) Value of elevated follicle-stimulating hormone levels and the differential diagnosis during the diagnostic subfertility work-up. Fertil Steril 79(3):489–490
22. Seifer D, Lambert-Messerlian G, Hogan J, Gardiner A, Blazar A, Berk C (1997) Day 3 serum inhibin-B is predictive of assisted reproductive technologies outcome. Fertil Steril 67(1):110–114
23. Hehenkamp W, Looman C, Themmen A, de Jong F, Te Velde E, Broekmans F (2006) Anti-Mullerian hormone levels in the spontaneous menstrual cycle do not show substantial fluctuation. J Clin Endocrinol Metab 91(10):4057–4063
24. Fanchin R, Taieb J, Lozano DH, Ducot B, Frydman R, Bouyer J (2005) High reproducibility of serum anti-Mullerian hormone measurements suggest a multi-staged follicular secretion and strengthens its role in the assessment of ovarian follicular status. Hum Reprod 20(4):923–927
25. Fanchin R, Schonauer LM, Righini C, Guibourdenche J, Frydman R, Taieb J (2003) Serum anti-Mullerian hormone is more strongly related to ovarian follicular status than serum inhibin B, estradiol, FSH and LH on day 3. Hum Reprod 18(2):323–327
26. Andolf E, Jorgensen C, Svalenius E, Sunden B (1987) Ultrasound measurement of the ovarian volume. Acta Obstet Gynecol Scand 66(5):387–389
27. Hendriks DJ, Kwee J, Mol BW, te Velde ER, Broekmans FJ (2007) Ultrasonography as a tool for the prediction of outcome in IVF patients: a comparative meta-analysis of ovarian volume and antral follicle count. Fertil Steril 87(4):764–775
28. Muttukrishna S, McGarrigle H, Wakim R, Khadum I, Ranieri DM, Serhal P (2005) Antral follicle count, anti-Mullerian hormone and inhibin B: predictors of ovarian response in assisted reproductive technology? BJOG 112(10):1384–1390
29. Balasch J, Gomez F, Casamitjana R, Carmona F, Rivera F, Vanrell JA (1992) Pituitary-ovarian suppression by the standard and half-doses of D-Trp-6-luteinizing hormone-releasing hormone depot. Hum Reprod 7(9):1230–1234
30. Feldberg D, Farhi J, Ashkenazi J, Dicker D, Shalev J, Ben-Rafael Z (1994) Minidose gonadotropin-releasing hormone agonist is the treatment of choice in poor responders with high follicle-stimulating hormone levels. Fertil Steril 62(2):343–346
31. Dirnfeld M, Fruchter O, Yshai D, Lissak A, Ahdut A, Abramovici H (1999) Cessation of gonadotropin-releasing hormone analogue (GnRH-a) upon down-regulation versus conventional long GnRH-a protocol in poor responders undergoing in vitro fertilization. Fertil Steril 72(3):406–411
32. Klinkert E, Broekmans F, Looman C, Habbema J, te Velde E (2005) Expected poor responders on the basis of an antral follicle count do not benefit from a higher starting dose of gonadotrophins in IVF treatment: a randomized controlled trial. Hum Reprod 20(3):611–615
33. Van Hooff MH, Alberda AT, Huisman GJ, Zeilmaker GH, Leerentveld RA (1993) Doubling the human menopausal gonadotrophin dose in the course of an in-vitro fertilization treatment cycle in low responders: a randomized study. Hum Reprod 8(3):369–373
34. Amsterdam A, Hanoch T, Dantes A, Tajima K, Strauss JF, Seger R (2002) Mechanisms of gonadotropin desensitization. Mol Cell Endocrinol 187(1–2):69–74
35. Van Blerkom J, Davis P (2001) Differential effects of repeated ovarian stimulation on cytoplasmic and spindle organization in metaphase II mouse oocytes matured in vivo and in vitro. Hum Reprod 16(4):757–764
36. Baart EB, Martini E, Eijkemans MJ et al (2007) Milder ovarian stimulation for in-vitro

fertilization reduces aneuploidy in the human preimplantation embryo: a randomized controlled trial. Hum Reprod 22(4):980–988

37. Sims JA, Seltman HJ, Muasher SJ (1994) Early follicular rise of serum progesterone concentration in response to a flare-up effect of gonadotropin-releasing hormone agonist impairs follicular recruitment for in-vitro fertilization. Hum Reprod 9(2):235–240
38. Scott RT, Navot D (1994) Enhancement of ovarian responsiveness with microdoses of gonadotropin-releasing hormone agonist during ovulation induction for in vitro fertilization. Fertil Steril 61(5):880–885
39. Schoolcraft W, Schlenker T, Gee M, Stevens J, Wagley L (1997) Improved controlled ovarian hyperstimulation in poor responder in vitro fertilization patients with a microdose follicle-stimulating hormone flare, growth hormone protocol. Fertil Steril 67(1):93–97
40. Sbracia M, Farina A, Poverini R, Morgia F, Schimberini M, Aragona C (2005) Short versus long gonadotropin-releasing hormone analogue suppression protocols for superovulation in patients > or = 40 years old undergoing intracytoplasmic sperm injection. Fertil Steril 84(3):644–648
41. Akman M, Halit E, Tosun S, Bayazit N, Esra A, Behceci M (2001) Comparison of agonist-flare-up protocol and antagonist multidose protocol in ovarian stimulation in poor responders: results of a prospective randomized trial. Hum Reprod 16(5):868–870
42. Fanchin R, Cunha-Filho JS, Schonauer LM, Kadoch IJ, Cohen-Bacri P, Frydman R (2003) Coordination of early antral follicles by luteal estradiol administration provides a basis for alternative controlled ovarian hyperstimulation regimens. Fertil Steril 79(2):316–321
43. Fanchin R, Castelo Branco A, Kadoch IJ et al (2004) Premenstrual administration of gonadotropin-releasing hormone antagonist coordinates early antral follicle sizes and sets up the basis for an innovative concept of controlled ovarian hyperstimulation. Fertil Steril 81(6):1554–1559
44. Dragisic KG, Davis OK, Fasouliotis SJ, Rosenwaks Z (2005) Use of a luteal estradiol patch and a gonadotropin-releasing hormone antagonist suppression protocol before gonadotropin stimulation for in vitro fertilization in poor responders. Fertil Steril 84(4):1023–1026
45. Shastri SM, Barbieri E, Kligman I, Schoyer KD, Davis OK, Rosenwaks Z (2011) Stimulation of the young poor responder: comparison of the luteal estradiol/gonadotropin-releasing hormone antagonist priming protocol versus oral contraceptive/microdose leuprolide. Fertil Steril 95:592–595
46. Frattarelli JL, Hill MJ, McWilliams GD, Miller KA, Bergh PA, Scott RT (2008) A luteal estradiol protocol for expected poor-responders improves embryo number and quality. Fertil Steril 89:1118–1122
47. D'Amato G, Caroppo E, Pasquadibisceglie A, Carone D, Vitti A, Vizziello GM (2004) A novel protocol of ovulation induction with delayed gonadotropin-releasing hormone antagonist administration combined with high-dose recombinant follicle-stimulating hormone and clomiphene citrate for poor responders and women over 35 years. Fertil Steril 81(6):1572–1577
48. Mitwally MF, Casper RF (2002) Aromatase inhibition improves ovarian response to follicle-stimulating hormone in poor responders. Fertil Steril 77(4):776–780
49. Schoolcraft WB, Surrey ES, Minjarez DA, Stevens JM, Gardner DK (2008) Management of poor responders: can outcomes be improved with a novel gonadotropin-releasing hormone antagonist/letrozole protocol? Fertil Steril 89(1):151–156
50. Poretsky L, Cataldo NA, Rosenwaks Z, Giudice LC (1999) The insulin-related ovarian regulatory system in health and disease. Endocr Rev 20:535–582
51. Hull KL, Harvey S (2001) Growth hormone: roles in female reproduction. J Endocrinol 168(1):1–23
52. Blumenfeld Z, Amit T, Barkey RJ, Lunenfeld B, Brandes JM (1991) Synergistic effect of growth hormone and gonadotropins in achieving conception in "clonidine-negative" patients with unexplained infertility. Ann N Y Acad Sci 626:250–265
53. Harper K, Proctor M, Hughes E (2003) Growth hormone for in vitro fertilization. Cochrane Database Syst Rev (3):CD000099

Chapter 16

Oocyte Retrieval and Quality Evaluation

Lucinda Veeck Gosden

Abstract

After more than 30 years of harvesting oocytes for in vitro fertilization, collection techniques are well established. Presented here are dependable methods for maintaining environmental stability during such procedures and general descriptions of the materials collected.

Key words Oocyte, Metaphase I, Metaphase II, Prophase I, Granulosa, Ooplasm, Oolemma, Zona pellucida

1 Oocyte Retrieval

During retrieval (harvests), the door between the laboratory and the OR remains open in order to allow the circulating nurse to deliver tubes of follicular fluid. Soon after the procedure, this door is closed and locked. Before the start of each procedure, an embryologist must check the patient's armband to verify her name. This step is mandatory; failure to appropriately identify the patient before beginning an oocyte collection procedure will result in employment termination.

The circulating nurse states the patient's name as she enters the laboratory. She must place the follicular fluid aspirates in the heating block under the hood, left to right in the order of aspiration. There should never be more than two or three tubes in the heating block. If three tubes have been collected and the embryologist is behind in searching for oocytes, the physician must be requested to slow the aspiration of follicles.

Each patient should be considered a potential carrier of infectious disease, even if appropriate tests have been negative. Viral testing results may be overlooked or reported in error. Because of the potential for exposure, it is mandatory to comply with OSHA regulations requiring gloves and masks when dealing with body

Zev Rosenwaks and Paul M. Wassarman (eds.), *Human Fertility: Methods and Protocols*, Methods in Molecular Biology, vol. 1154, DOI 10.1007/978-1-4939-0659-8_16, © Springer Science+Business Media New York 2014

fluids; goggles are recommended by current safety standards but not required because of the difficulty incurred during microscopic procedures.

Equipment: Laminar flow hood, heating block, dissecting microscope with stage warmer, inverted microscope with stage warmer, CO_2 incubator, video or computer system to display patient's name, intercom for communication, Pipet-Aid, slide warmer, Sharps container.

Supplies: Sterile 9″ pipettes and bulbs, pipette rack, 5 mL sterile pipettes, Falcon or Nunc 60 × 15 mm sterile petri dishes, Terumo 1 cc syringes, 26G needles, powderless latex gloves, specimen containers, HEPES-buffered follicle flush medium, HEPES-buffered oocyte wash medium, equilibrated mineral oil, prepared wash and insemination dishes, diamond etching pen, methanol burner, pipette rack.

2 Preparation for Collection of Oocytes

1. Familiarize yourself with the patients' names and the location of culture dishes for cases being retrieved. The patient's first name, last name, and the date must be displayed on a monitor within the laboratory and in the O.R. during oocyte collection. Check with physician *and verify patient's name on her armband*; check again in case of doubt. *Do not hesitate to satisfy yourself about the correct identification.*
2. Make certain that the heating block temperature is at 37 °C and that an appropriate number of tubes containing the culture media for oocyte washing are in the block. Use one 10 mL tube per case, plus 1–2 extra to cover any cases with large numbers of oocytes.
3. Place at least 20 Falcon or Nunc dishes on the slide warming unit. It is important not to stack dishes deeply since there will be little possibility for top dishes to warm properly. As the warmed dishes are discarded during the course of the retrieval, they may be replaced with new dishes so that a continuous supply of warmed dishes is available. Remember that the inside of dishes must remain sterile. Do not move hands or arms over an open dish or leave dishes open for long. Never leave dishes open when placed near the edge of the laminar flow hood. Place several specimen containers nearby to be used for discarding follicular aspirates.
4. Place one dish on the microscope stage warmer to be used for oocyte washing. Do not fill this dish with HEPES-buffered wash medium until just before the case begins.
5. Attach 26G needles to two tuberculin syringes and place under the hood. Do not remove plastic needle casings until use.

6. Flame polish two 9″ sterile Pasteur pipettes, attach pipette bulbs to both, and rest each on the pipette rack.
7. Make certain the thermometer in the heating block in the OR reads 37 °C. If it does not, delay the first case until the correct temperature is obtained; explain to the OR staff that optimal results are only realized with appropriate temperature conditions.

3 Collection of Oocytes

In conformance with general laboratory guidelines, never allow dishes from more than one patient to be placed under the hood at the same time.

While the first follicle is being aspirated, take a tube of HEPES-buffered wash medium from the heating block, remove the cap, and decant the contents into the sterile dish (labeled with patient name) on the microscope stage warmer. Pipette 5 mL of warmed, equilibrated mineral oil over this medium. Cover with the lid and push the dish aside until the first oocyte is found.

A stereomicroscope is utilized to search for oocytes in the follicular aspirate. As soon as the circulating nurse has placed the first aspirate in the heating block, take a prewarmed dish from the slide warmer and remove the lid. Decant approximately half of the follicular fluid into the dish, shake the remainder gently, and then decant the final volume into the same dish or the lid. Replace the cap on the tube and throw it in the Sharps container. Gently swirl the dish while looking for the presence of cumulus or other tissue masses of interest.

Once one or more oocytes is identified, use the first of the sterile Pasteur pipettes to pick up a small volume of the wash medium, then pick up the oocyte–corona–cumulus complex(es) (OCCs) in as little fluid as possible. Transfer these to the dish of HEPES-buffered wash medium and replace the lid. When it is determined that no more oocytes are to be found within the follicular aspirate (all aspirates checked twice), transfer the bloodied fluid to a specimen cup, cap, and discard the empty petri dish into the Sharps container (if not collecting this fluid for research purposes). When requested, save all aspirates from PGD patients for the PGD laboratory. Follow this procedure until all aspirates have been examined.

Physicians are usually anxious about the number of oocytes found, and they sometimes make decisions for or against flushing a follicle based on what is reported. It is recommended that the embryologist communicate this information as quickly as possible via the circulating nurse or intercom.

Once all oocytes have been collected, remove the plastic covers from the needles and proceed to reduce (dissect) the size of each cumulus mass, especially those which appear blood-stained or atretic (dark, clumped areas within the mass). While dissecting the cumulus, always keep the needles at a safe distance from the oocyte. Position needles at one pole of the oocyte, never on both sides. The cumulus may stick to the dish once dissection has been completed; detach the cumulus by moving a needle underneath the oocyte.

After dissection of the oocytes is completed, wash all oocytes thoroughly in a second dish of HEPES-buffered wash medium (if necessary). Move one of the patient's insemination dishes from the incubator to the microscope stage. Check the name on the bottom of the dish and read aloud—also check that the individual droplets within the dish are suitable for use without debris, contamination, or anything else that would compromise patient care. Use the second sterile Pasteur pipette to transfer the oocytes out of the wash dish into the center droplet of the insemination dish. Rinse oocytes several times in the center droplet and then in several droplets to the right of the name. Finally, place oocytes (1–8, depending on total number collected) into side droplets within the dish, starting with the one immediately to the left of the first letter of the name. If the dissection requires so much time that the oocytes are at risk of pH and osmolality changes, dissect a few oocytes at a time, transfer them to the insemination dish, and return the dish to the incubator. Return then to the wash dish for further dissections. While this procedure is in progress, one needs to keep several points in mind:

- Be aware of temperature and pH changes. Warmed mineral oil provides a buffer for additional time out of the incubator, but this time should not be overextended. When not being used, insemination dishes must be kept within a stable CO_2 environment; note the color of the medium as an indicator of pH.
- Speed in processing the oocyte is important. It is not wise to keep oocytes suspended in bloodied fluid for more than 1 or 2 min, since this is not consistent with their natural environment. If one falls behind (if there are more than three follicular aspirates in the heating block), politely request that the aspiration be stopped for a few minutes in order to finish processing the fluids.
- Evaporation quickly affects the osmolality of all media. While working with culture dishes, always make sure that all droplets are completely covered with oil.
- If follicular fluids are exceedingly bloody, replace the wash dish accordingly (prepare a second wash dish); do not continue with the same dish until the end of the retrieval.

- Do not keep oocytes in the dish of HEPES-buffered wash medium for prolonged periods of time. Never stop to chat or allow visitors or nurses to distract the process. Oocytes should be processed and put away at one's first opportunity.

Once the harvest is finished and all oocytes have been put away, record all relevant information. At this stage, evaluation for maturity is only a rough estimate, but it is important to comment on general appearance.

4 General Structure of the Oocyte

The diameter of the mature human oocyte is approximately 110–115 μm and is bounded by a plasma membrane called the *oolemma*. Surrounding the oocyte/oolemma is a glycoprotein envelope called the *zona pellucida*, a structure approximately 15–20 μm wide (becoming a bit thinner after fertilization) that protects the oocyte during transport and fertilization. Between the oolemma and the zona pellucida is the fluid-filled *perivitelline space*. The use of this term persists despite its inaccuracy when describing the oocytes of humans or most other mammals; it acknowledges the word *vitellus*, a term traditionally used to describe the yolky substance of a hen's egg which contains abundant nutrient reserves. The cytoplasm of the mammalian oocyte is usually referred to as the *ooplasm*, a more appropriate term for describing the living portion of the human gamete. The main organelles of the ooplasm are the mitochondria, the endoplasmic reticulum, and the Golgi system.

When fully capable of undergoing a normal fertilization process, the secondary oocyte is briefly arrested in its course of maturation at metaphase II of meiosis. Nuclear maturation is usually closely attended by a general maturation of the cytoplasm and is characterized by an increase in the number of organelles scattered throughout the ooplasm. The presence of a first polar body conveys that nuclear maturation has reached this stage.

Along with the zona pellucida and perivitelline space, the total diameter of the mature human oocyte is approximately 150 μm.

An oocyte incubated with spermatozoa before reaching metaphase II may incorporate a spermatozoon into its ooplasm and yet fail to initiate events leading to sperm decondensation; such an oocyte ultimately lacks a functional male pronucleus [1]. One study examining 518 non-fertilized oocytes demonstrated that 22 % had actually been penetrated by sperm, but without oocyte activation or pronuclear formation [2]. Many of these oocytes may have been immature when combined with spermatozoa.

Besides the requirement for nuclear maturation, it is believed that a brief period is necessary after extrusion of the first polar body for the oocyte to gain full cytoplasmic competence. An oocyte that is meiotically mature but slightly underdeveloped or overdeveloped with regard to its cytoplasm may be more apt to display one, three, or more pronuclei. With immature cytoplasm, the cortical granule numbers and response may be inadequate; with post-mature cytoplasm, cortical granule release may be inhibited due to the inward migration of the granules towards the interior of the cell. In either instance, there is evidence that the zona reaction is also often poorly functional when sperm–oocyte interaction is not appropriately timed with regard to oocyte nuclear and cytoplasmic maturity [3].

Oocytes collected for in vitro fertilization are generally surrounded by several layers of cells which define the *cumulus oophorus.* Cells of the cumulus are instrumental, via gap junctions, in nurturing the oocyte during growth and possibly in passing inhibiting factors (e.g., cAMP) necessary for deterring the resumption of meiosis [4]. The innermost layer of cells is called the *corona* or *coronal layer*. This layer expands and presents a radiant pattern as oocytes mature in response to exogenous hCG or a midcycle surge of LH. Near ovulation, as they loosen and expand, cumulus cells are observed to retract from the zona pellucida of the oocyte, presumably cutting off the previously important cellular–oocyte communication. It has been proposed that oocytes not associated with proliferative cellular changes near ovulation have very limited potential for implantation despite fertilization and apparently normal development in vitro [5].

In most mammalian species studied in vivo, the oocyte arrives at the site of fertilization in the ampulla of the fallopian tube still surrounded by the cumulus mass. The cumulus may play a role in assisting transport of the oocyte into the fallopian tube through fimbrial cilia–cumulus cell contact. Another possible use of the cumulus after oocyte maturation is that its radially arranged cells help to guide spermatozoa towards the oocyte just before fertilization; however, there is no hard evidence for this speculation. Breakup of the cumulus mass is brought about by dissolution of its mucoid hyaluronic acid matrix by enzymes released by the spermatozoa.

Follicular *membrana granulosa cells* disassociated from cumulus cells are found in follicular aspirates collected for in vitro fertilization. The number of cells collected will vary from follicle to follicle according to the extent of negative pressure exerted during the suction, the size of the needle, and the overall maturity of the follicle. As with cumulus cells, the correlation between morphological aspects of free granulosa cells and oocyte nuclear maturity is not exact, but mature-appearing cells (large, well-dispersed cells) are generally collected along with mature oocytes, and immature-appearing cells (smaller, tightly packed cells) along with immature oocytes. Follicular membrana granulosa cells may be assessed at the time of oocyte harvest to aid in the evaluation of follicular maturity.

They are subsequently often used during in vitro studies to examine metabolic activity or steroid synthesis.

The oocyte observed while its chromosomes are at metaphase I of maturation lacks a first polar body and requires some time in culture before attaining full meiotic competence [6]. Greater than 98 % of these oocytes will complete their journey towards metaphase II and first polar body extrusion.

Oocytes with chromosomes at prophase I of maturation exhibit a nucleus or germinal vesicle. More than 80 % of these will continue through metaphase I to metaphase II if isolated and incubated in an appropriate medium for 24 h.

5 Quality Evaluation: Assessment of Maturity

Traditionally, evaluation of oocyte maturity has been based upon the expansion and radiance of the cumulus–corona complex which surrounds the harvested oocyte [7, 8]. With this assessment, oocytes are rapidly categorized as mature (correlated to metaphase II of maturation) when they possess an expanded and luteinized cumulus matrix and a radiant or *sun-burst* corona radiata. A less expanded cumulus–corona complex denotes an intermediate stage of maturity (correlated to metaphase I of maturation), and absence of expanded cumulus is generally associated with immaturity (correlated to prophase I of maturation). While this type of analysis usually closely approximates the true nuclear status of the oocyte, it is too often imprecise and may lead to subsequent laboratory errors in the handling of gametes. In fact, nuclear maturation of the oocyte and cellular maturation of the cumulus are frequently disparate [8–12]. When disparity occurs, immature oocytes may be inseminated prematurely and fail to produce a favorable outcome. Besides fertilization failure, other detrimental side effects accompany combining sperm and eggs at suboptimal times; ovulation induction protocols may not be suitably appraised and male factor issues become difficult to interpret based on poor fertilization results.

Because of these pitfalls, techniques have been developed to more accurately assess the meiotic status of the oocyte. A systematic approach can be used to produce a *maturation score* by grading the size of the follicle, expansion of the cumulus mass, radiance of the corona cells, size/cohesiveness of associated membrana granulosa cells, and shape/color of the oocyte itself if visible within the mass of surrounding cellular investments. Alternatively, frank visualization of the oocyte and its germinal vesicle or first polar body can be attempted by spreading out the cumulus mass, or by removing it altogether with the aid of enzymes.

If clearly visible or denuded of cells, oocytes are classified according to the presence or absence of first polar bodies/germinal vesicles and are inseminated/injected accordingly:

Metaphase II (MII): First polar body present, no germinal vesicle; Inseminated or injected 3–5 h after collection

Metaphase I (MI): No first polar body, no germinal vesicle; Inseminated or injected 1–5 h after extrusion of the first polar body

Prophase I (PI): Germinal vesicle present; Inseminated or injected 26–29 h after collection

Our experience has been that oocytes collected at more advanced stages of in vivo maturation demonstrate the greatest ability to form two pronuclei after insemination [6, 13, 14]. Fertilization rates drop only slightly when oocytes require a period of 5–15 h in culture before extruding the first polar body, but fertilization is markedly reduced when more than 15 h pass before the maturational process is completed. The reason for this is likely related to sperm functionality as well as oocyte maturity since processed sperm may be more than 24 h old before being placed with an early MI or PI oocyte. Under these conditions, the precise cause of the lower incidence of fertilization of very immature oocytes is difficult to interpret [6].

Approximately 20–30 % of oocytes collected for in vitro fertilization are meiotically immature at the time of harvest from the ovary. This is undoubtedly due to the stimulation of multiple follicles during clinical ovulation induction, some large and well vascularized, and some small with late recruitment. If all oocytes are placed with sperm at the same time, a proportion slightly higher than this percentage will fail to become fertilized normally. Logically enough, when oocytes are placed with sperm only as they have reached full maturity, far better fertilization results are attained.

The incidence of abnormal fertilization (1 pronucleus, 3 or more pronuclei) is not different between MII oocytes and MI or PI oocytes that have matured in culture before insemination or injection [6, 13, 14]. Pregnancy potential after the transfer of preembryos developed from MII and MI oocytes is similar regardless of whether zero or 20 h has been required for maturation before insemination or injection [15]. Only preembryos developing from PI oocytes demonstrate a significantly reduced potential for implantation and live birth, although such births are certainly within the realm of possibility [6, 16–19].

6 The Metaphase II Oocyte (Figs. 1–3)

The MII oocyte is often termed *mature, ripened, or preovulatory*, vague descriptions that fail to specify the exact meiotic status of the gamete. This oocyte is at a resting stage of meiosis II after extrusion of the first polar body and direct passage to metaphase II. Chromosomes are divided between the oocyte and the polar body (23 chromosomes, 46 chromatids, $2n$ DNA in each), those in the oocyte being attached to spindle microtubules [1].

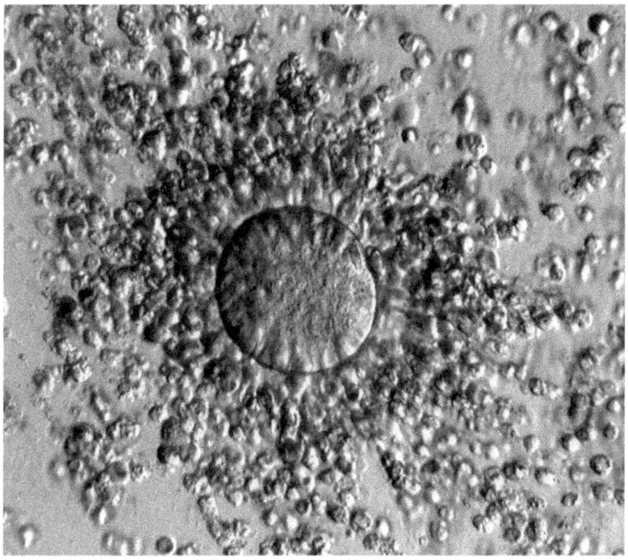

Fig. 1 Spreading out the cumulus enables one to visualize an oocyte more clearly. This oocyte exhibits homogeneously colored ooplasm and possesses a first polar body at 2 o'clock. Cells of the corona radiata are moderately expanded

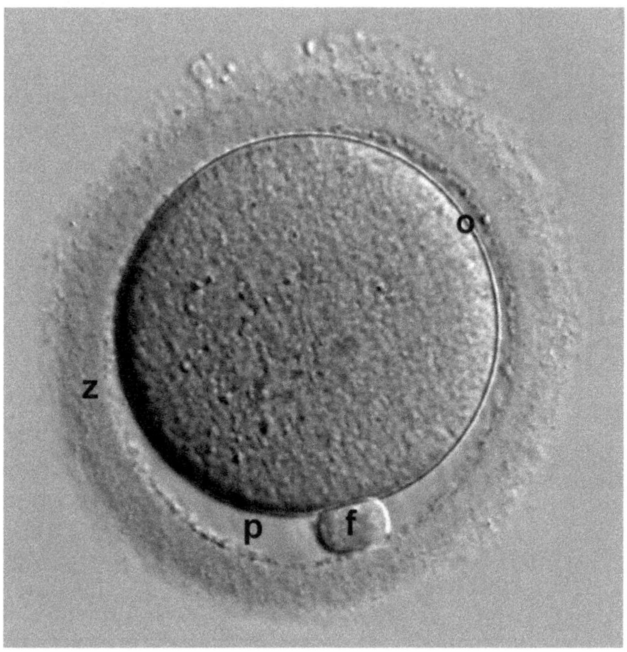

Fig. 2 Stripped of its cumulus cells, the fully mature human oocyte at metaphase II of meiosis is shown here (*o* oolemma, *p* perivitelline space, *f* first polar body, *z* zona pellucida)

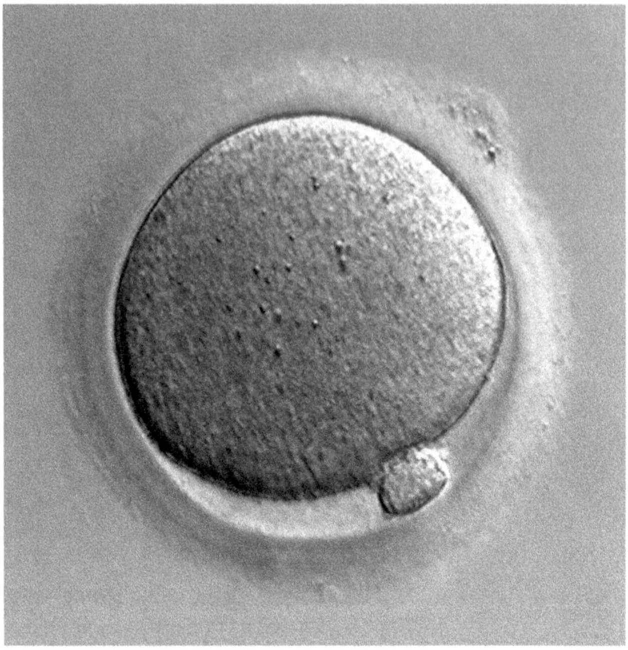

Fig. 3 An oocyte with chromosomes at metaphase II of maturation. Minute cytoplasmic inclusions speckle the oocyte surface. Minor defects are noted in the zona pellucida at 1 and 6 o'clock. The zona is relatively thin and compact

For a while after its formation, the first polar body remains connected to the oocyte by the meiotic spindle, forming a cytoplasmic bridge. Chromosomes within the first polar body may remain clumped together, may undergo a second meiotic division, or may scatter within the cytoplasm; generally a nucleus is not formed [1, 20]. The first polar body contains cortical granules because of its extrusion before sperm penetration and cortical granule release; in the oocyte, one to three layers of cortical granules are present at the periphery.

Under the microscope, the oocyte is characterized by its round, even shape and displays an ooplasm of light color and homogeneous granularity. It is usually associated with an expanded, luteinized cumulus and a *sun-burst* corona radiata. Membrana granulosa cells harvested along with the MII oocyte are loosely aggregated, with mature features [6, 8, 11, 19].

7 The Metaphase I Oocyte (Figs. 4–6)

The MI oocyte is considered *nearly mature* or *intermediate* in maturation. The oocyte has completed prophase of meiosis I; the germinal vesicle and its nucleolus have faded and disappeared. During this stage a spindle forms, and recombined maternal and paternal

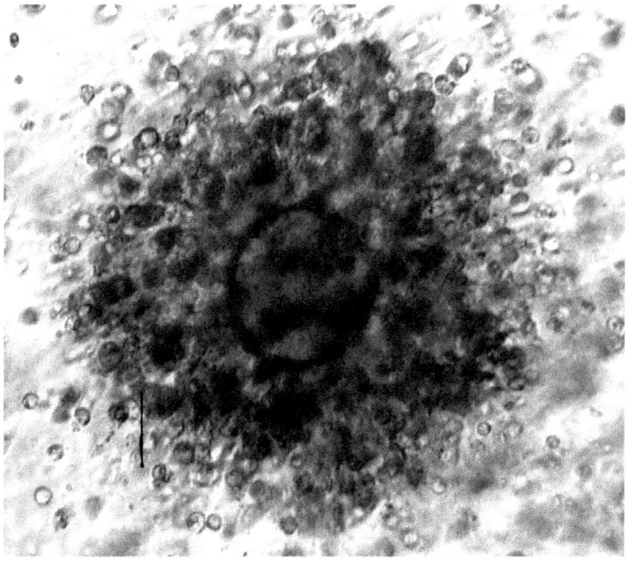

Fig. 4 A mature-appearing oocyte in its natural state near the time of ovulation. At ×100 magnification, the oocyte is seen surrounded by a radiant corona, many layers thick, and a less dense outer cumulus oophorus. Under the cover of these obscuring cells, the oocyte itself is difficult to evaluate, and therefore, precise maturational status is uncertain, but a first polar body cannot be visualized

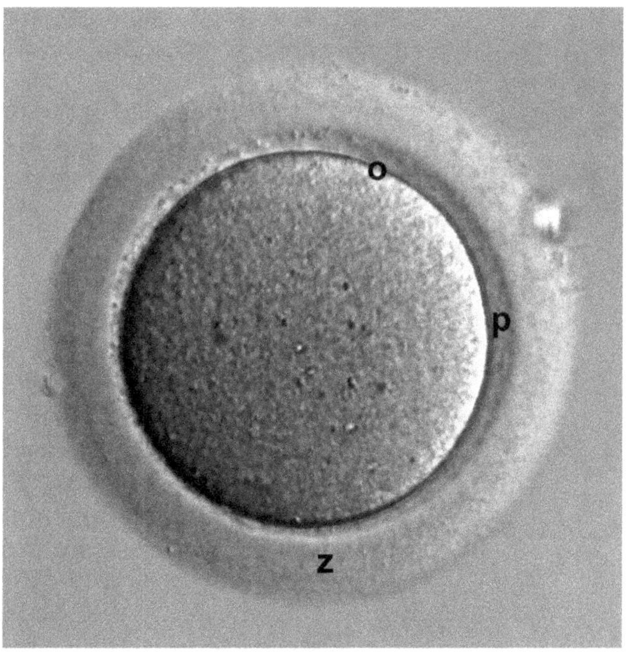

Fig. 5 The nearly mature human oocyte at metaphase I of meiosis (*o* oolemma, *p* perivitelline space, *z* zona pellucida, *note* no first polar body)

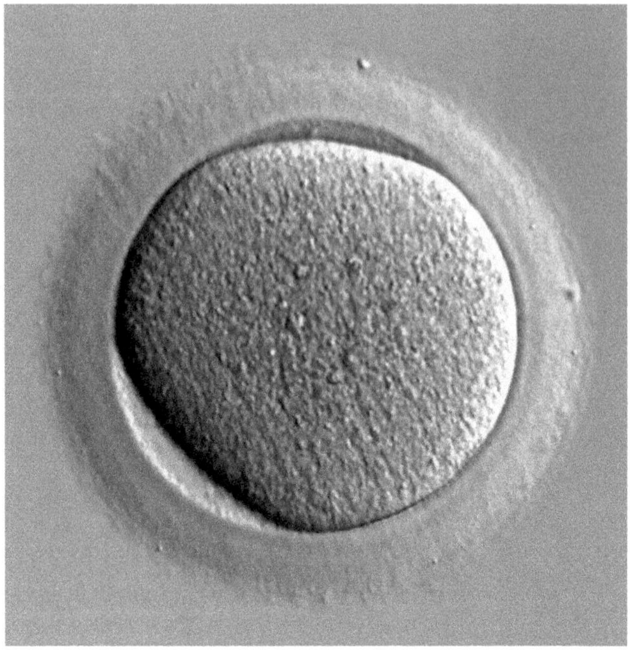

Fig. 6 Early metaphase I oocyte demonstrating *slightly darkened*, coarse ooplasm and an irregular shape at two poles. This oocyte has most likely undergone fairly recent germinal vesicle breakdown

chromosomes line up randomly towards the poles. Later, at telophase, whole chromosomes sort independently to oocyte or first polar body.

An MI oocyte requires 1–24 h in culture before reaching full maturity. Those needing less than 15 h are considered *late* in maturity, while those requiring more than 15 h are defined as *early* [6, 8, 11–14].

Under the microscope, the MI oocyte is characterized by the absence of both germinal vesicle and first polar body. A late MI oocyte is round and even in form, with homogeneously granular and light-colored ooplasm. Early MI oocytes may display minor central granularity. Mature-appearing cumulus cells are usually associated with late stages.

Because first polar body extrusion can occur at any time after harvest, it is necessary to examine the oocyte at regular intervals to determine the correct timing for insemination. If sperm are placed with the oocyte before nuclear and cytoplasmic maturation are complete, they generally fail to decondense within the ooplasm, or abnormal fertilization occurs. If insemination is delayed too long, in vitro aging may follow with similar undesired consequences [1, 6].

8 The Prophase I Oocyte (Figs. 7–9)

The PI oocyte is often termed *immature or unripened*. It possesses a tetraploid amount of DNA due to the presence of 46 double-stranded chromosomes. This oocyte begins to mature in response to gonadotropin surges and reduction in follicular maturation inhibiting factors. The germinal vesicle, which persisted throughout earlier growth phases, begins its progression to germinal vesicle breakdown (GVBD) and the oocyte enlarges. Most PI oocytes collected for in vitro fertilization have been stimulated to resume meiosis, are in the final stages of the first meiotic prophase, and have already reached full size. If a spermatozoon penetrates this immature oocyte, it will fail to promote activation since the oocyte is not meiotically mature, and its chromosomes will undergo premature condensation [21].

GVBD may occur within minutes or require up to several hours after harvest; the length of time appears to depend on how far maturational events have progressed within the follicle before collection. More than 80 % will succeed in passing through metaphase I of maturation to ultimately reach metaphase II [6].

The germinal vesicle, or nucleus, of the human oocyte is spherical and contains a large, refractile, exocentric nucleolus. Upon close examination, a second smaller nucleolus may be detected. The germinal vesicle is centrally located within the ooplasm of young PI oocytes and in those that exhibit developmental arrest.

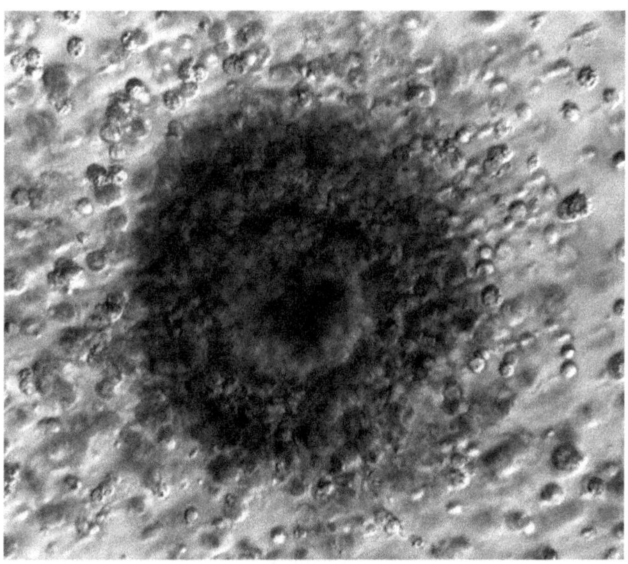

Fig. 7 An oocyte that appears undermature by the morphological aspects of the cumulus and corona cells. The coronal layer, although at least five cells deep, is compact rather than radiant, and completely obscures the oocyte. The oocyte itself appears *centrally darkened*, an observation often correlated to prophase I oocytes

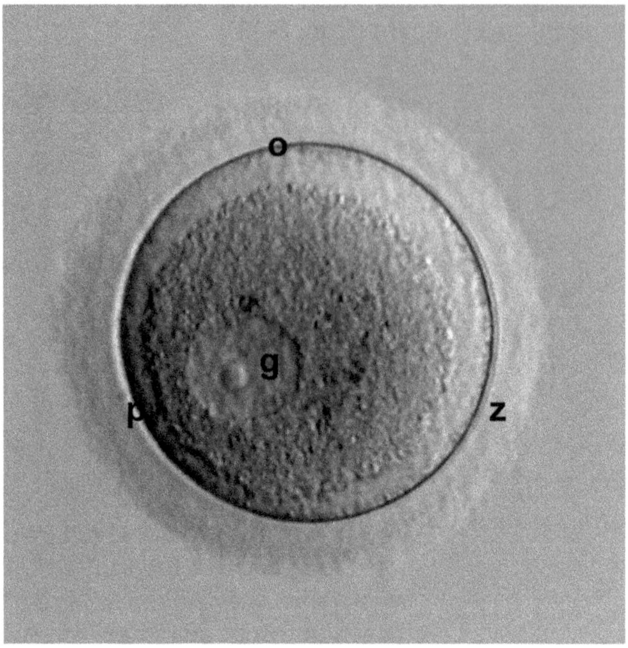

Fig. 8 The immature human oocyte at prophase I of meiosis (*g* germinal vesicle, *o* oolemma, *p* perivitelline space, *z* zona pellucida)

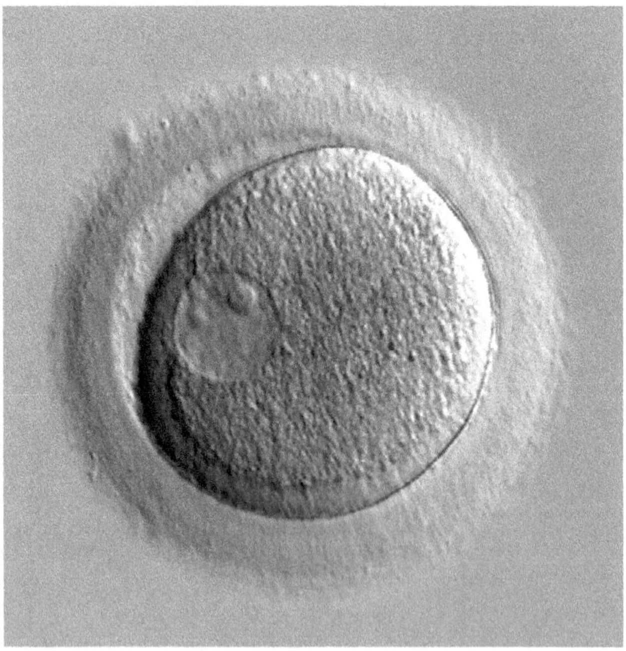

Fig. 9 A typical germinal vesicle-bearing oocyte showing coarse central granularity, a clear peripheral zone, and irregular shape. The germinal vesicle possesses a large, exocentric nucleolus and two smaller indistinct bodies. Because of the peripheral location of the germinal vesicle, it can be assumed that GVBD is imminent

It migrates to a more cortical position in healthy oocytes before GVBD. The dissolution of the germinal vesicle marks the first practical microscopic indication that meiosis has resumed. As the oocyte matures, defenses against polyspermy are established in the form of cortical granule accumulation and alignment at the oocyte periphery. These granules are sparse and discontinuous in immature oocytes [1].

Under the microscope, the PI oocyte is characterized by its distinct germinal vesicle and refractile nucleolus. An irregular shape, darkened center, and granular ooplasm are almost always displayed. Attached cumulus cells are usually compact and multilayered, but may be proliferative. Free follicular membrana granulosa cells within the immature follicle are usually small and appear in compact masses. PI oocytes with very mature characteristics of the cumulus (expanded appearance and very radiant corona) generally fail to undergo GVBD [11, 12].

9 The Zona Pellucida

The zona pellucida is the prime controller of interspecific fertilization and as such forbids non-species sperm from entering and contacting the oolemma. The human zona is a relatively thick (approximately 7–20 μm), transparent, acellular vestment that surrounds the oocyte and is separated from it by the perivitelline space. Under bright-field microscopy the healthy zona of an unfertilized oocyte appears clear and continuous, shows little variation in its thickness, and usually displays minor porosity and a slightly speckled aspect. Excessive porosity and breakup of the structure within a dispersed matrix of granular material can be associated with oocyte aging and deterioration. As the period of in vitro culture passes, the zona often acquires increased porosity and additional variation in shape and thickness.

10 Ooplasm

In our experience, approximately 8 % of oocytes collected for in vitro fertilization display a granular-appearing ooplasm at harvest. Fertilization rates are similar for clear and granular oocytes (91 % vs. 87 %) although triploid fertilization is marginally higher in the granular group.

11 Vacuoles

Ooplasmic vacuolization generally indicates a poor prognosis for optimal development of the oocyte. Vacuoles form by dilation and coalescence of vesicular elements of smooth endoplasmic reticulum, because of aggregation and fusion of preexisting smaller vacuoles, or may be the result of an aberrant endocytosis due to oolemma instability. Under the light microscope, vacuoles are noticeably membrane-bound, and appear vacant or empty. Although very small vacuolar structures (<5 μ) are relatively common and may not affect overall developmental potential, large vacuoles (>25 μ) probably represent defects that interfere with subsequent growth.

12 Abnormalities in Cytoplasmic Structure or Organization. From: Van Blerkom and Henry [22]

1. Dark cytoplasm, when examined under a dissecting microscope, and coarse cytoplasm when examined by DIC microscopy. This is highly associated with aneuploidy.
2. Extensive vesiculation indicated by the presence of highly irregular and coarse appearing cytoplasm.
3. Organelle and/or vesicle clustering, identified by the presence of a single, dark mass, usually centrally located. This is often associated with fragmented, pyknotic, and scattered chromosomes.
4. Organelle-free cytoplasm or partial cortical depletion of organelles. Very few of these were aneuploid.
5. Intracellular necrosis, indicated by small, membrane-limited vesicles containing dense inclusions.
6. A single massive accumulation of saccules of the smooth endoplasmic reticulum (SER), identified by the presence of a smooth, elliptical body of pronuclear dimensions. This is associated with aneuploidy.
7. Vacuoles presumed to be of endocytotic origin. This is not associated with aneuploidy.

13 Other Oocyte Abnormalities

1. Bilayered zona pellucida (ZP); excessively pigmented and thick ZP; perivitelline debris
2. Binovular or "double" oocytes: two oocytes surrounded by one ZP
3. Refractile body
4. Fragmented first polar body (PB); overly large PB; overly small PB
5. Severely misshapen oocyte

14 What to Do in the Event of

Medium contamination: Notify the laboratory director of any suspicious cloudiness or precipitation in culture media. Investigate further to see if potential contamination is restricted to one patient only, or affects others as well. If only one patient is affected, make new medium, dishes, and droplets, and proceed with the retrieval. If the potential contamination is noted after oocytes have already been placed in culture, prepare several wash dishes and wash all oocytes extensively before transferring to new droplets. In the unlikely event that the contamination affects several patients, a serious quality control problem has occurred and incubators require decontamination. The offending microbe will be identified by sending a sample to the hospital microbiology laboratory. Be sure to make notation on the patient's laboratory form if excessive tissue or epithelial cells (non-granulosa) are present in the follicular aspirates. These cells may be a source of potential bacterial contamination.

Known viral condition: If hepatitis, herpes, or another viral condition is known, use special safety precautions during all procedures requiring oocyte/sperm handling; wash hands immediately before and after procedures. Take extreme care to avoid spillage and any form of contact with human fluids. Warn other embryologists verbally and document condition on the patient laboratory form by means of a red circular label. Only those embryologists who have been immunized with HEPTAVAX* should perform procedures on hepatitis positive patients (it is recommended that each member of the laboratory staff be immunized).

Excessive blood clotting: Extensive dissection of each clot is required.

Obvious oocyte immaturity: A long incubation period extending to the next day may be necessary before insemination. Notify the laboratory director or supervisor for evaluation.

Obvious oocyte atresia: Do not inseminate.

No oocytes found during harvest: All aspirates must be checked by a second individual (or more) before discarding the fluid. The attending physician must be notified immediately.

References

1. Sathananthan AH, Trounson AO, Wood C (1986) Atlas of fine structure of human sperm penetration, eggs, and embryos cultured in vitro. Praeger, New York, p 2 (penetration of immature oocytes); p 4 (chromosomes and spindle microtubules); p 10 (cortical granules); p 42, p 126 (oocyte ageing); p 90 (polar body nucleus)
2. Van Blerkom J, Davis PW, Merriam J (1994) A retrospective analysis of unfertilized and presumed parthenogenetically activated human oocytes demonstrates a high frequency of sperm penetration. Hum Reprod 9:2381
3. Edwards RG (1980) Conception in the human female. Academic, New York, p 634

4. Dekel N, Beers WH (1980) Development of the rat oocyte in vitro: inhibition and induction of maturation in the presence or absence of the cumulus oophorus. Dev Biol 75:247
5. Gregory L, Booth AD, Wells C et al (1994) A study of the cumulus-corona cell complex in in-vitro fertilization and embryo transfer; a prognostic indicator of the failure of implantation. Hum Reprod 9:1308
6. Veeck LL (1988) Oocyte assessment and biological performance. Ann N Y Acad Sci 541:259
7. Veeck LL (1985) Extracorporeal maturation, Norfolk, Virginia. Ann N Y Acad Sci 442:357
8. Veeck LL (1986) The morphologic estimation of mature oocytes and their preparation for insemination. In: Jones HW Jr, Jones GS, Hodgen GD, Rosenwaks Z (eds) In vitro fertilization—Norfolk. Williams & Wilkins, Baltimore, p 81
9. Hammitt DG, Syrop CH, Van Voorhis BJ et al (1993) Maturational asynchrony between oocyte cumulus-coronal morphology and nuclear maturity in gonadotropin-releasing hormone agonist stimulations. Fertil Steril 59:375
10. Laufer N, Tarlatzis BC, DeCherney AH et al (1984) Asynchrony between human cumulus-corona cell complex and oocyte maturation after human menopausal gonadotropin treatment for in vitro fertilization. Fertil Steril 42:366
11. Veeck LL (1986) Atlas of the human oocyte and early conceptus, vol 1. Williams & Wilkins, Baltimore, p 7, p 127 (mature granulosa); p 57, p 68 (oocyte classification); p 74 (disparity between cumulus and nucleus)
12. Veeck LL (1991) Atlas of the human oocyte and early conceptus, vol 2. Williams & Wilkins, Baltimore, p 3, p 13, p 27 (membrana granulosa); p 13 (oocyte classification); p 27 (disparity between cumulus and nucleus)
13. Veeck LL (1990) Pregnancy rate and pregnancy outcome associated with laboratory evaluation of spermatozoa, oocytes, and preembryos. In: Mashiach S, Ben-Rafael Z, Laufer N, Schenker JG (eds) Advances in assisted reproductive technologies. Plenum Press, New York, p 745
14. Veeck LL (1990) The morphological assessment of human oocytes and early concepti. In: Keel BA, Webster BW (eds) Handbook of the laboratory diagnosis and treatment of infertility. CRC Press, Boca Raton, p 353
15. Coetzee K, Windt ML (1996) Fertilization and pregnancy using metaphase I oocytes in an intracytoplasmic sperm injection program. J Assist Reprod Genet 13:768
16. Barnes FL, Crombie A, Gardner DK et al (1995) Blastocyst development and birth after in-vitro maturation of human primary oocytes, intracytoplasmic sperm injection, and assisted hatching. Hum Reprod 10:3243
17. Cha KY, Koo JJ, Choi DH et al (1991) Pregnancy after in vitro fertilization of human follicular oocytes collected from non-stimulated cycles, their culture in vitro, and their transfer in a donor oocyte program. Fertil Steril 55:109
18. Liu J, Katz E, Garcia JE et al (1997) Successful in vitro maturation of human oocytes not exposed to human chorionic gonadotropin during ovulation induction, resulting in pregnancy. Fertil Steril 67:566
19. Veeck LL, Wortham JWE, Witmyer J et al (1983) Maturation and fertilization of morphologically immature human oocytes in a program of in vitro fertilization. Fertil Steril 39:594
20. Austin CR (1961) The mammalian egg. Blackwell Press, Oxford, 75
21. Planchot M, Crozet N (1992) Fertilization abnormalities in human in vitro fertilization. Hum Reprod 7 suppl 1:89
22. Van Blerkom J, Henry G (1992) Oocyte dysmorphism and aneuploidy in meiotically mature human oocytes after ovarian stimulation. Hum Reprod 7:379–390

Chapter 17

Sperm Retrieval and Quality Evaluation

Peter J. Stahl, Peter N. Schlegel, and Marc Goldstein

Abstract

Technical refinements in sperm retrieval methods and the application of advanced reproductive technologies (ART) using surgically retrieved sperm have enabled biological paternity in azoospermic men who were considered untreatable 20 years ago. Achievement of optimal reproductive outcomes in these patients benefits greatly from a multistep, interdisciplinary process of sperm acquisition that involves reproductive endocrinologists, urologists, or other specialists in male subfertility, and laboratory personnel with expertise in characterizing and isolating sperm from surgically retrieved specimens. The critical steps in this process are discussed in this chapter.

Key words Microsurgical epididymal sperm aspiration, MESA, Male factor, Sperm retrieval, Testicular fine needle aspiration, Testicular large needle aspiration, Testicular sperm extraction, TESE, Microdissection testicular sperm extraction, mTESE, Azoospermia, Percutaneous epididymal sperm aspiration, PESA, Testicular core needle biopsy, Testicular percBiopsy

1 Introduction

The clinical evaluation of azoospermia is reviewed, which provides critical information about the nature of the azoospermia (obstructive vs. nonobstructive), determines adjuvant treatments to enhance sperm production prior to retrieval, and guides selection of the optimal method for sperm retrieval. Methods for sperm retrieval from the epididymis and testis are discussed, including percutaneous epididymal sperm aspiration (PESA), microsurgical epididymal sperm aspiration (MESA), percutaneous testicular fine (FNA) and large needle aspiration (LNA), percutaneous core needle testicular biopsy (percBiopsy), conventional testicular sperm extraction (TESE), and microdissection TESE (mTESE). Recommendations and methodology for the intra-procedural assessment of sample quality to guide the nature and extent of sperm retrieval procedures are provided.

Zev Rosenwaks and Paul M. Wassarman (eds.), *Human Fertility: Methods and Protocols*, Methods in Molecular Biology, vol. 1154, DOI 10.1007/978-1-4939-0659-8_17, © Springer Science+Business Media New York 2014

2 Clinical Evaluation of Azoospermic Men

Proper and thorough evaluation of the azoospermic man is essential for determining whether the infertility is treatable, the need for adjunctive fertility therapy (i.e., varicocelectomy or gonadotropin replacement), and the optimal method for surgical sperm retrieval. Azoospermia may be associated with occult health-threatening pathology, such as testicular cancer that can be missed without complete evaluation [1–4]. Although thorough discussion of the evaluation for azoospermia is beyond the scope of this chapter, several critical points that affect the choice of sperm retrieval procedure must be emphasized. The key aspects of the evaluation are determination of the nature of azoospermia (obstructive vs. nonobstructive) and assessment of the risk for testicular complications.

The history is important for identification of factors that may predispose to infertility. For example, a history of cryptorchidism or gonadotoxic exposure (such as chemotherapy or radiation for cancer treatment) suggests nonobstructive azoospermia (NOA), whereas a history of vasectomy or inguinal herniorrhaphy after prior proven fertility indicates obstructive azoospermia (OA). Moreover, the patient's medical history is the cornerstone of the risk assessment for complications after sperm retrieval procedures. Patients with bleeding disorders (either congenital or pharmacologically induced), low serum testosterone levels, and solitary testes may be at higher risk for adverse outcomes and should be managed accordingly.

The volume, consistency, and presence of each testis, epididymis, and vas deferens should be assessed by physical examination. Small soft testes suggest impaired sperm production; while indurated epididymides or absence of palpable vasa deferentia with normal volume testes suggests obstruction. Each testis should also be examined for indurated areas suggestive of intra-testicular mass. The spermatic cords should be inspected and palpated for varicoceles, which do not generally cause azoospermia but may adversely affect sperm production and are surgically correctable [5].

Laboratory testing should include a semen analysis that must confirm azoospermia by analysis of the pellet derived from centrifugation, and serum tests for testosterone (T) and follicle stimulating hormone (FSH) [6]. Normal serum hormone levels suggest but are not diagnostic of OA, particularly in the setting of normal testicular volumes. Although individual laboratory normal ranges are highly variable, it is important to recognize that young healthy males should have a serum FSH less than 7.6 IU/L. Low T and elevated levels of serum FSH typically suggest NOA. It is particularly important to identify the rare patient with unrecognized hypogonadotropic hypogonadism (low T, FSH, and LH) because spermatogenesis may be restored by treatment of the underlying disorder (i.e., dopamine agonist therapy for macroprolactinoma) or with gonadotropin replacement therapy in primary hypogonadotropic hypogonadism [7]. In contrast, men with low T and normal or elevated FSH levels suffer from NOA due to primary testicular failure. These patients are much more common and typically require surgical sperm retrieval for reproduction.

The laboratory evaluation of azoospermic men should also include genetic testing [6] that can be critically informative about both the prognosis for sperm retrieval and the risk of disease or subfertility in offspring conceived by ART. In OA, testing for CFTR mutations is indicated for men with non-palpable vasa deferentia or unexplained OA with low ejaculate volumes. Seventy percent of such men test will test positive [8]. In NOA patients, testing for karyotypic abnormalities and Y chromosome microdeletions is indicated.

Karyotypic abnormalities including 47XXY (Klinefelter's syndrome), are found in 14 % of men with NOA. Klinefelter's syndrome is extremely common with a prevalence of 1/500–1/1,000 men within the general population. Affected men often go undiagnosed and may initially present with infertility and NOA. Reproduction is possible in over 60 % of men with Klinefelter's syndrome using microdissection testicular sperm extraction followed by intracytoplasmic sperm injection (ICSI). It appears that these offspring are not at significantly increased risk of Klinefelter's syndrome or other cytogenetic abnormalities [9].

Y chromosome microdeletions are transmissible genetic lesions that are detected in up to 10 % of American men with NOA. The most common is the AZFc deletion that comprises 40 % of the microdeletions detected in azoospermic men. Sperm retrieval is possible in two-thirds of cases. The remaining 60 % of Y microdeletions detected in azoospermic men involve loss of the more centromeric AZFa and AZFb regions (or combinations of AZFb and c regions) of the Y chromosome. Unfortunately the latter microdeletions are associated with a near-zero chance of sperm retrieval, and must be identified for triage of affected patients to use of donor sperm or adoption [10]. Whenever a genetic anomaly is detected, genetic counseling is mandatory prior to utilization of ART.

3 Selection of the Optimal Sperm Retrieval Procedure

The sperm retrieval technique should be selected that minimizes trauma to the male reproductive tract and yields the highest quality sperm in a quantity sufficient to satisfy the patient's immediate and future reproductive goals. Sperm retrieval techniques vary by anatomic target (epididymis vs. testis) and the method by which sperm

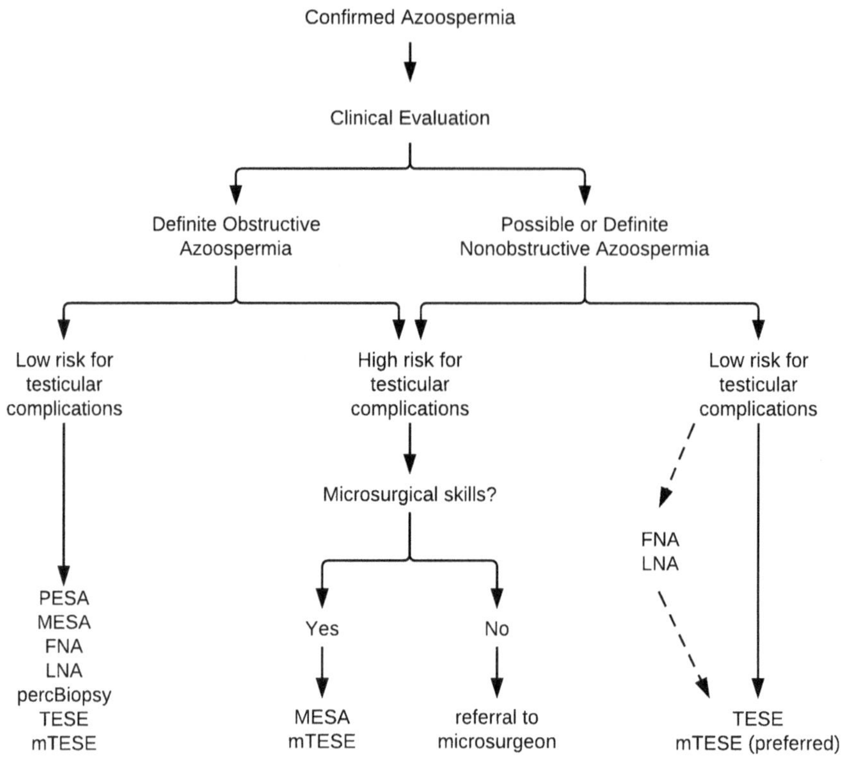

Fig. 1 Flowchart for determining optimal sperm retrieval method

acquisition is attempted (percutaneous vs. surgical). Surgical techniques furthermore vary by whether or not the procedure is assisted by intraoperative optical magnification (conventional vs. microsurgical). The optimal method for sperm retrieval depends upon the etiology of azoospermia, the patient-specific risk of testicular complications, the reproductive goals of the patient, the technical capabilities of the embryology laboratory, and the skill set and preferences of the clinician performing the sperm retrieval procedure (Fig. 1).

3.1 Sperm Retrieval Procedures in Patients with Obstructive Azoospermia

Obstructive azoospermia is characterized by diffuse distribution of normal or near-normal sperm production throughout the testes. Millions of sperm accumulate in the dilated portions of the vas deferens and epididymis on the testicular side of the anatomic point of obstruction. In most cases, motile sperm may be found in the proximal portions of the epididymis and even in the testes of men with OA [11].

Given the wide distribution of sperm throughout the testes and epididymides in men with OA, sperm retrieval is possible in most cases using any technique. However, each method is associated with different anesthetic requirements, safety profiles, required surgical skills, and sperm yields (Table 1). MESA is the retrieval technique that reliably yields sufficient sperm for both immediate

Table 1
Characteristics of sperm retrieval procedures for men with obstructive azoospermia

Procedure	Anesthetic	Required skills	Sperm retrieval rate (%)	Yield
MESA [30, 38, 39]	General (preferred), regional plus sedation, or local plus sedation	Scrotal exploration, microsurgery	95–100	Average of 15–95 × 10^6 total sperm with 15–42 % total motility, cryopreservation possible in 98-100 % of cases with an average of 5.3–7.6 vials per patient
PESA [15, 23–27]	Local with or without sedation	Spermatic cord block, percutaneous needle aspiration	80–100	Thousands to millions of sperm with variable motility (poorly reported in most studies), cryopreservation possible in 43–96 % of cases
Testicular FNA (21–23G) [29, 30]	Local with or without sedation	Spermatic cord block, percutaneous needle aspiration	52–100	Hundreds of thousands to millions of sperm with variable motility (poorly reported in most studies), cryopreservation possible in 38 % of cases in one study [29]
Testicular LNA (14–20 G) [15, 29, 32]	Local with or without sedation	Spermatic cord block, percutaneous needle aspiration	98–100	Hundreds of thousands to millions of sperm with variable motility (poorly reported in most studies), cryopreservation possible in 100 % of cases in one study [29]
Testicular percBiopsy [30, 35, 36]	Local with or without sedation	Spermatic cord block, percutaneous needle biopsy	82–100	Hundreds of thousands to millions of sperm with variable motility (poorly reported in most studies), often sufficient for cryopreservation (poorly reported)
TESE [54]	General, regional with or without sedation, or local with or without sedation	Scrotal exploration, excisional testis biopsy	100	Hundreds of thousands to millions of sperm in most cases (poorly reported in most studies), usually sufficient for cryopreservation (poorly reported)
mTESE [54]	General (preferred), regional plus sedation, or local plus sedation	Scrotal exploration, microsurgery	100	Hundreds of thousands to millions of sperm in most cases (poorly reported in most studies), usually sufficient for cryopreservation (poorly reported)

assisted conception and cryopreservation for future reproduction. Blind percutaneous procedures and testicular retrievals should be avoided in patients at high risk for testicular complications, including men with solitary testes, those with congenital or pharmacologically induced bleeding diatheses, and patients with low serum T who are at risk for post-procedural androgen deficiency. The safest approach in these patients is MESA performed by an experienced microsurgeon.

3.2 Sperm Retrieval Procedures in Patients with Nonobstructive Azoospermia

Nonobstructive azoospermia is characterized by severely deficient or absent sperm production such that sperm are not present in the ejaculate. However, testicular sperm are retrievable in more than half of men with NOA [12]. Histological evaluation of the testes of men with NOA commonly demonstrates a predominance of seminiferous tubules devoid of germ cells (Sertoli cell only pattern) or with germ cells that fail to mature to elongated spermatids (maturation arrest pattern). However, small foci of complete spermatogenesis are often present in these testes with more complete evaluation than is possible with a limited random biopsy.

The fundamental understanding that sperm production is randomly and heterogeneously distributed throughout the testes is the most important principle to consider when selecting a sperm retrieval procedure for an NOA patient. Regardless of the method by which retrieval is attempted, the objective is to maximize the likelihood of sampling the small sperm-containing areas of the testes. Several strategies have been employed, including removal of a large volume of randomly selected seminiferous tubular tissue in a single biopsy (TESE), removal of multiple random specimens from multiple areas of each testis (multiple site TESE), and percutaneous testicular needle aspiration (FNA or LNA) or core needle biopsy (percBiopsy) from multiple random sites within each testis. Most recently, mTESE has been developed wherein larger diameter, more opaque seminiferous tubules are visually identified and selectively sampled with the aid of optical magnification provided by an operating microscope [13].

The optimal method for sperm acquisition in men with NOA remains controversial. The relevant literature is comprised predominantly of retrospective, non-randomized observational studies. Sufficiently powered randomized prospective trials comparing sperm retrieval methods have not been performed in men with NOA. Nonetheless, significant evidence suggests that open surgical sperm retrieval is more effective than percutaneous procedures for sperm acquisition in men with NOA. The characteristics of the sperm retrieval procedures most commonly used in clinical practice for treatment of NOA are listed in Table 2.

Although several groups have reported SRRs for percutaneous procedures in men with NOA that are similar to the published SRRs for open surgical approaches [14, 15], most studies that have directly compared percutaneous and surgical retrieval have

Table 2
Characteristics of sperm retrieval methods used in men with nonobstructive azoospermia

Procedure	Anesthetic	Required skills	Sperm retrieval rate (%)	Yield
Testicular FNA (21–23 G) [14, 28, 54]	Local with or without sedation	Spermatic cord block, percutaneous needle aspiration	17–59	<10 to thousands of sperm, cryopreservation possible in one study in 42 % of cases when sperm were retrieved [14]
Testicular LNA (14–20 G) [14–16, 28, 32, 55]	Local with or without sedation	Spermatic cord block, percutaneous needle aspiration	26–60	<10 to thousands of sperm (poorly reported in most studies), cryopreservation possible in some cases (poorly reported)
TESE [16, 21, 41]	General, regional with or without sedation, or local with or without sedation	Scrotal exploration, excisional testis biopsy	17–70	<10 to thousands of sperm (poorly reported in most studies), cryopreservation possible in some cases (poorly reported)
mTESE [21, 41]	General (preferred), regional plus sedation, or local plus sedation	Scrotal exploration, microsurgery	33–77	<10 to thousands of sperm (poorly reported in most studies), cryopreservation possible in some cases (poorly reported)

demonstrated that open surgery is associated with higher sperm retrieval rates [16, 17]. Furthermore, several studies have shown that sperm may be immediately identified with open surgery in a significant proportion of cases when percutaneous retrieval has failed [18, 19]. For these reasons, the American Urological Association recommends TESE or mTESE as the procedures of choice for sperm retrieval in men with NOA [20].

Moreover, most comparisons of TESE with mTESE have demonstrated that mTESE yields higher SRRs, increased quantities of retrieved sperm from smaller tissue fragments, and a lower risk of testicular atrophy or fibrosis [21]. However, mTESE requires significantly longer operative times and microsurgical expertise that is not accessible at many centers.

4 Percutaneous Methods for Sperm Retrieval

Percutaneous sperm retrieval techniques are minimally invasive procedures that may be performed in the outpatient setting under local anesthesia. They are considerably more convenient, less costly, and more accessible than open surgical procedures because neither an operating room nor an anesthesiologist nor a trained surgeon is required. However, these procedures are performed blindly and pose a theoretical risk of injury to the testis, epididymis, and spermatic cord structures. In most reported series, however, significant complications are rare. Perhaps of more significance than the theoretical risk of complications is that percutaneous procedures generally yield lower quantities and quality of sperm with greater sample contamination with red blood cells than open surgical procedures, particularly when the open procedures are assisted by optical magnification. Therefore percutaneous sperm retrieval procedures are best suited to uncomplicated patients for whom cryopreservation of sperm for subsequent attempts at assisted reproduction is not a high priority.

4.1 Local Anesthesia for Percutaneous Sperm Retrieval Procedures

Successful administration of local anesthetic enables office-based, percutaneous sperm retrieval in appropriately selected cases. Anesthesia of the scrotum and scrotal contents is best achieved when a spermatic cord block is combined with infiltration of the scrotal skin with a local anesthetic, as the skin and scrotal contents are innervated separately (the skin by cutaneous branches of the pudendal nerve, and the scrotal contents by branches of the genitofemoral and ilioinguinal nerves as well as sympathetic and parasympathetic nerves within the spermatic cord). This approach is safe and provides highly satisfactory or satisfactory anesthesia in 92 % and 8 % of cases, respectively [22].

Suggested equipment and medications for spermatic cord block are topical anesthetic cream (i.e., 2.5 % lidocaine/2.5 % prilocaine cream, EMLA), local anesthetic without epinephrine (i.e., 0.25 % bupivacaine mixed 1:1 with 0.5 % lidocaine) buffered with $NaHCO_3$ (local anesthetic mixture mixed 9:1 with 8.4 % NaHCO3 solution), a 20 mL syringe with finger holes for one-handed aspiration and injection, a 25–30 G 1.5 cm hypodermic needle, sterile drapes, and antiseptic solution. Administration of an oral or sublingual benzodiazepine such as clonazepam 1 h prior to the procedure may facilitate patient relaxation and comfort. Topical anesthetic cream is applied to the scrotal skin for 15 min, after which the scrotum is washed with antiseptic solution and isolated with sterile drapes.

The spermatic cord is identified at the inguinal–scrotal junction after it has exited the external inguinal ring or within the high scrotum. It is grasped between the thumb and index or middle finger of the surgeon's non-dominant hand and isolated, making sure that the vas deferens is included. The needle is inserted perpendicularly to the spermatic cord and advanced until the surgeon is certain that the tip of the needle has reached the most posterior aspect. The spermatic cord is then infiltrated with 10 mL of buffered local anesthetic as the needle is slowly withdrawn. Manipulation

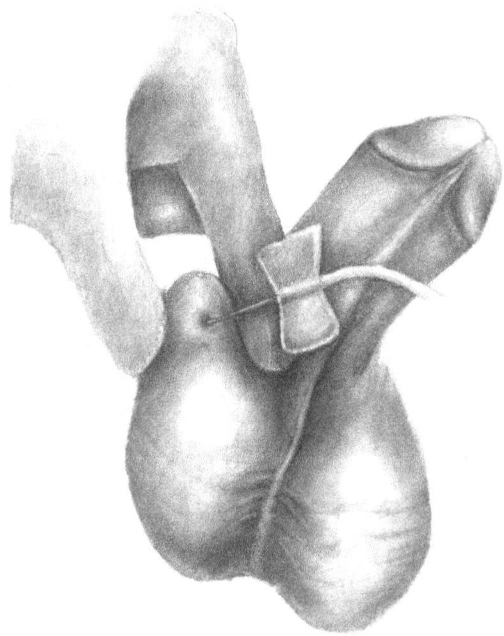

Fig. 2 Percutaneous epididymal sperm aspiration (PESA)

of the needle within the spermatic cord is minimized so as to decrease the chance of vascular or vasal injury. After completion of the spermatic cord block, the scrotal block is completed by subcutaneous infiltration of the scrotal skin at the intended site of the procedure with local anesthetic solution (e.g., 1 % lidocaine with epinephrine). It may be helpful to buffer the acidity of this solution with a 1:9 addition of bicarbonate solution to the anesthetic to limit the acidity of fluid that can have contact with the retrieved sperm as well as to limit the pain of injection of an acidic solution in the scrotum.

4.2 Percutaneous Epididymal Sperm Aspiration

Percutaneous epididymal sperm aspiration (PESA) (Fig. 2) is typically performed in the outpatient setting after administration of local anesthesia, with or without oral or intravenous sedation. It is a simple and safe method that is effective in approximately 81–100 % of men with OA [23–28]. PESA is only effective when the epididymis is dilated due to obstruction, and therefore should only be used in patients with OA. Equipment needed for the procedure includes two 1 mL syringes, a 21–28 G butterfly needle connected to silicone tubing, human tubule fluid (HTF) or another appropriate sperm transport solution, a sperm collection container, and a benchtop microscope.

The epididymis is stabilized between the thumb and forefinger of the surgeon's non-dominant hand. The butterfly needle is inserted through the skin into the caput epididymis. Constant and gentle negative pressure is applied by aspirating with a 1 mL syringe attached to the silicone tubing of the butterfly needle. The needle is manipulated within the epididymis until fluid is returned. Some investigators prefer to use a standard straight needle connected directly to a 1 mL syringe. It is important to note that in some cases the amount of fluid aspirated is so small that it is not easily visible. After aspiration, the epididymal aspirate is flushed with a small amount of sperm transport solution into the sperm collection container. A droplet of the collected fluid is placed on a slide and examined microscopically for the presence, quantity and motility of sperm. Multiple or bilateral punctures may be required to obtain sufficient sperm-containing fluid. We have generally avoided PESA because of the admixture of sperm with high DNA fragmentation with the best sperm that necessarily occurs with this procedure.

4.3 Testicular Sperm Aspiration (FNA and LNA)

Testicular sperm aspiration is generally considered the least involved and technically easiest procedure for acquiring sperm. It may be performed in the office setting under local anesthesia, requires little time, is well-tolerated, does not require advanced surgical or microsurgical training, and is generally safe. Testicular needle aspiration techniques are more versatile than PESA and may be used in patients with both OA and NOA, although the sperm retrieval rates (SRR) are lower in patients with NOA. Generally, these procedures may be subdivided into FNAs that use small bore needles (21–23 G) and LNAs that use large bore needles (14–20 G).

Reported SRRs in OA patients with FNA and LNA are 52–100 % [29–31] and 98–100 % [15, 29, 32], respectively. Reported SRRs in NOA for FNA and LNA are 7–59 % [14, 17, 28, 33, 35] and 26–60 % [15, 16, 28, 34], respectively. The wide range of reported SRRs with testicular sperm aspiration in NOA reflects significant inter-study variability in patient characteristics, needle aspiration technique, needle size, number of testicular punctures, and surgeon experience.

Required equipment includes a 10 or 20 mL syringe with finger holes for one-handed creation of negative pressure during aspiration, a smaller 1–5 mL flushing syringe filled with human tubule fluid (HTF) or another appropriate sperm transport solution, a 14–23 G hypodermic or butterfly needle connected to silicone tubing, a clamp for occlusion of the silicone tubing (if used), a sperm collection container, and a benchtop microscope.

The testis is stabilized with the surgeon's non-dominant hand within the taut scrotal skin after a spermatic cord and scrotal skin block. The needle is inserted through medial or lateral surface of the testis near the anterior mid-pole, where the testes are relatively avascular, and directed towards the contralateral upper pole while constant, negative pressure is applied (Fig. 3). It is important to avoid the midline of the lower pole because of the critical location of the

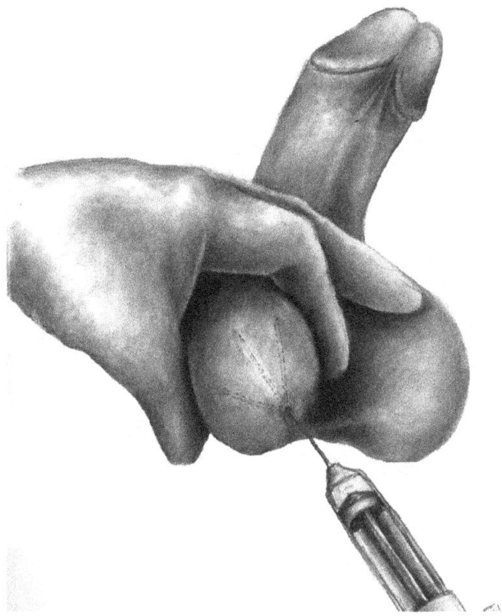

Fig. 3 Needle aspiration of testis

testicular blood supply in this region of the testis under the tunica albuginea. The needle may be gently redirected and manipulated within the testis without exiting the entry site through the tunica albuginea, so as to maximize the area of the testis sampled while minimizing intrusions through the vascular tunica albuginea. If a butterfly needle is used, the silicone tubing is clamped to maintain negative pressure prior to removal of the needle from the testis, after which the 20 mL syringe is exchanged for the smaller flushing syringe containing sperm transport solution.

The contents of the tubing are then flushed into the sperm collection container. If a butterfly needle is not used, the aspirated fluid and/or tissue fragments in the needle and syringe are simply flushed into the sperm collection container. A droplet of the collected fluid is placed on a slide and examined microscopically for the presence, quantity and quality (morphology and motility) of sperm. If significant tissue fragments were obtained, those fragments should be mechanically disrupted prior to microscopic evaluation of the specimen. Multiple punctures are often required.

4.4 Testicular Core Needle Biopsy (Testicular PercBiopsy)

Testicular percBiopsy (Fig. 4) is another minimally invasive percutaneous technique whereby sperm may be harvested from the testis in the office setting under local anesthesia. In contrast to TESA and PESA, testicular percBiopsy provides morphologically undisturbed testicular tissue fragments that are sufficient for standard histologic assessment of spermatogenesis. Most studies of testicular percBiopsy for sperm retrieval have focused upon men

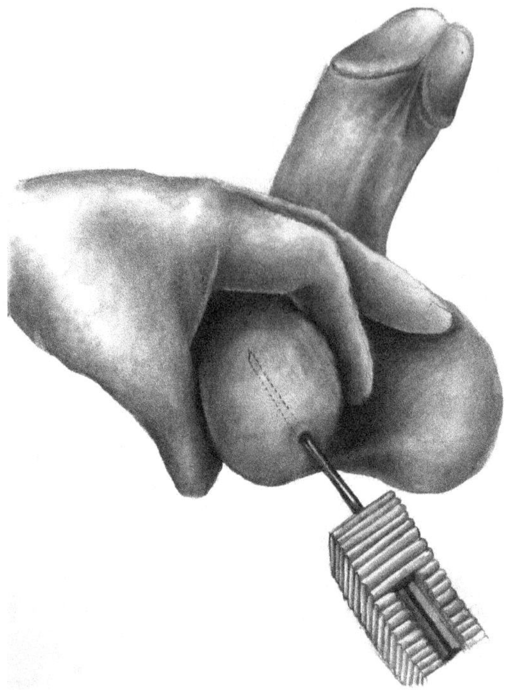

Fig. 4 Testicular core needle biopsy

with OA, in whom reported SRRs are 82–100 % [30, 35, 36]. One small study has also reported successful sperm retrieval in 60 % of 10 men with NOA using testicular percBiopsy [37].

Testicular percBiopsy is performed with an automatic spring-loaded needle with an outer cannula and an inner, notched rod in which a testicular tissue fragment is automatically cut, trapped and withdrawn from the testis upon firing (i.e., Trucut needle). A 14 G spring-loaded needle with a short 1 cm excursion is ideal. In addition to the spring-loaded biopsy needle, other required equipment includes a tissue collection container filled with human tubule fluid (HTF) or another appropriate sperm transport solution, a separate specimen container containing Bouin's solution or another appropriate fixative (if diagnostic biopsy is clinically indicated), and a benchtop microscope. A scalpel for mini-incision of the scrotal skin is optional.

The testis is stabilized with the surgeon's non-dominant hand within the taut scrotal skin after a spermatic cord and scrotal skin block. A mini-incision of the scrotal skin with a pointed scalpel blade avoids contamination of the sample with skin and facilitates accurate placement of the biopsy needle tip prior to firing. The needle tip is placed through the skin or skin mini-incision until the tip of the needle is in direct contact with the tunica albuginea of the testis. Gentle pressure is applied to maintain steady contact with the tunica albuginea and the spring-loaded biopsy needle is fired. The needle should be directed such that its path is through the relatively avascular medial or lateral anterior mid-pole towards the contralateral upper pole.

After each biopsy, the spring-loaded needle is removed and the tissue is placed in the proper specimen container. If diagnostic biopsy is being performed, manipulation of the biopsy specimen should be minimized. Core needle biopsies obtained for the purpose of sperm retrieval should be mechanically disrupted within the sperm transport media to liberate sperm from intact seminiferous tubules within the biopsy prior to microscopic analysis. Multiple core biopsies may be obtained through the same entry points in the skin and testis.

4.5 Post-procedural Care Following Percutaneous Sperm Retrieval

Patients who undergo percutaneous sperm retrieval procedures should be treated similarly to any patient who undergoes a minor surgical procedure. Immediate, appropriate post-procedural care, activity restrictions, and appropriate clinical follow-up are critically important to decrease the risk of complications and to ensure that any complications that do arise are expeditiously managed. Ice and compression should be applied to the scrotum immediately following any sperm retrieval procedure, including those performed percutaneously. Placement of an ice pack within an appropriately sized scrotal supporter for 30 min is usually sufficient. Patients should be advised to rest for 1–2 days, to minimize physical activity for 2 weeks, and to avoid heavy lifting for at least 4 weeks. Mild post-procedural scrotal swelling is expected, but significant swelling or scrotal pain should be promptly evaluated. Clinical follow-up including physical examination of the scrotum is mandatory and should occur within 2 weeks of the procedure. Post-procedural scrotal ultrasonography is not required in routine cases but is an excellent modality by which possible complications may be investigated.

5 Surgical Methods for Sperm Retrieval

Open surgical methods for sperm retrieval from the epididymis and testis are the most commonly performed sperm retrieval procedures and are considered by most experts to be the gold standard [21]. During these procedures, the testis or epididymis is surgically exposed and epididymal fluid or testicular tissue is extracted under direct vision, with or without optical magnification provided by an operating microscope.

5.1 Microsurgical Epididymal Sperm Aspiration

Microsurgical epididymal sperm aspiration (MESA) (Fig. 5) is the gold standard approach for epididymal sperm retrieval in OA because it yields millions of high quality, motile sperm with minimal specimen contamination with red blood cells [38, 39].

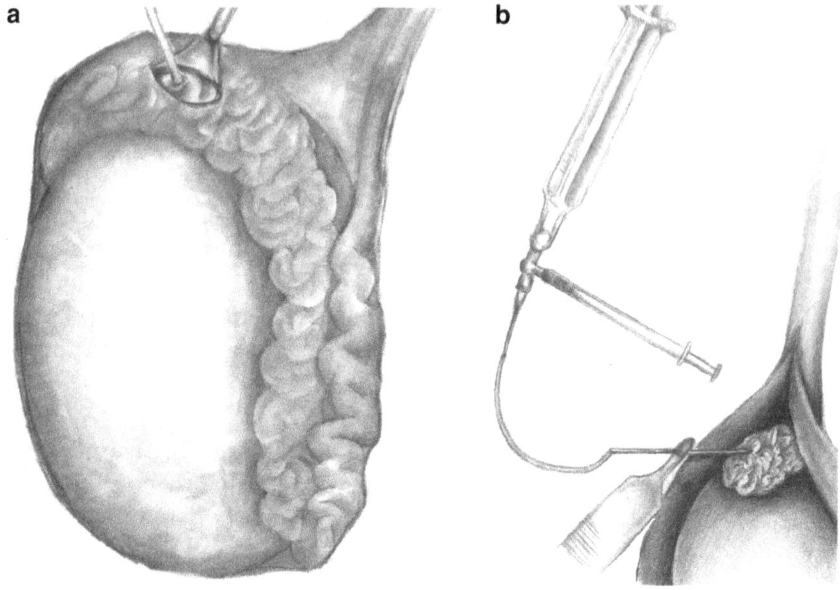

Fig. 5 (**a**) Microsurgical epididymal sperm aspiration (MESA). (**b**) View of MESA with procedural apparatus

General anesthesia, spinal or local anesthesia may be used. Required equipment includes a 6–25× operating microscope, surgical and microsurgical instruments including an ophthalmic micro-knife, a benchtop phase-contrast or light microscope, 5 μL glass pipettes, silicone tubing that attaches to both a syringe and the pipettes, human tubule fluid (HTF) or another appropriate sperm transport solution, and glass slides with coverslips.

After induction of general anesthesia the patient's scrotum is shaved, prepped, and draped in sterile fashion. The scrotal skin, dartos muscle, and tunica vaginalis are divided to expose the testis and epididymis. The surgeon grasps the testis with his or her non-dominant hand and stabilizes the epididymis between his or her thumb and index finger. The epididymis is inspected at 10–15× magnification to select dilated tubules containing golden, semi-translucent fluid (usually in the caput) for aspiration. Opaque, yellow tubules typically contain only sperm heads and debris and should be avoided.

The epididymal tubules are exposed by careful incision or excision of the overlying epididymal tunic at 15–20× magnification. Meticulous hemostasis is obtained with bipolar electrocautery to avoid contamination of the epididymal fluid with blood, avoiding the application of excess heat to the underlying epididymal tubules. The epididymal tubule is then sharply punctured with an ophthalmic micro-knife or aspirated using the quantitative micropuncture apparatus. A touch-preparation slide of epididymal fluid is the prepared by immediately and gently pressing a slide onto the incised epididymal tubule. The epididymal fluid on the slide is covered with a coverslip and examined under the benchtop microscope.

This procedure is repeated until an epididymal site containing microscopically adequate fluid is identified, moving toward the caput of the epididymis with each successive puncture that does not yield sufficient quality epididymal fluid. In nearly all patients, better sperm motility is found closer to the testis in the caput region of the epididymis. It is exceedingly rare to have no sperm motility in all areas of the obstructed epididymis. Such a finding should alert the surgeon that they are too far distal in the obstructed reproductive tract, except in the extremely rare patient with ciliary dysfunction. If necessary, the efferent ductules may be exposed by reflection of the caput epididymis away from the testis and individual tubules are aspirated with compression of the testis to facilitate outflow of intratesticular fluid. In some cases, it may be necessary to expose and aspirate the contralateral epididymis to optimize the yield of the procedure.

When adequate epididymal fluid is identified, it is collected atraumatically with glass micropipettes by simple capillary action or with a specialized micropuncture apparatus. The tip of a 5 μL micropipette (0.5 mm luminal diameter, 0.9 mm outer diameter) is placed adjacent to the effluxing epididymal tubule. Fluid is collected into multiple pipettes until its flow ceases or enough high-quality epididymal fluid has been collected. Ten to 20 microliters of epididymal fluid may be easily collected with this technique within 10–20 min [40]. The micropipettes are flushed with a small amount of HTF or another sperm transport solution into a collection container by connecting a 1 mL syringe to the pipette with medical grade silicone tubing (or tubing from a butterfly needle). If the micropuncture apparatus is used, the collecting tubing and micropipette are similarly back-flushed with sperm wash medium. The epididymal tubule is sealed with bipolar electrocautery and hemostasis is ensured. The epididymis and testis are returned to their normal anatomic position within the tunica vaginalis, which is closed with an absorbable suture. The dartos and skin are then closed. Bacitracin ointment is applied to the incision. Postoperatively, a scrotal supporter provides patient comfort and minimizes scrotal edema.

5.2 Conventional Testicular Sperm Extraction

Open excisional biopsy of seminiferous tubular tissue, referred to as TESE (Fig. 6) is the most commonly described and performed procedure for sperm retrieval. TESE may be performed in cases of both OA and NOA, though some clinicians prefer less-invasive techniques to minimize the morbidity of sperm retrieval in cases of OA. Several techniques have been described that vary in the approach by which the testis is surgically accessed and the number and location of excisional testicular biopsies that are taken.

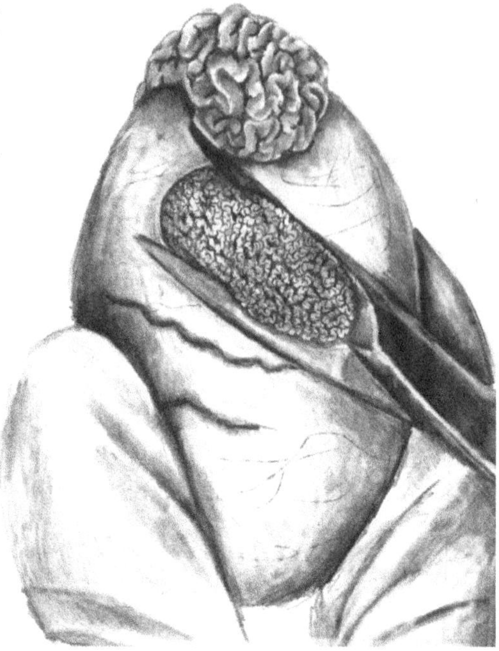

Fig. 6 Conventional testicular sperm extraction (TESE)

Reported SRRs with TESE in men with NOA range from 17 to 70 % [41]. The large range of reported outcomes reflects significant inter-study heterogeneity in the characteristics of the patient populations studied, methodological differences in TESE technique, and differences in surgical and laboratory expertise in finding sperm within testicular tissue. In a 2005 analysis of the published literature including observational studies, Colpi et al. reported a mean SRR of 52 % for TESE [42]. However, only a few studies have included adequate numbers of patients and have also reported histological features of men with NOA who underwent TESE. In these studies, reported SRRs are 41–58 % [43–47], with the higher SRR reported in a cohort predominantly composed of men with hypospermatogenesis, which is the most favorable histologic variant of NOA.

TESE is performed in an operating room under local, regional, or general anesthesia. Standard surgical instruments including a small curved scissor, and an electrocautery device, appropriate suture material, HTF buffered with HEPES and Plasmanate/other protein source or another sperm transport solution, Bouin's solution or another appropriate fixative (if diagnostic biopsy is indicated), and specimen collection containers are required. A benchtop microscope is highly recommended to direct the procedure.

The scrotal skin is incised transversely within a skin fold or longitudinally within the median raphe. The underlying dartos muscle is divided with cautery to expose the tunica vaginalis. At this point either a small incision may be made in the tunica vaginalis to fashion a window through which the testis may be accessed, or a large incision in the tunica vaginalis may be made through which the testis and epididymis may be completely delivered into the surgical field. If multiple biopsies are planned, the latter approach provides maximal exposure and facilitates access to all portions of the testis.

A small incision is made in the tunica albuginea of the testis in an avascular plane between the transverse subtunical blood vessels, which may be visually identified using optical magnification provided by loupes or an operating microscope [48]. Meticulous hemostasis is obtained with bipolar or monopolar electrocautery. The testis is gently squeezed to extrude seminiferous tubular tissue, which is sharply excised using a small sharp curved scissor (i.e., curved Iris scissor). If a diagnostic biopsy is indicated, the excisional biopsy is transferred directly from the scissor to a nonstick surface (such as a moistened piece of cut-out surgical drape) without grasping the tissue, in order to preserve the architecture of the seminiferous tubules to enable proper histologic assessment of spermatogenesis. Excised seminiferous tubular tissue for sperm retrieval is placed directly into a specimen collection container containing sperm transport solution. If a benchtop microscope is accessible in the operating room, the tissue is processed within sperm transport solution to mechanically disrupt the seminiferous tubules. A droplet of the testicular tissue suspension is examined under the microscope to direct the extent and duration of the procedure.

Multiple specimens may be taken from the same site by repeatedly squeezing the testis to extrude tissue. Moreover, multiple incisions in the tunica albuginea may be made to sample different areas of one or both testes. Indeed, observational evidence suggests higher SRRs with multiple biopsies taken from multiple sites in comparison to single site biopsies [49]. Furthermore, several publications have shown that sperm are found in a significant portion of cases in second or third biopsies when the first biopsy is negative [50].

The tunica albuginea is closed with delayed absorbable or nonabsorbable suture once the surgeon has decided to stop excising testicular tissue. Meticulous hemostasis must be obtained, after which the tunica vaginalis is closed. Long-acting local anesthetic without epinephrine (0.5 % bupivacaine) may be injected into the space between the tunica albuginea and tunica vaginalis after closure of the tunica vaginalis. The skin and dartos muscles are closed with absorbable suture after infiltration with long-acting local anesthetic containing epinephrine (0.5 % bupivacaine with epinephrine).

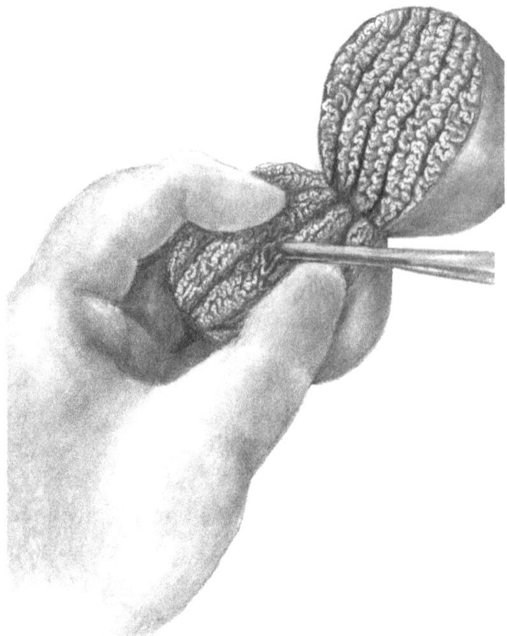

Fig. 7 Microdissection testicular sperm extraction (mTESE)

Postoperatively, an ice pack is left within a scrotal supporter. Activities are restricted to mostly rest for 1–2 days, minimization of vigorous activity for 2–3 weeks, and avoidance of heavy lifting for 4 weeks. Patients are provided with postoperative analgesics and reevaluated clinically within 4 weeks of surgery.

5.3 Microdissection Testicular Sperm Extraction

Microdissection testicular sperm extraction (mTESE) (Fig. 7) employs high power optical magnification to optimize the efficiency, safety and efficacy of TESE. This technique requires the coordinated efforts of a trained microsurgeon and an experienced embryology team. Reported SRRs are the highest of all sperm retrieval techniques in NOA (33–77 %) and most comparative studies have demonstrated superiority compared to TESE in SRR, sperm yield per mg of testicular tissue excised, and safety profile [12, 13, 21]. However, mTESE is significantly more time-consuming, more expensive, and less accessible due to the scarcity of trained andrological microsurgeons.

The initial surgical steps are identical to those used during conventional TESE to expose the testis. However, the operating microscope is brought into the field prior to incision of the testis. Optical magnification facilitates incision of the tunica albuginea in an avascular plane so as to minimize disruption of the blood supply to the testicular parenchyma. The tunica albuginea is incised in the testicular equatorial plane for 270°, thereby maximizing surgical exposure of the testis and allowing for manual eversion and microdissection of the upper and lower testicular poles. Wide incision of the tunica albuginea is important to allow clear access to all areas of bleeding, especially the rich plexus of vessels that are immediately underneath the tunica albuginea. Small areas of bleeding are selectively coagulated with a micro-bipolar forceps, thereby minimizing damage to adjacent testicular tissue.

Healthy-appearing seminiferous tubules that appear thicker and more opaque are identified by anatomic microdissection within the superficial and deep layers of the testis. Manipulation of the tissue must be done in an atraumatic fashion to avoid disruption of tissues from their blood supply, but is necessary to allow access to all areas of testicular tissue. Each selected tubule is carefully dissected out along as much of its length as possible. A microsurgical platform forceps (to allow grasping of the fine seminiferous tubules) is highly preferred to expedite the procedure. As they are extracted, tubules are mechanically disrupted and passed to an embryologist that is present in the operating room for immediate evaluation. By evaluating testicular samples in real time, the surgeon can stop the procedure as soon as the presence of sperm is confirmed, thereby minimizing operative time and removal of testicular tissue. Complete bilateral testicular microdissection may take up to 4 h, or initially more, early in a surgeon's experience. Once the procedure is completed, the tunica albuginea is reapproximated with delayed absorbable or non-absorbable suture. Closure of the tunica vaginalis, dartos muscle, and scrotal skin is performed as was described for TESE.

6 Intra-procedural Evaluation of Sperm Quality

Intra-procedural evaluation of sperm quality within extracted epididymal fluid and testicular tissue provides critical information that directs the duration and extent of sperm retrieval procedures. Immediate microscopic evaluation of extracted specimens is the cornerstone of the quality evaluation and relies upon several important factors including the intra-procedural handling and processing of extracted specimens, the method by which wet mount slides are prepared, and the expertise of the embryologist or andrologist performing the microscopic evaluation.

6.1 Preparation of Wet Mount Slides from Epididymal Fluid

The intra-procedural quality evaluation of epididymal fluid is easier when specimen contamination with red blood cells is minimized. During MESA, which yields significantly less contaminated fluid than PESA, a clean touch-prep slide can be made by simply touching the slide to the incised epididymal tubule. A droplet of HTF or another sperm transport solution is gently mixed with the epididymal fluid on the slide and the fluid mixture is covered with a coverslip. Touch-prep wet mount slides are the simplest, quickest slides to make and are useful for intraoperative decision-making during MESA.

Wet mount slides can also easily be prepared from epididymal fluid after it has been collected and deposited into the sperm collection containers during MESA or PESA. Such slides may be made with more precision than touch-prep slides and are preferred for quantitative determination of sperm concentration and percent motility. Ten microliters of sperm-containing fluid is placed on a glass slide and covered with a coverslip to achieve an optimal chamber depth of approximately 20 μM. Shallower chambers (less volume) for microscopic evaluation may encumber sperm motility, whereas sperm will move in and out of focus in deeper chambers (larger volume) and be more difficult to observe and count [51].

6.2 Preparation of Wet Mount Slides from Testicular Tissue

Extracted testicular tissue is immediately placed in 0.5–1 mL of HTF or another appropriate sperm transport solution within a small petri dish or conical tube. If the specimen is contaminated by blood, it may be gently irrigated with lactated Ringer's solution either during or immediately following extraction using a 10 mL syringe attached to an angiocatheter. The use of gauze to remove blood from testicular tissue should be avoided so as to avoid loss of seminiferous tubules that may become trapped within the gauze fibers.

The efficiency of finding sperm in extracted testicular tissue is enhanced by aggressive mechanical disruption of seminiferous tubular tissue to allow sperm to be released from the tubules and identified. Intra-procedural processing of testicular tissue is critically important in men with NOA to optimize the efficiency of sperm recognition, enabling the surgeon to stop removing testicular tissue as soon as adequate sperm are identified. The extracted tissue is minced with a small sharp scissor for several minutes. After mincing, the processed suspension of testicular tissue is aspirated into a small syringe and passed through sequentially smaller angiocatheters (16–24 G). This process is continued until the suspension is easily passed multiple times through a 24 G angiocatheter to assure that the testicular tissue suspension has been adequately disrupted. This method of testicular tissue processing can increase sperm yield up to 300-fold [52]. Wet mount slides are prepared by placing 10 μL of the testicular tissue suspension on a glass slide and covering with a coverslip.

6.3 Microscopic Assessment of Epididymal and Testicular Wet Mount Slides

Microscopic evaluation is best performed with a phase-contrast microscope, but may also be performed with a standard light microscope. The wet mount slide is viewed under 200–400× magnification. The presence and quantity of other cell types including red blood cells and germ cells are noted. Sperm concentration, total motility, progressive motility, and morphology are assessed by standard World Health Organization (WHO) methodology. Testicular sperm are expected to have abnormal motility, and epididymal sperm will commonly have retained cytoplasmic droplets. Immediate, real-time reporting of extracted sperm quality provides critical information to the operating surgeon that may be used for intra-procedural decision-making. In some cases, reassessment of motility after 30–60 min of incubation in HTF or another sperm transport solution, or stimulation of motility with pentoxyfilline, may aid sperm selection for intracytoplasmic sperm injection (ICSI).

6.4 Real-Time Interpretation of Sample Quality to Direct Sperm Retrieval

Standardized criteria that establish quantitative and qualitative guidelines for surgically retrieved epididymal or testicular sperm do not exist. For obstructed patients, enough sperm should be harvested from every patient so that the sample may be divided into multiple aliquots that contain at least several thousand motile sperm with strictly normal morphology. This approach maximizes the opportunity for both immediate assisted reproduction and cryopreservation. The number of sample aliquots required depends upon the reproductive goals of the couple and the number of anticipated IVF cycles that will be required to meet those goals.

During MESA, microsurgical exploration of the epididymis with selective sampling from multiple epididymal sites yields optimal results. Some experts advocate bilateral procedures unless the aspirate from the first side is optimal, whereas other experts believe that there is little advantage to exploring both epididymides. The typical yield in patients with an intact epididymis is 80–100 million total sperm with 30–50 % motility [41]. For men with caput-only (as in patients with congenital absence of the vas), maximal sperm motility is typically only 5–10 %. In obstructed patients, abundant high quality sperm are available for ICSI after thawing of cryopreserved samples even when the percentage of sperm with strictly normal morphology is less than 1 %.

For patients with NOA, obtaining at least 1,000 testicular sperm, preferably with twitching motility and nearly normal morphology (no major head defects and full length sperm tail) is desired. However, the number of sperm obtained must be balanced with the extent of the procedure and the potential effects on sperm production that may result from extensive testicular dissection. If significant dissection is required to find more sperm, but adequate sperm to inject all eggs have been retrieved, the procedure should be halted. It should be noted that testicular sperm never have strictly normal morphology [53], and that pregnancy is possible even using significantly abnormal-appearing sperm from the testes of men with NOA.

7 Conclusion

Biological paternity is now possible for azoospermic men using sperm retrieval procedures and ICSI. The efficacy and safety of sperm retrieval can be by informed selection of an appropriate sperm retrieval procedure, proper technical performance of the procedure that is selected, and intra-procedural microscopic assessment of the retrieved specimen to direct the extent of the procedure. In this chapter we have provided an overview of these clinical processes.

References

1. Kolettis PN, Sabanegh ES (2001) Significant medical pathology discovered during a male infertility evaluation. J Urol 166:178–180
2. Hotaling JM, Walsh TJ (2009) Male infertility: a risk factor for testicular cancer. Nat Rev Urol 6:550–556
3. Walsh TJ, Croughan MS, Schembri M et al (2009) Increased risk of testicular germ cell cancer among infertile men. Arch Intern Med 169:351–356
4. Raman JD, Nobert CF, Goldstein M (2005) Increased incidence of testicular cancer in men presenting with infertility and abnormal semen analysis. J Urol 174:1819–1822, discussion 1822
5. Matthews GJ, Matthews ED, Goldstein M (1998) Induction of spermatogenesis and achievement of pregnancy after microsurgical varicocelectomy in men with azoospermia and severe oligoasthenospermia. Fertil Steril 70:71–75
6. Male Infertility Best Practice Policy Committee of the American Urological Association, Practice Committee of the American Society for Reproductive Medicine (2006) Report on optimal evaluation of the infertile male. Fertil Steril 86:S202–S209
7. Sokol RZ (2009) Endocrinology of male infertility: evaluation and treatment. Semin Reprod Med 27:149–158
8. Chillon M, Casals T, Mercier B et al (1995) Mutations in the cystic fibrosis gene in patients with congenital absence of the vas deferens. N Engl J Med 332:1475–1480
9. Paduch DA, Bolyakov A, Cohen P et al (2009) Reproduction in men with Klinefelter syndrome: the past, the present, and the future. Semin Reprod Med 27:137–148
10. Stahl PJ, Masson P, Mielnik A et al (2010) A decade of experience emphasizes that testing for Y microdeletions is essential in American men with azoospermia and severe oligozoospermia. Fertil Steril 94:1753–1756
11. Jow WW, Steckel J, Schlegel PN et al (1993) Motile sperm in human testis biopsy specimens. J Androl 14:194–198
12. Schlegel PN (2009) Nonobstructive azoospermia: a revolutionary surgical approach and results. Semin Reprod Med 27:165–170
13. Schlegel PN (1999) Testicular sperm extraction: microdissection improves sperm yield with minimal tissue excision. Hum Reprod 14:131–135
14. Lewin A, Reubinoff B, Porat-Katz A et al (1999) Testicular fine needle aspiration: the alternative method for sperm retrieval in non-obstructive azoospermia. Hum Reprod 14:1785–1790
15. Levine LA, Dimitriou RJ, Fakouri B (2003) Testicular and epididymal percutaneous sperm aspiration in men with either obstructive or non-obstructive azoospermia. Urology 62:328–332
16. Hauser R, Yogev L, Paz G et al (2006) Comparison of efficacy of two techniques for testicular sperm retrieval in nonobstructive azoospermia: multifocal testicular sperm extraction versus multifocal testicular sperm aspiration. J Androl 27:28–33
17. Friedler S, Raziel A, Strassburger D et al (1997) Testicular sperm retrieval by percutaneous fine needle sperm aspiration compared with testicular sperm extraction by open biopsy in men with non-obstructive azoospermia. Hum Reprod 12:1488–1493
18. Westlander G, Hamberger L, Hanson C et al (1999) Diagnostic epididymal and testicular sperm recovery and genetic aspects in azoospermic men. Hum Reprod 14:118–122
19. Khadra AA, Abdulhadi I, Ghunain S et al (2003) Efficiency of percutaneous testicular sperm aspiration as a mode of sperm collection

for intracytoplasmic sperm injection in nonobstructive azoospermia. J Urol 169:603–605
20. Practice Committee of American Society for Reproductive Medicine in collaboration with Society for Male Reproduction and Urology (2008) The management of infertility due to obstructive azoospermia. Fertil Steril 90:S121–S124
21. Carpi A, Sabanegh E, Mechanick J (2009) Controversies in the management of nonobstructive azoospermia. Fertil Steril 91:963–970
22. Issa MM, Hsiao K, Bassel YS et al (2004) Spermatic cord anesthesia block for scrotal procedures in outpatient clinic setting. J Urol 172:2358–2361
23. Meniru GI, Gorgy A, Podsiadly BT et al (1997) Results of percutaneous epididymal sperm aspiration and intracytoplasmic sperm injection in two major groups of patients with obstructive azoospermia. Hum Reprod 12:2443–2446
24. Shrivastav P, Nadkarni P, Wensvoort S et al (1994) Percutaneous epididymal sperm aspiration for obstructive azoospermia. Hum Reprod 9:2058–2061
25. Glina S, Fragoso JB, Martins FG et al (2003) Percutaneous epididymal sperm aspiration (PESA) in men with obstructive azoospermia. Int Braz J Urol 29:141–145, discussion 145–6
26. Craft I, Tsirigotis M, Bennett V et al (1995) Percutaneous epididymal sperm aspiration and intracytoplasmic sperm injection in the management of infertility due to obstructive azoospermia. Fertil Steril 63:1038–1042
27. Lin YH, Huang LW, Seow KM et al (2004) Intentional cryopreservation of epididymal spermatozoa from percutaneous aspiration for dissociated intracytoplasmic sperm injection cycles. Acta Obstet Gynecol Scand 83:745–750
28. Rosenlund B, Westlander G, Wood M et al (1998) Sperm retrieval and fertilization in repeated percutaneous epididymal sperm aspiration. Hum Reprod 13:2805–2807
29. Fahmy I, Kamal A, Aboulghar M et al (2004) Percutaneous aspiration biopsy using an intravenous catheter for testicular sperm retrieval in patients with obstructive azoospermia. Reprod Biomed Online 9:102–105
30. Sheynkin YR, Ye Z, Menendez S et al (1998) Controlled comparison of percutaneous and microsurgical sperm retrieval in men with obstructive azoospermia. Hum Reprod 13:3086–3089
31. Semiao-Francisco L, Braga DP, Figueira Rde C et al (2010) Assisted reproductive technology outcomes in azoospermic men: 10 years of experience with surgical sperm retrieval. Aging Male 13:44–50
32. Belker AM, Sherins RJ, Dennison-Lagos L et al (1998) Percutaneous testicular sperm aspiration: a convenient and effective office procedure to retrieve sperm for in vitro fertilization with intracytoplasmic sperm injection. J Urol 160:2058–2062
33. Tournaye H (1999) Surgical sperm recovery for intracytoplasmic sperm injection: Which method is to be preferred? Hum Reprod 14(Suppl 1):71–81
34. Ezeh UI, Moore HD, Cooke ID (1998) A prospective study of multiple needle biopsies versus a single open biopsy for testicular sperm extraction in men with non-obstructive azoospermia. Hum Reprod 13:3075–3080
35. Hovatta O, Moilanen J, von Smitten K et al (1995) Testicular needle biopsy, open biopsy, epididymal aspiration and intracytoplasmic sperm injection in obstructive azoospermia. Hum Reprod 10:2595–2599
36. Rajfer J, Binder S (1989) Use of biopty gun for transcutaneous testicular biopsies. J Urol 142:1021–1022
37. Tuuri T, Moilanen J, Kaukoranta S et al (1999) Testicular biopty gun needle biopsy in collecting spermatozoa for intracytoplasmic injection, cryopreservation and histology. Hum Reprod 14:1274–1278
38. Janzen N, Goldstein M, Schlegel PN et al (2000) Use of electively cryopreserved microsurgically aspirated epididymal sperm with IVF and intracytoplasmic sperm injection for obstructive azoospermia. Fertil Steril 74:696–701
39. Schroeder-Printzen I, Zumbe J, Bispink L et al (2000) Microsurgical epididymal sperm aspiration: Aspirate analysis and straws available after cryopreservation in patients with non-reconstructable obstructive azoospermia. MESA/TESE group giessen. Hum Reprod 15: 2531–2535
40. Matthews GJ, Goldstein M (1996) A simplified method of epididymal sperm aspiration. Urology 47:123–125
41. Donoso P, Tournaye H, Devroey P (2007) Which is the best sperm retrieval technique for non-obstructive azoospermia? A systematic review. Hum Reprod Update 13:539–549
42. Colpi GM, Piediferro G, Nerva F et al (2005) Sperm retrieval for intra-cytoplasmic sperm injection in non-obstructive azoospermia. Minerva Urol Nefrol 57:99–107
43. Vernaeve V, Verheyen G, Goossens A et al (2006) How successful is repeat testicular sperm extraction in patients with azoospermia? Hum Reprod 21:1551–1554
44. Friedler S, Raziel A, Schachter M et al (2002) Outcome of first and repeated testicular sperm

extraction and ICSI in patients with non-obstructive azoospermia. Hum Reprod 17:2356–2361

45. Seo JT, Ko WJ (2001) Predictive factors of successful testicular sperm recovery in non-obstructive azoospermia patients. Int J Androl 24:306–310
46. Su LM, Palermo GD, Goldstein M et al (1999) Testicular sperm extraction with intracytoplasmic sperm injection for nonobstructive azoospermia: testicular histology can predict success of sperm retrieval. J Urol 161: 112–116
47. Gil-Salom M, Romero J, Minguez Y et al (1998) Testicular sperm extraction and intracytoplasmic sperm injection: a chance of fertility in nonobstructive azoospermia. J Urol 160:2063–2067
48. Dardashti K, Williams RH, Goldstein M (2000) Microsurgical testis biopsy: a novel technique for retrieval of testicular tissue. J Urol 163:1206–1207
49. Amer M, Haggar SE, Moustafa T et al (1999) Testicular sperm extraction: impact of testicular histology on outcome, number of biopsies to be performed and optimal time for repetition. Hum Reprod 14:3030–3034
50. Hauser R, Botchan A, Amit A et al (1998) Multiple testicular sampling in non-obstructive azoospermia – is it necessary? Hum Reprod 13:3081–3085
51. World Health Organization (2010) WHO laboratory manual for the examination and processing of human semen. World Health Organisation, Geneva, p 271
52. Ostad M, Liotta D, Ye Z et al (1998) Testicular sperm extraction for nonobstructive azoospermia: results of a multibiopsy approach with optimized tissue dispersion. Urology 52: 692–696
53. Steele EK, McClure N, Lewis S (2000) A comparison of the morphology of testicular, epididymal, and ejaculated sperm from fertile men and men with obstructive azoospermia. Fertil Steril 73:1099–1103
54. Okada H, Dobashi M, Yamazaki T et al (2002) Conventional versus microdissection testicular sperm extraction for nonobstructive azoospermia. J Urol 168:1063–1067
55. Carpi A, Menchini Fabris FG, Palego P et al (2005) Fine-needle and large-needle percutaneous aspiration biopsy of testicles in men with nonobstructive azoospermia: safety and diagnostic performance. Fertil Steril 83:1029–1033

Chapter 18

Treatment of Male Infertility

Gianpiero D. Palermo, Justin Kocent, Devin Monahan, Queenie V. Neri, and Zev Rosenwaks

Abstract

Major difficulties exist in the accurate and meaningful diagnosis of male reproductive dysfunction, and our understanding of the epidemiology and etiology of male infertility has proven quite complex.

The numerous spermatozoa produced in mammals and other species provides some degree of protection against adverse environmental conditions represented by physical and chemical factors that can reduce reproductive function and increase gonadal damage even resulting in testicular cancer or congenital malformations. The wide fluctuations of sperm production in men, both geographical and temporal, may reflect disparate environmental exposures, occurring on differing genetic backgrounds, in varying psychosocial conditions, and leading to the diversified observed outcomes.

Sperm analysis is still the cornerstone in diagnosis of male factor infertility, indeed, individually compromised semen paramaters while adequately address therapeutic practices is progressively flanked by additional tests. Administration of drugs, IUI, correction of varicocele, and, to a certain extent, IVF although they may not be capable of restoring fertility itself often result in childbearing.

Key words ICSI, IVF, Assisted fertilization, Spermatogenesis, Andrology, Male infertility, Surgically retrieved spermatozoa

1 Introduction

In contrast with general medicine, "infertility" belongs to a special category of human ailments. Where a specific disease is present, diagnosis and treatment are closely linked and the successful outcome involves a close binomial relationship between the patient and the medical practitioner. In human infertility, the treatment is focused on the two partners, and the discomfort generated by human infertility is usually not affecting their general health but is linked for the most part to the performance of the female and the male gametes per se.

Infertility is commonly defined as the failure of conception after 12 months of unprotected intercourse [1]. It affects about 14 % of individuals in reproductive age and in about half is

Zev Rosenwaks and Paul M. Wassarman (eds.), *Human Fertility: Methods and Protocols*, Methods in Molecular Biology, vol. 1154, DOI 10.1007/978-1-4939-0659-8_18,

related to the male partner resulting in considerable distress for the couple [2].

While the diagnosis of male factor infertility is the most straightforward the etiology is still unclear. In fact, the underlying cause of oligo-astheno-terato-zoospermia (OAT) is still unknown rendering the effectiveness of any conventional treatment for male infertility extremely doubtful. Reports have drawn attention to genetic defects in spermatozoa as a cause of male infertility [3]; however, the availability in the last 20 years of micromanipulation techniques has made it feasible to treat infertility related to issues in the male partner in ways that previously would have been deemed hopeless. These new therapeutic options have opened fresh perspectives, providing ground for critical analysis of the available conventional diagnostic and therapeutic approaches that involve either drugs, varicocele surgery, intrauterine insemination (IUI), IVF, and ICSI.

According to varying surveys, it appears now that the male partner is the culprit in some 25–50 % of infertility cases. Since the focus on human semen quality and its relationship to male fertility began in the late 1940s [4, 5], numerous studies have probed the fluctuations in sperm concentration among human ejacuates in relation to the ability to reproduce [6]. Apart from physiological variations, three elements that appear to potentially affect human sperm production include environmental, genetic, and psychosocial factors. While each of these can act individually they can often compound their effect producing the male infertile state.

In the general population, about 25 % of couples do not achieve pregnancy within 1 year, 15 % seek treatment for infertility, and less than 5 % remain childless. Nonetheless, despite advances in the diagnostic workup of infertile men, in about 50 % of men with compromised spermatogenesis the cause is unrecognized [7–9]. It has been demonstrated that DNA damage in human spermatozoa has been linked with poor semen performance, resulting in poor fertilization rates, impaired embryo development, increased pregnancy loss and a possible health consequences in the offspring, not excluding cancer [10–13]. However, at this stage the origin and role of sperm DNA fragmentation in the male gamete, as currently measured, is still controversial.

The introduction of intracytoplasmic sperm injection (ICSI) nearly 20 years ago offered a way to bypass many of these problems [14]. Within 3 years of its introduction methods such as microdroplet IVF [15], partial zona dissection [16], and sub-zonal sperm insertion [17] became obsolete. Emerging half a century earlier, the field of clinical andrology, had long left most afflicted men either childless or facing donor insemination.

In this treatise, we first outline the criteria currently utilized to evaluate couples with male factor infertility. We identify the etiology of this condition by focusing on different aspects related to sperm production, genetics, and environmental factors. We also describe the most recent approaches and their clinical outcomes to the treatment of male infertility.

2 Screening the Male Patient

Semen analysis still provides the fundamental information on which clinicians base their initial diagnosis while gaining information on the fertility potential of a male patient. The diagnosis of male infertility is based upon the semen profile, constructed according to recognized guidelines [18–20]. This measures the volume of the ejaculate, the concentration of spermatozoa, their motility, and their morphology. Marked inter-ejaculates variability is a recurrent phenomemon [21, 22], in addition many recorded parameters of the spermatogenic profile are subjective, and so assessments have been inconsistent across laboratories [23–25]. Although WHO has published a range of 'normal' values, these are not evidence-based, either in terms of their diagnostic value, or in terms of their relationship to the normal ejaculate. As a consequence, many couples with 'unexplained' infertility eventually prove to have defective spermatozoa when appropriately sensitive assays (such as acrosome reaction, antisperm antibodies, PLCζ) are used, yet some couples with subnormal semen parameters prove to have normal sperm function [26, 27]. Given this evidence, it would perhaps be most approapriate for laboratories to establish their own definition of 'normal' based on their individual experience in screening ejaculates.

Since many of the male diagnostic categories are of a descriptive nature (e.g., idiopathic oligozoospermia) or of questionable therapeutic relevance (e.g., male accessory gland infection) and genetic and epigenetic causes of infertility for a man to reproduce, this classification is now in need of review [28, 29]. The relative frequency of the major diagnostic categories obtained from 8,500 couples from 33 centers in 25 countries [30] showed that the largest seminal abnormalities were of unknown cause. Other common male problems included varicocele, and male accessory gland infection, and systemic, iatrogenic, and endocrine conditions were also very frequent.

The position of andrology in the male infertility workup has been strengthened by the introduction of testicular sperm extraction (TESE) or microsurgical epididymal sperm aspiration (MESA) following appropriate differential diagnosis of infertility causes (obstructive/non-obstructive azoospermia, congenital bilateral absence of the vas deferens) and in some cases, testicular biopsies. The effectiveness of treating varicoceles in terms of pregnancy outcome is still a matter of discussion. However, it seems clear that spermatogenesis can be, at least partially restored, after microsurgical varicocelectomy in some men with cryptozoospermia or secretory azoospermia [31].

While obtaining an inferility history it is important to attempt to identify any risk factors (e.g., cryptorchidism, environmental hazards), that may be of aid in their andrological assessment, unfortunately, the overzealous enthusiasm of the reproductive specialists may drive the excessive use of unconfirmed diagnostic tests. However, these tests that claim to predict fertility status or to indicate the appropriate ART procedure risk to emotionally and financially drain childless men.

A number of epidemiological factors that may have a bearing on a couple's fertility include age, though the impact of male age is less certain than that of the female. Smoking by both partners is highly relevant as well, there being evidence that smokers have lower sperm concentrations than nonsmokers [32–34]. Occupational, environmental, and genetic factors may also be highly relevant. In regard to the latter, there is no doubt that recent advances in assisted conception technology have increased our understanding of the etiology of male infertility, particularly by drawing attention to the major contribution of specific genes [3, 29, 35–38].

3 Genetic and Epigenetic Aspects of Male Infertility

Human male germ-cell development begins in early embryogenesis, but the first mature spermatozoa appear only at puberty (ca. 12 years old). Genetic disorders disrupting this male-specific cell differentiation and maturation can be reflected at the chromosomal or molecular DNA level.

In the case of of chromosomal anomalies, these can be numeral, structural, or both. Aneuploidy leading to male infertility may involve the sex chromosomes (e.g., an additional X-chromosome in Klinefelter's syndrome) or autosomes (e.g., trisomy 21). Structural chromosome anomalies (small deletions, translocations, inversions) can lead to male infertility and these may involve both sex or autosomal chromosomes. Chromosomal rearrangements such as reciprocal translocations can give rise to abnormal meiotic chromosome pairing, and in turn disruption of spermatogenesis. For instance, evidence has accumulated since the 1970s that disruption through meiosis and recombination follows a significant rearrangement of the Y-chromosome. In fact, deletions in the Yq region can be associated with azoospermia, suggesting a spermatogenesis locus in the region designated AZF (Azoospermia Factor) where a later study has shown three such loci (AZFa, AZFb, AZFc) in Yq [29].

Our comprehension concerning these genes involved in human spermatogenesis is still poor and mostly deduced from animal studies, but such extrapolation should be treated with caution due to the various spectrum of altered spermatogenesis/spermiogenesis seen in humans. Genes involved in spermatogenesis may be expressed functionally in the germ line, during the development of male gonads, or in testicular somatic cells, but those expressed specifically in the germ line are assumed to be the most relevant for regulation of germ cell maturation. This appears to be the case for RBM, SPGY, and DAZ genes, even if the number of candidate genes that may regulate male fertility continues to increase, and disruption of any gene that is involved in male gonad development (e.g., the androgen receptor gene mutation) leads to sterility with intersexual phenotypes. Recessive and dominant mutations in genes generally controlling somatic cells may indirectly induce infertility as a consequence of other problems, as seen for instance in men with Kartagener's syndrome, cystic fibrosis, or myotonic dystrophy.

There is probably a link between genetic and environmental factors. Exposure to DNA and chromosome-damaging agents is of great concern, and such damage in germ cells may contribute to a legacy of genetically based male infertility projecting on future generations. Abnormal chromosomal number is an important genetic hazard, in fact ionizing radiation and a variety of chemical compounds have been shown to increase aneuploidy in mammals [39] or by inducing chromosomal breakage. Alternatively, chemicals may act on the spindle and so affect the process of chromosomal segregation. Physical rearrangements of chromosomes can give rise to abnormal meiotic chromosome pairing, which in turn can result in infertility as reflected in Klinefelter Syndrome. Structural rearrangements, particularly reciprocal translocations, have been associated with infertility as seen in Robertsonian translocation. More than a dozen chemicals have been shown to induce transmissible translocations in the mouse. Some such as ethylene oxide, induce mutations in the germ cell genome [40], but thus far no environmental agent has been shown to produce heritable germ-line mutations which bear on the fertility of the next generation. The current lack of evidence in humans regarding germ-line mutations is due to various factors namely the small number of human genes currently available to estimate these mutations.

While the benefits of ICSI are clear, the success of this approach may have also reinforced the perception that the sperm cell is simply a vehicle for the transport of the paternal genome. Therefore, if any motile spermatozoa are available it has become quite common to expect normal fertilization and embryogenesis, regardless of the potential intrinsic defects which may be associated with subfertile spermatozoa [41]. In addition to a haploid set of chromosomes, the human sperm cell must provide the oocyte with an activation stimulus, a functional centrosome, proper packaging and coding of the genome, and, possibly, other factors such as mRNAs and microRNAs (miRNAs) that may regulate transcription or complement the maternal transcriptome within the cleaving embryo [42–46].

During spermatogenesis, sperm chromatin undergoes a dramatic reorganization, including protamine replacement of histones, histone modifications and DNA methylation [47, 48]. Protamines are small arginine-rich proteins whose positive charge allows a tight binding of sperm nuclear DNA and form cysteine disulphide bonds that facilitate a higher order chromatin packaging than histones [49]. The thiol-rich nature and so the tendency of eutherian protamine to form –S-S– cross-links after spermiation in the epididymis is also believed to be responsible for to nuclear compaction and DNA stability. Others have claimed that –S-S– cross-links has nothing to do with chromatin compaction and when in fact it almost certainly relates to a role in stiffening/rigidification of the nucleus that may help it penetrate the relatively tough zona [50] as seen in other vertebrates that have dense nuclei void of cysteine in the protamine [51]. In fact, human spermatozoa have the lowest content of protamine-related cysteine among mammals—ca. 8.5 mols% v. about 12.5 in the rat. Protamine replacement takes place in a stepwise fashion in which histones are initially replaced by testis-specific histone variants, which are then replaced by transition proteins, followed by their replacement with protamines [52].

Modifications of the N-terminal region of histones confer another epigenetic regulatory mechanism on gene expression. Generally, methylation of the histones is associated with silencing of the gene, while acetylation is associated with transcription [53]. These and other histone modifications offer a short-term mechanism for regulation of gene expression and mark DNA imprinting control regions during spermatogenesis [54, 55]. While the significance of this epigenetic process is well known, the relevance and implications for patients with severe male factor infertility is intriguing.

Methylation of the genome is associated with repression of gene expression. Methylation is carried out by DNA methyltransferases, which transfer a methyl group to deoxycytosines found in CpG islands, and regulate transcription [47, 56]. It has recently been demonstrated that the methylation pattern in the testis is different from somatic tissue, this being diminished eightfold [57]. Whether this hypomethylation is present in mature spermatozoa, or whether it represents an epigenetic process to prepare spermatozoa for increased transcription following pronuclear development is still not known.

Some genes are differentially methylated according to the gender of their lineage, a process known as imprinting. For example, insulin-like growth factor (IGF)-II and H19 are reciprocally methylated in maternal and paternal loci, respectively, resulting in paternal expression of IGF-II and maternal of H19. Imprinted genes are not only regaulated by DNA methylation, similarly histone modifications are likely involved in regulating their transcription [56]. In oocytes, gene imprinting occurs later than in spermatogenesis, and is so labile that may be affected by culture conditions or medication used during ovulation induction [58]. Conversely, the impact on the imprinting of embryonic genome by using less mature spermatozoa as those obtained through testicular biopsy are not clear. However, it is generally accepted that altered parental imprinting marks have been shown to be associated with compromised embryo development [59, 60].

Imprinted genes are usually found in clusters, such as those found on chromosome 11p15 linked to Beckwith–Wiedemann syndrome (BWS), and the cluster on chromosome 15q11–13 associated with Angelman syndrome (AS) [61]. Since both AS and BWS have been reported to be more represented in offspring from ART [62, 63], and since ICSI often involves the use of immature or dysfunctional spermatozoa, the role of abnormal DNA methylation and altered imprinting profiles in the male gamete of infertile men is of intense interest [42, 64]. A relatively recent report failed to evidence no difference in global methylation patterns of spermatozoa from men with abnormal protamine expression; however, more sensitive techniques are needed to definitively answer the question [65] even if a subsequent important study has indicated a higher rate of abnormal methylation of imprinted genes in infertile men [66].

Recent studies have demonstrated that mature spermatozoa contain 10–20 fg/cell RNA [67]. While this level of RNA is trivial compared with the large quantity in other cells, it is nevertheless surprising given the extensive removal of excess cytoplasm during spermiogenesis and the transcriptionally quiescent state of the cell [68]. The mRNAs found in the RNA pool lack intronic sequences and contain poly-A tails, indicating they are normally and fully processed following transcription [69].

The role, if any, that the sperm mRNA plays in embryogenesis is controversial. Krawetz and others have demonstrated that a stable pool of mRNA appears to consist of transcripts involved in spermatogenesis, and may be used to profile transcription and translation events during spermatogenesis [44, 70–72].

In addition to mRNA, in human spermatozoa small antisense RNAs have been identified that remain in the embryo throughout early embryogenesis [72, 73]. These small RNAs include PIWI-interacting RNAs (piRNAs) associated with MIWI, a germ-line-specific member of the argonaute family of proteins [74–76].

Both piRNAs and MIWI are present in the mature spermatozoa, a finding that raises the exciting possibility that paternal miRNAs may have a role in regulating gene expression in the embryo [47, 68]. While this hypothesis is still being evaluated, a recent study [77] has raised serious questions as to whether such a role exists. In one study, many of the miRNAs in spermatozoa are also present in oocytes at higher concentrations, and co-injection of oocytes with spermatozoa and anti-miRNAs to suppress any potential function of the miRNAs, did not seem to affect embryogenesis [77].

4 Assisted Fertilization

With the first successful human delivery in 1978, in vitro fertilization (IVF) opened a floodgate of possibilities for infertile patients. The decade that followed was a time of exciting research that set the framework for widespread clinical implementation of this novel procedure as a field of young and ambitious scientists raced to improve efficiency and answer unsolved questions. Human IVF provides a benchmark in our history, and marked the first time a species consciously combated it's own reproductive deficiencies and took propagation into their hand.

Notwithstanding its historic significance the practicioners of IVF had hurdles to overcome. For instance, patients with marginal semen characteristics and poor spermatozoa experienced limited successes with traditional IVF. Suboptimal functioning capacity, i.e., diminished concentration, motility, and/or morphology present a more complex obstacle for spermatozoa attempting to penetrate the thick glycoprotein layer—the zona pellucida (ZP)—that constitutes the egg coat. Traditional means of overcoming such hurdles were limited and dealt primarily with increasing sperm concentration and at enhancing their selection. Embryologists opted to increase the sperm concentration within the inseminating suspension or if extremely few spermatozoa with impaired motility were available lower the actual volume of the insemination medium to enhance comingling of the two gametes. Simple procedures included the "swim-up" method that selected only highly motile cells in the upper fraction. Other methods utilized multilayer density gradients to select highly motile spermatozoa that also exhibited higher penetration capacity [78, 79]. While these selection procedures optimized the ability to treat IVF patients with moderately compromised semen characteristics, often described by moderate oligo-asthenospermia, they fell short in dealing with more severe issues of sperm dysfunction.

The focus then shifted to efforts to assist the spermatozoon to penetrate the zona pellucida. Early attempts involving complete removal of the zona resulted in polyspermy or impaired embryonic development. These were followed by attempts to overcome the zona as a barrier either by "softening" it enzymatically with trypsin or pronase [80, 81], or penetrating it chemically via localized or "pinpoint" exposure to acidified Tyrode's solution prior to sperm exposure. The latter technique became known as zona-drilling (ZD) [82] but the low pH solution involved proved to possibly damage the oocyte [83]. A variant of ZD opened a fissure in the zona via mechanical means—dubbed partial zona dissection (PZD)—utilized a larger but virtual opening than did ZD, thus mitigating the rate of polyspermy [16]. The PZD procedure seemed to lend itself to a situation in patients exhibiting some oligozoospermia and pseudo-asthenozoospermia, though the sperm required to be capacitated, undergo acrosome reaction and display a progressive motility. Still a large obstacle persisted in the form of sperm cells that exhibit motility in place required further assistance to properly interact with oolemma. This procedure was replaced by a more refined procedure in which spermatozoa were brought with a pipette through the zona and deposited beneath the zona into the perivitelline space—sub-zonal insemination (SUZI). This provided some success in patients with more impaired sperm motility while controlling the incidence of polyspermy but yielded fertilization rates still relatively low, not least because of requiring spermatozoa to undergo a complete acrosome reaction—a necessary precursor to fusion with the oolemma [84, 85]. These preliminary efforts to artificially assist sperm penetration soon became obsolete with the introduction of a microsurgical method for insertion of spermatozoa directly into the egg [14].

5 ICSI and Clinical Results

Intracytoplasmic sperm injection (ICSI), a procedure that involves the insertion of a single spermatozoon into the cytoplasm of an oocyte was once considered a theoretical option while procedures such as SUZI were being applied. ICSI bypasses both the zona pellucida and sperm defects in the male gamete that compromise its ability to fertilize. Documented trials on mammalian gametes date this approach as far back as 1966 [86]. However, initial efforts with human gametes yielded a high incidence of oocyte damage and an unsatisfactory level of embryo implantation [87, 88]. However, this was investigated further at the Free University in Brussels, and a clinical study was after promising fertilization results from a murine model, and four human pregnancies were reported in 1992 [14].

While ICSI is currently employed widely, its crowning achievements arguably lie in the establishment of pregnancies with male germ cells from azoospermic men. Until advanced micromanipulation techniques were pioneered, such couples relied on donor sperm or adoption. A complete absence of spermatozoa in the ejaculate occurs in 1 % of all men and 10–15 % of all infertile men, and diagnosis of this condition is subcategorized into obstructive

(OA) and non-obstructive (NOA) azoospermia. The obstructive form can result from CBAVD, trauma, infection, vasectomy, or failed vasectomy reversal. Sperm retrieval and subsequent ICSI may be achieved via epididymal aspirations including microscopic (MESA), percutaneous (PESA), and fine needle (FNA) [89, 90]. The non-obstructed form of azoospermia due to a failure of spermatogenesis requires surgical intervention to obtain spermatozoa for subsequent injection. Testicular sperm extraction (TESE) and the now-refined micro-TESE (mTESE) retrieve seminiferous tubules for search with the latter achieving a higher probability of sperm retrieval and maintaining a greater anatomical integrity of the testicle [91].

Aside from primary indicators, ICSI is the preferred option for male patients experiencing various sperm-related problems that include ejaculatory dysfunction, retrograde ejaculation, and complications stemming from paraplegia. Post-ejaculatory urinalysis is often utilized when a patient experiences low-volume or absent ejaculation. Conditions linked to ultralow ejaculate volumes include retrograde ejaculation, lack of emission, ejaculatory duct obstruction, hypogonadism, and CBAVD. The quest for spermatozoa is carried out following a high-speed centrifugation and examination of a pellet under oil at 400× magnification with an inverted microscope.

The direct and reliable ICSI outcome lends itself well to methods requiring blastomere biopsies on embryos such as preimplantation genetic diagnosis (PGD). Such genetic testing is utilized to avoid transmission of serious recurrent disorders by identifying single gene mutations from biopsied blastomeres after PCR, whole-genome amplification, or SNP arrays. Furthermore, the sex of the embryo as well as chromosomal translocation can be identified using fluorescent in situ hybridization, or FISH. Also, selecting only one isolated spermatozoon for contact with an oocyte significantly decreases the chance of viral transmission with HIV, HBV, and HCV amongst others. In fact, viruses that are freely present in semen and accompanying cells yield a reduction when exposed to sperm gradient preparation, which further decreases the probability of transmission [92].

For the past 18 years (September 1993 to December 2010), a total of 11,038 IVF and 19,983 ICSI cycles (Fig. 1a) have been performed at our center where ICSI is currently utilized in approximately 73 % in all cycles (Fig. 1b). The resulting positive pregnancies was in about 40 % of all cases, based on the presence of at least one fetal heartbeat.

Approximately 90 % ($n=17,938$) of all ICSI cycles were performed using ejaculated spermatozoa, and the remainder involved specimens ($n=2,045$) that were surgically retrieved from the epididymis or testis (Fig. 2). In cycles utilizing ejaculated spermatozoa, a total of 153,596 MII were oocytes injected, resulting in a

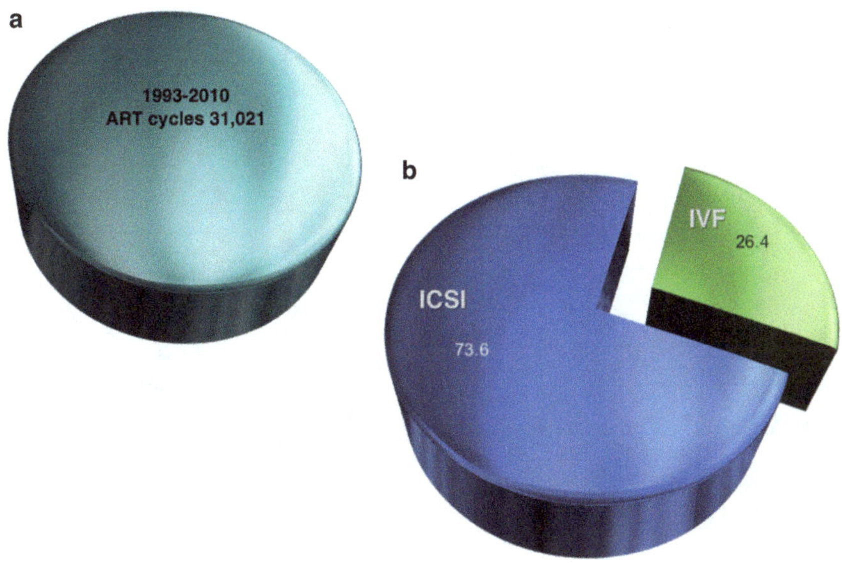

Fig. 1 ART cycles performed at Cornell in the past 18 years (**a**) and the proportion of cycles utilizing ICSI (**b**)

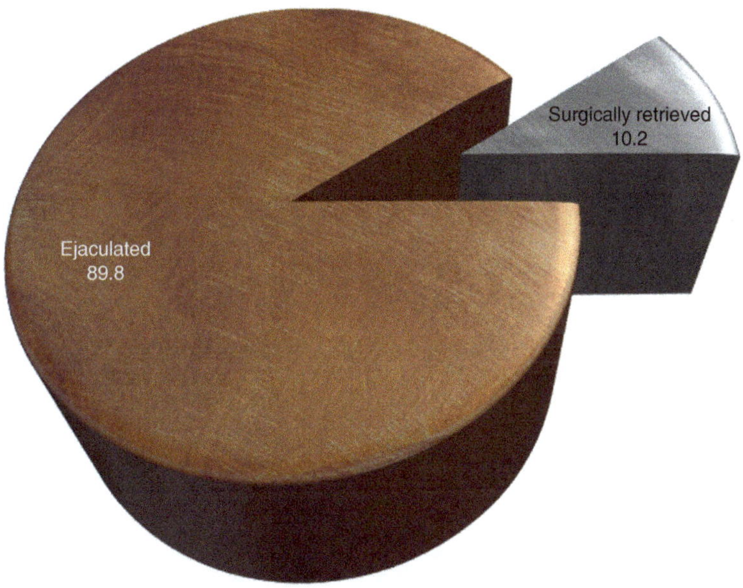

Fig. 2 A total of 20,629 ICSI cycles performed from September 1993 to December 2010 according to the semen source

survival rate of 94.5 % (Fig. 3a). Of those that survived 114,370 oocytes fertilized normally, with only 1PN in 3,615 and 3PN in 5,110 and the remaining 22,089 being unactivated (Fig. 3b).

Among men with severe oligozoospermia (Table 1), fertilization reached 66.1 % with a clinical pregnancy rate of 44.0 %, one thus comparable to normozoospermic cycles.

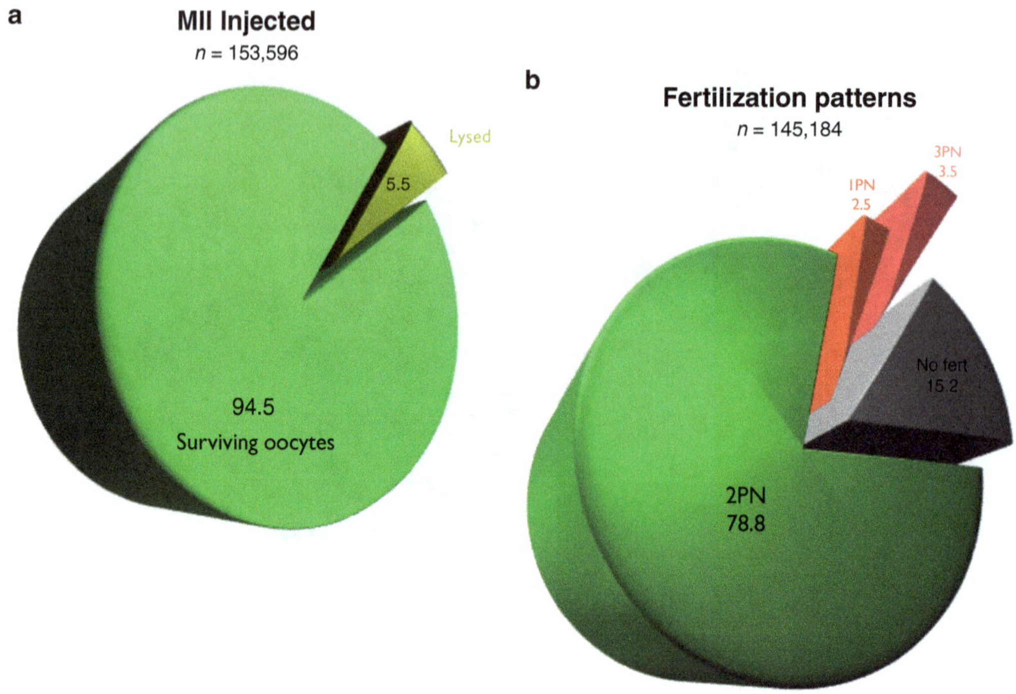

Fig. 3 Survival (**a**) and fertilization (**b**) characteristics of ejaculated spermatozoa

Table 1
ICSI outcome of men with severe oligozoospermia ($\leq 1 \times 10^6$/ml)

No. of cycles	1,372
Mean density (10^6/ml ± SD)	0.3 ± 0.3
Mean motility (% ± SD)	17.3 ± 19
Mean morphology (% ± SD)	0.6 ± 1
Fertilization (%)	8,597/13,006 (66.1)
Clinical pregnancy (+FHB %)	604 (44.0)

In cases where no spermatozoa were identified in a counting chamber ($n = 179$), specimens were further processed to finally obtain an average concentration of $0.009 \pm 0.002 \times 10^3$/ml and motility of 13.3 % ± 16. These cryptozoospermic cases had a fertilization rate of 59.6 % (1,082/1,815) and clinical pregnancy of 38.0 % (68/179).

A total of 902 cycles were performed with epididymal spermatozoa and 1,143 cycles with testicular spermatozoa (Table 2). Both fertilization (71.5 %) and clinical pregnancy rates (52.5 %)

Table 2
ICSI outcome and surgical sampling

	MESA	TESA
No. of cycles	902	1,143
Mean density (10^6/ml ± SD)	28.5 ± 35	0.4 ± 2
Mean motility (% ± SD)	8.9 ± 14	3.1 ± 7
Mean morphology (% ± SD)	1.5 ± 3	0
Fertilization (%)	6,167/8,629 (71.5)*	5,964/10,482 (56.9)*
Clinical pregnancy (+FHB %)	475 (52.7)†	466 (40.8)†

*†χ^2, 2 × 2, 1 df, Sperm source on fertilization and clinical pregnancy rates, $P < 0.001$

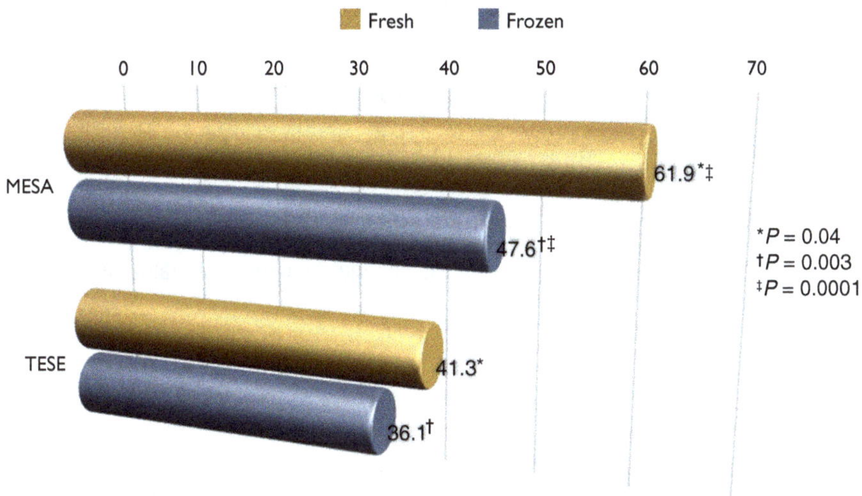

Fig. 4 Clinical pregnancies derived from surgically retrieved specimens either fresh or cryopreserved

were higher ($P < 0.001$) with epididymal versus testicular samples (TESE fertilization 56.9 % and clinical pregnancy rate 40.8 %). In assessing whether or not cryopreservation affected the clinical outcome, we observed that fresh MESA had a higher clinical pregnancy than TESE ($P = 0.04$) and a similar result was observed when specimens were cryopreserved (frozen MESA 47.6 versus frozen TESE 36.1 %; $P = 0.003$). When we looked within a particular surgical retrieval method, the clinical pregnancy rate with TESE was comparable regardless of the sample being fresh or cryopreserved. However, fresh MESA showed a higher pregnancy rate in comparison to its frozen counterpart (61.9 vs 47.6 %, $P = 0.0001$) (Fig. 4).

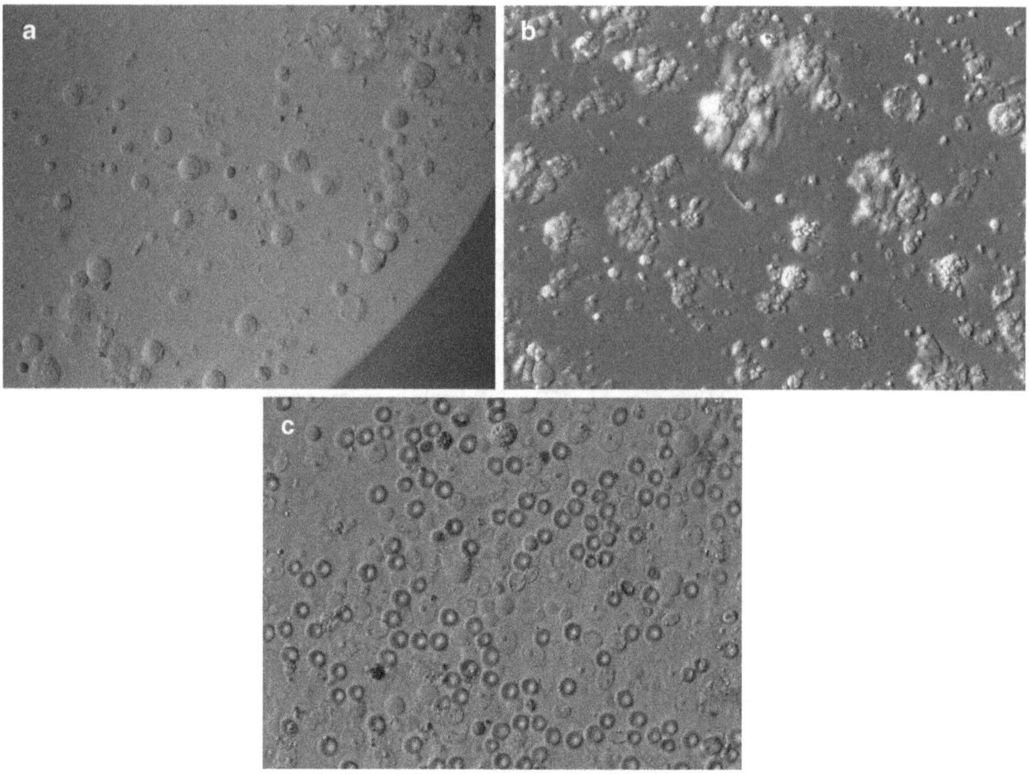

Fig. 5 Testicular tissue dissection with presence of spermatozoa (**a**), spermatozoon-like structure (**b**), and complete azoospermia (**c**)

In 8,293 patients where a viable fetal heart was observed by ultrasound, 898 of these later miscarried, aborted, or had an ectopic. The ongoing pregnancy rate was 36.1 % per retrieval (7,444/19,226), and 38.7 % per replacement procedure (7,444/10,676). So far, we have 8,695 neonates born from 6,537 deliveries.

In a small series, we have performed testicular biopsies with patients diagnosed with Klinefelter's Syndrome. A total of 88 couples had 131 testicular biopsies. We were able to retrieve spermatozoa in 72.5 % ($n=95$) that resulted in a fertilization rate of 52.3 % (570/1,087). Thirty-five women (36.8 %) have delivered thus far, generating 50 healthy babies.

The degree of spermatogenic failure often varies in NOA cases and consequently sufficient spermatozoa for ICSI can only be identified in 40–60 % of the patients. In those cases where few sperm cells are present it is often necessary to search the biopsy for an extended period of time (Fig. 5). We assessed whether such an extended search for spermatozoa in these NOA patients has an effect on ICSI outcome. The average search time for routine TESE cases was no more than 30 min (control), mostly in relation to the oocyte cohort.

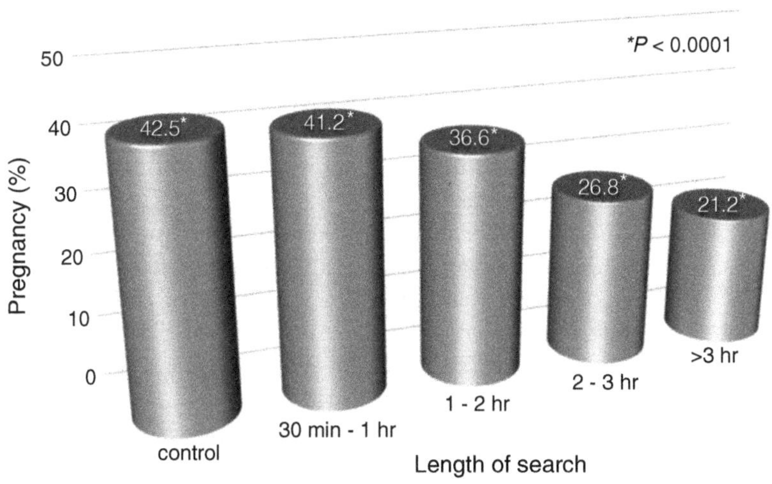

Fig. 6 Pregnancy outcome as a function of search time

The extensive searches, often carried out by several embryologists, were divided into four groups based on time: 30 min to 1 h, 1–2 h, 2–3 h, and >3 h, and compared to clinical outcome. A total of 739 NOA men who underwent 1,087 ICSI cycles were included in this study. The mean ages of the female and male patients were 37.2 and 35.4, respectively. Of the 1,087 cycles included in this study, 225 (26.1 %) required an extended search. The length of sperm hunt ranged from 30 min to as long as 10 h with a mean of 82 min. The average number of embryologists involved in the searches was 4 ± 2. Pentoxifylline was used in almost all of the extended search cycles and in about 57 % of the control cycles.

The control and the extended search groups had similar patient profiles and number of oocytes retrieved. The fertilization rate was 44.0 % in the search group and 57.1 % in the control group ($P<0.002$), delivery rates were 34.3 % and 46.8 %, respectively ($P<0.001$). Fertilization and pregnancy rates were plotted according to the length of time spent for each search (30 min–1 h, 1–2 h, 2–3 h, >3 h). The fertilization rates for these five groups were 54.2, 46.3, 28.0, and 25.4 %, respectively ($R^2=0.9315$; $P<0.001$). A progressive decrease in pregnancy rate with lengthening search times was also observed ($P<0.0001$; Fig. 6). Pregnancy loss rates were comparable between all the extended search groups and control.

It appears that the length of time required to extensively search a testicular tissue sample and to perform ICSI on all the oocyte cohorts, is inversely related to fertilization and pregnancy outcomes. In spite of the time-dependent clinical performance, searching for precious spermatozoa is still warranted even after several hours of hunting. Though labor-intensive and time-consuming, this procedure still grants many couples the opportunity to conceive.

6 Conclusions

Infertility is a common and distressing condition, and problems in the male partner are one of the common causes.

The information generated by conventional semen analysis has historically classified patients into categories lacking knowledge of causality and leaving conventional therapy as somewhat empirical and at times ineffective (Table 3). A single condition such as oligozoospermia may involve a multitude of different etiologies. It is not until we resolve the causes of male infertility at a molecular level that we shall be able to achieve the holy grail of diagnosis, treatment, and prevention.

The epidemiology of male reproduction has been evolving quite rapidly in recent years. A better understanding of spermatogenesis and its genetic control has led to the formulation of new hypotheses on the role of the DNA packing and small RNAs.

Apart from a few exceptions, central hypogonadism and some post-testicular forms, the only available treatment option for the large majority of male infertility situations is medically assisted reproductive technologies, represented by in vitro fertilization [14, 93] or

Table 3
Diagnostic categories for the infertile male based on WHO criteria

No demonstrable cause
Idiopathic oligozoospermia
Idiopathic asthenozoospermia
Idiopathic teratozoospermia
Idiopathic azoospermia
Obstructive azoospermia
Isolated seminal plasma abnormalities
Sexual or ejaculatory dysfunction

Systematic causes
Endrocrine
Iatrogenic
Congenital abnormalities
Acquired testicular damage
Varicocele
Immunological infertility
Male accessory gland infection

in the presence of extreme OAT and various forms of azoospermia, ICSI as the preferred method [14]. ICSI is the most effective means of treating couples with male factor infertility and previous ART fertilization failures. This is consistent with the spermatozoa collected from the epididymis and from the testis achieving comparable fertilization and consistent pregnancy outcomes.

However, ART is a symptomatic therapy which does not address the underlying cause for infertility with the risk of transmitting both identified and concealed genetic anomalies. An increased incidence in chromosomal anomalies and possibly neonatal malformations have been reported especially when the indication for utilizing ART is severe male factor infertility [94, 95]. Apart from the mentioned health consequences of the offspring fathered by a man with severe spermatogenic failure, including the inability to reproduce such as sons of men with Y deletions. In fact, there is still very little known about the long term health conditions of both the infertile man and their offspring [35, 96]. A higher incidence of sperm aneuploidy in infertile men with secretory azoospermia [97–100] translates to a higher frequency of gonosomal abnormalities in the male progeny [101, 102], possibly because of meiotic defects surfacing during male germ line maturation [103].

A different issue arises from the DNA fragmentation observed in suboptimal spermatozoa that could affect embryo development and in genotypically normal offspring, become a source of epigenetic disorders [104]. However, such fragmentation does not seem to preclude establishment of a pregnancy [105, 106], the success of which can be attributed to a corrective role exerted by the selection of the best individual spermatozoon during the ICSI procedure [107–109].

Since some ART procedures have been implicated in various adverse outcomes of offspring, basic research is required to elucidate the biological mechanisms underlying the genetic and epigenetic effects of assisted reproduction. In addition, it is important for clinicians to precisely record the assisted reproduction procedures including the stimulation protocol, method of embryo culture, culture media used, and timing of embryo transfer. As it is not yet possible to evaluate precisely the consequences of assisted reproduction on imprinting, long-term, large-scale epidemiological follow-up studies that could estimate the magnitude of the risks posed by assisted reproduction on human pregnancies are paramount.

Acknowledgments

We are very appreciative to all clinicians and scientists at the The Ronald O. Perelman and Claudia Cohen Center for Reproductive Medicine and the Urology Department. We are thankful to Dr. J. Michael Bedford for his critical review and for Dr. Alessia Uccelli for listing the references.

References

1. Collins JA, Milner RA, Rowe TC (1991) The effect of treatment on pregnancy among couples with unexplained infertility. Int J Fertil 36(3):140–141, 145–152
2. Wright J et al (1991) Psychosocial distress and infertility: men and women respond differently. Fertil Steril 55(1):100–108
3. Reijo R et al (1995) Diverse spermatogenic defects in humans caused by Y chromosome deletions encompassing a novel RNA-binding protein gene. Nat Genet 10(4):383–393
4. MacLeod J, Wang Y (1979) Male fertility potential in terms of semen quality: a review of the past, a study of the present. Fertil Steril 31(2):103–116
5. Swan SH, Elkin EP, Fenster L (1997) Have sperm densities declined? A reanalysis of global trend data. Environ Health Perspect 105(11):1228–1232
6. Spira A (1994) Has sperm quality decreased during the last 50 years? Rev Epidemiol Sante Publique 42(6):563–564
7. Dohle GR et al (2005) EAU guidelines on male infertility. Eur Urol 48(5):703–711
8. Forti G, Krausz C (1998) Clinical review 100: evaluation and treatment of the infertile couple. J Clin Endocrinol Metab 83(12):4177–4188
9. Guzick DS et al (2001) Sperm morphology, motility, and concentration in fertile and infertile men. N Engl J Med 345(19):1388–1393
10. Borini A et al (2006) Sperm DNA fragmentation: paternal effect on early post-implantation embryo development in ART. Hum Reprod 21(11):2876–2881
11. Cohen-Bacrie P et al (2009) Correlation between DNA damage and sperm parameters: a prospective study of 1,633 patients. Fertil Steril 91(5):1801–1805
12. Ji BT et al (1997) Paternal cigarette smoking and the risk of childhood cancer among offspring of nonsmoking mothers. J Natl Cancer Inst 89(3):238–244
13. Lewis SE, Aitken RJ (2005) DNA damage to spermatozoa has impacts on fertilization and pregnancy. Cell Tissue Res 322(1):33–41
14. Palermo G et al (1992) Pregnancies after intracytoplasmic injection of single spermatozoon into an oocyte. Lancet 340(8810):17–18
15. Cohen J et al (1984) Male infertility successfully treated by in-vitro fertilisation. Lancet 1(8388):1239–1240
16. Cohen J et al (1988) Implantation of embryos after partial opening of oocyte zona pellucida to facilitate sperm penetration. Lancet 2(8603):162
17. Ng SC et al (1988) Pregnancy after transfer of sperm under zona. Lancet 2(8614):790
18. WHO (1992) WHO laboratory manual for the examination and processing of human semen and serpm-cervical mucus interaction, 3rd edn. World Health Organization, Geneva
19. WHO (1999) WHO laboratory manual for the examination and processing of human semen and serpm-cervical mucus interaction, 4th edn. World Health Organization, Geneva
20. WHO (2010) WHO laboratory manual for the examination and processing of human semen, 5th edn. World Health Organization, Geneva
21. Mallidis C, Howard EJ, Baker HW (1991) Variation of semen quality in normal men. Int J Androl 14(2):99–107
22. Schwartz D et al (1979) Within-subject variability of human semen in regard to sperm count, volume, total number of spermatozoa and length of abstinence. J Reprod Fertil 57(2):391–395
23. Cooper TG et al (1992) Internal quality control of semen analysis. Fertil Steril 58(1):172–178
24. Matson PL (1995) External quality assessment for semen analysis and sperm antibody detection: results of a pilot scheme. Hum Reprod 10(3):620–625
25. Neuwinger J, Behre HM, Nieschlag E (1990) External quality control in the andrology laboratory: an experimental multicenter trial. Fertil Steril 54(2):308–314
26. Glazener CM et al (1987) The value of artificial insemination with husband's semen in infertility due to failure of postcoital sperm-mucus penetration – controlled trial of treatment. Br J Obstet Gynaecol 94(8):774–778
27. Irvine DS et al (1986) Failure of high intrauterine insemination of husband's semen. Lancet 2(8513):972–973
28. Najmabadi H et al (1996) Substantial prevalence of microdeletions of the Y-chromosome in infertile men with idiopathic azoospermia and oligozoospermia detected using a sequence-tagged site-based mapping strategy. J Clin Endocrinol Metab 81(4):1347–1352
29. Vogt HJ (1996) [Sperm intolerance as a possible cause for infertility?]. J Hautarzt 47(4):312–313

30. Comhaire FH, Vermeulen L, Schoonjans F (1987) Reassessment of the accuracy of traditional sperm characteristics and adenosine triphosphate (ATP) in estimating the fertilizing potential of human semen in vivo. Int J Androl 10(5):653–662
31. Matthews GJ, Matthews ED, Goldstein M (1998) Induction of spermatogenesis and achievement of pregnancy after microsurgical varicocelectomy in men with azoospermia and severe oligoasthenospermia. Fertil Steril 70(1):71–75
32. Joffe M, Li Z (1994) Association of time to pregnancy and the outcome of pregnancy. Fertil Steril 62(1):71–75
33. Vine MF (1996) Smoking and male reproduction: a review. Int J Androl 19(6):323–337
34. Vine MF et al (1994) Cigarette smoking and sperm density: a meta-analysis. Fertil Steril 61(1):35–43
35. Kent-First MG et al (1996) The incidence and possible relevance of Y-linked microdeletions in babies born after intracytoplasmic sperm injection and their infertile fathers. Mol Hum Reprod 2(12):943–950
36. McLachlan RI et al (1998) Genetic disorders and spermatogenesis. Reprod Fertil Dev 10(1):97–104
37. Simoni M et al (1997) Screening for deletions of the Y chromosome involving the DAZ (Deleted in AZoospermia) gene in azoospermia and severe oligozoospermia. Fertil Steril 67(3):542–547
38. Repping S et al (2002) Recombination between palindromes P5 and P1 on the human Y chromosome causes massive deletions and spermatogenic failure. Am J Hum Genet 71(4):906–922
39. Dellarco VL et al (1995) Mutagenesis and human genetic disease: an introduction. Environ Mol Mutagen 25(Suppl 26):2–6
40. Rhomberg L et al (1990) Quantitative estimation of the genetic risk associated with the induction of heritable translocations at low-dose exposure: ethylene oxide as an example. Environ Mol Mutagen 16(2):104–125
41. Tal J et al (2005) ICSI outcome in patients with transient azoospermia with initially motile or immotile sperm in the ejaculate. Hum Reprod 20(9):2584–2589
42. Emery BR, Carrell DT (2006) The effect of epigenetic sperm abnormalities on early embryogenesis. Asian J Androl 8(2):131–142
43. Menezo YJ (2006) Paternal and maternal factors in preimplantation embryogenesis: interaction with the biochemical environment. Reprod Biomed Online 12(5):616–621
44. Miller D, Ostermeier GC (2006) Towards a better understanding of RNA carriage by ejaculate spermatozoa. Hum Reprod Update 12(6):757–767
45. Patrizio P et al (2007) Molecular methods for selection of the ideal oocyte. Reprod Biomed Online 15(3):346–353
46. Yanagimachi R (2005) Intracytoplasmic injection of spermatozoa and spermatogenic cells: its biology and applications in humans and animals. Reprod Biomed Online 10(2):247–288
47. Biermann K, Steger K (2007) Epigenetics in male germ cells. J Androl 28(4):466–480
48. Oliva R (2006) Protamines and male infertility. Hum Reprod Update 12(4):417–435
49. Ward WS, Coffey DS (1991) DNA packaging and organization in mammalian spermatozoa: comparison with somatic cells. Biol Reprod 44(4):569–574
50. Calvin HI, Bedford JM (1971) Formation of disulphide bonds in the nucleus and accessory structures of mammalian spermatozoa during maturation in the epididymis. J Reprod Fertil Suppl 13(p Suppl 13):65–75
51. Bedford JM, Calvin HI (1974) The occurrence and possible functional significance of -S-S- crosslinks in sperm heads, with particular reference to eutherian mammals. J Exp Zool 188(2):137–155
52. Carrell DT, Emery BR, Hammoud S (2007) Altered protamine expression and diminished spermatogenesis: what is the link? Hum Reprod Update 13(3):313–327
53. Li J et al (2002) Involvement of histone methylation and phosphorylation in regulation of transcription by thyroid hormone receptor. Mol Cell Biol 22(16):5688–5697
54. Delaval K et al (2007) Differential histone modifications mark mouse imprinting control regions during spermatogenesis. EMBO J 26(3):720–729
55. Reik W (2007) Stability and flexibility of epigenetic gene regulation in mammalian development. Nature 447(7143):425–432
56. Reik W, Dean W (2001) DNA methylation and mammalian epigenetics. Electrophoresis 22(14):2838–2843
57. Oakes CC et al (2007) A unique configuration of genome-wide DNA methylation patterns in the testis. Proc Natl Acad Sci U S A 104(1):228–233
58. Sato A et al (2007) Aberrant DNA methylation of imprinted loci in superovulated oocytes. Hum Reprod 22(1):26–35
59. Deng T et al (2007) Disruption of imprinting and aberrant embryo development in

completely inbred embryonic stem cell-derived mice. Dev Growth Differ 49(7):603–610
60. Mitalipov SM (2006) Genomic imprinting in primate embryos and embryonic stem cells. Reprod Fertil Dev 18(8):817–821
61. Schaefer CB et al (2007) Epigenetic decisions in mammalian germ cells. Science 316(5823):398–399
62. Allen C, Reardon W (2005) Assisted reproduction technology and defects of genomic imprinting. BJOG 112(12):1589–1594
63. Chang AS et al (2005) Association between Beckwith-Wiedemann syndrome and assisted reproductive technology: a case series of 19 patients. Fertil Steril 83(2):349–354
64. Li Z et al (2006) Correlation of sperm DNA damage with IVF and ICSI outcomes: a systematic review and meta-analysis. J Assist Reprod Genet 23(9–10):367–376
65. Aoki VW, Emery BR, Carrell DT (2006) Global sperm deoxyribonucleic acid methylation is unaffected in protamine-deficient infertile males. Fertil Steril 86(5):1541–1543
66. Kobayashi H et al (2007) Aberrant DNA methylation of imprinted loci in sperm from oligospermic patients. Hum Mol Genet 16(21):2542–2551
67. Krawetz SA et al (2005) In silico and wet-bench identification of nuclear matrix attachment regions. Methods Mol Med 108:439–458
68. Boerke A, Dieleman SJ, Gadella BM (2007) A possible role for sperm RNA in early embryo development. Theriogenology 68(Suppl 1):S147–S155
69. Martins RP, Krawetz SA (2005) Towards understanding the epigenetics of transcription by chromatin structure and the nuclear matrix. Gene Ther Mol Biol 9(B):229–246
70. Krawetz SA (2005) Paternal contribution: new insights and future challenges. Nat Rev Genet 6(8):633–642
71. Miller D, Ostermeier GC, Krawetz SA (2005) The controversy, potential and roles of spermatozoal RNA. Trends Mol Med 11(4): 156–163
72. Ostermeier GC et al (2005) Toward using stable spermatozoal RNAs for prognostic assessment of male factor fertility. Fertil Steril 83(6):1687–1694
73. Ostermeier GC et al (2005) A suite of novel human spermatozoal RNAs. J Androl 26(1):70–74
74. Girard A et al (2006) A germline-specific class of small RNAs binds mammalian Piwi proteins. Nature 442(7099):199–202
75. Grivna ST, Pyhtila B, Lin H (2006) MIWI associates with translational machinery and PIWI-interacting RNAs (piRNAs) in regulating spermatogenesis. Proc Natl Acad Sci U S A 103(36):13415–13420
76. Kim VN (2006) Small RNAs just got bigger: piwi-interacting RNAs (piRNAs) in mammalian testes. Genes Dev 20(15):1993–1997
77. Amanai M et al (2006) Injection of mammalian metaphase II oocytes with short interfering RNAs to dissect meiotic and early mitotic events. Biol Reprod 75(6):891–898
78. Mahadevan MM, Trounson AO (1984) The influence of seminal characteristics on the success rate of human in vitro fertilization. Fertil Steril 42(3):400–405
79. Mortimer D (1994) Sperm recovery techniques to maximize fertilizing capacity. Reprod Fertil Dev 6(1):25–31
80. Gordon JW et al (1988) Fertilization of human oocytes by sperm from infertile males after zona pellucida drilling. Fertil Steril 50(1):68–73
81. Kiessling AA et al (1988) Fertilization in trypsin-treated oocytes. Ann N Y Acad Sci 541:614–620
82. Gordon JW, Talansky BE (1986) Assisted fertilization by zona drilling: a mouse model for correction of oligospermia. J Exp Zool 239(3):347–354
83. Santella L et al (1992) Is the human oocyte plasma membrane polarized? Hum Reprod 7(7):999–1003
84. Palermo G et al (1993) Sperm characteristics and outcome of human assisted fertilization by subzonal insemination and intracytoplasmic sperm injection. Fertil Steril 59(4): 826–835
85. Palermo G et al (1992) Induction of acrosome reaction in human spermatozoa used for subzonal insemination. Hum Reprod 7(2):248–254
86. Lin TP (1966) Microinjection of mouse eggs. Science 151(708):333–337
87. Lanzendorf SE et al (1988) A preclinical evaluation of pronuclear formation by microinjection of human spermatozoa into human oocytes. Fertil Steril 49(5):835–842
88. Veeck LL (1988) Oocyte assessment and biological performance. Ann N Y Acad Sci 541:259–274
89. Schlegel PN, Li PS (1998) Microdissection TESE: sperm retrieval in non-obstructive azoospermia. Hum Reprod Update 4(4):439
90. Schlegel PN et al (1997) Testicular sperm extraction with intracytoplasmic sperm injection

for nonobstructive azoospermia. Urology 49(3):435–440

91. Schlegel PN (2009) Nonobstructive azoospermia: a revolutionary surgical approach and results. Semin Reprod Med 27(2):165–170
92. Vitorino RL et al (2011) Systematic review of the effectiveness and safety of assisted reproduction techniques in couples serodiscordant for human immunodeficiency virus where the man is positive. Fertil Steril 95(5):1684–1690
93. Nyboe Andersen A et al (2009) Assisted reproductive technology and intrauterine inseminations in Europe, 2005: results generated from European registers by ESHRE: ESHRE. The European IVF Monitoring Programme (EIM), for the European Society of Human Reproduction and Embryology (ESHRE). Hum Reprod 24(6):1267–1287
94. Bonduelle M et al (2002) Prenatal testing in ICSI pregnancies: incidence of chromosomal anomalies in 1586 karyotypes and relation to sperm parameters. Hum Reprod 17(10):2600–2614
95. Sutcliffe AG, Ludwig M (2007) Outcome of assisted reproduction. Lancet 370(9584):351–359
96. Katagiri Y et al (2004) Y chromosome assessment and its implications for the development of ICSI children. Reprod Biomed Online 8(3):307–318
97. Colombero LT et al (1999) Incidence of sperm aneuploidy in relation to semen characteristics and assisted reproductive outcome. Fertil Steril 72(1):90–96
98. Palermo GD et al (1995) Intracytoplasmic sperm injection: a novel treatment for all forms of male factor infertility. Fertil Steril 63(6):1231–1240
99. Palermo GD et al (2002) Chromosome analysis of epididymal and testicular sperm in azoospermic patients undergoing ICSI. Hum Reprod 17(3):570–575
100. Takeuchi T et al (2006) The value of the sperm aneuplolidy assay in men undergoing assisted fertilization technology. Fertil Steril 86(supplement 1):S104–S105
101. Bonduelle M et al (2004) Medical follow-up study of 5-year-old ICSI children. Reprod Biomed Online 9(1):91–101
102. Bonduelle M et al (1996) Prospective follow-up study of 423 children born after intracytoplasmic sperm injection. Hum Reprod 11(7):1558–1564
103. Ferguson KA et al (2007) Abnormal meiotic recombination in infertile men and its association with sperm aneuploidy. Hum Mol Genet 16(23):2870–2879
104. Aitken RJ, De Iuliis GN (2007) Origins and consequences of DNA damage in male germ cells. Reprod Biomed Online 14(6):727–733
105. Feliciano M et al (2004) Assays of sperm nuclear status do not correlate with ICSI success. Hum Reprod 19(supplement 1):i52
106. Neri QV et al (2004) Does DNA fragmentation in spermatozoa affect ICSI offspring development? Hum Reprod 19(supplement 1):i107
107. Bungum M et al (2004) The predictive value of sperm chromatin structure assay (SCSA) parameters for the outcome of intrauterine insemination. IVF and ICSI. Hum Reprod 19(6):1401–1408
108. Chen C et al (2011) Kinetic characteristics and DNA integrity of human spermatozoa. Hum Reprod 19(Supplement 1):i30
109. Virro MR, Larson-Cook KL, Evenson DP (2004) Sperm chromatin structure assay (SCSA) parameters are related to fertilization, blastocyst development, and ongoing pregnancy in in vitro fertilization and intracytoplasmic sperm injection cycles. Fertil Steril 81(5):1289–1295

Chapter 19

Techniques for Slow Cryopreservation of Embryos

Lucinda Veeck Gosden

Abstract

The slow cryopreservation of embryos has been used for nearly three decades as a means of storing surplus conceptuses from single IVF (in vitro fertilization) cycles. Doing so has allowed caregivers to maximize pregnancy rates without wastage of precious biological materials. Very detailed methods are described here using a popular biological freezing unit manufactured by Planer PLC (Middlesex, UK). Culture media preparation and tranfer protocols, including replacement in both natural and stimulated cycles, are included.

Key words Cryopreservation, Thawing, Prouclear stage oocyte, Embryo, Blastocyst, Slow freezing

1 Introduction

As published in the 2008 CDC Report, it was demonstrated that, in women over the age of 35 years old, implantation rates after thawing and transfer of frozen embryos was remarkably similar, even somewhat better than fresh transfer in women over age 41, when compared to fresh embryo transfer in the same age groups. In younger women and donor oocyte recipients, while implantation rates were still considerably higher transferring fresh embryos, the differences were less pronounced than a decade ago [1] (Centers for Disease Control, *Assisted reproductive technology fertility clinic report*, 2008, Figs. 1–3). It is evident that cryopreservation strategies have become mainstream in most IVF programs and far more effective in terms of assisting couples in their quest for a child. For example, it was shown that when the cumulative effect of adding thawed pregnancies (only from cycles failing to become pregnant following fresh transfer) to fresh pregnancies was examined, ultimate delivery outcomes are significantly enhanced [2, 3]. Additionally, patients at risk of ovarian hyperstimulation syndrome (OHSS) have been managed more effectively by freezing all conceptuses upfront, thereby reducing, although not eliminating, the likelihood of adverse clinical symptoms following pregnancy [4]. Some programs still decline to freeze a single conceptus unless the

Zev Rosenwaks and Paul M. Wassarman (eds.), *Human Fertility: Methods and Protocols*, Methods in Molecular Biology, vol. 1154, DOI 10.1007/978-1-4939-0659-8_19, © Springer Science+Business Media New York 2014

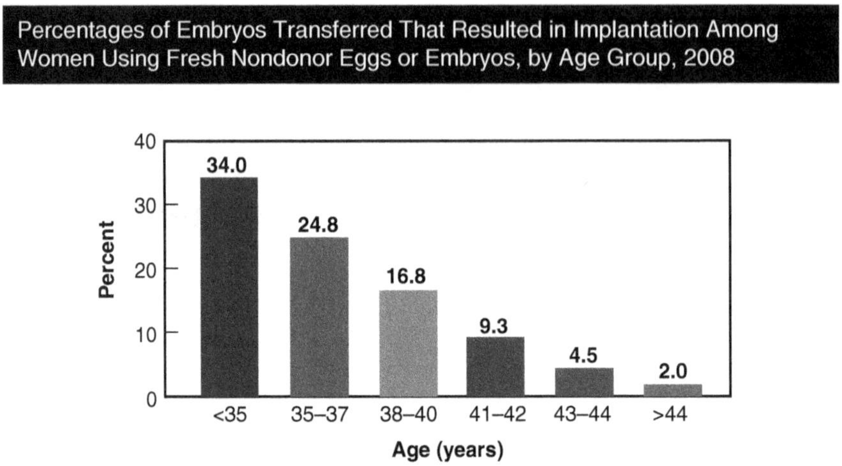

Fig. 1 Implantation rate versus age using fresh nondonor embryos. Centers for Disease Control, *Assisted reproductive technology fertility clinic report*, 2008

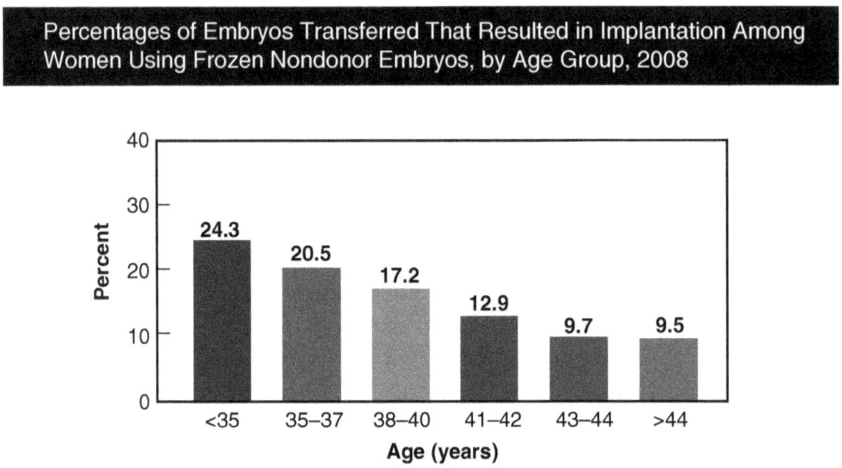

Fig. 2 Implantation rate versus age using frozen–thawed embryos in nondonor cycles. Centers for Disease Control, *Assisted reproductive technology fertility clinic report*, 2008

patient has others in cryostorage or will not be receiving a fresh transfer. Since cryopreservation results have improved tremendously with single embryo transfer after freezing and thawing, this strategy is now less often observed. There are even occasional reports of good results after routinely freezing all conceptuses upfront, thereby eliminating fresh transfer altogether [5, 6].

1.1 Freezing Considerations

The primary goal in establishing a freezing program is to do as little damage as possible while exposing gametes and embryos to non-physiologic low temperatures. Protocols in use today are essentially "freeze-dry" techniques that involve dehydrating the cell to prevent intracellular ice from forming.

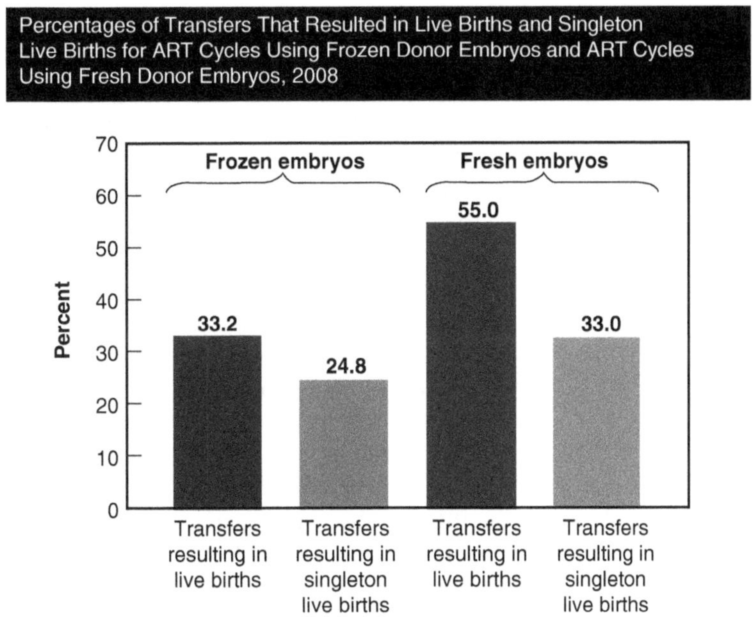

Fig. 3 Comparison of live birth rates after fresh versus frozen–thawed embryo transfer in oocyte recipients. Centers for Disease Control, *Assisted reproductive technology fertility clinic report*, 2008

The formation of intracellular ice crystals can mechanically damage oocytes and conceptuses by disrupting and displacing organelles, or slicing through membranes. This is why freezing techniques are based on the necessity for cryoprotective agents and controlled ice formation at critical temperatures. It has been shown that when human cells are placed into a medium that contains an intracellular cryoprotective agent, intracellular water readily exits the cell as a result of the higher extracellular concentration of cryoprotectant. This causes some cell shrinkage until osmotic equilibrium is reached by the slower diffusion of the cryoprotectant into the cell [7]. Once equilibrium is reached, the cell resumes a normal appearance. The rate of permeation of cryoprotectant and water is dependent on temperature; equilibrium is achieved faster at higher temperatures. For this reason, oocytes and embryos are usually added to cryoprotective media at room temperature. However, since some cryoprotectants like dimethylsulfoxide (DMSO) are toxic at elevated concentrations, these are often used at lower temperatures to reduce adverse effects.

Cryoprotectants are also beneficial in their ability to lower the freezing point of a solution. Solutions may remain unfrozen at −5 °C to −15 °C because of "supercooling" (cooling to well below the freezing point without extracellular ice formation). When solutions supercool, cells do not dehydrate appropriately since there is no increase in osmotic pressure from the formation of extracellular ice crystals. To prevent supercooling, an ice crystal is introduced in a controlled fashion in process call "seeding." This contributes to intracellular dehydration as water leaves the cell to achieve equilibrium with the extracellular environment [7, 8]. If the rate of cooling is too rapid, water cannot pass quickly enough from the cell, and as temperature continues to drop, it reaches a point when the intracellular solute concentration is not high enough to prevent the formation of ice crystals. Mammalian oocytes and embryos, which possess relatively small ratios of surface area to unit volume and hold substantial intracellular water, have until recently with the improvement of vitrification techniques, been cooled at slow rates (0.1–1 °C/min) to permit adequate dehydration.

Membrane permeability by cryoprotectants differs between the oocyte, embryo, and blastocyst. As such, it has been found that different cryoprotective agents are more suitable for certain stages of gamete and embryo development. DMSO and 1,2 propanediol (PROH) are frequently used for freezing early cleavage-stage embryos, and propylene glycol (glycerol) is commonly used for blastocysts. All three intracellular agents have fairly small molecules that permeate cell membranes easily. In addition to these, there are several extracellular substances that help dehydrate and protect cells. The most frequently used extracellular substance is sucrose, which possesses large, non-permeating molecules and exerts an osmotic effect to aid in accelerated cell dehydration. Sucrose cannot be used alone but is often used in conjunction with standard permeating, intracellular cryoprotectants.

1.2 Thawing Considerations

If cooling is terminated at relatively high temperatures (≥–30 °C), the cell carries more intracellular ice than if cooled longer to lower temperatures (≤–80 °C). In order to protect the cell, thawing must be carried out rapidly to induce rapid ice dispersal. Conversely, samples cooled to ≤–80 °C should be thawed more slowly to allow for gradual rehydration [9]. If water reenters the cell too rapidly, it may swell or burst. It is common to expose thawed specimens to progressively lower dilutions of cryoprotectant to gently and slowly remove it from the cell.

1.3 Vitrification and Ultrarapid Freezing

The idea behind vitrification is to protect the cell by completely avoiding all ice crystal formation. To accomplish this, cryoprotective solutes must be increased to 40 % (wt/vol) or higher. DMSO is frequently used, but PROH, glycerol, and other agents have been tested. Because at room temperature high concentrations of cryoprotectant are toxic, embryos are exposed to them at 0 °C. Samples may be plunged directly into liquid nitrogen without needing to introduce a seed; the viscosity is so great that solutions solidify into glasslike states. Vitrified specimens must be thawed rapidly in ice water, a process that can be troublesome [10–12]. Nonetheless, the use of vitrification has been gaining in popularity for many years now; some IVF programs have essentially dropped slow freezing methods in favor of a more simplified and rapid vitrification model [13–15].

Using ultrarapid freezing, specimens are exposed for short intervals (2–3 min) to relatively high concentrations of DMSO (3.5 M) and sucrose (0.25 M) followed by direct plunging into LN_2. Specimens are then thawed rapidly in a 37 °C waterbath and cryoprotectant removed in a single step [16]. Using these simple and rapid techniques, a number of children have been born. Gordts et al. as early as 1990 [17], reported four pregnancies after ultrarapid freezing of pronuclear stage oocytes. In this study, much better survival was noted for pronuclear oocytes as compared to cleaved embryos, a finding that has been reported by other investigators as well [18]. In contradiction to this, Lai et al. report an 83 % survival rate (at least one blastomere intact) and 16 % birthrate for ultrarapidly frozen cleaved embryos [19]. Mitochondrial distribution and overall subcellular structure are described as being normal after using this method of freezing [20].

Technically, vitrification and ultrarapid freezing are very similar techniques. If no ice crystals form during the latter process, it is actually a vitrification procedure [21].

2 Materials

Freezing–thawing media for prezygotes/embryos/blastocysts
(prepared weekly or bimonthly)

Date prepared/prepared by: ____________/____________
Base water resistivity (>18 MΩ/cm required): ____________

Lot numbers:

Component	Supplier	Lot number and date
1,2 Propanediol	Sigma; P1009	
Sucrose	Sigma; S1888	
Glycerol	Sigma; G9012	

Solution I (1,000 mL)

Component	Measurement	Supplier	Lot number and date
HHI medium (stands for "Harvest, Hatching, ICSI")	700 mL	Laboratory Stock	This is simply a HEPES-supplemented medium based on Stage I sequential culture medium
Plasmanate	300 mL (30.0 %)	NHS; 120-42P	

Solution II (300 mL; 0.2 M sucrose)

Component	Measurement	Supplier
Solution I	300 mL	Just prepared
Sucrose	20.55 g	Sigma; #S1888

Solution III (50 mL; 0.1 M sucrose)

Component	Measurement	Supplier
Solution I	50 mL	Just prepared
Sucrose	1.71 g	Sigma; #S1888

Solution IV (150 mL; 0.4 M sucrose)

Component	Measurement	Supplier
Solution I	150 mL	Just prepared
Sucrose	20.55 g	Sigma; #S1888

Formulae for non-sucrose freeze–thaw solutions

#		Solution I	PROH
1.	1.5 M	44.3 mL	5.7 mL
2.	1.0 M	46.2 mL	3.8 mL
3.	0.5 M	48.1 mL	1.9 mL
4.	0.0 M	50.0 mL	–

Formulae for sucrose freeze–thaw solutions

#		Solution II	PROH
5.	1.5MS.2	44.3 mL	5.7 mL
6.	1.0MS.2	46.2 mL	3.8 mL
7.	0.5MS.2	48.1 mL	1.9 mL
8.	0.0MS.2	50.0 mL	–

Formulae for blastocyst freeze–thaw solutions with glycerol

#		Solution I, II, III, or IV	Glycerol
9.	Bl Cryo 5 % G	47.5 mL Solution I	2.5 mL
10.	Bl Cryo 10 % GS.2	45.0 mL Solution II	5.0 mL
11.	Bl Thaw 10 %GS.4	45.0 mL Solution IV	5.0 mL
12.	Bl Thaw 5 %GS.4	47.5 mL Solution IV	2.5 mL

(continued)

#		Solution I, II, III, or IV	Glycerol
13.	Bl Thaw S.4	50.0 mL Solution IV	–
14.	Bl Thaw S.2	50.0 mL Solution II	–
15.	Bl Thaw S.1	50.0 mL Solution III	–

2.1 Computerized Programs: Planer Biological Freezer

Prezygote freezing

Action	Ramp #	Program#1
Start temp		22.0 °C
Start		Chamber
Rate	01	-1.00 °C
Temp	01	-6.5 °C
Temp=	01	Chamber
Hold	02	5 min
Temp	02	-6.5 °C
Rate	03	-0.50 °C
Temp	03	-80.0 °C
Temp=	03	Chamber
Seeding		Manual
Seed temp		-6.5 °C
Seed=		Chamber
Soak time		5 min
Trigger		Temp

Embryo freezing

Action	Ramp #	Program#3
Start temp		22.0 °C
Start		Chamber
Rate	01	-2.00 °C
Temp	01	-7.0 °C
Temp=	01	Chamber
Hold	02	5 min
Temp	02	-7.0 °C
Rate	03	-0.30 °C
Temp	03	-30.0 °C
Temp=	03	Chamber

(continued)

Action	Ramp #	Program#3
Seeding		Manual
Seed temp		-7.0 °C
Seed=		Chamber
Soak time		5 min
Trigger		Temp

Blastocyst freezing

Action	Ramp #	Program#4
Start temp		22.0 °C
Start		Chamber
Rate	01	-2.00 °C
Temp	01	-7.0 °C
Temp=	01	Chamber
Hold	02	10 min
Temp	02	-7.0 °C
Rate	03	-0.30 °C
Temp	03	-38.0 °C
Temp=	03	Chamber
Seeding		Manual
Seed temp		-7.0 °C
Seed=		Chamber
Soak time		5 min
Trigger		Temp

Prezygote thawing

Action	Ramp #	Program#6
Start temp		-100.0 °C
Start		Chamber
Hold	01	5 min
Temp	01	-100.0 °C
Rate	02	8.00 °C
Temp	02	22.0 °C
Temp=	02	Chamber
Hold	03	5 min

(continued)

Action	Ramp #	Program#6
Temp	03	22.0 °C
Seeding		No seeding
Trigger		Temp

3 Methods

3.1 Prezygotes (Figs. 4–6)

Freezing success with this stage has spanned more than three decades and has resulted in thousands of births. It is thought that the prezygote's lack of a spindle in large part explains its excellent survival and potential for implantation. Being single-celled, it is easy to determine whether or not a prezygote has survived thawing; when its membrane is not intact, the cell appears flattened and usually dark in color. Left in culture for 15–24 h, the healthy thawed pronucleate oocyte enters into syngamy, completes the fertilization process, and proceeds to the first cleavage. Cell division is the true indicator of survival after thaw; fewer than 5 % of prezygotes appearing healthy after thaw fail to follow this pattern.

Despite the good results achieved after freezing at this stage, there are also certain disadvantages. Because prezygotes are frozen before cleavage occurs, there are no standard morphological parameters to aid in selection; consequently, prezygotes with poor

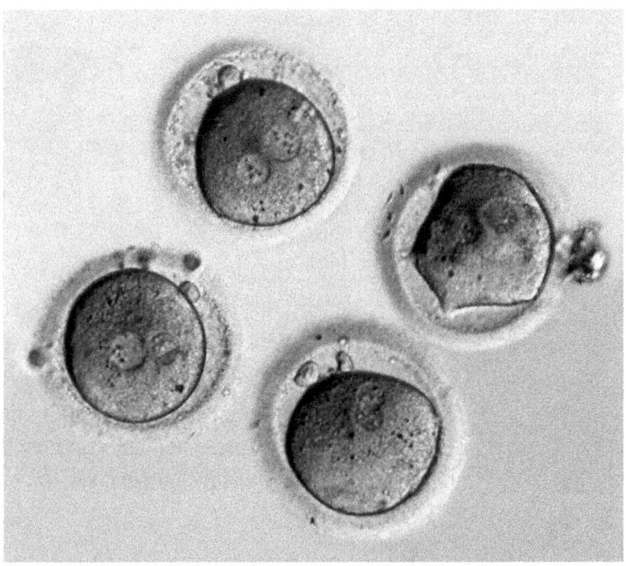

Fig. 4 Four prezygotes photographed immediately after thawing. These were frozen using 1,2 propanediol as the cryoprotectant, without sucrose. Although one prezygote is slightly misshapen, it regained its round shape within a short period of time. All four display intact zonae and oolemmae

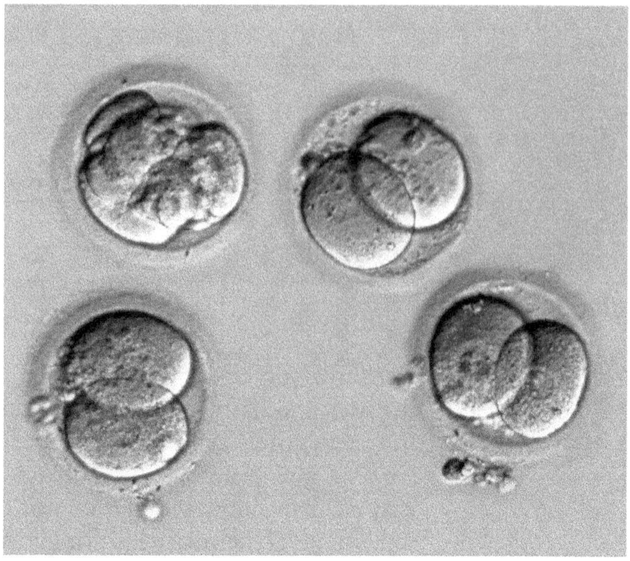

Fig. 5 The same conceptuses after 24 h of culture. Three have developed to the 2-cell stage, and one is cleaving to four cells. Two gestational sacs were identified by ultrasound and one male child delivered

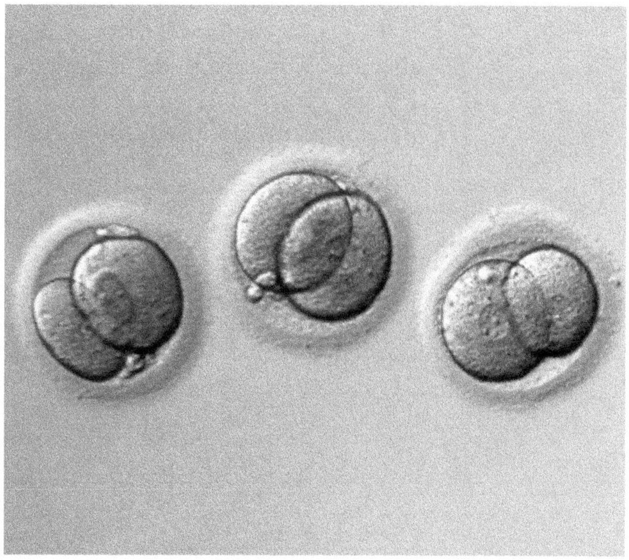

Fig. 6 Three conceptuses developing from frozen–thawed prezygotes stored for 482 days. One of these implanted in the uterus of a 38-year-old mother and led to the birth of a healthy female child

developmental potential are sometimes frozen. It is disappointing to freeze a large number of prezygotes for a woman only to find that the ones not frozen arrest in culture or exhibit abnormal morphological characteristics on days 2 or 3. In such cases, the patient might have been better served if consideration of freezing had been delayed 2–5 days.

It is important to freeze the prezygote before breakdown of pronuclei since waiting too long negatively impacts results. This urgency to begin freezing may be inconvenient for some programs without adequate staffing.

The morphology of thawed prezygotes is generally similar to their pre-freeze appearances, but occasionally the cytoplasm is clearer and organelle accumulation around pronuclear structures is reduced. After thawing, nucleoli are often seen scattered within pronuclei despite their alignment at pronuclear junctions before freezing. Interestingly, two pronuclei have been observed several times to coalesce into one large pronucleus during these procedures (personal observation).

3.1.1 Detailed Prezygote (Pronuclear Stage) Slow Freezing and Thawing Methods: Specific Techniques Using a Biological Freezing Unit (Planer Kryo 10-1.7 II or III)

Before loading the cryovial, wash prezygotes in 1.5 M PROH to avoid transferring oil. To do this, first use an automatic pipettor with sterile tip or a flamed glass pipette to transfer prezygotes to 1.5 M PROH within a sterile dish. Change the pipettor tip/glass pipette and avoid oil layer when loading conceptus into cryovial.

1. Evaluate the morphology of the pronuclear specimen. Document.
2. Appropriately label a Nunc 1.8 mL sterile cryovial. Prepare a dated QC sample cryovial in the same manner.
3. Pipette 0.3 mL freezing medium (1.5 M PROH) into cryovials using sterile technique. Freezing medium should be at room temperature.
4. Place prezygotes into cryovials, carrying as little culture medium as possible. Take care to place prezygotes at the bottoms of the vials and do not allow bubbles of air to enter. Visually check each cryovial under higher magnification after loading to assure that the specimens are properly located inside. Cryovials must be maintained in a level, upright position at all times and caps must be well tightened. Equilibrate cryovials at room temperature for 30 min (40 min maximum from the time the first cryovial loaded); during this time, securely load cryovials onto freezing canes (canes must be also be at room temperature).
5. Approximately 3 min prior to beginning the freezing run, turn Planer main switch to the "ON" position and open valve of LN_2 tank.
6. The Planer will display *Run*, *Print*, *Program* on the screen. Press *Run*.
7. The Planer will display *Enter Passcode* on the screen. Enter "3333."
8. Choose program number and name by pressing the < and > keys. Choose the pronuclear freezing program and press *Enter*.

9. When the proper temperature is reached, an audible alarm will sound that indicates the unit is at room temperature. Carefully load freezing canes/cryovials and QC sample cryovial (thermosensor must be immersed in freezing medium) into Planer and set clock alarm for 26 min. Press *Run*. The freezing program will now be under computerized control until the time of manual seeding. Temperature will drop within the unit at a rate of −1.0 °C/min until a temperature of −6.5 °C is reached. When −6.5 °C is reached, the display will read *Soaking*. Immediately record the date, patient's names, and display temperature of the chamber on a QC log sheet.
10. After 5 min of soaking, an audible alarm will sound to indicate that manual seeding may be performed. Check the temperature of the chamber and ensure that it is at a temperature of at least −6.0 °C. If not, delay seeding and wait for the chamber to reach this temperature. If this does not occur, press *Run* in order for program to continue as if seeding had been performed. When the appropriate control temperature is reached, press *Run* again to put the unit into manual override (pause) in order to perform manual seeding. Record this temperature in the QC log sheet before proceeding with seeding. In this case, you will have to perform the seeding and then wait for 5 min before pressing *Run* again to resume program. If chamber is at the proper temperature when the audible alarm, record this display temperature on the QC log sheet.
11. Perform manual seeding:
    (a) Freeze sponge forceps in liquid nitrogen by immersing forceps in filled dewar.
    (b) Quickly raise freezing cane from unit and grasp cryovial at the level of the meniscus of medium. Watch for ice crystal formation within the cryovial. When crystals are visualized, immediately return the cryovial to Planer. Check the cryovial for ice crystal formation after approximately 30 s (medium should appear "slushy"). If needed, reseed to achieve crystal formation. This is a critical step in the process.
12. Press *Run*. Under computerized control, the Planer will hold at same temperature for an additional five minutes before proceeding with program. After hold period, temperature will drop at a rate of −0.5 °C/min until a temperature of −80 °C is reached. Machine should be checked at regular intervals to ensure that operation is satisfactory. Approximately $2^1/_2$ h are required to complete the program.
13. An audible alarm sounds at the completion of the freezing program and the display reads *Remove Sample*. At this point, the cryovials should be removed from the Planer and immediately plunged into LN_2 within styrofoam/plastic container.

14. Cryovials should be snapped into properly prelabeled storage canes, label side up, while constantly under LN_2. Storage canes are then fitted into cryosleeves. Storage canes and cryosleeves must be at LN_2 temperature before touching cryovials.
15. Transfer canes/cryosleeves to appropriate canisters within designated storage tanks.
16. Press *Run* to end program. The display will read *Do Not Switch Off* while the unit stabilizes to room temperature. This should take 4–5 min. When the display reads *Ready To Restart*, turn off main switch and LN_2 valve. Attach the patient's chart recording to the back of the cryopreservation form.

3.2 Prezygote Thawing

1. Turn Planer main switch to the "ON" position and open valve of LN_2 tank.
2. The Planer will display *Run*, *Print*, *Program* on the screen. Press *Run*.
3. The Planer will display *Enter Passcode* on the screen. Enter "3333."
4. Choose program number and name by pressing the < and > keys. Choose the prezygote thawing program and press *Enter*.
5. An audible alarm will sound when the unit is at −10 °C.
6. Quickly transfer patient cryovial from storage receptacle and snap onto freezing cane within the Planer unit.
7. Press *Run*. Under programmed control, the Planer will hold for 5 min and then warm at a rate of +8.0 °C/min until room temperature is reached. An audible alarm will sound at the end of the thawing process which takes approximately 20 min. Wait for an additional 5 min after alarm sounds before removing cryovial from unit (this delay will allow the actual sample temperature to reach the warmer chamber temperature).
8. At the finish of the thawing program, remove cryovial from freezing cane and allow to equilibrate at room temperature for 5 min. During this time, prepare a 4-well culture dish with 1.0 M PROH, 0.75 M PROH, 0.5 M PROH, and 0.25 M PROH dilutions (the 0.75 M dilution is prepared by adding one part 1.0 M to one part 0.5 M; the 0.25 M dilution is prepared by adding one part 0.5 M to one part 0.0 M). Prepare one Falcon 3037 organ culture dish with 0.0 M PROH dilution. Label wells and dish appropriately. All dilutions should be maintained at room temperature.
9. Press *Run* to end program. The display will read *Do Not Switch Off* while the unit stabilizes to room temperature. This should take 1–2 min. When the display reads *Ready To Restart*, turn off main switch and LN_2 valve.

10. Empty contents of cryovial into petri dish and locate thawed prezygote. If it cannot be found in the first scan, refill cryovial with freezing medium (1.5 M PROH at room temperature), gently shake vial, and pour contents into fresh dish. Repeat this process up to 10 times if prezygote cannot be located. One should be able to tell almost immediately whether or not the prezygote has survived the freeze–thaw process by judging ooplasmic color and the integrity of the zona pellucida. Occasionally, only an empty zona pellucida is located.
11. Transfer prezygote into 1.0 M PROH medium. Wait for 3 min.
12. Transfer prezygote into 0.75 M PROH medium. Wait for 3 min.
13. Transfer prezygote into 0.5 M PROH medium. Wait for 3 min.
14. Transfer prezygote into 0.25 M PROH medium. Wait for 3 min.
15. Transfer prezygote into 0.0 M PROH medium in an organ culture dish. Wait for 3 min.
16. Prepare a fire-polished sterile pipette and locate patient's culture dish.
17. Transfer prezygote into fresh medium droplet, wash through four outside droplets, and place in clean droplet. Label droplet appropriately. Incubate overnight.
18. Record survival information on the thawing form; note how many conceptuses remain in storage.

3.3 Frozen Prezygote Replacement Protocols

1. Natural cycle replacement
   (a) Regular ovulatory cycle with normal luteal phase progesterone
   (b) Thaw day of ovulation or next (day after LH peak and/or day of E_2 drop)
      - Transfer day after thaw
      - No progesterone unless indicated or previous failure without it
      - Medrol and tetracycline starting on day of LH surge for 4 days
2. Programmed cycle replacement with GnRHa
   (a) Luteal suppression 0.2 mg GnRHa; 0.1 mg on predetermined day 1 until day 15
   (b) Transdermal estrogen patches:

      0.1 mg every other day, days 1–4

      0.2 mg every other day, days 5–8

0.3–0.4 mg every other day, days 9–14 (depending on estrogen levels)

0.2 mg every other day, days 15—pregnancy testing (if pregnant, until E2 rises at approximately 7 weeks)

(c) Tetracycline and Medrol starting on day 15 for 4 days

(d) Thaw day 16, transfer day 17

(e) 50 mg progesterone starting on day 15, daily until end of 12th week of gestation, gradually weaned down starting weeks 9–10 (depending on serum levels)

Progesterone: *Natural cycle, if required*: 200 mg micronized vaginal suppositories two or three times a day; continued until negative pregnancy test or through week 12 if pregnant (gradually weaned down starting week 9–10, depending on serum levels).

Programmed cycle: 50 mg/day IM beginning day 15; continued until negative pregnancy test or through week 12 if pregnant (gradually weaned down starting week 9–10, depending on serum levels)

Estrogen patches: Programmed cycle: Climara (comes in 0.1 mg patches; gives high serum E_2 levels)

Medrol: 16 mg/day for 4 days starting day of LH surge (natural cycle) or day 15 (programmed cycle)

Tetracycline: 250 mg four times per day for 4 days starting day of LH surge (natural cycle) or day 15 (programmed cycle)

3.4 Embryos (Figs. 7–17)

The first reported birth following cryopreservation and thaw developed from a frozen embryo [22]. Like pronuclear oocytes, cleaving embryos do well after thawing and contribute to pregnancy at acceptable rates. Almost any cleavage stage can be frozen successfully, from 2-cell to blastocyst. Freezing the embryo is convenient because, unlike the prezygote, there are no time restraints. In addition, both morphology and growth rate are known, allowing for selection of potentially viable conceptuses. It is becoming more common to choose the best embryos for fresh transfer and freeze all others with good morphology only after fresh selection has been made.

Sometimes survival is difficult to evaluate because not all blastomeres survive the rigors of freezing and thawing. Dying blastomeres may be present amongst the living ones, but can be removed easily during assisted hatching procedures. Generally, an embryo possessing >50 % viable blastomeres upon thaw is considered a survivor. There is no convincing evidence that the loss of one or two blastomeres is overtly detrimental to very early developing embryos [23–26]. Despite this, it has been reported that fully intact human embryos demonstrate a higher implantation rate than do partially intact ones [27].

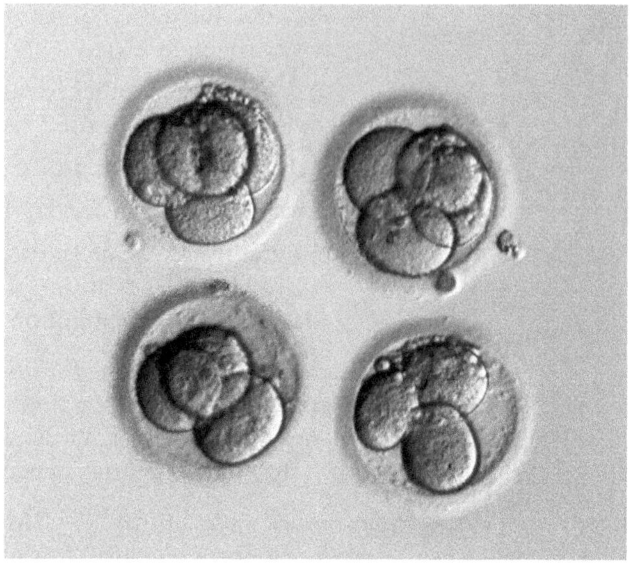

Fig. 7 In this example, two of four embryos, stored for 861 days, survived thawing with all blastomeres intact. The remaining two conceptuses lost a single blastomere during the process. Degenerative cells were removed before intrauterine transfer to a 35-year-old woman. A healthy female child was delivered

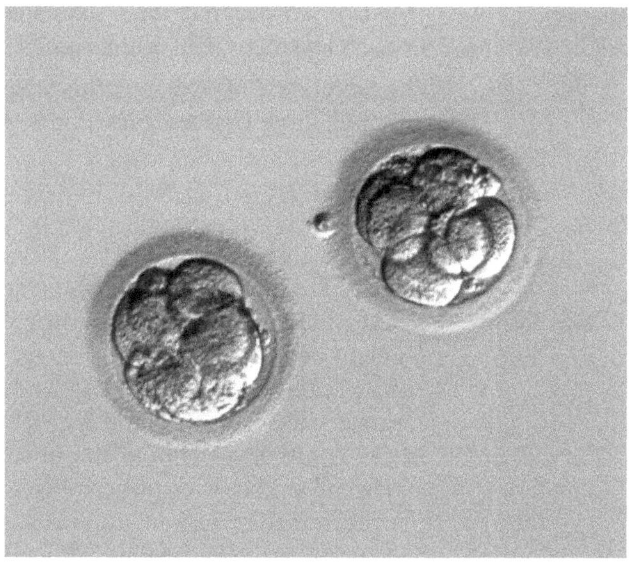

Fig. 8 A photomicrograph of two day 3 embryos as they appeared just before freezing

3.4.1 Detailed Embryo Slow Freezing and Thawing Methods: Specific Techniques Using a Biological Freezing Unit (Planer Kryo 10-1.7 II or III)

1. Evaluate the morphology of the embryo. Document.
2. Appropriately label a Nunc 1.8 mL sterile cryovial. Prepare a dated QC sample cryovial in the same manner.
3. Pipette 0.3 mL freezing medium (1.5 M solution with sucrose) into cryovials using sterile technique. Freezing medium should be at room temperature.

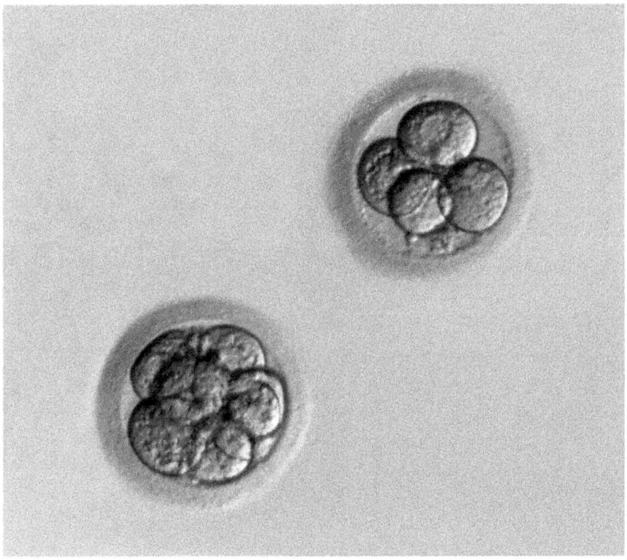

Fig. 9 After cryostorage for 144 days, both embryos survived, although two blastomeres were damaged and microsurgically removed in the upper right conceptus. The 38-year-old patient failed to achieve pregnancy after the transfer of these embryos, but returned for second thawing attempt of prezygotes and subsequently conceived

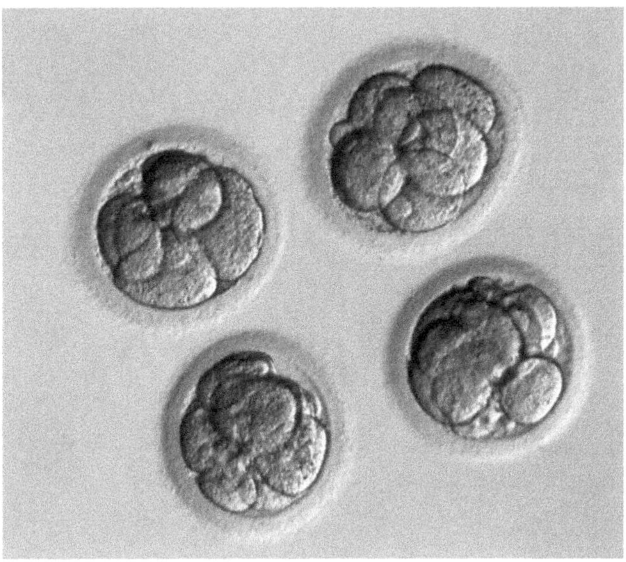

Fig. 10 Four embryos photographed just before freezing on day 3 after harvest

4. Take embryos through non-sucrose dilutions as follows: 0.5 M solution (no sucrose) for 5 min; 1.0 M (no sucrose) for 5 min; 1.5 M (no sucrose) for 10 min; 1.5 M with sucrose for 10 s. Change pipette before loading cryovials to avoid transferring oil.

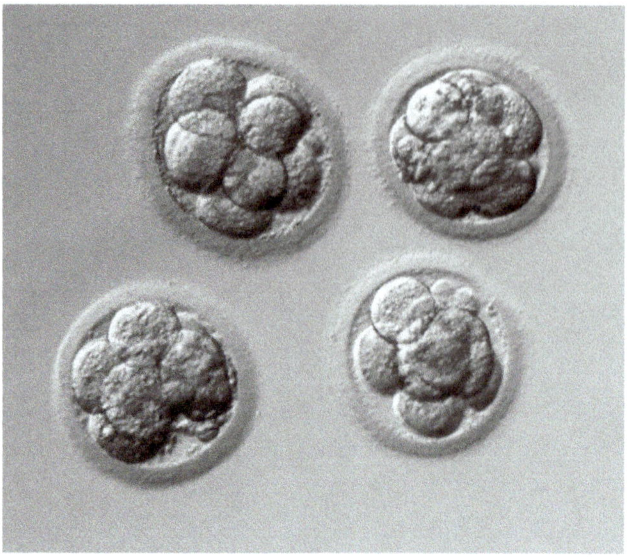

Fig. 11 The same four embryos as in Fig. 10, photographed a few hours after thawing 94 days later. All tolerated the freezing and thawing procedures well. After intrauterine transfer, the 35-year-old mother became pregnant from this cohort and subsequently delivered a healthy male infant

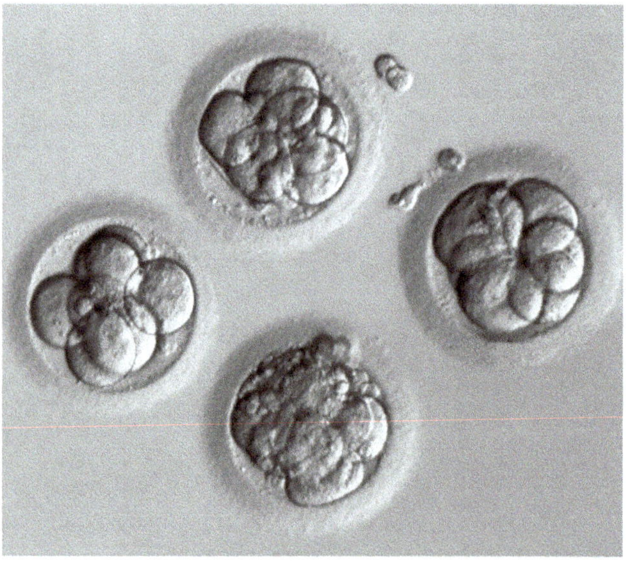

Fig. 12 Four embryos photographed on day 3 after harvest, just before freezing was carried out

5. Place embryos into cryovials, carrying as little culture medium as possible. Take care to place embryos at the bottoms of the vials and do not allow bubbles of air to enter. Visually check each cryovial under higher magnification after loading to assure

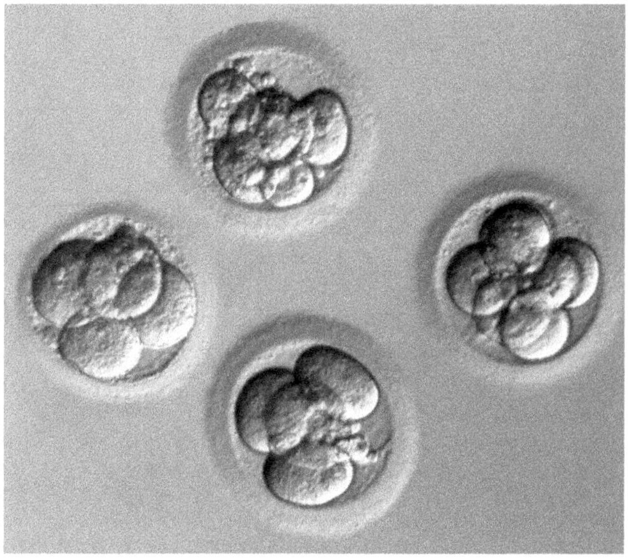

Fig. 13 The same four embryos photographed a few hours after thawing 68 days later. Two of these had degenerative blastomeres and fragments removed before the photograph was taken. The 37-year-old patient became pregnant after the transfer of these conceptuses, showing two sacs and one fetal heart by ultrasound. Miscarriage followed

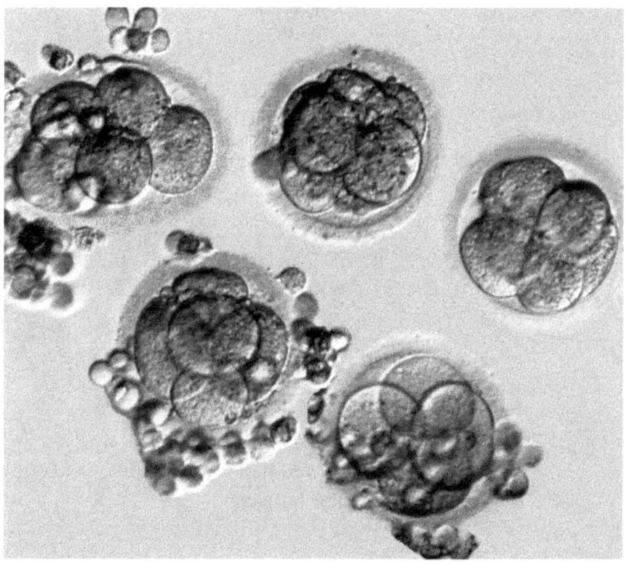

Fig. 14 Five conceptuses photographed before freezing on day 3 after harvest

that the specimens are properly located inside. Cryovials must be maintained in a level, upright position at all times and caps must be well tightened. Equilibrate cryovials at room temperature for 15 min (25 min maximum from the time the first cryovial

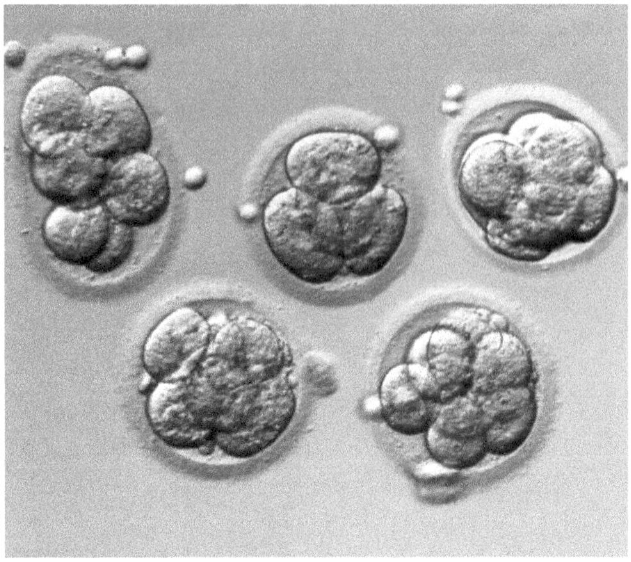

Fig. 15 Upon thawing 68 days later, all demonstrated evidence of survival. A preclinical pregnancy and loss was established in the 38-year-old mother after transfer

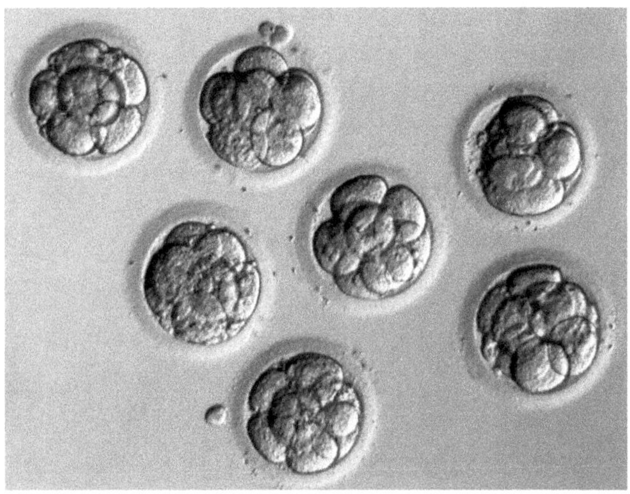

Fig. 16 Seven healthy-appearing conceptuses photographed before freezing on day 3 after harvest

loaded); during this time, securely load cryovials onto freezing canes (canes must be also be at room temperature).

6. Approximately 3 min prior to beginning the freezing run, turn Planer main switch to the "ON" position and open valve of LN_2 tank.
7. The Planer will display *Run, Print, Program* on the screen. Press *Run*.

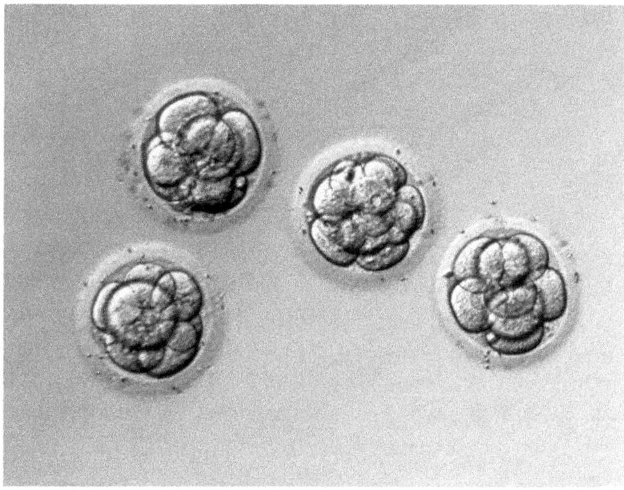

Fig. 17 Upon thawing four of these conceptuses 122 days later, all survived. Unfortunately, no pregnancy was established in the 38-year-old woman producing these conceptuses

8. The Planer will display *Enter Passcode* on the screen. Enter "3333."
9. Choose program number and name by pressing the < and > keys. Choose the embryo freezing program and press *Enter*.
10. When the proper temperature is reached, an audible alarm will sound that indicates the unit is at room temperature. Carefully load freezing canes/cryovials and QC sample cryovial (thermosensor must be immersed in freezing medium) into Planer and set clock alarm for 14 min. Press *Run*. The freezing program will now be under computerized control until the time of manual seeding. Temperature will drop within the unit at a rate of −2.0 °C/min until a temperature of −7.0 °C is reached. When −7.0 °C is reached, the display will read *Soaking*. Immediately record the date, patient's names, and display temperature of the chamber on a QC log sheet.
11. After 5 min of soaking, an audible alarm will sound to indicate that manual seeding may be performed. Check the temperature of the chamber and ensure that it is at a temperature of at least −6.5 °C. If not, delay seeding and wait for chamber to reach this temperature. If this does not occur, press *Run* in order for program to continue as if seeding had been performed. When the appropriate control temperature is reached, press *Run* again to put the unit into manual override (pause) in order to perform manual seeding. Record this temperature in the QC log sheet before proceeding with seeding. In this case, you will have to perform the seeding and then wait for 5 min before pressing *Run* again to resume program. If the

chamber is at the proper temperature when the audible alarm sounds, record this display temperature on the QC log sheet.

12. Perform manual seeding:
    (a) Freeze sponge forceps in liquid nitrogen by immersing forceps in filled dewer.
    (b) Quickly raise freezing cane from unit and grasp cryovial at the level of the meniscus of medium. Watch for ice crystal formation within the cryovial. When crystal is visualized, immediately return cryovial to Planer. Check cryovial for ice crystal formation after approximately 30 s (medium should appear "slushy"). If needed, reseed to achieve crystal formation. This is a critical step in the process.
13. Press *Run*. Under computerized control, the Planer will hold at same temperature for an additional 5 min before proceeding with program. After hold period, temperature will drop at a rate of −0.3 °C/min until a temperature of −30 °C is reached. Machine should be checked at regular intervals to ensure that operation is satisfactory. Approximately $1^{1}/_{4}$ h are required to complete the program.
14. An audible alarm sounds at the completion of the freezing program and the display reads *Remove Sample*. At this point, the cryovials should be removed from the Planer and immediately plunged into LN_2 within styrofoam/plastic container.
15. Cryovials should be snapped into properly prelabeled storage canes, label side up, while constantly under LN_2. Storage canes are then fitted into cryosleeves. Storage canes and cryosleeves must be at LN_2 temperature before touching cryovials.
16. Transfer canes/cryosleeves to appropriate canisters within designated storage tanks.
17. Press *Run* to end program. The display will read *Do Not Switch Off* while the unit stabilizes to room temperature. This should take 4–5 min. When the display reads *Ready To Restart*, turn off main switch and LN_2 valve. Attach the patient's chart recording to the back of the cryopreservation form.

3.5 Embryo Thawing (Waterbath Method)

1. Ensure that the 30 °C waterbath is at the appropriate temperature.
2. Remove cryovial from the liquid nitrogen storage tank and warm in a 30 °C waterbath for approximately 30 s with gentle agitation.
3. Allow cryovial to sit at room temperature for 1 min. During this time, prepare a 4-well culture dish with 1.0 M PROH (sucrose), 0.75 M PROH (sucrose), 0.5 M PROH (sucrose), and 0.25 M PROH (sucrose) (the 0.75 M dilution is prepared by adding one part 1.0 M to one part 0.5 M; the 0.25 M dilution

is prepared by adding one part 0.5 M to one part 0.0 M). Prepare two Falcon 3037 organ culture dishes, one with 0.0 M PROH (sucrose) and one with 0.0 M PROH (no sucrose) dilutions. Label wells and dishes appropriately.

4. Empty contents of cryovial into petri dish and locate specimen. If not located in the first scan, refill cryovial with 1.5 M PROH (sucrose) at room temperature, gently shake vial, and pour contents into fresh dish. Repeat this process up to 10 times if specimen cannot be located. One should be able to tell almost immediately whether or not the specimen has survived the freeze–thaw process by judging both blastomere color and integrity of the zona pellucida. Survival is defined as greater than 50 % of the blastomeres remaining intact and exhibiting a healthy appearance.
5. Transfer embryo into 1.0 M (sucrose) medium. Allow to equilibrate for 3 min.
6. Transfer embryo into 0.75 M (sucrose) medium. Allow to equilibrate for 3 min.
7. Transfer embryo into 0.5 M medium (sucrose). Allow to equilibrate for 3 min.
8. Transfer embryo into 0.25 M medium (sucrose). Allow to equilibrate for 3 min.
9. Transfer embryo into 0.0 M medium (sucrose) in the organ culture dish. Wait for 3 min.
10. Transfer embryo to the 0.0 M (no sucrose) medium in the second organ culture dish. Wait for 3 min.
11. During this last interval, prepare a fire-polished sterile pipette and locate patient's culture dish.
12. Transfer embryo into fresh medium droplet, wash through four outside droplets, and place in clean droplet. Label droplet appropriately. Incubate until intrauterine transfer.
13. Record survival information on the thawing form; note how many frozen conceptuses remain in storage.

3.6 Frozen Embryo Replacement Protocols

1. Natural cycle replacement
   (a) Regular ovulatory cycle with normal luteal phase progesterone
   (b) Thaw day after ovulation or next (2 days after LH peak and/or day after E_2 drop)
   - Transfer day of thaw
   - No progesterone unless indicated or previous failure without it
   - Medrol and tetracycline starting on day of LH surge for 4 days

2. Programmed cycle replacement with GnRHa
   (a) Luteal suppression 0.2 mg GnRHa; 0.1 mg on predetermined day 1 until day 15
   (b) Transdermal estrogen patches:
      0.1 mg every other day, days 1–4
      0.2 mg every other day, days 5–8
      0.3–0.4 mg every other day, days 9–14 (depending on estrogen levels)
      0.2 mg every other day, days 15—pregnancy testing (if pregnant, until E_2 rises at approximately 7 weeks)
   (c) Tetracycline and Medrol on day 15 for 4 days
   (d) Thaw and transfer day 17
   (e) 50 mg progesterone on day 15, daily until end of 12th week of gestation, gradually weaned down starting weeks 9–10 (depending on serum levels)

Progesterone: *Natural cycle, if required*: 200 mg micronized vaginal suppositories two or three times a day; continued until negative pregnancy test or through week 12 if pregnant (gradually weaned down starting week 9–10, depending on serum levels).

Programmed cycle: 50 mg/day IM beginning day 15; continued until negative pregnancy test or through week 12 if pregnant (gradually weaned down starting week 9–10, depending on serum levels)

Estrogen patches: programmed cycle: climara (comes in 0.1 mg patches; gives high serum E_2 levels)

Medrol: 16 mg/day for 4 days starting day of LH surge (natural cycle) or day 15 (programmed cycle)

Tetracycline: 250 mg four times per day for 4 days starting day of LH surge (natural cycle) or day 15 (programmed cycle)

3.7 Blastocysts (Figs. 18–21)

Blastocysts have generated a great deal of interest over the last decade. It may well be that we are walking a road towards routine culture to this stage, and as a consequence, to routine freezing of expanded blastocysts. Most groups report successful freezing with subsequent live births, some of these using coculture systems to support embryo growth [3, 28–31].

Often, blastocysts are frozen following the fresh transfer of day 3 conceptuses; after intrauterine transfer, remaining viable embryos are examined each day for 2 or 3 additional days to evaluate their suitability for freezing. This has been termed the post-transfer observation period. Blastocysts forming on either day 5 or day 6 are cryopreserved for future use. Only rarely and under special

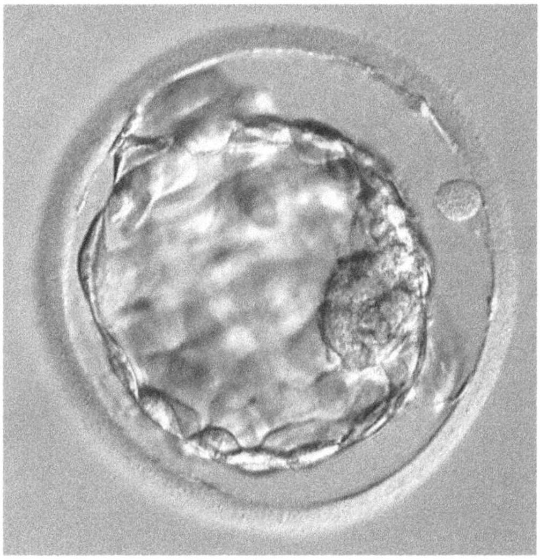

Fig. 18 Blastocyst immediately after thawing; note contraction

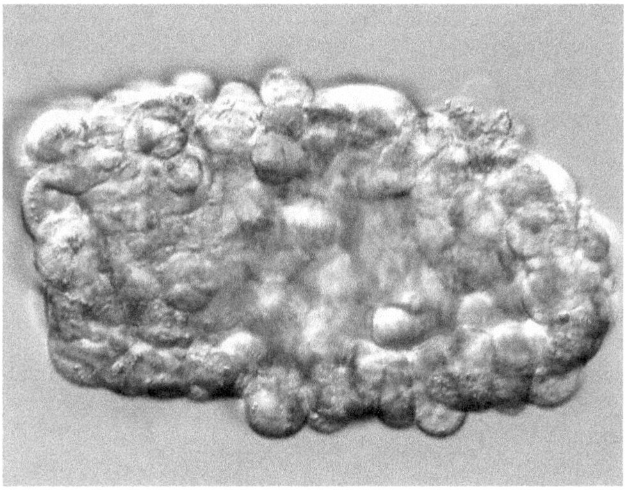

Fig. 19 Same blastocyst, expanded and then recontracted, after hatching in culture within 30 min after thawing

circumstances have later blastocysts been frozen since pregnancy rates have been disappointing after freezing day 7/8 specimens. Despite poor results beyond day 6, no differences were found between day 5 and day 6 blastocysts. While it is intuitive to assume that embryos reaching the blastocyst stage faster (day 5) might be "healthier" than their day 6 counterparts, our own data suggest that the rate of development within this window may not be crucial to subsequent post-thaw success as long as blastocysts have reached the same stage of growth [3].

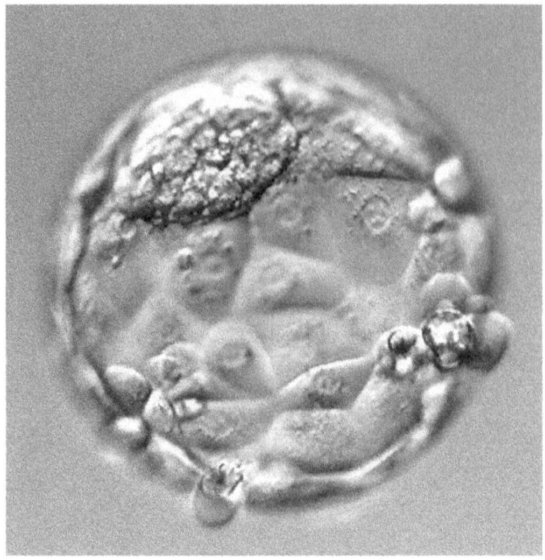

Fig. 20 Expanded blastocyst from a 39-year-old woman just before freezing

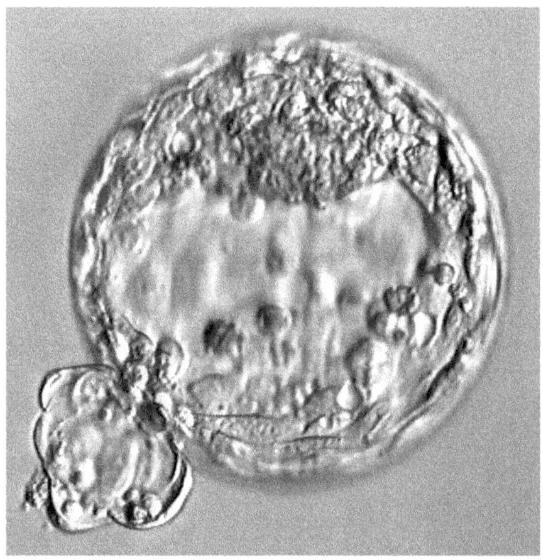

Fig. 21 Same blastocyst after thawing 141 days later. Membrane blebbing through the zona pellucida is seen, likely a result of subtle zona pellucida damage during thaw. This single blastocyst was transferred and a healthy male child was delivered

3.7.1 Detailed Blastocyst Slow Freezing and Thawing Methods: Specific Techniques Using a Biological Freezing Unit (Planer Kryo 10-1.7 II or III)

1. Evaluate the morphology of the blastocyst. Document.
2. Appropriately label a Nunc 1.8 mL sterile cryovial. Prepare a dated QC sample cryovial in the same manner.
3. Pipette 0.3 mL freezing medium (10 % glycerol/0.2 M sucrose) into cryovials using sterile technique. Freezing medium should be at room temperature.

4. Place blastocysts in 5 % glycerol (no sucrose) for 10 min. Change pipette before loading cryovials to avoid transferring oil.
5. Transfer to 10 % glycerol/0.2 M sucrose for 10 min.
6. Place blastocysts into cryovials, carrying as little medium as possible. Take care to place them at the bottoms of the vials and do not allow bubbles of air to enter. Visually check each cryovial under higher magnification after loading to assure that the specimens are properly located inside. Cryovials must be maintained in a level, upright position at all times and caps must be well-tightened. Securely load cryovials onto freezing canes (canes must be also be at room temperature).
7. Approximately 3 min prior to beginning the freezing run, turn Planer main switch to the "ON" position and open valve of LN_2 tank.
8. The Planer will display *Run*, *Print*, *Program* on the screen. Press *Run*.
9. The Planer will display *Enter Passcode* on the screen. Enter "3333."
10. Choose program number and name by pressing the < and > keys. Choose the blastocyst freezing program and press *Enter*.
11. When the proper temperature is reached, an audible alarm will sound that indicates the unit is at room temperature. Carefully load freezing canes/cryovials and QC sample cryovial (thermosensor must be immersed in freezing medium) into Planer and set clock alarm for 14 min. Press *Run*. The freezing program will now be under computerized control until the time of manual seeding. Temperature will drop within the unit at a rate of –2.0 °C/min until a temperature of –7.0 °C is reached. When –7.0 °C is reached, the display will read *Soaking*. Immediately record the date, patient's name, and display temperature of the chamber on a QC log sheet.
12. After 5 min of soaking, an audible alarm will sound to indicate that manual seeding may be performed. Check the temperature of the chamber and ensure that it is at a temperature of at least –6.5 °C. If not, delay seeding and wait for chamber to reach this temperature. If this does not occur, press *Run* in order for program to continue as if seeding had been performed. When the appropriate control temperature is reached, press *Run* again to put the unit into manual override (pause) in order to perform manual seeding. Record this temperature on the QC log sheet before proceeding with seeding. In this case, you will have to perform the seeding and then wait for 5 min before pressing *Run* again to resume program. If the chamber is at the proper temperature when the audible alarm sounds, record this display temperature on the QC log sheet.

13. Perform manual seeding:
    (a) Freeze sponge forceps in liquid nitrogen by immersing forceps in filled dewer.
    (b) Quickly raise freezing cane from unit and grasp cryovial at the level of the meniscus of medium. Watch for ice crystal formation within the cryovial. When crystal is visualized, immediately return cryovial to Planer. Check cryovial for ice crystal formation after approximately 30 s (medium should appear "slushy"). If needed, reseed to achieve crystal formation. Remember, this is a critical step in the process.
14. Press *Run*. Under computerized control, the Planer will hold at same temperature for an additional 10 min before proceeding with program. After hold period, temperature will drop at a rate of −0.3 °C/min until a temperature of −38 °C is reached. Machine should be checked at regular intervals to ensure that operation is satisfactory. Approximately $1^1/_2$ h are required to complete the program.
15. An audible alarm sounds at the completion of the freezing program and the display reads *Remove Sample.* At this point, the cryovials should be removed from the Planer and immediately plunged into LN_2 within styrofoam/plastic container.
16. Cryovials should be snapped into properly prelabeled storage canes, label side up, while constantly under LN_2. Storage canes are then fitted into cryosleeves. Storage canes and cryosleeves must be at LN_2 temperature before touching cryovials.
17. Transfer canes/cryosleeves to appropriate canisters within designated storage tanks.
18. Press *Run* to end program. The display will read *Do Not Switch Off* while the unit stabilizes to room temperature. This should take 4–5 min. When the display reads *Ready To Restart*, turn off main switch and LN_2 valve. Attach patient's chart recording to the back of the cryopreservation form.

3.8 Blastocyst Thawing

1. Ensure that the 30 °C waterbath is at the appropriate temperature.
2. Remove cryovial from the liquid nitrogen storage tank and hold at room temperature for one minute; follow by warming in a 30 °C waterbath for approximately 30 s with gentle agitation.
3. Allow cryovial to sit at room temperature for 1 min. During this time, prepare a 4-well culture dish with 10 % glycerol/0.4 M sucrose, 5 % glycerol/0.4 M sucrose, 0.4 M sucrose, and 0.2 M sucrose. Prepare one Falcon 3037 organ culture dish with the 0.1 M sucrose dilution. Label wells and dish appropriately.

4. Empty contents of cryovial into petri dish and locate specimen. If not located in the first scan, refill cryovial with 10 % glycerol/0.4 M sucrose at room temperature, gently shake vial, and pour contents into fresh dish. Repeat this process up to 10 times if specimen cannot be located. One should be able to tell whether or not the specimen has likely survived the freeze–thaw process by judging color and cellular aspects.
5. Transfer into 10 % glycerol/0.4 M sucrose medium. Allow to equilibrate for 1 min.
6. Transfer into 5 % glycerol/0.4 M medium. Allow to equilibrate for 3 min.
7. Transfer into 0.4 M sucrose medium. Allow to equilibrate for 3 min.
8. Transfer into 0.2 M sucrose medium. Allow to equilibrate for 2 min.
9. Transfer into 0.1 M sucrose medium in the organ culture dish. Wait for 1 min.
10. Prepare a fire-polished sterile pipette and locate patient's culture dish.
11. Transfer into fresh medium droplet, wash through four outside droplets, and place in clean droplet. Label droplet appropriately. Incubate until intrauterine transfer.
12. Record survival information on the thawing form; note how many frozen conceptuses remain in storage.

3.9 Frozen Blastocyst Replacement Protocols

1. Natural cycle replacement
   (a) Regular ovulatory cycle with normal luteal phase progesterone.
   (b) Day 5 blastocysts: Thaw 3 days after ovulation or next (4 days after LH peak and/or 4 days after E2 drop); culture overnight before transfer.
   (c) Day 6 blastocysts: Thaw 4 days after ovulation or next (5 days after LH peak and/or 5 days after E2 drop); transfer day of thaw.
   - No progesterone unless indicated or previous failure without it or luteal phase defect.
   - Medrol and tetracycline starting on day of LH surge for 4 days.
2. Programmed cycle replacement with GnRHa
   (a) Luteal suppression 0.2 mg GnRHa; 0.1 mg on predetermined day 1 until day 15
   (b) Transdermal estrogen patches:

   0.1 mg every other day, days 1–4

0.2 mg every other day, days 5–8

0.3–0.4 mg every other day, days 9–14 (depending on estrogen levels)

0.2 mg every other day, days 15—pregnancy testing (if pregnant, until E_2 rises at approximately 7 weeks)

(c) Tetracycline and Medrol on day 15 for 4 days

(d) Thaw and transfer day 20 (day 6 blastocysts) or thaw day 19 and transfer day 20 (day 5 blastocysts)

(e) 50 mg progesterone on day 15, daily until end of 12th week of gestation, gradually weaned down starting weeks 9–10 (depending on serum levels)

Progesterone: *Natural cycle, if required*: 200 mg micronized vaginal suppositories two or three times a day; continued until negative pregnancy test or through week 12 if pregnant (gradually weaned down starting week 9–10, depending on serum levels).

Programmed cycle: 50 mg/day IM beginning day 15; continued until negative pregnancy test or through week 12 if pregnant (gradually weaned down starting week 9–10, depending on serum levels)

Estrogen patches: Programmed cycle: Climara (comes in 0.1 mg patches; gives high serum E_2 levels)

Medrol: 16 mg/day for 4 days starting day of LH surge (natural cycle) or day 15 (programmed cycle)

Tetracycline: 250 mg four times per day for 4 days starting day of LH surge (natural cycle) or day 15 (programmed cycle)

4 Summary

Nearly 30 years ago, the world's first pregnancy from a frozen and thawed embryo was reported [22]. Since then, most IVF programs have embraced cryopreservation technologies as a method for augmenting pregnancy from a single ovarian stimulation cycle. As ovulation induction protocols have improved, allowing the recruitment of multiple healthy oocytes, so has the need grown to responsibly manage their numbers. It is not unusual today to collect in excess of ten mature oocytes from a woman, often substantially more. Before freezing techniques were routinely used in the laboratory, a woman producing so many gametes would be forced either to limit the number inseminated or risk having to discard healthy embryos. These concerns are not an issue today because cryopreservation has simply become routine in nearly all IVF programs.

However, routine cryopreservation programs do not exist without some difficulties. Cryoprotective media are only as good as the components they are made from, and various groups report problems with lot numbers of purchased media, reagents, or protein supplements. Cryoprotective agents must be precisely measured if making one's own solutions. Biological freezers must be well-maintained and regularly undergo stringent quality control measures lest future thaws prove less than satisfactory. By the time discrete problems involving freezing and thawing are first recognized, many women may have had their embryos frozen under adverse conditions, requiring months of agonizing thaw attempts with poor results. Add these concerns to the issues of divorcing couples (ownership problems), non-payment of storage fees, prolonged storage without decision from parents regarding disposition, frozen embryo abandonment, or unforeseen natural hazard in the laboratory, and one begins to realize the great responsibility of supervising the care of frozen biological tissues.

References

1. Centers for Disease Control and Prevention (2008) Assisted reproductive technology success rates: national summary and fertility clinic reports. US Department of Health and Human Services, Atlanta Georgia
2. Veeck LL, Amundson CH, Brothman LJ et al (1993) Significantly enhanced pregnancy rates through cryopreservation and thaw of pronuclear stage oocytes. Fertil Steril 59:1202
3. Veeck LL, Bodine R, Clarke RN et al (2004) High pregnancy rates can be achieved after freezing and thawing human blastocysts. Fertil Steril 82:1418–1427
4. Queenan JT Jr, Veeck LL, Toner JP et al (1997) Cryopreservation of all prezygotes in patients at risk of severe hyperstimulation does not eliminate the syndrome, but the chances of pregnancy are excellent with subsequent frozen-thaw transfers. Hum Reprod 12:1573
5. Zhou F, Lin XN, Tong XM et al (2009) A frozen-thawed embryo transfer program improves the embryo utilization rate. Chin Med J (Engl) 122:1974
6. Tatpati LL, Hudson SB, Gera PS et al (2010) High cumulative live births in oocyte donation cycles with cryopreservation of all embryos. Gynecol Obstet Invest 70:76
7. Mazur P (1984) Freezing of living cells: mechanisms and implications. Am J Physiol 247(3 Pt 1):C125
8. Whittingham DG (1977) Some factors affecting embryo storage in laboratory animals. Ciba Found Symp 52:97
9. Schneider U (1986) Cryobiological principles of embryo freezing. J In Vitro Fert Embryo Transf 3:3
10. Quinn P, Kerin JFP (1986) Experience with the cryopreservation of human embryos using the mouse as a model to establish successful techniques. J In Vitro Fert Embryo Transf 3:40
11. Friedler S, Shen E, Lamb EJ (1987) Cryopreservation of mouse 2-cell embryos and ova by vitrification: methodologic studies. Fertil Steril 48:306
12. Friedler S, Giudice LC, Lamb EJ (1988) Cryopreservation of embryos and ova. Fertil Steril 49:743
13. Zhu D, Zhang J, Cao S et al (2011) Vitrified-warmed blastocyst transfer cycles yield higher pregnancy and implantation rates compared with fresh blastocyst transfer cycles—time for a new embryo transfer strategy? Fertil Steril 95:1691
14. Desai N, Abdelhafez F, Bedaiwy MA et al (2010) Clinical pregnancy and live births after transfer of embryos vitrified on day 3. Reprod Biomed Online 20:808
15. Lin TK, Su JT, Lee FK et al (2010) Cryotop vitrification as compared to conventional slow freezing for human embryos at the cleavage stage: survival and outcomes. Taiwan J Obstet Gynecol 49:272
16. Trounson A, Sjoblom P (1988) Cleavage and development of human embryos in vitro after ultrarapid freezing and thawing. Fertil Steril 50:373
17. Gordts S, Roziers P, Campo R et al (1990) Survival and pregnancy outcome after ultrarapid freezing of human embryos. Fertil Steril 53:469
18. Ciriminna R, Schillaci R, Cefalu E et al (1994) What is the most idoneous developmental

stage for embryo-freezing? Acta Eur Fertil 25:173
19. Lai AC, Lin BP, Chang CC et al (1996) Pregnancies after transfer of ultrarapidly frozen human embryos. J Assist Reprod Genet 13:625
20. Noto V, Campo R, Roziers P et al (1993) Mitochondrial distribution after fast embryo freezing. Hum Reprod 8:2115
21. Quinn P (1995) Cryopreservation of embryos and oocytes. In: Keye WR, Chang RJ, Rebar RW, Soules MR (eds) Infertility: evaluation and treatment. WB Sanunders, Philadelphia, PA, p 821
22. Trounson AO, Mohr L (1983) Human pregnancy following cryopreservation, thawing and transfer of an eight-cell embryo. Nature 305:707
23. Willadsen SM (1980) The viability of early cleavage stages containing half the normal number of blastomeres in the sheep. J Reprod Fertil 59:357
24. Veiga A, Calderon G, Barri PN et al (1987) Pregnancy after the replacement of a frozen-thawed embryo with less than 50 % intact blastomeres. Hum Reprod 2:321
25. Hartshorne GM, Wick K, Elder K et al (1990) Effect of cell number at freezing upon survival and viability of cleaving embryos generated from stimulated IVF cycles. Hum Reprod 5:857
26. Rulicke T, Autenried P (1995) Potential of two-cell mouse embryos to develop to term despite partial damage after cryopreservation. Lab Anim 29:320
27. Van den Abbeel E, Camus M, Van Waesberghe L et al (1997) Viability of partially damaged human embryos after cryopreservation. Hum Reprod 12:2006
28. Cohen J, Simons RF, Edwards RG et al (1985) Pregnancies following the frozen storage of expanding human blastocysts. J In Vitro Fert Embryo Transf 2:59
29. Hartshorne GM, Elder K, Crow J et al (1991) The influence of in vitro development upon post-thaw survival and implantation of cryopreserved human blastocysts. Hum Reprod 6:136
30. Kaufman RA, Menezo Y, Hazout A et al (1995) Cocultured blastocyst cryopreservation: experience of more than 500 transfer cycles. Fertil Steril 64:1125
31. Menezo YJ, Ben Khalifa M (1995) Cytogenetic and cryobiology of human cocultured embryos: a 3-year experience. J Assist Reprod Genet 12:35

Chapter 20

Cryopreservation of Eggs

Zsolt Peter Nagy, Liesl Nel-Themaat, Ching-Chien Chang, Daniel B. Shapiro, and Diana Patricia Berna

Abstract

Oocyte cryopreservation is playing an increasingly important role in the field of human infertility treatment. The ability to store viable oocytes for later use has given many women the option to delay childbearing in order to pursue other ventures in life, without the concern of losing the opportunity to have a family. Furthermore, oocyte cryopreservation is very valuable for diseased patients who have to undergo treatments that may compromise fertility. Also, infertility patients who produce large numbers of oocytes during a retrieval cycle now have the option of storing some eggs prior to fertilization, thereby reducing the number of embryos that have to be managed. Lastly, oocyte cryopreservation enables egg donation programs that are independent of fresh donations, which makes it possible for numerous recipients to benefit from a single donor.

Traditionally, slow freezing was the only method available for oocyte cryopreservation. However, recent years have shown that ultrarapid cooling of oocytes results in higher survival and developmental rates. Thus, vitrification is today's preferred method of oocyte cryopreservation and therefore the only technique described.

In this chapter, we present two reliable methods of oocyte vitrification that have been in use for several years and that have been experimentally validated. Since no single vitrification method is clearly superior to the rest, other systems are also briefly described to give the reader options when deciding which methods to utilize in their practice.

Key words Oocyte cryopreservation, Vitrification, Egg freezing, Open pulled straw, Cryotop

1 Introduction

Oocyte cryopreservation is an integral part of infertility treatment. The ability to store oocytes long-term not only enables the delay of childbearing for social reasons, but also allows patients undergoing treatment for various diseases known to diminish fertility to save oocytes for use later in life. More recently, frozen donor egg banks have started to play an enormous role in helping patients with poor egg quality, genetic mutations, and social infertility to obtain donor oocytes. This increased availability of donor oocytes has reduced the need for fresh-cycle donors, who are sometimes difficult to find.

Zev Rosenwaks and Paul M. Wassarman (eds.), *Human Fertility: Methods and Protocols*, Methods in Molecular Biology, vol. 1154, DOI 10.1007/978-1-4939-0659-8_20, © Springer Science+Business Media New York 2014

According to 2010 data from the Society for Assisted Reproduction Technologies (SART), more than 18,000 embryo transfers (ET) were performed (fresh and frozen combined), and although the data does not distinguish between fresh and cryopreserved donor oocytes, it can be assumed that a significant proportion of these were from cryopreserved oocytes (www.sart.org). In the USA, more than 50 % of reproductive medicine clinics were offering oocyte cryopreservation in 2009 [1]. This proportion is also expected to increase dramatically as more clinics become efficient in donor oocyte cryopreservation.

Traditionally, oocyte cryopreservation is considered relatively inefficient. Research using animal models has highlighted the principal challenges of mammalian egg freezing, which apply to the human oocyte as well. Slow freezing protocols, which have been used most frequently, typically result in low survival rates due to the large volume-to-surface ratio of the large egg and the sensitivity of the meiotic spindle to freezing techniques. In recent years, however, oocyte vitrification has shown great promise with high survival rates and developmental rates. Furthermore, the technique is more practical than many slow-cooling protocols due to the short incubation and cooling times generally required.

Several different protocols have been described, using unique media formulations and carrier devices. These mostly utilize similar principles of cryoprotectant exposure, dehydration, rapid cooling, warming, and rehydration. To date, however, no single one appears to be clearly superior to the others. Therefore, in this chapter, we will describe two of the most commonly used systems, the Open Pulled Straw (OPS) system, which has been in use for nearly 13 years and the Cryotop method, which has been in use for 7 years. The effectiveness of both these systems has been documented. We will then proceed to touch on some of the other devices and commercially available media that have been successfully used to cryopreserve human oocytes. Deciding which system to use would ultimately be the choice of individual programs based on personal preference and their own research. Our goal is to simply provide a detailed description of two dependable protocols and then introduce the equally effective alternatives.

There are basically two different categories that will be discussed: (1) mini-straw systems, where oocytes are vitrified in a small column of vitrification medium in a fine straw-like device, and (2) minimum volume vitrification (MVV) systems, where the oocyte is vitrified with only a thin layer of medium on the carrier surface or on a medium film. The protocols and devices will thus be presented in these two categories. We would like to note that the discussed systems are not all-inclusive, and a recent review listing a more extensive list of the available carriers used in human and animal models for vitrification of embryos and oocytes is available [2].

One consideration when choosing devices is whether to use an "open" or "closed" system. Concerns about contamination from direct exposure of the gametes to liquid nitrogen (LN_2) have initiated the development of several closed systems, in which the gametes are never exposed to LN_2 directly due to a tightly sealed outer sleeve, or sealing of the straw after loading the gametes and before submersion into LN_2. Some of the original open systems have thus been modified to create new, closed systems. Modification of these carriers do not typically significantly change the procedures that can be used with these specific devices, and therefore, the techniques for open and closed systems will be discussed simultaneously.

2 Materials

2.1 Mini-straw Systems: The Open Pulled Straw (OPS) System

The OPS (Fig. 1a) system was first described by Vajta et al. in 1998 as a purpose-driven method of embryo and oocyte vitrification in the bovine model [3]. It consists of a finely pulled plastic straw in which oocytes are vitrified suspended in a small column of vitrification medium. The pulled straw can be inserted into a larger outer straw and sealed for protection against cross-contamination to create a closed system, but this is optional. The OPS system was used during studies that resulted in the first human birth after oocyte vitrification [4] and the first bovine birth after cryopreservation of immature oocytes [5]. Today this system is manufactured by RVT (Cairns, Australia) and is widely used for oocyte vitrification, as it continues to result in high live birth rates. Dr. Gabor Vajta kindly provided the OPS protocols, and additional information can be found at www.wix.com/opsstraws/home.

2.1.1 Equipment

1. Stereomicroscope equipped with heated stage for embryo visualization and manipulation.
2. Slide warmer or heated pad next to microscope for warming dishes.
3. Open Pulled Straws.
4. 60 mm Petri dish.
5. Four-well Nunc dish (Nalge Nunc International. Rochester, NY, USA).
6. Calibrated pipettes and sterile tips (10, 100, and 1,000 μL).
7. Styrofoam box or special liquid nitrogen holder, deep enough for total submersion of straws (approx. 15 cm deep).
8. Liquid Nitrogen.
9. Canes, goblets, canisters, and liquid nitrogen storage tank.
10. Stopwatch.
11. Marker pen.

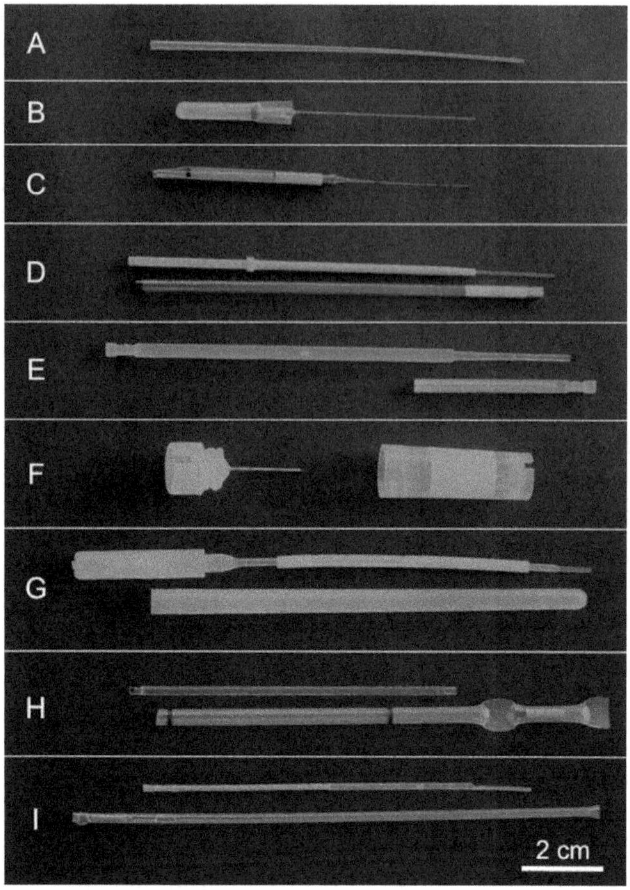

Fig. 1 A selection of mini-straw (**a**–**c**) and minimum volume vitrification (**d**–**i**) system carriers that have been used for human oocyte vitrification: Open pulled straw (**a**), Cryopette (**b**), CryoTip (**c**), Cryotop (**d**), Cryolock (**e**), Cryoloop (**f**), McGill Cryoleaf (**g**), Rapid-i (**h**, note the outer sleeve has been cut shorter than it would normally be), and High Security Vitrification system (**i**)

2.1.2 Media

Holding Medium (HM)

HEPES-buffered Tissue Culture Medium (TCM) 199 (other handling media that are routinely used for oocyte or embryo handling may also be used), supplemented with 20 % Serum Substitute Supplement (Irvine Scientific, Santa Ana, CA, USA) or protein supplement of choice.

Vitrification Solution 1 (VS1)

HM supplemented with 7.5 % (v/v) ethylene glycol (EG, Sigma-Aldrich, St, Louis, MO, USA; Cat # E9129) and 7.5 % (v/v) dimethylsulfoxide (DMSO, Sigma-Aldrich, St. Louis, MO, USA; Cat # D2650, 5 mL ampoules. Use each ampoule within 3 weeks of opening).

Vitrification Solution 2 (VS2)

HM supplemented with 16 % (v/v) EG, 16 % (v/v) DMSO, and 0.5 M sucrose.

Warming Solution 1 (WS1)

HM supplemented with 1.0 M sucrose.

Warming Solution 2 (WS2)

HM supplemented with 0.5 M sucrose.

2.1.3 Storage

All prepared media can be stored at 4 °C for up to 3 months. *See* **Note 1** for volumes required and for storage tips.

2.1.4 Other Fine Straw Devices

Several other fine straw devices exist that can be used with described media, or in some cases have media formulated especially for use with the specific system. Below are brief descriptions of a select few, but several others exist [2].

2.1.5 Cryopette Vitrification System

The Cryopette (Fig. 1b) is manufactured by Origio MidAtlanta Devices, Inc. (Mount Laurel, NJ, USA) and the system is sold with a custom cryobath, sealing and labeling devices, cutting tools and storing equipment. It may also be used with existing tools, but the starter kit that contains the above-mentioned components is encouraged. The carrier consists of a small plastic bulb attached to a fine tube, which can be used as a one-piece pipette and vitrification carrier. Vitrification in LN_2 provides a cooling rate of 23, 700 °C/min, and warming in water at 37 °C a warming rate of 34,480 °C/min. To aspirate the oocytes into the tube, the fingers are used to squeeze the bulb to create suction. After loading, the open end of the pipette is sealed using the special sealer before vitrification. This ensures a closed system and prevents any contact with LN_2. While Origio manufactures and recommends the Medicult vitrification and warming media kits, the Cryopette can be used with any other media formulations, such as described for the OPS system above. Detailed information on the system is available on the company Web site at www.origio.com. When this chapter was prepared, no publications on human oocyte vitrification using the device existed.

2.1.6 CryoTip

Like the Cryopette, the CryoTip (Fig. 1c) is a finely pulled plastic straw that can be sealed for a closed vitrification system. It is manufactured by Irvine Scientific (Santa Ana, CA, USA) and is attached to a metal cap that slides over the tip for protection and to ensure that the straw will sink in LN_2. There is also a connector that attaches to the CryoTip for aspiration. The device is designed for use with Irvine Scientific's closed vitrification system media, which are similar in formulation to media described for the OPS and Cryotop systems. Information is available on the company Web site at www.irvinesci.com.

A recent publication found that preservation of human oocyte structure after vitrification and warming using the CryoTip was inferior to that obtained using a MVV system (Cryotop) [6], although a live birth using the device after vitrification and warming of human oocytes have been reported [7]. The authors suggested results were likely due to a lower cooling rate/warming rate and longer exposure time to the cryoprotectants in the CryoTip method [6]. Thus, although the system offers the advantage of lowering contamination potential, survival rates appear to be compromised.

2.2 Minimum Volume Vitrification: The Cryotop Method

The Cryotop method is a minimum vitrification volume vitrification approach, where the oocytes are vitrified in <0.1 μL of vitrification medium [8]. This approach increases the cooling and warming rate (~22,800 and 42,100 °C/min, respectively) while allowing a significant reduced amount of cryoprotectants in the vitrification media [9]. The Cryotop device (Fig. 1d) is manufactured by Kitazo Supply Co. (Fujinomiya, Japan) and consists of a very fine polypropylene strip attached to a hard plastic handle. It also comes with a protective outer sleeve for storage. The vitrification and warming procedure described below was modified from [9]. Figure 3 is a diagram of the original protocol. More details regarding the Cryotop device and method is available at www.kitazato-biopharma.com.

2.2.1 Equipment

1. Stereomicroscope equipped with heated stage for embryo visualization and manipulation.
2. Slide warmer or heated pad next to microscope for warming dishes.
3. Cryotops.
4. 30 mm Petri dish.
5. Four-well Nunc dish.
6. Calibrated pipettes and sterile tips (10, 100, and 1,000 μL).
7. Styrofoam box or special LN_2 holder, deep enough for total submersion of Cryotops (approx. 15 cm deep).
8. Liquid Nitrogen.
9. Canes, goblets, canisters, and liquid nitrogen storage tank.
10. Stopwatch.
11. Marker pen.

2.2.2 Media

The Cryotop method can be used with a variety of vitrification and warming media, and Kitazo Supply Co. also sells media specially formulated for use with the device. Media formulations described below are from the original publication of the Cryotop method [9]. Later publications using the system modified the media slightly by increasing the serum supplement to 20 %, and adding DMSO to equal proportions of EG [8, 10] (*see* **Note 5**).

Basic Medium (BM)

HEPES-buffered Tissue Culture Medium (TCM) 199 (other handling media that are routinely used for oocyte or embryo handling may also be used), supplemented with 10 % Serum Substitute or protein supplement of choice.

Equilibration Solution (ES)

BM supplemented with 7.5 % mol/L EG. (Optional: 7.5 % (v/v) DMSO, *see* **Note 5**).

Vitrification Solution (VS)

BM supplemented with 15 % (v/v) EG and 0.5 M sucrose. (Optional: 15 % (v/v) DMSO, *see* **Note 5**).

Warming Solution (WS)

BM supplemented with 1.0 M sucrose.

Diluent Solution (DS)

BM supplemented with 0.5 M sucrose.

2.2.3 Storage

All prepared media can be stored at 4 °C for up to 3 months. *See* **Note 6** for aliquot and storage tips.

2.2.4 Other MVV Systems

Several other systems were developed in recent years, and the following list briefly describes some of the most common carriers. See ref. [2] for a more extensive list of available vitrification devices.

2.2.5 Cryolock

Similar to the Cryotop, the Cryolock (Fig. 1e) is a MVV system that has a thin plastic tip onto which embryos or oocytes are loaded with a minute volume of medium, and a cap to protect the delicate tip after vitrification directly in LN_2. It can be used with any of the available vitrification media kits, and has been tested extensively using the Cryotop method. The device is manufactured by BioDiseño (Bogota, Columbia) and more information is available at www.biodiseno.com. Live births following oocyte vitrification using the Cryolock have been reported [11, 12] and high survival and clinical outcomes were obtained using the Cryotop method in combination with the Cryolock carrier [11].

2.2.6 Cryoloop

The CryoLoop (Fig. 1f) was originally designed by Hampton Research (Aliso Viejo, CA, USA) for cryocrystallography applications, but was adopted for embryo and oocyte vitrification mainly in animal models several years ago. Although the original product has not been endorsed by the company for this purpose, similar devices, such as the Vitroloop (Vitrolife) and Vitriloop (SciTech Invention, Chandler, AZ, USA), are now produced by other manufacturers specifically for embryo and oocyte vitrification. The original device consists of a cryovial with a metal rod attached to the inside of the screw cap. At the tip of the metal rod, a fine nylon loop is mounted, where the embryos or oocytes are loaded with a thin film of media. Good survival and developmental rates have been achieved using the Cryoloop for human oocyte vitrification [11, 13–16], and since the device has been used for a several years, more data is available than for many of the newer carriers.

2.2.7 McGill Cryoleaf Open Vitrification System

The Cryoleaf (Fig. 1g) is also manufactured by Origio and is an open system that consists of a polypropylene rod with a thick handle on the one end and a thin, flat tip on the opposite end onto which embryos/oocytes are loaded before direct submersion into LN_2. A colored protective tube that slides back and forth over the shaft and tip of the rod can be locked in position after loading to protect the delicate loading surface. An additional thick outer cap that snaps onto the rod handle is placed over the rod and colored tube after vitrification for additional protection. The manufacturer recommends the use of their Medicult vitrification media kit with the McGill Cryoleaf, although other media, such as described for the Cryotop method, should also give good results. Since the device has been around for a few years, more clinical data is available than for many of the other carrier systems, especially for in vitro matured oocytes [17–19]. More information and additional references are available on the company Web site (www.origio.com).

2.2.8 Rapid-i

The Rapid-i (Fig. 1h) is a carrier device for that consists of a clear plastic rod with a flat tip containing a 50 nL loading hole into which the embryos or oocytes are loaded with a small volume of medium. The inner rod fits into an outer sleeve straw that is sealed before submersion into LN_2, ensuring a closed system. A recent study that compared sealing the outer sleeve before and after vitrification using mouse embryos found a reduced subsequent embryo cell number with pre-sealing, likely due to a relatively low cooling rate (<500 °C/min) when insulating the rod in air before submersion into LN_2. With post-sealing, vitrification is achieved by exposure to super-cooled air that collects in the pre-cooled outer sleeve, resulting in a cooling rate of approximately 1,200 °C/min and encouraging survival and developmental rates [20]. The Rapid-i is manufactured by Vitrolife (Englewood, CO, USA) and specially formulated media can also be purchased for use with the device, although it is not necessary.

At the time when this chapter was prepared, the Vitrolife Web site stated that in the USA, the Rapid-i has been approved for vitrification of 4–8 cell stage embryos only (www.vitrolife.com) and data on human oocyte vitrification could not be found.

2.2.9 High Security Vitrification (HSV) System

The HSV vitrification system (Fig. 1i) is manufactured by Cryo Bio System (IMV International Corporation USA, Maple Grove, MN, USA) and is a closed system consisting of a capillary tube with a pre-formed gutter attached to a color-coded handling rod, which fits into an ionomeric resin straw. The outer straw is sealed before submersion into LN_2. The kit also comes with a plastic insertion device for easy insertion and removal of the inner tube from the outer sleeve straw. There is no specific media for the carrier, and commercially available vitrification and warming solutions or "home-made" solutions can be used. In a recent study, the molecular profiles of oocytes previously vitrified and warmed using the HSV system were biologically comparable to those of freshly isolated gametes [21]. More research is needed to determine the physiological survival of oocytes following vitrification and warming with this device. Information on the HSV system is available at www.cryobiosystem.com.

3 Methods

3.1 The Open Pulled Straw (OPS) System

All procedures (except the first warming step) for the OPS system should be performed at room temperature (RT) of 25–27 °C. Therefore, the room, media, dishes, and handling devices should be at this temperature range. Warming in WS1 should be performed at 37 °C, and this solution should be heated accordingly. Figure 2 is a diagram of the vitrification and warming steps used in the OPS system.

3.1.1 Vitrification of Oocytes

Preparation

Oocyte retrieval and denuding is performed using conventional methods. Vitrification of oocytes should be performed 2–6 h after retrieval (38–43 h after hCG administration), preferably at 2 h post-retrieval and immediately after removal of the granulosa and cumulus cells. For each batch of oocytes, a separate 60 mm petri dish should be prepared stepwise throughout the procedure (Fig. 2). A cylinder can be inserted into the LN_2 to pre-cool the outer straw (if used) with the metal weight at the bottom position and the open end up.

Vitrification Procedure

1. Make a 50 μL drop of HM at the 12 o'clock position and one 50 μL drop of VS1 directly adjacent and then one in the center below the first two drops. Place denuded oocytes into the HM drop, wait one minute and then merge the HM drop with the adjacent VS1 drop. After 3 min, merge with the second VS1 drop below and wait for 3 min.

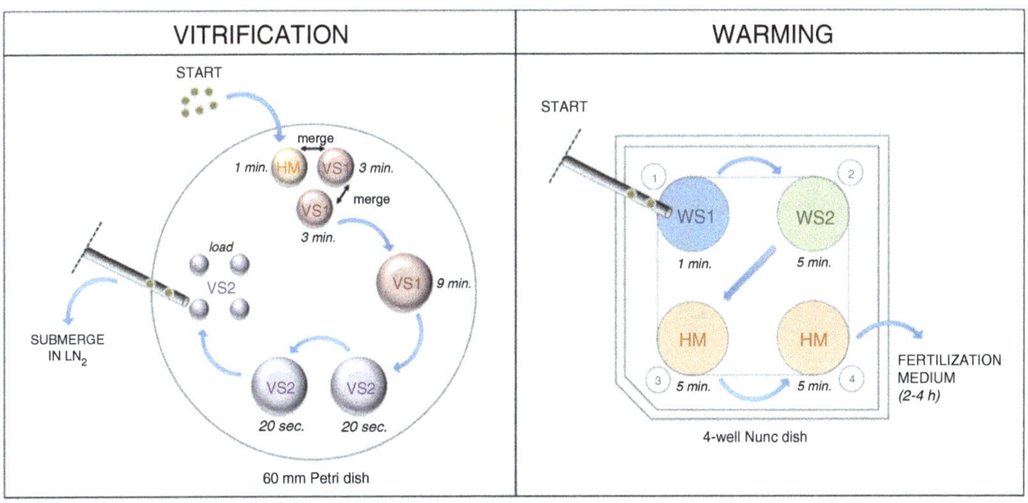

Fig. 2 Diagram of the Open Pulled Straw vitrification and warming procedures

2. While waiting, make a 100 μL drop of VS1 in the 3 pm position of the dish. When the 3 min are up, transfer the oocytes into the VS1 drop near the surface, let them gently sink, and equilibrate undisturbed for 9 min. After equilibration, oocytes should have sunk to the bottom of the drop and no shrinkage should visible. If equilibration is not complete after 9 min, additional equilibration time (3–6 min) should be allowed.
3. In the meantime, make two 100 μL drops of VS2 at the 6 o'clock position. After equilibration, place oocytes into the one on the right and pipette up and down for approximately 20 s, then place the oocytes in the second VS2 drop and pipette up and down for 20 s to ensure proper mixing.
4. Pick up the oocytes in vitrification groups (typically 2–3 vitrified per straw) in approximately 1 μL of VS2 and place the drop at the 9 o'clock position. Load the oocytes into the OPS straw by touching the open end to the drop containing the oocytes and let the media and oocytes move into the straw by capillary force (*see* **Note 2** for additional tips).
5. Immediately submerge the entire straw tip into LN_2 while firmly holding the opposite end. Start at a 30° angle tip facing down and quickly rotate from the holding end until the tip is vertical in the LN_2. The OPS can now be inserted into the precooled outer straw (kept submerged in LN_2) and heat-sealed using flamed, flat forceps should a closed system be desired.

3.1.2 Warming of Oocytes

Preparation

Warming is performed in a 4-well Nunc dish (Fig. 2). Media should be warmed to RT (25–27 °C) before starting the procedure. Place 1,000 μL of WS1 into the first well of the dish and heat to 37 °C before starting the warming procedure. Immediately before warming, place 100 μL of WS2 at RT into well 2.

1. After careful selection of the appropriate straw, the sealed end of the outer straw (if used) is cut open while the rest of the straw is kept under LN_2.
2. Use fine forceps to remove the OPS from the outer straw and quickly (should take 3 s) submerge the OPS tip directly into the WS1, curve facing up (*see* **Note 3**). As soon as WS1 starts entering the tip, seal the opposite open with forefinger. Embryos will be expelled into SW1 drop from the pressure of the warming air in the OPS (*see* **Note 4**).
3. Recover the oocytes within 1 min and transfer into WS2 with 10–12 μL WS1 and gentle pipette oocytes with solution up and down for gradual mixing. Expel residual medium from pipette tip over oocytes in the well. Let equilibrate for 5 min.

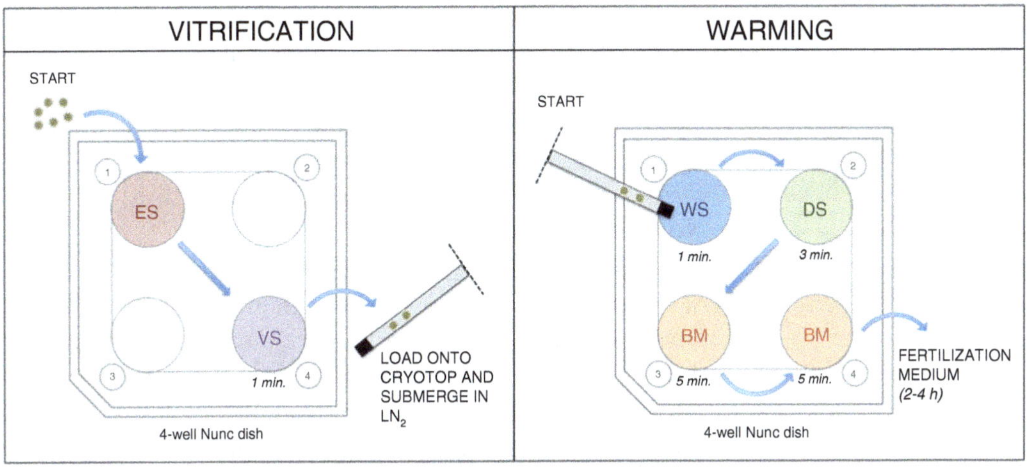

Fig. 3 Diagram of the Cryotop vitrification and warming procedures

4. While waiting, aliquot 100 µL of RT HM into well 3 and 4. Transfer oocytes, again mixing gently, into well 3 and wait for 5 min.
5. Transfer oocytes into well 4 for 5 min.

After the final wash, transfer oocytes into the fertilization or culture dish and return to incubator for 2–4 h before insemination.

3.2 *Cryotop Method*

All procedures (except the first warming step) for the Cryotop method should be performed at room temperature (RT) of ~22 °C. Therefore, the room, media, dishes, and handling devices should be at this temperature. Warming in WS should be performed at 37 °C and this solution can be heated on a slide warmer accordingly.

3.2.1 *Vitrification of Oocytes*

Preparation

Oocyte retrieval and denuding is performed using conventional methods. Vitrification of oocytes should be performed 2–6 h after retrieval (38–43 h after hCG administration), preferably at 2 h post retrieval and immediately after removal of the granulosa and cumulus cells. For each batch of oocytes, a 4-well Nunc dish should be prepared and warmed to RT as described below (*see* Fig. 3, **Note 6**).

Label the appropriate number of Cryotops, remove the outer sleeves and lay them out for easy access during vitrification. Before loading oocytes on each Cryotop, an outer sleeve should be clamped with large metal forceps so that it can be placed over the Cryotop while submerged in LN_2 with one hand while the other hand is holding unto the Cryotop. Fill a Styrofoam box or other container with LN_2 and place the labeled cane and goblet into the LN2 to have it ready for inserting each Cryotop after vitrification.

Procedure

1. After collection and processing, equilibrate denuded oocytes for 5–15 min in ES until no signs of shrinkage can be detected under light microscopy and oocytes have sunken to the bottom of the dish (*see* **Notes 7, 8**).
2. Pick up oocytes individually or in groups of 2–3 and transfer to the vitrification solution well (No. 4 of dish, *see* Fig. 3, **Notes 8**).
3. After 30 s of rinsing, after the oocyte(s) remain on the bottom of the dish, pick up the oocyte(s) in an extremely small volume (<0.1 μL) of VS and transfer onto the polypropylene strip of a labeled Cryotop, about 2 mM behind the black tip (*see* **Notes 9, 10**).
4. Immediately submerge the entire strip into LN_2 vertically.
5. Carefully slide the outer sleeve over the Cryotop without disturbing the polypropylene strip and ensure that is stays in the proper position.
6. Transfer the Cryotop to a labeled cane and goblet before transferring the next oocyte(s) to VS.

3.2.2 Warming of Oocytes

Preparation

Approximately 3 mL of WS is heated to 37 °C beforehand in a 35 mm petri dish. A diluent/washing dish can be prepared by aliquoting about 0.5 mL diluent DS at RT in the 1st well of in a 4-well Nunc dish, and BM (RT) in the 3rd and 4th well for washing. Fill a container with LN_2 and insert the cane and goblet holding the appropriate Cryotop(s) for easy access.

Procedure

1. Place the dish containing WS at 37 °C under the microscope and make sure the bottom of the dish is in focus.
2. Select the appropriate Cryotop using metal forceps and hold the end between fingers while keeping the end with the oocytes submerged in LN_2.
3. Remove the protective sleeve using metal forceps while keeping the tip submerged under LN_2.
4. Rapidly submerse the polypropylene tip into WS while observing under the microscope and locate the oocytes on the strip (*see* **Notes 11, 12**). Aspirate them in and out of the pipette tip for gentle mixing and to ensure they do not float out of view. They should be in the warming dish for no more than 1 min.
5. Transfer the oocytes into DS at RT for 3 min (well no.1 of 4-well dish, *see* Fig. 2, **Note 13**).
6. Wash oocytes in the first well of BM (well no. 3) for 5 min, and then in the second well of BM (well no. 4 of the dish) for 5 min.
7. Transfer oocytes to a regular culture dish and place in the incubator until further use.

4 Notes

4.1 OPS Method Notes

4.1.1 Media Preparation

1. For each vitrification procedure, the following volumes will be needed:

   2,500 μL HM.
   200 μL VM1.
   200 μL VM2.
   1,000 μL WM1.
   1,000 μL WM2.

   It is thus useful to aliquot appropriate volumes and store at 4 °C for up to 3 months.

4.1.2 Loading of Oocytes into Straws

2. For effective loading into the straw, keep the OPS straw with the narrow end curving down at a 45–60° angle. Touch the 1 μL drop containing the oocytes on one side of the drop. Capillary force will pull a square-shaped column of VS2 containing the oocytes into the straw. Keep the tip in the drop until the column stops moving and the bottom end of the column is completely flat (not curved), which will ensure proper mixing during warming.

4.1.3 Warming of Oocytes

3. After removal of the OPS straw from the outer straw, the timing from removal to immersion into WS1 is crucial to prevent damage to oocytes. Too quickly may result in bubbles or zona pellucida damage, while taking too long will allow ice crystal formation and cell damage. The process should take about 3 s.
4. Ensure that the entire VS2 column in the top of the OPS is submerged under WS1 to ensure even warming. Once submerged, the column will initially start moving upwards into the straw. At that point, immediately seal the opposite end with the index fingertip to ensure that the oocytes will be expelled into the WS1 well. If they do not move out, attach a pipette or syringe with needle on the opposite end to expel. If a bubble forms between the WS1 and VS2 column, inappropriate loading was performed and survival rates may be compromised.

4.2 Cryotop Method Notes

4.2.1 Preparation

5. DMSO can be added to ES and VS in equal proportion (15 % each v/v) to the EG for increased permeabilization and higher efficiency [8], but was omitted from the current, original formulation [9]. Some subsequent protocols also used 20 % serum supplement instead of 10 % [10].
6. Media can be prepared in advance and stored in 2–3 mL aliquots at 4 °C up to 2 months. A properly labeled 4-well Nunc dish can be prepared and warmed to RT before start of the procedure. Place approximately 0.5 mL of ES in the 1st well and 0.5 mL of VS in the 4th well for a good workflow.

For large groups of oocytes (>10), it is useful to prepare more than one well of ES (well no. 1 and 2 of the 4-well dish, with VS in well no. 4).

4.2.2 Equilibration and Vitrification

7. For large groups of oocytes, batches of no more than 10 (depending on comfort level of the technician) should be equilibrated together in the 1st well to prevent over exposure to cryoprotectants. After approximately 7 min, the second batch can be transferred into the 2nd well of ES. This will ensure that by the time the first batch is vitrified, the second batch will be fully equilibrated and ready for vitrification.
8. When oocytes are placed into ES, they should be placed at the surface where they will initially float and slightly shrink. As they equilibrate, they will descent to the bottom and re-expand. When they are transferred to VS solution, oocytes should be pipetted in and out several times to ensure even equilibration without losing them, as they tend to float off before being thoroughly equilibrated. When they remain on the bottom of the dish, they can be loaded in a small volume of medium unto the polypropylene strip.
9. The side on which the oocytes are loaded should be marked on the Cryotop to ensure the oocytes can be located upon warming and to prevent damaging them by warming them beneath the strip.
10. Once the oocytes are transferred to the polypropylene strip, the drop of medium can be aspirated leaving the oocytes on only a small layer of medium and a halo can be seen around them. They should then be submerged into LN_2 immediately to prevent drying. The entire VS equilibration and loading process and submersion into LN_2 should take no longer than 1 min.

4.2.3 Warming

11. Care should be taken that the oocytes are warmed in such a way that they will be on top of the polypropylene strip, as they may be damaged if squeezed between the strip and the bottom of the dish. Upon warming, oocytes can be loosened from the polypropylene tip of the Cryotop by gently scraping the bottom of the strip against the bottom of the dish.
12. The oocytes tend to float and be difficult to see until they have equilibrated with the warming medium. Therefore, the tip of the Cryotop should be placed into the warming solution in view so that the oocytes can be observed the moment they are warmed and the observer can follow their location in the dish. If they float off, they should be aspirated and placed at the bottom of the dish.
13. After 1 min, when transferring the oocytes to DS, they should be placed at the bottom of the dish (they tend to sink naturally)

with some extra WS that can be expelled over the oocytes to ensure gradual rehydration. The same technique should be followed when transferring the oocytes from DS to the BM for washing.

Acknowledgements

The authors would like to express their gratitude to the following individuals who shared valuable information regarding the various devices and protocols: Barry Behr, Ri-Cheng Chian, Vladimir Isachenko, Mitch Schiewe, Pierre Vanderzwalmen, and Gabor Vajta.

References

1. Rudick B, Opper N, Paulson R, Bendikson K, Chung K (2010) The status of oocyte cryopreservation in the United States. Fertil Steril 94:2642–2646
2. Saragusty J, Arav A (2011) Current progress in oocyte and embryo cryopreservation by slow freezing and vitrification. Reproduction 141:1–19
3. Vajta G, Holm P, Kuwayama M, Booth PJ, Jacobsen H, Greve T, Callesen H (1998) Open pulled straw (OPS) vitrification: a new way to reduce cryoinjuries of bovine ova and embryos. Mol Reprod Dev 51:53–58
4. Kuleshova L, Gianaroli L, Magli C, Ferraretti A, Trounson A (1999) Birth following vitrification of a small number of human oocytes: case report. Hum Reprod 14:3077–3079
5. Vieira AD, Mezzalira A, Barbieri DP, Lehmkuhl RC, Rubin MI, Vajta G (2002) Calves born after open pulled straw vitrification of immature bovine oocytes. Cryobiology 45:91–94
6. Bonetti A, Cervi M, Tomei F, Marchini M, Ortolani F, Manno M (2011) Ultrastructural evaluation of human metaphase II oocytes after vitrification: closed versus open devices. Fertil Steril 95:928–935
7. Oakes MB, Gomes CM, Fioravanti J, Serafini P, Motta EL, Smith GD (2008) A case of oocyte and embryo vitrification resulting in clinical pregnancy. Fertil Steril 90(2013):e5–e8
8. Kuwayama M (2007) Highly efficient vitrification for cryopreservation of human oocytes and embryos: the Cryotop method. Theriogenology 67:73–80
9. Kuwayama M, Vajta G, Kato O, Leibo SP (2005) Highly efficient vitrification method for cryopreservation of human oocytes. Reprod Biomed Online 11:300–308
10. Trokoudes KM, Pavlides C, Zhang X (2011) Comparison outcome of fresh and vitrified donor oocytes in an egg-sharing donation program. Fertil Steril 95:1996–2000
11. Garcia JI, Noriega-Portella L, Noriega-Hoces L (2011) Efficacy of oocyte vitrification combined with blastocyst stage transfer in an egg donation program. Hum Reprod 26:782–790
12. Selvaraj P, Selvaraj K, Srinivasan K (2010) First successful birth of twins in India following the transfer of vitrified oocytes. J Hum Reprod Sci 3:44–48
13. Keskintepe L, Agca Y, Sher G, Keskintepe M, Maassarani G (2009) High survival rate of metaphase II human oocytes after first polar body biopsy and vitrification: determining the effect of previtrification conditions. Fertil Steril 92:1706–1715
14. Sher G, Keskintepe L, Mukaida T, Keskintepe M, Ginsburg M, Agca Y, Maassarani G, Bayrak A (2008) Selective vitrification of euploid oocytes markedly improves survival, fertilization and pregnancy-generating potential. Reprod Biomed Online 17:524–529
15. Nottola SA, Coticchio G, Sciajno R, Gambardella A, Maione M, Scaravelli G, Bianchi S, Macchiarelli G, Borini A (2009) Ultrastructural markers of quality in human mature oocytes vitrified using cryoleaf and cryoloop. Reprod Biomed Online 19(Suppl 3):17–27
16. Liebermann J, Tucker MJ (2002) Effect of carrier system on the yield of human oocytes and embryos as assessed by survival and developmental potential after vitrification. Reproduction 124:483–489
17. Chian RC, Huang JY, Gilbert L, Son WY, Holzer H, Cui SJ, Buckett WM, Tulandi T, Tan SL (2009) Obstetric outcomes following

vitrification of in vitro and in vivo matured oocytes. Fertil Steril 91:2391–2398

18. Chian RC, Gilbert L, Huang JY, Demirtas E, Holzer H, Benjamin A, Buckett WM, Tulandi T, Tan SL (2009) Live birth after vitrification of in vitro matured human oocytes. Fertil Steril 91:372–376

19. Cao Y, Xing Q, Zhang ZG, Wei ZL, Zhou P, Cong L (2009) Cryopreservation of immature and in-vitro matured human oocytes by vitrification. Reprod Biomed Online 19:369–373

20. Larman MG, Gardner DK (2011) Vitrification of mouse embryos with super-cooled air. Fertil Steril 95:1462–1466

21. Di Pietro C, Vento M, Guglielmino MR, Borzi P, Santonocito M, Ragusa M, Barbagallo D, Duro LR, Majorana A, De Palma A, Garofalo MR, Minutolo E, Scollo P, Purrello M (2010) Molecular profiling of human oocytes after vitrification strongly suggests that they are biologically comparable with freshly isolated gametes. Fertil Steril 94:2804–2807

Chapter 21

Ovarian Tissue Cryopreservation and Transplantation: A Realistic, Effective Technology for Fertility Preservation

Dror Meirow, Hila Ra'anani, and Hannah Biderman

Abstract

It is clear that ovarian tissue cryopreservation can serve a very important role in providing fertility preservation. To date, more than 30 live human births have resulted from the transplantation of cryopreserved ovarian tissue, proving the successful implementation of this technique. The ideal conditions for ovarian tissue cryopreservation have yet to be determined, indicating the crucial need for more research in this field. Nonetheless, it is recommended that ovarian tissue cryopreservation be offered as an option to women before undergoing chemotherapy treatment in instances where there is no time to delay for hormonal stimulation and oocyte retrieval.

Key words Ovarian tissue cryopreservation, Fertility, Transplantation

1 Introduction

Ovarian tissue cryopreservation and transplantation is a means of preserving fertility potential that has been contemplated by physicians and scientists for many years. This procedure offers women and prepubertal girls the possibility of attempting to preserve their fertility prior to acute insult to the ovary that can result in permanent sterility. One of the most devastating consequences of radiation and chemotherapy as a cancer treatment is the subsequent diminishment or complete loss of fertility in young women who have yet to finish building their families. Anticancer treatments, especially pelvic radiation and high-dose alkylating agents such as cyclophosphamide, have been shown to cause ovarian atrophy, harm the ovarian blood vasculature, and diminish the primordial follicle pool, which directly represents future fertility potential [1]. As a result, attempts to preserve fertility prior to treatment are imperative. Physicians are finding that more and more young female cancer patients are seeking referrals to reproductive endocrinologists and it is important that the safety and efficacy of different fertility preservation options be analyzed and fully understood.

Zev Rosenwaks and Paul M. Wassarman (eds.), *Human Fertility: Methods and Protocols*, Methods in Molecular Biology, vol. 1154, DOI 10.1007/978-1-4939-0659-8_21,

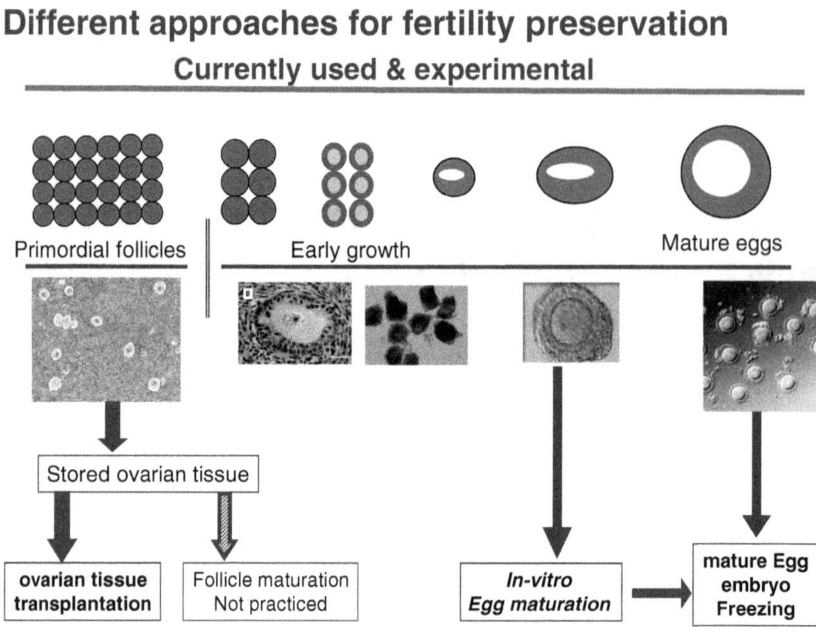

Fig. 1 Different approaches for fertility preservation that are currently practiced or experimental use different types of follicles. For egg and embryo freezing, mature eggs (IVF) or immature eggs are collected (IVM), whereas when freezing ovarian tissue primordial follicles are predominantly stored. Follicle maturation is currently used in research projects only

The options available to preserve fertility in cancer patients include embryo cryopreservation, oocyte cryopreservation, and ovarian tissue cryopreservation (Fig. 1). Embryo cryopreservation is considered the "gold standard" of fertility treatments and it is the only recognized, non-experimental procedure according to the Ethics Committee of the American Society for Reproductive Medicine [2]. However, it is not an option for prepubertal girls or women without a partner or sperm donor. Ovarian tissue cryopreservation is an alternative option for this population and allows for fertility preservation, especially in women who do not have the option of delaying cancer treatment in order to undergo hormonal stimulation or in those who have already initiated chemotherapy treatment. Ovarian tissue cryopreservation has also been indicated in patients with nonmalignant diseases who are treated with chemotherapy [3].

Ongoing advances in cryopreservation techniques have allowed for effective cryopreservation and subsequent grafting of ovarian tissue. Ovarian tissue can be harvested by a simple laparoscopic surgery [4], or during a laparotomy if it is carried out for oncologic therapy. The successful reestablishment of fertility from ovarian transplantation is aided by the relative resistance of ovarian follicles to ischemia since they are found in a largely avascular environment. Additionally, the location of primordial follicles in the ovarian cortex allows them to benefit quickly from revascularization. Both heterotopic and orthotopic transplantation sites have been used for grafting procedures. However, with all freezing protocols and transplantation techniques, significant numbers of follicles are lost. Endocrine studies and sonography allow for accurate patient monitoring and follow-up. In cancer patients, techniques to study the presence of minimal residual disease (MRD) in extracted ovarian tissue samples are mandatory and have improved in order to ensure that cancerous cells are not grafted back into the patient.

2 History and Basic Research

Attempts at ovarian grafting began over a century ago and there is a long history of ovarian grafting experimentation. Ovarian tissue transplantation has been performed both in the attempt to restore fertility and also as a potential treatment for menopausal symptoms, especially prior to the advent of hormonal therapy [5]. Very early works claimed successful pregnancies as a result of both ovarian tissue autografts and allografts, and it was initially thought that the ovary had immunological privilege. This has subsequently been found not to be the case and any ovarian allograft requires immunosuppression [6]. Ovarian tissue has been transplanted back to the original site: to the ovary or to nearby areas in the pelvis such as peritoneal pouches and round ligament; an orthotopic transplantation Whereas, in the past, when the goal was to treat menopause rather than restore fertility, ovarian tissue had been transplanted to heterotopic sites that include the rectus muscle, arm or abdominal skin, omentum, and cervix [7].

Early animal studies have shown the successful transplantation of frozen–thawed ovarian tissue without the need for vascular reanastomosis. In 1994, Gosden et al. report a successful pregnancy in a lamb after cryopreserving ovarian cortical slices and then transplanting the tissue back. The combination of tissue cryopreservation followed by subsequent autogenic transplantation proved the functionality of ovarian tissue grafting as a means of preserving future fertility potential [8]. Later, the same group reported the long term survival of the grafts [9]. Additional animal studies carried out since the mid 1990s have shown the success of ovarian tissue transplantation following cryopreservation [10–12]. The success of these animal studies was largely due to the rapid revascularization of the grafted tissue. Most importantly, the stockpile of primordial follicles found in the ovarian cortex was found to benefit quickly from the growth of blood vessels into the graft. Rodent studies show that within 48 h of transplantation, there is a significant network of blood vasculature that invades the ovarian cortex [13]. Human ovarian tissue grafted to immune deficient mice showed that both host and graft vessels contributed sequentially to graft revascularization. Angiogenesis initiated reperfusion started from day 5 post transplantation [14]. In monkeys it was shown that freshly grafted heterotopic ovarian tissue enabled endocrine recovery, oocyte collection and subsequent successful fertilization and pregnancy; this was accomplished without surgical reanastomosis of major blood vessels [15].

Human ovarian tissue xenografts into SCID mice have been performed to prove the efficacy of human ovarian tissue transplantation and to evaluate results. Follicular growth and development were successfully observed. Follicles were documented to mature to the antral stage [16] and oocytes were retrieved in the metaphase II stage [17, 18]. The follicles in the SCID mice were found to be capable of ovulation as proven by documented progesterone production and corpora lutea formation [18, 19].

After the first reports on fertility restoration post transplantation of frozen–thawed ovarian tissue in sheep, ovarian tissue was harvested and stored in cancer patients prior to potentially sterilizing chemotherapy treatments. The focus was on two main issues—the safety [4] and the technique [20]. Although huge improvements have been achieved in the last 15 years, these are still the main issues.

3 Indications and Patients' Selection

Initially, the patient is evaluated to ascertain her risk for fertility loss. Factors used to evaluate patients include the diagnosis and staging of the cancer; sterilization risk based on the patient's age, the drugs used, and the treatment protocol implemented, and the patient's ovarian reserve, which is dependent on the patient's age, endocrine profile, AMH levels, and antral follicle count. Also of importance is whether the patient was previously exposed to chemotherapy. The consultation carefully determines the patient's current medical status. estimating the risk of fertility preservation procedure which includes: any surgical or anesthesiology complications, whether the patient is anemic or thrombocytopenic, analysis of the tumor mass, pressure, and effusion of the thorax, heart, liver, and kidney functions and if the patient is in a hypercoagulable state [21].

Patients that are in good health, currently have good ovarian reserve, and are at high risk for sterilization post-chemotherapy are deemed good candidates for fertility preservation techniques. Of the fertility preservation options available, embryo cryopreservation is not indicated for prepubescent girls or females without a partner or sperm donor. Advances in oocyte cryopreservation techniques using vitrification protocols, have shown equivalent success rates to protocols using fresh oocytes [22]. In many instances patients are not afforded the time necessary to undergo ovarian stimulation. In addition, ovarian tissue cryopreservation can be used in the prepubescent female population. For these patients, storing ovarian tissue is an alternative, successful method to preserve fertility and unlike embryo or oocyte cryopreservation, this technique can be performed immediately. However, ovarian tissue cryopreservation is not indicated for older populations and many institutions will not perform this procedure on patients over the age of 38.

Furthermore, ovarian tissue cryopreservation, which mainly preserves primordial follicles, can be performed in patients who have recently been exposed to chemotherapy (within the last 6 months), unlike IVF (in vitro fertilization) and IVM (in vitro maturation). This is mainly due to the possibly mutagenic effects of chemotherapy on growing follicles. In animal studies, chemotherapy has been shown to have deleterious effects on reproductive outcome; high abortion and malformation rates have been observed, but only in growing follicles dependent on the stage of oocyte development during exposure [23]. Chemotherapy has been shown to induce condensation, fragmentation, and numerical chromosomal abnormalities in growing and mature follicles [24, 25]. These damaging effects are not present in primordial follicles even after long period of chemotherapy exposure. In humans there have been no observable increases in miscarriage rates or teratogenic effects post cessation of anticancer therapy [26]. This is believed to be the result of different susceptibilities of growing oocytes as it takes an oocyte approximately 6 months to develop and the current clinical recommendation is to wait longer than 6 weeks after cancer treatment before attempting pregnancy.

Studies have also indicated that previous exposure to chemotherapy decreases the likelihood of retrieving oocytes after ovarian stimulation and it has been reported that within a timeframe of 2–3 months post chemotherapy exposure, no oocytes were successfully retrieved, while primordial follicles were simultaneously present in cryopreserved ovarian tissue [21]. Ovarian tissue cryopreservation therefore, has the potential to serve a significant role in preserving fertility in young cancer patients who must undergo anticancer therapy immediately and functions as an option to preserve tissue without the risk of exposing growing follicles to the deleterious effects of anticancer therapy.

Ovarian tissue cryopreservation is not exclusively indicated for cancer patients but may rather be utilized for both malignant and nonmalignant diseases. Patients now undergo hematopoietic stem cell transplantation to treat hematological and autoimmune diseases and as a result are treated with high doses of ionizing radiation or alkylating agents which results in diminished fertility outcomes. High doses of chemotherapy are used to treat benign diseases such as microscopic polyangiitis, sickle cell anemia, and alkylating agents are used for rheumatologic diseases such as systemic lupus erythematosus [27]. Other surgical indications to store ovarian tissue may include benign disease such as recurrent ovarian endometriosis [28]. Additionally, patients with genetic diseases such as mosaic Turner syndrome may benefit from ovarian tissue cryopreservation techniques [29].

4 Ovarian Tissue Harvesting and Clinical Workup

When indicated for cancer patients, ovarian tissue harvesting and cryopreservation (OTCP) is performed as soon as possible, preferably prior to the start of anticancer therapy. In some instances, patients who have already undergone non sterilizing chemotherapy will undergo OTCP prior to high-dose chemotherapy. Patients exposed to non-sterilizing doses of chemotherapy will still have regular menses and hormone profiles. Exceptions include patients pretreated with GnRH-a or those exposed to chemotherapy within 3 months of OTCP. For these patients, hormone profiles will not represent ovarian reserve, and the decision to perform OTCP will solely depend upon clinical judgment, the patient's age and if their chemotherapy regime was not considered to be within the sterilizing range [21].

Tissue collection is performed via a laparoscopic procedure and the extent of ovarian tissue extraction should be determined by the likelihood of subsequent ovarian failure (Fig. 2). In some cases a unilateral oophorectomy may be performed [30]. Alternatively, a partial oophorectomy is performed, removing half to two-thirds of the ovary [31, 32]. Patients often prefer this partial tissue removal to a full oophorectomy. Techniques used to remove small biopsy

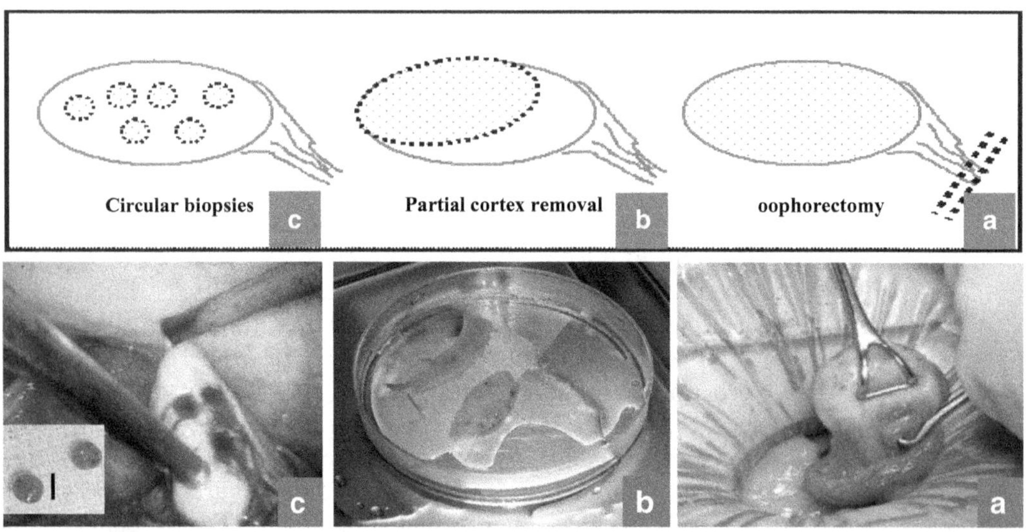

Fig. 2 Different approaches for tissue harvesting. Unilateral oophorectomy (**a**), partial cortical harvesting (1/2-2/3 of ovarian cortex) (**b**), small biopsies using specially developed biopter (**c**). **a** and **b** with permission from Meirow and Silber. C from [20]

segments from the ovaries (5-mm disk of 2–3 mm thickness) [33] are not commonly used today. Many follicles are lost during the freezing–thawing procedure and due to ischemic damage, and hence, it is important to remove a large enough sample in order to store many follicles.

Once ovarian tissue has been harvested, it is immediately place in the appropriate medium and transferred to the laboratory, where it is prepped for cryopreservation. If necessary, samples can be transported (even for hours) to centers that provide cryopreservation services in which case the sample is rinsed with PBS, transferred to a sterile tube containing IVF medium, and placed in an aluminum container in order to protect the sample, especially if it transported by airfreight [30].

Once the sample arrives at the laboratory it is transferred to a petri dish containing medium and placed on ice in order to maintain the temperature at 4 °C. When the operating theatre is adjacent to the laboratory, immature eggs are collected from the tissue for IVM and the tissue is kept at room temperature. The sample is then cleaned from the medulla and prepped for cryopreservation by dividing the cortical tissue into 5 × 10 mm slices [21]. If possible, during tissue preparation oocyte cumulus complexes from antral follicles are also collected. This "combined procedure" provides a possible second means of restoring patients' fertility after cryopreservation in addition to the transplantation of cryopreserved tissue. With advances in maturation techniques, the simultaneous cryopreservation of oocytes in addition to the ovarian tissue may allow for subsequent fertilization [34].

Prior to cryopreservation, one piece of cortical tissue is fixed and embedded in paraffin wax in preparation for hematoxylin–eosin staining to verify the presence of primordial follicles and for early detection of malignant cells [21, 35]. In addition, a piece of the medulla that is cleaned off is kept and evaluated for the presence of cancerous cells. It is prudent to separately store an additional small piece of cortical tissue in a separate vial for additional tests that may be needed in the future. In cases where patients request transplantation, this sample is often used to assess freezing, storage conditions, and thawing technique prior to transplantation and for minimal residual disease (MRD) monitoring.

5 Cryopreservation Techniques (Freezing–Thawing Procedures)

The cortex is predominantly populated by primordial follicles, which consist of primary oocytes surrounded by pregranulosa cells. Additionally, a small percentage of follicles found within the cortex will have already begun development and are considered to be primary follicles. Stromal cells and collagen bundles are also found within the ovarian cortex [36]. As a result there is a wide variation

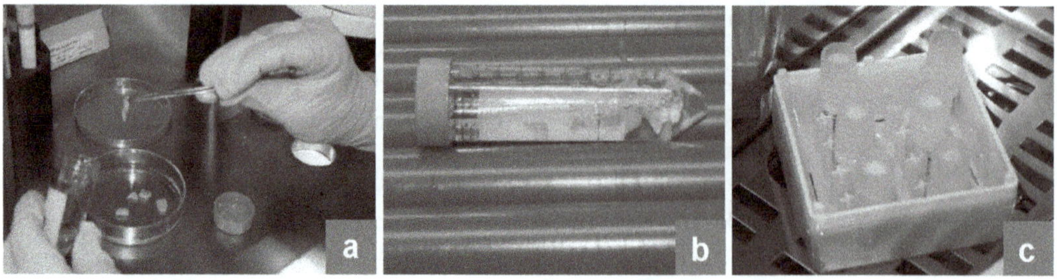

Fig. 3 Tissue Preparation for slow freezing. (**a**) The tissue is cut into pieces and inserted into the medium with cryoprotectants. (**b**) The tissue is rolled in the medium containing cryoprotectant to allow for the penetration and equilibrium of cryoprotectants in the tissue. (**c**) The tissue is placed in cryovials. **a**–**c** with permission from Meirow and Dor

in the density and cell types present in the cortex and this lack of uniformity has been found to increase as patients age [37]. This poses a serious problem since optimal cryopreservation techniques for one cell type may not be effective for others. The disparity in cell size, density, and membrane hydraulic permeability found between the cells of the ovarian cortex means that the rate of dehydration necessary for cryopreservation as well as the ideal rate of cooling varies for each cell type [38, 39].

The focus of ovarian tissue cryopreservation is the successful freezing and thawing of all cell subtypes, promoting survival of the maximal number of primordial follicles as they represent fertility reserve (Fig. 3). Techniques to optimize primordial follicle survival and reduce ischemic damage have been studied and debated [40, 41]. Successful ovarian tissue cryopreservation has relied on traditional slow program freezing techniques. The distribution of primordial follicles within the cortex allows for the freezing of very thin strips of cortical tissue. This is crucial since the increased surface area allows for better penetration of cryoprotectants (Fig. 4). The ideal cooling rates should minimize injury to primordial follicles and should take into account that slow cooling rates will prevent the formation of ice crystals but faster cooling rates prevent cells from being overexposed to cryoprotectants [42]. Several different types of transport medium exist; these include Leibovitz-based medium, histidine–tryptophan–ketoglutarate solution (HTK, an organ transport medium) and human serum albumin-containing medium [36]. The three cryoprotectants used for human tissue include dimethyl sulfoxide (DMSO), ethylene glycol (EG), and propanediol (PROH). Sucrose or other non-permeable osmolytes have been for use as osmotic buffers in order to aid in dehydration but in some instances prolonged exposure to these agents has been shown to hinder oocyte survival rates [43]. Our freezing protocol uses 1.5 M DMSO in an oocyte wash buffer, 15 % synthetic serum substitute supplement, and 0.1 M sucrose [44]. Manual seeding is often conducted in order to induce ice

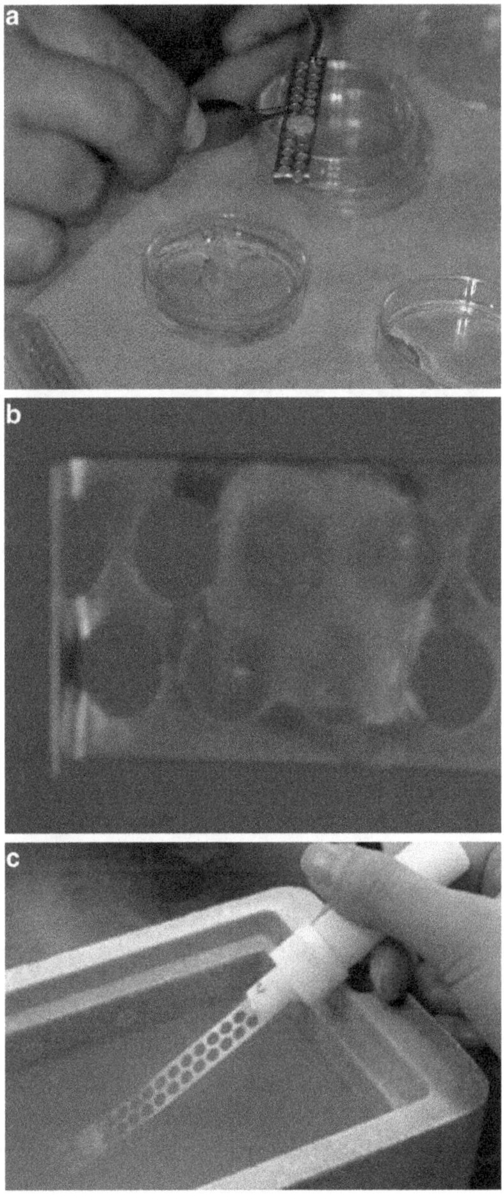

Fig. 4 Preparation of ovarian tissue for vitrification. **a–c** with permission from Meirow and Silber

crystal formation. High follicular survival rates of 84 and 74 % have been reported with implementation of the following slow-freeze protocol: immersion of thin cortical slices (<1 mm) in a 4 °C solution of 1.5 M EG or DMSO with 0.1 M sucrose followed by rapid rewarming [8].

Studies using vitrification techniques on ovarian tissue have been promising [45]. This technique allows for the rapid freezing of the heterogeneous cells of the ovarian cortex without the formation of ice crystals, creating a solidified amorphous state with the addition of high doses of cryoprotectants. Cryoprotectant toxicity may be avoided by using a combination of different cryoprotectants [46, 47]. In a side-by-side comparison of slow-freeze and vitrification protocols, follicle preservation was approximately equivalent; however, ovarian stroma is better preserved by vitrification techniques as assessed by electron microscopy [47]. Tissue vitrification techniques have not yet been optimized and to date, there have been no live births reported with this cryopreservation technique. The next necessary step to prove the efficacy of this methodology would be the report of successful pregnancies and live births.

Thawing procedures vary based on the freezing protocol. After slow-freeze, samples are removed from the liquid nitrogen and are allowed to sit at room temperature for 30 s before being placed in a warm bath (37 °C) for 2 min. The tissue is then placed into a dish and exposed to a series of thawing solutions that allow for the quick washing out of the cryoprotectant. In procedures that used PrOH as the cryoprotectant, the thawing protocol calls for 5 min in thawing solution 1 (1.0 M ProH and 0.2 M sucrose) followed by 5 min in solution 2 (0.5 M PrOH and 0.2 M sucrose) followed by 10 min in solution 3 (0.2 M sucrose) and lastly, 10 min in solution 4 (Cryo-PBS). Alternatively, for protocols that used EG as the cryoprotectant, the thawing procedure calls for several step down washes in solutions containing EG and sucrose [47]. Our ovarian tissue thawing protocol was previously described [44]: Vials are removed from the liquid nitrogen, held in the air at room temperature for one and a half minutes, then placed in a 37 °C water bath and stirred for 2 min to melt the medium in the tube. The contents of the vials (medium and tissue) are washed in a dish containing the first-step thawing solution: oocyte wash buffer, synthetic serum substitute supplement 15 %, 1.0 M DMSO and 0.1 M sucrose. The tissue is transferred to washing vials using prepared thawing solutions for graded dilution of the cryoprotectant. The first stage is carried out by rolling the tissue (1 Hz) in a large volume of oocyte wash buffer (50-mL vials), 0.1 M sucrose and 1.0 M DMSO for 5 min. The procedure and solutions for the following second dilution stage (0.5 M DMSO) and third dilution stage (without DMSO) are identical to the first stage except for the DMSO concentration; 5 min for each stage.

The thawing procedure following vitrification involves removing cryotubes from the liquid nitrogen and placing them into warming solution (37 °C) (HBSS and HAS with 0.5 M sucrose). Afterwards, samples are placed in a petri dish and undergo two more rounds of 5 min washes in HBSS/HAS with 0.25 M then 0.125 M sucrose. Lastly, tissue is placed in HBSS/HAS solution for 5 min [47].

6 Transplantation

Prior to auto-transplantation, evaluation of ovarian functional status is mandatory: ovarian failure evidenced by amenorrhea, high gonadotropin levels, low levels of ovarian steroids, low AMH levels as well as the absence of antral follicles on ovarian sonography must be confirmed. Alternatively, some patients may not undergo complete ovarian failure but rather experience an extensive loss of primordial follicle stockpiles while still maintaining occasional, irregular menses post treatment. Although these patients show significant hormonal disturbances with high FSH levels, they may still be able to conceive without the need of ovarian tissue transplantation. It is imperative that consultation with the oncological team occurs before transplantation in order to assure that reimplantation of ovarian tissue is only performed when the patient is free from disease [5] Permission must be granted from the institutional review board since ovarian tissue transplantation is still considered to be an experimental technique. The patient's general health post treatment and risk of relapse must be considered. A general algorithm has been suggested to increase the safety of cryopreservation-transplantation procedures which includes both clinical evaluation and laboratory studies [5] (Fig. 5). Pelvic sonography is a necessary first step to ensure that no macroscopic ovarian pathology exists in order to prevent unnecessary surgeries or complications. Prior to transplantation, a small piece of frozen cortical tissue is initially thawed, histologically stained, and analyzed for the presence of follicles after the freeze–thaw protocol and to ensure the absence of any malignant cells.

Concern has been raised that cryopreserved ovarian tissue may harbor markers for malignant cells, putting the patient at risk of disease recurrence post-transplantation. In leukemia patients,

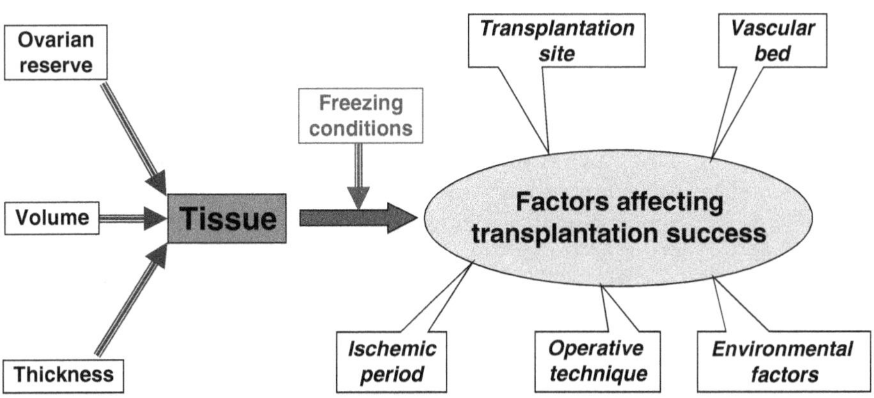

Fig. 5 Many factors are involved in successful ovarian tissue transplantation. These include the tissue stored, freezing conditions, and transplantation (See text). With permission from D Meirow

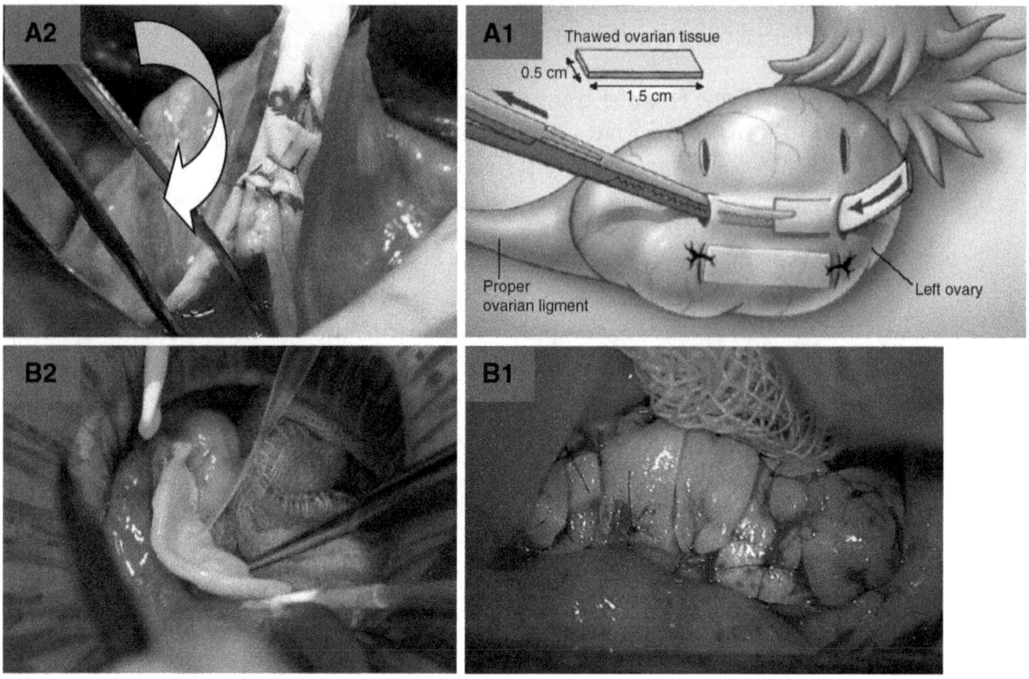

Fig. 6 Orthotopic Surgical grafting of ovarian tissue to (**a**) parietal sac (laparoscopy) Demeestere (**b**) round ligament (laparotomy). Meirow & Silber with permission

malignant cells present in the bloodstream are a serious risk and effective techniques to discover the presence of these cells is imperative in order to ensure the safety of transplantation of frozen–thawed tissue. Studies have shown that when cryopreserved tissue of leukemia patients is analyzed for malignant cells, the RT-PCR techniques are able to detect the presence of minimal residual disease even when the disease cannot be ascertained by histological analysis alone [48–50]. Techniques to uncover minimal residual disease even in the setting of clinical remission include specific immunohistochemical stains, polymerase chain reaction (PCR) amplification and fusion transcripts detection (BCL-ABL for example), rearranged immunoglobulin or T cell receptor genes, and flow cytometric detection of aberrant immunophenotypes [5]. It is imperative that these techniques be optimized and implemented in all patient samples prior to transplantation.

6.1 Transplantation Site

When deciding on a graft site there are several clinical points that should be considered which include: the ease of the surgical procedure, access for oocyte collection, the volume of tissue to be transplanted, and if the aim is to achieve a natural conception, only orthotopic sites should be selected (Fig. 6). Factors that determine transplantation success include the vascular bed, as a rich vascular bed is crucial in order to minimize ischemic injury to the graft post transplantation. In cases where ovarian tissue was grafted

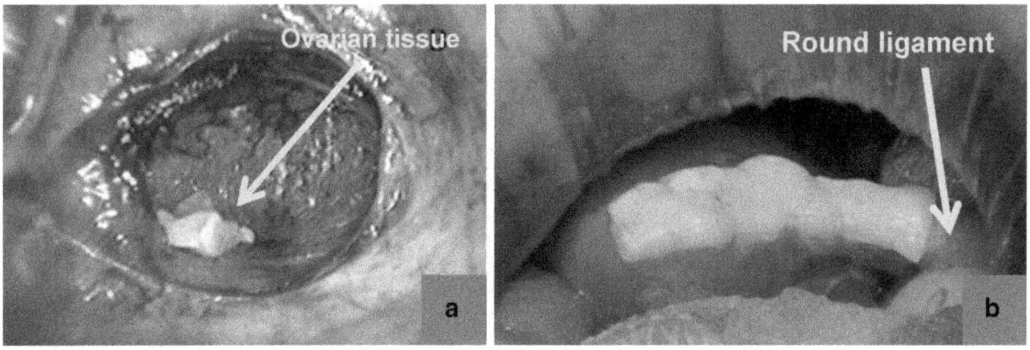

Fig. 7 Transplantation of ovarian tissue back to the ovary. (**a**) Insertion of thawed tissue fragments into the ovary (A1) First successful transplant Meirow & Dor et.al. 2005 scheme [23]. (A2) Transplantation 2011, the patient is currently pregnant Meirow & Dor with permission. (**b**) Transplantation of ovarian tissue after existing ovarian cortex was removed. (B1) Donnez et al. (B2) Meirow & Silber with permission

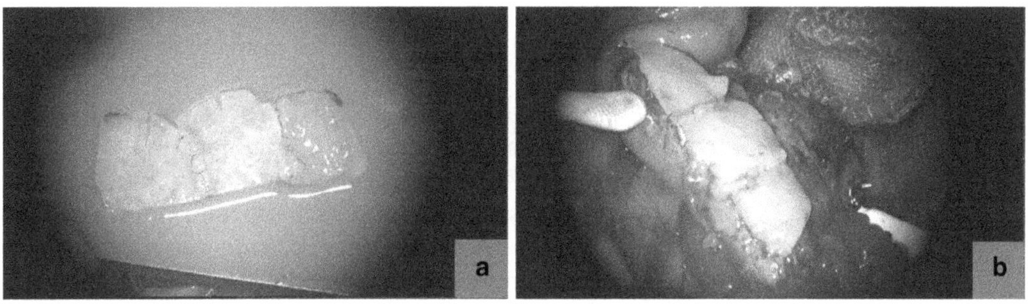

Fig. 8 Transplantation of vitrified ovarian tissue (pre-transplantation—after suturing three cortical fragments) (**a**) and post transplantation (**b**) Meirow & Silber with permission

orthotopically to the pelvic peritoneum or ovary and to a subcutaneous heterotopic site simultaneously, the results were more promising in the orthotopic site. More follicles developed at the orthotopic site while at the heterotopic site there was reduced follicular growth, limited oocyte recovery rates and fertilization rates [51, 52]. All reports of successful pregnancies have been achieved by returning ovarian tissue to the ovaries directly or orthotopically in the pelvis near the ovarian site [32].

6.2 Transplantation

A key consideration in ovarian transplantation is the maintenance of blood supply to the grafted tissue. Ischemic damage is the major cause of follicular loss therefore the amount of time that it takes for the graft to be adequately perfused plays a significant role in graft survival. There are two surgical techniques for orthotopic transplantation to the ovary (Fig. 7). In the first technique, the non-functioning ovarian cortex is removed and cortical pieces are grafted onto the ovary after removal of the cortex [27, 53] (Fig. 8). In the second method, cavities are formed using blunt dissection beneath the ovarian cortex. Thawed ovarian tissue is then placed within the cavities, and the ovarian peritoneum is sutured [30, 31].

Alternatively, transplantation to the peritoneal wall is indicated if the ovary is considered small or if there is a poor vasculature bed available. In this case, a peritoneal window is created and ovarian tissue is placed into the window and then sutured or covered [27, 54].

Several reports have touted the advantage of a two-step laparoscopic approach to create angiogenic granulation tissue during wound healing several days before autotransplantation of the grafted tissue [27, 54]. Piver et al. have modified this two-step orthotopic transplantation by also performing a double transplantation in order to increase the vascular bed [55]. Revel et al. have reported the use of ovarian micro-organs as an alternative transplantation methodology. Micro-organs are tissue fragments of microscopic thickness (300–350 μM) that still maintain the architecture of the organ. Crucially, ovarian micro-organs can function without vascularization but at the same time are capable of upregulating angiogenic factors, thus leading to improved graft longevity. After two failed transplantation attempts using more traditional ovarian cortex pieces, Revel et al. grafted ovarian micro-organs into a patient that successfully lead to pregnancy and live birth [56].

Although it has been suggested that the restoration of ovarian function post transplantation lasts for only a short time, it has been observed that in most patients the transplant functions for a few years even in patients that received chemotherapy prior to ovarian tissue harvesting. However, in older patients the grafts function only for a shorter period of time. Attempts have been made to use additional techniques and environmental factors as a means to improve graft revascularization, shorten the ischemic period, and therefore improve overall graft survival. Ovarian xenografts in SCID mice showed increased follicular survival rates when the antioxidant, vitamin E was administered [57]. The question of adding gonadotropins to the transplants has been studied in animal models as a way of upregulating angiogenesis [58]. It is difficult to define what the effects of gonadotropin administration are in human transplants since, there is still no standard against which grafting procedures can be measured. In our experience, we have administered estrogen and progesterone tablets both before and after grafting procedures as a means of avoiding elevated gonadotropin levels [44]. Alternatively, in other transplantations we performed, we did not administer medications. We could not observe any differences in transplantation results relating to hormone supplementation during transplantation.

7 Patient Follow-up and Pregnancies

Patients should be monitored after transplantation in order to assess post procedural ovarian response. Hormonal profiling of gonadotropin levels, AMH, and sex steroids should be done although these hormone profiles often do not exhibit expected patterns. Follicular development and endometrial growth can be assessed by follow-up sonography. The time needed for endocrine recovery will vary depending on whether or not the tissue transplanted had been exposed to chemotherapy.

To date, more than 30 healthy births have been reported after autotransplantation of cryopreserved ovarian tissue. These women underwent chemotherapy, subsequently suffered from premature ovarian failure or significantly low ovarian reserve and later requested re-implantation of their ovarian tissue. Donnez et al. reported the first successful pregnancy and live birth of a healthy girl after transplantation of cryopreserved ovarian tissue in a 25 year old that was treated for Hodgkin's lymphoma and had significantly low ovarian reserve [27]. In 2005, we reported the first proof of successful transplantation of frozen–thawed ovarian tissue in a woman that suffered from premature ovarian failure. A live birth was reported after oocyte retrieval from grafted cryopreserved ovarian tissue. The patient was diagnosed with primary mediastinal B cell non-Hodgkin's lymphoma at the age of 28. Ovarian tissue was cryopreserved between first and second-line chemotherapy. In 2003, four of seven pieces of cortical tissue were reimplanted and 9 months post reimplantation, spontaneous follicular development was observed. The patient underwent IVF, leading to the birth of a healthy daughter [31]. This patient conceived naturally again a few months post-delivery but aborted. Thereafter, the patient's hormone profile showed high FSH and LH levels, indicating that the return of ovarian function post-transplantation did not last more than 2 years. Silber described a series of successful transplantation cases of fresh ovarian tissue between monozygotic twins that resulted in the birth of healthy children. Although the tissue used was fresh and not cryopreserved, these cases clearly indicated the high success rate of the transplantation procedure.

During the last few years successful births after the cryopreservation of ovarian tissue were also reported by, Anderson, Demeestere, Piver, Sanchez-Serrano, Revel [30, 54–56, 59]. A summary of the first 13 live births after ovarian transplantation reported that age at tissue collection was 19–36, 40 % of patients had previous chemotherapy, all patients showed recovery of endocrine function, 50 % of pregnancies were after IVF and 50 % were spontaneous pregnancies [32]. All babies were normal. Today there are reports on multiple live births following ovarian tissue transplantation [60].

It remains unclear exactly how many patients worldwide have, in total, undergone cryopreservation-transplantation attempts, since reported results are indications of success. We have reported our results from one center indicating high transplantation success rates. Out of seven patients that underwent transplantation of ovarian tissue, all of them resumed menstruation and endocrine function. In five of these patients, grafts survived for a number of years, while in one patient for only 6 months (tissue harvesting done at age 39), and one patient has recently undergone a second transplantation. In all patients IVF was performed, mature eggs were retrieved and fertilized, resulting in six pregnancies (in 4 women). These results are highly encouraging and indicate high success rates of ovarian tissue transplantation (unpublished data).

8 Conclusions

The recent reports of successful pregnancies and deliveries resulting from ovarian tissue cryopreservation and grafting provide great hope to women that are threatened with the loss of their future fertility due to the need to start chemotherapy or radiation treatment immediately. For these women, this technique allows for the immediate cryopreservation of tissue prior to the sterilizing and potentially teratogenic effects of chemotherapy. More research is necessary in order to optimize protocols for both cryopreservation and tissue grafting. Additionally, further improvements in the detection of minimal residual disease are necessary in order to ensure that no malignant cells are transferred back to cancer-free patients. Of concern, is the fact that ovarian tissue grafts seem to provide only a limited timeframe of restored ovarian function. Nevertheless, it is clear that the recent advances in ovarian tissue cryopreservation techniques with demonstration of pregnancies and live births show the real potential of this treatment option for women, especially for prepubertal patients and women who do not have the option to delay cancer therapy for IVF hormonal stimulation.

References

1. Meirow D (2000) Reproduction post-chemotherapy in young cancer patients. Mol Cell Endocrinol 169(1–2):123–131
2. ECotACSfRM (2005) Fertility preservation and reproduction in cancer patients. Fertil Steril 83:1622–1628
3. Donnez J, Dolmans MM (2010) Cryopreservation and transplantation of ovarian tissue. Clin Obstet Gynecol 53(4):787–796
4. Meirow D, Schenker JG (1998) Cryopreservation and transplantation of ovarian tissue as a means of preserving female fertility. Harefuah 134(6):461–464
5. Meirow D (2010) Transplantation of cryopreserved ovarian tissues. Fertil Cryopreserv. In: Fertility Preservation, ed Ri Cheng Chian and Patrick Quinn. Published by Cambridge University Press, 218–232
6. Goldman MB (1974) Immunogenetic factors in the survival of ovarian transplants. Transplantation 17(5):518–523
7. Nugent D, Meirow D, Brook PF, Aubard Y, Gosden RG (1997) Transplantation in reproductive medicine: previous experience, present knowledge and future prospects. Hum Reprod Update 3(3):267–280
8. Gosden RG, Baird DT, Wade JC, Webb R (1994) Restoration of fertility to oophorectomized sheep by ovarian autografts stored at −196 degrees C. Hum Reprod 9(4): 597–603
9. Baird DT, Webb R, Campbell BK, Harkness LM, Gosden RG (1999) Long-term ovarian function in sheep after ovariectomy and transplantation of autografts stored at −196 C. Endocrinology 140(1):462–471

10. Candy CJ, Wood MJ, Whittingham DG (1995) Follicular development in cryopreserved marmoset ovarian tissue after transplantation. Hum Reprod 10(9):2334–2338
11. Salle B, Demirci B, Franck M, Rudigoz RC, Guerin JF, Lornage J (2002) Normal pregnancies and live births after autograft of frozen-thawed hemi-ovaries into ewes. Fertil Steril 77(2):403–408
12. Almodin CG, Minguetti-Camara VC, Meister H et al (2004) Recovery of fertility after grafting of cryopreserved germinative tissue in female rabbits following radiotherapy. Hum Reprod 19(6):1287–1293
13. Dissen GA, Lara HE, Fahrenbach WH, Costa ME, Ojeda SR (1994) Immature rat ovaries become revascularized rapidly after autotransplantation and show a gonadotropin-dependent increase in angiogenic factor gene expression. Endocrinology 134(3):1146–1154
14. Van Eyck AS, Bouzin C, Feron O, Romeu L, Van Langendonckt A, Donnez J, Dolmans MM (2010) Both host and graft vessels contribute to revascularization of xenografted human ovarian tissue in a murine model. Fertil Steril 93(5):1676–1685
15. Lee DM, Yeoman RR, Battaglia DE et al (2004) Live birth after ovarian tissue transplant. Nature 428(6979):137–138
16. Oktay KNH, Mullan J, Gosden RG (1998) Development of human primordial follicles to antral stages in SCID/hpg mice stimulated with follicle stimulating hormone. Hum Reprod 13:1133–1138
17. Gook DA, Edgar DH, Borg J, Archer J, Lutjen PJ, McBain JC (2003) Oocyte maturation, follicle rupture and luteinization in human cryopreserved ovarian tissue following xenografting. Hum Reprod 18(9):1772–1781
18. Gook DA, Edgar DH, Borg J, Archer J, McBain JC (2005) Diagnostic assessment of the developmental potential of human cryopreserved ovarian tissue from multiple patients using xenografting. Hum Reprod 20(1):72–78
19. Kim SS, Soules MR, Battaglia DE (2002) Follicular development, ovulation, and corpus luteum formation in cryopreserved human ovarian tissue after xenotransplantation. Fertil Steril 78(1):77–82
20. Meirow D, Fasouliotis SJ, Nugent D, Schenker JG, Gosden RG, Rutherford AJ (1999) A laparoscopic technique for obtaining ovarian cortical biopsy specimens for fertility conservation in patients with cancer. Fertil Steril 71(5): 948–951
21. Meirow D, Baum M, Yaron R et al (2007) Ovarian tissue cryopreservation in hematologic malignancy: ten years' experience. Leuk Lymphoma 48(8):1569–1576
22. Cobo A, Meseguer M, Remohi J, Pellicer A (2010) Use of cryo-banked oocytes in an ovum donation programme: a prospective, randomized, controlled, clinical trial. Hum Reprod 25(9):2239–2246
23. Meirow D, Schiff E (2005) Impact of maternal exposure to chemotherapy on embryos and offspring J National Cancer Institute 34:21–25
24. Kujjo LL, Chang EA, Pereira RJ et al (2011) Chemotherapy-induced late transgenerational effects in mice. PLoS One 6(3):e17877
25. Bar-Joseph H, Ben-Aharon I, Rizel S, Stemmer SM, Tzabari M, Shalgi R (2010) Doxorubicin-induced apoptosis in germinal vesicle (GV) oocytes. Reprod Toxicol 30(4):566–572
26. Winther JF, Boice JD Jr, Mulvihill JJ et al (2004) Chromosomal abnormalities among offspring of childhood-cancer survivors in Denmark: a population-based study. Am J Hum Genet 74(6):1282–1285
27. Donnez J, Dolmans MM, Demylle D et al (2004) Livebirth after orthotopic transplantation of cryopreserved ovarian tissue. Lancet 364(9443):1405–1410
28. Donnez J, Squifflet J, Dolmans M (2011) Ovarian tissue transplantation. In: Donnez J, Samuel Kim S (eds) Principles and practice of fertility preservation. Cambridge University Press, New York, pp 357–366
29. Borgstrom B, Hreinsson J, Rasmussen C et al (2009) Fertility preservation in girls with turner syndrome: prognostic signs of the presence of ovarian follicles. J Clin Endocrinol Metab 94(1):74–80
30. Andersen CY, Rosendahl M, Byskov AG et al (2008) Two successful pregnancies following autotransplantation of frozen/thawed ovarian tissue. Hum Reprod 23(10):2266–2272
31. Meirow D, Levron J, Eldar-Geva T et al (2005) Pregnancy after transplantation of cryopreserved ovarian tissue in a patient with ovarian failure after chemotherapy. N Engl J Med 353(3):318–321
32. Donnez J, Silber S, Andersen CY et al (2010) Children born after autotransplantation of cryopreserved ovarian tissue. a review of 13 live births. Ann Med 43(6):437–450
33. Meirow D, Fasouliotis SJ, Nugent D, Schenker JG, Gosden RG, Rutherford AJ

(1999) A laparoscopic technique for obtaining ovarian cortical biopsies for fertility conservation in cancer patients. Fertil Steril 71:948–951
34. Revel A, Koler M, Simon A, Lewin A, Laufer N, Safran A (2003) Oocyte collection during cryopreservation of the ovarian cortex. Fertil Steril 79(5):1237–1239
35. Seshadri T, Gook D, Lade S et al (2006) Lack of evidence of disease contamination in ovarian tissue harvested for cryopreservation from patients with Hodgkin lymphoma and analysis of factors predictive of oocyte yield. Br J Cancer 94(7):1007–1010
36. Gook D, Edgar DH (2011) Ovarian tissue cryopreservation. In: Donnez J, Samuel Kim S (eds) Principles and practice of fertility preservation. Cambridge University Press, New York, pp 342–356
37. Schmidt KL, Ernst E, Byskov AG, Nyboe Andersen A, Yding Andersen C (2003) Survival of primordial follicles following prolonged transportation of ovarian tissue prior to cryopreservation. Hum Reprod 18(12):2654–2659
38. Mazur P (1970) Cryobiology: the freezing of biological systems. Science 168(3934):939–949
39. Hunter J, Bernard A, Fuller B, McGrath J, Shaw RW (1992) Plasma membrane water permeabilities of human oocytes: the temperature dependence of water movement in individual cells. J Cell Physiol 150(1):175–179
40. Newton H, Fisher J, Arnold JR, Pegg DE, Faddy MJ, Gosden RG (1998) Permeation of human ovarian tissue with cryoprotective agents in preparation for cryopreservation. Hum Reprod 13(2):376–380
41. Oktay K, Nugent D, Newton H, Salha O, Chatterjee P, Gosden RG (1997) Isolation and characterization of primordial follicles from fresh and cryopreserved human ovarian tissue. Fertil Steril 67(3):481–486
42. Picton HM, Kim SS, Gosden RG (2000) Cryopreservation of gonadal tissue and cells. Br Med Bull 56(3):603–615
43. Gook DA, Edgar DH, Stern C (1999) Effect of cooling rate and dehydration regimen on the histological appearance of human ovarian cortex following cryopreservation in 1, 2-propanediol. Hum Reprod 14(8):2061–2068
44. Meirow D, Levron J, Eldar-Geva T et al (2007) Monitoring the ovaries after autotransplantation of cryopreserved ovarian tissue: endocrine studies, in vitro fertilization cycles, and live birth. Fertil Steril 87(2):418 e7–418 e15
45. Kagawa N, Silber S, Kuwayama M (2009) Successful vitrification of bovine and human ovarian tissue. Reprod Biomed Online 18(4):568–577
46. Fahy GM, Wowk B, Wu J, Paynter S (2004) Improved vitrification solutions based on the predictability of vitrification solution toxicity. Cryobiology 48(1):22–35
47. Keros V, Xella S, Hultenby K et al (2009) Vitrification versus controlled-rate freezing in cryopreservation of human ovarian tissue. Hum Reprod 24(7):1670–1683
48. Meirow D, Hardan I, Dor J et al (2008) Searching for evidence of disease and malignant cell contamination in ovarian tissue stored from hematologic cancer patients. Hum Reprod 23(5):1007–1013
49. Dolmans MM, Marinescu C, Saussoy P et al (2010) Reimplantation of cryopreserved ovarian tissue from patients with acute lymphoblastic leukemia is potentially unsafe. Blood 116(16):2908–2914
50. Rosendahl MAM, Ralfkiaer E et al (2010) Evidence of residual disease in cryopreserved ovarian cortex from female patients with leukemia. Fertil Steril 94:2186–2190
51. Demeestere I, Simon P, Emiliani S, Delbaere A, Englert Y (2006) Options to preserve fertility before oncological treatment: cryopreservation of ovarian tissue and its clinical application. Acta Clin Belg 61(5):259–263
52. Demeestere I, Simon P, Emiliani S, Delbaere A, Englert Y (2009) Orthotopic and heterotopic ovarian tissue transplantation. Hum Reprod Update 15(6):649–665
53. Silber SJ, Lenahan KM, Levine DJ et al (2005) Ovarian transplantation between monozygotic twins discordant for premature ovarian failure. N Engl J Med 353(1):58–63
54. Demeestere I, Simon P, Buxant F et al (2006) Ovarian function and spontaneous pregnancy after combined heterotopic and orthotopic cryopreserved ovarian tissue transplantation in a patient previously treated with bone marrow transplantation: case report. Hum Reprod 21(8):2010–2014
55. Piver P, Amiot C, Agnani G et al (2009) Two pregnancies obtained after a new technique of autotranplantation of cryopreserved ovarian tissue. Hum Reprod 24:i150-035
56. Revel A, Laufer N, Ben Meir A, Lebovich M, Mitrani E (2011) Micro-organ ovarian transplantation enables pregnancy: a case report. Hum Reprod 26(5):1097–1103
57. Nugent D, Newton H, Gallivan L, Gosden RG (1998) Protective effect of vitamin E on ischaemia-reperfusion injury in ovarian grafts. J Reprod Fertil 114(2):341–346

58. Maltaris T, Beckmann MW, Binder H et al (2007) The effect of a GnRH agonist on cryopreserved human ovarian grafts in severe combined immunodeficient mice. Reproduction 133(2):503–509
59. Sanchez-Serrano M, Crespo J, Mirabet V et al (2010) Twins born after transplantation of ovarian cortical tissue and oocyte vitrification. Fertil Steril 93(1):268 e11–268 e13
60. Ernst E, Bergholdt S, Jorgensen JS, Andersen CY (2010) The first woman to give birth to two children following transplantation of frozen/thawed ovarian tissue. Hum Reprod 25(5):1280–1281

Chapter 22

Detection of Monogenic Disorders and Chromosome Aberrations by Preimplantation Genetic Diagnosis

Kangpu Xu and David Reichman

Abstract

This chapter highlights the methodologies of single cell genetic diagnosis along with the strengths and weaknesses of existing techniques.

Key words PCR, Genetic diagnosis, PGD, PGS, Monogenic disorders

1 Introduction

Several groundbreaking genetic analysis tools were developed in the 1980s. Polymerase chain reaction (PCR) arose at this time and became a transformative method for molecular biology and biochemistry [1]. It was via this technique that the first PGD studies were performed, in which Handyside and colleagues successfully amplified regions of the Y chromosome from single cells and thereby prevented transmission of X-linked disorders [2]. Preimplantation genetic diagnosis is now in its third decade, and great strides have been made to evolve and refine the approach and technology underlying the technique [3]. Almost all single gene disorders can now be identified in preimplantation embryos, and PGD has increasingly become an important tool in reproductive medicine. Application of this technology has steadily increased since the human genome project came to completion and as revelation of genetic disorders has deepened.

2 Preimplantation Genetic Diagnosis for Mendelian Disorders

2.1 Single Cell DNA Amplification by PCR

Each single human cell, such as a blastomere from a preimplantation embryo, contains approximately 6–7 pg of DNA. Preimplantation genetic diagnosis must be sensitive enough to detect one base pair

Zev Rosenwaks and Paul M. Wassarman (eds.), *Human Fertility: Methods and Protocols*, Methods in Molecular Biology, vol. 1154, DOI 10.1007/978-1-4939-0659-8_22, © Springer Science+Business Media New York 2014

nucleotide changes within the context of the entire complex genome. Direct mutation analysis with this small amount of material is not possible. Using PCR, a few DNA molecules can be amplified to millions of copies of defined sequence in a matter of a few hours, thus making PCR an invaluable tool for PGD. However, risk of exogenous DNA contamination in PCR is high because of the exceptionally low copy numbers of target DNA in a single cell. Tremendous efforts were made to optimize PGD after its introduction, such as specific cell loading protocols, blank controls, and testing of specific sequences. Gel electrophoresis can be used for verification of PCR amplification and to rule out contamination; contamination is now infrequent in experienced laboratories.

Allelic drop-out (ADO) is a major risk that can lead to misdiagnosis in PCR-PGD. ADO refers to the amplification failure of one of the two alleles in heterozygous cells during PCR. This can occur when the quantity of input DNA is extremely low, such as in the case of PGD tests utilizing single cells. The frequency of ADO has been reported to range from 10 to 30 % and is estimated to occur in 5–10 % of tests. ADO was historically believed to be responsible for a large proportion of misdiagnosis cases, but has been less of an issue with current sophisticated protocols. Recently, improvements in *Taq* enzyme, the critical DNA polymerase for PCR, have contributed to lower rates of ADO, with better coverage and more efficient amplification.

2.2 Mutation Analysis by Enzyme Digestion, Heteroduplex Analysis, Fluorescence Fragment Analysis, and Mini-sequencing

Following amplification of target DNA via PCR, digestion of the amplified product with site-specific restriction endonucleases can reveal genetic sequence alterations. A myriad of restriction enzymes recognize unique and specific base pair sequences; if a genetic sequence alteration creates or abolishes a restriction enzyme recognition site, mutant variants can be discerned from the wild type via discrete band patterning on gel electrophoresis. One such example is the sickle cell mutation (Hb S, GAG → GTG), whereby the mutant allele can be identified via use of the restriction enzyme DdeI [4, 5]. Another example, representing diagnosis via restriction enzyme BanII of the E1322 mutation leading to Limb-Girdle muscular dystrophy 2B, is shown in Fig. 1.

Heteroduplex analysis is another classical method to detect small deletions or insertions. The formation of heteroduplexes from annealing complementary DNA strands containing base mismatches can also be used as a diagnostic tool for the detection of deletion/insertion mutations in a number of disease-causing alleles; this is due to the differential migration of heteroduplexes versus homoduplexes on acrylamide gels. Heteroduplexes can be formed intentionally by co-incubation of known fragments and unknown PCR products. This method was first reported for PGD in a landmark paper (1992) for detecting a common three base pair

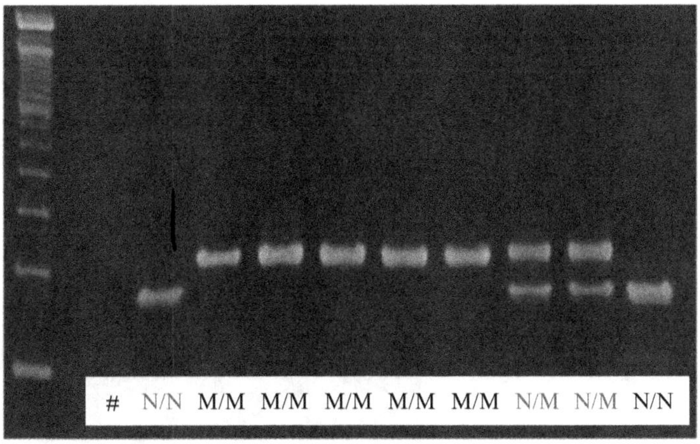

Fig. 1 Restriction enzyme analysis. Restriction analysis for detection of a mutation, E1322, Limb-Girdle Muscular Dystrophy 2B (LGMD2B, OMIM #253601). Restriction enzyme BanII is used. From *left*, DNA size marker, *Lanes 1* and *9*, (N/N, complete digestion), *Lane 2* to *6*, homozygote affected (M/M, no digestion); *Lanes 7* and *8*, heterozygotes (N/M)

deletion (ΔF508) in exon 10 of the CFTR gene (Fig. 2a) [6]. These procedures have been successfully used for PGD of many disorders/mutations due to their reliability, accuracy and sensitivity; such techniques can still be effectively applied for PGD today. Nevertheless, not all mutations can be recognized by existing restriction enzymes, and such techniques are quite labor intensive.

Fluorescent, real-time PCR is a more versatile and less labor-intensive methodology, allowing for automation of several tedious steps, and has largely replaced gel-based DNA band size assessment (Fig. 2b) [7–9]. Furthermore, a more universal, primer-specific mutation detection method, mini-sequencing, can be used for detection of nearly any mutation, and multiplex is possible (Fig. 3) [10, 11]. In mini-sequencing, a labeled detection primer positioned immediately to the 3′ end of the nucleotide sequence to be analyzed is included in the PCR reaction. In the case of a homozygous genotype, a signal is obtained in only one of the reactions and for a heterozygote genotype, signals are obtained in both reactions. The main benefit of the mini-sequencing strategy is the use of a mutation analysis protocol based on a common procedure, irrespective of the mutations involved. Mutation detection by mini-sequencing is not limited to point mutations and the method is generally applicable to both large and small deletions.

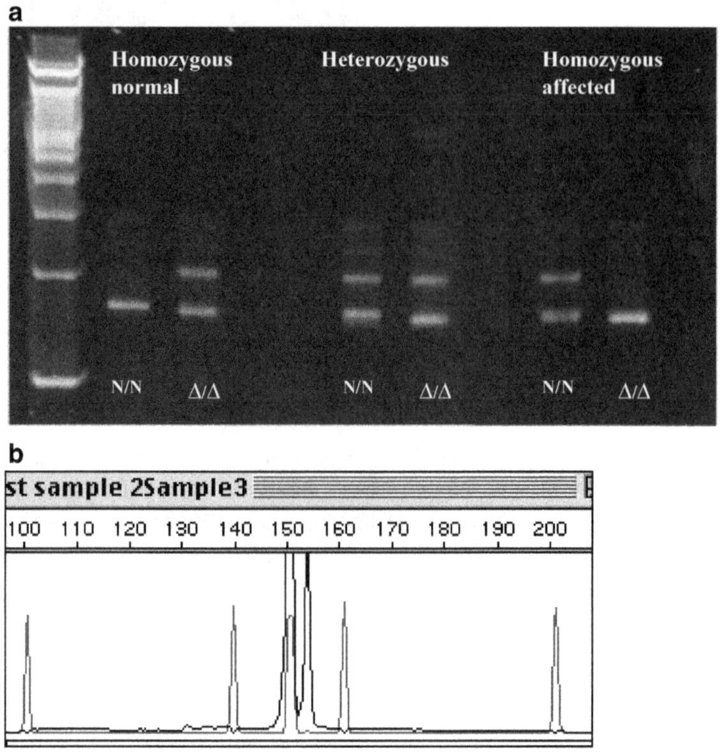

Fig. 2 CFTR mutation, delta F508, detected by heteroduplex analysis (**a**) and fluorescent fragment analysis (**b**). Heteroduplexes can be formed intentionally by co-incubation of known fragments (N/N and Δ/Δ) and unknown PCR products. Shown here are examples of heteroduplex analysis for CFTR gene, ΔF508 mutation on exon 10. The far left lane is a DNA ladder. The *first two lanes* represent a homozygous normal pattern, the *middle two lanes* represent a heterozygote, and the *last two lanes* represent an affected homozygote. Fluorescent fragment analysis for a heterozygote of delta F508 mutation, showing two peaks (*black*), 151 and 154bp clearly separated

2.3 Inclusion of Linkage Markers

PGD for single gene disorders has become a sensitive and accurate technique, with thousands of disparate mutations able to be diagnosed. Analysis for multiple genetic defects, or for multiple alleles simultaneously, poses additional challenges. Both members of a couple may be carriers for the same disease but with different allelic variants, leading to affected compound heterozygous offspring. In the case of autosomal dominant disorders, such as Huntington's disease, including linkage markers is critical to avoid false-negatives when ADO occur (Fig. 4). With the completion of the Human Genome Project, more detailed sequence data became available, allowing for inclusion of flanking linkage markers to be included in

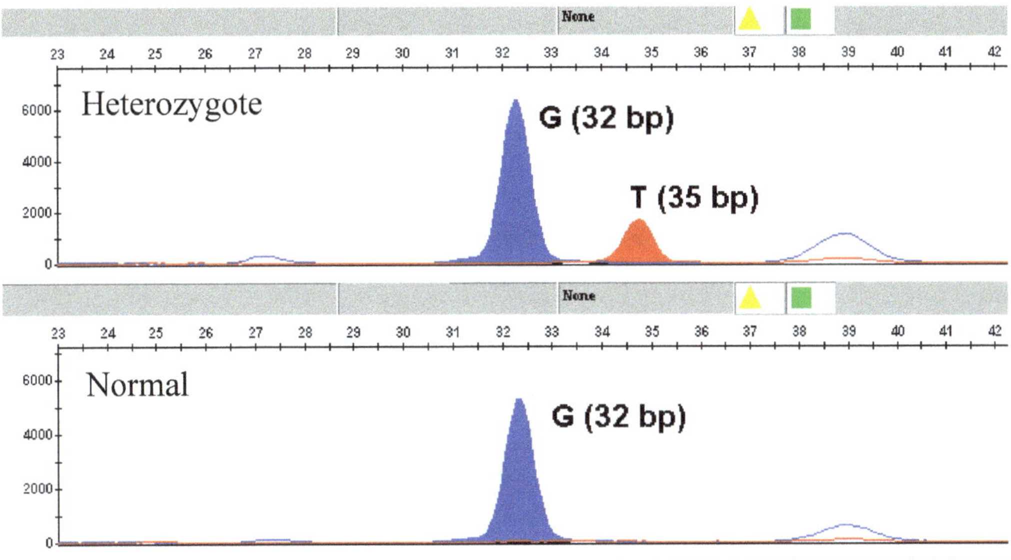

Fig. 3 Mini-sequencing analysis for CF G542X mutation. G542X is a G to T substitution at nucleotide 1756 on exon 11, Gly to Stop at 542

PCR [12]. Informative allelic markers for PGD patients are now routinely used, in addition to the mutations tested, allowing for assessment of whether the maternal or paternal allele has been inherited. Such linkage analysis allows for diagnosis to be made even when allelic drop-out occurs. In fact, with the use of informative linkage markers, one may obtain information regarding mutation transmission without knowing the exact details of the mutation; thus, even unknown mutations may be detected via such techniques.

2.4 Expanding PGD Indications

Primer design for each mutation or disorder can be accomplished in a relatively short amount of time. Online detailed information is readily available from public or pay-for-service websites. For example, OMIM (Online Mendelian Inheritance in Man) is a comprehensive, authoritative compendium of human genes and genetic phenotypes that is freely available and updated daily (http://www.ncbi.nlm.nih.gov/omim). The Human Gene Mutation Database (http://archive.uwcm.ac.uk) reports that 90 % of disease-causing mutations in humans are attributable to micro-lesions such as single-base substitutions and small deletions and/or insertions, rather than gross lesions (repeat expansions, complex gene rearrangements, gross insertions and/or deletions). Therefore, it is vital that mutation detection methods are sensitive to the content of nucleotide variations at the single-base level.

Armed with the above-mentioned mutation analysis methods, and encouraged by improvements in accuracy, the PGD

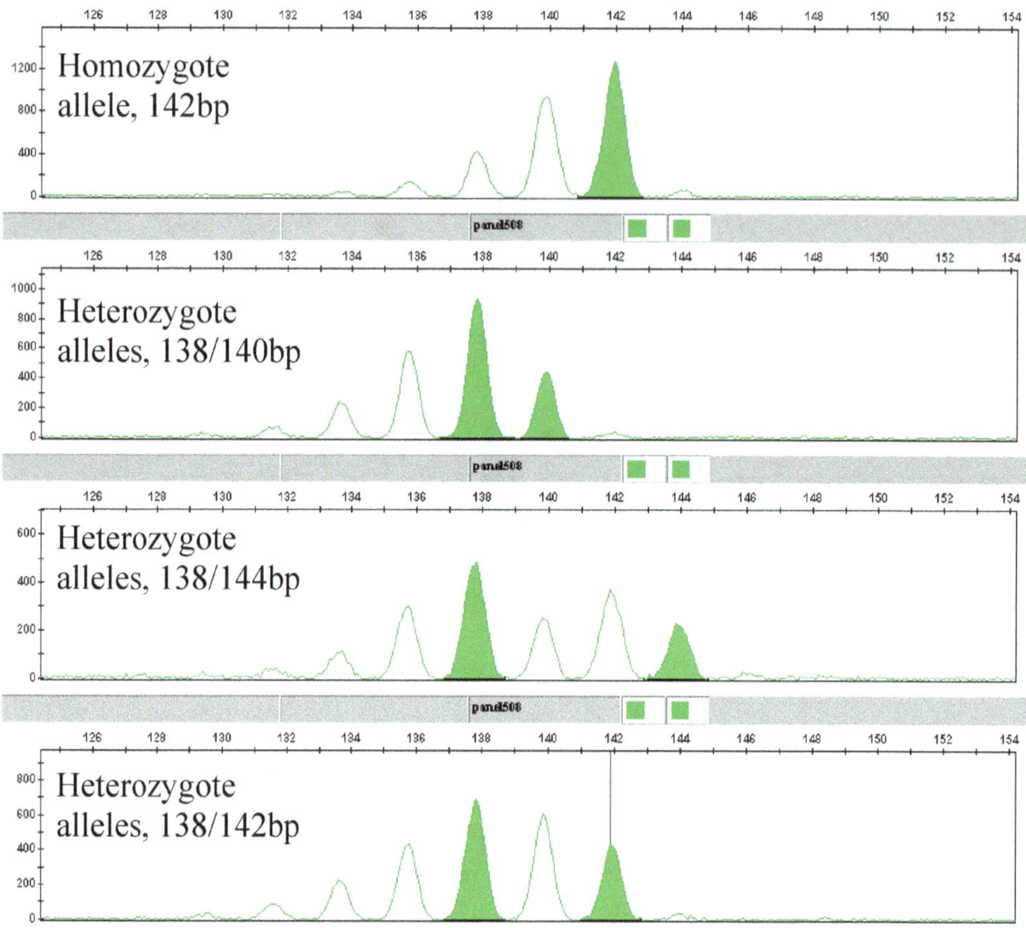

Fig. 4 Use of STR linkage markers for Huntington's disease. Informative STR D4S127 can be used. HD (#143100) is caused by an expanded trinucleotide repeat (CAG)n, encoding glutamine, in the gene encoding Huntington's on chromosome 4p16.3

community has quickly enriched its inventory of tools for the diagnosis of various genetic disorders. The application of PGD to single gene disorders has rapidly expanded, and the number of genetic disorders diagnosed is now too numerous to enumerate. Major landmarks were the ability to accurately diagnosis cystic fibrosis and sickle cell anemia [5, 13]. In addition to such diseases, PGD can also be performed for mutations predisposing to cancer, such as BRCA 1 and 2 mutations [14]. While these represent disorders of Mendelian inheritance, HLA matching via PGD represents a unique genetic tool. Using IVF to create embryos that are subsequently HLA-matched but unaffected by the disorder, siblings suffering from disease can be cured [15–17]. This can be achieved via multiplex PCR with amplification targets on all possible alleles of HLA-A, HLA-B, HLA-C, and HLA-DRB regions, and linked single tandem repeat (STR) markers selected to cover the extended HLA complex, combined with analysis for the single gene disorder in question [17]. Worldwide, thousands of such cases have been attempted, with hundreds of babies born. Recently, one large series of PGD for HLA was published in a single PGD unit, which included 327 cycles. Embryo transfer was performed in 212 cycles, with a 34.9 % clinical pregnancy rate per transfer and 59 healthy and HLA-compatible children born [18].

2.5 Preimplantation Chromosome Testing: From a Few Chromosomes to all 23 Pairs

Early studies in human preimplantation development revealed that lower human reproductive efficiency is largely explained by the high frequency of aneuploidy in gametes. Particularly, the incidence of aneuploidy is closely related to increasing female age [19]. This observation has been repeatedly confirmed in numerous studies by karyotyping, and more recently by high-resolution chromosome analysis [20–23]. Selection of euploid embryos for transfer decreases risk of miscarriage after IVF and allows for prevention of trisomy 13, 18, and 21 that may lead to live birth of children with specific syndromes. Florescent in situ hybridization (FISH), developed in the early 1980s, was the first cytogenetic technique to be applied towards screening for numerical chromosomal abnormalities [24]. The technique involves hybridizing fluorescent tagged probes using DNA sequences specific for individual chromosomes. Griffin and his colleagues in the UK successfully applied X and Y DNA probes to single blastomeres in 1991, followed by the Cornell team in the USA in 1993 [25, 26]. Fixation of the nucleus onto glass slides is accomplished via two methods: Tarkowski's fixation using acetic acid or otherwise Tween-20. The quality of fixation remains the most difficult step in FISH-based single cell tests; accurate screening of a single cell with FISH requires a high degree of technical proficiency.

With the development of various fluorophores, multi-color FISH allowed up to 5 probes to be included in one single hybridization run, which enabled the detection of additional chromosomes beyond X and Y (Fig. 5). Because FISH signals can be washed away, another round of FISH can be applied with different chromosomal probes, allowing for even more chromosomes to be analyzed. Procedures for up to 3 rounds of FISH, including 12–14 FISH probes were established in many PGD laboratories. While FISH was almost exclusively used for several years, it requires nuanced technical skill, is limited by intrinsic limitations to hybridization efficiency, and the predictive value of the test is diminished with increasing number of chromosomes analyzed. A number of randomized controlled trials using a FISH-based approach to preimplantation genetic screening for chromosomal aneuploidy failed to reveal a benefit, and in certain instances was associated with decreased implantation [27, 28]. A few more powerful FISH based systems that would allow for screening of all 24 chromosomes,

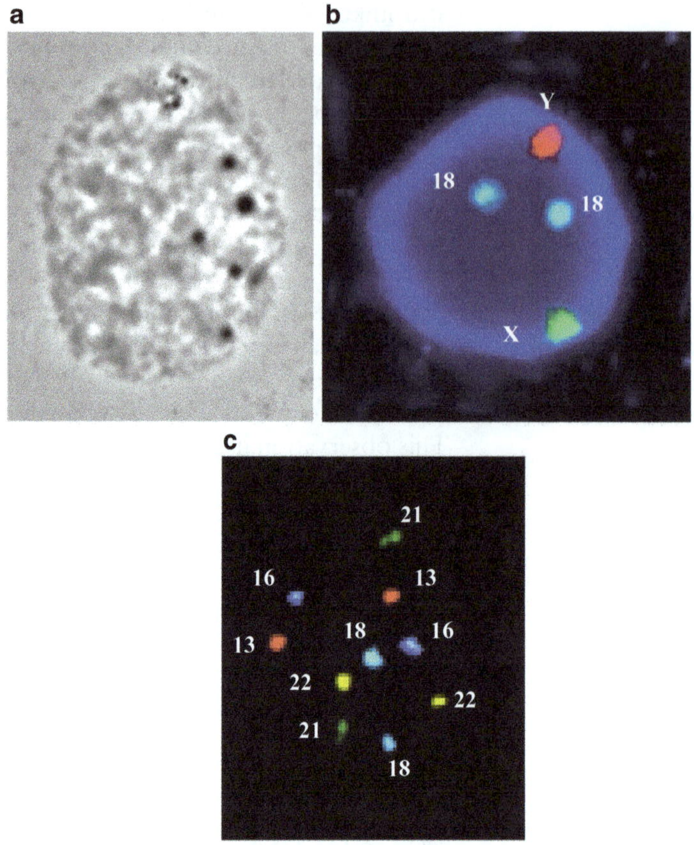

Fig. 5 FISH for aneuploidy screening. (**a**) Fixed nucleus from D3 blastomere. (**b**) FISH for sex determination. Chromosome X: *Green* and Y: *Orange*; 18: Aqua (control). (**c**) Five FISH probes for aneuploidy screening. Chromosome 13: *Red*, 16: *Blue*; 18: Aqua; 21: *Green* and 22: *Yellow*

such as SKY (spectral karyotype) were also introduced to single cells, but with little clinical success [29]. Twenty-four chromosome FISH using oligonucleotide probes has also been reported, but has never become a mainstream modality.

Comparative genomic hybridization (CGH), first reported in 1992, represented a major technical breakthrough for PGD when it was first applied to single cells [30–32]. Specific technical improvements allowed reliable clinical application of this technology [33, 34]. Initially, cryopreservation for biopsied embryos was needed due to the time required for completing the hybridization procedure; short-CGH methods were subsequently established that allowed for fresh embryo transfer after day 3 biopsy [35]. Modification of classic CGH using metaphase spreads to an array based modality made it a

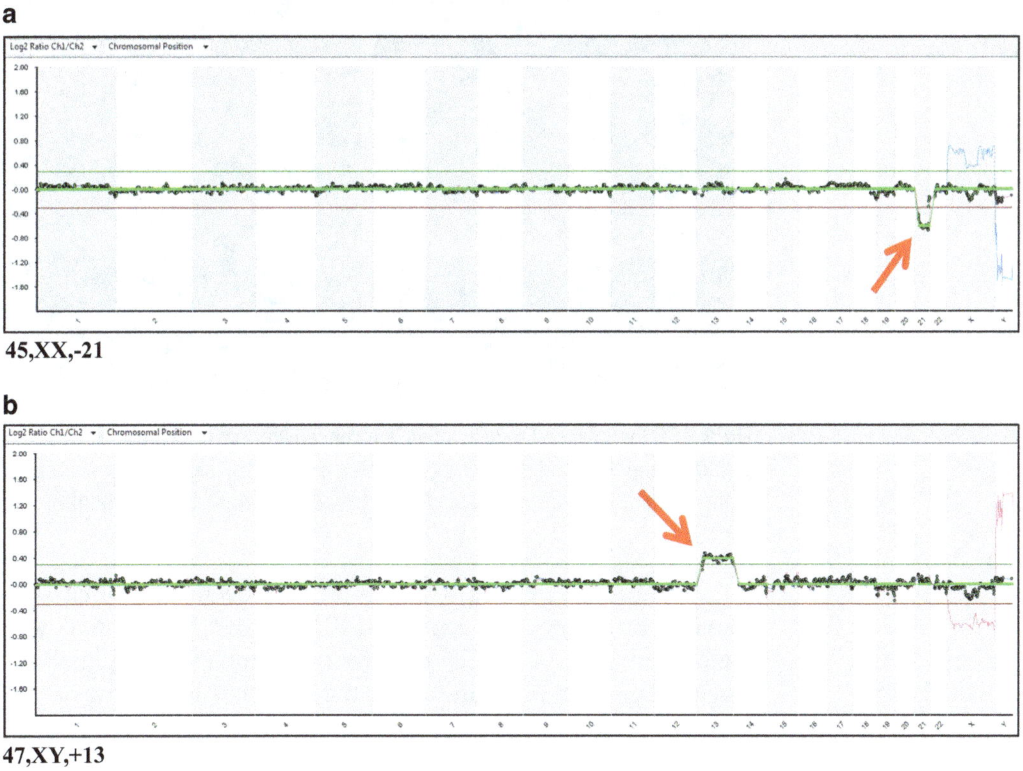

Fig. 6 Array CGH for aneuploidy screening with 24Sure V3 Chip, BlueGnome-Illumina. (**a**) 45,XX,-21. (**b**) 47,XY,+13. *Red arrows* indicate a loss of chromosome 21 (*panel A*) and a gain of chromosome 13 (*panel B*)

more practical and efficient approach for chromosomal screening and it has now become a prevailing modality for preimplantation genetic screening [36, 37]. In array CGH, the tested genomic DNA sample and a reference genomic DNA sample are labeled with different fluorescent dyes and co-hybridized to a glass slide on which a large number of small DNA fragments of known chromosomal location are arrayed. Gain or loss of genetic material as compared to control DNA can thus be determined (Fig. 6).

In addition to array CGH, single nucleotide polymorphism array (SNP array) technology can also be used for chromosomal screening, in which a number of oligonucleotide probes are used; because such oligonucleotide probes interrogate both copy number and SNP sites, genotype as well as DNA copy number information can be obtained. SNP array has successfully been used for PGD of IVF embryos, and has the added advantage of being able to detect uniparental disomy (UPD) [38]. While both aCGH and SNP array platforms serve as accurate diagnostic platforms for PGS, results from aCGH appear to be more consistent [39].

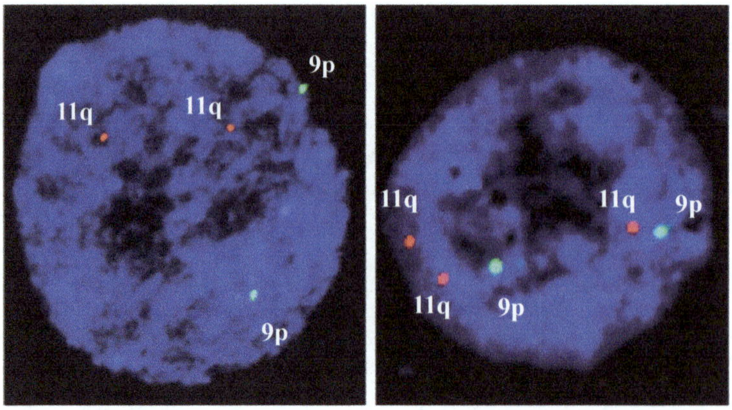

Fig. 7 FISH for balanced translocation. A balanced embryo (*left*) showing two *green* (subtelomeric 9p) and two *orange* (subtelomeric 11q) signals. An unbalanced embryo (*right*) for a carrier of balanced translocation, 46,XY,t(9;11)(p22;q22)

Quantitative PCR (qPCR) has recently been used for chromosomal copy number assessment [40]. In qPCR, selected target regions on each of the 24 chromosomes are PCR amplified in a strictly controlled reaction so that the number of amplicons at specific PCR cycles can then be compared with controls, thus allowing copy number of each chromosome to be calculated. When, for example, an extra chromosome 21 is present in the specimen, less PCR cycles are needed to reach a specific DNA yield as compared to a control sample which has two copies of chromosome 21. Although qPCR has the advantage of speed (the whole process can be completed in as little as 4–6 h) with limited targets (2–4 per chromosome), the resolution for chromosomal analysis is very low.

2.6 Chromosome Translocations and Deletions

Carriers for balanced, reciprocal, and Robertsonian translocations have high risks for producing abnormal gametes that have missing or additional chromosomal material. Detection of unbalanced chromosomal complements via PGD increases the likelihood of successful, normal pregnancy. Patient breakpoint-specific probes were designed and first implemented for translocation analysis in the late 1990s [41]. Although this technique had the advantage of detecting and distinguishing normal chromosomes from unbalanced and/or balanced chromosomes, it was prohibitively expensive for individual patient use. The discovery of chromosome specific telomeric probes made diagnosis of translocations in IVF embryos more universally applicable for PGD (Fig. 7) [42]. When FISH probes are available, deletions and pericentric inversions can also be detected [43]. However, in cases of paracentric inversions, this strategy may not work—such inversions may not lead to a gain or loss of the terminal region, where FISH probes are useful. Other methodology, such as PCR-based or array-based approaches have also been successfully applied [44]. Similar to the use of telomeric FISH probes, the limitation of array-based approaches is that they are unable to differentiate between normal chromosomal complements and balanced rearrangements. The resolution of current array platforms is relatively lower than telomeric FISH (typically at 5–10 MB), and thus limits the ability to detect cryptic translocations. Therefore, FISH may still be required for these patients.

2.7 Acquisition of DNA: Polar Body, Day 3, and Day 5 Analysis

Until the mid-2000s, the majority of PGD and PGS cases were performed using individual blastomeres from day 3 embryos. One or two nucleated cells (blastomeres) are removed 3 days after fertilization, at which time normally developing embryos typically contain 6–9 cells. An inverted microscope equipped with biopsy tools for micromanipulation is required. Typically, a holding pipet is used to fix the embryo in place, and a small amount of acidified Tyrode's solution is used to make a small hole in the zona pellucida; alternatively, a laser is applied to cut a slit in the zona. Controlled negative pressure is applied with a biopsy pipet to remove the blastomere (or polar body in the case of oocyte sampling) from the embryo (Fig. 8). Similar steps are needed for trophectoderm (TE) biopsy, except that the connection between TE cells must then be divided, typically with a laser (Fig. 9) [45].

Blastomere biopsy must occur before compaction begins, as individual blastomere removal becomes impossible after sibling blastomeres become tightly adherent to each other. The chromosomal complement of the embryo is then inferred from the analysis of the one or two removed cells. Mosaicism, however, is very common in cleavage stage embryos, as has been demonstrated in a number of published studies [20, 46–51]. It is known that mosaicism occurs in women of all ages, including young patients [52, 53]. The prevalence of mosaicism in D3 cleavage embryos is perhaps best summarized in a systematic review by van Echten-Arends and colleagues using FISH with 5–15 chromosomes in 815 embryos, in which 73 % of embryos were mosaic [50]. These numbers may be overestimated given that the embryos analyzed were de-selected for transfer, but on the other hand, the detection of mosaicism would increase if all 24 chromosomes had been analyzed.

The prevalence of mitotic nondisjunction in individual blastomeres on day 3 of development limits the ability of day 3 biopsy to unambiguously prognosticate regarding the potential for normal development and implantation. Whether "self-correction" occurs in mosaic embryos remains unclear [54, 55]. Moreover, it has been argued that removal of 1/8th (1 cell removed from 8 cell embryo) or 1/4th (2 cells removed from 8 cell embryo) of an embryo's genetic complement may adversely affect its subsequent development to the blastocyst stage [56, 57]. While the pendulum has

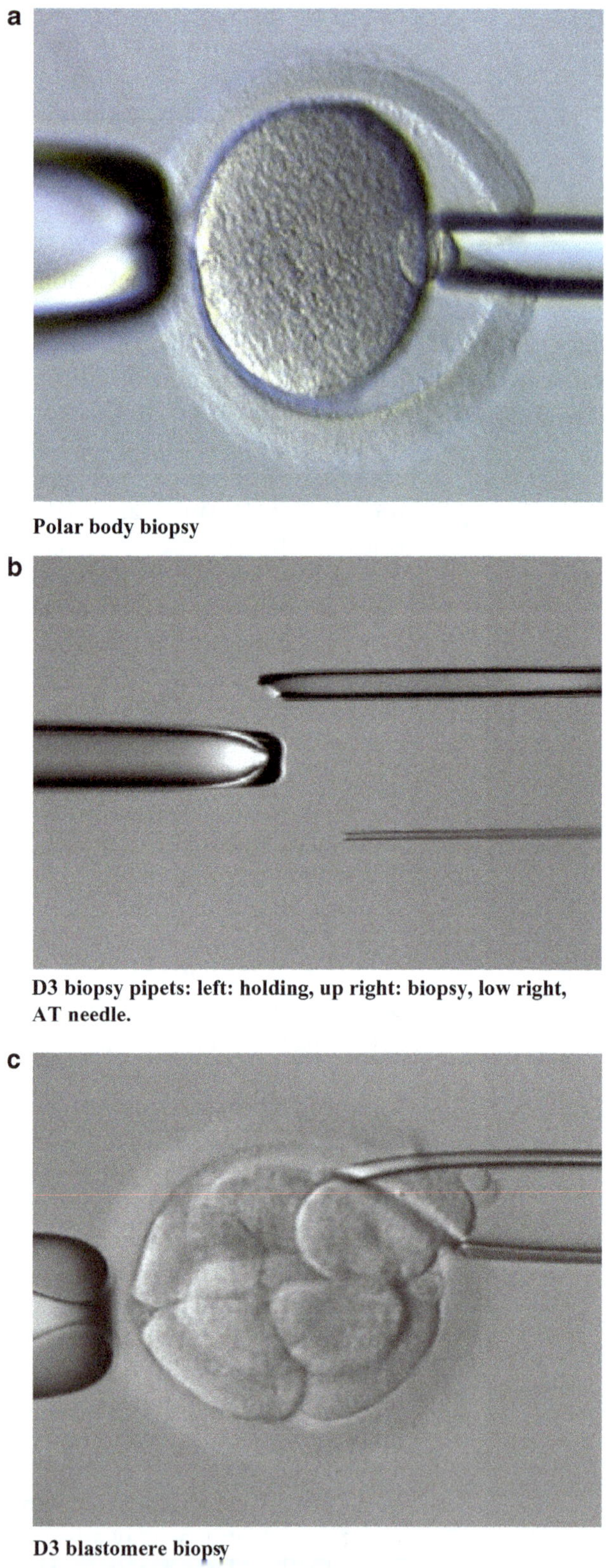

Fig. 8 Polar body and D3 biopsies. (**a**) Polar body biopsy. (**b**) D3 biopsy pipets: *left*: holding, *up right*: biopsy, *low right*, AT needle. (**c**) D3 blastomere biopsy

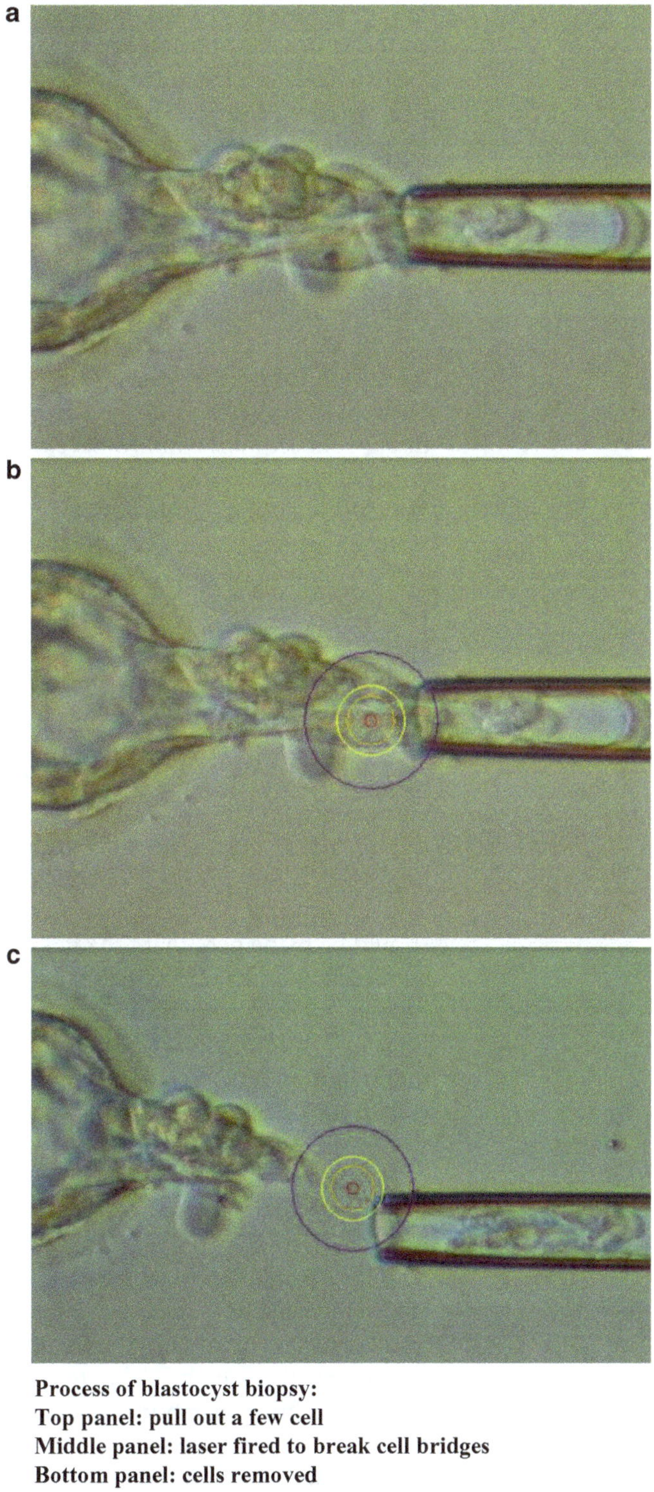

Fig. 9 Blastocyst (trophectoderm) biopsy. Process of blastocyst biopsy: (**a**) *Top panel*: pulling out a few cells. (**b**) *Middle panel*: laser fired to break cell bridges. (**c**) *Bottom panel*: cells removed

begun to therefore swing away from day 3 biopsy, it is important to recognize that thousands of babies have been born from day 3 biopsy PGD, and reported pregnancy rates from large retrospective cohorts still reveal robust performance [58]. With current techniques, day 3 biopsy almost universally allows for transfer to occur on day 5 or 6 during the cycle in which the biopsy is performed, without requiring freezing of embryos.

Biopsy at the blastocyst stage on day 5 or 6 of development allows for sampling of several cells from the trophectoderm, and therefore offers several potential advantages over day 3 biopsy. Whereas day 3 biopsy requires removal of cells that will ultimately contribute to the embryo proper, biopsy on day 5 allows for sampling of cells that give rise to the placenta, and thus is less likely to disrupt subsequent embryo development. Moreover, multiple cells are available for analysis, thus reducing the likelihood of diagnostic errors. Of course, biopsy on day 5 requires an efficient culture system with high rates of blastulation; with improved laboratory culture conditions and evolving technology (i.e., time-lapse kinetics) for prediction of blastulation, however, ideal cases for day 5 biopsy can now be selected without significant risk of being left with no embryos for biopsy.

While mosaicism does exist in blastocysts, it is present to a much lesser extent than in cleavage stage embryos, and poses less of a diagnostic challenge; the exact frequency of blastocyst mosaicism is unknown [59]. There seems to be no difference between the inner cell mass and trophectoderm in terms of mosaicism, suggesting that there is no selection mechanism for selective aggregation of normal cells to the inner cell mass [60–63]. A caution should be made here that the rates of mosaicism derived from FISH studies are much more accurate than those from array studies, because the latter usually can only provide a rough estimation due to software limitations [64]. As in day 3 embryos, mosaicism is much more frequent in poor quality embryos [65, 66]. The reduction in mosaicism and lower prevalence of aneuploidy in blastocysts as opposed to cleavage stage embryos led to the notion that embryos may have a self-correction mechanism [54, 55]. Supporting evidence also can be found in segmental aberrations, wherein the same embryos that showed segmental deletions/duplications on day 3 are likely to have normal karyotypes when retested on D5/6 (Xu et al. unpublished).

While less commonly used in the USA, polar body biopsy remains an alternate modality to day 3 or day 5 biopsy. In this technique, the first polar body is removed from metaphase II oocytes after retrieval, and the second polar body is removed after fertilization has occurred. Correction of meiotic errors frequently occurs, as does recombination, and thus, analysis of only one polar body does not afford sufficient information for accurate diagnosis [67–70]. Genes closer to the telomeric region of the chromosome are more likely to undergo recombination, and thus provide further diagnostic challenges. Polar body analysis is more limited than the aforementioned embryo biopsy techniques given that it provides only maternal genetic information; polar body biopsy is thus not useful if the mother is an affected individual (homozygous) for a recessive trait, given that all her oocytes will carry the mutation. Moreover, polar body analysis is further limited by its inability to capture post-zygotic errors of mitosis, and will miss events such as "self-correction" which occur during compaction and blastulation [71]. Nevertheless, a recent small scale observational study suggested polar body biopsy and array CGH analysis accurately predicted most aneuploidies in cleavage stage embryos [72]. It remains to be seen what role polar body biopsy will play in genetic diagnosis as embryo biopsy techniques continue to evolve and improve.

3 Detection of Aneuploidy at Preimplantation Stages: Clinical Controversy

In the last two decades, technologies related to genomics and genetic testing have evolved at an astoundingly fast pace. As the techniques have improved, so to have the indications for PGD begun to expand, along with patient demand for use of the technology. While analysis for single gene disorders, HLA typing, and translocations have become widely accepted and recognized as effective diagnostic tools for clinical use, routine use of preimplantation genetic screening for unselected patients remains one of the most controversial issues in the field of reproductive medicine today. There is little disagreement within the scientific and medical community regarding the relationship between aneuploidy, implantation failure, and pregnancy loss [22, 73]. Theoretically, aneuploidy testing allows for selection of euploid embryos for transfer, thus allowing for the selection of embryos with the highest implantation potential and the lowest risk of subsequent miscarriage. PGS now accounts for more than half of all PGD cases performed around the world in the last decade, and in some US centers, as many as one-third of all IVF patients undergo PGS [74].

Several early non-randomized studies demonstrated a beneficial role of PGS [47, 52, 75]. Overall, the reported benefits include reducing the risk of having live-birth of trisomic offspring, reducing risk of miscarriages, reducing risk of multiple pregnancies (via the transfer of fewer embryos), and improving the efficiency of IVF when the number of good quality embryos exceeds the number of embryos recommended for transfer. Of all the arguments in favor of PGS, reduction in the rate of miscarriages is supported by the most robust clinical data, with an approximately fourfold reduction in the rate of miscarriage with either polar body or cleavage stage biopsy (depending on the age of the patient) [76].

For the most part, however, randomized controlled trial data has not supported the widespread implementation of this fast-evolving technology. In one of the first large randomized trials evaluating day 3 biopsy in unselected IVF patients, no beneficial effects were observed, and in fact patients undergoing biopsy experienced lower pregnancy rates [77]. This study was met with much subsequent criticism against its alleged technical shortcomings and experimental design limitations [78, 79]. A 2011 Cochrane meta-analysis of 9 randomized controlled trials regarding preimplantation genetic screening, however, again failed to show a benefit in unselected patients as well as patients of advanced maternal age, and in fact confirmed a decrease in live birth rates [80]. Indeed, the Society for Assisted Reproductive Technology (SART) and the American Society for Reproductive Medicine (ASRM) cautioned in 2008 that there was insufficient data to suggest PGS for advanced maternal age, multiple prior IVF failures, or prior miscarriages [81]. Proponents of PGS argue that the studies analyzed in such meta-analyses were limited by their use of day 3 biopsy, which may be associated with diminished implantation potential, and FISH, which is limited by hybridization inefficiency and interpretation.

Over the last 5 years, the switch from day 3 blastomere biopsy to day 5 trophectoderm biopsy, coupled with the commercial launch of microarray technology specifically designed for PGS, has led to renewed excitement regarding screening of embryos for chromosomal aneuploidy. Recent, large retrospective studies, along with randomized controlled trial data, have revealed improved pregnancy rates and lower miscarriage rates with 24 chromosome comprehensive screening, albeit in highly selected patient groups with robust ovarian reserve [33, 40, 58, 82, 83]. While some practitioners have come to feel that only proven euploid embryos should be transferred, such testing involves added embryo manipulation and embryo cryopreservation (in most cases), and a philosophical divide between those in favor and those against universal screening has grown. Interpretation of the data is confounded by disparate control groups; in one recent study, for example, single euploid blastocyst transfer was compared against transfer of 2 blastocysts that were not subjected to biopsy, as opposed to one [84].

While blastocyst biopsy has conventionally required cryopreservation of embryos and transfer in a subsequent thaw cycle, recent advancements in the efficiency of DNA processing has allowed for fresh embryo transfer on day 6 after day 5 biopsy in select clinics; while this avoids subjecting the embryo to cryoprotectants and the stress of freezing, and prevents the patient from having to wait for transfer, it also subjects the embryo to a more narrow endometrial window of implantation [83, 85]. As the technology continues to evolve, it will remain to be determined whether fresh transfer late on day 5 or day 6 versus transfer in a thaw cycle subsequent to cryopreservation will be associated with the most favorable pregnancy rates.

4 Future PGD and Research Directions: From Dual Sampling to Sequencing

After two decades of intensive research and clinical practice, personalized PGD has entered a completely new era [86]. At present, whole chromosome copy number alternations in embryos are mostly identified by microarray platforms at relatively low resolution, similar to the resolution (10–20 MB) achievable by conventional karyotyping [63]. The diagnosis for various disorders at the level of base substitutions and short insertions/deletions is achieved for each individual couple with the specific protocols outlined in the previous sections. In many PGD cases, however, both diagnosis for single gene disorders and chromosomal copy numbers are required, and unique protocols designed for each individual couple are becoming less practical and cost-effective. Reliable, simple and universal procedures which will be able to detect most, if not all, disorders are now the focus of the field; karyomapping is one such method which may have the potential to achieve multiple diagnostic goals in tandem [87].

Several laboratories have already begun to offer combined tests for diagnosis of both monogenic disorders and copy number variation. Multiplex PCR, utilizing multiple primer pairs (double up to several thousand pairs) can amplify multiple genomic targets, with analysis of mutations along with enough single nucleotide polymorphisms (SNPs) or single tandem repeats (STRs) to allow for detection of embryo genotype in a single assay. Alternatively (and more effectively), whole-genome amplification (WGA) is used to pre-amplify a minute amount of DNA from a single or few cells. There are several effective WGA procedures that have been developed over the last two decades. Based on the DNA polymerases used, two types of WGA are available. The first type uses Taq polymerase and different primers, such as random primers (PEP) or degenerative primers (DOP) [88, 89]. The second type, first described in 2000, takes advantage of an isothermal DNA polymerase, phi 29, and the entire amplification can be accomplished in as short as 2–3 h [90]. DNA from single cells can thus be efficiently amplified and accurately used for SNP array analysis [91]. The information provided can be used to predict chromosomal ploidy, and when parental SNP information is available, accuracy is increased [38, 92].

4.1 Next-Generation Sequencing (NGS)

Next-generation Sequencing (NGS), a term to differentiate it from classic "Sanger" sequencing and whole-genome sequencing (WGS), describes the system that sequences all DNA base pairs, and may well be the future of both PGD and PGS. Although different NGS platforms utilize slightly different methods, most technologies involve similar DNA preparation. When a sufficient quantity of DNA is available via WGA or multiplex PCR, the steps of NGS involves library preparation, DNA fragmentation and labeling, sequencing, data analysis and reporting. All of these steps can be accomplished in as little as 12 h if lower resolution diagnosis is provided, or a couple of days, when deeper sequencing is needed. Currently, the duration of the diagnostic procedure in its entirety precludes day 5 afternoon transfer following day 5 morning biopsy.

NGS for genetic testing has begun to play an increasing role in noninvasive prenatal diagnosis [93]. Early studies designed for single cells have demonstrated its potential for PGD, wherein accurate diagnosis for trisomy 21 can be made with a half-million sequencing reads on a bench-top NGS platform (Fig. 10) [94]. Given that NGS renders so much data, creating data-analysis pipelines remains the most challenging part of implementing the technology [95]. Although the birth of healthy babies from NGS PGD has been reported, the routine and wide use of NGS for clinical PGS is still a few years away. Two NGS strategies are currently being developed in parallel; an amplicon approach (or targeted sequencing) versus whole-genome sequencing. The former focuses on sequencing known genes or known genomic regions but with deeper coverage, while the latter covers the entire genome with some tolerance for loss of a small proportion of genetic information. Each approach requires its own data analysis pipeline and has unique advantages and limitations. As the technology evolves, so too will the efficiency of sequencing, with a corresponding reduction in price and increase in accuracy; as an example, single cell DNA can now be sequenced on micro-well displacement amplification systems (MIDAS) [96].

While NGS has the potential to provide information regarding all 3 billion base pairs of the human genome at an increasingly affordable cost, access to such data may pose complex ethical and medicolegal issues. Families without infertility may request the technology so as to have offspring with select genetic qualities or may desire to cull embryos based on predisposition to adult onset diseases. Similar to the use of IVF for social gender selection, PGD with NGS may well offer fertile couples other nonmedical reasons to pursue the technology. Moreover, while the human genome project greatly enhanced our knowledge of genetic disorders, new disorders are continually being discovered, and knowledge of these disorders may lag our ability to detect them via sequencing data. While solving some diagnostic problems, NGS will undoubtedly give rise to a completely new set of issues, such as incidental findings of unknown significance. Recommendations regarding how to deal with incidental findings have recently been published from the American College of Medical Genetics as well as the President's Bioethics committee [97, 98].

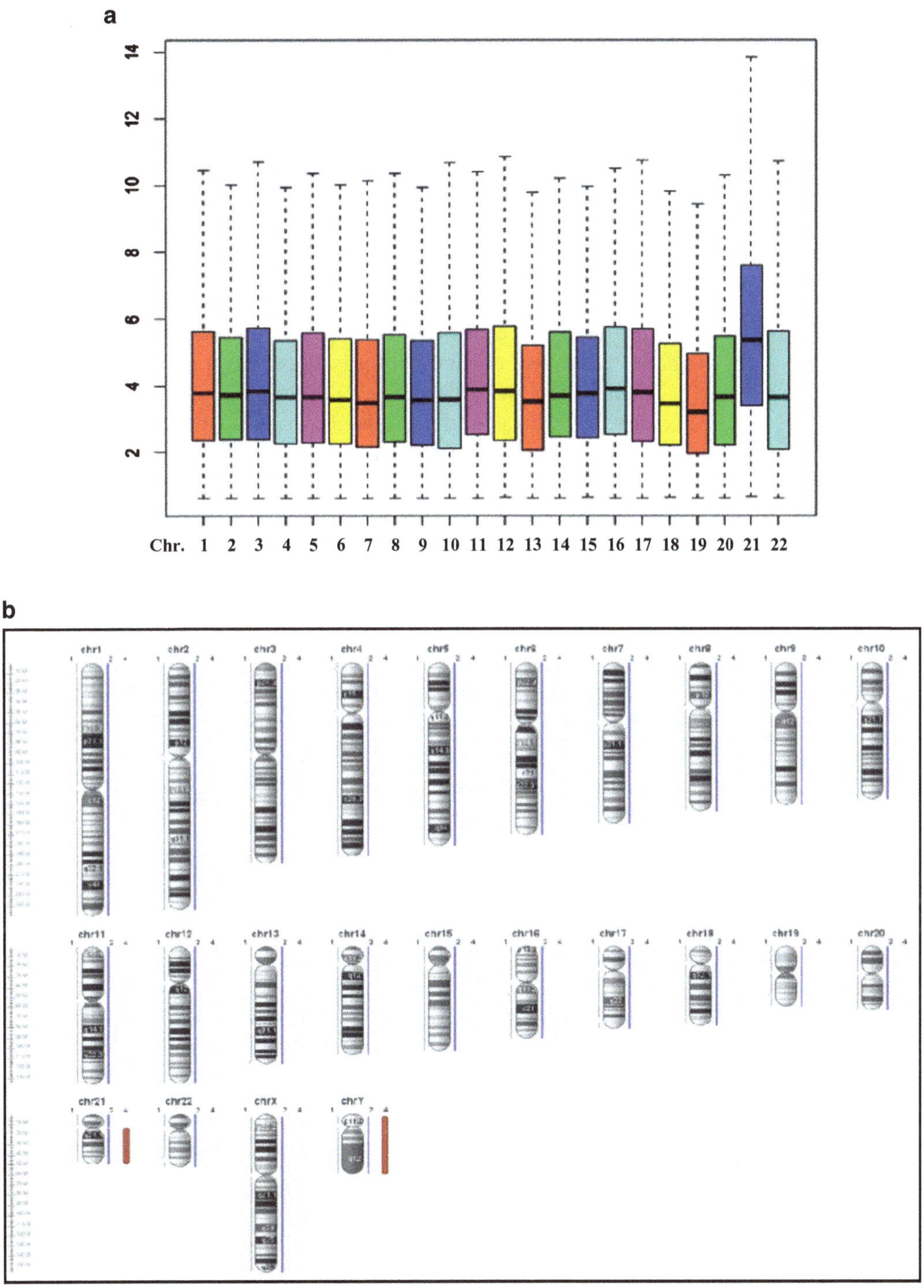

Fig. 10 Aneuploidy detection for single or a few cells by next-generation sequencing on two platforms. (**a**) Single cell: 47,XY,+21 on a MiSeq NGS sequencer (Illumina). The *Y* axis indicates the copy ratio. (**b**) Single cell NGS, 45,XX,-21; on Ion Torrent PGM (Life Technology, CA). The *red line* on the *right side* of chromosome 21 indicating a loss of one chromosome 21, whereas the *red line* on the *right side* of Y indicating no Y chromosome sequence detected

5 Final Remarks

Technically, PGD is now a mature treatment modality. Accumulated evidence indicates that PGD is a safe procedure based on several studies when day 3 biopsy is combined with day 5 fresh embryo transfer, and blastocyst biopsy is considered less invasive than day 3 biopsy because a smaller proportion of cells is removed [99, 100]. Nevertheless, attention on following up children born as a result of this technology in the long term must continue. It should be pointed out that the aforementioned established procedures may be yet be replaced by novel strategies, such as testing of blastocoele fluid, or analysis of metabolites in the spent medium [101].

There are also many other important aspects including legal, regulatory, ethical, and social issues that will affect the implementation and widespread use of this technology. It remains to be seen if PGD will be widely implemented for reducing the overall burden on society for treating genetic disorders, and if so, who will ultimately shoulder the cost of such screening [102, 103].

Finally, as effective as PGD is for preventing transmission of genetic disorders, PGD is ultimately limited to its inability to fundamentally change biology. When women enter the later stages of reproductive aging, what PGD/PGS can offer is limited. Further study regarding gamete and embryo biology will hopefully provide further insight into the processes underlying reproductive aging. Other new technologies, such as regenerative reproductive approaches involving novel genetic therapies may be the next frontier to be explored.

References

1. Mullis KB, Faloona FA (1987) Specific synthesis of DNA in vitro via a polymerase-catalyzed chain reaction. Meth Enzymol 155:335–350
2. Handyside AH, Kontogianni EH, Hardy K, Winston RM (1990) Pregnancies from biopsied human preimplantation embryos sexed by Y-specific DNA amplification. Nature 344:768–770
3. Handyside AH, Xu K (2012) Preimplantation genetic diagnosis comes of age. Semin Reprod Med 30:255–258
4. Kan YW, Dozy AM (1978) Antenatal diagnosis of sickle-cell anaemia by D.N.A. analysis of amniotic-fluid cells. Lancet 2:910–912
5. Xu K, Shi ZM, Veeck LL, Hughes MR, Rosenwaks Z (1999) First unaffected pregnancy using preimplantation genetic diagnosis for sickle cell anemia. JAMA 281:1701–1706
6. Handyside AH, Lesko JG, Tarin JJ, Winston RM, Hughes MR (1992) Birth of a normal girl after in vitro fertilization and preimplantation diagnostic testing for cystic fibrosis. New Engl J Med 327:905–909
7. Findlay I, Ray P, Quirke P, Rutherford A, Lilford R (1995) Allelic drop-out and preferential amplification in single cells and human blastomeres: implications for preimplantation diagnosis of sex and cystic fibrosis. Hum Reprod 10:1609–1618
8. Sermon K, De Vos A, Van de Velde H, Seneca S, Lissens W, Joris H et al (1998) Fluorescent PCR and automated fragment analysis for the clinical application of preimplantation genetic diagnosis of myotonic dystrophy (Steinert's disease). Mol Hum Reprod 4:791–796
9. Van de Velde H, Sermon K, De Vos A, Lissens W, Joris H, Vandervorst M et al (1999) Fluorescent PCR and automated fragment analysis in preimplantation genetic diagnosis for 21-hydroxylase deficiency in congenital adrenal hyperplasia. Mol Hum Reprod 5:691–696

10. Blake DL, Dean NL, Knight C, Tan SL, Ao A (2001) Direct comparison of detection systems used for the development of single-cell genetic tests in preimplantation genetic diagnosis. J Assist Reprod Genet 18:557–565
11. Fiorentino F, Magli MC, Podini D, Ferraretti AP, Nuccitelli A, Vitale N et al (2003) The minisequencing method: an alternative strategy for preimplantation genetic diagnosis of single gene disorders. Mol Hum Reprod 9:399–410
12. Dreesen JC, Jacobs LJ, Bras M, Herbergs J, Dumoulin JC, Geraedts JP et al (2000) Multiplex PCR of polymorphic markers flanking the CFTR gene; a general approach for preimplantation genetic diagnosis of cystic fibrosis. Mol Hum Reprod 6:391–396
13. Sermon K, Goossens V, Seneca S, Lissens W, De Vos A, Vandervorst M et al (1998) Preimplantation diagnosis for Huntington's disease (HD): clinical application and analysis of the HD expansion in affected embryos. Prenat Diagn 18:1427–1436
14. Offit K, Sagi M, Hurley K (2006) Preimplantation genetic diagnosis for cancer syndromes: a new challenge for preventive medicine. JAMA 296:2727–2730
15. Verlinsky Y, Rechitsky S, Schoolcraft W, Strom C, Kuliev A (2001) Preimplantation diagnosis for Fanconi anemia combined with HLA matching. JAMA 285:3130–3133
16. Verlinsky Y, Rechitsky S, Sharapova T, Morris R, Taranissi M, Kuliev A (2004) Preimplantation HLA testing. JAMA 291: 2079–2085
17. Fiorentino F, Biricik A, Karadayi H, Berkil H, Karlikaya G, Sertyel S et al (2004) Development and clinical application of a strategy for preimplantation genetic diagnosis of single gene disorders combined with HLA matching. Mol Hum Reprod 10:445–460
18. Kahraman S, Beyazyurek C, Ekmekci CG (2011) Seven years of experience of preimplantation HLA typing: a clinical overview of 327 cycles. Reprod Biomed Online 23:363–371
19. Angell RR, Aitken RJ, van Look PF, Lumsden MA, Templeton AA (1983) Chromosome abnormalities in human embryos after in vitro fertilization. Nature 303:336–338
20. Vanneste E, Voet T, Le Caignec C, Ampe M, Konings P, Melotte C et al (2009) Chromosome instability is common in human cleavage-stage embryos. Nat Med 15: 577–583
21. Lu S, Zong C, Fan W, Yang M, Li J, Chapman AR et al (2012) Probing meiotic recombination and aneuploidy of single sperm cells by whole-genome sequencing. Science 338:1627–1630
22. Hassold T, Hunt P (2009) Maternal age and chromosomally abnormal pregnancies: what we know and what we wish we knew. Curr Opin Pediatr 21:703–708
23. Pellestor F, Andreo B, Arnal F, Humeau C, Demaille J (2003) Maternal aging and chromosomal abnormalities: new data drawn from in vitro unfertilized human oocytes. Hum Genet 112:195–203
24. Langer-Safer PR, Levine M, Ward DC (1982) Immunological method for mapping genes on Drosophila polytene chromosomes. Proc Natl Acad Sci U S A 79:4381–4385
25. Griffin DK, Handyside AH, Penketh RJ, Winston RM, Delhanty JD (1991) Fluorescent in-situ hybridization to interphase nuclei of human preimplantation embryos with X and Y chromosome specific probes. Hum Reprod 6:101–105
26. Munne S, Lee A, Rosenwaks Z, Grifo J, Cohen J (1993) Diagnosis of major chromosome aneuploidies in human preimplantation embryos. Hum Reprod 8:2185–2191
27. Meyer LR, Klipstein S, Hazlett WD, Nasta T, Mangan P, Karande VC (2009) A prospective randomized controlled trial of preimplantation genetic screening in the "good prognosis" patient. Fertil Steril 91:1731–1738
28. Staessen C, Verpoest W, Donoso P, Haentjens P, Van der Elst J, Liebaers I et al (2008) Preimplantation genetic screening does not improve delivery rate in women under the age of 36 following single-embryo transfer. Hum Reprod 23:2818–2825
29. Schrock E, du Manoir S, Veldman T, Schoell B, Wienberg J, Ferguson-Smith MA et al (1996) Multicolor spectral karyotyping of human chromosomes. Science 273:494–497
30. Kallioniemi A, Kallioniemi OP, Sudar D, Rutovitz D, Gray JW, Waldman F et al (1992) Comparative genomic hybridization for molecular cytogenetic analysis of solid tumors. Science 258:818–821
31. Wilton L, Voullaire L, Sargeant P, Williamson R, McBain J (2003) Preimplantation aneuploidy screening using comparative genomic hybridization or fluorescence in situ hybridization of embryos from patients with recurrent implantation failure. Fertil Steril 80:860–868
32. Wells D, Escudero T, Levy B, Hirschhorn K, Delhanty JD, Munne S (2002) First clinical application of comparative genomic hybridization and polar body testing for preimplantation genetic diagnosis of aneuploidy. Fertil Steril 78:543–549
33. Sher G, Keskintepe L, Keskintepe M, Maassarani G, Tortoriello D, Brody S (2009)

Genetic analysis of human embryos by metaphase comparative genomic hybridization (mCGH) improves efficiency of IVF by increasing embryo implantation rate and reducing multiple pregnancies and spontaneous miscarriages. Fertil Steril 92:1886–1894

34. Gutierrez-Mateo C, Wells D, Benet J, Sanchez-Garcia JF, Bermudez MG, Belil I et al (2004) Reliability of comparative genomic hybridization to detect chromosome abnormalities in first polar bodies and metaphase II oocytes. Hum Reprod 19:2118–2125
35. Rius M, Obradors A, Daina G, Cuzzi J, Marques L, Calderon G et al (2010) Reliability of short comparative genomic hybridization in fibroblasts and blastomeres for a comprehensive aneuploidy screening: first clinical application. Hum Reprod 25:1824–1835
36. Hellani A, Abu-Amero K, Azouri J, El-Akoum S (2008) Successful pregnancies after application of array-comparative genomic hybridization in PGS-aneuploidy screening. Reprod Biomed Online 17:841–847
37. Fragouli E, Alfarawati S, Daphnis DD, Goodall NN, Mania A, Griffiths T et al (2011) Cytogenetic analysis of human blastocysts with the use of FISH, CGH and aCGH: scientific data and technical evaluation. Hum Reprod 26:480–490
38. Johnson DS, Gemelos G, Baner J, Ryan A, Cinnioglu C, Banjevic M et al (2010) Preclinical validation of a microarray method for full molecular karyotyping of blastomeres in a 24-h protocol. Hum Reprod 25: 1066–1075
39. Pinto D, Darvishi K, Shi X, Rajan D, Rigler D, Fitzgerald T et al (2011) Comprehensive assessment of array-based platforms and calling algorithms for detection of copy number variants. Nat Biotechnol 29:512–520
40. Treff NR, Tao X, Ferry KM, Su J, Taylor D, Scott RT Jr (2012) Development and validation of an accurate quantitative real-time polymerase chain reaction-based assay for human blastocyst comprehensive chromosomal aneuploidy screening. Fertil Steril 97:819–824
41. Cassel MJ, Munne S, Fung J, Weier HU (1997) Carrier-specific breakpoint-spanning DNA probes: an approach to preimplantation genetic diagnosis in interphase cells. Hum Reprod 12:2019–2027
42. National Institutes of Health and Institute of Molecular Medicine (1996) A complete set of human telomeric probes and their clinical application. Nat Genet 14:86–89
43. Iwarsson E, Ahrlund-Richter L, Inzunza J, Rosenlund B, Fridstrom M, Hillensjo T et al (1998) Preimplantation genetic diagnosis of a large pericentric inversion of chromosome 5. Mol Hum Reprod 4:719–723
44. Vanneste E, Melotte C, Voet T, Robberecht C, Debrock S, Pexsters A et al (2011) PGD for a complex chromosomal rearrangement by array comparative genomic hybridization. Hum Reprod 26:941–949
45. McArthur SJ, Leigh D, Marshall JT, de Boer KA, Jansen RP (2005) Pregnancies and live births after trophectoderm biopsy and preimplantation genetic testing of human blastocysts. Fertil Steril 84:1628–1636
46. Munne S, Weier HU, Grifo J, Cohen J (1994) Chromosome mosaicism in human embryos. Biol Reprod 51:373–379
47. Delhanty JD, Harper JC, Ao A, Handyside AH, Winston RM (1997) Multicolour FISH detects frequent chromosomal mosaicism and chaotic division in normal preimplantation embryos from fertile patients. Hum Genet 99:755–760
48. Magli MC, Jones GM, Gras L, Gianaroli L, Korman I, Trounson AO (2000) Chromosome mosaicism in day 3 aneuploid embryos that develop to morphologically normal blastocysts in vitro. Hum Reprod 15:1781–1786
49. Ziebe S, Lundin K, Loft A, Bergh C, Nyboe Andersen A, Selleskog U et al (2003) FISH analysis for chromosomes 13, 16, 18, 21, 22, X and Y in all blastomeres of IVF pre-embryos from 144 randomly selected donated human oocytes and impact on pre-embryo morphology. Hum Reprod 18:2575–2581
50. van Echten-Arends J, Mastenbroek S, Sikkema-Raddatz B, Korevaar JC, Heineman MJ, van der Veen F et al (2011) Chromosomal mosaicism in human preimplantation embryos: a systematic review. Hum Reprod Update 17:620–627
51. Mertzanidou A, Wilton L, Cheng J, Spits C, Vanneste E, Moreau Y et al (2013) Microarray analysis reveals abnormal chromosomal complements in over 70% of 14 normally developing human embryos. Hum Reprod 28:256–264
52. Baart EB, Martini E, van den Berg I, Macklon NS, Galjaard RJ, Fauser BC et al (2006) Preimplantation genetic screening reveals a high incidence of aneuploidy and mosaicism in embryos from young women undergoing IVF. Hum Reprod 21:223–233
53. Daphnis DD, Delhanty JD, Jerkovic S, Geyer J, Craft I, Harper JC (2005) Detailed FISH analysis of day 5 human embryos reveals the mechanisms leading to mosaic aneuploidy. Hum Reprod 20:129–137
54. Munne S, Velilla E, Colls P, Garcia Bermudez M, Vemuri MC, Steuerwald N et al (2005)

Self-correction of chromosomally abnormal embryos in culture and implications for stem cell production. Fertil Steril 84:1328–1334

55. Barbash-Hazan S, Frumkin T, Malcov M, Yaron Y, Cohen T, Azem F et al (2009) Preimplantation aneuploid embryos undergo self-correction in correlation with their developmental potential. Fertil Steril 92:890–896
56. Scott RT Jr, Upham KM, Forman EJ, Zhao T, Treff NR (2013) Cleavage-stage biopsy significantly impairs human embryonic implantation potential while blastocyst biopsy does not: a randomized and paired clinical trial. Fertil Steril 100:624–630
57. Goossens V, De Rycke M, De Vos A, Staessen C, Michiels A, Verpoest W et al (2008) Diagnostic efficiency, embryonic development and clinical outcome after the biopsy of one or two blastomeres for preimplantation genetic diagnosis. Hum Reprod 23:481–492
58. Harton GL, Munne S, Surrey M, Grifo J, Kaplan B, McCulloh DH et al (2013) Diminished effect of maternal age on implantation after preimplantation genetic diagnosis with array comparative genomic hybridization. Fertil Steril 100(6):1695–1703
59. Dekel-Naftali M, Aviram-Goldring A, Litmanovitch T, Shamash J, Yonath H, Hourvitz A et al (2013) Chromosomal integrity of human preimplantation embryos at different days post fertilization. J Assist Reprod Genet 30:633–648
60. Evsikov S, Verlinsky Y (1998) Mosaicism in the inner cell mass of human blastocysts. Hum Reprod 13:3151–3155
61. Bielanska M, Jin S, Bernier M, Tan SL, Ao A (2005) Diploid-aneuploid mosaicism in human embryos cultured to the blastocyst stage. Fertil Steril 84:336–342
62. Clouston HJ, Herbert M, Fenwick J, Murdoch AP, Wolstenholme J (2002) Cytogenetic analysis of human blastocysts. Prenat Diagn 22:1143–1152
63. Johnson DS, Cinnioglu C, Ross R, Filby A, Gemelos G, Hill M et al (2010) Comprehensive analysis of karyotypic mosaicism between trophectoderm and inner cell mass. Mol Hum Reprod 16:944–949
64. Mamas T, Gordon A, Brown A, Harper J, Sengupta S (2012) Detection of aneuploidy by array comparative genomic hybridization using cell lines to mimic a mosaic trophectoderm biopsy. Fertil Steril 97:943–947
65. Derhaag JG, Coonen E, Bras M, Bergers Janssen JM, Ignoul-Vanvuchelen R, Geraedts JP et al (2003) Chromosomally abnormal cells are not selected for the extra-embryonic compartment of the human preimplantation embryo at the blastocyst stage. Hum Reprod 18:2565–2574
66. Ruangvutilert P, Delhanty JD, Serhal P, Simopoulou M, Rodeck CH, Harper JC (2000) FISH analysis on day 5 post-insemination of human arrested and blastocyst stage embryos. Prenat Diagn 20:552–560
67. Verlinsky Y, Rechitsky S, Cieslak J, Ivakhnenko V, Wolf G, Lifchez A et al (1997) Preimplantation diagnosis of single gene disorders by two-step oocyte genetic analysis using first and second polar body. Biochem Mol Med 62:182–187
68. Forman EJ, Treff NR, Stevens JM, Garnsey HM, Katz-Jaffe MG, Scott RT Jr et al (2013) Embryos whose polar bodies contain isolated reciprocal chromosome aneuploidy are almost always euploid. Hum Reprod 28:502–508
69. Scott RT Jr, Treff NR, Stevens J, Forman EJ, Hong KH, Katz-Jaffe MG et al (2012) Delivery of a chromosomally normal child from an oocyte with reciprocal aneuploid polar bodies. J Assist Reprod Genet 29:533–537
70. Kuliev A, Zlatopolsky Z, Kirillova I, Spivakova J, Cieslak Janzen J (2011) Meiosis errors in over 20,000 oocytes studied in the practice of preimplantation aneuploidy testing. Reprod Biomed Online 22:2–8
71. Capalbo A, Bono S, Spizzichino L, Biricik A, Baldi M, Colamaria S et al (2013) Sequential comprehensive chromosome analysis on polar bodies, blastomeres and trophoblast: insights into female meiotic errors and chromosomal segregation in the preimplantation window of embryo development. Hum Reprod 28: 509–518
72. Christopikou D, Tsorva E, Economou K, Shelley P, Davies S, Mastrominas M et al (2013) Polar body analysis by array comparative genomic hybridization accurately predicts aneuploidies of maternal meiotic origin in cleavage stage embryos of women of advanced maternal age. Hum Reprod 28: 1426–1434
73. Warburton D (2005) Biological aging and the etiology of aneuploidy. Cytogenet Genome Res 111:266–272
74. Harper JC, Wilton L, Traeger-Synodinos J, Goossens V, Moutou C, SenGupta SB et al (2012) The ESHRE PGD Consortium: 10 years of data collection. Hum Reprod Update 18:234–247
75. Gianaroli L, Magli MC, Ferraretti AP, Munne S (1999) Preimplantation diagnosis for aneuploidies in patients undergoing in vitro fertilization with a poor prognosis: identification of the categories for which it should be proposed. Fertil Steril 72:837–844

76. Verlinsky Y, Cohen J, Munne S, Gianaroli L, Simpson JL, Ferraretti AP et al (2004) Over a decade of experience with preimplantation genetic diagnosis: a multicenter report. Fertil Steril 82:292–294
77. Mastenbroek S, Twisk M, van Echten-Arends J, Sikkema-Raddatz B, Korevaar JC, Verhoeve HR et al (2007) In vitro fertilization with preimplantation genetic screening. New Engl J Med 357:9–17
78. Cohen J, Grifo JA (2007) Multicentre trial of preimplantation genetic screening reported in the New England Journal of Medicine: an in-depth look at the findings. Reprod Biomed Online 15:365–366
79. Cohen J, Wells D, Munne S (2007) Removal of 2 cells from cleavage stage embryos is likely to reduce the efficacy of chromosomal tests that are used to enhance implantation rates. Fertil Steril 87:496–503
80. Mastenbroek S, Twisk M, van der Veen F, Repping S (2011) Preimplantation genetic screening: a systematic review and meta-analysis of RCTs. Hum Reprod Update 17:454–466
81. Practice Committee of Society for Assisted Reproductive T, Practice Committee of American Society for Reproductive M (2008) Preimplantation genetic testing: a Practice Committee opinion. Fertil Steril 90: S136–S143
82. Schoolcraft WB, Fragouli E, Stevens J, Munne S, Katz-Jaffe MG, Wells D (2010) Clinical application of comprehensive chromosomal screening at the blastocyst stage. Fertil Steril 94:1700–1706
83. Scott RT Jr, Upham KM, Forman EJ, Hong KH, Scott KL, Taylor D et al (2013) Blastocyst biopsy with comprehensive chromosome screening and fresh embryo transfer significantly increases in vitro fertilization implantation and delivery rates: a randomized controlled trial. Fertil Steril 100:697–703
84. Forman EJ, Hong KH, Ferry KM, Tao X, Taylor D, Levy B et al (2013) In vitro fertilization with single euploid blastocyst transfer: a randomized controlled trial. Fertil Steril 100:100–7 e1
85. Treff NR, Scott RT Jr (2013) Four-hour quantitative real-time polymerase chain reaction-based comprehensive chromosome screening and accumulating evidence of accuracy, safety, predictive value, and clinical efficacy. Fertil Steril 99:1049–1053
86. Lander ES (2011) Initial impact of the sequencing of the human genome. Nature 470:187–197
87. Handyside AH, Harton GL, Mariani B, Thornhill AR, Affara N, Shaw MA et al (2010) Karyomapping: a universal method for genome wide analysis of genetic disease based on mapping crossovers between parental haplotypes. J Med Genet 47:651–658
88. Zhang L, Cui X, Schmitt K, Hubert R, Navidi W, Arnheim N (1992) Whole genome amplification from a single cell: implications for genetic analysis. Proc Natl Acad Sci U S A 89:5847–5851
89. Telenius H, Carter NP, Bebb CE, Nordenskjold M, Ponder BA, Tunnacliffe A (1992) Degenerate oligonucleotide-primed PCR: general amplification of target DNA by a single degenerate primer. Genomics 13:718–725
90. Dean FB, Hosono S, Fang L, Wu X, Faruqi AF, Bray-Ward P et al (2002) Comprehensive human genome amplification using multiple displacement amplification. Proc Natl Acad Sci U S A 99:5261–5266
91. Ling J, Zhuang G, Tazon-Vega B, Zhang C, Cao B, Rosenwaks Z et al (2009) Evaluation of genome coverage and fidelity of multiple displacement amplification from single cells by SNP array. Mol Hum Reprod 15:739–747
92. van Uum CM, Stevens SJ, Dreesen JC, Drusedau M, Smeets HJ, Hollanders-Crombach B et al (2012) SNP array-based copy number and genotype analyses for preimplantation genetic diagnosis of human unbalanced translocations. Eur J Hum Genet 20:938–944
93. Lo YM, Chan KC, Sun H, Chen EZ, Jiang P, Lun FM et al (2010) Maternal plasma DNA sequencing reveals the genome-wide genetic and mutational profile of the fetus. Sci Transl Med 2:61ra91
94. Xu KPVA, Zhang CH, Jiang H, Zhang XQ, Rosenwaks Z (2011) Whole genome deep sequencing from single cells for preimplantation genetic diagnosis. Fertil Steril 96:S220
95. Yin X, Tan K, Vajta G, Jiang H, Tan Y, Zhang C et al (2013) Massively parallel sequencing for chromosomal abnormality testing in trophectoderm cells of human blastocysts. Biol Reprod 88:69
96. Gole J, Gore A, Richards A, Chiu YJ, Fung HL, Bushman D et al (2013) Massively parallel polymerase cloning and genome sequencing of single cells using nanoliter microwells. Nat Biotechnol 31:1126–1132
97. Green RC, Berg JS, Grody WW, Kalia SS, Korf BR, Martin CL et al (2013) ACMG recommendations for reporting of incidental findings in clinical exome and genome sequencing. Genet Med 15:565–574

98. Presidential Commission for the Study of Bioethical Issues (2012). Privacy and progress in whole genome sequencing. http://www.bioethics.gov
99. Nekkebroeck J, Bonduelle M, Desmyttere S, Van den Broeck W, Ponjaert-Kristoffersen I (2008) Mental and psychomotor development of 2-year-old children born after preimplantation genetic diagnosis/screening. Hum Reprod 23:1560–1566
100. Nekkebroeck J, Bonduelle M, Desmyttere S, Van den Broeck W, Ponjaert-Kristoffersen I (2008) Socio-emotional and language development of 2-year-old children born after PGD/PGS, and parental well-being. Hum Reprod 23:1849–1857
101. Palini S, Galluzzi L, De Stefani S, Bianchi M, Wells D, Magnani M et al (2013) Genomic DNA in human blastocoele fluid. Reprod Biomed Online 26:603–610
102. Tur-Kaspa I, Aljadeff G, Rechitsky S, Grotjan HE, Verlinsky Y (2010) PGD for all cystic fibrosis carrier couples: novel strategy for preventive medicine and cost analysis. Reprod Biomed Online 21: 186–195
103. Davis LB, Champion SJ, Fair SO, Baker VL, Garber AM (2010) A cost-benefit analysis of preimplantation genetic diagnosis for carrier couples of cystic fibrosis. Fertil Steril 93: 1793–1804

Chapter 23

Embryo Culture and Selection: Morphological Criteria

Aparna Hegde and Barry Behr

Abstract

In this chapter, we have outlined the various morphological criteria for selection of the best embryo at each important milestone encountered in the progress from the oocyte to the blastocyst. As Gerris et al. stated, a combination of one, two, or even three selection points should lead to a more accurate selection of the best embryo, as no one criterion is better than the other. An embryo that fails to meet the entire set of selection criteria must be avoided as culture cannot correct an impaired embryo.

Key words Embryo grading, Oocyte quality, Pronuclear morphology, Early cleavage, Day 2 scoring, Day 3 scoring, Day 5/6 scoring, Cumulative scoring

1 Introduction

The overarching goal of embryo culture in the use of assisted reproductive technology (ART) is the transfer of only one embryo. The choice of the "right one" has been a matter of considerable research and debate since the inception of in vitro fertilization [1]. However, the development of a comprehensive and definitive selection protocol for the most viable embryo with the best implantation potential is still elusive. Though biochemical methods for embryo selection have been described [2–4], they are time-consuming and impractical for most busy ART laboratories. Hence, the most widely used criteria for selecting the best embryos for transfer have been based on cell number and morphology [5]. Assessment of morphology is quick and imperfect but has been shown to have some predictive value [6]. All the morphological criteria, though, have limitations which principally speak to the fact that each of these criteria relies on a single static observation of the embryos. However, embryo development is a dynamic process with a very strict set of clocks governing division and initiation of key events from gene activation through compaction and blastulation [7]. This implies that optimal evaluation of an embryo's potential would necessitate multiple assessments of the embryo at each

Zev Rosenwaks and Paul M. Wassarman (eds.), *Human Fertility: Methods and Protocols*, Methods in Molecular Biology, vol. 1154, DOI 10.1007/978-1-4939-0659-8_23, © Springer Science+Business Media New York 2014

developmental checkpoint as it makes its transition from a single-celled zygote after fertilization of the oocyte to a blastocyst and from maternal to embryonic genome activation, initiation of protein synthesis, and cell differentiation. Thus, conducting non-invasive evaluation of the preimplantation period in a systematic, cumulative fashion should provide for the most predictive information regarding embryo quality [8].

1.1 Oocyte Quality (Day 0)

Greater than 50 % of all oocytes retrieved in human ART cycles have potential problems [9]. The developmental fate of the embryo is largely dictated by the quality of the oocyte [10] and there is increasing evidence that oocyte quality profoundly affects fertilization and subsequent development [11]. Sirard et al. described five levels of oocyte developmental competence: the ability to resume meiosis, to cleave upon fertilization, to develop into a blastocyst, to induce pregnancy, and to generate a healthy offspring [12]. Cytoplasmic changes accompanying oocyte growth, including mRNA transcription, protein translation, and posttranslational modification [13, 14], are necessary for meiotic maturation [15], activation of the zygotic genome [16], and blastocyst formation [17]. Any dysfunction or dislocation of oocyte components such as meiotic spindle, cortical granules or mitochondria can decrease oocyte viability and negatively impact embryo quality [18–20]. Disturbances or asynchrony in cytoplasmic [21–23] and nuclear maturation [24, 25] of the oocyte may both result in different morphological abnormalities. The causative factors could be reduced blood supply of the follicle during controlled hyperstimulation resulting in oxygen deficiency and reduced viability [26] or abnormal oocyte spindle or chromosomal defects [27]. This cytoplasmic defect can lead to impairment in the formation of the polar axes which are necessary for morphogenesis throughout the preimplantation and postimplantation stages [28]. Disproportionate segregation of mitochondria and proteins at the first cleavage division may also result in cleavage arrest or a limited implantation potential [4].

The use of morphological criteria for the assessment of oocyte quality may not be as precise as the use of cellular and molecular predictors of oocyte quality. It is also not possible to get any useful information on the day of follicle aspiration because cumulus cells impair evaluation in IVF oocytes. Historically, the evaluation of oocyte maturity in conventional IVF has been based on the expansion and radiance of the cumulus-corona complex [29]: expanded cumulus matrix and a “sun-burst” corona radiata denote preovulatory (metaphase II) oocytes, less-expanded complex denotes intermediate maturity and the absence of an expanded cumulus may indicate oocyte immaturity (prophase I). However, this method is unreliable in case of disparity of maturation process in oocyte and cumulus as can happen when there is a drop in estradiol levels during controlled hyperstimulation [30].

The various morphological parameters that help assess oocyte quality are:

1. Cumulus-oocyte complex: Compactness and thickness of the cumulus investment and brightness of the cytoplasm [31–33].
2. Cytoplasm: Granularity of the cytoplasm (large or small granules, homogenous or clustering distribution of granules in center or in periphery of oocyte), coloration, regions of organelle clustering (vacuoles, endoplasmic reticulum) [23, 34, 35].
3. Polar body: Shape (round or ovoid), size (large or small), surface (smooth or rough), cytoplasm (intact or fragmented) [25, 36].
4. Zona pellucida: Thickness and structure [37].
5. Perivitelline space: Size (normal or increased), the presence or absence of grain [21, 38].
6. Meiotic spindle: Location and reflection [39–41].

ICSI facilitates morphological assessment of the oocyte due to the inherent requirement for the removal of cumulus and corona cells [8]. Only oocytes at metaphase II are suitable for ICSI. The fertilization rate of oocytes at metaphase I is very low, and hence, they are used only when a few MII oocytes are available [42] Oocytes at prophase I and giant MII oocytes (which mostly result in digynic triploidy) should be discarded [43–45]. It has been recognized that MII oocytes of good morphology should have a clear, moderately granulate cytoplasm, a small perivitelline space, an intact first polar body, and a colorless zona pellucida [25, 38, 45].

1.1.1 Cytoplasm and Polar Body

Though oocytes of normal cytoplasmic structure have been found to undergo the best uterine implantation [35], De Sutter et al. [38], Balaban et al. [35], and Hammah et al. [46] did not find oocyte morphology to be correlated with fertilization rate or embryo quality. However, only a 3 % pregnancy rate has been reported with oocytes with cytoplasmic abnormalities as compared to 24 % in patients in which normal oocytes were used [34]. Poor pregnancy rates have also been reported in patients with oocytes with granular cytoplasm (12.8 %) [23]. Also a high frequency of aneuploidy has been found in the embryos arising from oocytes that had cytoplasmic abnormalities [23, 35, 45]. A higher rate of aneuploidy has been found in dysmorphic oocytes [23, 47, 48], and this could account for the significant decrease in preclinical pregnancy found in patients in whom none of the replaced embryos was dysmorphic in origin (20 % versus 58.3 % in oocytes with multiple dysmorphism) [49].

Based on the cytoplasm characteristics, Balaban et al. classified human oocytes into five categories:

1. Normal oocytes.
2. Oocytes with extracytoplasmic abnormalities (dark zona pellucida, large perivitelline space with debris, fragmentation of

the polar body and abnormal consistency of the oolemma and the zona pellucida).

3. Oocytes with intracytoplasmic abnormalities (dark or granular cytoplasm and cytoplasmic fragments).
4. Oocytes with shape abnormalities.
5. Oocytes with multiple abnormalities [35].

1.1.2 Polar Body

Studies on in vitro maturation of oocytes have suggested the prognostic relevance of first polar body morphology [50]. It is thought that ageing in vivo in MII before aspiration may lead to degeneration of the first polar body, and that the resultant overmaturity of such oocytes may further contribute to a diminished developmental potential [24, 51].

Ebner et al. have proved that an intact first polar body showing a smooth surface has a positive prognostic value in terms of fertilization and embryo quality [25, 51]. Xia et al. [45] showed that oocytes with a fragmented first polar body developed more poorly (55.1 %) after fertilization than those with a normal polar body (60.3 %). An enlarged polar body has also been found to be related to poorer rates of fertilization, cleavage, and top-quality embryos [52]. Ciotti et al. [36], however, found that identification of first polar body fragmentation did not seem to interfere with ICSI outcomes.

1.1.3 Perivitelline Space and Zona Pellucida

The characteristics of the perivitelline space that have been found to be associated with poor outcomes are a larger space (37.5 % poor development of the oocyte as compared to 60.3 % for normal sized space) [48] and large grains (59 % worse development after fertilization as compared to 71.1 % in those without grains [21]. Also oocytes with thickness of the zona pellucida less than 18.6 μm are fertilized best in vitro [53]. Bertrand et al. [54] suggested that presence of a thick zona pellucida (22 μm and thicker) could be an indication for intracytoplasmic sperm injection for infertile patients.

1.1.4 Meiotic Spindle

The meiotic spindle has a significant influence on the correct alignment of chromosomes in the oocyte and their segregation during meiosis. Parameters of the meiotic spindle (location and refraction) may be used to determine the quality of oocytes. Nowadays, meiotic spindle can be examined and its location can be determined by means of polarizing microscopy which avoids the damage confocal microscope causes [53].

The birefringence of meiotic spindle can be studied with the help of a PolScope microscope. It has been estimated that the oocytes with a birefringent spindle have higher developmental potential after fertilization in vitro or intracytoplasmic sperm injection than oocytes without a birefringent spindle [39, 40, 55, 56]. Moon et al. [10] estimated by PolScope that the location of meiotic

Table 1
Metaphase II oocyte morphology scoring system (MOMS) [58]

	Points
Extracytoplasmic features	
Abnormal polar body	2.0
Large perivitelline space	1.4
Cytoplasmic features	
Granular cytoplasm	1.4
Centrally located granular area	2.7
Vacuoles	2.1

spindle could vary in an oocyte. Therefore, it can be damaged when "blind" intracytoplasmic sperm injection is made into human oocytes. However, no relationship was found between deviation of the meiotic spindle from the polar body within oocytes and oocyte developmental competence [39, 40]. Battaglia et al. [57] reported that the number of oocytes with spindle abnormalities (abnormal placement of tubulin) increased with increasing women's age (40 years and older). This abnormality is associated with displacement of one or more chromosomes from the metaphase plate during the second meiotic division and thus can cause aneuploidy [57]. The routine use of a PolScope in the IVF lab is challenging as temperature and orientation of the oocyte during the spindle observation can significantly affect the interpretation of the results.

Rienzi et al. derived a MII (metaphase II) oocyte morphological score (MOMS) by identifying relationships among oocyte appearance, fertilization status, PN score, and day-2 embryo quality [58]. Morphological examination of 1,191 metaphase II oocytes showed that presence of vacuoles, abnormal first polar body and large perivitelline space were related to lower fertilization rate. Pronuclear morphology was adversely affected by the presence of a large perivitelline space, diffuse cytoplasmic granularity, and/or centrally located granular area. The latter characteristic was also found to be negatively related to day-2 embryo quality. According to the odds ratios obtained for each oocyte morphotype, an MII oocyte morphologic score (MOMS) was calculated (Table 1). A significant relationship was found between MOMS and female age, female basal follicle-stimulating hormone (FSH), and clinical outcome.

1.2 Zygote Quality (Day 1)

It has been suggested that major abnormalities of pronuclear development (i.e., pronuclei of unequal sizes, at a distance or not centrally located within the zygote cytoplasm) are incompatible with normally progressive development [59]. Asynchrony in the formation and polarization of pronuclear nucleoli (the active sites

Table 2
Definition of presumptive abnormal patterns of pronuclear stage morphology: Tesarik and Greco score [61]

Pattern	Description
1	Big difference (>3) in the number of NPB in both pronuclei
2	Small number (<7) of NPB without polarization in at least one pronucleus
3	Large number (>7) of NPB without polarization in at least one pronucleus
4	Very small number (<3) of NPB in at least one pronucleus
5	Polarized distribution of NPB in one pronucleus and non-polarized in the other

NPB nucleolar precursor bodies

of rRNA synthesis) has been described to severely impair further development of the preimplantation embryo [60–62] (Table 2). Conversely, good-quality embryos were found to arise from oocytes that had more uniform timing from injection in ICSI to pronuclear abuttal [63]. A number of studies have confirmed that zygotes exhibiting unequal pronuclear sizes are associated with an increased risk of embryonic anomalies and also mosaicism [64–66]. The sperm derived centriole and associated microtubule organizing region function to reposition the pronuclei into apposition [67–69] and visual changes in the pronuclei may thus correlate with functional abnormalities in chromosome segregation [59]. The resetting of a new axis after fertilization is governed by cytoplasmic contraction waves organized by the sperm centrosome [70]. When the cytoplasmic waves are shorter, embryos are unable to achieve optimal pronuclear rotation [63] leading to poor morphology, for example uneven cleavage or fragmentation [71]. Nucleolar precursor body (NPB) alignment has been related to the chromatin rotation that occurs in developing pronuclei [63, 72]. Chromatin rotation plays an important role in establishing embryonic axis, an essential part of cell determination for preimplantation embryos, and therefore, abnormalities in nucleolar precursor body alignment may potentially have severe consequences [70].

1.2.1 Pronuclear Scoring Systems

A number of non-invasive pronuclear scoring systems have been proposed by different ART laboratories which rely upon one static observation of simple morphological parameters under phase microscopy utilizing either Hoffman modulation contrast (HMC) or differential-interference contrast (DIC) optics. It is routinely performed at 16–18 h after insemination with both standard IVF and ICSI. All pronuclear scoring systems attempt to classify zygotes based on the following characteristics of the two pronuclei: symmetry (equal versus unequal), position (in apposition versus at a distance), and location (central versus non-central).

The nucleoli are scored based on number (3–7), symmetry (equal versus unequal sizes), and location (polarized or aligned versus non-polarized or non-aligned). Other parameters that may be factored in are polar body morphology and alignment, cytoplasmic morphology (presence or absence of halo), and finally a second embryo evaluation at 25–27 h post insemination for pronuclear morphology (presence versus absence of pronuclei) and/or early cell division [59]. There are many studies that attempt to relate these parameters with IVF outcome.

Scott and Smith were the first to propose a scoring system for assessing pronuclear morphology in 1998 [73] which they simplified later, allowing for a single observation to be performed and eliminating the scoring of entry into the first cleavage division [62, 74].

In 2000, Scott et al. introduced a revised score which included only parameters that could be assessed at the time of fertilization assessment (17–18 h post insemination) classifying the zygotes into five basic groups denoted as z score [62]. The five groups were further condensed into four basic categories (*see* Fig. 1) [74]: the Z-1 zygotes

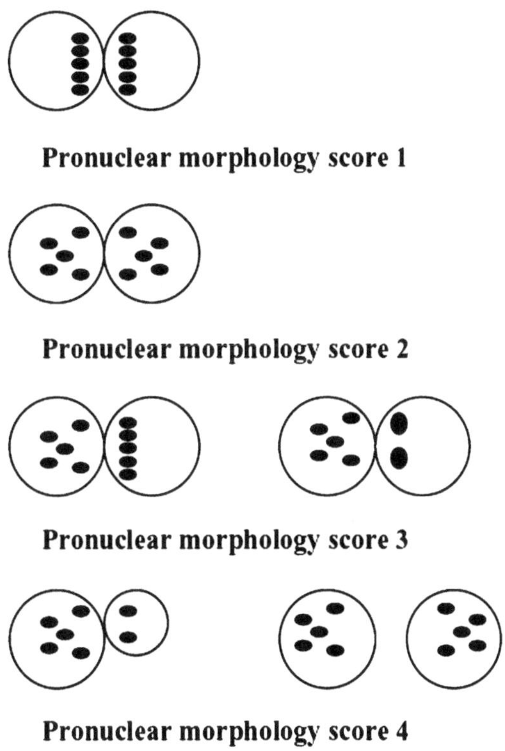

Fig. 1 Representation of the revised Pronuclear Morphology Score (PNMS) by Scott [91] describing the size and location of pronuclei and nucleoli. PNMS 1, 2, 3 and 4 are also referred to as Z1, Z2, Z3, and Z4 respectively. PNMS 1 and 2 are considered normal patterns. PNMS 2 will have equal pronuclear sizes, and PNMS 4 (*left*) will have unequal pronuclear sizes [184]

exhibited polarized NPBs, with each PN having the same number; Z-2 zygotes exhibited equal numbers and sizes of NPBs which were equally scattered in the two nuclei; Z-3 zygotes had either equal numbers of NPBs in each PN but with polarization only evident in one, or unequal numbers or sizes of NPBs between the two PNs; Z-4 zygotes had unequal numbers of NPBs with or without PN alignment.

Scott et al. used the revised score to show that zygotes displaying equality between the nuclei had 49.5 % blastocyst formation and those with unequal sizes, numbers, or distribution of nucleoli had 28 % blastocyst formation [62]. They also showed a 1.6-fold in implantation rates when the *z*-scoring system was included in embryo selection [74]. In recent years they have asserted that the ratio of the NPBs per nucleus is also very important for continued viability of the embryos, but this parameter has not yet been included in the scoring system [75]. Several other studies have refined or modified the concept of inter-pronuclear synchrony, NPB number, and polarity [59, 61, 72, 76–79].

Kahraman et al. evaluated the relationship between pronuclei morphology scoring (PNMS) and the chromosomal complement of embryos in couples with severe male infertility undergoing ICSI [80]. Pre-embryos were classified into eight categories based on the alignment, size, linear or irregular distribution of pronuclear bodies (PNB), position and clarity of cytoplasmic halo, and abutting of the pronucleus. These categories were subdivided into groups I and II according to the similarity and distribution of PNB. They found that more group II pre-embryos with markedly different morphology from group I were formed after ICSI with testicular sperm than with fresh ejaculated sperm (32.1 versus 22.7 %). The rate of chromosomal abnormality was higher in embryos developed from group II pre-embryos (52.2 %) than in embryos developed from group I prezygotes (37.6 %) and the rate of blastocyst formation was lower (50.45 vs 28.2 %). Thus, they proved that pronuclear morphology was impacted upon by the sperm source in ICSI.

Gianaroli et al. proposed another PNMS system based on three parameters, the location and position of pronuclei (A through E), size and polarity of nucleoli (1 through 4), and pronuclear orientation in relation to the second polar body (α, β, and γ) [81]. Upon completion of chromosome analysis, it was determined that four configurations had a greater chance of producing euploid embryos (A1α, A2β, A3β, and A3α) which correlated with better embryonic development [59].

1.2.2 Cytoplasmic Appearance at Pronuclear Stage

In humans and other mammals, immediately prior to pronuclear formation, there is a microtubule-mediated withdrawal of the mitochondria and other cytoplasmic components from the cortex towards the center of the oocyte [82] leaving a clear halo [63]. The physiological relevance of this halo, which is seen in two-thirds of all oocytes, is not clear. No agreement currently exists regarding the utility of using the halo as a marker of embryo quality [83]. Salumets et al. found that presence of halo correlates significantly with embryo quality but not with outcome [84]. Others reported increased blastocyst formation [10] and implantation [4]. Stalf et al. observed a significantly higher pregnancy rate (44 %) in halo-positive women compared with those who were halo-negative (28.3 %) [4]. Interestingly, Zollner et al. found that a halo of extreme dimensions might have a detrimental effect on blastocyst development [78]. More studies are required before any conclusion can be reached regarding the suitability of assessing for cytoplasmic halo as a non-invasive test regarding the embryo quality on day 1.

Evaluation of pronuclear morphology scoring systems by various ART facilities have had inconsistent results in their usefulness in predicting embryonic implantation potential [59]. This is probably due to the multifactorial situations in which nearly all of these studies were performed, namely the use of different culture media, stimulation protocols and differences in timing of fertilization assessments [73]. Given the dynamic nature of PN formation (Tesarik and Kopecny [61]), including NPB distribution, migration, coalescence, and dissolution, the timing of evaluation is of critical importance [zygotes formed from ICSI reveal their PNs approximately 4 h earlier than those formed by routine insemination [83, 85]. Accurate assessment of the three-dimensional disposition and number of NPBs is difficult as it requires the visual memorization of the spatial organization of the structures through multiple focal planes and this memorization must be done rapidly to avoid prolonged exposure of the zygote to light, and to temperature and pH shifts [83].

Pronuclear scoring systems have been shown to have strong correlations with blastocyst development potential [62, 74, 86] and implantation rates [61, 62, 73, 77, 86–91]. Chromosomal status has also been found to be associated with zygote pronuclear morphology, an increased prevalence of genetically abnormal embryos has been observed when the pronuclear patterns are abnormal [66, 80, 86, 92, 93]. The incidence of chromosomally normal embryos is highest with equal-sized nuclei and equal numbers of NPB that are aligned [80, 81, 93–95] Gamiz et al. showed a collation with age, with the PN score positively correlating with chromosomal normality only in patients <37 years of age [94]. Scott et al. reported that when only Z3 derived blastocysts were available for transfer on day 5, there were no pregnancies [62].

Indubitably, pronuclear scoring is a rapid, simple, early, noninvasive selection technique requiring one static observation of simple morphological parameters at 16–18 h post insemination that enables the laboratory to separate zygotes with implantation potential from those that clearly have a fundamental defect at a molecular and/or cellular level [10]. However, no standardized scoring system for zygote grading is currently in use. Hence, comparison of success rates between laboratories while controlling for embryo quality is a challenge [59]. Pronuclear morphology is also very fluid and there is considerable biological variation within normal development with different numbers and dimensions of nucleoli being compatible with development. The scoring is more time consuming than the cursory assessment of the zygote that is currently done for determination of fertilization. Further refinement of the scoring system is, however, occurring to allow for more accurate assessment of the NPBs in the nuclei in terms of number, size, and location which is showing correlations with both embryo morphology and implantation and delivery [75].

1.2.3 Time of First Cleavage

According to Bavister et al. ascertainment of strict developmental time points is one of the most critical factors for embryo selection [96]. Final oocyte maturation is triggered by the LH surge, a time point which should be used to temporally assess all the ensuing embryologic developmental hallmarks [7]. Strangely this time point is seldom used when assessing embryos. Any delay in a developmental event will result in the embryo being out of synchrony with the so-called oocyte clock [10]. A 4-cell embryo scored in the morning of day 2 is definitely not the same as one that was scored as 4-cell in the afternoon [3]. The first cleavage to the two-cell stage at 24–27 h after insemination or microinjection has been shown to be one such critical time point for selecting embryos for transfer [97, 98]. Early entry into the first mitotic division accounts for early completion of the final events of fertilization, alignment of the pronuclei, alignment of the chromosomes on the metaphase spindle, and finally the first mitotic division. Also early cleavage is a clearly visible event, whereas pronuclear morphology may vary during the dynamic process of syngamy [72]. Entry into first mitotic division by 22–24 h after insemination has been correlated with increased blastocyst formation [99, 100], increased implantation rates [98, 101–106], increased pregnancy rates [84, 97–99, 105, 106] and presence of a euploid set of chromosomes [103, 107]. Significantly, in the original embryo scoring system described by Scott and Smith [73], embryos that had already cleaved to the 2-cell stage by 25–26 h post insemination were assigned an additional score of 10. Terriou et al. [108] found that even early first cleavage was strongly associated with good embryo morphology on day 1 and multivariate analysis demonstrated that early embryo cleavage and embryo score on day had strong complementary predictive value for pregnancy.

Interestingly, first cleavage can occur earlier in ICSI zygotes compared to those obtained by IVF [103]. This is because the direct injection of a spermatozoan bypasses most of the fertilization steps and results in a shorter fertilization time [85]. Hence, Lundin et al. suggest that it may be better either to inseminate oocytes earlier or screen IVF zygotes later in order to obtain comparable data [103]. However, until more studies are available on ICSI embryos, checking embryos on day 1, at 22–24 h after insemination can be considered a simple and non-invasive procedure that may help to select embryos in which the time clock is advanced or at least not delayed [10].

1.3 Day 2 Scoring: Multinucleation and Blastomere Morphology

In the initial years of ART, grading systems for day 2 embryo based on the number (representative of cleavage speed), shape of the blastomeres, and the percentage of fragmentation were introduced [4, 29] as oocyte quality could not be estimated and there were no day 1 scoring systems available. Many studies have demonstrated a positive association of transfer of high-quality embryos as determined by the various scores with improved developmental potential, implantation rate, and pregnancy rate [8]. However, a major problem with most of these studies is that the embryos were grouped, and hence, it was not possible to determine which embryo actually implanted [4, 109]. To circumvent this problem, Guerif et al. [109] cultured more than 4,000 embryos individually in microdrops and sequentially evaluated them from day 1 to day 6 and found that early cleavage and cell number on day 2 were the most powerful parameters to predict the development of a good morphology blastocyst on day 5. The grading system used by Guerif et al. [109] was as follows:

Blastomeres were classified into three different groups based on cell number: <4, 4, and >4 cells. The degree of fragmentation was expressed as a percentage of the total oocyte volume occupied by anucleate cytoplasmic fragments. The rate of fragmentation was scored as follows: <20 % of the volume of the embryo, between 20 and 50 % of the volume of the embryo, >50 % of the volume of the embryo. Embryos with one or more multinucleated blastomeres were excluded from extended embryo culture. Embryos with pattern 0, early cleavage, four regular blastomeres, <20 % of fragmentation, and no multinucleated blastomeres were classified as "top quality" [109].

There are various day 2 scoring systems that have been proposed (*see* Table 3). They are either exclusive day 2 scores [110–112] or include grading that can be done on either day 2 or 3 [5] or grading that includes parameters observed on both day 2 and 3 [1]. Van Royen's "top quality embryo" includes (a lack or absence of??) multinucleation on day 2 (1999) in addition to day 3 parameters. However, Scott et al. [75] demonstrated that day 2 morphology when combined with pronucleate morphology is a stronger predictor of implantation than day 3 morphology or the ability to achieve the blastocyst stage. Parameters that were most consistently correlated with no delivery were lack of pronuclear symmetry and abnormal day 2 morphology parameters; multinucleation and uneven cell size (defined as >20 % difference in size/volume of the cells; [75].

Table 3
Classification of fragmentation as proposed by Alikani et al. [83]

Pattern	Description	Degree	% of fragmentation
Type 1	Minimal in volume, and fragments are associated with one blastomere	W1	0–5 %
Type 2	Localized fragments predominantly occupying the perivitelline space	W2	6–15 %
Type 3	Small, scattered fragments may be in the cleavage cavity or peripherally positioned	W3	16–25 %
Type 4	Large fragments distributed through the embryonic mass and are associated with asymmetric cells	W4	26–355
Type 5	Fragments appear neurotic, with characteristic granularity and cytoplasmic contraction within the intact blastomeres	W5	>35 %

1.3.1 Multinucleation

The larger dimensions of day 2 cells and also their better optical accessibility (less overlap) due to the smaller number of cells make day 2 an ideal time frame for nuclear observations. 30 % of embryos with multinucleation in the 2-cell stage do not show multinucleation in the 3–8 cell stage [113]. Multinucleation is a fairly common phenomenon and was found in 79.4 % and 74 % of all cycles by Van Royen et al. [114] and Jackson et al. [115], respectively. The observed multinucleation is actually an underestimation of the actual rate as the percentages reported in literature only reflect multinucleation at the interphase stage when the nucleus is visible [114].

Day 2 state of nucleation has proven to be highly significant in terms of fetal development and delivery [75].

Multinucleation indicates a breakdown of one or more cellular events. Multinucleation occurring during the first mitotic event is most likely due to chromosome segregation error and/or mitotic error in the first cleavage [10]. Kligman et al. [116] found that more than 70 % of embryos displaying multinucleation had aneuploidy when analyzed by FISH. The occurrence of multinucleation has been correlated with an increased rate of aneuploidy and chromosomal abnormalities [65, 116–118]. Multinucleation occurring in embryos while proceeding through the second and third mitotic event may be due to presence of karyokinesis without cytokinesis, accounting for about 30 % of multinucleated blastomeres or due to errors in chromosome segregation or other mitotic errors [117]. Van Royen et al. [114] hypothesized that multinucleation is due to developmental failure of the oocyte as they found that the incidence of multinucleation was positively correlated with factors such as shorter than average stimulations, higher than-average number of oocytes retrieved and higher than average FSH dose for stimulation. Multinucleation on day 2 has been linked to reduced in vitro development [107, 119, 120] and lower implantation rate [107, 120–123]. Multinucleation screening on day 2 coupled with day 3 morphology has enabled the use of single embryo transfers without any decrease in pregnancy rates compared to the use of more than one embryo and a single selection criteria [121, 123, 124]. Also they found that there is a significant relationship between multinucleation and other negative morphological characteristics of early cleaving embryos like fragmentation and cleavage rate and the lowest incidence of multinucleation coincides with minimal fragmentation and optimal cleavage rate [114].

1.4 Blastomere Morphology

In the 4-cell stage, both the tetrahedron shape of the embryo and the need for all the cells to be intact is a necessary part of normal embryonic development [10]. The second polar body which remains attached to the embryo until the blastocyst stage, is adherent to only one cell and its progeny ends up on the outside of the blastocyst at the junction between the inner cell mass (ICM) and the trophectoderm and determines the embryonic axis [10]. Screening for abnormal day-2 embryos (i.e., 4-cell embryos without the correct orientation, size of blastomeres or number) is a simple selection criterion for embryos with the best implantation potential.

1.4.1 Cell Number

Embryos progress through preimplantation development along a predictable timeline, with 95 % estimated to reach the 2-cell stage by 33.2 h, the 4-cell stage by 49.0 h, the 8-cell stage by 64.8 h and the 16-cell stage 80.7 h after insemination [125]. Optimal cleavage rates exist and those embryos that cleave either too quickly or too slowly are associated with compromised development [111, 112, 119]. Four cell embryos, even when showing minor fragmentation, should be preferred for transfer to good-quality 2-cell embryos [111]. At the 42–44 h post-hCG stage, even number of cells (two and four) rather than uneven (three, five and greater) is associated with positive outcome [75]. The 2-cell embryo proceeds through an intermediate 3-cell stage before rapidly cleaving again to form a 4-cell tetrahedron embryo [126–128]. An even-sized 3-cell embryo or a 3-cell embryo that is not progressing quickly to a 4-cell on day 2, is most likely abnormal or has cytokinetic delays [75] and is less likely to result in a normal delivery.

Guerif et al. [109] also found that the blastocyst formation rate was higher for 4-cell embryos on day 2 compared to faster cleaving (5–8 cell) or slower cleaving (2–3 cell) embryos. Dokras et al. [129], Shapiro et al. [130], and Langley et al. [131] have also seconded the contention that the presence of greater than 4 cells on day 2 has a negative impact on blastocyst formation. It may be that some blastomeres in the faster cleaving embryos may actually be large anucleate fragments [112] or that accelerated division might be an indicator of developmental instabilities that would affect the embryo's ability to develop to a blastocyst [109]. Fast developing embryos have been reported to exhibit higher levels of aneuploidy compared to synchronous embryos [132].

1.5 Cell Size and Symmetry

Blastomere size and symmetry are significant scoring parameters that should be incorporated into any gated/sequential embryo selection system [75]. Asymmetry in an embryo with an even number of blastomeres most likely arises from an uneven distribution of proteins, mRNA and various organelles, including mitochondria, between the two sister cells. Asymmetry in an embryo having an uneven number of blastomeres (i.e., 5 or 7 cells) most likely reflects an asynchrony of cell division rather than an uneven distribution of cytoplasm [83]. It must be remembered though that in normal human development, there is a degree of cleavage asynchrony [10], resulting in 3-, 5-, and 7-cell embryos [8]. This is a normal phenomenon and allows the correct spatial arrangement of the cells in the embryo as it divides [126]. Thus, at any one point not all cells would be of the same size [8]. However, in an embryo there can be only three sizes from the 3 to 8 cell stage and any deviation from this rule will mean that the embryo is fragmented or that one blastomere has arrested [10]. Blastomere cell size has been linked to fragmentation [133] where the degree of fragmentation was correlated with the extent of unevenness of blastomeres. This is disputed by Scott et al. [75] as uneven cell size on day 2 in their study appeared to be more inherent to the embryo and related to cleavage rather than to fragmentation.

Asymmetry has been shown to be associated with poor development potential of embryos [107, 123, 133], a markedly reduced implantation rate [107, 111], and a higher incidence of aneuploidy [111]. No deliveries were recorded when transferred embryos presented with unequal cell size on day 2, regardless of day 1, 3, or 5 morphology in the study by Scott et al. [75]. In another group of patients in the same study, only 12 of 132 clinical pregnancies (9 %) were from the transfer of at least one embryo with uneven cell size on day 2.

The relationship of spatial arrangement of the blastomeres to developmental competency of the embryos has yet to be established though one study failed to show any significance [134].

1.6 Fragmentation

Fragmentation is distinct from normal "blebs" that occur transiently during cell division [135]. It refers to the presence of extracellular cytoplasmic fragments not associated with the blastomeres per se [83] but with abnormalities in the link between nuclear and cytoplasmic cell division [83] that may be associated with apoptosis [136, 137] or abnormalities in chromosomal segregation [83]. It has been speculated that the fragments may provide a trigger for apoptosis if a certain level of developmentally important proteins is eliminated from their polarized domains [135]. The discordance between nuclear and cytoplasmic cell division may arise from intrinsic problems within the embryo, and/or from developmental abnormalities caused by poor culture conditions [83].

1.6.1 Scoring Systems for Fragmentation

Fragmentation has many phenotypes described which differ in the sizes of the fragments described, percentage of the volume of the embryo occupied by fragments, and the distribution of this anomaly among the blastomeres [138]. The simple scoring system for fragmentation is based on the volume of the embryo occupied by fragments (e.g., Score 0 = 0 %; score 1 = <10 %; score 2 = 10–25 %; Score 3 = >25 %) [139]. Alikani et al. [138] proposed a more detailed classification system that took into account the size and location of the fragments relative to the size and position of the nucleated cell (*see* Table 4).

Fragmentation in the embryo is significant as it has been observed that the rate of blastocyst formation decreases significantly with fragmentation on day 2 [109, 119, 140].

Several hypotheses have been suggested to explain the detrimental effects of fragments on embryo development [109]. Van Blerkom et al. [141] suggested that fragments might physically impede cell–cell interactions, interfering with compaction, cavitation, and blastocyst formation. In addition, ultrastructural observations of degeneration in blastomeres adjacent to fragments suggest that fragments might release toxic substances and therefore damage nearby cells [138, 142]. Alternatively, fragments might also reduce the volume of cytoplasm and deplete the embryos of essential organelles or polarized domains [135]. Heavily fragmented embryos are known to present a higher rate of chromosome abnormalities, in particular mosaicism [65, 143, 144]. In the case of moderate fragmentation, it has been suggested that more than the occurrence of fragments per se, the different temporal or spatial patterns of fragmentation have a more profound effect on embryo development [49, 135, 141]. On the other hand, certain phenotypes of minor fragmentation may disappear during in- vitro culture, either by lysis or by resorption [107, 141], thereby providing for a high quality, stage-appropriate embryo on the day of transfer. Interestingly, Alikani et al. [138] showed that microsurgical removal of small fragments can improve implantation potential of embryos by restoring spatial relationship of cells within the embryo and prevention of secondary degeneration, thus mimicking normal embryonic development in the presence of minor fragmentation. This approach to morphology improvement is controversial.

1.6.2 Day 3 Scoring

Cell number (denoting cleavage rate), blastomere morphology and fragmentation on day 3 have been correlated with developmental potential and the many different day 3 scoring systems in existence

Table 4
Examples of various scoring systems for embryo selection [83]

Author	Name	Day of assessment	System type	Scoring range	Best embryo	Cell number	Frag.	Symm	Compaction /expansion	Cytoplasm features	Cleavage rate	Multi nucleation
Cummins et al. [5]	Embryo quality score (EQS)/ embryo development rating (EDR)	2 or 3	Score	1–4	Score 4		+	+		+	+	
Puissant et al. [110]	Embryo scoring	2	Grade	1–4	Grade 4	+	+	+				
Veeck [29]	Morphological grading	3	Grade	I–IV	Grade I	+	+	+				
Steer et al. [145]	Cumulative embryo score (CES)	3	Grade	1–4	Grade 4	+	+	+				
Giorgetti et al. [111]	4 point embryo score	2	Point	1–4	4 point	+	+	+				
Ziebe et al. [112]	Embryo quality score	2	Symm or frag score with cell	Symm: 1.0–2.0 Frag: 2.1, 2.2, 3.0, 4.0	≥4 cell +2.1	+	+	+			+	
Van Royen et al. [1]	Top quality embryo	2 and 3	Score	Frag: A,B,C	Top Quality	+	+					+
Desai et al. [147]	Embryo quality store (D3EQ)	3	Score	1–10	Score 10	+	+	+	+/+	Pitting vacuole		

+ characteristic included in assessment system, *Frag* fragmentations, *Symm* symmetry have all been shown to have a correlation with implantation [145–148]. Positive association has been found between the number of cells in day 3 embryos (up to 8) with implantation rates following day 3 transfer [110, 145], rate of blastocyst formation [149] and pregnancy rates [149, 150] when compared to embryos with less than 8 cells.

Alikani et al. [119] also found that embryos with 7–9 cells on day 3 converted to blastocysts at a significantly higher rate than day 3 embryos with <7 cells or >9 cells [8]. Racowsky et al. [139] demonstrated that those embryos with exactly eight cells on day 3 had the highest implantation rates and the embryos with more than eight cells had a significantly lower implantation rate than those with eight cells (18.1 % versus 24.9 %). The later may be related to an increased incidence of aneuploidy [65]. The extent of fragmentation on day 3 has been found to be closely related to both implantation following day 3 transfer [139] and also the likelihood of progressing to blastocyst formation and subsequent implantation [83]. Pregnancies with greater than 50 % fragmentation have been found to have poor outcome [151].

Van Royen et al. [1, 123] defined a "top quality embryo" for transfer on day 3 using the following strict criteria, combining day 2 and 3 morphology: four or five blastomeres on day 2 and at least seven blastomeres on day 3 after fertilization, absence of multinucleated blastomeres and <20 % of fragments on day 2 and day 3 after fertilization. Gerris used the "top quality embryo" criteria and in a prospective randomized trial and found an implantation rate of 42.3 % and an ongoing pregnancy rate of 38.5 % for single top quality embryo transfers as compared to 48.1 % and 74 % respectively for double top quality embryo transfers [121]. However, the patients in the study were heavily selected: less than 25 % of patients during the time of the study had embryos of sufficient quality to have a single embryo transfer [3]. The same group [124] conducted a retrospective analysis of single embryo transfers using the strict top quality criteria and they found that the average number of embryos transferred decreased from 2.265 (in 1998) to 1.79 (in 2001) and the multiple pregnancy and twinning rates dropped from 33.6 and 29.5 % (in 1998) to 18.6 and 16.3 % (in 2001) respectively while still maintaining the average ongoing pregnancy rate (33.5 % per transfer).

1.6.3 Day 3–4 Scoring

Day 4 compacting/morula assessment has not been a part of standard approach to determination of embryo quality. However, two studies have examined the impact of embryo morphology parameters on day 4 such as compaction, degree of fragmentation, and cytoplasmic vacuolization on further embryonic development and contend that they could be useful markers for embryonic development [152, 153]. In addition to compaction, morula formation can also be a useful indicator of developmental competence [152, 154]. The degree of compaction on day 4 has been found to be associated with implantation potential [154], and hence, it is plausible that early compaction on day 3 may be an important positive predictor of embryonic development [83]. The day 3 score created by Desai et al. [147] included compaction as one of the parameters. Skiadas et al. [152] found that early compaction was significantly associated with implantation depending upon the degree of fragmentation: in optimal embryos on day 3 (≥ 8 cells and displaying <10 % fragmentation), early compaction was associated with a significantly higher implantation rate as compared to embryos with ≥10 % fragmentation.

A potentially informative scoring system for day 4 of embryonic development is the one suggested by Feil et al. [153]: Grade 1: early blastocyst, with cavitation or compacted embryo; Grade 2: grade 1 compacted morula with one or more morphological anomaly; Grade 3: partially compacted embryo with vacuoles or excessive fragmentation present, or embryo with eight cells or more and without any signs of compaction; grade 4: embryos with eight cells or more, with no signs of compaction and having vacuoles or excess fragments, or embryos with less than eight cells, and with no sign of compaction. In their study [153], Day 4 and 5 single embryo transfers resulted in similar ongoing pregnancy rates of 38.7 % and 32.1 %, respectively.

More studies are needed before this day 4 grading system can be accepted as a viable option to the accepted day 5/6 scoring system of Gardner et al. [155].

1.6.4 Day 5 or 6 Scoring of the Blastocyst

Embryo selection at either the pronucleate or cleavage stage may not be optimal as it can at best be considered an assessment of the oocyte and may result in transfer of embryos that are abnormal or arrest at later developmental stages [156–158]. There is also a paternal effect on development that is mainly evident after the 8-cell stage [159, 160]. A significant negative correlation exists between DNA damage in ejaculated spermatozoa and subsequent blastocyst development [161].

Consequently, extended culture till blastocyst stage and day 5 or 6 scoring of the blastocyst for the selection of the best embryo may be beneficial.

The blastocyst score needs to be sequential as blastocyst development is very dynamic and rapid. The blastocyst should have a defined blastocele, a distinct ICM protruding into the blastocele cavity, and a ring of evenly spaced and sized trophectoderm cells at approximately 154 h after hCG or 112–114 h after insemination [10].

If compaction, which normally begins at the 8–16 cell stage, starts too early, it can lead to the formation of a trophoblastic vesicle, where all the cells are allocated on the outside leaving none for the inner cell mass (ICM; [162]. As compaction progresses, refractile bodies in the cells begin to fuse in defined areas to form the blastocele. If these vesicles form too early, there is the appearance of a paving-stone formation which has a negative impact on blastocyst formation and implantation [10]. If the blastocysts are very expanded by this stage, they may have lowered viability [163]. Also blastocysts that have finger-like projections across the blastocele which do not break down with time, should be avoided [10].

Inner Cell Mass dimensions have also been found to be highly indicative of blastocyst implantation potential [164]. Richter et al. [164] found that day 5 expanded blastocysts with ICMs of >4,500 μm^2 implanted at a higher rate than those with smaller ICMs (55 % versus 31 %). Day 5 expanded blastocysts with slightly oval ICMs implanted at a higher rate (58 %) compared with those with either rounder ICMs (7 %) or more elongated ICMs (33 %). Implantation rates were highest (71 %) for embryos with both optimal ICM size and shape [164]. These criteria are not used routinely in most ART units.

A number of scoring systems for blastocysts have been proposed that take into account some of the above concerns and show correlations with implantation [129, 155, 165]. However, the score devised by Gardner et al. [155] is most commonly used and will be described here. It was determined in the mouse model that total cell number, ICM cell number and glycolysis had the strongest correlation with embryo viability. Blastocyst formation and hatching, however, were poorly correlated with pregnancy outcome [3]. An alphanumeric scoring system was therefore developed that takes into account three aspects of blastocyst morphology: degree of expansion, ICM development and trophectoderm development (*see* Fig. 2; [166]. In a retrospective analysis of the blastocyst score and subsequent implantation rates it was determined that when a patient received two top scoring blastocysts, i.e., 3AA or higher, an implantation rate of 70 % was attained [155]. Conversely, when a patient received slow or low scoring blastocysts, i.e., <3AA, the implantation rate fell to 28 % [155].

Interestingly, Richter et al. [164] found that blastocyst diameter and trophectoderm cell numbers were unrelated to implantation rates. In their study, pregnancy rates were higher for day 5 transfers of optimally shaped ICMs compared with day 5 transfers of optimally sized ICMs. They have suggested that this lack of association between trophectoderm cell numbers and implantation rates casts doubts on the appropriateness of inclusion of the trophectoderm grade in the scoring system defined by Gardner et al. [155]. However, this lack of association is surprising, given that it is the trophectoderm that forms the initial connection to the uterine wall and develops into the placenta and associated tissues supporting embryonic development [164]. It may be that the sample sizes used in the study were insufficient to detect existing

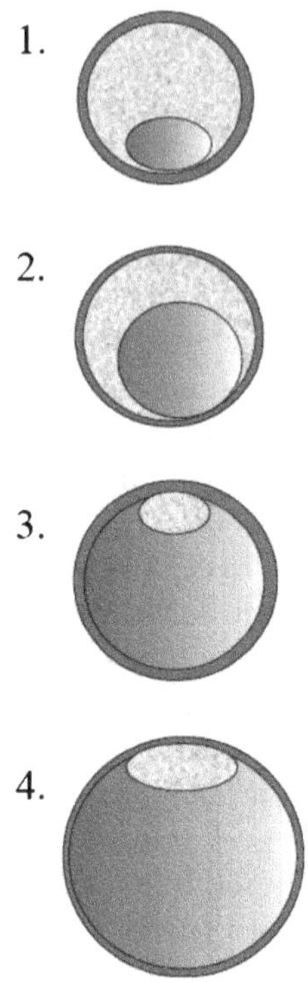

Fig. 2 Blastocyst scoring system used to select embryos for transfer from Gardner and Schoolcraft [166]. Initially blastocysts are given a numerical score from 1 to 6 based upon their degree of expansion and hatching status; (1) early blastocyst; the blastocele being less than half the volume of the embryo; (*2*) blastocyst; the blastocele being greater than or equal to half of the volume of the embryo; (*3*) full blastocyst; the blastocele completely fills the embryo; (*4*) expanded blastocyst; the blastocele volume is now larger than that of the early embryo and the zona is thinning; (*5*) hatching blastocyst; the trophectoderm has started to herniated though the zona; (*6*) hatched blastocyst; the blastocyst has completely escaped from the zona. The initial phase of the assessment can be performed on a stereo microscope. The second step in scoring the blastocysts should be performed on an inverted microscope. For blastocysts graded as 3–6 (i.e., full blastocysts onwards) the development of the inner cell mass (ICM) and trophectoderm can then be assessed; ICM Grading. (A) Tightly packed, many cells. (B) Loosely grouped, several cells. (C) Very few cells. Trophectoderm Grading. (C) Many cells forming a tightly knit epithelium. (B) Few cells. (C) Very few cells forming a loose epithelium

relationships. It is also possible that some as yet undetermined characteristic of the trophectoderm layer could be indicative of blastocyst viability and implantation potential [164].

1.6.5 Sequential Multiday Scoring Systems

Several numerical continuous scoring systems have been proposed in which multiple multiday parameters are used to select the best embryo as opposed to a single evaluation performed shortly before transfer.

Racowsky et al. [167] have classified these systems as follows: morphological observations and the allotment of scores [1, 111, 147, 168, 169]; the application of logistic regression analysis [108, 109, 170–172]; class probability tree analysis [173]; a case based reasoning system [174]; decision tree data mining [175] and automated pattern analysis [176].

The Graduated Embryo Score (GES) was proposed by Fisch et al. [177] to evaluate parameters from the first 3 days of development (pronuclear morphology, early cleavage and day 3 morphology) following a study using 1,245 embryos (*see* Table 4). Fisch et al. [177] found that embryos scoring 90–100 had 64 % blastocyst formation; 70–85 had 31 % blastocyst formation and 30–65 had 11 % blastocyst formation. In patients with at least one transferred embryo scoring ≥70, the pregnancy rate was 59 % compared with 34 % if all embryos scored <70. Among embryos scoring 70–100, an implantation rate of 39 % was seen, compared with 24 % among embryos scoring 0–65. In later studies they further validated their score and found prospectively that transfer of one or more embryos with GES ≥70 predicts pregnancy and implantation rates better than a single morphological evaluation on day 3 (168; (Table 5). They also found that day 3 transfers using GES and sHLA-G improves predictive accuracy of ART outcome allowing single embryo transfers, with age ≤37 years being an important qualifier [178].

Neuber et al. [179] examined morphological parameters of 1,550 individually cultured embryos, on all 5 days of development,

Table 5
Graduated Embryo Score of Fisch et al. [168]

Evaluation	Hours after insemination	Developmental milestone	Score
1	16–18	Nucleoli aligned along pronuclear axis	20
2	25–27	Cleavage regular and symmetrical Fragmentation[a] Absent <20 % >20 %	30 30 25 0
3	64–67	Cell number and grade[b] 7CI, 8CI, 8CII, 9CI 7CII, 9CII, 10CI, 11CI, Compacting I	20 10
Total score			100

[a]If the embryo was not cleaved at 25–27 h, grading of fragmentation should occur at the 64–67 h evaluation if the embryo reached the 7-cell stage and had <20 % fragmentation

[b]Grade I = symmetrical blastomeres and absent fragmentation. Grade II = slightly uneven blastomeres and <20 % fragmentation. Grade III = uneven blastomeres and >20 % fragmentation. Grade A embryos are seven or more cells with <20 % fragmentation

Fisch. Graduated embryo score (GES). Fertil Steril 2003

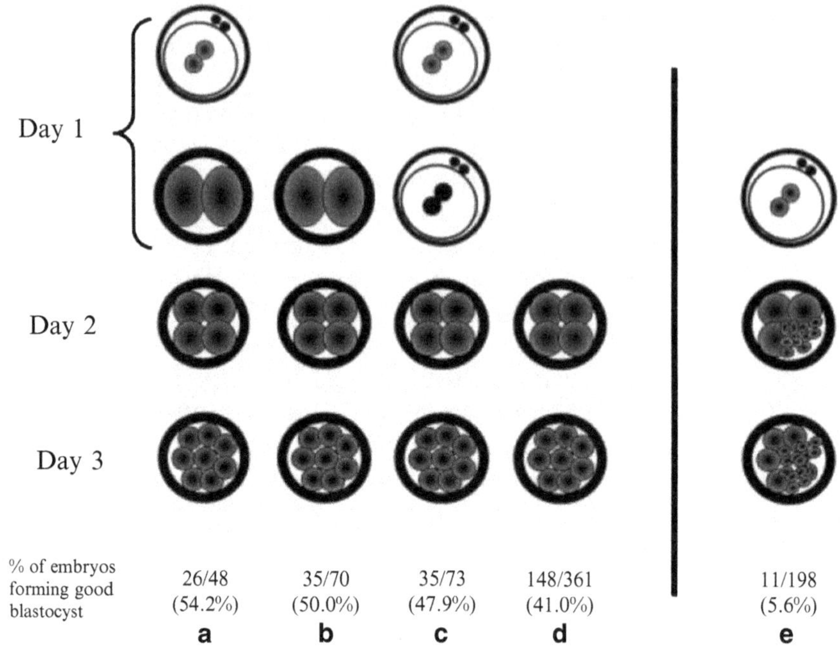

Fig. 3 Sequential assessment of individual embryos and their ability to reach the blastocyst stage on day 5. Percentage development to the blastocyst stage is calculated from the number of blastocysts divided by the number of embryos showing the sequential cleavage characteristics. (**a**) All embryos showing PN alignment, early cleavage, good quality >4- and >7-cell development. (**b**) All embryos showing early cleavage, good quality >4- and >7-cell development. (**c**) All embryos showing PN alignment, PN breakdown and good quality >4- and >7-cell development. (**d**) All embryos showing good quality >4- and >7-cell development only. (**e**) All embryos showing intact PN, poor quality >4- and >7-cell development [177]

and proposed a multistep Sequential Embryo Assessment scoring system. Figure 3 shows a number of scenarios for sequential assessment predicting blastocyst development [179]. They found a significant positive relationship between early-cleaving 2-cell embryos and subsequent good quality ≥4-cell, ≥7-cell and blastocyst development. Combining all parameters, they found that a developing embryo showing PN symmetry with early cleavage, and subsequent good ≥4-cell and ≥7-cell cleavage, has a one in two chance of developing into a good-quality blastocyst [179]. Neuber et al. [180] later developed a computer algorithm based on the sequential assessment tool which took into account quantifiable visual characteristics of multiple stages of development, beginning with pronuclear alignment, the length of the first cell cycle, and cleavage stages (cell number and morphological appearance) up to day 3 of insemination.

Rienzi et al. [181] used a combined pronuclear, cleavage stage and day 3 scoring system to state that in a selected population of good prognosis patients, implantation potential of day 3 and day 5 embryos is equal. Using a multiple-step scoring system, Rienzi

[41] later reported 77 % blastocyst formation on day 5 for embryos with normal pronuclear stage and early cleavage on day 1; 4- to 5-cells with equal blastomere size, <10 % fragmentation, and no multinucleation on day 2; and >6 cells with equal blastomere size, <10 % fragmentation, and no multinucleation on day 3. However, the proposed values that defined these scores were attributed quite arbitrarily, since they were only based on observed frequencies [109].

Gardner and Sakkas [3] have also proposed a multiday scoring system combining morphological parameters which evaluated development at 18–19 h (pronuclear, cytoplasm, NPB, and polar body morphology), 25–26 h (early cleavage and nuclear membrane breakdown), 42–44 h (multinucleation, cell number, and fragmentation), 66–68 h (multinucleation, cell number, and fragmentation), 94–96 h (compaction, signs of blastocele formation), and 106–108 h (trophectoderm, ICM, and blastocele cavity) post insemination/ICSI. The maximum score for a perfect embryo over 5 days would be 180. This score needs to be prospectively studied for its efficacy.

1.7 Key Issues Regarding Sequential Assessment

Though the idea of sequentially assessing embryos seems promising, conflicting results exist overall and there is currently no consensus on (1) the optimum day(s) for evaluation; (2) the optimum set of variables that should be used in a predictive model; (3) the statistical methods used for data analyses and interpretation; (4) the scoring system to be used for embryo selection; and (5) model performance benchmarks for future comparison. Answers to these questions are necessary so that quantitative comparisons can be made about the relative efficacies of the different evaluation protocols [83, 167].

There are several concerns regarding the various morphological grading systems:

1. Most studies evaluating embryo viability using pronuclear stage [61, 62], early cleaving 2-cell embryos [98, 102, 105] and culture until the blastocyst stage [155] have involved grouped embryos. Since the developmental fate of each embryo was not traceable to a viable implantation and there is no proof that the predicted top quality embryo was the one that implanted out of all the embryos transferred, the conclusions made must be interpreted with caution [83].
2. Early development especially the pronuclear stage is very dynamic and evaluation of early stage embryo requires testing to be accomplished within strict time windows. In a busy laboratory, adherence to such strict timelines is difficult. Moreover, evaluations are usually accomplished during regular work hours that may not coincide with the optimal or most predictive times to evaluate an embryo.

3. A "numerical scoring system" for selection is frequently based on assignment of seemingly arbitrarily weighted values [83] without multivariate analyses or Spearman's correlation being performed [181, 182]. And hence, it is plausible that the obtained score does not precisely predict the implantation potential. Also categorical grouping of continuous variables (for example, fragmentation) may lead to contradictory results.
4. Most of the studies on which the scoring systems are based are retrospective with very few reporting prospective assessment with demonstrated improvement in selection [105, 168]. Rigorous prospective testing of any specific algorithm is necessary to prove that its application does improve embryo selection [83].
5. Many studies have involved datasets involving both day 3 and day 5 transfers [75, 168], an approach likely to reduce utility of the analyses due to various confounders relating to patient selection bias, culture influences (e.g., probable improved overall quality of embryos cultured to day 5, possible loss of developmentally competent day 3 embryos not supported by extended culture conditions), and variances in uterine receptivity [83].
6. Studies based on multiple embryo transfers may introduce further confounding factors due to the possible inter-embryo cooperation/interaction whereby a poorer quality embryo may either increase or decrease the likelihood of a better quality embryo implanting, or, conversely a better quality embryo may enhance the independent implantation potential of one considered of poorer quality [83, 183].
7. Precision and consistency of scoring systems among embryologists (particularly for embryos of marginal quality) within and across laboratories must be taken into consideration [83, 182] as it may reduce the accuracy of the system.
8. The potential benefits of the multi-day scoring systems must be weighed against the possible detrimental effects of environmental perturbations (light exposure, temperature and pH shifts etc.) caused due to increased time taken for conducting the strict evaluations.

Racowsky et al. [167] have called into question the need for a sequential system of assessment. They assert that logically an embryo that shows normal development on day 3 is likely to have exhibited normal development earlier in culture [83]. They have claimed that unpublished preliminary data from their group, utilizing multivariate analysis of a dataset of single embryo transfers of embryos cultured individually, suggest that no additional benefit is accrued from early cleavage and day 2 assessments over that obtained exclusively from evaluation on day 3. They found that there is considerable overlap in the morphological appearance of those embryos that successfully implant versus those that fail. Guerif et al. [109] have also advocated the need to search for additional criteria, including the ability of the blastocyst to develop. They prospectively and sequentially evaluated (pronuclear morphology on day 1, early cleavage, cell number and fragmentation rate on day 2) in 4,042 individually grown embryos and found that the combination of all four parameters allowed the prediction of blastocyst development with an area under the receiver operating characteristics curve of 0.688, which represents a fairly low prediction of embryo viability.

However, it is plausible that early developmental insults resulting in morphological defects that can be seen on day 0 or 1 may not be visually apparent when the embryo is examined on day 3 or 5, but may manifest later in the form of lowered implantation potential and reduced pregnancy rate. More stringent studies comparing sequential assessment with single assessment are required before any definitive conclusions can be made.

2 Summary

A universally accepted, accurate, non-invasive, easy, simple, and quick grading system to select the best possible embryo with maximum implantation potential still eludes the field of assisted reproductive technology. A significant number of human preimplantation embryos undergo deviant development in vitro, failing to follow the expected normal developmental timeline. Hence, it would appear prudent to incorporate as much information gathered at the pronuclear and cleavage stages into the final assessment as possible. This realization has prompted the hypothesis that multiple evaluations through early preimplantation development may improve selection compared with a single evaluation, performed shortly before transfer. Such an evaluation would logically take into account every developmental hurdle in the preimplantation development period of the embryo, thus improving predictive accuracy. Though various scoring systems have been proposed, there is still debate over various parameters that form a part of these scoring systems, the optimal day(s) or time for evaluation, predictive accuracy of the various scores and inter-grader consistency of evaluation and also optimal research design to test the scores. The grouping of the embryos also clouds the reliability of the results of the various studies as it is not possible to determine the exact developmental fate of the selected "best embryo" [8].

In conclusion, a major part of the 'art" in assisted reproductive technology is selecting embryos for transfer [168]. The search for the perfect scoring system for determination of embryo quality continues. Until it is found, a universally accepted sequential scoring system that does a multifactorial analysis of various morphological parameters at each milestone of embryonic development should serve as a viable alternative.

References

1. Van Royen E et al (1999) Characterization of a top quality embryo, a step towards single-embryo transfer. Hum Reprod 14: 2345–2349
2. Mendoza C et al (2002) Follicular fluid markers of oocyte developmental potential. Hum Reprod 17:1017–1022
3. Gardner D et al (2003) Assessment of embryo viability: the ability to select a single embryo for transfer – a review. Placenta 24:S5–S12
4. Ebner T et al (2003) Selection based on morphological assessment of oocytes and embryos at different stages of preimplantation development: a review. Hum Reprod 9: 251–262
5. Cummins J et al (1986) A formula for scoring human embryo growth rates in vitro fertilization: its value in predicting pregnancy and in comparison with visual estimates of embryo quality. J In Vitro Fert Embryo Transf 3(5): 284–295
6. Sakkas D et al (2005) Noninvasive methods to assess embryo quality. Curr Opin Obstet Gynecol 17:283–288
7. Johnson M et al (2000) Egg timers: how is developmental time measured in the early vertebrate embryo? BioEssays 22:57–63
8. Behr B et al (2011) Cumulative morphological assessment of embryo quality. In: Agarwal A, Varghese A, Peter Nagy Z (eds) Practical manual of in vitro fertilization: advanced methods and novel devices. Humana, New York, NY
9. Vlaisavljevic V et al (2001) Is there any benefit from the culture of a single oocyte to a blastocyst-stage embryo in unstimulated cycles. Hum Reprod 16:3279–2383
10. Scott L (2003) The biological basis of non-invasive strategies for selection of human oocytes and embryos. Hum Reprod 9(3): 237–249
11. Wang Q et al (2007) Evaluation of oocyte quality: morphological, cellular and molecular predictors. Reprod Fertil Dev 2007(19):1–12
12. Sirard M et al (2006) Contribution of the oocyte to embryo quality. Theriogenology 65:126–136
13. Sirard M et al (1989) Timing of nuclear progression and protein synthesis necessary for meiotic maturation of bovine oocytes. Biol Reprod 40:1257–1263
14. Kastrop P et al (1991) Protein synthesis and phosphorylation patterns of bovine oocytes maturing in vivo. Mol Reprod Dev 29: 271–275
15. Hunter A et al (1987) Stage-dependent effects of inhibiting ribonucleic acids and protein synthesis on meiotic maturation of bovine oocytes In vitro. J Dairy Sci 70:1646–1651
16. Barnes F et al (1991) Embryonic transcription in in vitro cultured bovine embryos. Mol Reprod Dev 29:117–123
17. De Sousa P et al (1998) Temporal patterns of embryonic gene expression and their dependence on oogenetic factors. Theriogenology 49:115–128
18. Sun Q et al (2001) Translocation of active mitochondria during pig oocyte maturation, fertilization and early embryo development in vitro. Reproduction 122:155–163
19. Coticchio G et al (2004) What criteria for the definition of oocyte quality? Ann N Y Acad Sci 1034:132–144
20. Combelles C et al (2005) Assessment and optimization of oocyte quality during assisted reproductive technology treatment. Semin Reprod Med 23:277–284
21. Hassan-Ali H et al (1998) Perivitelline space granularity: a sign of human menopausal gonadotrophin overdose in intracytoplasmic sperm injection. Hum Reprod 13: 3425–3430
22. Loutradis D et al (1999) Oocyte morphology correlates with embryo quality and pregnancy rate after intracytoplasmic sperm injection. Fertil Steril 72:240–244
23. Kahraman S et al (2000) Relationship between granular cytoplasm of oocytes and pregnancy outcome following intracytoplasmic sperm injection. Hum Reprod 15:2390–2393
24. Eichenlaub-Ritter U et al (1995) Recurrent failure in polar body formation and premature chromosome condensation in oocytes from a human patient: indicators of asynchrony in

nuclear and cytoplasmic maturation. Hum Reprod 9:2343–2349
25. Ebner T et al (2000) Prognostic value of first polar body morphology on fertilization rate and embryo quality in intracytoplasmic sperm injection. Hum Reprod 15:427–430
26. Chui D et al (1997) Follicular vascularity – the predictive value of transvaginal power Doppler ultrasonography in an in-vitro fertilization programme: a preliminary study. Hum Reprod 12:191–196
27. Van Blerkom J et al (1997) The developmental potential of the human oocyte is related to the dissolved oxygen content of follicular fluid: association with vascular endothelial growth factor levels and perifollicular blood flow characteristics. Hum Reprod 12:1047–1055
28. Edwards R et al (2000) The role of embryonic polarities in preimplantation growth and implantation of mammalian embryos. Hum Reprod 15(Suppl 6):1–8
29. Veeck L (1990) The morphological assessment of human oocytes and early concepti. In: Keel BA, Webster BW (eds) Handbook of the laboratory diagnosis and treatment of infertility. CRC, Boca Raton, pp 353–369
30. Laufer N et al (1984) Asynchrony between human cumulus-corona cell complex and oocyte maturation after human menopausal gonadotropin treatment for in vitro fertilization. Fertil Steril 42:366–372
31. Blondin P et al (1995) Oocyte and follicular morphology as determining characteristics for developmental competence in bovine oocytes. Mol Reprod Dev 41(1):54–62
32. Warriach HM et al (2004) Thickness of cumulus cell layer is a significant factor in meiotic competence of buffalo oocytes. J Vet Sci 5(3):247–251
33. Nagano M et al (2006) ATP content and maturational/developmental ability of bovine oocytes with various cytoplasmic morphologies. Zygote 14(4):299–304
34. Serhal P et al (1997) Oocyte morphology predicts outcome of intracytoplasmic sperm injection. Hum Reprod 12:1267–1270
35. Balaban B et al (1998) Oocyte morphology does not affect fertilization rate, embryo quality and implantation rate after intracytoplasmic sperm injection. Hum Reprod 13(12): 3431–3433
36. Ciotti P et al (2004) First polar body morphology before ICSI is not related to embryo quality or pregnancy rate. Hum Reprod 19(10): 2334–2339
37. Gabrielsen A et al (2001) The impact of the zona pellucida thickness variation of human embryos on pregnancy outcome in relation to suboptimal embryo development. A prospective randomized controlled study. Hum Reprod 16(10):2166–2170
38. De Sutter P et al (1996) Oocyte morphology does not correlate with fertilization rate and embryo quality after intracytoplasmic sperm injection. Hum Reprod 11:595–597
39. Wang W et al (2001) Developmental ability of human oocytes with or without birefringent spindles imaged by Polscope before insemination. Hum Reprod 16:1464–1468
40. Moon J et al (2003) Visualization of the metaphase II meiotic spindle in living human oocytes using the Polscope enables the prediction of embryonic developmental competence after ICSI. Hum Reprod 18:817–820
41. Rienzi L (2005) Significance of morphological attributes of the early embryo. Reprod Biomed Online 10:669–681
42. De Vos A et al (1999) Invitro matured metaphase-I oocytes have a lower fertilization rate but similar embryo quality as mature metaphase-II oocytes after intracytoplasmic sperm injection. Hum Reprod 14:1859–1863
43. Balakier H et al (2002) Morphological and cytogenetic analysis of human giant oocytes and giant embryos. Hum Reprod 17: 2394–2401
44. Rosenbusch B et al (2002) Cytogenetic analysis of giant oocytes and zygotes to assess their relevance for the development of digynic triploidy. Hum Reprod 17:2388–2393
45. Xia P (1997) Intracytoplasmic sperm injection: correlation of oocyte grade based on polar body, perivitelline space and cytoplasmic inclusions with fertilization rate and embryo quality. Hum Reprod 12:1750–1755
46. Hammah S (2005) Oocyte and embryo quality: is their morphology a good criterion? J Gynecol Obstet Biol Reprod 34(7):5S38–5S41
47. Plachot M et al (1988) Chromosome analysis of ovocytes and human embryos collected after fertilization in vitro. A model of natural selection against aneuploidy. Rev Fr Gynecol Obstet 83(10):613–617
48. Van Blerkom J et al (1992) Oocyte dysmorphism and aneuploidy in meiotically – mature human oocytes after ovarian stimulation. Hum Reprod 7:379–390
49. Alikani M et al (1995) Intracytoplasmic sperm injection in dysmorphic human oocytes. Zygote 3:283–288
50. Mikkelson A et al (2001) Morphology of in-vitro matured oocytes: impact on fertility potential and embryo quality. Hum Reprod 16:1714–1718

51. Ebner T et al (1999) Elective embryo transfer selected on the basis of first polar body morphology is associated with increased rates of implantation and pregnancy. Fertil Steril 72: 599–603
52. Navarro P et al (2009) Relationship between first polar body morphology before intracytoplasmic sperm injection and fertilization rate, cleavage rate, and embryo quality. Int J Gynaecol Obstet 104(3):226–229
53. Lasiene K et al (2009) Morphological criteria of assessment of oocyte quality. Medicina (Kaunas) 45(7):505–515
54. Bertrand E et al (1995) Does zona pellucida thickness influence the fertilization rate? Hum Reprod 10(5):1189–1193
55. Shen Y et al (2006) Light retardance by human oocyte spindle is positively related to pronuclear score after ICSI. Reprod Biomed Online 12(6):737–751
56. Fang C et al (2007) Visualization of meiotic spindle and subsequent embryonic development in in vitro and in vivo matured human oocytes. J Assist Reprod Genet 24(11):495–499
57. Battaglia D et al (1996) Influence of maternal age on meiotic spindle in oocytes from naturally cycling women. Hum Reprod 11:2217–2222
58. Rienze L et al (2008) Significance of metaphase II human oocyte morphology on ICSI outcome. Fertil Steril 90(5):1692–1700
59. James A (2007) Human pronuclei as a mode of predicting viability. In: Elder K, Cohen J (eds) Human preimplantation embryo selection. Informa UK Ltd, London, pp 31–40
60. Van Blerkom J (1990) Occurrence and developmental consequences of abberant cellular organization in meiotically mature oocytes after exogeneous ovarian hyperstimulation. J Electron Microsc Tech 16:324–346
61. Tesarik J et al (1999) The probability of abnormal preimplantation development can be predicted by a single static observation on pronuclear stage morphology. Hum Reprod 14:1318–1323
62. Scott L et al (2000) The morphology of human pronuclear embryos is positively related to blastocyst development and implantation. Hum Reprod 15:2394–2403
63. Payne D et al (1997) Preliminary observations on polar body extrusion and pronuclear formation in human oocytes using time-lapse video cinematography. Hum Reprod 12:532–541
64. Manor D et al (1996) Undocumented embryos: do not trash them, FISH them. Hum Reprod 11:2502–2506
65. Munne A et al (1998) Chromosome abnormalities in human embryos. Hum Reprod Update 4:842–855
66. Sadowy S et al (1998) Impaired development of zygotes with uneven pronuclear size. Zygote 63:137–141
67. Wright G et al (1990) Observations on the morphology of pronuclei and nucleoli in human zygotes and implications for cryopreservation. Hum Reprod 5:109–115
68. Schatten G (1994) The centrosome and its mode of inheritance: the reduction of the centrosome during gametogenesis and its restoration during fertilization. Dev Biol 165: 299–335
69. Asch R et al (1995) The stages at which human fertilization arrests microtubule and chromosomal configurations in inseminated oocytes which failed to complete fertilization and development in humans. Hum Reprod 10:1897–1906
70. Edwards R et al (1997) Oocyte polarity and cell determination in early mammalian embryos. Mol Hum Reprod 3:863–905
71. Garello C et al (1999) Pronuclear orientation, polar body placement, and embryo quality after intracytoplasmic sperm injection and in-vitro fertilization: further evidence for polarity in human oocytes? Hum Reprod 14: 2588–2595
72. Tesarik J et al (1989) Development of human male pronucleus: ultrastructure and timing. Gamete Res 24:135–149
73. Scott L et al (1998) The successful use of pronuclear embryo transfers the day following oocyte retrieval. Hum Reprod 13:1003–1013
74. Scott L (2003) Pronuclear scoring as a predictor of embryo development. Reprod Biomed Online 6:57–60
75. Scott L et al (2007) Morphological parameters of early cleavage – stage embryos that correlate with fetal development and delivery: prospective and applied data for increased pregnancy rates. Hum Reprod 22(1):230–240
76. Montag M et al (2001) Evaluation of pronuclear morphology as the only selection criterion for further embryo culture and transfer: results of a prospective multicentre study. Hum Reprod 16:2384–2389
77. Ludwig M et al (2000) Clinical use of a pronuclear stage score following intracytoplasmic sperm injection: impact on pregnancy rates under the conditions of the German embryo protection law. Hum Reprod 15:325–329
78. Zollner U et al (2002) The use of a detailed zygote score after IVF/ICSI to obtain good quality blastocysts: the German experience. Hum Reprod 17:1327–1333
79. Senn A et al (2006) Morphological scoring of human pronuclear zygotes for prediction of pregnancy outcome. Hum Reprod 21:234–239

80. Kaharman S et al (2002) Pronuclear morphology scoring and chromosomal status of embryos in severe male infertility. Hum Reprod 17:3193–3200
81. Gianaroli L et al (2003) Pronuclear morphology scoring and chromosomal status of embryos in severe male infertility. Fertil Steril 80:341–349
82. Bavister B et al (2000) Mitochondrial distribution and function in oocytes and early embryos. Hum Reprod 15:189–198
83. Ceyhan S et al (2009) Biennial review of infertility. Humana, New York, pp 143–184
84. Salumets A et al (2001) The predictive value of pronuclear morphology of zygotes in the assessment of human embryo quality. Hum Reprod 16:2177–2181
85. Nagy Z et al (1998) Timing of oocyte activation, pronucleus formation, and cleavage in humans after intracytoplasmic sperm injection (ICSI) with testicular spermatozoa and after ICSI or invitro fertilization on sibling oocytes with ejaculated spermatozoa. Hum Reprod 13:1606–1612
86. Balaban B et al (2001) The effects of pronuclear morphology on embryo quality parameters and blastocyst transfer outcome. Hum Reprod 16:2357–2361
87. Ludwig M et al (2000) Experience with the elective transfer of two embryos under the conditions of the German embryo protection law: results of a retrospective data analysis of 2573 transfer cycles. Hum Reprod 15: 319–324
88. Tesarik J et al (2000) Embryos with high implantation potential after intracytoplasmic sperm injection can be recognized by a simple, non-invasive examination of pronuclear morphology. Hum Reprod 15:1396–1399
89. Tesarik J et al (2002) Paternal effects acting during the first cell cycle of human preimplantation development after ICSI. Hum Reprod 17:184–187
90. Wittemer C et al (2000) Zygote evaluation: an efficient tool for embryo selection. Hum Reprod 15:2591–2597
91. Scott L (2002) Embryologic strategies for overcoming recurrent ART treatment failure. Hum Fertil 5:206–214
92. Balaban B et al (2004) Pronuclear morphology predicts embryo development and chromosome constitution. Reprod Biomed Online 8(6):695–700
93. Gamiz P et al (2003) The effect of pronuclear morphology on early development and chromosomal abnormalities in cleavage-stage embryos. Hum Reprod 18:2413–2419
94. Chen C et al (2003) The relationship of pronuclear state morphology and chromosome status at cleavage stage. J Assist Reprod Genet 20(10):413–420
95. Edirisinghe W et al (2005) Association of pronuclear Z score with rates of aneuploidy in in vitro-fertilised embryos. Reprod Fertil Dev 17(5):529–534
96. Bavister B (1995) Culture of preimplantation embryos: facts and artifacts. Hum Reprod Update 1:91
97. Sakkas D et al (1998) Early cleavage of human embryos to the two-cell stage after intracytoplasmic sperm injection as an indicator of embryo viability. Hum Reprod 13:182–187
98. Shoukir Y et al (1997) Early cleavage of in-vitro fertilized human embryos to the 2-cell stage: a novel indicator of embryo quality and viability. Hum Reprod 12:153–156
99. Van Montfoort A et al (2004) Early cleavage is a valuable addition to existing embryo selection parameters: a study using single embryo transfers. Hum Reprod 19:2103–2108
100. Fenwick J et al (2002) Time from insemination to first cleavage predicts developmental competence of human preimplantation embryos in vitro. Hum Reprod 17:407–412
101. Edwards R et al (1980) Establishing full term human pregnancies using cleaving embryos grown in vitro. Br J Obstet Gynecol 87: 737–757
102. Bos-Mikish A et al (2001) Early cleavage of human embryos: an effective method for predicting successful IVF/ICSI outcome. Hum Reprod 16:2658–2661
103. Lundin K et al (2001) Early embryo cleavage is a strong indicator of embryo quality in human IVF. Hum Reprod 16:2652–2657
104. Petersen C et al (2001) Embryo selection by the first cleavage parameter between 25 and 27 h after ICSI. J Assist Reprod Genet 18:209–212
105. Sakkas D et al (2001) Assessment of early cleaving in vitro fertilized human embryos at the 2- cell stage before transfer improves embryo selection. Fertil Steril 76:1150–1156
106. Giorgetti C et al (2007) Early cleavage: an additional predictor of high implantation rate following elective single embryo transfer. Reprod Biomed Online 14(1):85–91
107. Hardarson T et al (2001) Human embryos with unevenly sized blastomeres have lower pregnancy and implantation rates: indications for aneuploidy and multinucleation. Hum Reprod 16:313–318
108. Terriou P et al (2007) Relationship between even early cleavage and day 2 embryo score

and assessment of their predictive value for pregnancy. Reprod Biomed Online 14(3): 294–299
109. Guerif F (2007) Limited value of morphological assessment at days 1 and 2 to predict blastocyst developmental potential: a prospective study based on 4042 embryos. Hum Reprod 22:1973–1981
110. Puissant F et al (1987) Embryo scoring as a prognostic tool in IVF treatment. Hum Reprod 2:705–708
111. Giorgetti C et al (1995) Embryo score to predict implantation after in – vitro fertilization: based on 957 single embryo transfers. Hum Reprod 10(9):2427–2431
112. Ziebe S et al (1997) Embryo morphology or cleavage stage: how to select the best embryos for transfer after in-vitro fertilization. Hum Reprod 10(9):2427–2431
113. Staessen C et al (1998) Genetic constitution of multinuclear blastomeres and their derivative daughter blastomeres. Hum Reprod 13:1625–1631
114. Van Royen E et al (2003) Multinucleation in cleavage stage embryos. Hum Reprod 18(5): 1062–1069
115. Jackson K et al (1998) Multinucleation in normally fertilized embryos is associated with an accelerated ovulation induction response and lower implantation and pregnancy rates in in vitro fertilization – embryo transfer cycles. Fertil Steril 70:60–66
116. Kligman I et al (1996) The presence of multinucleated blastomeres in human embryos is correlated with chromosomal abnormalities. Hum Reprod 11:1492–1498
117. Pickering S et al (1995) An analysis of multinucleated blastomere formation in human embryos. Hum Reprod 10:1912–1922
118. Laverge H et al (1997) Triple colour fluorescent in-situ- hybridization for chromosomes X, Y and 1 on spare human embryos. Hum Reprod 12:1492–1498
119. Alikani M et al (2000) Cleavage anomalies in early human embryos and survival after prolonged culture in-vitro. Hum Reprod 15: 2634–2643
120. Levy R et al (1998) Laser scanning confocal imaging of abnormal arrested human preimplantation embryos. J Assist Reprod Genet 15:485–495
121. Gerris J et al (1999) Prevention of twin pregnancy after in-vitro fertilization or intracytoplasmic sperm injection based on strict embryo criteria: a prospective randomized clinical trial. Hum Reprod 14:2581–2587
122. Pelinck M et al (1998) Embryos cultured in vitro with multinucleated blastomeres have poor implantation potential in-vitro fertilization and intracytoplasmic sperm injection. Hum Reprod 13:960–963
123. Van Royen E et al (2001) Calculating the implantation potential of day 3 embryos in women younger than 38 years of age: a new model. Hum Reprod 16(2):326–332
124. Gerris J et al (2002) Elective single day 3 embryo transfer halves the twinning rate without decrease in the ongoing pregnancy rate of an IVF/ICSI programme. Hum Reprod 17:2626–2631
125. Edwards R et al (1981) The growth of human preimplantation embryos in vitro. Am J Obstet Gynecol 141(4):408–416
126. Roux C et al (1995) Morphometric parameters of living human in-vitro fertilized embryos: importance of the asynchronous division process. Hum Reprod 10: 1201–1207
127. Hiragi T et al (2005) Mechanisms of first cleavage specification in the mouse egg: is our body plan set at day 0? Cell Cycle 4:661–664
128. Gardener R et al (2006) An investigation of the origin and significance of bilateral symmetry of the pronuclear zygote in the mouse. Hum Reprod 21:492–502
129. Dokras A et al (1993) Human blastocyst grading: an indicator of developmental potential. Hum Reprod 8:2119–2127
130. Shapiro B et al (2000) Predictive value of 72-hr blastomere cell number on blastocyst development and success of subsequent transfer based on the degree of blastocyst development. Fertil Steril 73:582–586
131. Langley M et al (2001) Extended embryo culture in human assisted reproduction treatments. Hum Reprod 16(5):902–908
132. Magli M et al (1998) Incidence of chromosomal abnormalities from a morphologically normal cohort of embryos in poor-prognosis patients. J Assist Reprod Genet 15(5): 297–301
133. Hnida C et al (2004) Computer-controlled, multilevel, morphometric analysis of blastomere size as biomarker of fragmentation and multinuclearity in human embryos. Hum Reprod 19:288–293
134. Carrell D et al (2009) A simplified coculture system using homologous attached cumulus tissue results in improved human embryo morphology and pregnancy rates during in vitro fertilization. J Assist Reprod Genet 16(7): 344–349
135. Antczak M et al (1999) Temporal and spatial aspects of fragmentation in early human embryos: possible effects on developmental competence and association with the differential

elimination of regulatory proteins from polarized domains. Hum Reprod 14: 429–447

136. Juriscova A et al (1996) Programmed cell death and human embryo fragmentation. Mol Hum Reprod 2(2):93–98
137. Perez G et al (1999) Fragmentation and death (a.k.a. apoptosis) of ovulated oocytes. Mol Hum Reprod 5(5):414–420
138. Alikani M et al (1999) Human embryo fragmentation in vitro and its implications for pregnancy and implantation. Fertil Steril 71: 836–842
139. Racowsky C et al (2003) Human embryo fragmentation in vitro and its implications for pregnancy and implantation. Fertil Steril 71: 836–842
140. Hardy K et al (2003) Maintenance of the inner cell mass in human blastocysts from fragmented embryos. Biol Reprod 68(40): 1165–1169
141. Van Blerkom J et al (2001) A microscopic and biochemical study of fragmentation phenotypes in stage-appropriate human embryos. Hum Reprod 16:719–729
142. Sathanathan H et al (1988) The effects of cooling human oocytes. Hum Reprod 3: 968–977
143. Plachot M et al (1987) Chromosome analysis of ovocytes and human embryos collected after fertilization in vitro. A model of natural selection against aneuploidy. Rev Fr Gynecol Obstet 83(910):613–617
144. Pellestor F et al (1994) Relationship between morphology and chromosomal constitution in human preimplantation embryo. Mol Reprod Dev 39(2):141–146
145. Steer C et al (1992) The cumulative embryo score: a predictive embryo scoring technique to select the optimal number of embryos to transfer in an in-vitro fertilization and embryo transfer programme. Hum Reprod 7:117–119
146. Tan S et al (1992) Cumulative conception and live birth rates after in-vitro fertilization. Lancet 339:1390–1394
147. Desai N et al (2000) Morphological evaluation of human embryos and derivation of an embryo quality scoring system specific for day 3 embryos: a preliminary study. Hum Reprod 15(10):2190–2196
148. Dawson K et al (1995) Delaying transfer to the third day post-insemination, to select nonarrested embryos, increases development to the fetal heart stage. Hum Reprod 10: 177–182
149. Jones G et al (1998) Factors affecting the success of human blastocyst development and pregnancy following in vitro fertilization and embryo transfer. Fertil Steril 70(6): 1022–1029
150. Carillo A et al (1998) Improved clinical outcomes for in vitro fertilization with delay of embryo transfer from 48 to 72 hours after oocyte retrieval: use of glucose and phosphate free media. Fertil Steril 69:329–334
151. Ebner T et al (2001) Embryo fragmentation in vitro and its impact on treatment and pregnancy outcome. Fertil Steril 76:281–285
152. Skiadas CC et al (2006) Early compaction on day 3 may be associated with increased implantation potential. Fertil Steril 86(5):1386–1391
153. Fiel D et al (2008) Day 4 embryo selection is equal to Day 5 using a new embryo scoring system validated in single embryo transfers. Hum Reprod 23(7):1505–1510
154. della Ragione T et al (2007) Developmental stage on day-5 and fragmentation rate on day-3 can influence the implantation potential of top-quality blastocysts in IVF cycles with single embryo transfer. Reprod Biol Endocrinol 5:2
155. Gardner R et al (2000) Flow of cells from polar to mural trophectoderm is polarized in the mouse blastocyst. Hum Reprod 15:694–701
156. Rijinder P et al (1998) The predictive value of day 3 embryo morphology regarding blastocyst formation, pregnancy and implantation rate after day 5 transfer following in-vitro fertilization or intracytoplasmic sperm injection. Hum Reprod 13:2869–2873
157. Graham J et al (2000) Day 3 morphology is a poor predictor of blastocyst quality in extended culture. Fertil Steril 74:495–497
158. Milki A et al (2002) Accuracy of day 3 criteria for selecting the best embryo. Fertil Steril 77:1191–1195
159. Janny L et al (1994) Evidence for a strong paternal effect on human preimplantation embryo development and blastocyst formation. Mol Reprod Dev 38:36–42
160. Miller J et al (2001) The effect of intracytoplasmic sperm injection and semen parameters on blastocyst development in vitro. Hum Reprod 16:918–924
161. Seli E et al (2002) The extent of nuclear DNA damage in ejaculated spermatozoa impacts on blastocyst development after IVF. Fertil Steril 78(Suppl 1):S61
162. Johnson M et al (1981) The foundation of two distinct cell lineages within mouse morula. Cell 24:71–80
163. Racowsky C et al (2000) The number of eight-cell embryos is a key determinant for selecting day 3 or day 5 transfer. Fertil Steril 73:558–564

164. Richter K et al (2001) Quantitative grading of a human blastocyst: optimal inner cell mass size and shape. Fertil Steril 76(6): 1157–1167
165. Balaban B et al (2000) Blastocyst quality affects the success of blastocyst-stage embryo transfer. Fertil Steril 74:282–287
166. Gardner D et al (1999) In-vitro culture of human blastocysts. In: Jansen R, Mortimer D (eds) Towards reproductive certainty: fertility and genetics beyond 1999. Parthenon, Carnforth, pp 378–388
167. Racowsky C et al (2009) Is there an advantage in scoring early embryos on more than one day? Hum Reprod 24(9):2104–2113
168. Fisch J et al (2003) The graduated embryo score predicts the outcome of assisted reproductive technologies better then a single day 3 evaluation and achieves results associated with blastocyst transfer from day 3 embryo transfer. Fertil Steril 80(6):1352–1358
169. Sjoblom P et al (2006) Prediction of embryo developmental potential and pregnancy based on early stage morphological characteristics. Fertil Steril 86:848–861
170. Terriou P et al (2001) Embryo score is a better predictor of pregnancy than the number of transferred embryos or female age. Fertil Steril 75:525–531
171. Holte J et al (2007) Construction of an evidence-based integrated morphology cleavage embryo score for implantation potential of embryos scored and transferred on Day 2 after oocyte retrieval. Hum Reprod 22: 548–557
172. Vergouw C et al (2008) Metabolomic profiling by near-infrared spectroscopy as a tool to assess embryo viability: a novel, non-invasive method for embryo selection. Hum Reprod 23:1499–1504
173. Saith R et al (1998) Relationships between the developmental potential of human in-vitro fertilization embryos and features describing the embryo, oocyte and follicle. Hum Reprod 4:121–134
174. Jurisica I et al (1998) Case-based reasoning in IVF: prediction and knowledge mining. Artif Intell Med 12:1–24
175. Trimarchi J et al (2003) Comparing data mining and logistic regression for predicting IVF outcome. Fertil Steril 80:S100
176. Patrizi G et al (2004) Pattern recognition methods in human-assisted reproduction. Int Trans Oper Res 11:365–379
177. Fisch J et al (2001) The Graduated Embryo Score (GES) predicts blastocyst formation and pregnancy rate from cleavage-stage embryos. Hum Reprod 16(9):1970–1975
178. Fisch J et al (2006) Graduated Embryo score and soluble human leukocyte antigen-expression improve assisted reproductive technology outcomes and suggest a basis for elective single-embryo transfer. Fertil Steril 87(4):757–763
179. Neuber E et al (2003) Sequential assessment of individually cultured human embryos as an indicator of subsequent good quality blastocyst development. Hum Reprod 18:1307–1312
180. Neuber E et al (2006) Sequential embryo assessment investigator – driven morphological assessment at selecting a good quality blastocyst. Fertil Steril 85(3):794–796
181. Rienzi L et al (2002) Day 3 embryo transfer with combined evaluation at the pronuclear and embryo stages compares favourably with day 5 blastocyst transfer. Hum Reprod 17: 1852–1855
182. Carrell D et al (1999) A simplified coculture system using homologous attached cumulus tissue results in improved human embryo morphology and pregnancy rates during in vitro fertilization. J Assist Reprod Genet 16(7):344–349
183. Hunault C et al (2002) A prediction model for selecting patients undergoing in vitro fertilization for selecting patients undergoing in vitro fertilization for elective single embryo transfer. Fertil Steril 77(40):725–732
184. James A et al (2006) The limited importance of pronuclear scoring of human zygotes. Hum Reprod 21(6):1599–1604

Chapter 24

Embryo Selection Using Metabolomics

D. Sakkas

Abstract

Faced with an increasing demand to select one embryo to transfer back to patients, a number of techniques are being developed to assist in discriminating differences within the cohort of a patient's embryos. A new and emerging technology which allows us to measure the profile of different metabolites in embryo culture media and formulate a viability score correlated to implantation potential is metabolomics.

Key words Metabolomics, Embryo selection, Near infra red spectrometry, Algorithm, Embryo viability

1 Introduction

In 1978, the culmination of many years research in both animal and human reproductive associated technologies led to the first in vitro fertilization (IVF) birth in the world [1]. Further improvements in the early 1980s by groups in Australia led to births after drug-induced superovulation in the mother, the world's first frozen embryo baby [2] and the first donor egg baby [3].

Since the first births the use of assisted reproductive technologies (ART) continues to increase annually, with close to half a million treatment cycles being initiated in the USA and Europe alone [4, 5]. This trend has been driven by improved success rates, better access to care, and the long timeline of some other treatment options. Currently, between 1 and 4 % of all children born in the USA and Europe are from ART related conceptions.

1.1 Multiple Embryo Transfers

The ability to perform controlled ovarian hyperstimulation in women undergoing IVF treatment has created a conundrum in ART. Whereas on the one hand it has improved pregnancy rates, it has also led to a subsequent increase in multiple pregnancy rates. The improved success rates are attained, in many cases, only through the simultaneous transfer of multiple embryos. For the <35 age group in 2008 in the USA, an average of 2.2 embryos per patient were transferred leading to a 41.3 % delivery rate per

Zev Rosenwaks and Paul M. Wassarman (eds.), *Human Fertility: Methods and Protocols*, Methods in Molecular Biology, vol. 1154, DOI 10.1007/978-1-4939-0659-8_24, © Springer Science+Business Media New York 2014

initiated IVF cycle and an overall multiple birth rate of 35.2 %. The live birth results for older age groups decrease dramatically even though more embryos are transferred. The relative risk for multiple births in the 35–40 age group however remains above 25 % [4].

The risks related to multiple gestations are well documented and include preterm delivery, low birth weight, and a dramatic increase in the relative risk for cerebral palsy (reviewed by ref. [6]). These complications lead to a higher incidence of medical, perinatal, and neonatal complications and a tenfold increase in health-care costs compared to a singleton delivery [7]. The ability to decrease the prevalence of multiple gestations while maintaining or improving overall pregnancy rates has become one of the most significant goals of current IVF research.

1.2 Single Embryo Transfer (SET)

The documented dangers associated with multiple pregnancies have been allayed in many countries, in particular northern Europe, by legal restrictions on the number of embryos that can be transferred in a single IVF cycle. For example, in most Scandinavian countries and Belgium the government has set a legal limit of SET for specific patient groups. In other parts of the world, where no legal restrictions exist, the onus is on the individual clinic (as well as the couple) to decrease the number of embryos transferred so that a balance can be achieved between the risks associated with multiple gestations and sustained pregnancy rates. This approach has worked well in countries such as Finland and Australia which enjoy overall rates of SET greater than 60 %. It is evident that clinics in the USA and other countries, currently lacking legislation, will eventually be compelled via legal, financial and/or moral obligation to restrict the number of embryos transferred in order to minimize the risk of multiple gestations.

2 Methods for Implementing SET in the Laboratory

Currently there is no accurate objective methodology which will enable an embryologist to estimate the reproductive potential of individual embryos within a cohort of embryos. Great progress has been made using the existing selection techniques, which largely encompasses morphological evaluation, but many limitations remain. Morphological assessment has been the major tool of the embryologist for selecting which embryo(s) to replace. Since the early years of IVF it was noted that day 2–3 embryos cleaving faster and those of limited fragmentation were more likely to lead to a pregnancy [8, 9]. Morphological assessment systems have evolved significantly over the past decade and in addition to the classical parameters of cell number and fragmentation, numerous other characteristics have been examined including: pronuclear morphology, early cleavage to the 2-cell stage, top quality embryos on successive days and various forms of sequential assessment of embryos (*see* reviews by refs. [10–12]). For all the improvements in cleavage embryo assessment the ability to culture and assess blastocyst stage embryos has probably led to the greatest improvement in our ability to select embryos on the basis of morphology [13].

Faced with the inevitability that the worldwide IVF community, will, in the future, have to select only one or two embryos for transfer, we will be forced to make certain changes to the current practice of routine IVF. The first may be to rely on less aggressive stimulation protocols hence generating a lower number of eggs at collection [14, 15]. The second will be to improve the selection process for defining the quality of individual embryos so that the ones we choose for transfer are more likely to implant.

The inherent ease for the laboratory to assess various morphological markers makes it the preferred assessment technique to transfer embryos. Even with the adoption of more complex forms of assessment it will still remain as one of the main tools we have in our armory for assessment. For a new technique to be acceptable in a laboratory setting, however, it must satisfy a number of criteria.

It should be an analytical technique that:

1. Predicts embryo viability without damaging the embryo (invasively or noninvasively);
2. Is relatively rapid and performed in the clinic; and,
3. Provides consistent and accurate results.

3 Embryo Metabolism and Metabolomics

The hypothesis that there is a relationship between metabolic parameters and embryo viability has been upheld in numerous studies over the past 30 years. In 1980, Renard et al. [16] observed that Day 10 cattle blastocysts which had an elevated glucose uptake developed better, both in culture and in vivo after transfer than those blastocysts with a lower glucose uptake. In 1987, using noninvasive microfluorescence, Gardner and Leese [17] measured glucose uptake by individual Day-4 mouse blastocysts prior to transfer to recipient females. Those embryos that went to term had a significantly higher glucose uptake in culture than those embryos that failed to develop after transfer. Lane and Gardner [18] also showed that the glycolytic rate of mouse blastocysts could be used to select embryos for transfer prospectively. Morphologically identical mouse blastocysts with equivalent diameters were identified as viable using metabolic criteria prior to transfer and had a fetal development rate of 80 %. In contrast, embryos that exhibited an abnormal metabolic profile compared to in vivo developed controls, developed at a rate of only 6 %. This data provides clear evidence that metabolic function is linked to embryo viability and is independent of the information provided by morphology.

There are three approaches that allow researchers to obtain metabolic information about embryos: (1) Analysis of carbohydrate utilization, (2) Analysis of amino acids, and (3) Analysis of the embryonic metabolome. The first two approaches provide more specific information about candidate metabolites or the activity of specific metabolic pathways.

The third option provides a systematic analysis of the inventory of metabolites that represent the functional phenotype at the cellular level. Depending upon the technology employed to analyze the metabolome [19], one does not necessarily obtain identification of specific metabolites, but an algorithm is created that relates to cell function.

4 Metabolomics

A new and emerging technology which allows us to measure factors in embryo culture media is metabolomics. The complete array of small-molecule metabolites that are found within a biological system constitutes the metabolome and reflects its functional phenotype [20]. Metabolomics is the systematic study of this dynamic inventory of metabolites, as small molecular biomarkers representing the functional phenotype in a biological system. Using various forms of spectral and analytical approaches, metabolomics attempts to determine metabolites associated with physiologic and pathologic states [21]. A number of proof of principal studies of individual embryos and their implantation outcome indicated that embryos that result in pregnancy are different in their metabolomic profile compared to embryos that failed to implant [22].

Investigation of the metabolome of embryos, as detected in the culture media they grow in using targeted spectroscopic analysis and bioinformatics can therefore divulge these differences. In a study by Seli et al. [22], these differences were found to be detectable in the culture media using both Raman and Near Infrared (NIR) spectroscopy. In his study, a total of 69 Day 3 spent embryo culture media samples from 30 patients with known outcome (0 or 100 % sustained implantation rates) were evaluated using Raman and/or NIR spectroscopy. A regression formula was developed to calculate a relative "embryo viability score"—relating to embryo reproductive potential—which correlated implantation outcome to both Raman and NIR spectral metabolomic profiles. Both Raman and NIR spectroscopic analysis of the spent culture media of embryos with proven reproductive potential demonstrated significantly higher viability scores than those that failed to implant. Interestingly, when human embryos of similar morphology were examined using the same NIR spectral profile, their viability scores varied remarkably in relation to morphology indicating that the metabolome of embryos that looked similar differed significantly. This observation was in agreement with the study of Katz-Jaffe et al. [23, 24], who revealed that the proteome of individual human blastocysts of the same grade differed between embryos, again indicating that embryo morphology is not completely linked to its physiology.

The algorithm established was subsequently used to predict the likelihood of pregnancy from blinded embryo culture media samples. When the algorithm developed was used to blindly test a subgroup of 16 Day 3 embryo samples collected at a different center and cultured using a different type of commercial media, viability scores of embryos with proven reproductive potential were significantly higher compared to embryos that failed to implant [25]. A larger analysis of single embryo transfer cycles was also undertaken whereby NIR spectral analysis of frozen Day 2 and 3 embryo culture media samples was performed blinded to outcome. Individual metabolic profiles were established from 7 μl of the samples with each measurement taking less than 1 minute. Statistical analysis performed on the metabolic profiles established a viability score (as generated above) that was significantly different ($P<0.001$) between the pregnant and nonpregnant patients. A cutoff value for predicting pregnancy was taken at >0.3. When this cutoff was used to examine embryos of excellent and good morphology that underwent single embryo transfer a significant difference was found in the establishment of pregnancy [26].

More recent data has used a semiquantitative approach to apply this technology. A series of studies have been undertaken whereby algorithms for predicting outcomes of Day 2, 3, and 5 SETs were developed and tested blindly against samples unrelated to those used to develop the algorithms. The data again confirmed that as the Viability Scores™ generated by the algorithms increased the tendency for the assessed embryo to implant and display fetal cardiac activity also increased [27]. Subsequent data has also shown that this pattern is independent of morphology as the ability of the Viability Score™ to relate to FCA is maintained within groups of embryos with the same morphology.

5 The Pros and Cons of Metabolomics

Although the series of preliminary studies [25–29] showed a benefit of this technique they were largely based on retrospective studies and performed in a single research laboratory as distinct from a real clinical setting. Recently a number of in house clinical studies have been conducted using either a prototype or commercial version of the Molecular Biometrics Inc. NIR system (ViaMetrics) showing inconsistent results. These studied aimed to test the second and third parameters established above. The largest of these studies were performed as Randomized Clinical Trials after SET using a prototype instrument. All studies compared standard Morphological techniques for embryo selection versus using the NIR system to rank embryos within a cohort that had good morphology and were being selected for either transfer or cryopreservation.

In the Gothenburg study [30] both day 2 and day 5 SETs were performed. Although not significant, the results indicated a possible benefit of embryo selection through addition of NIR on day 2 transfer. Of 83 and 87 SETs in the Morphology alone and ViaMetrics (NIR) groups the pregnancy rates were 26.5 % and 31 %, respectively. In the same study no benefit for selection of day 5 SET was observed, whereby of 80 and 77 SETs in the Morphology alone and ViaMetrics (NIR) groups the pregnancy rates were 45 % and 39 %, respectively. Interestingly, significant differences were observed in this study in respect to the overall morphology of the blastocysts selected in the morphology alone group. The benefits of selecting a single good quality blastocyst on day 5 have also been found to be beneficial in many other studies and remains as one of the best selection criteria to date [31].

In another randomized trial examining Day 3 SETs no difference at all was observed between the two groups [32]. Concurrently, two trials were conducted using the commercial platform where two or three embryos were selected for transfer on the basis of Morphology alone or Morphology plus ViaMetrics (NIR) both showed improvements in the Viametrics assisted selection group. Sfontouris et al. [33] performed a study in which for every two patients where embryos were selected based on morphology, one patient had a transfer based on ViaMetrics (NIR) embryo selection. They reported implantation rates (IR) with fetal cardiac activity that were significantly ($P=0.04$) improved in the Viametrics group [$N=39$ patients, IR = 33/102 (32.4 %)] compared to the Morphology alone selection group [55/257 (21.4 %)].

One of the underlying problems encountered in the NIR system was that the threshold of signal distinguishing between a viable and nonviable embryo was highly sensitive to signal noise. NIR spectroscopy systems and the algorithms generated from them can create models that inadvertently conceal problems in a particular platform. In doing so, a method that has been established and cross-validated on a larger scale can still be problematic, as the variation can still lie within the technical platform itself. This would then mean that any algorithm created on one group of instruments may suffer when used on other instruments as their noise thresholds and behavior may differ significantly enough to mask the true analytical information obtained in each spectra. The commercial version of the NIR instrument was unfortunately withdrawn due to the wide variability in performance between clinics. This is not dissimilar to the situation faced by aneuploidy screening of embryos, whereby chromosomal screening of preimplantation embryos has always thought to be a strong theory, and however, using FISH proved to be inadequate [34], while it now appears that modern comprehensive screening techniques are providing more consistent results [35].

6 Conclusion and Future

It is very clear that markers do exist in the spent embryo culture media indicative of viability. The major benefits of a noninvasive type of technology is the fact that the technology can be used on spent media and the time taken to assess the samples is very short, making it possible to perform the analysis just prior to ET. So far, NIR spectroscopy, when tested in stringent clinical trials, has failed to consistently improve the chance of selecting a single embryo for a viable pregnancy and the NIR technology appears to need further development before being used as an objective marker of embryo viability. Therefore, although metabolomic assessment met the first two key criteria necessary to be used routinely in clinical IVF, it currently fails to consistently meet the final criterion of providing consistent and accurate results. Further refinement of NIR and Raman based spectroscopy techniques will hopefully lead to the widespread adoption of this methodology in the near future.

References

1. Steptoe PC, Edwards RG (1978) Birth after the reimplantation of a human embryo [letter]. Lancet 2(8085):366
2. Trounson A, Mohr L (1983) Human pregnancy following cryopreservation, thawing and transfer of an eight-cell embryo. Nature 305(5936):707–709
3. Lutjen P, Trounson A, Leeton J, Findlay J, Wood C, Renou P (1984) The establishment and maintenance of pregnancy using in vitro fertilization and embryo donation in a patient with primary ovarian failure. Nature 307(5947): 174–175
4. Society for Assisted Reproduction Technology (SART) 2009 data. https://www.sartcorsonline.com/rptCSR_PublicMultYear.aspx?ClinicPKID=0
5. De MJ, Goossens V, Bhattacharya S, Castilla JA, Ferraretti AP, Korsak V et al (2010) Assisted reproductive technology in Europe, 2006: results generated from European registers by ESHRE. Hum Reprod 25(8):1851–1862
6. Adashi EY, Barri PN, Berkowitz R, Braude P, Bryan E, Carr J et al (2003) Infertility therapy-associated multiple pregnancies (births): an ongoing epidemic. Reprod Biomed Online 7(5):515–542
7. Ledger WL, Anumba D, Marlow N, Thomas CM, Wilson EC (2006) The costs to the NHS of multiple births after IVF treatment in the UK. BJOG 113(1):21–25
8. Cummins J, Breen T, Harrison K, Shaw J, Wilson L, Hennessey J (1986) A formula for scoring human embryo growth rates in in vitro fertilization: its value in predicting pregnancy and in comparison with visual estimates of embryo quality. J In Vitro Fert Embryo Transf 3:284–295
9. Edwards R, Fishel S, Cohen J (1984) Factors influencing the success of in vitro fertilization for alleviating human infertility. J In Vitro Fert Embryo Transf 1:3–23
10. De Neubourg D, Gerris J (2003) Single embryo transfer—state of the art. Reprod Biomed Online 7(6):615–622
11. Sakkas D, Gardner DK (2004) Evaluation of embryo quality. In: Gardner DK, Weissman A, Howles C, Shoham Z (eds) Assisted reproduction

technology laboratory and clinical perspectives, 2nd edn. Martin Dunitz Press, London, pp 211–234
12. Sakkas D, Gardner DK (2005) Noninvasive methods to assess embryo quality. Curr Opin Obstet Gynecol 17(3):283–288
13. Gardner DK, Surrey E, Minjarez D, Leitz A, Stevens J, Schoolcraft WB (2004) Single blastocyst transfer: a prospective randomized trial. Fertil Steril 81(3):551–555
14. Inge GB, Brinsden PR, Elder KT (2005) Oocyte number per live birth in IVF: were Steptoe and Edwards less wasteful? Hum Reprod 20(3):588–592
15. Patrizio P, Sakkas D (2009) From oocyte to baby: a clinical evaluation of the biological efficiency of in vitro fertilization. Fertil Steril 91(4):1061–1066
16. Renard JP, Philippon A, Menezo Y (1980) In-vitro uptake of glucose by bovine blastocysts. J Reprod Fertil 58(1):161–164
17. Gardner DK, Leese HJ (1987) Assessment of embryo viability prior to transfer by the noninvasive measurement of glucose uptake. J Exp Zool 242(1):103–105
18. Lane M, Gardner DK (1996) Selection of viable mouse blastocysts prior to transfer using a metabolic criterion. Hum Reprod 11(9):1975–1978
19. Botros L, Sakkas D, Seli E (2008) Metabolomics and its application for non-invasive embryo assessment in IVF. Mol Hum Reprod 14(12):679–690
20. Oliver SG, Winson MK, Kell DB, Baganz F (1998) Systematic functional analysis of the yeast genome. Trends Biotechnol 16(9):373–378
21. Ellis DI, Goodacre R (2006) Metabolic fingerprinting in disease diagnosis: biomedical applications of infrared and Raman spectroscopy. Analyst 131(8):875–885
22. Seli E, Sakkas D, Scott R, Kwok SH, Rosendahl S, Burns DH (2007) Non-invasive metabolomic profiling of embryo culture media using Raman and near infrared spectroscopy correlates with reproductive potential of embryos in women undergoing in vitro fertilization. Fertil Steril 88(5):1350–1357
23. Katz-Jaffe MG, Schoolcraft WB, Gardner DK (2006) Analysis of protein expression (secretome) by human and mouse preimplantation embryos. Fertil Steril 86(3):678–685
24. Katz-Jaffe MG, Gardner DK, Schoolcraft WB (2006) Proteomic analysis of individual human embryos to identify novel biomarkers of development and viability. Fertil Steril 85(1):101–107
25. Scott R, Seli E, Miller K, Sakkas D, Scott K, Burns DH (2008) Noninvasive metabolomic profiling of human embryo culture media using Raman spectroscopy predicts embryonic reproductive potential: a prospective blinded pilot study. Fertil Steril 90(1):77–83
26. Seli E, Vergouw CG, Morita H, Botros L, Roos P, Lambalk CB et al (2010) Noninvasive metabolomic profiling as an adjunct to morphology for noninvasive embryo assessment in women undergoing single embryo transfer. Fertil Steril 94(2):535–542
27. Ahlstrom A, Wikland M, Rogberg L, Barnett JS, Tucker M, Hardarson T (2011) Cross-validation and predictive value of near-infrared spectroscopy algorithms for day-5 blastocyst transfer. Reprod Biomed Online 22(5):477–484
28. Vergouw CG, Botros LL, Roos P, Lens JW, Schats R, Hompes PG et al (2008) Metabolomic profiling by near-infrared spectroscopy as a tool to assess embryo viability: a novel, non-invasive method for embryo selection. Hum Reprod 23(7):1499–1504
29. Seli E, Bruce C, Botros L, Henson M, Roos P, Judge K et al (2011) Receiver operating characteristic (ROC) analysis of day 5 morphology grading and metabolomic Viability Score on predicting implantation outcome. J Assist Reprod Genet 28(2):137–144
30. Hardarson T, Ahlstrom A, Rogberg L, Botros L, Hillensjo T, Westlander G et al (2012) Non-invasive metabolomic profiling of Day 2 and 5 embryo culture media: a prospective randomized trial. Hum Reprod 27:89–96
31. Blake DA, Farquhar CM, Johnson N, Proctor M (2007) Cleavage stage versus blastocyst stage embryo transfer in assisted conception. Cochrane Database Syst Rev (4):CD002118
32. Vergouw CG. Metabolomic profiling of culture media by NIR spectroscopy as an adjunct to morphology for selection of a single Day 3 embryo to transfer: a double blind randomised trial (unpublished data)
33. Sfontouris IA, Lainas GT, Sakkas D, Zorzovilis IZ, Petsas GK, Lainas TG (2013) Non-invasive metabolomic analysis using a commercial NIR instrument for embryo selection. J Hum Reprod Sci 6(2):133–139
34. Mastenbroek S, Twisk M, van Echten-Arends J, Sikkema-Raddatz B, Korevaar JC, Verhoeve HR et al (2007) In vitro fertilization with preimplantation genetic screening. N Engl J Med 357(1):9–17
35. Wells D, Alfarawati S, Fragouli E (2008) Use of comprehensive chromosomal screening for embryo assessment: microarrays and CGH. Mol Hum Reprod 14(12):703–710

Chapter 25

Embryo Transfer

Anate Aelion Brauer and Glenn Schattman

Abstract

Embryo transfer is a critical step in the overall success of in vitro fertilization (IVF). Despite its apparent simplicity, it is an integral part of the IVF cycle that can be difficult to teach and perform well. In this chapter we describe the procedure of embryo transfer and discuss various modifications that have been evaluated in an effort to improve the probability of embryo implantation including ultrasound guidance, catheter design, transfer technique, as well as enrichment of the embryo transfer media.

Key words Embryo transfer, Trial transfer, Ultrasound guided transfer, Embryo transfer catheter, Embryo transfer media

1 Introduction

Since the first successful pregnancy from an in vitro fertilized oocyte in humans almost 30 years ago, significant improvements in the in vitro fertilization (IVF) procedure have been realized. This first IVF pregnancy used natural cycle monitoring with oocyte recovery during open abdominal surgery. Since then, gonadotropins as well as gonadotropin releasing hormone agonists and antagonists have significantly improved the success rates of IVF by increasing the number of oocytes collected in a single cycle. Improvements in ultrasound technology have facilitated the retrieval of oocytes during an outpatient, trans-vaginal follicle puncture and oocyte aspiration and advancements in our understanding of the developmental requirements of the human embryo have improved our ability to culture embryos in vitro. The procedure of embryo transfer, placing the embryo(s) into the uterine cavity has remained relatively unchanged. While embryo quality and uterine receptivity are important factors in implantation, the transfer technique itself is also recognized as a key player in overall success. A successful embryo transfer should deliver the embryos atraumatically to the point in the endometrial lining where implantation is most likely to occur. Complications that may lead to a

Zev Rosenwaks and Paul M. Wassarman (eds.), *Human Fertility: Methods and Protocols*, Methods in Molecular Biology, vol. 1154, DOI 10.1007/978-1-4939-0659-8_25, © Springer Science+Business Media New York 2014

suboptimal transfer may include endometrial trauma, induction of uterine contractions, damage to the embryos, or deposition of embryos in a suboptimal location. Several steps associated with embryo transfer have been studied in an effort to minimize such complications and increase the efficiency of this final step of the IVF procedure. Here, we describe our embryo transfer protocol.

2 Materials

1. Procedure room with table capable of lithotomy position for GYN procedures.
2. Sterile speculum.
3. Sterile sponge stick.
4. Sterile lint free gauze.
5. Embryo culture media or other suitable sterile liquid for cleaning the ectocervix.
6. Trial transfer catheter (identical to transfer catheter).
7. Transfer catheter.
8. Inverted dissection microscope.
9. +/– Ultrasound with transabdominal transducer.
10. Ultrasound technician.

3 Methods

The procedure of embryo transfer may be divided into four distinct sections:

1. Preparation of the patient
2. Preparation of the catheter
3. Transfer of the embryos
4. Post transfer care

3.1 Preparation of the Patient

1. A trial transfer which illustrates the path of the cervical canal as well as the depth of the uterine cavity from external os to fundus should be performed 1 or more months prior to the patient undergoing oocyte retrieval and be available as a reference during the ET if needed (*see* **Note 1**).
2. The trial transfer also identifies patients in whom the transfer may be difficult due to a stenotic or tortuous cervical canal and would benefit from ultrasound guidance, dilation and/or anesthesia (*see* **Note 2**).
3. Fluid or mucous in the endometrial cavity identified at the time of oocyte retrieval or immediately prior to embryo transfer

should be aspirated with an atraumatic catheter such as a soft embryo transfer catheter. Absence of retained fluid should be confirmed prior to ET.

4. Anesthesia for embryo transfer is not routinely administered and should not be necessary. Anesthesia may be used if the patient has a history of a difficult trial transfer or difficult prior embryo transfer. Occasional use of a benzodiazepine is recommended in situations where patient anxiety may interfere with the procedure. Rarely a patient may require complete relaxation in order to negotiate a difficult cervical canal or due to extreme anxiety.
5. Antibiotics are not routinely administered at the time of ET as their use has not been shown to increase pregnancy rate or decrease the rare occurrence of infection [1, 2]. In patients with a history of prior pelvic infection or tubal disease due to an infectious etiology, antibiotics are administered prior to embryo transfer.
6. The number of embryos to be transferred should be discussed with the patient and her partner (if applicable) prior to the ET procedure. The decision should be based on objective criteria based on the clinic's own data and patient specific information including, but not limited to age, embryo quality, and prior history [3]. Limiting the risk of high-order multiples while maximizing the probability of a singleton pregnancy should be the priority.
7. Patient identification is confirmed at two time points by the practitioner doing the transfer as well as the embryologist loading the catheter: Upon entrance to the operating room and again just prior to transfer of the embryos.
8. The patient is placed in a dorsal lithotomy position.
9. A sterile speculum is placed in the vagina with good visualization of the cervix. Optimal placement of the speculum is important in obtaining the proper cervico-uterine angle for transfer. A variety of speculum sizes should be available in the procedure room.
10. The external cervical os is cleaned with sterile gauze moistened with saline or transfer medium. Additional mucus from the cervical canal can be aspirated using a sterile tuberculin syringe although there does not seem to be a benefit in removing cervical mucous prior to embryo transfer (*see* **Note 3**).
11. At this point we perform another trial transfer. The trial transfer catheter should be the same type of catheter that will be used for the actual transfer. Differences in catheter shape, thickness and flexibility compared to the trial transfer catheter as well as

differences in the patient's uterine position following ovarian stimulation make this an important step, as the actual transfer catheter may have to be shaped to accommodate the patient's anatomy (*see* **Note 4**).

12. The catheter should only be advanced up to the anticipated location for the embryo transfer. Typically 10–15 mm from the uterine fundus determined at the trial transfer.
13. Occasionally, we identify a patient who has a change in the position of their uterus, possibly related to stimulation and ovarian enlargement.
14. Once the trial transfer is completed, the catheter can be removed and the embryos can be loaded into a new, sterile embryo transfer catheter by the embryologist.

3.2 Preparation of the Catheter

1. Transfer catheters fall into two main categories: Soft or firm.
2. Firm catheters have the advantage of being rigid with a curve molded into the tip of the catheter and are potentially easier to negotiate a tortuous cervical canal.
3. Soft catheters are trickier to use, especially in patients with extreme angles in their cervix or lower uterine segment. A longer learning curve is required for their proper use.
4. Some catheters are 2 piece catheters with a soft inner and firm outer sheath which allow for more flexibility with transfer technique and molding of the catheter.
5. We use a soft, 2 piece catheter such as a Wallace® catheter for almost all of our embryo transfer procedures and only use the firm catheters for the rare occasion when we are unable to negotiate the soft catheter into the uterine cavity. Comparative studies suggest that soft catheters and atraumatic placement improve the likelihood for implantation (*see* **Note 5**).
6. Most transfers can be performed with a straight catheter as the flexible inner catheter will follow the natural curves of the cervix and uterus.
7. If needed, the outer sheath of the soft catheter may be gently curved to negotiate cervical bends and align the inner catheter to the uterine axis. The degree to which the catheter is curved is decided on by the operator at the time of trial transfer and is communicated to the embryologist. The outer catheter is then molded by the embryologist prior to embryo loading.
8. The catheter is then flushed with media to remove all air from the column. The embryo(s) are then aspirated into the catheter with a minimal volume of culture media (20–40 μl) so that the embryo(s) are loaded towards the tip of the catheter. Some labs include a small air column, proximal to and/or distal to

the embryos which allows for visualization on ultrasound of the air bubble in the endometrial cavity. We use our standard embryo culture media when loading embryos.

3.3 Transfer of the Embryos

1. The overwhelming majority of embryo transfers at our institution utilize a method often described as the "clinical touch" method. This method uses prior trial embryo transfer "roadmaps" to guide the catheter into the correct position in the uterine cavity. While a gentle touch is required to make sure the catheter is in the correct location, the transfer is atraumatic and does not "touch" the fundus to confirm location of the catheter within the uterine cavity. The embryos are deposited at ~10–15 mm shy of the fundus using the trial transfer measurements as a guide and the embryos are deposited in a similar location within the uterus with both the "clinical touch" and ultrasound guided methods.
2. At our institution ultrasound guidance during embryo transfer is reserved for those patients who we suspect will have a difficult transfer based on experience gleaned from the trial transfer done prior to stimulation or a history of a difficult embryo transfer. Ultrasound can be helpful in a uterus distorted by fibroids or those with a cesarean section scar in which the catheter may be misdirected. Ultrasound guidance is typically done using a 3.5–5 MHz 2-D transabdominal transducer from a mid-sagittal or oblique plane of the uterus and cervix (*see* **Note 6**).
3. Deposition of the embryos should occur in the mid to lower midportion of the uterus, avoiding transfers high in the fundus [4, 5]. It is our practice to deposit embryos 10–15 mm from the fundus.
4. After depositing the embryos into the uterine cavity, the operator should apply and maintain pressure on the plunger of the syringe to reduce the possibility that the embryos will be aspirated back into the cavity until the catheter is withdrawn from the cervix.
5. Some practitioners wait for up to 30 s before withdrawing the catheter from the uterine cavity.
6. The placement as well as the removal of the transfer catheter should be done slowly and gently.
7. After the transfer has been performed, the catheter is handed to the embryologist to be flushed and inspected for retained embryos. Retained embryos should be reloaded into a new catheter and immediately transferred back into the uterus.

3.4 Post transfer Care

1. The patients are transferred to the recovery room where they may remain on bed rest for up to 30 min. Recommendations regarding physical activity after transfer vary, although studies have shown that bed rest of >20–30 min after transfer has not improved pregnancy rates [6, 7].

4 Notes

1. During the trial transfer the uterine cavity is sounded, passing the catheter to the uterine fundus to measure the total length of the cervix and uterine cavity. Notes are made regarding the axis and curve of the catheter required to navigate the cervical canal. In addition to the practical benefits of a trial transfer, one RCT showed improvement in pregnancy rates when using trial transfer compared to touching the uterine fundus at the time of transfer [8].
2. Difficulty with an embryo transfer can be anticipated by performing a trial transfer prior to the start of an IVF cycle. If difficulty with a trial transfer is encountered and it is not possible to get a catheter into the uterine cavity, the patient can undergo cervical dilation and hysteroscopy under ultrasound guidance prior to undergoing ovarian stimulation. Cervical dilation up to the diameter of the intended transfer catheter can also be gently performed at the time of oocyte retrieval.
3. Cervical mucus may interfere with delivery of the embryos into the uterine cavity by plugging the tip of the catheter. Additionally, embryos may adhere to the mucus around the catheter and may become displaced upon catheter removal. Studies looking at effect of cervical mucus removal have yielded mixed results [9, 10].
4. Difficulties that may be encountered/and strategies that may help overcome them in preparation for the actual ET are as follows:
   (a) Coiling of the soft catheter inside the cervical canal: Soft catheters may become coiled inside the cervical canal. If this is suspected, a simple test of either rotating the catheter 360° resulting in recoil or withdrawing the catheter by 1–2 cm and feeling for a difference in resistance will usually confirm this suspicion.
   (b) Cervical stenosis: Cervical dilation using ultrasound guidance can temporarily alleviate the difficulty of negotiating the cervical canal in a patient with stenosis. Techniques of cervical dilation may include mechanical dilation [11, 12], osmotic dilation with laminaria [13], placement of a Malecot catheter [14], and operative hysteroscopy with shaving. We prefer mechanical cervical dilation. While timing of dilation is still debatable, most practitioners at our institution perform any necessary cervical dilation at the time of egg retrieval.
   (c) Extreme utero-cervical angle: Another cause of failure of the catheter to pass the internal cervical os is a lack of achieving proper alignment between the ET catheter and cervico-uterine angle. Curving the ET catheter according

to the cervico-uterine angle may help negotiate this angle. Gentle maneuvering of the vaginal speculum may also facilitate introduction of the catheter.

In cases of very anteflexed uteri, straightening of the angle can be achieved by filling the bladder before ET.

In cases of severely flexed uteri which may potentially require a tenaculum on the cervix for traction, a cervical stitch cut long can be placed at the time of oocyte retrieval and then used for relatively atraumatic traction at ET. We do not generally recommend use of a tenaculum at the time of transfer as it may initiate uterine contractions.

If placement is still difficult, one may use a malleable stylet through the outer sheath and negotiate the cervical canal before introducing the soft catheter.

5. Data suggests that the use of a soft catheter may be associated with a higher pregnancy rate than the use of a firm one [15–17].
6. Ultrasound guidance may also be used as a tool in dealing with difficult transfers. At our institution, use of ultrasound guidance is reserved for potentially difficult transfers as anticipated by the trial transfer. The bladder should be full enough to cover the uterine fundus to aid with visualization. Proponents of ultrasound guidance site several advantages:
   (a) Using ultrasound ensures that the catheter is placed at a specified distance from the fundus, avoiding fundal contact;
   (b) U/S guidance may be helpful in uteri distorted by fibroids or patients with a prior cesarean section scar
   (c) The full bladder required for U/S guidance may help straighten the cervico-uterine angle in patients with a severely anteverted uterus
   (d) U/S guidance may be helpful in a training institution, as it offers real time feedback for both the trainee and instructor
   (e) U/S guidance provides the patient reassurance that the embryos are being delivered appropriately.
   (f) A 2010 Cochrane review of 17 RCTs comparing ultrasound and clinical touch method concluded that ultrasound increased ongoing clinical pregnancy rates (441/1,254 vs. 350/1,218, odds ratio 1.38, 95 % CI 1.16–1.64, $p<0.0003$) [18]. Two additional meta-analyses of RCTs have led to similar conclusions [19–20]. Most studies included, however, are limited by markedly different embryo transfer techniques, lack of standardization of operator experience, low power and inadequate randomization.

Disadvantages of ultrasound may include the need for a second operator, a longer procedure time, and patient discomfort secondary to a full bladder and pressure from the probe.

References

1. Mains L, Voorhis B (2010) Optimizing the technique of embryo transfer. Fertil Steril 94:785–790
2. Brook N, Khalaf Y, Coomarasamy A, Edgeworth J, Braude P (2006) A randomized controlled trial of prophylactic antibiotics (co-amoxiclav) prior to embryo transfer. Hum Reprod 23:2911–2915
3. Practice Committee of Society of Assisted Reproductive Technology, Practice Committee or American Society for Reproductive Medicine (2008) Guidelines of number of embryos transferred. Fertil Steril 90:163–164
4. Coroleu B, Barri PN, Carreras O, Martinez F, Parriego M, Mereter L (2002) The influence of the depth of the embryo replacement into the uterine cavity on implantation rates after IVF: a controlled, ultrasound-guided study. Hum Reprod 17:341–346
5. Franco JG, Bai G, Prosinecki V, Abrunhosa F, Ferreira GC, Bastos M (2004) Best site for ET: the upper or lower half of endometrial cavity? Hum Reprod 19:1785–1790
6. Amarin ZO, Obeidat BR (2004) Bed rest versus free mobilization following ET: a prospective randomized study. BJOG 111:1273–1276
7. Purcell KJ, Schembri M, Telles TL, Fujimoto VY, Cedars MI (2007) Bed rest after embryo transfer: a randomized controlled trial. Fertil Steril 87:322–326
8. Mansour R, Aboulghar M, Serour G (1990) Dummy embryo transfer: a technique that minimizes the problems of embryo transfer and improves the pregnancy rate in human in vitro fertilization. Fertil Steril 54:678–681
9. Eskandar MA, Abou-setta AM, El-Amin M, Almushait MA, Sobande AA (2007) Removal of cervical mucus prior to embryo transfer improves pregnancy rates in women undergoing assisted reproduction. Reprod Biomed Online 14:308–313
10. Visschers BA, Bots RS, Peeters MF, Mol BW, van Dessel JH (2007) Removal of cervical mucus: effect on pregnancy rates in IVF/ICSI. Reprod Biomed Online 15:310–315
11. Prapas N, Prapas Y, Panagiotidis Y, Prapa S, Vanderzwalmen P, Makedos G (2004) Cervical dilatation has a positive impact on the outcome of IVF in randomly assigned cases having two previous difficult embryo transfers. Hum Reprod 19:1791–1795
12. Groutz A, Yovel I, Lessing J, Azem F, Wolf Y, Amit A (1997) Cervical dilatation during ovum pick-up in patients with cervical stenosis: effect on pregnancy outcome in an in vitro fertilization-embryo transfer program. Fertil Steril 67:909–911
13. Serhal P, Ranieri D, Khadum I, Wakim R (2003) Cervical dilatation with hygroscopic rods prior to ovarian stimulation facilitates embryo transfer. Hum Reprod 18:2618–2620
14. Yanushpolsky E, Ginsburg S, Fox J, Stewart E (2000) Transcervical placement of a Malecot catheter after hysteroscopic evaluation provides for easier entry into the endometrial cavity for women with histories of difficult intrauterine inseminations and/or embryo transfers: a prospective case series. Fertil Steril 73:402–405
15. Abou-Setta AM, Al-Inany HG, Mansour RT, Serour GI, Aboulghar MA (2005) Soft versus firm embryo transfer catheters for assisted reproduction: a systematic review and meta-analysis. Hum Reprod 20:3114–3121
16. Buckett W (2006) A review and meta-analysis of prospective trials comparing different catheters used for embryo transfer. Fertil Steril 85:728–734
17. El Shawarby SA, Ravhon A, Skull J et al (2008) A prospective randomized controlled trial of Wallace and Rocket embryo transfer catheters. Reprod Biomed Online 17:549–552
18. Brown JA, Buckingham K, About-Setta A, Buckett W (2010) Ultrasound versus "clinical touch" for catheter guidance during embryo transfer in women [review]. Cochrane Database Syst Rev (1):CD006107
19. Abou-Setta AM, Mansour RT, Al-Inany HG, Aboulghar MM, Aboulghar MA, Serour GI (2007) Among women undergoing embryo transfer, is the probability of pregnancy and live birth improved with ultrasound guidance over clinical touch alone? A systemic review and meta-analysis of prospective randomized trials. Fertil Steril 88:333–341
20. Buckett WM (2003) A meta-analysis of ultrasound-guided versus clinical touch embryo transfer. Fertil Steril 80:1037–1041

Chapter 26

Safety of Intracytoplasmic Sperm Injection

Gianpiero D. Palermo, Queenie V. Neri, and Zev Rosenwaks

Abstract

Early follow-up studies of IVF children showed that the frequency of birth anomalies resembled those arising with natural conception. More detailed analyses confirmed these findings, reinforcing the concept of the preimplantation period as teratologically "safe." The use of intracytoplasmic sperm injection (ICSI) to achieve fertilization introduced another variable.

ICSI's safety has often been criticized because the fertilizing spermatozoon neither binds to the zona pellucida nor fuses with oolemma. Bypassing these physiologic steps together with the arbitrary selection of the spermatozoon has been reason for concern. Thus far, ICSI offspring undergoing adolescence and beyond has provided sufficient information to reassure these qualms. In fact, the health of the offspring generated through ICSI, once taken into consideration the gestational order, the age and the genetic makeup of the couples are generally reassuring.

Key words ICSI, IVF, Children follow-up, Assisted reproductive technologies, Multiple gestation, Birthweight, Gestational age

1 Safety of ART

It is likely that well over five million babies have been born worldwide using assisted reproductive technologies (ART) [1]. The most important aspect of ART is the generation of healthy offspring. Since IVF's early days [2, 3], the long-term health of children born as a result of this procedure has been closely scrutinized. Though assisted reproduction treatments have rapidly become largely standardized worldwide, the greatest variation in ART procedures can be found in the use of different embryo culture media and media protein supplements; the adoption of a variety of different procedures for the collection, storage, and insemination of gametes; and the generation and storage of conceptuses.

Early follow-up studies of IVF children showed that the frequency of birth anomalies resembled those arising with natural conception [3]. More detailed analyses confirmed these findings, reinforcing the concept of the preimplantation period as

Zev Rosenwaks and Paul M. Wassarman (eds.), *Human Fertility: Methods and Protocols*, Methods in Molecular Biology, vol. 1154, DOI 10.1007/978-1-4939-0659-8_26, © Springer Science+Business Media New York 2014

teratologically "safe." The use of intracytoplasmic sperm injection (ICSI) [4] to achieve fertilization introduced another variable. ICSI's safety has often been criticized because the fertilizing spermatozoon neither binds to the zona pellucida nor fuses with the oolemma. Calcium spiking and sperm decondensation in ICSI fertilizations are also peculiar, and the claim of the persistence of acrosomal vesicle may delay the decondensation of early paternal pronucleus [5]. Nevertheless, despite these apparent risk factors, birth anomaly rates were only slightly higher in children conceived via ICSI in follow-up studies [6].

At present 2–4 % of children born in developed countries are conceived through ART [7, 8], with ICSI being the most effective means of treating couples with male factor infertility and previous fertilization failures. In fact, even spermatozoa collected from the epididymis and from the testis achieve comparable fertilization and consistent pregnancy rates when used in conjunction with this technique.

2 Genetic Disease Concern

As more and more men with severe oligospermia or azoospermia become fathers via ART, the occasional association of these conditions with karyotypic anomalies or Y-chromosome microdeletions has raised concerns as to whether ICSI facilitates the transmission of genetic defects to these patients' offspring. For instance, a higher incidence of sperm aneuploidy in infertile men with secretory azoospermia [9–12] is expected to produce a higher frequency of gonosomal abnormalities in the male progeny [13, 14], possibly because of meiotic defects surfacing during male germ line maturation [15]. These offspring anomalies can also arise *de novo*, as was indicated from an in-house genetic assessment of 5-year-old ICSI children with a 3.6 % incidence of gonosomal abnormalities [16] compared with 0.15 % incidence in a naturally conceived population [17]. In fact, based on current data and in agreement with the literature, it appears that a deletion in the AZF region of the Y chromosome presents in the male progeny alongside a compromised ability to produce spermatozoa [10, 18].

Expansion in the triplet repeat constituting the androgen receptor (*AR*) gene has also been linked with spermatogenetic defects [19]. However, our study confirmed that its maternal transmission of the AR gene protects the male progeny generated by infertile men from inheriting a dysfunctional AR gene, and therefore from the gene's eventual deleterious effect on spermatogenesis [20].

A different issue arises from the DNA fragmentation observed in suboptimal spermatozoa, which could affect embryo development and become a source of epigenetic disorders in phenotypically normal offspring [21]. However, such fragmentation does not seem to preclude the establishment of a pregnancy [22, 23], which can possibly be attributed to the corrective role of single sperm selection in the ICSI procedure [24, 25].

3 Follow-Up of Children Born After ART

Shorter gestations and lower birth weights have been noted in assisted reproduction pregnancies, irrespective of the conception method [26, 27] utilized. The basis for this could lie with the very factors that are involved in the parents' infertility [28], the fact that a singleton may result from early in utero loss of a twin [29], or that Caesarian deliveries of ART babies are often performed at the slightest sign of fetal distress [30]. However, when we stratified our data according to maternal age, singleton ICSI children had the same mean gestational age and birth weight as those conceived naturally [16]. Similarly, in the study of Lin *et al.* [31], levels of neonatal distress, NICU admission, and congenital malformations were comparable in both groups.

One unsettling investigation on the health of assisted reproduction offspring reported a twofold increase in major malformations [32]. However, this report failed to correct for maternal age, other patient characteristics [33], the presence of male factor infertility [34], and in its abnormality classification methods [35].

Follow-up studies of ICSI children, beginning in the mid-1990s, have revealed an incidence of malformations within the expected range for the general population of New York State [36]. Another series comparing singleton ICSI children that were matched for maternal age with naturally conceived children displayed no difference in neonatal outcomes [31]. A recent series investigating the outcomes of neonates generated by different assisted conception procedures, ICSI versus IVF with conventional fertilization, provided further confirmation of the expected rate of malformation [37].

Bowen and colleagues [38] evaluated the medical and developmental state of 1-year-old children born after ICSI or IVF as well as after natural conceptions. They found that most 1-year-old ICSI children were healthy and were developing normally, as measured by the Bayley Scales of Infant Development. However, about 17 % displayed an increase in learning difficulties compared with those conceived by IVF or naturally. A later report dismissing this concern in 2-year-old ICSI toddlers [39] inspired a follow-up study in 3 year olds. This age was chosen because mental development scores collected at 2.5 years are more predictive of mental performance in later life [40]. Moreover, in screening for a larger number of children using parent-administered questionnaires, it was found that the great majority of the 3-year-old children analyzed in the ICSI and IVF groups had normal cognitive abilities, socio-emotional development, and motor skill scores. This could not be confirmed for children from high-order gestations, however, as more frequent cognitive delays are observed in this group than in children resulting from singleton pregnancies, regardless of the conception method [41]. Interestingly, ICSI children whose fathers' spermatozoa were harvested surgically appeared to score better than those conceived with ejaculated spermatozoa [42]. In a German follow-up study of 5- and 10-year-old children, researchers found that the child's capacity in two different cognitive styles, sequential versus simultaneous mental processing, as measured by the Kaufman Assessment of Battery for Children, fell within the normal range for both IVF and ICSI children [43]. However, children born after standard IVF had better simultaneous mental processing than ICSI children ($P < 0.005$). In a different follow-up of 10-year-old children, it was found that ICSI and their naturally conceived counterparts had similar motor skills and IQ [44].

Because of the confounding role of multiple gestations, it was then decided to assess singleton births at 5 years of age as part of a multicenter international investigation. Although this study's results revealed no differences in the full scale IQ between ICSI and naturally conceived children, parents of naturally conceived children surprisingly displayed higher levels of distress and dysfunctional child interactions, and children appeared to be more difficult compared with those conceived by ICSI parents ($P < 0.05$) [45]. This study also revealed that ICSI children were indeed characterized by a lower birth weight ($P < 0.05$) and also by a higher proportion of major malformations ($P < 0.05$) than their naturally conceived counterparts. The higher proportion of affected children was concentrated in the Belgian surveys [13]. Another study comprising 1,500 children from five European countries failed to show a higher incidence of major malformation in the assisted reproduction group compared with the naturally conceived group despite a clear difference in their parental age [46]. The same Brussels group evidenced a malformation rate of 10.0 % versus 3.4 % in the same cohort of naturally conceived children once they reached 8 years of age [47]. These three additional malformations consisted of an inguinal hernia and two nevus flammeus that required only minor surgical corrections. Interestingly, when the same ICSI progeny were reassessed for physical and intellectual performances at 10 years of age, their state matched that of a similar natural conception group [44]. In another follow-up study, ICSI, IVF, and naturally conceived neonates were assessed for perinatal outcome, health, growth, and medical care needs. While the authors observed that ICSI and IVF children's developmental outcome were comparable, ICSI children had poorer perinatal outcome in comparison to their naturally conceived counterparts ($P < 0.05$) [48]. A recent survey on neonates demonstrated that

ICSI children born after the use of surgically retrieved spermatozoa had, however, a higher incidence of cardiac defects in comparison to ICSI using ejaculated sperm and those that are spontaneously conceived [49].

4 Epigenetic Concerns

Questions have been posed about the effect of embryo culture media on the health of assisted reproduction children and whether the extended embryo culture can induce imprinting disorders [50]. Some reports have claimed an association of Angelman syndrome with assisted reproduction; however, assessment of the q11–13 locus of chromosome 15 has not revealed any methylation abnormalities. Although the inability to detect any methylation abnormalities in this locus may be attributed to some assay variability [51], a clear link between the occurrence of Angelman syndrome and Prader–Willi syndrome with assisted conception has not been confirmed [52]. Thus, some studies claimed that the association of imprinting disorders and assisted reproduction is tentative and that use of ART carries an absolute risk that may be negligible [53]. Moreover, even in the loss of methylation/imprinting, parental alleles can retain their parent-of-origin-dependent specificity, possibly through other epigenetic phenomenon such as histone modifications [54]. The quantification of the relative expression of some key developmental imprinted genes using fetal adnexa of assisted reproduction offspring revealed imbalances insufficient to cause any clinical manifestation [55].

The fact that nonsymptomatic anomalies of gene expression are indeed present and detectable in placental tissues [56] of assisted reproduction offspring underscores the need to continue monitoring the influence of reproductive techniques on the conceptus. These variations in gene expression have also been noted in a study of a transgenic mouse strain characterized by its high sensitivity to oxidative insults, in which assisted conception procedures such as *in vitro* insemination or ICSI did not affect the phenotype in comparison with natural mating [57]. Although this study excludes the conception methods as being potentially responsible for altering offspring epigenetics, other common steps used in the assisted reproduction protocols need to be evaluated. While imprinting imbalances may be a consequence of simple embryo manipulations such as embryo transfer procedures, their incidence seems to be most dependent upon the *in vitro* culture used. *In vitro* culture appears to exert more serious imprinting imbalances, characterized by loss of paternal methylation of H19 and consequent reduction in expression in the same allele of IGF2 [58]. H19 and IGF2's interdependence results from the binding of the repressor factor CTCF (CCCTC-binding factor zinc finger protein, human transcriptional repressor mRNA) to the unmethylated differential methylation regions (DMR), which then prevents the H19/IGF2 common enhancers from activating the IGF2 promoter [59, 60]. This contrasts with the normal scenario of a paternal allele, whose H19 is methylated and silenced, whereas the reciprocally linked IGF2 gene is actively transcribed. The Rivera study [58] reported that downregulation of H19 in the assisted reproduction population occurred without a concurrent overexpression of IGF2. Although this could be an occasional finding, it may point to a subliminal deleterious effect of the *in vitro* culture conditions on the preferentially expressed maternal genes H19 and CDKN1C. The observed downregulation of these genes was clearly isolated and insufficient to induce an upregulation of the preferentially expressed paternal genes, and therefore, was incapable of projecting their negative effect on the offspring phenotype. This minor imbalance is confirmed by the difference in the measurement of whole genome analysis and the distribution of gene expression. The two conception methods seem to share a comparable, analogous gene pattern that is somewhat different from naturally conceived children, also seen in the above-mentioned experimental work [57].

A survey of 1,000 British families raising children afflicted with transient neonatal diabetes mellitus, Prader–Willi syndrome, Angelman syndrome, or Beckwith–Wiedemann syndrome showed that only the latter, where H19 downregulation plays a pivotal role, appeared to maintain an association with assisted reproduction [52].

That the maternal genes H19 and CDKN1C are involved with imprinting disorders in assisted reproduction children independent of the insemination method suggests that the spermatozoon or the fertilization technique itself is not the main culprit. This, however, was not confirmed during an analysis of moderate and severe oligozoospermic men who displayed a loss of methylation of the H19 gene [61]. Other authors observed methylation imbalances of paternal genes such as HUMARA and SNPRN during seminiferous tubule spermatogenesis where a partial methylation of HUMARA in non-obstructive azoospermia was detected [62]. A later study established a complete loss of methylation of the paternal H19 and a concurrent hypermethylation of paternally expressed MEST in men with spermatozoa concentration of $\leq 10 \times 10^6$/ml [63]. These imprinting defects are identifiable in individuals with disrupted spermatogenesis and worsen in conjunction with the severity of the spermatogenic failure. Erroneous methylation of these genes could be due to the erasure of methylation marks during germline development or could be acquired *de novo*, either during the formation of the spermatozoa in the seminiferous tubules or through abnormal epididymal function [64]. This suggests that genomic imprinting defects might underlie infertility by itself [63] again carrying a significant risk of causing biallelic inactivation of the IGF2 gene in the human embryo consequent to a biallelic expression of the linked H19. However, different patterns of sperm methylation may occur in the same patient, allowing spermatozoa with severe imprinting errors to coexist with sperm with normal imprinting marks. This fact suggests that misregulation leading to error in the establishment of imprinting marks is not necessarily widespread throughout the seminiferous tubules [63]. Accordingly, abnormal methylation patterns of imprinted genes in cases with disrupted spermatogenesis might be derived from different imprinting marks in the adult testicular germinal stem cell pool, from defects occurring during progenitor diploid germ cell expansion, or during haploid germ cell differentiation [63]. In fact, by analyzing methylation characteristics of SNRPN and H19 genes, other investigators did not confirm the finding even in spermatogenetically compromised seminiferous tubules [65]. This is supported by the observation of these disorders in association with standard *in vitro* insemination, where spermatogenesis is presumably not compromised, and points at factors in play at conception or during early embryo development that may be attributable to the exposure to culture medium [58, 66, 67].

On the other hand, since imprinting defects occur more frequently as a function of the number of assisted reproduction cycles, hormonal treatments rather than laboratory procedures may be to blame [53, 68, 69]. This suggests that the main risk of epigenetic diseases after assisted reproduction might be oocyte-related. This is consistent with the finding that the only identifiable common factor in 12 Beckwith–Wiedemann syndrome (BWS) children conceived by different reproductive techniques was ovarian stimulation [53, 65, 70, 71]. In fact, in a case report of two BWS children born to the same parents, one child was conceived naturally (after exposure to ovarian stimulation) and one child via standard IVF. This case suggests that both infertility and ovarian stimulation are possible risk factors for imprinting disorders [72].

Since some assisted reproduction procedures have been implicated in various adverse outcomes of the offspring, basic research is required to elucidate the biological mechanisms underlying the genetic and epigenetic effects of assisted reproduction. In addition, it is important for clinicians to precisely record the assisted reproduction procedures including the stimulation protocol, method of embryo culture, culture media used, and timing of embryo transfer. As it is not yet possible to precisely evaluate the consequences of assisted reproduction on imprinting, long-term, large-scale epidemiological projects and follow-up studies that could estimate the magnitude of the risks posed by assisted reproduction on human pregnancies are sorely needed.

While multiple case reports and case series have suggested an association between imprinting disorders and ART, recent cohort studies have failed to confirm the association [73]. Taking all of these studies into consideration, the current data suggest an association between imprinting disorders and ART that carries a low absolute incidence risk.

5 Prepubertal and Pubertal Development

A British follow-up study of young ART children identified that IVF and ICSI children were comparable to naturally conceived adolescents counterparts in terms of head circumference, height, and weight at any time point analyzed from birth, 5 years, 7–9 years, and 10–12 years [74]. Results from another survey, completed by parents, suggested that 5- and 7-year-old ART children had higher risks of respiratory problems characterized by asthma, wheezing, or taking anti-asthmatic medication when compared to children conceived naturally or after ovulation induction [75]. When Leydig cell function of pubertal boys aged 14 years resulting from ICSI was assessed, either through venous puncture or saliva sampling, testosterone concentrations were comparable to naturally conceived boys [76, 77]. In another series of investigations, the same researchers monitored pubertal development by Tanner stage (breast, genital, and pubic hair development) and age at menarche, in singleton born ICSI boys and girls, and did not observe any obvious differences in sexual development in ICSI adolescents compared to their 14-year-old spontaneously conceived counterparts, except for less pronounced breast development in ICSI females [78]. Finally, adiposity and body fat distribution were also measured in these 14-year-old adolescents. No differences in body composition measurements were found between ICSI and the control cohort. However, in boys with more advanced pubertal stages, there was a higher sum of peripheral skinfolds. In addition, the peripheral adiposity and body fat percentage of ICSI girls were significantly higher than their spontaneously conceived counterparts [78].

Overall, studies of children ranging from newborn to 14 years of age [43, 44, 47, 48, 74–77] have been reassuring in terms of perinatal outcome, IQ, and physical development [79]. Further follow-up on ICSI teenagers into adulthood should be continued to evaluate the reproductive capacity of these youngsters.

6 Counseling Infertile Couples About Risks

Once the ART option is offered to a couple in our practice, choosing a specific method of insemination is a complex decision because of the looming chances of failed fertilization due to subtle factors conferred by both gametes. This pressure on infertility specialists may induce them to bypass the academic approach and directly target a microinsemination method, as ICSI may spare the couple the emotional and financial costs related to fertilization failure. While ICSI was developed to address specific defects of the male gamete, it is now mastered worldwide and is effective in a variety of contexts.

ICSI is equally effective with fresh or frozen sperm specimens as well as ejaculated or surgically retrieved spermatozoa. Further, the procedure's utility is independent of the characteristics of the semen parameters or the presence of antisperm antibodies.

Moreover, ICSI's dependability has broadened its initial use from a technique capable of overriding the dysfunctionality of spermatozoa to one that may partly compensate for problems with the egg. Indeed, ICSI has allowed successful fertilizations when only few and/or abnormal oocytes were available because it allows direct visualization of oocyte maturation by stripping cumulus cells from the oocytes. In fact, the availability of ICSI has been instrumental in some European countries, including Italy and Germany, in circumventing restrictive legislation that limits the number of oocytes inseminated or embryos to be replaced.

ICSI has also been instrumental in fertilizing cryopreserved oocytes which otherwise are resistant to natural sperm penetration due to zona pellucida hardening resulting from premature exocytosis of cortical granules. Furthermore, ICSI is the preferred conception method during the application of preimplantation genetic diagnosis since it avoids contamination of the zona with sperm DNA, thus increasing the number of embryos available for screening.

Moreover, ICSI has an impact in the arena of HIV infection. While many patients infected with HIV-1 show interest in beginning a family, most serodiscordant couples are concerned with the possibility of both horizontal and vertical transmission of the virus. In such cases, intrauterine insemination with spermatozoa processed by double gradient centrifugation followed by swim up has been the suggested method of treating serodiscordant couples with an HIV-1 infected male partner. The use of ICSI in this context has been proposed by several groups because of its negligible oocyte exposure to semen, thereby reducing the risk of viral transmission.

ICSI has been so effective, in fact, that we can now use any spermatozoa collected from different sources within the male reproductive system, can be less concerned about the presence of antisperm antibodies and sperm preparation method utilized, and even disregard morphology, acrosome presence, abnormal chromatin packing, or the lack of a complete flagellum or display poor motility. ICSI's use seems destined only to increase, particularly with the introduction of streamlined and automated approaches to *in vitro* insemination.

It is because of ICSI that we reap most of the benefits related to this entirely different way of generating a conceptus. In fact, the removal of the cumulus cells to allow sperm injection offers a window to assess the maturity of oocytes and pinpoint the exact timing of fertilization's beginning. This also provides us with information on the efficacy of specific ovarian stimulation protocols, which may be particularly helpful as we strive to reduce the stress to the ovary and minimize the number of oocytes retrieved while optimizing their quality. Abnormal ICSI fertilizations have helped us to shed light on the inheritance of the centrosome, as previously elucidated.

ICSI has been invaluable in understanding the mechanism of oocyte activation, whether due to ooplasmic asynchrony or a lack of spermatozoal cytosolic factor. The use of ICSI has generated a great deal of concern since it may increase the risk of transmitting genetic diseases to the offspring, as confirmed by studies reporting on the higher incidence of sex chromosomal aneuploidy of paternal origin and structural *de novo* chromosomal abnormalities in children conceived after ICSI compared to the general population. These reports indicated that men with spermatogenic impairment due to a primitive testiculopathy have an increased sperm aneuploidy rate, which negatively correlates with the main sperm parameters despite a normal somatic karyotype. This suggests that the *noxae* acting at the testicular level not only impairs spermatogenesis but also affects the molecular mechanisms involved in chromosomal segregation.

In spite of these unsettling implications, ICSI offspring resulting from a singleton pregnancy are in general reassuringly healthy. Normal, healthy offspring also result when surgically retrieved samples or when scarce sperm cells with "prohibitive" morphology are injected.

It is our belief that during consultation with infertile couples presenting with borderline semen parameters, where there are dubious chances of fertilizing a conceptus, or when it is foreseen that the individuals may not be able to emotionally handle an unexpected fertilization failure, ICSI should not be denied. As in all aspects of medicine the counseling begins with the description of the side effects of medications, the surgical risks linked to egg retrieval, and the possibility of transmission of genetic disorders (even if currently unknown to the prospective parents) related to the oocyte or the sperm. Nonetheless, the chances of a *de novo* appearance of a disorder resulting from the syngamy of the two parental gametes should also be discussed. The concerns related to the health of ICSI offspring are not linked to the procedure itself, but mostly to the genetic or epigenetic conditions of the parents that is later expressed through the union of their gametes.

In summary, the most important factor that can lead to adverse outcomes in offspring conceived by IVF or ICSI is the occurrence of high-order pregnancies. However, the introduction of single embryo transfer has reduced this considerably. Although perinatal outcomes such as prematurity, low birth weight, perinatal mortality, and increased incidence of malformations have been linked to the techniques of IVF and ICSI, infertility itself seems to be the larger issue that leads to negative clinical outcomes. Overall, no significant long-term neurodevelopmental differences have been found in connection with ART, though the risks associated with childhood cancer and future fertility still require further investigation.

Acknowledgements

We are very appreciative to all clinicians and scientists at The Ronald O. Perelman & Claudia Cohen Center for Reproductive Medicine and the Urology Department.

References

1. Adamson GD, de Mouzon J, Lancaster P, Nygren KG, Sullivan E et al (2006) World collaborative report on in vitro fertilization, 2000. Fertil Steril 85:1586–1622
2. Steptoe PC, Edwards RG (1978) Birth after the reimplantation of a human embryo. Lancet 2:366
3. Steptoe PC, Edwards RG, Walters DE (1986) Observations on 767 clinical pregnancies and 500 births after human in-vitro fertilization. Hum Reprod 1:89–94
4. Palermo G, Joris H, Devroey P, Van Steirteghem AC (1992) Pregnancies after intracytoplasmic injection of single spermatozoon into an oocyte. Lancet 340:17–18
5. Ludwig M, Schroder AK, Diedrich K (2001) Impact of intracytoplasmic sperm injection on the activation and fertilization process of oocytes. Reprod Biomed Online 3:230–240
6. Ludwig M, Katalinic A (2002) Malformation rate in fetuses and children conceived after ICSI: results of a prospective cohort study. Reprod Biomed Online 5:171–178
7. Fauser BC, Devroey P, Macklon NS (2005) Multiple birth resulting from ovarian stimulation for subfertility treatment. Lancet 365:1807–1816
8. Klemetti R, Gissler M, Hemminki E (2002) Comparison of perinatal health of children born from IVF in Finland in the early and late 1990s. Hum Reprod 17:2192–2198
9. Colombero LT, Moomjy M, Sills ES, Rosenwaks Z, Palermo GD (1999) The role of structural integrity of the fertilising spermatozoon in early human embryogenesis. Zygote 7:157–163
10. Katagiri Y, Neri QV, Takeuchi T, Schlegel PN, Megid WA et al (2004) Y chromosome assessment and its implications for the development of ICSI children. Reprod Biomed Online 8:307–318
11. Palermo GD, Cohen J, Alikani M, Adler A, Rosenwaks Z (1995) Development and implementation of intracytoplasmic sperm injection (ICSI). Reprod Fertil Dev 7:211–217, discussion 217–218
12. Palermo GD, Colombero LT, Hariprashad JJ, Schlegel PN, Rosenwaks Z (2002) Chromosome analysis of epididymal and testicular sperm in azoospermic patients undergoing ICSI. Hum Reprod 17:570–575
13. Bonduelle M, Bergh C, Niklasson A, Palermo GD, Wennerholm UB (2004) Medical follow-up study of 5-year-old ICSI children. Reprod Biomed Online 9:91–101
14. Bonduelle M, Legein J, Buysse A, Van Assche E, Wisanto A et al (1996) Prospective follow-up study of 423 children born after intracytoplasmic sperm injection. Hum Reprod 11:1558–1564
15. Ferguson KA, Wong EC, Chow V, Nigro M, Ma S (2007) Abnormal meiotic recombination in infertile men and its association with sperm aneuploidy. Hum Mol Genet 16:2870–2879
16. Palermo GD, Neri QV, Takeuchi T, Squires J, Moy F et al (2008) Genetic and epigenetic characteristics of ICSI children. Reprod Biomed Online 17:820–833
17. Hassold T, Abruzzo M, Adkins K, Griffin D, Merrill M et al (1996) Human aneuploidy: incidence, origin, and etiology. Environ Mol Mutagen 28:167–175
18. Kent-First MG, Kol S, Muallem A, Blazer S, Itskovitz-Eldor J (1996) Infertility in intracytoplasmic-sperm-injection-derived sons. Lancet 348:332
19. Dowsing AT, Yong EL, Clark M, McLachlan RI, de Kretser DM et al (1999) Linkage between male infertility and trinucleotide repeat expansion in the androgen-receptor gene. Lancet 354:640–643
20. Katagiri Y, Neri QV, Takeuchi T, Moy F, Sills ES et al (2006) Androgen receptor CAG poly-

morphism (Xq11-12) status and human spermatogenesis: a prospective analysis of infertile males and their offspring conceived by intracytoplasmic sperm injection. Int J Mol Med 18:405–413

21. Aitken RJ, De Iuliis GN (2007) Origins and consequences of DNA damage in male germ cells. Reprod Biomed Online 14:727–733
22. Feliciano M, Neri QV, Kent-First M, Wehbe SA, Evenson D et al (2004) Assays of sperm nuclear status do not correlate with ICSI success. Hum Reprod 19:i52
23. Neri QV, Evenson D, Wehbe SA, Kent-First M, Squires J et al (2004) Does DNA fragmentation in spermatozoa affect ICSI offspring development? Hum Reprod 19:i107
24. Bungum M, Humaidan P, Spano M, Jepson K, Bungum L et al (2004) The predictive value of sperm chromatin structure assay (SCSA) parameters for the outcome of intrauterine insemination, IVF and ICSI. Hum Reprod 19:1401–1408
25. Virro MR, Larson-Cook KL, Evenson DP (2004) Sperm chromatin structure assay (SCSA) parameters are related to fertilization, blastocyst development, and ongoing pregnancy in in vitro fertilization and intracytoplasmic sperm injection cycles. Fertil Steril 81:1289–1295
26. Schieve LA, Ferre C, Peterson HB, Macaluso M, Reynolds MA et al (2004) Perinatal outcome among singleton infants conceived through assisted reproductive technology in the United States. Obstet Gynecol 103:1144–1153
27. Schieve LA, Meikle SF, Ferre C, Peterson HB, Jeng G et al (2002) Low and very low birth weight in infants conceived with use of assisted reproductive technology. N Engl J Med 346:731–737
28. Kovalevsky G, Coutifaris C (2002) Low and very low birth weight after use of assisted reproductive technology. N Engl J Med 347:1451–1452, author reply 1451–1452
29. Davies MJ (2002) Low and very low birth weight after use of assisted reproductive technology. N Engl J Med 347:1451–1452, author reply 1451–1452
30. Shwayder JM (2007) Liability in high-risk obstetrics. Obstet Gynecol Clin North Am 34: 617–625
31. Lin K, Kang HJ, Neri QV, Rosenwaks Z, Palermo GD (2003) Obstetrical and perinatal outcomes of ICSI versus natural singleton pregnancies. Fertil Steril 80:S109
32. Hansen M, Kurinczuk JJ, Bower C, Webb S (2002) The risk of major birth defects after intracytoplasmic sperm injection and in vitro fertilization. N Engl J Med 346:725–730
33. Steinkampf MP, Grifo J (2002) Major birth defects after assisted reproduction. N Engl J Med 347:1449–1451, author reply 1449–1451
34. Sills ES, Palermo GD (2002) Major birth defects after assisted reproduction. N Engl J Med 347:1449–1451, author reply 1449–1451
35. Sutcliffe AG, Bonduelle M, Taylor BW (2002) Major birth defects after assisted reproduction. N Engl J Med 347:1449–1451, author reply 1449–1451
36. Palermo GD, Colombero LT, Schattman GL, Davis OK, Rosenwaks Z (1996) Evolution of pregnancies and initial follow-up of newborns delivered after intracytoplasmic sperm injection. JAMA 276:1893–1897
37. Neri QV, Takeuchi T, Kang HJ, Lin K, Wang A et al (2006) Genetic assessment and development of children that result from assisted reproductive technology. Clin Obstet Gynecol 49:134–137
38. Bowen JR, Gibson FL, Leslie GI, Saunders DM (1998) Medical and developmental outcome at 1 year for children conceived by intracytoplasmic sperm injection. Lancet 351: 1529–1534
39. Bonduelle M, Joris H, Hofmans K, Liebaers I, Van Steirteghem A (1998) Mental development of 201 ICSI children at 2 years of age. Lancet 351:1553
40. Mccall RB, Hogarty PS, Hurlburt N (1972) Transitions in infant sensorimotor development and prediction of childhood IQ. Am Psychol 27:728–748
41. Neri QV, Squires J, Williams-Pitman M, Wang A, Rosenwaks Z et al (2003) Multiple gestation and not semen character is responsible for developmental abnormalities in 3 and 5 year old ART children. Hum Reprod 18:xviii102
42. Neri QV, Williams-Pitman M, Rosenwaks Z, Squires J, Palermo GD (2007) Cognitive and behavioral developmental profiles of ICSI children conceived from surgically retrieved spermatozoa. Fertil Steril 88:S39
43. Goldbeck L, Gagsteiger F, Mindermann I, Strobele S, Izat Y (2009) Cognitive development of singletons conceived by intracytoplasmic sperm injection or in vitro fertilization at age 5 and 10 years. J Pediatr Psychol 34: 774–781
44. Leunens L, Celestin-Westreich S, Bonduelle M, Liebaers I, Ponjaert-Kristoffersen I (2008) Follow-up of cognitive and motor development of 10-year-old singleton children born after ICSI compared with spontaneously conceived children. Hum Reprod 23:105–111

45. Ponjaert-Kristoffersen I, Tjus T, Nekkebroeck J, Squires J, Verte D et al (2004) Psychological follow-up study of 5-year-old ICSI children. Hum Reprod 19:2791–2797
46. Bonduelle M, Wennerholm UB, Loft A, Tarlatzis BC, Peters C et al (2005) A multi-centre cohort study of the physical health of 5-year-old children conceived after intracytoplasmic sperm injection, in vitro fertilization and natural conception. Hum Reprod 20:413–419
47. Belva F, Henriet S, Liebaers I, Van Steirteghem A, Celestin-Westreich S et al (2007) Medical outcome of 8-year-old singleton ICSI children (born > or = 32 weeks' gestation) and a spontaneously conceived comparison group. Hum Reprod 22:506–515
48. Knoester M, Helmerhorst FM, Vandenbroucke JP, van der Westerlaken LA, Walther FJ et al (2008) Perinatal outcome, health, growth, and medical care utilization of 5- to 8-year-old intracytoplasmic sperm injection singletons. Fertil Steril 89:1133–1146
49. Fedder J, Loft A, Parner ET, Rasmussen S, Pinborg A (2013) Neonatal outcome and congenital malformations in children born after ICSI with testicular or epididymal sperm: a controlled national cohort study. Hum Reprod 28:230–240
50. Edwards RG (2002) Molecular genetics analyses in basic and clinical embryology. Reprod Biomed Online 5:220–221
51. White HE, Durston VJ, Harvey JF, Cross NC (2006) Quantitative analysis of SNRPN(correction of SRNPN) gene methylation by pyrosequencing as a diagnostic test for Prader-Willi syndrome and Angelman syndrome. Clin Chem 52:1005–1013
52. Sutcliffe AG, Peters CJ, Bowdin S, Temple K, Reardon W et al (2006) Assisted reproductive therapies and imprinting disorders–a preliminary British survey. Hum Reprod 21:1009–1011
53. Shiota K, Yamada S (2005) Assisted reproductive technologies and birth defects. Congenit Anom (Kyoto) 45:39–43
54. Delaval K, Govin J, Cerqueira F, Rousseaux S, Khochbin S et al (2007) Differential histone modifications mark mouse imprinting control regions during spermatogenesis. EMBO J 26:720–729
55. Kang HJ, Katagiri Y, Neri QV, Baergen R, Rosenwaks Z et al (2007) Altered gene expression patterns in ART offspring. Fertil Steril 82:S292
56. McMinn J, Wei M, Sadovsky Y, Thaker HM, Tycko B (2006) Imprinting of PEG1/MEST isoform 2 in human placenta. Placenta 27: 119–126
57. Wilson TJ, Lacham-Kaplan O, Gould J, Holloway A, Bertoncello I et al (2007) Comparison of mice born after intracytoplasmic sperm injection with in vitro fertilization and natural mating. Mol Reprod Dev 74: 512–519
58. Rivera RM, Stein P, Weaver JR, Mager J, Schultz RM et al (2008) Manipulations of mouse embryos prior to implantation result in aberrant expression of imprinted genes on day 9.5 of development. Hum Mol Genet 17:1–14
59. Reik W, Walter J (2001) Genomic imprinting: parental influence on the genome. Nat Rev Genet 2:21–32
60. Takai D, Gonzales FA, Tsai YC, Thayer MJ, Jones PA (2001) Large scale mapping of methylcytosines in CTCF-binding sites in the human H19 promoter and aberrant hypomethylation in human bladder cancer. Hum Mol Genet 10:2619–2626
61. Marques CJ, Carvalho F, Sousa M, Barros A (2004) Genomic imprinting in disruptive spermatogenesis. Lancet 363:1700–1702
62. Dasoula A, Georgiou I, Kontogianni E, Sofikitis N, Syrrou M (2007) Methylation status of the SNRPN and HUMARA genes in testicular biopsy samples. Fertil Steril 87:805–809
63. Marques CJ, Costa P, Vaz B, Carvalho F, Fernandes S et al (2008) Abnormal methylation of imprinted genes in human sperm is associated with oligozoospermia. Mol Hum Reprod 14:67–74
64. Ariel M, Cedar H, McCarrey J (1994) Developmental changes in methylation of spermatogenesis-specific genes include reprogramming in the epididymis. Nat Genet 7: 59–63
65. Hartmann S, Bergmann M, Bohle RM, Weidner W, Steger K (2006) Genetic imprinting during impaired spermatogenesis. Mol Hum Reprod 12:407–411
66. Maher ER, Brueton LA, Bowdin SC, Luharia A, Cooper W et al (2003) Beckwith-Wiedemann syndrome and assisted reproduction technology (ART). J Med Genet 40:62–64
67. Paoloni-Giacobino A (2007) Genetic and epigenetic risks of ART. Fertil Steril 88:761–762, author reply 762
68. Ludwig M, Katalinic A, Gross S, Sutcliffe A, Varon R et al (2005) Increased prevalence of imprinting defects in patients with Angelman syndrome born to subfertile couples. J Med Genet 42:289–291
69. Niemitz EL, Feinberg AP (2004) Epigenetics and assisted reproductive technology: a call for investigation. Am J Hum Genet 74:599–609

70. Chang AS, Moley KH, Wangler M, Feinberg AP, Debaun MR (2005) Association between Beckwith-Wiedemann syndrome and assisted reproductive technology: a case series of 19 patients. Fertil Steril 83:349–354
71. Gosden R, Trasler J, Lucifero D, Faddy M (2003) Rare congenital disorders, imprinted genes, and assisted reproductive technology. Lancet 361:1975–1977
72. Strawn EY, Jr., Bick D, Swanson A (2010) Is it the patient or the IVF? Beckwith-Wiedemann syndrome in both spontaneous and assisted reproductive conceptions. Fertil Steril 294:754 e1–2
73. Odom LN, Segars J (2010) Imprinting disorders and assisted reproductive technology. Curr Opin Endocrinol Diabetes Obes 17:517–522
74. Basatemur E, Shevlin M, Sutcliffe A (2010) Growth of children conceived by IVF and ICSI up to 12years of age. Reprod Biomed Online 20:144–149
75. Carson C, Sacker A, Kelly Y, Redshaw M, Kurinczuk JJ et al (2013) Asthma in children born after infertility treatment: findings from the UK Millennium Cohort Study. Hum Reprod 28:471–479
76. Belva F, Bonduelle M, Painter RC, Schiettecatte J, Devroey P et al (2010) Serum inhibin B concentrations in pubertal boys conceived by ICSI: first results. Hum Reprod 25:2811–2814
77. Belva F, Bonduelle M, Schiettecatte J, Tournaye H, Painter RC et al (2011) Salivary testosterone concentrations in pubertal ICSI boys compared with spontaneously conceived boys. Hum Reprod 26:438–441
78. Belva F, Painter R, Bonduelle M, Roelants M, Devroey P et al (2012) Are ICSI adolescents at risk for increased adiposity? Hum Reprod 27:257–264
79. Basatemur E, Sutcliffe A (2008) Follow-up of children born after ART. Placenta 29(Suppl B): 135–140

Chapter 27

Human Germ Cell Differentiation from Pluripotent Embryonic Stem Cells and Induced Pluripotent Stem Cells

Jose V. Medrano, Carlos Simon, and Renee Reijo Pera

Abstract

Although 10–15 % of couples are infertile, little is known of the diverse, underlying pathologies in men and women with poor germ cell production; furthermore, for those with few or no high-quality germ cells, there are few options available for treatment. Thus, over the last decade, concerted efforts have been aimed at developing a biological system to probe the fundamentals of human egg and sperm production via pluripotent stem cell cells with the hopes of informing clinical decisions and ultimately providing alternative methods for therapy which may include developing a source of germ cells ultimately for reproductive purposes.

Key words Pluripotent embryonic stem cells, Induced pluripotent stem cells, Germ cell differentiation, Feeder-free culture of hESCs and iPSCs, Lentiviral transduction and differentiation, Flow cytometry, FISH, Synaptonemal complex formation

1 Introduction

Historically, the quality of life of infertile couples has been greatly diminished by the loss of opportunity to conceive; however, in recent years, novel clinical interventions such as intracytoplasmic sperm injection have dramatically changed the outlook for some couples, particularly those with severe forms of infertility [1]. In parallel with clinical successes, there have also been groundbreaking scientific advances including sequencing of the human genome, derivation of human embryonic stem cell (hESC) lines, and reprogramming of adult human somatic cells to pluripotency [2–6]. Together, these advances now allow us to overcome two historically insurmountable limitations in studies of human germ cell (egg and sperm) differentiation: the *inaccessibility* of early human development to exploration and manipulation and the *genetic-intractability* of the genome during development.

From a basic science point of view, there are several unique aspects to human germ cell development that merit investment in

Zev Rosenwaks and Paul M. Wassarman (eds.), *Human Fertility: Methods and Protocols*, Methods in Molecular Biology, vol. 1154, DOI 10.1007/978-1-4939-0659-8_27, © Springer Science+Business Media New York 2014

understanding human development per se: First, genes and dosages required for human germ cell development differ from those of mice, including both autosomal and sex chromosomal genes and dosages [7–13]. Second, humans are rare among species in that infertility is common, with half of all cases linked to faulty germ cell development [14]. Moreover, pathologies associated with meiotic errors are numerous in humans relative to other species, with errors in meiotic chromosome segregation occurring in as many as 5–30 % of human germ cells depending on sex and age [15]. This is in contrast to frequencies of ca. 1/10,000 cells in yeast, 1/1,000 cells in flies, and 1/100 cells in mice. With recent advances, we now have the ability to incorporate new strategies in order to examine the specifics of human germ cell development. This will allow us to derive full benefit from the wealth of data from model systems such as the fly and the mouse to begin to understand the complex genetics of human germ cell formation and differentiation. In seeking to understand germ cell biology, we also acknowledge that the ability to contribute to the germ line is a fundamental property that distinguishes pluripotent stem cells. For example, Han and colleagues recently demonstrated that by addition of a 5th factor to the commonly used 4 factor mixture for reprogramming, ability to contribute to the germ line was significantly increased [16]. This was in spite of the fact that iPSCs derived from 4 factor-reprogramming were indistinguishable from standard iPSCs or mESCs in gene expression and markers of pluripotency. Thus, recent research allows us to address fundamental questions regarding our germ line origins, function, and pathology and lays the groundwork for designing rational therapeutics and diagnostics to inform clinical decisions based on data obtained from model organisms, human genetic studies and direct experimental analysis of human germ cells. It also contributes to the related field of pluripotent stem cell biology and regenerative medicine.

In spite of the obvious need for information regarding human germ cell development, most of our knowledge of germ cell development derives from studies of model organisms over the last decades [17]. Yet, it is often difficult to translate data from model systems to human germ cell development, given the lack of comparative data on human development. Thus, it is critical that we shift the current research paradigm and use the vast array of tools and knowledge derived from other systems to focus on human germ cell development. To accomplish this, we can build upon knowledge of mechanisms that act to form and differentiate germ cells in *Drosophila* and the mouse [18–21]. In addition, we can combine our knowledge of the genetics of human infertility, especially as it is related to decades-old questions regarding the sex chromosomes, with fundamental principles of stem cell biology to develop an innovative scientific platform that is capable of addressing questions regarding gene function via classical genetic means (natural mutations, complementation, suppressor analysis, and epistasis) on a human genome background [9, 22–31]. We can also shift the paradigm of clinical practice by ultimately providing a platform for pharmaceutical, genetic, and environmental testing and contributing to a vision of the day when we may intervene to preserve and/or restore fertility of men and women.

Previous studies have resulted in the development of hESC-based tools that allowed us to conduct the first direct genetic analysis of human germ cell development. Studies indicated that germ cell formation, maintenance and differentiation through meiosis could be modulated via expression of *DAZ*, *DAZL*, and *BOL* [32, 33]. However, clearly the ability to interrogate gene function is limited to simple genetics with hESCs and the use of morpholinos or siRNA; thus, subsequent studies focused on new approaches based on iPSCs (induced pluripotent stem cells) [34]. Classically, pluripotent stem cells have been derived from the inner cell mass, are capable of contributing to all three germ layers, as well as the germ line, and are characterized by specific cell markers, epigenetic and genetic status, and growth requirements [2, 35–37]. However, more recently, cells with extensive similarities to hESCs have been derived via reprogramming of skin fibroblasts [4–6]. By introducing a cocktail of transcription factors, somatic cells can be reprogrammed to an ESC-like fate. Although molecular details of reprogramming have not been well described, human iPSCs may provide an optimal system to probe the genetics of human infertility and ultimately obtain functional germ cells. Although many obstacles remain, much progress has been made in differentiating germ cells from both induced pluripotent stem cells and human embryonic stem cells in the last 5–10 years. Here, we describe the state of the art in germ line differentiation from pluripotent stem cells.

2 Materials

Below we outline the methods used most commonly and successfully to differentiate pluripotent human stem cell lines to the germ cell lineage; note that we include differentiation via overexpression of germ cell-specific genes (*DAZ*, *DAZL*, *BOULE*, and *VASA*). Methods can be modified for expression of other genes, testing of signaling factors or peptides, or spontaneous differentiation; methods are based on previous studies [32, 34, 38–41] and unpublished data (Medrano-Plaza, Simon, and Reijo Pera).

2.1 Reagents and Chemicals

1. Powder porcine skin gelatin (Sigma).
2. Inactivated mouse embryonic fibroblasts (MEF) in passage 3–5.
3. Established human embryonic stem cell (hESC)/human induced pluripotent stem cell (iPSC) lines in a passage number between 15 and 55.

4. MEF medium: DMEM high glucose (Invitrogen) supplemented with 10 % fetal bovine serum (Hyclone), 2 mM L-glutamine (Invitrogen), and 1 % Penicillin/Streptomycin (Invitrogen).
5. hESC maintenance medium: DMEM/F12 supplemented with 20 % KnockOut serum replacer, 2 mM L-glutamine, 0.1 mM nonessential amino acids, 10 ng/mL basic FGF (Invitrogen), and 0.1 mM 2-mercaptoethanol (Millipore).
6. Powder collagenase type IV (Invitrogen).
7. Matrigel growth factor reduced (BD Biosciences).
8. High titer DAZ2, DAZL, BOULE, and VASA expression lentivirus supernatants (if overexpression is desired).
9. High titer pLVGV (VASA-GFP) lentivirus supernatant.
10. Polybrene (Millipore).
11. PBS-Ca^{2+}-Mg^{2+} (Invitrogen).
12. Differentiation medium: DMEM/F12 supplemented with 20 % fetal bovine serum (Hyclone), 2 mM L-glutamine, 0.1 mM nonessential amino acids (Invitrogen), and 0.1 mM 2-mercaptoethanol (Millipore).
13. Blasticidin (Invitrogen).
14. Geneticin (Invitrogen).
15. Trypsin 0.05 % (Invitrogen).
16. Absolute ethanol (Panreac).
17. DNA staining buffer (0.2 mg/mL RNase A (Sigma) and 0.02 mg/mL propidium iodide (PI) (Invitrogen) in BD Perm/Wash buffer (BD Biosciences)).
18. Hypoextraction buffer (30 mM Tris (Sigma), 50 mM Sucrose (Sigma), 17 mM Citric acid (Sigma), 5 mM EDTA (Invitrogen), two tablets of Complete Mini (Roche); pH 8.2).
19. Paraformaldehyde (PFA) 4 % (Sigma).
20. Photo-Flo (KODAK).
21. Blocking solution (4 % goat serum (Sigma) in 1 % BSA (Sigma), 0.1 % Tween-20 (Sigma) in PBS (Invitrogen)).
22. Rabbit polyclonal anti-SCP3 primary antibody (Novus Biologicals).
23. Mouse polyclonal anti-CENPA antibody (Abcam).
24. Goat anti-rabbit Alexa 594 secondary antibody (Invitrogen).
25. Goat anti-mouse Alexa 488 secondary antibody (Invitrogen).
26. ProLong Gold antifade reagent with DAPI (Invitrogen).
27. Carnoy's fixative (1:3 acetic acid–methanol).

28. Cytoplasm wash buffer (50 μg/mL pepsin in 0.01 N HCl).
29. FISH hybridization mixture 1× (7 μL CEP hybridization buffer, 1 μL CEP 18 (D18Z1) SpectrumOrange probe, 1 μL CEP 16 (D16Z3) SpectrumGreen probe, 1 μL CEP 16 (DYZ1) SpectrumAqua probe) (Vysis, Abbott Molecular).
30. Rubber cement.
31. Formamide.
32. 20× saline–sodium citrate (SSC) buffer (Sigma).

2.2 Equipment and Consumables

1. T175cm² culture flasks (Falcon, BD Biosciences).
2. 6-well plates (Falcon, BD Biosciences).
1. Micropipettes and plastic tips (2.5, 10, 20, 200, and 1,000 μL).
2. Serological pipettes (2, 5, 10, 25, and 50 mL).
3. Culture incubators with humidity and CO_2% control.
4. Plastic or rubber cell scrapers (e.g., "policeman").
5. Biosafety level 2 (BL2) work area with a culture laminar flow hood.
6. Stereomicroscope.
7. Cell strainer, 30 μm (BD Biosciences).
8. Hemocytometer.
9. Inverted microscope with phase-contrast optics.
10. Fluorescence activated cell sorting system.
11. Cytospin centrifuge with adapters and Cytospin slides (Thermo Scientific).
12. Staining Coplin cuvettes.
13. Water bath.
14. Thermoplate.
15. Fluorescence microscope.
16. Cell counter.

2.3 Other Essential Laboratory Equipment and Supplies

1. Autoclave.
2. Precision balances.
3. Cell culture laminar flow hood.
4. Centrifuges.
5. Forceps.
6. Informatics equipment.
7. Filtration apparatus.
8. First aid kits.
9. Heating blocks and warming plates.

10. Ice bucket.
11. Lab furniture low VOC emission.
12. Liquid nitrogen tank.
13. Refrigerator (4 °C) and freezer (−20 and −80 °C).
14. pH meter.
15. Timer and stopwatch.
16. Vortex mixers.
17. Purified water supply or molecular grade bottled water.
18. Biohazard disposal bags and containers.
19. Disk filter units.
20. Gloves.
21. Protective goggles.
22. Lab coats.
23. Laboratory glassware.
24. Microscope slides and coverslips.
25. Test tubes and racks.

3 Methods

3.1 Cell Culture of hESCs and iPSCs, Transduction, and Differentiation

3.1.1 Preparation of Conditioned Medium for Feeder-Free Culture of hESCs and iPSCs

1. Coat T175 cm^2 culture flasks with 5 mL of 0.1 % gelatin (diluted in water) and incubate for 30 min at 37 °C.
2. Remove gelatin from flasks and add 20 mL of MEF medium to each flask.
3. Gently thaw inactivated MEFs and resuspend them in 10 mL MEF medium.
4. Centrifuge the cells at 200 × *g* for 5 min at room temperature, discard supernatant and resuspend cell pellet in MEF medium in order that cell concentration is 3,000,000 cells/mL.
5. Seed 1 mL of the fibroblasts cell suspension per flask, in order that the cell density reaches approximately 17,000 cells/cm^2. Incubate the flasks overnight at 37 °C, 5 % CO_2.
6. The next day, remove MEF medium from flasks, wash with PBS, and add 40 mL of hESC medium.
7. During the next 5 days, collect conditioned hESC medium from flasks and replace it by fresh hESC medium every 24 h. The conditioned hESC medium can be frozen down at −20 °C and stored for up to 6 months.
8. When ready to use, thaw conditioned medium, filter it through a 0.22 μm pore diameter membrane system and add βFGF to a final concentration of 10 ng/mL

3.1.2 Feeder-Free Culture of hESCs and iPSCs for the Lentiviral Transduction

1. To prepare Matrigel-coated plates, slowly thaw a 1 mL Matrigel aliquot at 4 °C to avoid the formation of gel and dilute it in 49 mL of cold DMEM/F12 (dilution 1:50). Then add 1 mL of Matrigel solution to coat each well of 6-well plates.
2. Incubate the plates for 1 h at room temperature or overnight at 4 °C (Matrigel coated plates can be stored at 4 °C up to 2 weeks before use).
3. To passage hESC/iPSC colonies to the feeder-free culture on Matrigel, aspirate medium from a 6-well plate with ready to passage colonies and add 1 mL of collagenase type IV (1 mg/mL diluted in DMEM/F12) per well. Incubate plates at 37 °C for 10–15 min. Check colony morphology every 5 min during collagenase incubation in order to stop reaction when the edge of the colonies start to detach.
4. Remove the collagenase and add 2 mL of fresh hESC medium per well.
5. Gently detach colonies with a cell scraper and collect them into 15 mL centrifuge tubes. Dissociate the cells into small clusters of 100–500 cells with gentle pipetting.
6. Let the cells to settle down for 5 min in order to remove the supernatant with fibroblasts, and resuspend the cell pellet with colonies in conditioned medium.
7. Remove Matrigel from coated plates, wash with PBS twice and add 3 mL of the clump suspension in conditioned medium per well (optimal colony density should be 60–80 colonies per well of a 6-well plate). Return plates to the incubator. Shake the plates left to right and back to front to distribute cell clumps inside the wells.
8. Change conditioned medium of plates every day and remove spontaneously differentiated areas mechanically during the next 2–4 days. When the confluence inside the wells is around 50 % the cells are ready for the lentiviral transduction.

3.1.3 Lentiviral Transduction and Differentiation

Note: Because we are manipulating biohazardous virus, this part of the protocol should be performed in a Biosafety Level 2 (BL2) work area.

1. For the ectopic expression of germ cell related RNA-binding proteins, thaw a vial of high titer lentiviral supernatant of DAZ2, DAZL, BOULE, and VASA. For the Analysis of the potential of pluripotent cell lines to form germ cells via VASA-GFP FACS, thaw a vial of high titer lentiviral supernatant of pLVGV.
2. Remove conditioned medium from wells and add 250 μL of each lentiviral supernatant to a final volume of 1 mL per well

(DAZ2 gene is located at the Y chromosome and is not present in female cell lines, so in these cell lines infection should be done with 250 μL of each VASA, DAZL, and BOULE viral supernatant, replacing the 250 μL of DAZ2 supernatant by conditioned medium). For the pLVGV transduction, use directly 1 mL of viral supernatant per well.

3. Add polybrene to a final concentration of 8 μg/mL.
4. Incubate plates at 37 °C, 5 % CO_2 for 6 h.
5. Add 2 mL of conditioned medium to each well (dilution of viral supernatant 1:3) and incubate overnight at 37 °C, 5 % CO_2.
6. The next day, remove viral supernatants from wells and wash the at least twice with 4 mL of PBS per well in order to completely remove viruses. Once you have completely removed the virus supernatants, you can move your plates back to the standard tissue culture lab.
7. To let the cells recover from the lentiviral infection, add 3 mL of fresh conditioned medium per well and incubate overnight at 37 °C, 5 % CO_2.
8. The next day, start selection of resistant infected cells by adding blasticidin to a final concentration of 2 μg/mL in conditioned medium during the next 3 days with media changes every day in the wells transduced with the ectopic expression lentiviruses. For the wells transduced with pLVGV, the selection antibiotic is 200 μg/mL geneticin during 7 days, followed by an additional passage of cells without antibiotic in order to let them recover from the treatment. Because the efficiency of transduction in hESCs and iPSCs is low, it is normal that a significant portion of cells start to die and detach starting on day 2 of selection and during the next 3–5 days. Also, these selection treatments are based in our titration experiments with our cell lines, but maybe different cell lines can show a different response to the antibiotics, so we recommend to perform new titration experiments if different cells lines are going to be tested.
9. Once finished the selection period, replace the conditioned medium of wells by differentiation medium and keep the cells on this medium during the next 2 weeks with media changes every 2–3 days. During the spontaneous differentiation, cells will change their typical morphology of small round cells growing in colonies and start to form different cell types with mesenchymal, fibroblast and epithelial morphology (Fig. 1). Check daily the morphology change and GFP fluorescence of differentiating cells in order to detect any significant change.

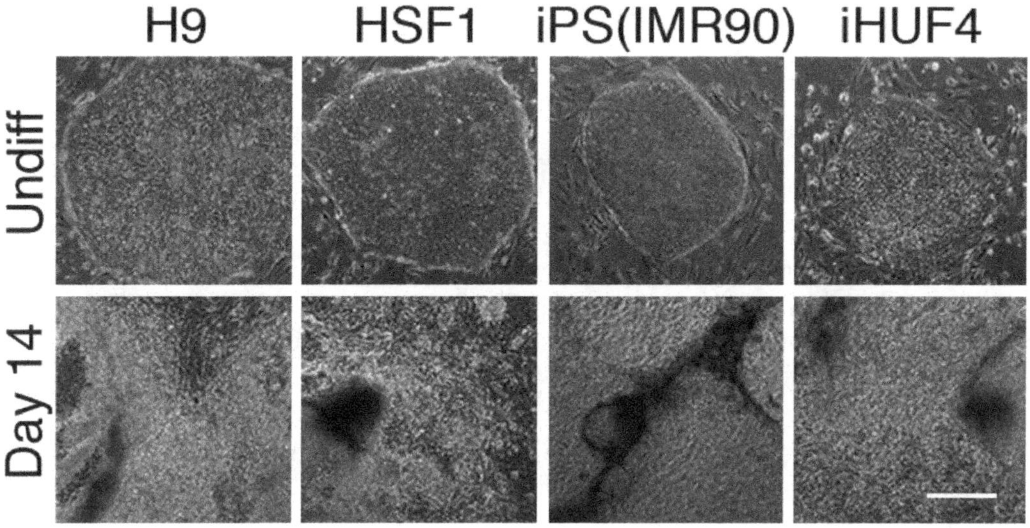

Fig. 1 Morphology of undifferentiated H9, HSF1, iPS(IMR90), and iHUF4 cell colonies routinely cultured on MEFs and cells differentiated as adherent culture for 14 days. After 14 days of differentiation, cultures appear confluent, with cell morphology distinct from the undifferentiated cells for all the cell lines (from Panula et al. [34])

3.2 Analysis of the Potential of Pluripotent Cell Lines to Form Germ Cells Via VASA-GFP Fluorescence Activated Cell Sorting (FACS)

1. Remove differentiation medium and wash the wells with PBS-$Ca2^{+}$-Mg^{2+} twice.
2. Add 500 μL of trypsin 0.05 % to each well and incubate for 10 min at 37 °C.
3. Stop reaction by adding 2 mL of differentiation medium and harvest the cells still attached to the bottom of the well by flushing.
4. Filter cell clumps through a 30 μm cell strainer in order to obtain a single cell suspension without any cell clump. Collect cells in 15 mL centrifuge tubes.
5. Centrifuge cells at 200 × *g* for 5 min at room temperature.
6. Discard supernatant, count cells and resuspend cell pellet with differentiation medium to a final concentration of 0.5–1 million of cells/mL.
7. Put tubes on ice during the next steps in order to maintain cells alive during the flow cytometer set up (Fig. 2).
8. Before starting the FACS of VASA-GFP positive and negative cells, it is necessary to apply several gates to the cell suspension in order to improve the sorting:
   (a) Set the voltage and gain of the forward and side scatter in order to visualize the cell population in a dot plot representing size (linear forward scatter) versus complexity

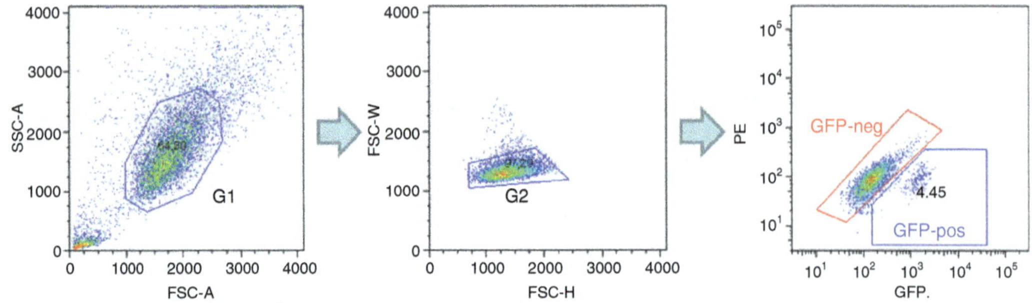

Fig. 2 Schematic representation of the setup protocol for the VASA-GFP flow cytometry sorting

(linear side scatter). In this graph, set a gate (G1) that contains most of cells excluding too small events (cell debris).

(b) Create a second plot showing the G1 population in the linear forward-high scatter versus the linear forward-wide scatter and set the gate G2 in the area that contains the events in the linear population in order to exclude possible doublets (upper side) and cell debris (lower side). This G2 gate contains the working cell population to sort.

(c) Create a plot showing the G2 population in the logarithmic red detector versus the logarithmic green detector. In the negative control population transduced with the empty lentiviral backbone, autofluorescence in the red and green detectors increases linearly with the size and complexity of cells. However, when a subset of cells in the whole population expresses GFP under the control of the VASA promoter, a cloud of events will move out of the linear autofluorescence zone to the green fluorescence. Alternative, you can just create a gate with the linear forward scatter versus the logarithmic green detector and compare the transduced cells with their non-GFP transduced controls to discriminate GFP positive cells.

9. Set a gate that comprises the linear autofluorescence area (GFP negative population) and another one for the cloud of events out of the linear area moved to the green detector (GFP positive population), and sort them in different collection tubes. In our hands, different cell lines have different potential with an expected efficiency that ranges between 1 and 10 % of VASA-GFP positive cells. The percentage of GFP positive cells is indicative of that since the GFP is expressed under control of the germ line specific VASA promoter. Different cell lines could have different potential to form germ cells. Additionally, sorted cells can be moved back to culture conditions, or be subjected to gene and protein expression analysis, as well as other analytic techniques in order to confirm their germinal identity.

3.3 Flow Cytometry Sorting of Cells by Their DNA Content

1. Remove differentiation medium and wash the wells with PBS-Ca2^{+}-Mg^{2+} twice.
2. Add 500 μL of trypsin 0.05 % to each well and incubate for 10 min at 37 °C.
3. Stop reaction by adding 2 mL of differentiation medium and harvest the cells still attached to the bottom of the well by flushing.
4. Filter cell clumps through a 30 μm cell strainer in order to obtain a single cell suspension without any cell clump. Collect the cells in 15 mL centrifuge tubes.
5. Centrifuge the cells at 200 × *g* for 5 min at room temperature.
6. Discard supernatant, count the cells, and resuspend cell pellet in ice-cold ETOH 70 %.
7. Incubate the cells for 45–60 min at room temperature in order to fix them.
8. Centrifuge the cells at 200 × *g* for 5 min at room temperature.
9. Discard supernatant, wash with PBS, and repeat the centrifugation.
10. Discard supernatant and resuspend the cells in DNA staining buffer (0.2 mg/mL RNase A and 0.02 mg/mL propidium iodide in BD Perm/Wash buffer containing saponin and FCS) at a final concentration of 0.5–1 million of cells/mL.
11. Incubate the cells protected from light at 37 °C for 30 min. Keep the cells protected from light also during the cytometer set up (Fig. 3).
12. Before starting the fluorescence activated cell sorting (FACS) of VASA-GFP positive and negative cells, it is necessary to apply several gates to the cell suspension in order to improve the sorting:
    (a) Set the voltage and gain of the forward and side scatter in order to visualize the cell population in a dot plot representing size (linear forward scatter) versus complexity (linear side scatter). In this graph, set a gate (G1) that contains most of cells excluding too small events (cell debris).
    (b) Create a second plot showing the Gl population in the linear PI detector versus the linear PI-wide scatter and set the gate G2 in the area that contains the events in the linear population in order to exclude possible doublets (upper side) and cell debris (lower side). This G2 gate contains the working cell population to sort.
13. Create a histogram showing the G2 population in the linear PI detector. Typical diploid cycle shows two peaks representing the G_1 (2N, 1C) and G_2 (2N, 2C) phases with a valley representing the cells in mitosis/S phase.

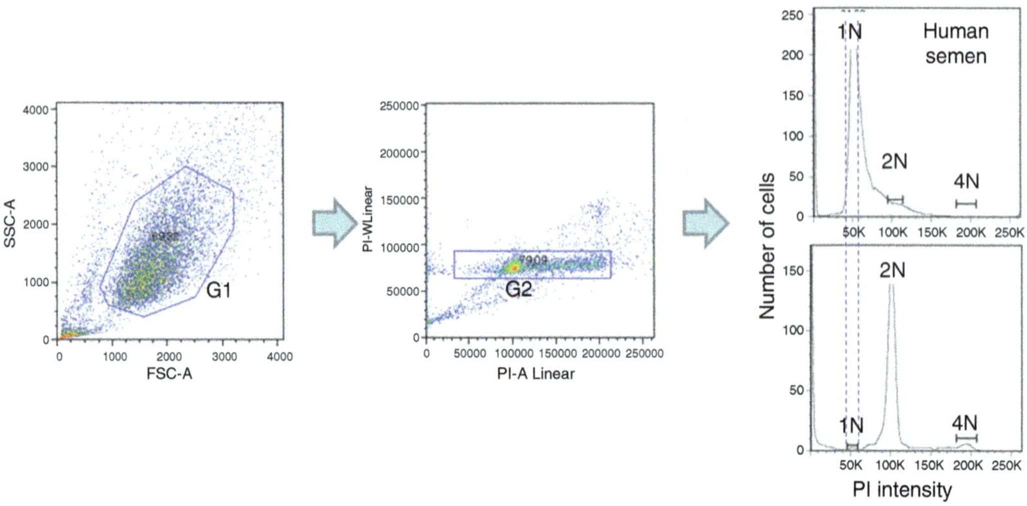

Fig. 3 Schematic representation of the setup protocol for the flow cytometry sorting of cells by their DNA content

14. Using a PI-labeled sperm sample as a positive 1N control, set the gates that comprise the putative 1N and 2N cells and sort them into different collection tubes. Expected efficiency of haploid cells is between 0.5 and 2 % in our experiments performed with different hESC/iPSC lines. Sorted cells can be subjected to further analysis to determine their ploidy.

3.4 Meiotic Progression Analysis by FISH and Synaptonemal Complex Formation

3.4.1 Synaptonemal Complex Formation Analysis by Nuclear SCP3 Staining Pattern

1. Collect and gently resuspend by pipetting up and down single cells in 1 mL of fresh prepared hypoextraction buffer (30 mM Tris, 50 mM Sucrose, 17 mM Citric acid, 5 mM EDTA, two tablets of Complete Mini (Roche); pH 8.2) at a concentration of 100,000 cells/mL.
2. Place tubes containing cell suspensions on ice and incubate for 5 min.
3. Load 100 μL of the cell suspension onto a Cytospin slide and centrifuge at 800 rpm for 3 min in a Cytospin centrifuge.
4. Remove the Cytospin adapter. Immediately place the slides into a humid chamber and fix the cells in 4 % PFA for 15 min at room temperature.
5. Drip off the PFA and rinse once with PBS.
6. Permeabilize the cells with 0.04 % Photo-Flo (KODAK) in PBS for 5 min at room temperature.
7. Drip off the Photo-Flo and incubate the slides in blocking solution (4 % goat serum in 1 % BSA, 0.1 % Tween-20 in PBS) for 30 min at room temperature.
8. Incubate with primary antibodies Rabbit polyclonal anti-SCP3 primary antibody (1:1,000 dilution) and Mouse polyclonal

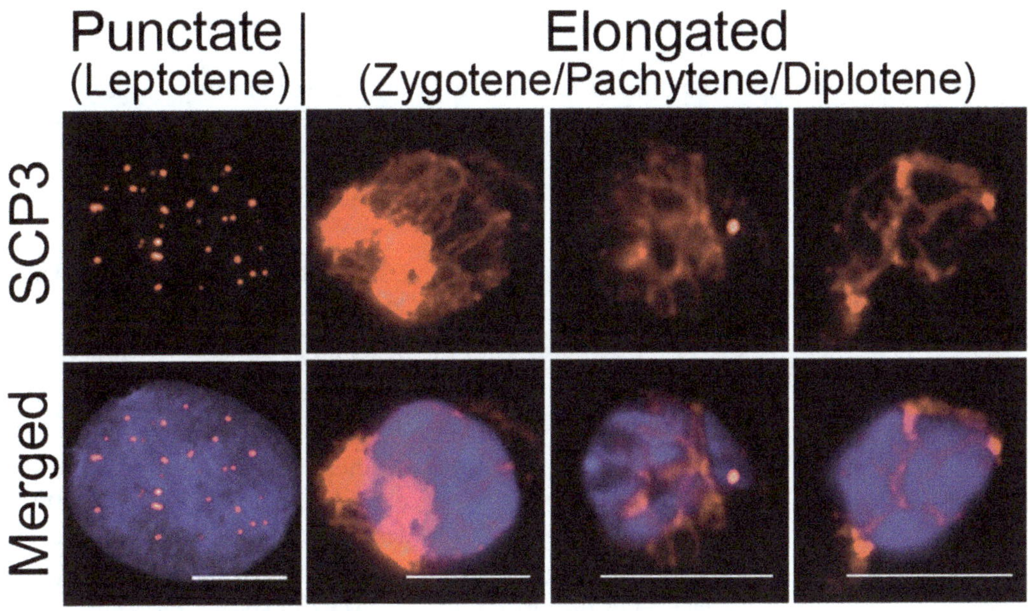

Fig. 4 Representative SCP3 staining patterns for the different stages of meiotic prophase I. Punctuated SCP3 signal indicates that cells are in an early meiotic stage (leptotene stage of prophase I). On the other hand, an elongated SCP3 staining pattern inside the nucleus is indicative of progression through zygotene, pachytene, or diplotene meiotic prophase I stages (Kee et al. [32])

anti-CENPA antibody (1:500 dilution) for 2 h at room temperature in the same blocking solution.

9. Wash the slides with PBS + 0.1 % Tween-20 for 5 min at room temperature, three times.
10. Incubate with secondary antibodies Goat anti-rabbit Alexa 594 and Goat anti-mouse Alexa 488 for 1 h protected from light.
11. Wash the slides with PBS + 0.1 % Tween-20 for 5 min at room temperature, three times.
12. Mount coverslips with ProLong Gold antifade reagent with DAPI. Dry overnight.
13. Analyze the synaptonemal complex formation under fluorescence microscope. CENPA (green) would be helpful to co-localize chromosomic centromeres inside the nucleus. Punctuated SCP3 signal (red) indicates that cells are in an early meiotic stage (leptotene stage of prophase I). In the other hand, an elongated SCP3 staining pattern inside the nucleus is indicative of progression through zygotene, pachytene, or diplotene meiotic prophase I stages (Kee et al., 2009) (Fig. 4).

3.4.2 Fluorescent In Situ Hybridization (FISH)

1. Load 100 μL from a cell suspension of 100,000–200,000 cells/mL onto a Cytospin slide and centrifuge at 800 rpm for 3 min in a Cytospin centrifuge.

2. Remove the Cytospin adapter and immediately fix cells in glacial Carnoy's fixative solution (1:3 acetic acid–methanol). Let the fixative dry for 5 min at room temperature. At this point, slides can be stored at −20 °C to resume the protocol later.
3. Incubate the slides in 2× SSC buffer for 30–60 min at room temperature.
4. Treat the slides with cytoplasm wash buffer (50 μg/mL pepsin in 0.01 N HCl) for 5 min at 37 °C.
5. Wash the slides with PBS for 5 min at room temperature twice.
6. Optional: Postfix nucleus by incubation of the slides in 1 % formaldehyde and wash with PBS.
7. Dehydrate the samples by incubating slides in increasing concentration baths of ice cold ETOH 70, 80 and 100 %, 2 min each. Air-dry the slides once the last incubation is over.
8. Add 2 μL of FISH hybridization mixture 1× (7 μL CEP hybridization buffer, 1 μL CEP 18 (D18Z1) SpectrumOrange probe, 1 μL CEP 16 (D16Z3) SpectrumGreen probe, 1 μL CEP 16 (DYZ1) SpectrumAqua probe) to each coverslip upside down and then mount them to each slide. Seal it with rubber cement to prevent evaporation.
9. Denature samples and probes together by incubating them on a thermoplate at 85 °C for 5 min protected from light. Alternatively, probes and samples can be denatured separately by incubating them at 76 °C for 5 min in 70 % formamide/2× SSC buffer.
10. Transfer the slides immediately to humidified chambers and incubate them at 37 °C overnight protected from light.
11. The next day, carefully remove rubber cement with forceps.
12. Place the slides in a denaturing wash of 50 % formamide/2× SSC buffer at 43 °C for 10 min.
13. Repeat the **step 12** twice. Usually, coverslips fall in the second wash. If not, carefully remove them from the slides with the forceps.
14. Wash again the slides in 2× SSC buffer at 43 °C for 10 min.
15. Optional: Wash the slides in 2× SSC buffer + 0.1 % NP-40 at 43 °C for 10 min.
16. Wash the slides in 2× SSC buffer for 5 min at room temperature.
17. Mount coverslips with ProLong Gold antifade reagent with DAPI. Dry overnight.
18. Analyze cell ploidy under a fluorescence microscope, using a cell counter to determine the percentage of haploid cells.

References

1. Cedars M (2005) In: Cedars MI (ed) Infertility practical pathways in obstetrics and gynecology. McGraw-Hill, p xv–xviii
2. Thomson JA, Itskovitz-Eldor J, Shapiro SS, Waknitz MA, Swiergiel JJ, Marshall VS, Jones JM (1998) Embryonic stem cell lines derived from human blastocysts. Science 282: 1145–1147
3. International Human Genome Sequencing Consortium (2001) Initial sequencing and analysis of the human genome. Nature 409:860–921
4. Park I, Zhao R, West J, Yabuuchi A, Huo H, Ince T, Lerou P, Lensch M, Daley G (2008) Reprogramming of human somatic cells to pluripotency with defined factors. Nature 451:141–146
5. Yu J, Vodyanik M, Smuga-Otto K, Antosiewicz-Bourget J, France J, Tian S, Nie J, Jonsdottir G, Ruotti V, Stewart R, Slukvin I, Thomson J (2007) Induced pluripotent stem cell lines derived from human somatic cells. Science 318:1917–1920
6. Takahashi K, Tanabe K, Ohnuki M, Narita M, Ichisaka T, Tomoda K, Yamanaka S (2007) Induction of pluripotent stem cells from adult human fibroblasts by defined factors. Cell 131:861–872
7. Reijo R, Alagappan RK, Patrizio P, Page DC (1996) Severe oligospermia resulting from deletions of the *Azoospermia Factor* gene on the Y chromosome. Lancet 347:1290–1293
8. Reijo R, Lee TY, Salo P, Alagappan R, Brown LG, Rosenberg M, Rozen S, Jaffe T, Straus D, Hovatta O et al (1995) Diverse spermatogenic defects in humans caused by Y chromosome deletions encompassing a novel RNA-binding protein gene. Nat Genet 10:383–393
9. Vogt PH, Edelmann A, Kirsch S, Henegariu O, Hirschmann P, Kiesewetter F, Kohn FM, Schill WB, Farah S, Ramos C, Hartmann M, Hartschuh W, Meschede D, Behre HM, Castel A, Nieschlag E, Weidner W, Grone H-J, Jung A, Engel W, Haidl G (1996) Human Y chromosome azoospermia factors (AZF) mapped to different subregions in Yq11. Hum Mol Genet 5:933–943
10. Skaletsky H, Kuroda-Kawaguchi T, Minx PJ, Cordum HS, Hillier L, Brown LG, Repping S, Pyntikova T, Ali J, Bieri T, Chinwalla A, Delehaunty A, Delehaunty K, Du H, Fewell G, Fulton L, Fulton R, Graves T, Hou SF, Latrielle P, Leonard S, Mardis E, Maupin R, McPherson J, Miner T, Nash W, Nguyen C, Ozersky P, Pepin K, Rock S, Rohlfing T, Scott K, Schultz B, Strong C, Tin-Wollam A, Yang SP, Waterston RH, Wilson RK, Rozen S, Page DC (2003) The male-specific region of the human Y chromosome is a mosaic of discrete sequence classes. Nature 423:825–837
11. Zinn AR, Page DC, Fisher EMC (1993) Turner syndrome: the case of the missing sex chromosome. Trends Genet 9:90–93
12. Hendry AP, Wenburg JK, Bentzen P, Volk EC, Quinn TP (2000) Rapid evolution of reproductive isolation in the wild: evidence from introduced salmon. Science 290:516–518
13. Swanson WJ, Vacquier VD (2002) The rapid evolution of reproductive proteins. Nat Rev Genet 3:137–144
14. Menken J, Larsen U (1994) Estimating the incidence and prevalence and analyzing the correlates of infertility and sterility. Ann N Y Acad Sci 709:249–265
15. Hunt PA, Hassold TJ (2002) Sex matters in meiosis. Science 296:2181–2183
16. Han J, Yuan P, Yang H, Zhang J, Soh B, Li P, Lim S, Cao S, Tay J, Orlov Y, Lufkin T, Ng H-H, Tam W-L, Lim B (2010) Tbx3 improves the germ-line competency of induced pluripotent stem cells. Nature 463:1096–1100
17. Matzuk M, Lamb D (2002) Genetic dissection of mammalian fertility pathways. Nat Cell Biol 4S:41–49
18. Fox MS, Clark AT, El Majdoubi M, Vigne JL, Urano J, Hostetler CE, Griswold MD, Weiner RI, Reijo Pera RA (2007) Intermolecular interactions of homologs of germ plasm components in mammalian germ cells. Dev Biol 301:417–431
19. Jaruzelska J, Kotecki M, Kusz K, Spik A, Firpo M, Reijo RA (2003) Conservation of a Pumilio-Nanos complex from Drosophila germ plasm to human germ cells. Dev Genes Evol 213:120–126
20. Houston DW, King ML (2000) Germ plasm and molecular determinants of germ cell fate. Curr Top Dev Biol 50:155–181
21. Lehmann R, Ephrussi A (1994) Germ plasm formation and germ cell determination in Drosophila. Ciba Found Symp 182:282–296
22. Nicholas C, Chavez S, Baker V, Reijo Pera RA (2009) Instructing an embryonic stem cell-derived oocyte fate: lessons from endogenous oogenesis. Endocr Rev 30:264–283
23. Judson R, Babiarz J, Venere M, Blelloch R (2009) Embryonic stem cell-specific microRNAs promote induced pluripotency. Nat Biotechnol 27(5):459–461
24. Bucay N, Yebra M, Cirulli V, Afrikanova I, Kaido T, Hayek A, Montgomery A (2009)

A novel approach for the derivation of putative primordial germ cells and Sertoli cells from human embryonic stem cells. Stem Cells 27: 68–77

25. Muller R, Lengerke C (2009) Patient-specific pluripotent stem cells: promises and challenges. Nat Rev Endocrinol 5:195–203
26. Kossack N, Meneses J, Shefi S, Nguyen H, Chavez S, Nicholas C, Gromoll J, Turek P, Reijo-Pera RA (2009) Isolation and characterization of pluripotent human spermatogonial stem cell-derived cells. Stem Cells 27:138–149
27. Yu J, Thomson J (2008) Pluripotent stem cell lines. Genes Dev 22:1987–1997
28. Kuroda-Kawaguchi T, Skaletsky H, Brown LG, Minx PJ, Cordum HS, Waterston RH, Wilson RK, Silber S, Oates R, Rozen S, Page DC (2001) The AZFc region of the Y chromosome features massive palindromes and uniform recurrent deletions in infertile men. Nat Genet 29:279–286
29. Sun C, Skaletsky H, Rozen S, Gromoll J, Nieschlag E, Oates R, Page DC (2000) Deletion of *azoospermia factor a* (*AZFa*) region of human Y chromosome caused by recombination between HERV15 proviruses. Hum Mol Genet 9:2291–2296
30. Saxena R, Vries JW, Repping S, Alagappan RK, Skaletsky H, Brown LG, Ma P, Chen E, Hoovers JM, Page DC (2000) Four *DAZ* genes in two clusters found in the *AZFc* region of the human Y chromosome. Genomics 67: 256–267
31. Namiki M (2000) Genetic aspects of male infertility. World J Surg 24:1176–1179
32. Kee K, Angeles V, Flores M, Nguyen H, Reijo Pera RA (2009) Human DAZL, DAZ and BOULE genes modulate primordial germ cell and haploid gamete formation. Nature 462: 222–225
33. Kee K, Gonsalves J, Clark A, Reijo Pera RA (2006) Bone morphogenetic proteins induce germ cell differentiation from human embryonic stem cells. Stem Cells Dev 15:831–837
34. Panula S, Medrano J, Kee K, Bergstrom R, Nguyen H, Byers B, Wilson K, Wu J, Simon C, Hovatta O, Reijo Pera RA (2010) Human germ cell differentiation from fetal- and adult-derived induced pluripotent stem cells. Hum Mol Genet 20:752–762
35. Chavez S, Meneses J, Nguyen H, Kim S, Reijo Pera RA (2008) Characterization of six new human embryonic stem cell lines (HSF-7, -8, -9. -10, -12 and -13) derived in minimal animal-component conditions. Stem Cells Dev 17: 535–546
36. International Stem Cell Consortium (2007) Characterization of human embryonic stem cell lines by the International Stem Cell Initiative. Nat Biotechnol 25:803–816
37. Reijo-Pera R, Dejonge C, Bossert N, Yao M, Yang JH, Asadi N, Wong W, Wong C, Firpo M (2008) Gene expression profiles of human inner cell mass cells and embryonic stem cells. Differentiation 78:18–23
38. Julaton V, Reijo Pera RA (2011) NANOS3 function in human germ cell development. Hum Mol Genet 20:2238–2250
39. Kee K, Flores M, Cedars M, Reijo Pera RA (2010) Human primordial germ cell formation is diminished by exposure to environmental toxicants acting through the AHR signaling pathway. Toxicol Sci 117:218–224
40. Clark AT, Bodnar MS, Fox MS, Rodriquez RT, Abeyta MJ, Firpo MT, Reijo Pera RA (2004) Spontaneous differentiation of germ cells from human embryonic stem cells in vitro. Hum Mol Genet 13:727–739
41. Clark AT, Rodriguez R, Bodnar M, Abeyta M, Turek P, Firpo M, Reijo RA (2004) Human STELLAR, NANOG, and GDF3 genes are expressed in pluripotent cells and map to chromosome 12p13, a hot-spot for teratocarcinoma. Stem Cells 22:169–179

Index

A

B

C

Zev Rosenwaks and Paul M. Wassarman (eds.), *Human Fertility: Methods and Protocols*, Methods in Molecular Biology, vol. 1154, DOI 10.1007/978-1-4939-0659-8, © Springer Science+Business Media New York 2014

www.ingramcontent.com/pod-product-compliance
Lightning Source LLC
LaVergne TN
LVHW082013200926
848481LV00010B/2849

* 9 7 8 1 4 9 3 9 4 1 1 7 9 *